Springer Series in

CHEMICAL PHYSICS 80

Springer Series in

CHEMICAL PHYSICS

The purpose of this series is to provide comprehensive up-to-date monographs in both well established disciplines and emerging research areas within the broad fields of chemical physics and physical chemistry. The books deal with both fundamental science and applications, and may have either a theoretical or an experimental emphasis. They are aimed primarily at researchers and graduate students in chemical physics and related fields.

65 **Fluorescence Correlation Spectroscopy**
Theory and Applications
Editors: R. Rigler and E.S. Elson

66 **Ultrafast Phenomena XII**
Editors: T. Elsaesser, S. Mukamel, M.M. Murnane, and N.F. Scherer

67 **Single Molecule Spectroscopy**
Nobel Conference Lectures
Editors: R. Rigler, M. Orrit, T. Basché

68 **Nonequilibrium Nondissipative Thermodynamics**
With Application to Low-Pressure Diamond Synthesis
By J.-T. Wang

69 **Selective Spectroscopy of Single Molecules**
By I.S. Osad'ko

70 **Chemistry of Nanomolecular Systems**
Towards the Realization of Molecular Devices
Editors: T. Nakamura, T. Matsumoto, H. Tada, K.-I. Sugiura

71 **Ultrafast Phenomena XIII**
Editors: D. Miller, M.M. Murnane, N.R. Scherer, and A.M. Weiner

72 **Physical Chemistry of Polymer Rheology**
By J. Furukawa

73 **Organometallic Conjugation**
Structures, Reactions and Functions of d–d and d–π Conjugated Systems
Editors: A. Nakamura, N. Ueyama, and K. Yamaguchi

74 **Surface and Interface Analysis**
An Electrochmists Toolbox
By R. Holze

75 **Basic Principles in Applied Catalysis**
By M. Baerns

76 **The Chemical Bond**
A Fundamental Quantum-Mechanical Picture
By T. Shida

77 **Heterogeneous Kinetics**
Theory of Ziegler-Natta-Kaminsky Polymerization
By T. Keii

78 **Nuclear Fusion Research**
Understanding Plasma-Surface Interactions
Editors: R.E.H. Clark and D.H. Reiter

79 **Ultrafast Phenomena XIV**
Editors: T. Kobayashi, T. Okada, T. Kobayashi, K.A. Nelson, S. De Silvestri

80 **X-Ray Diffraction by Macromolecules**
By N. Kasai and M. Kakudo

N. Kasai M. Kakudo

X-Ray Diffraction by Macromolecules

With 351 Figures and 56 Tables

Kodansha Springer

Nobutami Kasai
Professor Emeritus
Osaka University
and
Kobe Women's University

Masao Kakudo
Professor Emeritus
Osaka University
and
Former President
Himeji Institute of Technology

Series Editors:
Professor A. W. Castleman, Jr.
Department of Chemistry, The Pennsylvania State University
152 Davey Laboratory, University Park, PA 16802, USA

Professor J.P. Toennies
Max-Planck-Institut für Strömungsforschung
Bunsenstr. 10, 37073 Göttingen, Germany

Professor W. Zinth
Universität München, Institut für Medizinische Optik
Öttingerstr. 67, 80538 München, Germany

ISSN 0172-6218

ISBN 4-06-207405-2 Kodansha Ltd., Tokyo

ISBN-10 3-540-25317-3 Springer Berlin Heidelberg New York
ISBN-13 978-3-540-25317-4 Springer Berlin Heidelberg New York

Library of Congress Control Number: 2005923756

Springer is a part of Science + Business Media.

springeronline.com

Printed in Japan

Cover concept: eStudio Calamar Steinen
Cover production: *design & production* GmbH, Heidelberg

Printed on acid-free paper SPIN: 11406860 57/3141/di - 5 4 3 2 1 0

Preface

More than thirty years have past since the publication of *X-Ray Diffraction by Polymers* by the authors (original edition in Japanese published by Maruzen, Tokyo, 1968, and English edition by Kodansha-Elsevier, Tokyo-Amsterdam, 1972).

Since then, accelerated by the very rapid and the remarkable development of electronic computers, both hard- and software as well as new experimental techniques, marvelous developments have been achieved on X-ray sources, X-ray detectors, X-ray apparata in experimental technique and methods of identification of crystalline materials, methods of structure solution and structure analysis of small and macromolecular crystals X-ray diffraction data, presentation of the results obtained, *i.e.* crystal and molecular structures, and the crystallographic databases related to them.

Today, we can see the words, "'mail-in' crystallography" in scientific journals, referring to sending sample(s) with appropriate information to an institution by mail, making it possible to obtain the structure of a complex biological molecule by mail within days ('mail-in' service). However, the authors believe that there are many scientists who are not satisfied with the results obtained by conventional analysis and wish to try to find a way to obtain more detailed structural information on macromolecules or high polymers by themselves based on the fundamentals of X-ray diffraction.

The present volume is divided into three parts as in earlier editions: fundamental, experimental and analytical. In the fundamental part, X-ray small-angle scattering is more precisely described in Chapter 6. In the experimental part, recently developed devices and the latest version of X-ray instruments equipped with these detectors are described. On the other hand, for the basic understanding of X-ray diffraction, descriptions and usages of rather old X-ray instruments are also given. In the analytical section, in addition to the structure analysis of high polymers, a new introduction has been added on the crystallization and structure determination of biological macromolecules in Chapter 13.

In this way, it is hoped that whichever section the reader turns to, depending on the research field, knowledge and experience, a contribution will be made. That is to say, the volume is intended as an intermediate textbook bridging the gap between beginners and specialist workers. Explanatory daigrams have been planned as far as possible to provide an intuitive understanding, and in the description of the equipment and methods it is shown how these are adapted to suit the aims of the analysis. The procedure adopted in the analytical part is to advance from the simple to the complex, starting with analyses of crystalline diffraction spots, amorphous haloes, broadening of diffraction spots, and overall background scattering, then concentrating of X-ray diffraction pattern, and proceeding eventually to composite analytical methods constructed from these individual analyses. Also, many pages have been allotted to examples from original works in order to facilitate the practical application of the analysis for the less experienced. It is hoped that these examples will serve as a further step beyond the level of most primers, but without obscuring the forest for the trees.

Most of the revised manuscripts were finished at the end of 1994. However, the homes of both authors were severely damaged by the heavy earthquake which hit Kobe and nearby areas in Japan on 17 January, 1995, causing delay in the publication of the monograph. Last year minimum revisions were again made on the manuscripts and pictures of the latest instruments were included in the experimental part of the work.

The painstaking work of the staff of Kodansha Ltd. in preparing the text is gratefully acknowledged. One of the authors (N.K.) deeply thanks the late Mrs. Maria Hiroko KASAI for her assistance in the initial stages.

Kobe
February 2005

Nobutami KASAI

Masao KAKUDO

Contents

Part III Analytical

Part I

Fundamental

X-ray diffraction patterns of substances in various states are tabulated overleaf. Before going on to the analysis of these patterns, this part deals first with X-rays (Chapter 1) and X-ray diffraction by matter (Chapter 2), and then with how these patterns (except for the patterns obtained by the dynamical scattering of X-rays, which are outside the scope of this volume) are obtained by means of X-ray diffraction (Chapters 2 to 5). The essence of small-angle scattering is described in Chapter 6. Chapter 7 briefly summarizes the structure of high polymer substances.

State of specimens (dirfracting matter) and X-ray diffraction patterns

Crystalline (including liquid crystal)	**Single crystal** [†1] Inorganic, Organic Organo-metallic molecules, Metal and Macro-molecules	**LARGE single crystal** (*perfect* or *ideal crystal*)	**DYNAMICAL Sacttering** ***Pendellösung Fringe*** and *other patterns* 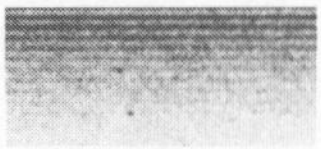 (a)	**Scattering intensity** $I \propto \lvert F \rvert$ F : Structure factor Perfection of crystal
		SMALL single crystal (< 10 μm) (*mosaic crystal*) 	**KINEMATICAL Scattering** ***Oscillation photograph*** 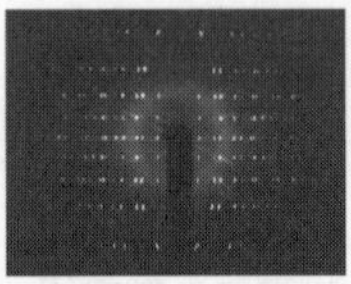 (b$_1$) (*Cyliydrical film*) ***Precession photograph*** 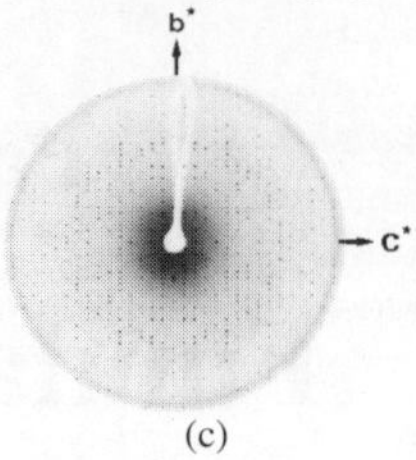 (c) (Flat film)	**Scattering intensity** $I \propto \lvert F \rvert^2$ F : Structure factor **Crystal structure** Molecular structure Distortion of atomic arrangement Thermal motion, Statistical disorder, etc.
	Poly-crystalline Inorganic, Organic, Organo-metallic molecules, Metal, Macro-molecules Liquid crystals, and Gels	**Oriented** 	***Fiber diagram.*** 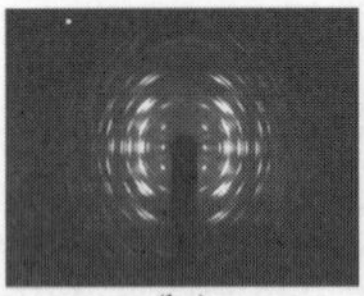 (b$_2$) (*Cylindrical film*) 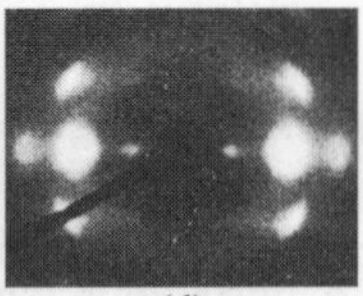 (d) (Flat film) 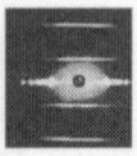 (e) (*Cyl. film*)	**Crystal slructure** Molecular structure **Preferred orientation of crystallites** (type and degree) **Size and shape and/or Imperfections of crystallites** ***Higher order structure*** (XSAS) [†2]  (j)  (k$_1$)  (k$_2$)
		Un-oriented 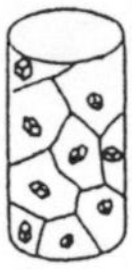	***Debye-Scherrer rings*** 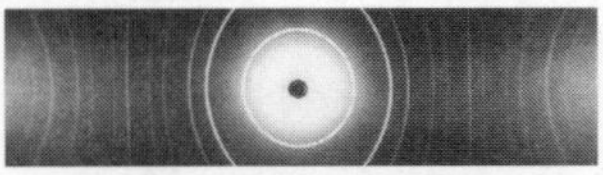 (f$_1$) Crystallite size : **moderate**  (f$_2$) Crystallite size : **very large** (*Cylindrical film*)	**Identification of crystals** **Crystal structure** *Rietveld method. etc.* **Size and shape and/or Imperfections of crystallites** **Crystallinity** ***Higher order structure*** (XSAS) [†2]  (l)
Non - crystalline	**Solid** ***Amorphous,*** and ***Glassy***	**Orineted** **Un-oriented**	***Amorphous halo***  (h) *Un-oriented* (Flat film) 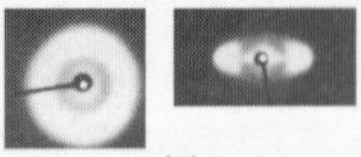 (g) *Oriented* (Flat film)	**Structural information** Short range order (*Radial distribution function*) ***Higher order structure*** (XSAS) [†2]  (m)  (n)
	Liquid ***Neat,*** and ***Solution***		***Halo***  (i) (Flat film)	**Molecular structure** (Small molecule) **Molecular shape in solution** (XSAS) [†2]  (o)

†1 Lamellar crystal (so-called polymer single crystal) is not included.
†2 X-ray small-angle scattering

1. Essential Properties of X-Rays

1.1 X-Rays as Electromagnetic Waves

X-rays, like light and ultraviolet radiation, are electromagnetic waves. Since the accompanying magnetlc field plays no role in X-rays diffraction, an X-ray may, for the purposes of this book, be pictured as an electric field oscillating in a single plane, where the plane is perpendicular to the direction of propagation (*i.e.* a polarized transverse wave), as shown in Fig. 1.1. X-ray wavelengths are generally considered to lie between 0.1 Å and about 100 Å [1 Å (Ångström) = 0.1 nm].

It is also possible to think of electromagnetic radiation as consisting of particle-like photons advancing with the velocity of light, and having energy $h\nu$, where ν is the frequency of oscillation and h is Planck's constant (6.626×10^{-34} Js). It is the wave-like properties of X-rays, however, which are of greater importance for diffraction phenomena. Fig. 1.1 shows the X-ray as a vibration in a single plane, but it is often helpful to regard an X-ray as having a wavefront with considerable lateral extension, like a wave on the sea shore.

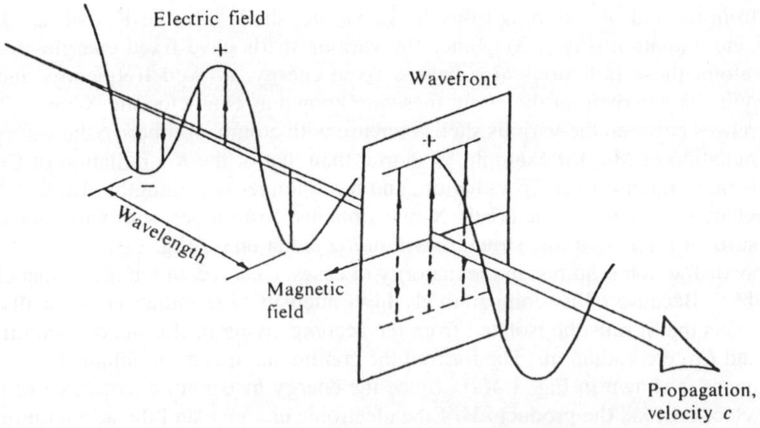

Fig. 1.1 X-ray propagation; polarized transverse electromagnetic wave.

1.2 Generation of X-Rays

1.2.1 X-ray tube

X-rays are produced when thermal electrons released from a heated filament under vacuum are accelerated towards, and eventually strike, a metal anode (anticathode) at a high potential (Fig. 1.2). The most important process for the generation of X-rays is the deceleration of these electrons as they penetrate matter. According to Maxwellian theory each change in velocity of a charged particle produces a continuous γ-spectrum. For electrons with energies of 5–100 keV, this so-called "Bremsspektrum" has the wavelength of X-rays. If, however, a bombarding electron should eject an electron from, say, the K shell of an atom, the resulting vacancy will be filled by an electron falling from a higher energy (L, M, etc.)

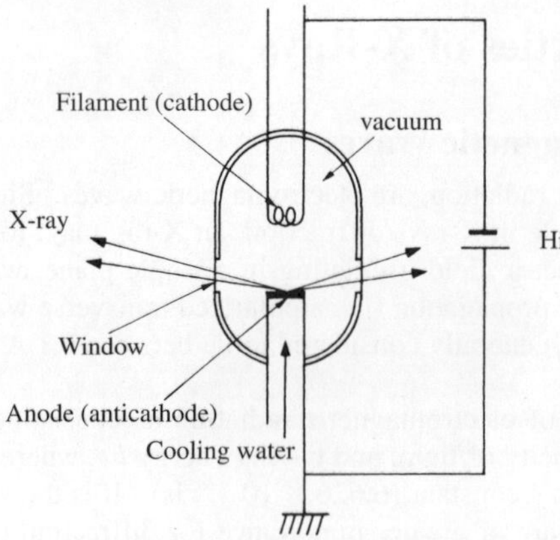

Fig. 1.2 Principle of the X-ray tube (cf. Fig. 8.1).

shell, the energy difference being radiated in the form of an X-ray photon. The X-rays resulting from the fall of electrons from the L, M, etc. shells into the K shell are known as $K\alpha$, $K\beta$, etc. radiation (Fig. 1.3). Since the various shells have fixed energies for a given type of atom, these radiations also have a fixed energy, a fixed frequency, and a fixed wavelength characteristic of the atom; these are known as characteristic X-rays. The energy differences between the various shells increase with atomic number so the wavelength of the $K\alpha$ radiation of Mo, for example, is shorter than that of the $K\alpha$ radiation of Cu. $K\alpha$ radiation is more intense than $K\beta$ radiation, and has a longer wavelength. Table 1.1[1)] shows the wavelengths of $K\alpha_1$ characteristic X-rays obtained from tubes with various anode metals: the suffix 1 refers to a fine structure within $K\alpha$ radiation (*cf* Fig. 1.3).

X-ray diffraction requires, in the majority of cases, a source which is as monochromatic as possible. Because of its comparatively high intensity, $K\alpha$ radiation is usually chosen, which means that it must be isolated from the accompanying background of continuous radiation and $K\beta$, etc. radiation. The form of the continuous spectrum obtained with different tube voltages is shown in Fig. 1.4(a). Since the energy $h\nu$ cannot exceed that of the bombarding electrons, *i.e.* the product eV of the electronic charge e and the accelerating voltage V, we have

$$h\nu_{\max} = eV = hc/\lambda_{\min} \quad \text{and}$$

$$\lambda_{\min} = 12.4/V \; (V \text{ in kV } (= 1000 \text{ volt units})) \tag{1.1}$$

Thus the radiation shows a continuous distribution from $\lambda_{\min}$ toward longer wavelengths.

The characteristic X-rays are generally superimposed on this continuous spectrum, as shown in Fig. 1.4(b), but the same wavelengths limit applies: unless $\lambda_{\min}$ is less than the wavelength of, say, the $K\alpha$ line, this line cannot appear. It follows that there are critical

Table 1.1 $K\alpha$ wavelength for various anode (anticathode) metals[1)]

Anode	Cr	Fe	Co	Cu	Mo	Ag
Wavelength, $K\alpha_1$ [Å]	2.289877	1.936306	1.789188	1.54059292(45)	0.70931713(41)	0.55942177(76)

[Reproduced with permission from *Internatinal Tables for Crystallography*, Vol. C, 2nd ed., (A. J. C. Wilson, E. Prince eds.), International Union of Crystallography, D. Reidel Pub., Dordrecht (abbreviated as IUCr, hereafter) (1999)]

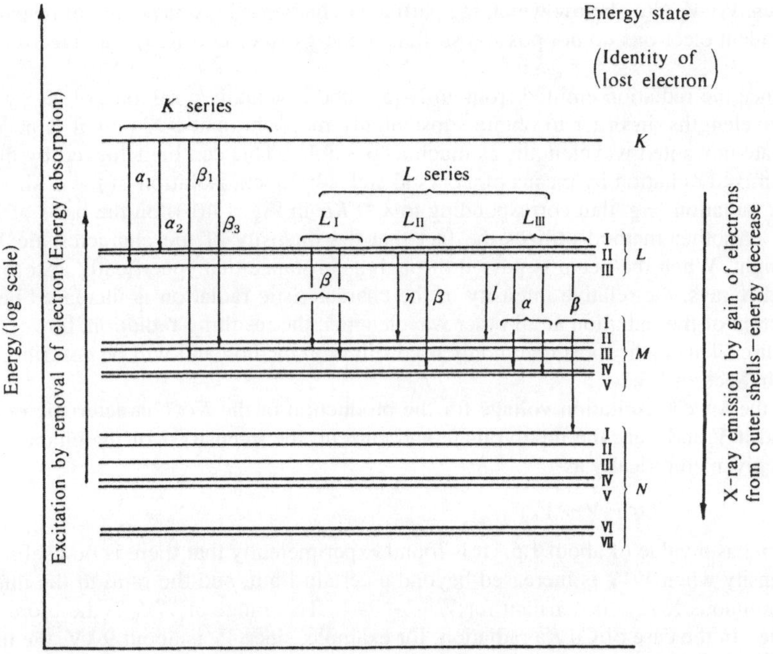

Fig. 1.3 Orbital electron energies and characteristic X-rays.

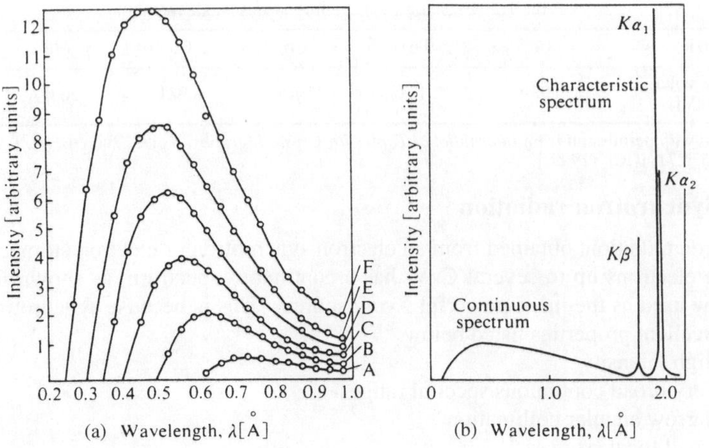

Fig. 1.4 Typical X-ray spectral intensity distributions.
(a) Continuous spectrum for various tube voltages(Tungsten anode).[2] A: 20 kV, B: 25 kV, C: 30 kV, D: 35 kV, E: 40 kV, F: 50 kV.
(b) Characteristic X-rays superimposed on the continuous spectrum. Intensity ratio of continuous spectrum to characteristic is about 1 : 100 for Cu radiation at tube voltage 40 kV.
$I(K\alpha_1) : I(K\alpha_2) : I(K\beta) \fallingdotseq 100 : 50 : (20\text{–}30)$.
[Reproduced from C. Urey, *Phys. Rev.*, **11**, 401, Am. Inst. Phys. (1918)]

voltages, V_K, V_L, etc., below which the particular charateristic X-rays cannot be excited, *i.e.* the incident electrons do not possess sufficient energy to eject a K, L, etc. electron from its orbit.

Since the radiation emitted from an X-ray tube is actually a mixture of X-rays of various wavelengths, in order to obtain substantially monochromatic X-rays it is necessary to eliminate unwanted wavelengths as much as possible. This can be achieved by diffraction of the mixed radiation by means of a crystal and subsequent isolation of the desired characteristic radiation (*e.g.* that corresponding to $\lambda = K\alpha$ in Fig. 1.4(b)) on the basis of its orientation. Another method utilizes the fact that the intensity of the characteristic X-rays is very high. When the beam is passed through a substance that specifically absorbs the unwanted X-rays, the relative intensity of the characteristic radiation is increased because of absorption of the radiation at all other wavelengths; the resulting radiation, however, is still a mixture. This is the most commonly used filtration method and will be described in more detail in Section 1.3.2.

If the lowest excitation voltage for the production of the $K\alpha$ characteristic radiation is V_K, and if V and i are the tube voltage and current, the intensity I of the characteristic X-rays is given empirically as

$$I \propto i\,(V - V_K)^n \tag{1.2}$$

where n has a value of about 1.5. It is found experimentally that there is no useful increase in intensity when V/V_K is increased beyond a certain limit, and the ratio to the intensity of the continuous X-rays is constant for $V/V_K = 3$–4. This range of V/V_K is therefore the most suitable. In the case of Cu *K*a radiation, for example, since V_K is about 9 kV, the most suitable effective (r.m.s.) voltage should be less than 36 kV and the most suitable peak voltage less than 45 kVp. Some important V_K values are given in Table 1.2.[1)]

Table 1.2 Excitation voltage for various anode metals[1)]

Anode	Cr	Fe	Co	Cu	Mo	Ag
Excitation voltage, V_K [kV]	5.989	7.111	7.709	8.981	20.00	25.60

[Reproduced with permission from *International Tables for Crystallography*, Vol C. 2nd ed., (A. J. C. Wilson, E. Prince eds.), p.71, IUCr. (1999)]

1.2.2 Synchrotron radiation

Synchrotron radiation, obtained from an electron synchrotron or electron storage ring by accelerating electrons up to several GeV, has a continuous spectrum, as shown in Fig. 1.5[3)], and is now used as the more powerful X-ray source. This is because synchrotron radiation has the excellent properties listed below.[4a)]

1. High intensity
2. Very broad continuous spectral range
3. Narrow angular collimation
4. Small focal size
5. High degree of linear/circular polarization
6. Regularly pulsed time structure
7. Ultra-high vacuum environment
8. Computability of properties

However the use of synchrotron radiation is limited to laboratories with available synchrotron sources (Fig. 1.6).[4b)]

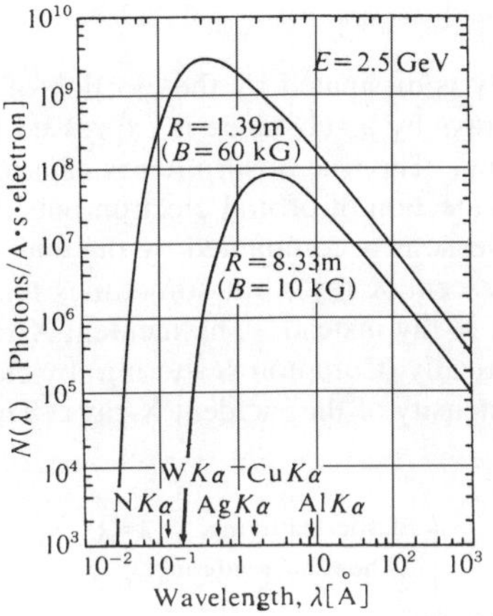

Fig. 1.5 Intensity distribution of a synchrotron radiation at an accelerating voltage $E = 2.5$ GeV.[3]
R: radius of orbital [m] and *B*: strength of superconducting bending magnet [kG].
[Reproduced with permission from K. Kohara, S. Kikuta, *X-Ray Diffraction Techniques*, 61, Tokyo Univ. Press (1979)]

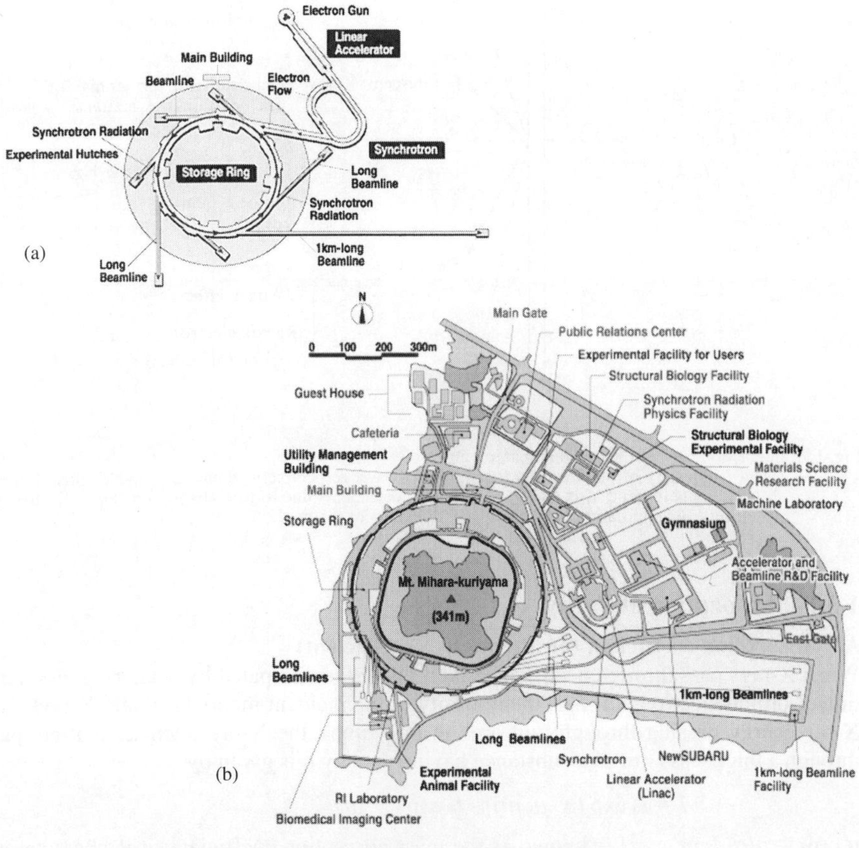

Fig. 1.6 SPring 8, Harima, Hyogo, Japan[4b]. (a) Plan of the SPring 8.
(b) Plan of the Storage Ring, the Beam lines and other facilities.
[Reproduced with permission from SPring 8 (2003)]

1.3 Properties and Effects of X-Rays

When X-rays pass through a substance their energy is dissipated by the ejection of orbital electrons and by scattering. The absorption of X-rays by a substance is expressed by true absorption and apparent absorption due to scattering. The true absorption is caused by an atom as a result of absorption of X-ray energy and ejection of orbital electron outside(photoelectric effect). Ordinarily this photoelectron release is accompanied by the emission of X-ray characteristic of the atom (Fig. 1.3) (fluorescent X-ray), but sometimes an Auger electron is ejected without emitting the fluorescent X-ray instead. The incident X-rays are scattered elastically (Thomson scattering) or inelastically (Compton scattering) by electrons in the substance, which results in decrease of the intensity of the incident X-rays. These effects are summarized in Fig. 1.7.

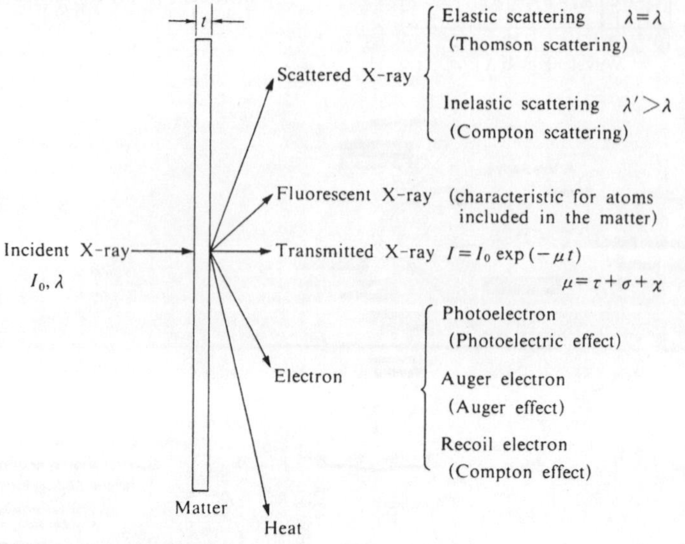

Fig. 1.7 Interaction between X-ray and matter.
I_0 and I: intensity of incident and transmitted X-rays, respectively; λ and λ': wavelengths of X-rays; μ: absorption coefficient; τ, σ and χ: absorption coefficients due to true absorption (photoelectron effect), scattering and annihilation, respectively; t: thickness of the matter.

1.3.1 Absorption of X-rays

A. Absorption coefficient (or attenuation coefficient)

When X-rays pass through a substance their energy is dissipated by scattering and the ejection of orbital electrons. If I_0 is the intensity of the incident monochromatic X-rays (*i.e.* the X-ray energy passing through unit area in unit time), the X-ray intensity I after passage through a thickness t cm of a substance having density ρ is given by

$$I = I_0 \exp(-\mu_m \rho t) = I_0 \exp(-\mu t) \tag{1.3}$$

μ_m (or $= \mu/\rho$) [$cm^2 \cdot g^{-1}$] is known as the mass absorption coefficient: it is characteristic of the substance and also varies with wavelength for a given substance. Reference books quote values for the so-called linear absorption coefficient μ rather than the mass absorption coefficient μ_m (above); this is the reciprocal of the thickness of substance which attenuates the beam by a factor e, and is equal to $\mu_m\rho[cm^{-1}]$ in Eq. 1.3. Since μ_m is often required in X-ray work, a number of values are given in Table 1 of the Appendix.

If the chemical constitution $A_{n_1}B_{n_2}C_{n_3}...$ of a substance is known, the absorption by this substance can be found as follows. The mass absorption coefficients μ_{mA}, μ_{mB}, μ_{mC}, etc. of the various elements for the wavelength of the X-rays used are first found from the table, and the mass absorption coefficient of the substance is then calculated by adding these in accordance with the weight ratios of the elements present. Thus if A_A, A_B, A_C, etc. are the atomic weights of the elements A, B, C, etc.,

$$\mu_m = \frac{n_1 A_A \mu_{m_A} + n_2 A_B \mu_{m_B} + \cdots}{n_1 A_A + n_2 A_B + \cdots} \tag{1.4}$$

B. Discontinuous nature of absorption

It is evident from Table 1 of the Appendix that the variation of absorption with increasing atomic weight is not monotonic for a given wavelength, but shows sudden changes, and that discontinuous variations are also to be expected for a given element with varying wavelengths. This is shown much more clearly by the curves of experimentally determined absorption coefficients in Fig. 1.8.[5] The discontinuities are known as absorption edges. The abrupt changes in absorption are due to the loss of energy associated with the generation of characteristic X-rays by the atoms of the substance. The wavelengths λ_K, λ_L, etc. of the absorption edges thus correspond to the excitation voltages V_K, V_L, etc. of the characteristic X-rays (Table 1.2). One therefore speaks of *K*, *L*, etc. absorption edges which correspond to the *K*, *L*, etc. characteristic radiation of the element.

The absorption curve of X-rays by atom shows a fine structure in the range down to several hundreds eV from the absorption edge. This is called extended X-ray absorption fine structure (EXAFS), which can be used to obtain information about the spatial arrangement of atoms neighboring the atom.

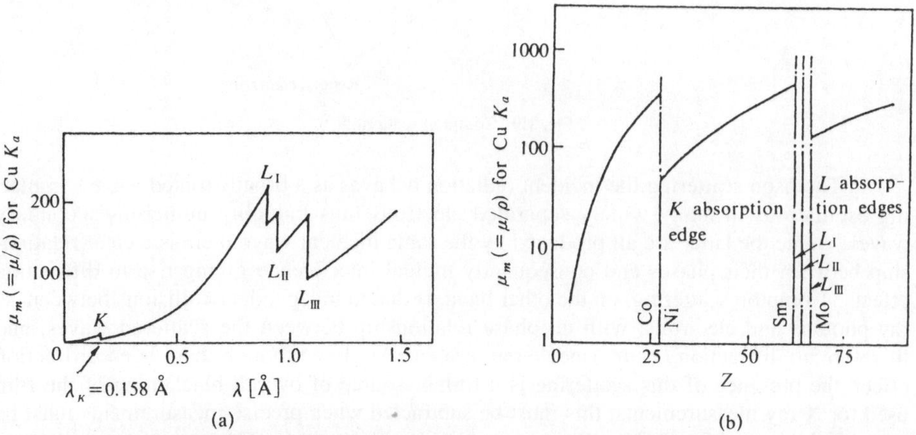

Fig. 1.8 Experimental absorption coefficients.[5]
(a) Mass absorption coefficients μ_m of Pt *vs.* wavelength λ.
(b) Mass absorption coefficients μ_m for Cu $K\alpha$ ($\lambda = 1.54$ Å) *vs.* atomic number *Z*.
[Reproduced with permission from A. Guinier, *Théorie et Technique Radiocristallographie* p.14, Dunod (1960)]

1.3.2 X-ray scattering

A. Thomson scattering

Since X-rays are electromagnetic waves, it is not surprising that they cause vigorous vibration of the shell electrons of the atoms of substances through which they pass. When electrically charged particles such as electrons are accelerated, secondary radiation is always emitted. Because the secondary emission is stimulated by the oscillating electric field of the incident X-rays, it is synchronous with it, consisting therefore of X-rays having the same frequency and wavelength as the incident X-rays. This scattering without change in wavelength is known as Thomson scattering and is the main type of scattering involved in X-ray diffraction (coherent scattering).

B. Compton scattering

X-rays also behave as particle-like photons having momentum $h\nu/c$: momentum is transferred when these particles collide inelastically with electrons, and the radiation is converted into X-rays with lower energy (longer wavelength). Since the energy change varies with the collision conditions, these scattered X-rays have a continuous spectrum at wavelengths greater than that of the incident X-rays. This scattering is known as Compton scattering and is of historical importance as experimental evidence for the dual wave/particle nature of X-rays. Fig. 1.9 shows the geometry of such a collision. Since these scattered X-rays do not give rise to diffraction, this type of scattering is of minor importance in the present volume.

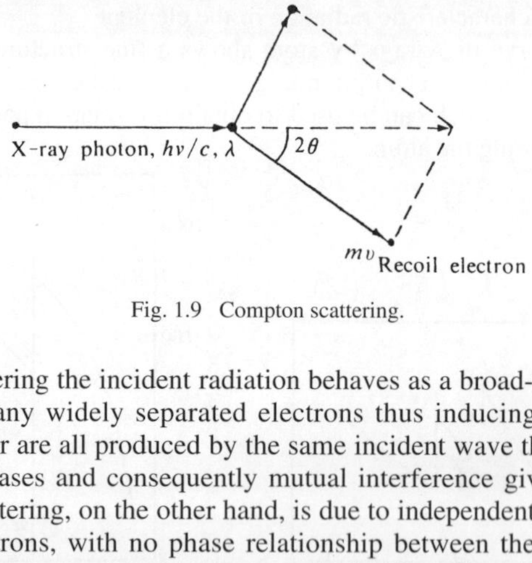

Fig. 1.9 Compton scattering.

In Thomson scattering the incident radiation behaves as a broad-fronted wave stimulating oscillations in many widely separated electrons thus inducing numerous secondary waves. Since the latter are all produced by the same incident wave there is a clear relationship between their phases and consequently mutual interference giving rise to diffraction effects. Compton scattering, on the other hand, is due to independent collisions between X-ray photons and electrons, with no phase relationship between the scattered waves, and therefore no diffraction occurs (incoherent scattering). Even though there is no diffraction effect, the presence of this scattering is a further source of overall blackening of the film used for X-ray measurements; this must be subtracted when precise measurements must be made of the intensities of the diffracted waves in calculating the Thomson scattering intensities. Special care is necessary in measuring the diffuse halos from amorphous regions. The intensities will be discussed quantitatively in the section on the scattering of X-rays by electrons.

1.3.3 X-ray refraction

It can be shown theoretically that all substances have refractive indices extremely close to unity for short wavelength electromagnetic waves such as X-rays. If the frequency of the incident radiation is considerably greater than the critical absorption frequencies of the scattering atoms, the refractive index n for X-rays of wavelength λ, be given as[6a)]

$$n = 1 - \delta$$

$$\delta = \frac{e^2}{2\pi mc^2 m_H A} \frac{Z}{A} \rho\lambda^2 \tag{1.5}$$

where e and m are charge and mass of the electron, c the velocity of light, m_{H} the mass of the hydrogen atom, Z the total number of electrons in the atom, A the atomic weight of the scattering atoms, and ρ the density of the medium. For Mo $K\alpha$, $\lambda = 0.70926$ Å, and $\delta = 1.36 \times 10^{-6}\ \rho Z/A$. Since Z/A is always about one-half, the value is usually so small $(10^{-5} - 10^{-6})$ that it can be neglected in practice. This means that lenses cannot be constructed for X-rays which thus, unlike light, cannot readily be made to converge or diverge. However, if the X-ray hits the surface of a solid material at a very minute angle, the total reflection of the X-ray may be observed. The critical angle is[6b)]

$$\theta_c = \sqrt{2(1-n)} = \sqrt{2\delta} \tag{1.6}$$

θ_c has a value of the magnitude of 10 – 30 min; the longer the wavelength of the X-ray the larger the θ_c. Total reflection from a curved surface may be used to focus the X-ray beam (Section 8.2.5D).

1.3.4 Effects used for the detection of X-rays

A. Photographic effects and photostimulated luminescence

1) Photographic effect. The exposure of photographic film to X-rays is widely used in fields ranging from medical diagnosis to the inspection of various materials in industry, and is of course a principal means for the observation of X-ray diffraction patterns. Here we shall simply describe the characteristics of film specially produced for X-ray applications.

Owing to the high penetrating power of X-rays, more than 50% of the radiation having wavelengths about that of Cu $K\alpha$ passes through a normal photographic film with little effect on the silver grains. The bases of X-ray films are therefore coated fairly thickly on both sides with a high-sensitivity emulsion. Films of this type, which are intended primarily for recording X-ray transmission patterns, are very sensitive, but their nonlinear energy/sensitivity characteristic curve renders them far from ideal for the measurement of X-ray intensities (cf. Section 8.2.2.C.a).

2) Photostimulated luminescence.[7)] Photostimulable phosphor such as $BaFBr : Eu^{2+}$ shows "photostimulated luminescence" (PSL). That is, when the phosphor is exposed to X-rays it stores a fraction of the absorbed X-ray energy temporarily in the form of quasistable color centers, and when they are later stimulated by visible light, they emit PSL with an intensity proportional to the number of X-ray photons absorbed (*cf.* Section 8.2.2. B.1)).

B. Ionization

When X-rays are passed through a gas energetic photons ionize it by stripping electrons from the gas atoms in their path. These electrons and the still energetic X-ray photons lose further energy by subsequent collision with other atoms and the formation of numerous ion pairs. This ionizing ability of X-rays can be exploited to measure the radiation dose. The X-rays are passed into a sealed chamber containing a gas (usually Ar, Ne, or Xe). The electrons produced are captured by an electrode of high potential (about 1,000–2,000 V) inside the chamber, and an electric current flows. This current is produced by a number of electrons that corresponds to the energy of the incident photons, and is therefore a measure of the radiation dose. If the body of the ionization chamber is the negative electrode and the positive electrode is a fine wire in the center of the chamber, the ejected electrons are accelerated towards the wire and ionize many more atoms on their way, so that an avalanche of electrons converges on the electrode and produces a large instantaneous current. The amount of ionization, hence the intensity, may be measured after amplification of these electric pulses. This is the principle of the ionic chamber and the proportional and Geiger-Müller (GM) counters. Earlier, GM counters were used in most X-ray diffraction apparata. Now however, proportional counters are widely used because they have the following advantages over GM counters: the pulse height observed is proportional to the energy of the incident X-ray photons, the dead time is much shorter, and the counting loss is very small up to a high counting rate of $10^4 - 10^5$ cps (*cf.* Section 8.2.2.A and B).

When X-rays are irradiated to a lithium drifted silicon or germanium semiconductor, Si(Li) or Ge(Li), or high-purity germanium semiconductors the energetic photons absorbed by the semiconductor ejects photoelectrons of high speed, which form numerous ion pairs, electrons and positive holes. The number of ion pairs formed is proportional to the energy of X-ray photons. The electrons and positive holes produced are swept toward the anode and cathode respectively under a high potential, and the flow of an electric pulse current can be detected.[8] In order to stabilize the semiconductor, increase the mobility of electrons and positive holes, and minirnize the dark current due to the thermal formation of the ion pairs the semiconductor must be kept at low temperatures (*cf* Section 8.2.2.A.b).

C. Fluorescence

Although X-rays are not directly visible they can be observed indirectly as a yellow to violet fluorescence with the aid of a fluorescent screen. The substances most commonly used are CaS, ZnS, and $CaWO_3$, as these give a strong yellow fluorescence. If a fluorescent screen is fixed to the back of a film, the X-rays passing through it produces fluorescence, and the film emulsion is affected by this as well as directly by the X-rays. Screens giving violet fluorescence are widely used in this way to increase sensitivity because of the high actinic value of fluorescence in this wavelength region.

The widely used scintillation counters utilize fluorescence (scintillation) to detect X-rays and other radiation by means of scintillators and photomultipliers (Section 8.2.2.A).

The X-ray image-orthicon used in the X-ray television for the real time observation of X-ray topographs or intensity data collection of single crystals is another example of the application of this fluorescence effect (*cf.* Section 8.2.2.C.d).

1.3.5 Other effects

A. Chemical effects

Because of the high energy of X-rays, they are capable of decomposing substances by breakdown of chemical bonds, and also of synthesizing substances by the formation of new bonds. It has long been known, for example, that a solution of iodoform in chloroform turns violet upon X-irradiation due to the liberation of iodine.

The decomposition and polymerization of macromolecular substances by radiation is widely used in industry at the present time. X-rays resemble both ultraviolet (UV) and β

radiation in producing many characteristic chemical effects, but as these fall outside the scope of this book they will not be detailed here. However, if we admit the possibility of changes in the sample itself during the course of X-ray diffraction studies, this presents important problems: one must always bear in mind that X-rays can have pronounced chemical effects.

B. Biological damage

Living organisms are complex assemblages of organic compounds with proteins as the main constituents, and are therefore susceptible to the chemical action of X-rays. Moreover, since the living body is the seat of a delicate system of chemical reactions, even slight chemical damage due to X-rays can disturb the entire physiological function of the organism. The risk of burns due to X-rays is well known, but also, because of malfunctions that can result from unknown metabolic changes, the investigator must always beware of the X-ray beam.

References

1. *International Tables for Crystallpgraphy*, Vol. C, 2nd ed., (A.J.C. Wilson, E. Prince eds.), Mathematical, Physical and Chemical Tables, pp.206-211, IUCr, Kluwer Academic Pub., Dordrecht (1999).
2. C. Urey, *Phys, Rev.*, **11**, 401 (1918).
3. K. Kohra, S. Kikuta, *X-Ray Diffraction Techniques*, p. 61, Tokyo Univ. Press (1979) (in Japanese).
4. a) *Synchrotron Radiation* (Phys. Soc. Jpn. ed.), Baifukan (1986).
   b) URL. http://www.spring8.or.jp/e/general.info/overview/
5. A. Guinier, *Thérie et Technique de la Radiocristallographie*, 3rd ed., p.14, Dunod Editeur, Paris (1964).
6. a) R.W. James, *The Optical Principles of the Diffraction of X-Rays*, p.54, G. Bell & Sons, London (1948).
   b) In ref. 6a) p. 172.
7. a) M. Sonoda, M. Takano, J. Miyahara, H. Kato. *Radiology*, **148**, 833 (1983).
   b) Y. Amemiya, J. Miyahara. *Nature*, **336**, 89 (1988).
   c) J. Miyahara, K. Takahashi, Y. Amemiya, N. Kamiya, Y. Satow, *Nuclear Instr. Methods*, **A246**, 572 (1986).
8. A. F. H. Muggleton, *Nuclear Instr. Methods*, **101**, 113 (1972).

Captions to and sources of the X-ray diffraction patterns cited in "State of specimens (diffracting matter) and X-ray diffraction patterns"

(a) Pendellösung fringes appeared in the traverse pattern of the 020 diffraction of Taurine (2-Aminoethylsulfonic acid) large crystal [Dynamical scattering] (Flat dry plate)
[Reproduced with permission from I. Suzuki, *et al.*, *Sulfur Amino Acids*, **4**, 254, Research Society for Sulfur-containing Amino Acids (1981)] (in Japanese).

(b_1)Rotation photograph of a single crystal like Poly(ethylene oxide)-Urea complex (Rotation about the *c* axis, *cf.* Figs. 7-13, 10-8 (a)). (*Cylindrical film*)

(b_2)Fiber diagram of drawn Poly(ethylene oxide)-Urea complex (draw direction vertical, *cf.* Fig.10.8 (b)). (*Cylindrical film*)

(b_1) (b_2) : [Reproduced with permission from H. Tadokoro, *et al.*, *J. Polym. Sci.* **B2**, 363, John Wiley & Sons, Inc. (1964)].

(c) Precession photograph of Cytochrome *c'* from *Rhodospirillum rubrum, h0l* diffractions. (Flat film)
[Reproduced with permission from M. Yasui, *et al.*, *J. Mol. Biol.*, **177**, 845, Academic Press (1984)]

(d) Fiber diagram of Silk fibroin (*Bombyx mori*). (draw direction vertical). (Flat film)
[Reproduced with permission from Y. Takahashi, *Silk Polymers, Material Science and Biotechnology, ACS Symposium Ser.*, **544**, 168, Am. Chem. Soc. (1994)]

(e) Fiber diagram of Poly (β-propiolactone) II. (draw direction vertical). (*Cylindrical film*)
Discrete diffractions have been observed on the equator but *streaks* on every layer line.
[Reproduced with permission from K. Suehiro *et al.*, *J. Polym. Sci.*, **7**, 352, John Wiley & Sons, Inc. (1975)]

(f_1) Debye-Scherrer patterns of five randomly oriented crystals of fluorite. (*Cylindrical film*)
(f_2) Debye-Scherrer patterns of fluorite powder. (*Cylindrical film*)
(f_1), (f_2) : [Reproduced with permission from F. A. Bannister, The Powder Method in Mineralogical Research in *X-Ray Diffraction by Polycrystalline Materials* (H. P. Peiser, H. Rocksby, A. J. C. Wilson eds.), p.526, Chapman and Hall (1960)]
(g) Amorphous halo of poly (ethylene terephthalate) fibers showing preferred orientation. (Flat film)
[Reproduced with permission from W. O. Statton, Characterization of Polymers in *Handbook of X-rays for Diffraction, Emission, Absorption and Microscopy* (E. F. Kaelbe ed.), p.21-15, McGraw-Hill (1967)]
(h) X-ray diffraction halo of a linear eicosamer of a tripeptide, $(PPG)_{20}$. (Flat film)
[M. Kakudo *et al.* Unpublished data]
(i) X-ray diffraction halo of water. (Flat film)
[Reproduced with permission from D. L. Hukins, *X-Ray Diffraction by Disordered and Ordered Systems, Covering X-Ray Diffraction by Gases. Liquids and Solids and Indicating How to the Theory of Diffractions by These Different State of Matter is Related and How It Can be Used to Solve Structural Problems*, p.66, Pergamon Press (1981)]
(j) X-ray small-angle diffraction pattern of rat sciatic nerve. (Flat film)
[N. Kasai, M. Kakudo, Unpublished data]
(k_1) X-ray small-angle scattering patterns of polyethylene. (Flat film)
(k_2) X-ray small-angle scattering patterns of polyethylene. (Flat film)
(k_1), (k_2) : [Reproduced with permission from W. O. Statton, Characterization of Polymers in *Handbook of X-rays for Diffraction, Emission, Absorption and Microscopy* (E. F. Kaelbe ed.), p.21-15, McGraw-Hill (1967)]
(l) Schematic X-ray small-angle scattering pattern of amorphous solid. (Flat film)
(m) X-ray small-angle scattering pattern of a silica gel. (Flat film)
[N. Kasai, M. Kakudo, Unpublished data]
(n) X-ray small-angle scattering pattern of a vanadium sol (Würtz sol, in glass capillary). (Flat film)
[N. Kasai, M. Kakudo, Unpublished data]
(o) Schematic X-ray small-angle scattering pattern of a solution. (Flat film)

2. X-Ray Scattering, Interference and Diffraction

The preceding chapter dealt with the general properties of X-rays that are essential to diffraction. This chapter proceeds to give a more detailed account of X-ray scattering and diffraction, and presents the principles on which the practical observation and analysis of diffraction phenomena are based.

Section 1.3.2 detailed the various types of scattering that occur when X-rays strike an object (atom or electron). Fig.2.1 illustrates the scattering of X-rays by a single electron. The following conventions form the basis of the scattering theory presented in this chapter.

1. The wavelengths of both the incident and scattered X-rays are identical (Thomson scattering). Thus, in both the following theory and experimental practice, fixed wavelength is always used.

2. The scattered X-rays are propagated over the whole of space. When the distribution of intensities is anisotropic, a particular direction is identified by a suitable angular parameter, the direction of the incident radiation being taken as the datum. When only one scattering angle is involved, as in Fig.2.1, it may be adequately specified by its magnitude 2θ. In general, however, the direction in three-dimensional space with reference to the datum is expressed by a vector.

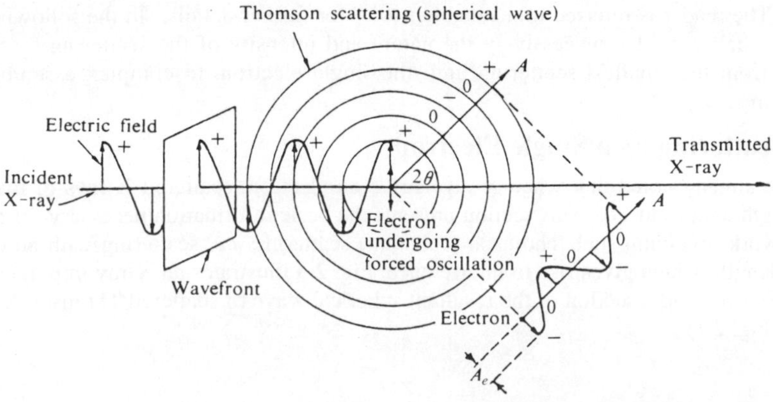

Fig. 2.1 Propagation of a spherical wave with amplitude A_e in direction A.

3. All the scattered X-rays arriving at a given observation point are parallel, just as the incident X-ray beam is taken to be parallel. This condition excludes observation points at a distance from two or more scattering points that is comparable with the separation between the points; in this case the scattering angle would differ for each point ($2\theta_1$, $2\theta_2$, etc., in Fig. 2.2). The observation distance should therefore be thought of as effectively infinite, so that $2\theta_1$, $2\theta_2$, etc., are all equal despite the divergence of the scattered rays in the immediate vicinity of the scattering points.

4. Scattered X-rays interfere with other scattered X-rays, but do not undergo any further interactions with other atoms. This condition is, in effect, an assumption that the X-rays from a given atom (which spread out over all space and so necessarily strike other atoms in the vicinity) will only induce secondary scattering interactions with other atoms to

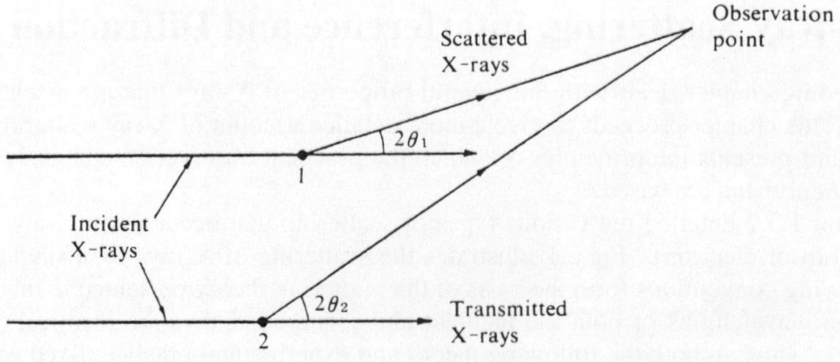

Fig. 2.2 The effect of a non-remote observation point.

a negligible extent (**Kinematical theory** of diffraction).

Condition 3. is widely known as the Fraunhofer approximation. Conditions 3. and 4. are particularly fundamental to the **kinematical theory** of diffractions as presented in this volume. With a few, rare exceptions, they may be taken as valid in practice for the applications treated. Condition 4. is vital to the **dynamical theory** of diffraction, which is not dealt with in this volume, although it is important in both X-ray and electron diffraction theory. The reader is referred to a suitable text[1–6] for further details. In the following sections we shall consider successively the nature and intensity of the scattering by objects ranging from the smallest scattering unit, the single electron, to complex assemblies of many atoms.

2.1 Scattering by a Single Electron

We have already noted that when X-rays strike an electron, scattered X-rays of the same wavelength are produced. This section presents the basic information necessary for experimental work involving such Thomson (coherent) scattering, *i.e.* scattering with no change in wavelength, which gives rise to interference. Fig. 2.3 illustrates an X-ray impinging on a single electron, and a section of the resultant spherical wave of scattered Thomson X-radiation (*cf.* Fig. 2.1).

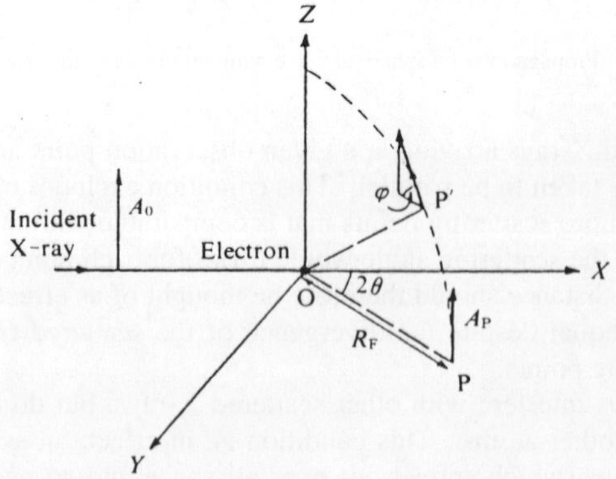

Fig. 2.3 X-ray coherent (Thomson) scattering by a free electron.

Let the electric field that accompanies the wave motion of the incident X-ray have an amplitude A_0 and a direction parallel to OZ. The corresponding amplitude of the scattered spherical wave at a point in the XY plane P, at a distance R_F from the electron, is

$$A_P = A_0 \frac{1}{R_F} \frac{e^2}{mc^2} \tag{2.1}$$

where m is the mass of the electron, e its charge, and c the velocity of light. The angle 2θ shown between OP and OX is at this stage immaterial, but see Eq.2.3 below.

Since the intensity of the wave, *i.e.* the energy passing through unit area per unit time, is proportional to the square of the amplitude of the wave, the intensities I_0 and I_P of the incident X-ray and the scattered wave at P will be

$$I_P = I_0 \frac{1}{R_F{}^2} \frac{e^4}{m^2c^4} \tag{2.2}$$

Equations 2.1 and 2.2, based on the given restrictions upon angle of incidence, observation direction and plane of vibration, in fact give the maximum observable intensity for an X-ray scattered by a single electron at a distance R_F from the electron. In general, however, when the point P Iies on the spherical surface of the wave outside the XY plane (point P′), where A_0 and $A_p{}'$ are no longer parallel, the component normal to the XY plane is reduced to $A_p \sin\varphi$, where φ is the angle between A_0 and OP′.

The above scattering calculation also rests on the assumption that there is a single polarized incident wave. An actual beam, on the other hand, is not uniquely polarized as a whole, but consists of waves with all possible directions of polarization within a plane perpendicular to the propagation direction. The amplitude A_e and intensity I_e of the scattered X-rays then vary with the scattering direction as follows:

$$\begin{aligned} I_e = A_e{}^2 &= I_0 \frac{1}{R_F{}^2} \frac{e^4}{m^2c^4} \frac{1+\cos^2 2\theta}{2} \\ &= 7.90\times10^{-26} \cdot I_0 \frac{1}{R_F{}^2} \frac{1+\cos^2 2\theta}{2} \end{aligned} \tag{2.3}$$

Thus the amplitude and intensity of the X-rays scattered in a particular direction when an X-ray beam strikes a single electron can be found from Eq. 2.3.

A few words should be added at this point about Compton (incoherent) scattering, which does not give rise to interference.

In the case where X-rays of wavelength λ give rise to Compton scattering of wavelength λ' we may put $\lambda' = \lambda + \Delta\lambda$, where $\Delta\lambda$ corresponds to the increase in wavelength of the center of the Compton spectrum. Where $\Delta\lambda$ is small we have

$$\Delta\lambda = \frac{2h}{mc} \sin^2\theta = 0.024(1-\cos 2\theta) \tag{2.4}$$

where 2θ is the scattering angle.

The values of $\Delta\lambda$ are in fact very small, and the intensity is only a few percent that of the Thomson scattering. Compton scattering does not cause diffraction, but is observed as a weak background scattering over the whole angular range, and may be disregarded for the present. However, since this scattering must be subtracted in the quantitative measurement of the diffuse halos from amorphous material, its intensity will be discussed in more detail later (Eq. 2.17).

2.2 Interference and Diffraction of Scattered X-Rays

2.2.1 The phenomena of interference and diffraction

Reference has already been made to interference, diffraction, and the phase of a wave. We now proceed to a detailed exposition of the phenomenon which is essential to this volume, *i.e.* the diffraction of X-rays.

Figure 2.4 represents part of a large tank filled with water, on the surface of which wave crests, induced by the lateral oscillations of a vertical plate (shown to the left of the diagram), are traveling from left to right. Upon encountering two corks, 1 and 2, which have been fixed so that they are partially submerged, the energy of the wave train is partially scattered in the form of circular ripples of lower amplitude (crest height) but having the same frequency and wavelength as the original waves. The encounter with 1 and 2 replaces the plane parallel wave train with a complex wave system resulting from the superimposition of the two circular waves upon the incident wave.

If we now regard the plane wave arriving from the left of the diagram as the wave front of a single X-ray wave (*cf.* Fig.1.1), 1 and 2 as scattering electrons, and the circular waves as spherical waves spreading out from the electrons, the situation in the above diagram provides an analogue of the phenomenon that occurs in the plane containing OP and the incident ray in Fig. 2.3.

The formation of a composite wave by superimposition of two or more waves is known as interference. The incident wave, confined initially to a single direction of propagation, is found (due to the scattering effect of 1 and 2 and the resultant spherical propagation) to be deflected so that it propagates in new and other directions. This latter phenomenon is known as diffraction.

The principles of diffraction can readily be understood with the aid of the diagram. In Fig. 2.4, the lines denote wave crests, with the troughs lying in the spaces between the lines. Since the incident wave reaches 1 and 2 simultaneously, the crests of the scattered waves traveling in direction A_0 (parallel with the incident wave) coincide, as do also the troughs. This is evident from the diagram, which represents the wave motion at a particular instant, by considering a section along this direction. When the crests and troughs of two waves do not coincide, however, there is a difference in phase between them. This phase difference is normally expressed either in wave numbers (the ratio of wave displacement or path difference to wavelength) or in terms of the angular difference between the associated circular motions. Fig. 2.5 (p.20) shows the relationship between these quantities for waves in direction C, where there is a significant difference in phase. The wave from 1 leads that from 2 by ΔC, so the phase difference is $\Delta C/\lambda$ in wave numbers, or $2\pi\Delta C/\lambda$ in radians. Since one period (the interval within which a wave completes one cycle of oscillation) usually appears in the equations of wave motion as the angular period 2π, phase difference in this volume will usually be expressed in terms of phase angles.

Let us consider the superimposition of the waves along the directions $A_{\rm I}$, $A_{\rm II}$, $B_{\rm I}$, and $B_{\rm II}$, as we have for directions A_0 and C. Both $A_{\rm I}$ and $A_{\rm II}$ are similar to A_0 in that crests and troughs coincide, *i.e.* there is effectively zero phase difference, and this results in waves with doubled amplitude, as shown for $A_{\rm I}$ in Fig. 2.4. The figure also indicates how the wave crests tend to become parallel at locations remote from the scattering electrons. The assumption that the scattered waves are parallel when viewed from a sufficiently remote observation point is known as the Fraunhofer approximation, mentioned at the start of this chapter. The lines joining such a point to 1 and 2 would be parallel, and would differ in

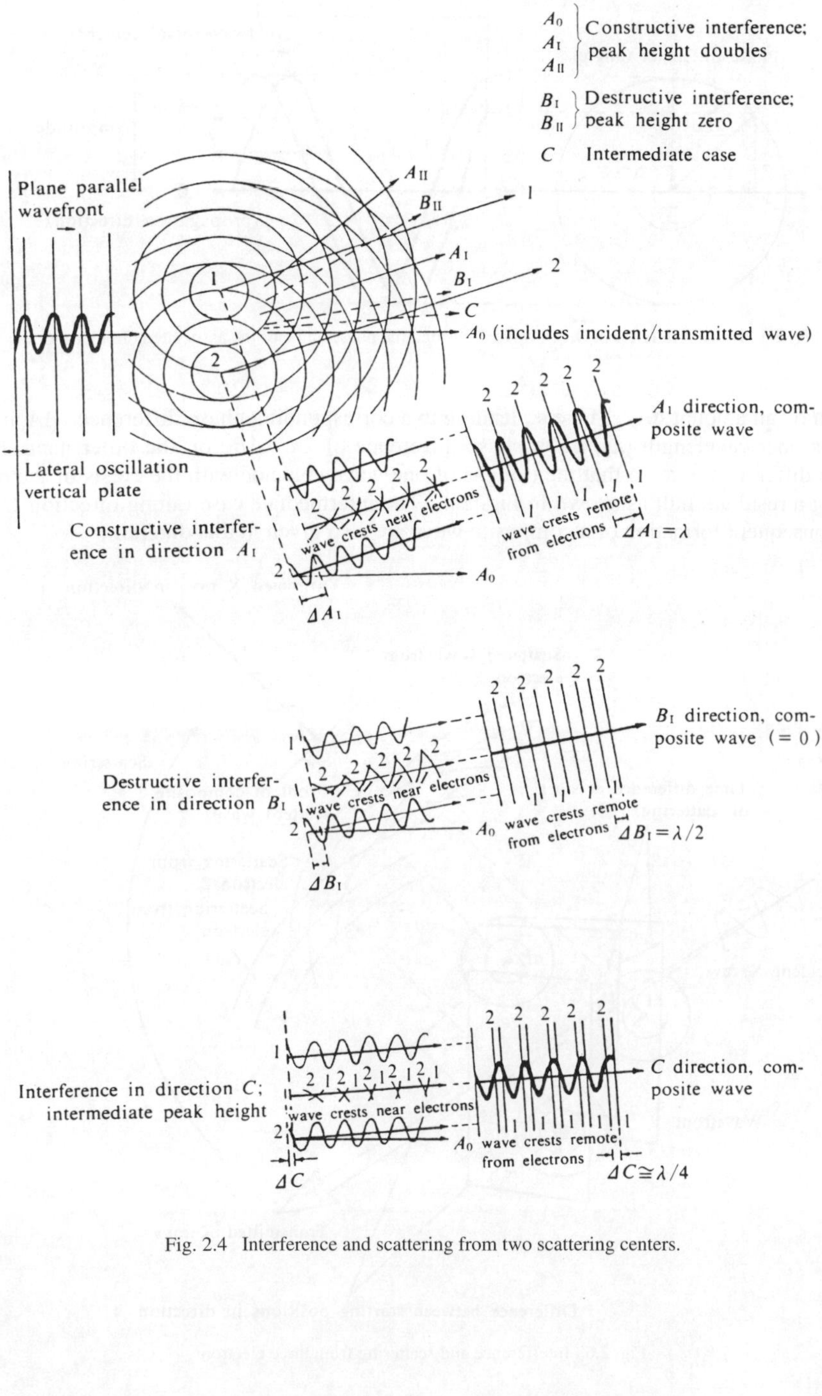

Fig. 2.4 Interference and scattering from two scattering centers.

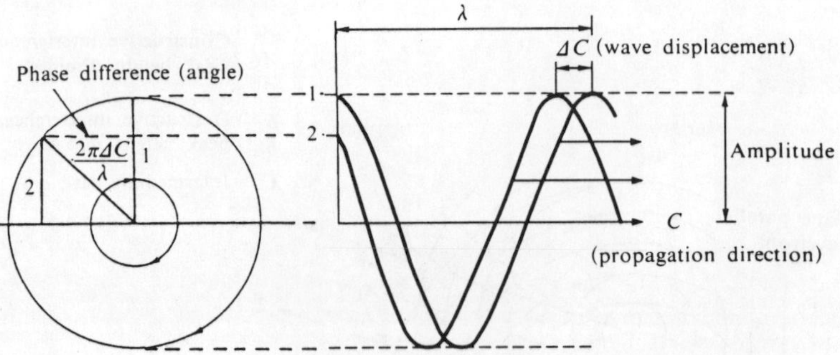

Fig. 2.5 Phase difference between wave trains and the phase angle of the associated circular motion.

length by an amount ΔA_{I}, ΔA_{II}, etc., leading to a corresponding phase difference. ΔA_{I} in fact equals one wavelength (λ), *i.e.* a phase difference of 2π. ΔB_{I}, on the other hand, has a phase difference of π, so that the troughs of one wave coincide with the crests of the other, giving a resultant null as shown in Fig. 2.4. The intermediate case (along direction *C*) and the consequent formation of a composite wave are also given in this diagram.

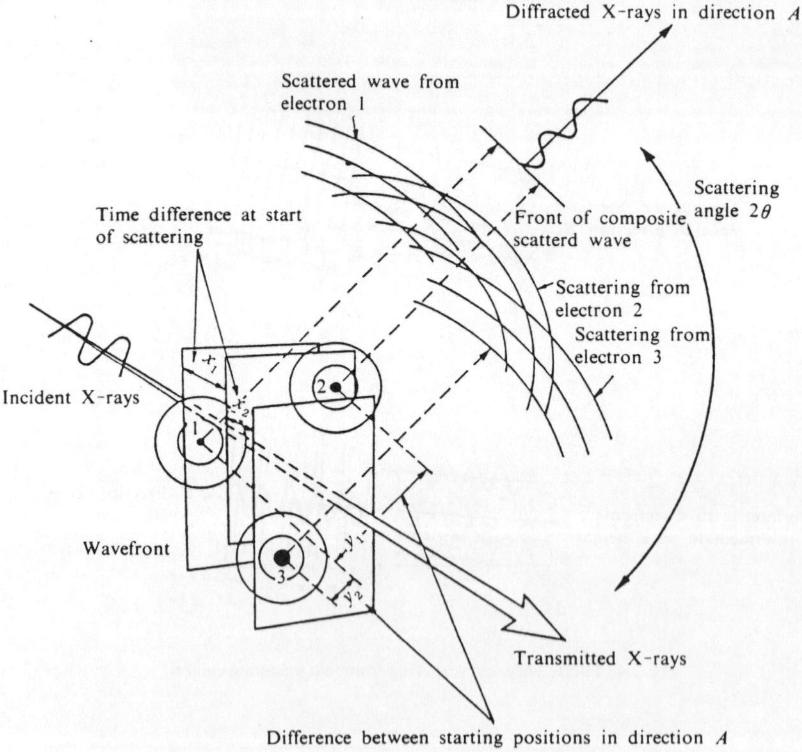

Fig. 2.6 Interference and scattering from three electrons.

Let us now examine the more complex case of Fig. 2.6. This is an extension to a more general situation where X-rays are incident in an arbitrary direction upon three electrons fixed in space. The incident wave front first reaches 1, simultaneously producing a scattered wave. After traveling a further distance x_1 the wave reaches 2, and another scattered wave is formed; finally, scattering occurs at 3 after a further distance x_2. We now consider the composite wave observed at a remote observation point in an arbitrary direction A. The waves from 2 and 3 lag the wave from 1 by x_1 and $x_1 + x_2$ initially, and are further retarded by y_1 and y_2 respectively, distances which depend upon the observation direction. The total path differences between the component waves are found from the sums of distances x and y, and the amplitude of the composite wave can, in principle, be found by the same kind of graphical treatment which yielded the amplitude of the scattered waves in various directions in Fig. 2.4 (p.19).

2.2.2 Basis for calculating the amplitudes of, and phase differences between, diffracted waves

However useful the above semi-graphical method may be in providing a picture of the physical processes which underlie X-ray diffraction, it obviously has serious limitations in applications to the actual problems of data computation. The following mathematical treatment confirms the preceding intuitive conclusions and permits a deeper understanding of the relationships between the variables involved, and a broadening of the range of possible applications. The details of the mathematical treatment may present some problems to the student, but it is sufficient at this stage merely to understand and memorize the principal results.

In wave theory, an electromagnetic wave is usually represented by a complex quantity

$$A = A_0 \exp(-2\pi i \bar{\nu}) = A_0 (\cos 2\pi \bar{\nu} - i \sin 2\pi i \bar{\nu}) \tag{2.5}$$

rather than by a simple sine wave with a real amplitude as shown in Fig. 1.1. In the above expression, A_0 is the maximum amplitude of the electric field, $\bar{\nu}$ is the distance from the source to the observation point in wave number units, and A is the amplitude of the electric field at the observation point. The intensity I of the wave, is given by the product of A and its complex conjugate A^*:

$$I = AA^* = A_0 \exp(-2\pi i \bar{\nu}) \cdot A_0 \exp(2\pi i \bar{\nu}) = A_0^2 \tag{2.6}$$

These are general expressions for wave motion. The time dependence of the wave motion does not appear explicitly because the mutual phase shifts of the waves produced by scattering and the resultant interference depend only upon the geometry of waves and objects, and are independent of time. However, if one is interested in the energy passing in unit time per cm^2 one must take into account that A is proportional to $\cos 2\pi \bar{\nu} t$, and hence A^2 is proportional to $\cos^2 2\pi \bar{\nu} t$. Since this (averaged over time) is equal to $\frac{1}{2}$, the energy is proportional to $\frac{1}{2}I$. In the mathematical treatment of diffraction problems throughout the remainder of this volume it is important to remember that the amplitude is generally a complex quantity involving the phase, and the intensity is AA^*, or $|A|^2$. Thus far, with the aim of making our explanations more readily intelligible, we have depicted X-rays as wavefronts or undulations. Now, on the other hand, we need to develop a simple mathematical formulation which will be adequate for calculating diffracted intensities, directions, etc., from complex scattering systems such as atoms, molecules, and crystals. For this purpose it is convenient to employ vectors to express the direction of the incident X-rays, the observation direction, and the position of the origin of scattering.

Figure 2.7 shows an analytical construction for finding the resultant amplitude of the waves scattered by more than two points; *cf.* Fig. 2.6 which indicated a qualitative treatment for the case with three electrons. The X-rays are scattered in a direction $\boldsymbol{s}$ ($|\boldsymbol{s}| = 1$) which makes an angle of 2θ with the direction of the incident beam $\boldsymbol{s}_0$ ($|\boldsymbol{s}_0| = 1$). Points 1, 2, etc., are at locations $\boldsymbol{r}_1$, $\boldsymbol{r}_2$, etc., with respect to the origin O (Fig. 2.7(a)). To obtain the amplitude of the diffracted wave, the path differences (x and y in Fig. 2.6) must first be calculated. Whereas in the former case we simply considered the path differences between 1, 2, and 3, we now follow the more general procedure of finding the path differences with respect to the origin O. Consider the general case of the nth point shown in Fig. 2.7(b), with position vector $\boldsymbol{r}_n$. If we define the vector $\boldsymbol{S}$ such that $\boldsymbol{S} = \boldsymbol{s} - \boldsymbol{s}_0$ (see diagram), then there will be no path difference between waves passing through n and those passing through n' (in direction $\boldsymbol{S}$ as shown). Thus the path difference between waves passing through n and those passing through the origin may be found by considering n as if it were at n'.

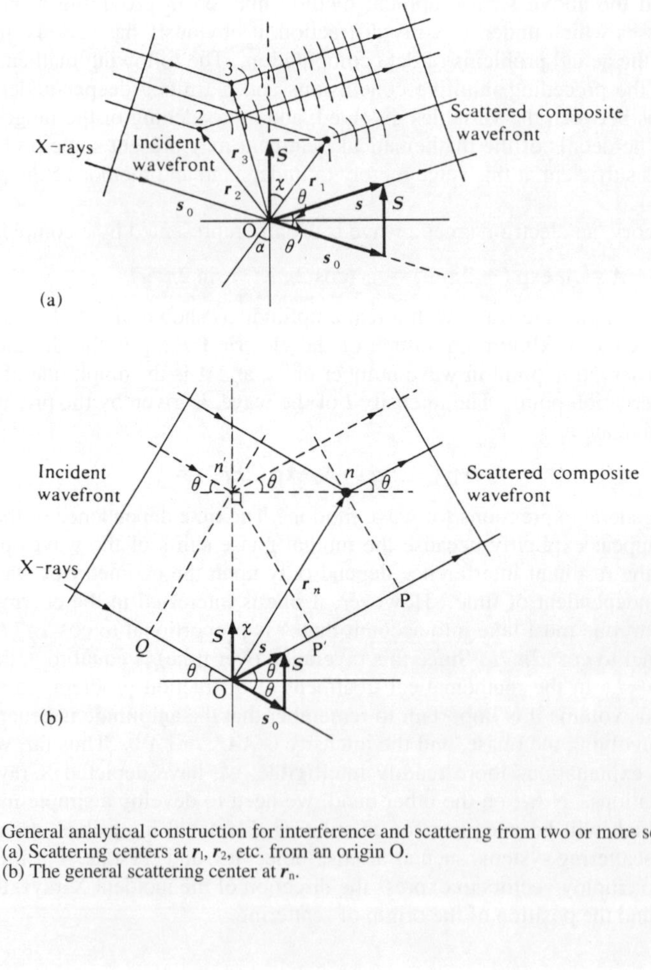

Fig. 2.7 General analytical construction for interference and scattering from two or more scattering centers. (a) Scattering centers at $\boldsymbol{r}_1$, $\boldsymbol{r}_2$, etc. from an origin O. (b) The general scattering center at $\boldsymbol{r}_n$.

If χ is the angle between $\boldsymbol{r}_n$ and $\boldsymbol{S}$, then the path difference is QO + OP′, that is $2r_n \cos\chi \sin\theta$. But since $\boldsymbol{S} = \boldsymbol{s} - \boldsymbol{s}_0$ (where both are unit vectors),

$$|\,\boldsymbol{S}\,| = |\,\boldsymbol{s} - \boldsymbol{s}_0\,| = S = 2\sin\theta \tag{2.7}$$

so the path difference may be written $Sr_n \cos\chi$ or (in vector terms) $\boldsymbol{S}\cdot\boldsymbol{r}_n$, the scalar product of $\boldsymbol{S}$ and $\boldsymbol{r}_n$.

The path differences with respect to the origin in Fig. 2.7(a), are therefore
for point 1, $(\boldsymbol{S}\cdot\boldsymbol{r}_1)$
and for point 2, $(\boldsymbol{S}\cdot\boldsymbol{r}_2)$
etc.,
and the difference between points 1 and 2 is

$$\begin{aligned}(\boldsymbol{S}\cdot\boldsymbol{r}_1) - (\boldsymbol{S}\cdot\boldsymbol{r}_2) &= (\boldsymbol{S}\cdot(\boldsymbol{r}_1 - \boldsymbol{r}_2)) \\ &= (\boldsymbol{S}\cdot(\text{displacement of 2 with respect to 1}))\end{aligned} \tag{2.8}$$

To calculate the amplitude, the phase differences of the waves scattered from 1 , 2, etc. with respect to the wave reaching the origin are first determined, and then substituted in the general wave equation, Eq. 2.5. Summation of these component amplitudes ultimately gives the composite total. Since $\bar{\nu}$ in Eq. 2.5 corresponds to path difference, we have

$$A_1 = A_0 \exp\{-2\pi i\,(\boldsymbol{S}\cdot\boldsymbol{r}_1)\},\quad A_2 = A_0 \exp\{-2\pi i\,(\boldsymbol{S}\cdot\boldsymbol{r}_2)\} \tag{2.9}$$

$\bar{\nu}$, it should be remembered, is in wave number units, so that we would expect a factor $1/\lambda$ to appear in association with $\boldsymbol{s}_0$, $\boldsymbol{s}$, and $\boldsymbol{S}$. To avoid this, *i.e.* to enable Eq. 2.9 to be cited in the above form, $|\,\boldsymbol{s}_0\,|$ and $|\,\boldsymbol{s}\,|$ are usually defined as $1/\lambda$ rather than as unity (as we have done hitherto). This does not, of course, affect the validity of the foregoing analysis. The amplitude A and intensity I of the composite wave in direction $\boldsymbol{s}$ for a total of N points is therefore

$$A = A_1 + A_2 + A_3 + \cdots + A_N$$

$$= A_0[\exp\{-2\pi i\,(\boldsymbol{S}\cdot\boldsymbol{r}_1)\} + \exp\{-2\pi i\,(\boldsymbol{S}\cdot\boldsymbol{r}_2)\} + \cdots + \exp\{-2\pi i\,(\boldsymbol{S}\cdot\boldsymbol{r}_n)\}] \tag{2.10}$$

$$= A_0 \sum_{n=1}^{N} \exp\{-2\pi i(\boldsymbol{S}\cdot\boldsymbol{r}_n)\} \tag{2.11}$$

$$I = AA^* = A_0{}^2\Big[\sum_{n=1}^{N} \exp\{-2\pi i(\boldsymbol{S}\cdot\boldsymbol{r}_n)\}\Big]\Big[\sum_{n=1}^{N} \exp\{2\pi i(\boldsymbol{S}\cdot\boldsymbol{r}_n)\}\Big] \tag{2.12}$$

With reference to Eq. 2.11, it is clear that any variation in either $\boldsymbol{S}$ (due to changes in incident direction $\boldsymbol{s}_0$ or observation direction $\boldsymbol{s}$) or in $\boldsymbol{r}$ (the spatial location of the scattering points) will lead to a change in the path differences between the various component waves, with a consequent change in the composite amplitude A. If the scattering points are electrons, A_p of Eq. 2.1 may be used for A_0.

2.2.3 The relationship between real and reciprocal space

From this point on, expressions of the form $(\boldsymbol{S}\cdot\boldsymbol{r})$ will often be encountered, and as mentioned earlier, these path differences, which are determined by the directions of the X-rays and the positions of the scattering points, will be expressed in wavelength units (wave numbers). Since $|\,\boldsymbol{S}\,|$ has the dimensions Å^{-1} and $|\,\boldsymbol{r}\,|$ has the dimensions Å, they are reciprocally related: $\boldsymbol{r}$ is called the physical-space or real-space vector, and the direction vector $\boldsymbol{S}$ is called the Fourier-space or (as distinct from real-space) reciprocal-space vector.

The set of vectors $\boldsymbol{r}$ denotes a physical structure in real space, whereas the diffraction due to scattering from such a structure is represented by the set of vectors $\boldsymbol{S}$ in reciprocal space. X-ray structural analysis may thus be regarded as the determination of the pattern in real space which corresponds to the directions and amplitudes of the scattered X-rays in reciprocal space.

We now proceed to develop an equation which, though rather unwieldy, is vital to an understanding of the relationship between X-ray diffraction and physical structures.

Referring to Fig. 2.7 again, if the number of scattering points is large, the amplitudes of the component waves, *i.e.* the scattering powers of the points, are simply multiplied by their phase factors (exponential factors), and the products added together as in Eq. 2.11. The discrete points denoted by the vectors $\boldsymbol{r}$ are, as a set, equivalent in their X-ray scattering powers to a substance whose scattering power has a continuous distribution $\rho(\boldsymbol{r})$. If the form of this function is known, the product of the function and the corresponding exponential phase factors for all values of $\boldsymbol{r}$ with respect to the origin can be integrated over real space without first being evaluated. Fig. 2.8 (a) and (b) illustrate the significance of the systems of real- and reciprocal-space coordinates in this situation. In Fig. 2.9 the combined amplitude A of the scattered waves from all points in the substance as observed from the direction $\boldsymbol{s}$ can be

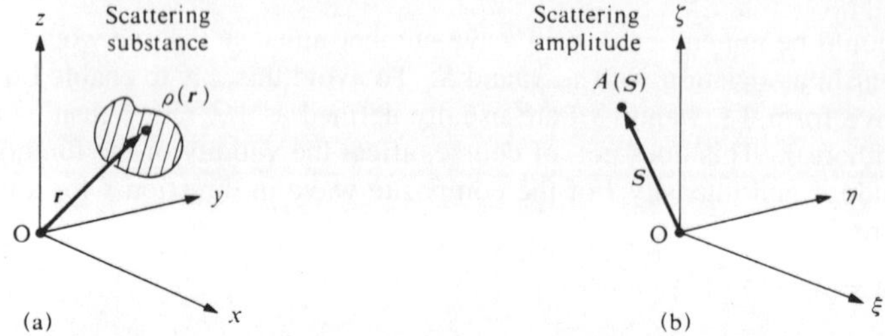

Fig. 2.8 Systems of coordinates in real and reciprocal space.
(a) Real space (coordinates have dimensions [L]).
(b) Reciprocal space (coordinates have dimensions [L^{-1}]).

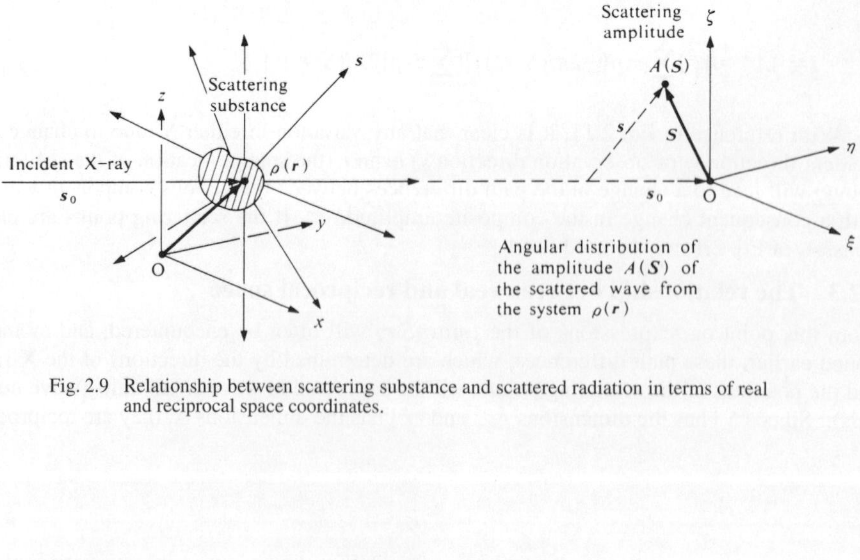

Fig. 2.9 Relationship between scattering substance and scattered radiation in terms of real and reciprocal space coordinates.

expressed as follows:

$$A(\boldsymbol{S}) = \int_0^\infty \rho(\boldsymbol{r}) \exp\{-2\pi i(\boldsymbol{S}\cdot\boldsymbol{r})\} \mathrm{d}v_r \tag{2.13}$$

The integration is carried out over the entire volume of the substance with a volume element $\mathrm{d}v_r$. This integral has the same form as the well-known Fourier integral (hence the term "Fourier space" above). Since the scattering points are generally electrons, $\rho(\boldsymbol{r})$ may be taken as the density of the electron distribution at $\boldsymbol{r}$. This is a single-valued continuous function in real space, and thus satisfies the conditions for Fourier transforms. We can therefore also write

$$\rho(\boldsymbol{r}) = \int_0^\infty A(\boldsymbol{S}) \exp\{2\pi i(\boldsymbol{S}\cdot\boldsymbol{r})\} \mathrm{d}v_s \tag{2.14}$$

The integration is carried out in this case over the whole of reciprocal space with a volume element $\mathrm{d}v_s$. Whereas Eq. 2.13 shows that the function $A(\boldsymbol{S})$ can be derived from the Fourier transform of $\rho(\boldsymbol{r})$, Eq. 2.14 (by means of the theory of Fourier integrals) shows that $\rho(\boldsymbol{r})$ can be obtained from the inverse Fourier transform of $A(\boldsymbol{S})$, *i.e.* that mutual transformation is possible. The physical significance of this is that the amplitudes of the diffraction pattern of a substance can be calculated from Eq. 2.13 if the electron density distribution is known, and the structure $\rho(\boldsymbol{r})$ of a substance can be calculated from Eq. 2.14 if the amplitudes of the diffracted X-rays are known for all $\boldsymbol{S}$. The structure of a substance and the X-ray scattering amplitude are therefore always related as shown by Eqs. 2.12 and 2.14 and Fig. 2.9.

2.3 Scattering of X-Rays by a Single Atom

2.3.1 Atomic scattering factor

In our initial discussion of Thomson scattering by a single electron, the electron was treated as a stationary point charge (*cf.* Section 2.1). Atomic electrons, however, are in constant orbital motion around the nucleus, so that the scattering of X-rays by a hydrogen atom with its single orbiting electron, or by other atoms with more numerous orbital electrons, is in fact rather different from the simple qualitative picture presented in Fig. 2.3. Whether the extranuclear electrons are regarded as particles circling in orbits or as a continuous electron cloud surrounding the nucleus, the atomic electron density should be expressed in terms of a time average of the density, for the probability of finding an electron in a given position varies continuously. Thus, in Eq. 2.13, taking the nucleus as the origin, the amplitude of the X-rays scattered from a single atom becomes

$$f(\boldsymbol{S}) = A_e \int \rho(\boldsymbol{r}) \exp\{-2\pi i(\boldsymbol{S}\cdot\boldsymbol{r})\} \mathrm{d}v \tag{2.13'}$$

with $\rho(\boldsymbol{r})$, the electron density surrounding the atom, expressed in polar coordinates with the origin at the nucleus. If $\psi(\boldsymbol{r})$ is the wave function of the atom, $\psi\psi^*$ is the instantaneous probability of finding an electron in an element of space $\mathrm{d}v$ in the vicinity of the atom. The $\psi(\boldsymbol{r})$ can be computed numerically by approximations of Hartree-Fock, Thomas-Fermi, or others, and hence the values of atomic scattering factors, f can be obtained as a function of $(\sin\theta)/\lambda$ (*i.e.* $\boldsymbol{S}$ in wave-number units) in good approximation.Table 2.A in the Appendix lists scattering factors, $f((\sin\theta)/\lambda)$ for free atoms.[7] Some of them are drawn in Fig. 2.10 against $(\sin\theta)/\lambda$.[8] Table 2.B in the Appendix gives the coefficients a_j, b_j, and c_j for fitting the scattering factors of Table 2.A to the analytic expression

$$f((\sin\theta)/\lambda) = \sum_{j=1}^{4} a_j \exp\{-b_j \sin^2\theta/\lambda^2\} + c_j \cdots \tag{2.15}$$

This analytic expression is a convenient form for entering the form into a computer.[9)]

f is variously known as the atomic scattering amplitude, atomic scattering factor, atomic scattering power, and atomic structure factor. Although "atomic scattering amplitude" is probably the more accurate term, "atomic scattering factor" is very widely used.

It is clear from the foregoing that the amplitude of the X-rays scattered from one atom is not the same as that from an equivalent number of electrons assumed to be at rest, but must include a further factor due to the motion of the electrons about the atom (the f factor). If A_e is the scattered amplitude by a single electron, then we must have $A = A_e{\cdot}f$, and since Eq. 2.3 gives $I_e = A_e^2$,

$$I = I_e f^2 \tag{2.16}$$

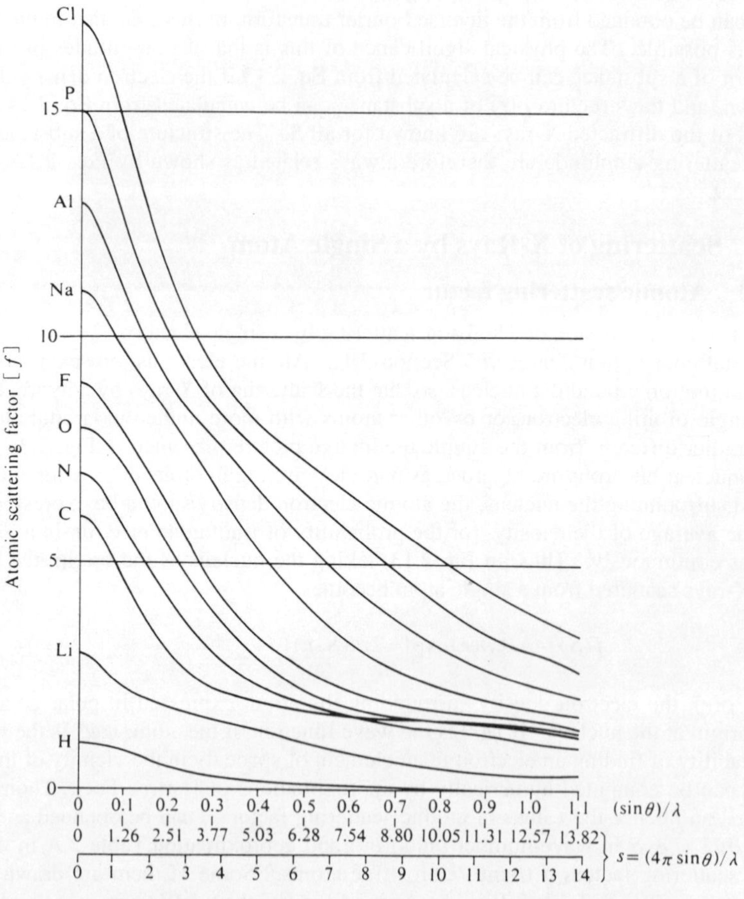

Fig. 2.10 Atomic scattering factors of some free atoms.
[Reproduced from L. Pauling, J. Sherman, *Z. Krist.*, **81**,24, Akademische Verlag (1932)]

2.3.2 Anomalous dispersion

The atomic scattering factor f derived here is independent of the wavelength λ of the incident X-rays only under the assumption that the bound electrons of the atom scatter as free electrons. However, if the wavelength is very close to an absorption edge the scattering power of a bound electron may be either greater or less than that of a free electron, and the phase of the scattered electron may be different, and should be taken into account. Thus the atomic scattering factor must be represented as a complex quantity:

$$f = f_0 + \Delta f' + i\Delta f'' \qquad (2.17)$$

where f_0 is equal to the usual f for $\lambda >> \lambda_k$ independent of λ, and dependent only on the scattering angle, viz. $(\sin\theta)/\lambda$. $\Delta f'$ is a small correction to the real part and $\Delta f''$ is the imaginary dispersion correction, which is non-zero (a few percent of $f_0 + \Delta f'$) only if $\lambda < \lambda_k$.†

This is so-called anomalous dispersion of an atom, which can be applied in the phase determination of structure factors (*cf.* Section 11.2.6) during the course of crystal structure analysis. Table 4 in the Appendix lists values of $\Delta f'$ and $\Delta f''$ for some characteristic radiations.[10] Fig. 2.11 shows changes in values of $\Delta f'$ and $\Delta f''$ of atoms for Cu $K\alpha$ radiation.[11]

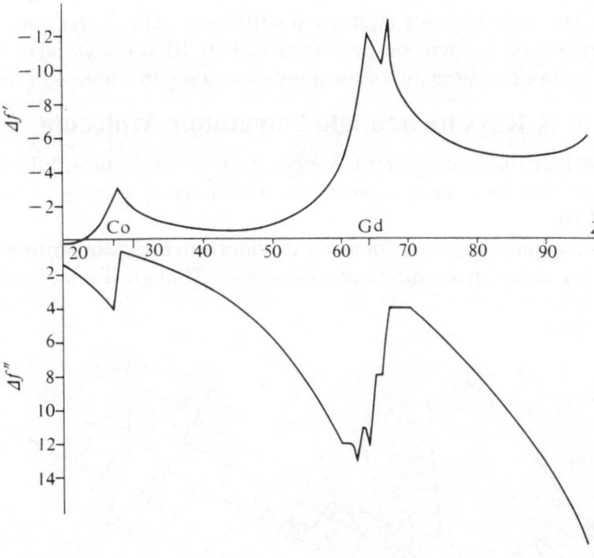

Fig. 2.11 $\Delta f'$ and $\Delta f''$ of atoms for Cu $K\alpha$ radiation.[11]
[Reproduced from C. H. Dauben, D. H. Templeton, *Acta Cryst.*, **8**, 841, IUCr. (1955)]

2.3.3 Compton scattering intensity

Reference has already made to the wavelength distribution of the scattered radiation in Compton (incoherent) scattering. The only information about this radiation that is required in practice, however, is its intensity distribution. Since Compton scattering does not lead to

† In some text f' and f'' are used instead of $\Delta f'$ and $\Delta f''$.

diffraction, there is no need to consider the positions at which scattering occurs and the phase difference due to the path differences between these points. The intensity thus depends only upon the scattering angle with respect to the direction of the incident X-rays. Since it results from an inelastic collision between an X-ray photon and an electron (*cf.* Fig. 1.6) the Compton scattering has an intensity of zero in the incident direction, *i.e.* at zero scattering angle. In general, the intensity I_{inco} of the Compton scattering from an atom having an atomic number Z is obtained by summation of the Compton scattering intensities for its Z electrons as follows:

$$I_{\text{inco}} = I_e R_{\text{BD}}[Z - \sum_m |f_{mm}|^2 - \sum_{m \neq n}\sum |f_{mn}|^2] \tag{2.18}$$

f_m is the scattering factor of the mth electron, and

$$f_{mn} = \int \psi_m^* \exp\{-2\pi i(\boldsymbol{S}\cdot\boldsymbol{r})\}\psi_n \, dv \tag{2.19}$$

f_{mn} is an exchange term due to interaction of the mth and nth electrons. R_{BD} is the Breit-Dirac electron recoil factor. Where the double summation term may be omitted in practice, and at zero scattering angle, since $\Sigma |f_{mm}|^2$ is equal to Z, I_{inco} is zero. As $(\sin\theta)/\lambda$ increases, I_{inco} also increases to give a continuous intensity distribution. The correction of diffraction intensities by subtraction of I_{inco} will be described in Part III (Analytical) (Section 13.1) Compton scattering factors for some important atoms are given in Table 4 of the Appendix.

2.4 Scattering of X-Rays by a Single Polyatomic Molecule

We now turn to consider the scattering of X-rays from a single molecule formed by N atoms held closely together by chemical bonds, as shown schematically in Fig. 2.12. (See also Fig.2.13 (a) (p.30)).

The molecules are assumed to be sufficiently distant from one another to avoid interference between scattered waves from different molecules. Though the molecule is moving

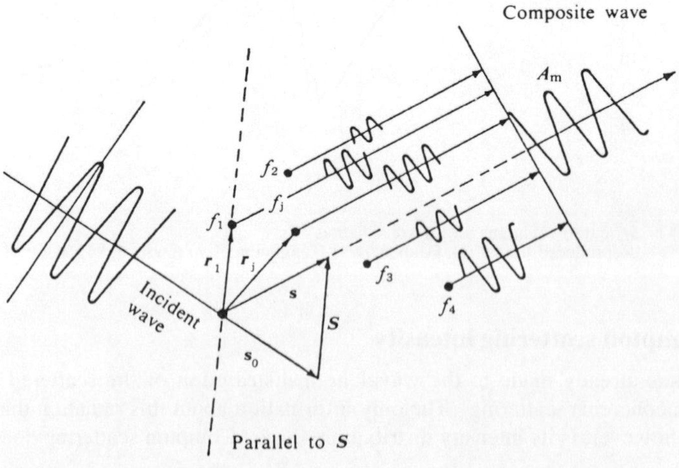

Fig. 2.12 Interference and scattering from the atoms within a single molecule.

and freely changing its orientation, we shall first combine the scattered waves from its individual atoms with respect to the position occupied by the molecule as a whole at a given instant. Let the scattering factors of the atoms in the molecule be $f_1, f_2, \ldots f_N$, and let the position vectors of these atoms with respect to a given origin be $\boldsymbol{r}_1, \boldsymbol{r}_2, \ldots \boldsymbol{r}_N$, so that the path differences of the scattered waves from the various atoms with respect to this origin are $(\boldsymbol{S}\cdot\boldsymbol{r}_1)$, $(\boldsymbol{S}\cdot\boldsymbol{r}_2)$, etc. The molecular amplitude A_m which results from the combination of waves having the above phase differences and amplitudes f_1, f_2, etc., is found with the aid of Eq. 2.13 (or Eq. 2.13′) (refer also to Section 2.3) as follows. The factor A_e, which is common to all the scattering equations derived in the following sections, is omitted. We find that

$$A_m = \sum_j^N f_j \exp\{-2\pi i(\boldsymbol{S}\cdot\boldsymbol{r}_j)\} \tag{2.20}$$

Thus, since $I_m = A_m A_m^*$, the intensity I_m is given by

$$\begin{aligned} I_m &= [\sum_j^N f_j \exp\{-2\pi i(\boldsymbol{S}\cdot\boldsymbol{r}_j)\}][\sum_j^N f_j \exp\{2\pi i(\boldsymbol{S}\cdot\boldsymbol{r}_j)\}] \\ &= \sum_j^N f_j^2 + \sum_j \sum_k^{j\neq k} f_j f_k\{-2\pi i(\boldsymbol{S}\cdot\boldsymbol{r}_{jk})\} \\ &= \sum_j^N f_j^2 + \sum_j \sum_k^{j\neq k} f_j f_k \cos 2\pi(\boldsymbol{S}\cdot\boldsymbol{r}_{jk}) \end{aligned} \tag{2.21†}$$

Since the molecule is in motion, however, the amplitude and intensity given by Eqs. 2.20 and 2.21 for the fixed molecule must be averaged over all orientations and positions of the molecule. For translational motion, though the $\boldsymbol{r}_j$ vectors are all continuously changing, the intensity I_m as given by Eq. 2.21 depends only on the vector $\boldsymbol{r}_{jk}$ between the jth and kth atoms, and so does not change provided that the structure of the molecule remains undistorted. For rotation, since $\boldsymbol{r}_{jk}$ is constantly changing with respect to $\boldsymbol{s}$ and $\boldsymbol{s}_0$, I_m must be averaged for the free rotations of the $\boldsymbol{r}_{jk}$ vectors with respect to $\boldsymbol{s}$ and $\boldsymbol{s}_0$ or with respect to (assuming that $|\boldsymbol{r}_{jk}|$ is constant), *i.e.* for all angles between the polar coordinate $\boldsymbol{r}_{jk}$ and $\boldsymbol{S}$, and for all angles of rotation about $\boldsymbol{S}$. This average is readily calculated:

$$<\cos 2\pi(\boldsymbol{S}\cdot\boldsymbol{r}_{jk})> = \frac{\sin 2\pi S r_{jk}}{2\pi S r_{jk}} \tag{2.22}$$

Thus the X-ray diffraction intensity for a moving polyatomic molecule is

$$<I_m(\boldsymbol{S})> = \sum_j^N f_j^2 + \sum_j \sum_k^{j\neq k} f_j f_k \frac{\sin 2\pi S r_{jk}}{2\pi S r_{jk}} \tag{2.23}$$

$$= J_a(S) + J_m(S) \tag{2.23′}$$

where $S = |\boldsymbol{S}|$, and $r = |\boldsymbol{r}|$. In Eq. 2.23′, the first term,

$$J_a(\mathrm{S}) = \sum_j^N f_j^2$$

is the scattering from individual atoms, and the second term,

† The appearance of a cosine term alone in place of the exponential term in Eq. 2.21 is due to the fact that when j and k are interchanged in $\boldsymbol{r}_{jk}$ the sign of the sine term changes so that jk and kj terms cancel to give zero.

$$J_m(S) = \sum_j^{j \neq k} \sum_k f_j f_k \frac{\sin 2\pi S r_{jk}}{2\pi S r_{jk}}$$

is the scattering dependant upon the molecular structure itself. If the structure of the molecule is known, the set of vectors $\boldsymbol{r}_{jk}$ for all the atoms in the molecule can be determined, whereupon multiplication by the atomic scattering factor for each atom and summation over the entire set of vectors will give a numerical value for the angular distribution of the scattering intensity of the molecule $J_m(s)$ $(s = 2\pi S)$.

Figure 2.13(b) shows the variation in scattering intensity with scattering angle for X-

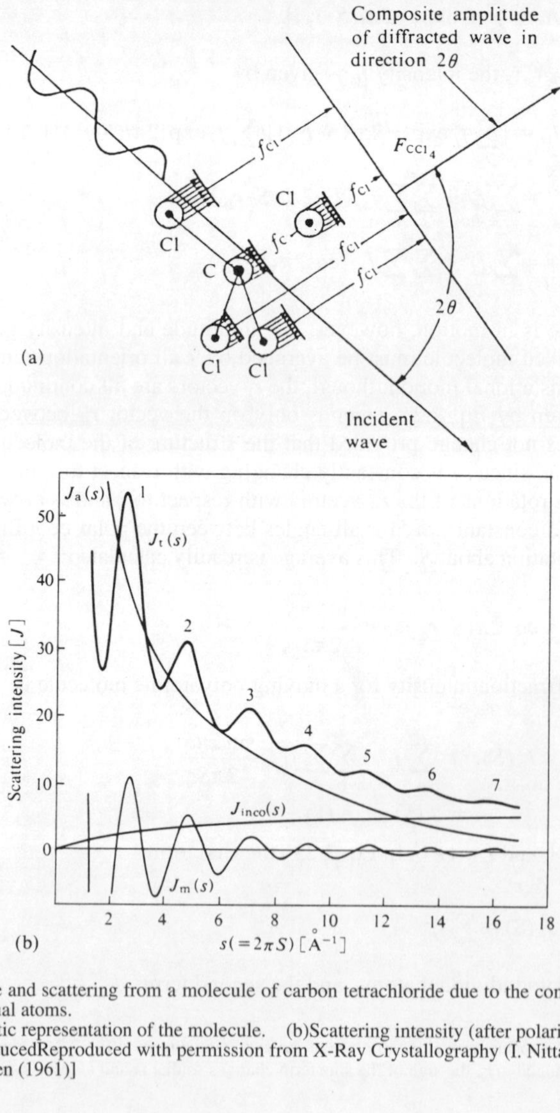

Fig. 2.13 Interference and scattering from a molecule of carbon tetrachloride due to the composite scattering from the individual atoms.
(a) Schematic representation of the molecule. (b)Scattering intensity (after polarization correction).[12]
(b):[ReproducedReproduced with permission from X-Ray Crystallography (I. Nitta ed.), Vol.II, p.423, Maruzen (1961)]

ray diffraction by carbon tetrachloride gas.

$$J_t(s) = J_a(s) + J_m(s) + J_{inco}(s)$$

where, $J_{inc}(s)$ is the incoherent scattering intensity (Eq. 2.17).

2.5 X-Ray Scattering from a Dense, Disordered Assemblage of Identical Atoms (a Monatomic Liquid)

We now turn to consider the amplitude and intensity of the X-rays scattered from a large number of identical atoms (*i.e.* all having the same *f*) separated by distances of the order of the atomic radius, as for example in liquid mercury.

Let the instantaneous positions of the atoms (which are in constant thermal motion) be as shown in Fig. 2.14(a). If $\boldsymbol{r}_j$ locates the *j*th atom, *N* is the total number of atom, and the directions of the incident and scattered radiation are $\boldsymbol{s}_0$ and $\boldsymbol{s}$, Eq. 2.20 applies, and gives the amplitude (and hence ultimately the intensity) of the scattered wave. Both the effect of phase differences with respect to the origin and of the individual scattering amplitudes of the atoms are included in this equation. Thus, exactly as in the case of the single polyatomic molecule, we have

$$A(\boldsymbol{S}) = \sum_j^N f_j \exp\{-2\pi i(\boldsymbol{S}\cdot\boldsymbol{r}_j)\} = f^2 \sum_j^N \exp\{-2\pi i(\boldsymbol{S}\cdot\boldsymbol{r}_j)\} \tag{2.24}$$

$$I = AA^*, \quad I(\boldsymbol{S}) = f^2 [\sum_j^N f_j \exp\{-2\pi i(\boldsymbol{S}\cdot\boldsymbol{r}_j)\} \sum_k^N \exp\{2\pi i(\boldsymbol{S}\cdot\boldsymbol{r}_k)\}]$$

$$= f^2 [\underset{j=k}{N} + \sum_j^{j\neq k} \sum_k \exp\{-2\pi i(\boldsymbol{S}\cdot(\boldsymbol{r}_j - \boldsymbol{r}_k))\}]$$

$$= f^2 [N + \sum_j^{j\neq k} \sum_k \cos 2\pi(\boldsymbol{S}\cdot\boldsymbol{r}_{jk})] \tag{2.25}$$

The problem here is to find a means of summing the second term of Eq. 2.25 over an almost infinite number of pairs of atoms. If the atoms are completely disordered (the diagram attempts to indicate this), their statistical distribution may be assumed to be independent of the particular choice of atom as origin, and statistical methods may be applied to the pairs of atoms $\boldsymbol{r}_{jk}$.

In Fig. 2.14(b), if the total volume is *V* and the total number of atoms is *N*, the average volume per atom is $V/N = v$. The probability of finding an atom in a volume element d*v* is therefore d*v*/*v*. Consider a particular atom. The probability that a second atom will be found within a volume element d*v* at a distance and direction $\boldsymbol{r}$ from this atom should, apparently, also be d*v*/*v*. In fact, however, this would only be true for values of $|\boldsymbol{r}|$ (or *r*) very much greater than the average inter-atomic distance. It certainly is not true for liquids at small values of $\boldsymbol{r}$; the atoms or molecules exhibit short-range order due to interactions operating over relatively short distances. In this case the probability of finding an atom in volume d*v* at $\boldsymbol{r}$ should be expressed as $\mathrm{d}P(\boldsymbol{r})$, where $P(\boldsymbol{r})$ is known as the radial density distribution function. We may therefore write $\mathrm{d}P(\boldsymbol{r}) = P(\boldsymbol{r})\mathrm{d}v/v$. $P(\boldsymbol{r})$ is a function which must clearly take the value zero for $\boldsymbol{r}$ less than twice the contact radius of the atoms, and which must tend to unity for sufficiently large $\boldsymbol{r}$. Fig. 2.14(c)[13] shows the actual form that this function takes in a representative case, intermediate values of $P(\boldsymbol{r})$ oscillating above and below unity. For this reason the double summation in Eq. 2.25 cannot be carried out by means of arithmetic addition, with equal weight given to each of the phase terms associated with a pair of atoms (*jk*). Instead a weight $(P(\boldsymbol{r}) - 1)/v$ must be assigned to the vectors $\boldsymbol{r}$

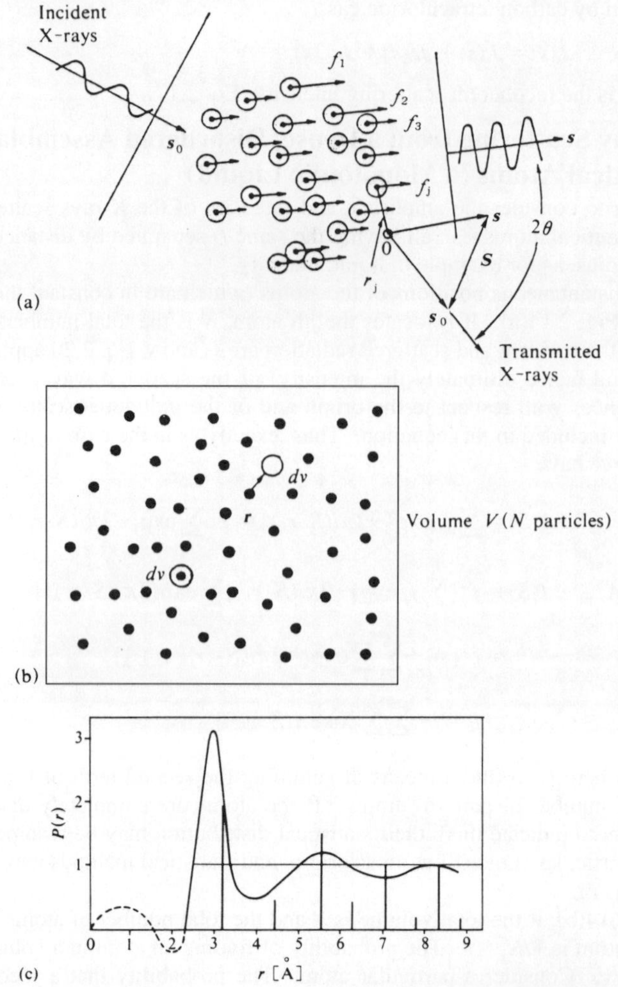

Fig. 2.14 Interference and scattering from a monatomic liquid. (a) Schematic representation of the monatomic liquid (b) Statistical assembly of N particles in volume V. (c) The form of a typical radial distribution function $P(r)$[13)]
(c): [Reproduced from P. Debye, H. Menke, *Phys. Z.*, **31**, 798, Verlag von S. Hirzel (1930)]

(*i.e.* $\boldsymbol{r}_{jk}$) of the pairs of atoms. Moreover, $\boldsymbol{r}_{jk}$ may be assumed to be continuously variable in the system and integration can be carried out instead of the summation. Thus outside the domain of the form scattering that occurs for $\boldsymbol{S}$ values which are not too small (see Eq. 5.14a), one obtains

$$\sum_{j}^{j\neq k}\sum_{k}\exp\{-2\pi i(\boldsymbol{S}\cdot\boldsymbol{r}_{jk})\}=\frac{1}{V}\int_0^{\infty}\{P(r)-1\}\exp\{-2\pi i(\boldsymbol{S}\cdot\boldsymbol{r})\}\mathrm{d}v \qquad (2.26)$$

If the system is isotropic, $\boldsymbol{r}$ is independent of direction, and may be replaced by r, and the exponential term is averaged over all angles between $\boldsymbol{r}$ and $\boldsymbol{S}$ and all angles of rotation about $\boldsymbol{S}$. Eq. 2.25 thus becomes

$$<I(\boldsymbol{S})> = Nf^2\left\{1+\frac{1}{v}\int_0^{\infty} 4\pi r^2\{P(r)-1\}\frac{\sin 2\pi Sr}{2\pi Sr}\,\mathrm{d}r\right\} \tag{2.27}$$

Substituting the atomic density $\rho(\boldsymbol{r})$, where $\rho(\boldsymbol{r}) = P(\boldsymbol{r})/v$, and the average density $\rho_0 = 1/v$, we obtain:

$$<I(\boldsymbol{S})> = Nf^2\left\{1+\int_0^{\infty} 4\pi r^2\{\rho(r)-\rho_0\}\frac{\sin 2\pi Sr}{2\pi Sr}\,\mathrm{d}r\right\} \tag{2.28}$$

Figure 2.15 is an example of the X-ray diffraction intensity curve for liquid mercury.

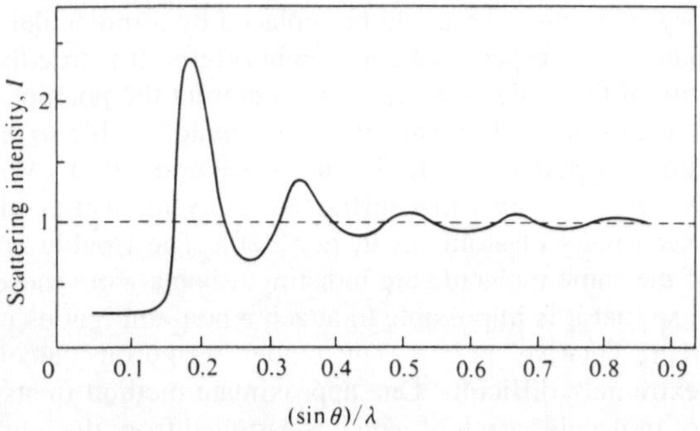

Fig. 2.15 X-ray scattering intensity distribution for liquid mercury.[13) [Reproduced from P. Debye, H. Menke, *Phys. Z.*, **31**, 798, Verlag von S. Hirzel (1930)]

Liquid and amorphous polymers such as polyethylene, since scattering by the hydrogen atoms is slight, may be regarded for practical purposes as consisting of carbon atoms alone, and Eq. 2.28 is applicable.

2.6 A Dense, Disordered Assemblage of Dissimilar Atoms

We now turn to consider scattering from a uniform mixture of dissimilar atoms, where the experimental conditions are otherwise as before. As in the previous calculations of the composite amplitude, the scattered amplitudes from all points are multiplied by the appropriate phase term, and the products summed. In this case, however, f cannot be isolated from the summation as a constant factor, because the atoms are no longer all alike. The combined amplitudes and intensities at any particular instant will be of the form

$$A(\boldsymbol{S}) = \sum_j^N f_j \exp\{-2\pi i(\boldsymbol{S}\cdot\boldsymbol{r}_j)\}$$

$$I(\boldsymbol{S}) = \sum_j^N f_j^2 + \sum_j^{j\neq k}\sum_k f_j f_k \exp\{-2\pi i(\boldsymbol{S}\cdot\boldsymbol{r}_{jk})\} \tag{2.29}$$

Since the set of atomic scattering factors f_j refers to a mixture of different atoms, $\sum_j^N f_j^2$ is statistically equivalent to the average $<f_j^2>$ of the values of f_j^2 of all the atoms, multiplied by the total number of atoms, N. The second term is a summation over $N(N-1)$ pairs of atoms, and the coefficient is the product of the scattering factors of any pair of atoms, so that it may be replaced by $<f_j>^2$. Replacing the vectors $\boldsymbol{r}_{jk}$ by scalar r, and applying the same statistical technique as for Eqs. 2.26 and 2.27, Eq. 2.29 becomes

$$<I(S)> = N\left[<f_j^2> + <f_j>^2 \frac{1}{v}\int_0^\infty 4\pi r^2\{P(r)-1\}\frac{\sin 2\pi Sr}{2\pi Sr}\mathrm{d}r\right] \tag{2.30}$$

This also can be expressed in terms of $\rho(r)$ (*cf.* Eq. 2.28). Examples of scattering substances of this type are molten alloys and glasses.

2.7 A Dense, Disordered Assemblage of Polyatomic Molecules (a Molecular Liquid)

By analogy with the cases which we have so far examined, it might be thought that the atomic scattering factor f of Section 2.5 could be replaced by a molecular scattering factor A_m (*cf.* Eq. 2.20) calculated with respect to a convenient origin. It is true that if we consider only the center of gravity of the molecules, *i.e.* if we represent the positions of the atoms in the molecules, etc., by means of single points, then we should be able to perform statistical calculations with the aid of a purely atomic density distribution $P(\boldsymbol{r})$. What this does not take into account, however, is the fact that diffracted X-ray intensities ultimately depend upon interference between pairs of atoms as in Eq. 2.20. The trouble arises because the vectors $\boldsymbol{r}_{jk}$ for atoms of the same molecule are indistinguishable from those between atoms of different molecules, so that it is impossible to attach a non-ambiguous meaning to a distribution $P(\boldsymbol{r})$ of our "representative" points. This renders a rigorous calculation of the second term in Eq. 2.20 extremely difficult. One approximate method treats the system as a statistical distribution of molecules, each of which is assumed from the outset to possess the average molecular scattering amplitude $<A_m>$ and intensity $<I_m>$ for all orientations of one molecule. From Eq. 2.18, using the averages given in Eq. 2.22, we obtain

$$A_m = \sum_j f_j \exp\{-2\pi i(\boldsymbol{S}\cdot\boldsymbol{r}_j)\}, \qquad |<A_m>| = \sum_j f_j \frac{\sin 2\pi S r_j}{2\pi S r_j} \tag{2.31}$$

$$<I_m> = <A_m{}^2> = \sum_j \sum_k^{j\neq k} f_j f_k \frac{\sin 2\pi S r_{jk}}{2\pi S r_{jk}} \tag{2.32}$$

If it were legitimate to treat the scattering from this system in the same way as that from a dense collection of fictitious atoms $<A_m>$, it would be possible to use Eq. 2.27. Molecules, however, unlike atoms, are not in general even approximately spherical, so that $|<A_m>|^2 \neq <A_m^2>$, and the system must therefore be treated as a collection of dissimilar atoms. Rather than Eq. 2.27 we should therefore use Eq. 2.30, with $<A_m^2>$ replacing $<f_j^2>$ and $|<A_m>|^2$ replacing $<f_j>^2$:

$$<I(S)> = N\left[<A_m{}^2> + |<A_m>|^2 \frac{1}{v}\int_0^\infty 4\pi r^2\{P(r)-1\}\frac{\sin 2\pi Sr}{2\pi Sr}\mathrm{d}r\right]. \tag{2.33}$$

where $P(r)$ here is not an atomic density distribution, but the appropriate molecular equivalent. If the structure of the molecule is known, evaluation of Eqs. 2.31 and 2.32 is possible, so that calculation of Eq. 2.33 is also possible provided the form of $P(r)$ can be inferred. The latter, however, is rarely the case, so that it is more common for the experimental values of $I(\boldsymbol{S})$ to be used in the direct Fourier transformation of Eq. 2.33 to find $P(r)$, which is important in connection with intermolecular forces in liquids. The form of $P(r)$ is discussed in more detail in Section 2.8 below.

Eq. 2.33 may be rewritten in the form

$$
\begin{aligned}
<I(S)> &= N[<A_m{}^2> + |<A_m>|^2 - |<A_m>|^2 \\
&\quad + |<A_m>|^2 \frac{1}{v}\int_0^\infty 4\pi r^2\{P(r)-1\}\frac{\sin 2\pi Sr}{2\pi Sr}\mathrm{d}r \\
&= N[<A_m{}^2> + |<A_m>|^2] \\
&\quad + N|<A_m>|^2\left[1+\frac{1}{v}\int_0^\infty 4\pi r^2\{P(r)-1\}\frac{\sin 2\pi Sr}{2\pi Sr}\mathrm{d}r\right]
\end{aligned}
\tag{2.34}
$$

The first and second terms of Eq. 2.34 are independent of the molecular shape factor, and hence of $P(r)$. The third term has exactly the same form as the right hand side of Eq. 2.27, with $|<A_m>|^2$ in the place of f^2. We are justified, therefore, in regarding the scattering intensity for a collection of molecules as equivalent to that for a collection of hypothetical atoms whose scattering factor is equal to an appropriately calculated average scattering factor for the molecules.

2.8 Scattering of X-Rays by Amorphous Solids

Use of the term "amorphous" implies the absence of regularity in the arrangement of the atoms, ions, or molecules of a substance. In amorphous solids, however, the spatial arrangement is not completely disordered, normally being subject to certain restraints. A completely disordered arrangement may be provisionally defined as one which would have $P(r)=1$ for all r which are larger than twice the contact radius (as in an ideal gas). The scattering intensity for such a system would thus be the same as that for diffraction by a gas, *i.e.* Nf^2 or $N<A_m^2>$, since the integral terms in Eqs. 2.27 and 2.34 are zero.

The above considerations apply strictly only to a completely disordered system, whereas we have already noted that virtually no amorphous substance will be completely without order. For example, let us consider a totally random collection of linear macromolecules. Irrespective of the particular molecular skeleton, its atoms will inevitably be linked by a succession of chemical bonds with fixed lengths and angles. Thus the $P(\boldsymbol{r})$ for the system will be highest for $\boldsymbol{r}$ corresponding to the bond length $\boldsymbol{r}_0$ of the most common bond type in the system, and $P(\boldsymbol{r})$ will oscillate in the vicinity of $\boldsymbol{r}_0$. Not only in this special case, but also in most other amorphous substances, it is found that $P(\boldsymbol{r})$ oscillates around a certain mean value. The corresponding form of $P(\boldsymbol{r})$ for various degrees of molecular disorder, including a schematic representation of the macromolecule, is shown in Fig. 2.16.

The X-ray patterns which are characteristic of each structure will be considered in later sections (*cf.* Fig. 13.1). Here, only one typical example of the X-ray diffraction intensities for an amorphous substance (a glass) is given in Fig. 2.17.[14)] The diffuse scattering, resulting in a typical diffuse halo, is mainly due to the predominant contribution of the third term in Eq. 2.34, which in turn results from the oscillations of the function $P(\boldsymbol{r})$, which correspond to the length of the most common Si–O bond (*ca.* 1.6 Å) and the O–O and Si–Si separations. The function differs in this respect from that of liquid (*cf.* Section 2.5).

Although X-ray diffraction from amorphous substances is of vital importance in connection with polymers, the discussion so far has been confined to the basic essentials. X-ray scattering from the amorphous component of a real polymer, however, is essentially similar to that for the $P(\boldsymbol{r})$ discussed above.

We now consider the case in which $P(\boldsymbol{r})$ is isotropic. In Fig. 2.14 the direction of $\boldsymbol{r}$ is immaterial, and $\boldsymbol{r}$ can be replaced by scalar r. The volume element dv of the second point then becomes the volume of a spherical shell between radii r and $r+\mathrm{d}r$. The probability of finding the second point in this volume element is therefore $4\pi r^2P(r)\mathrm{d}r$. If the substance

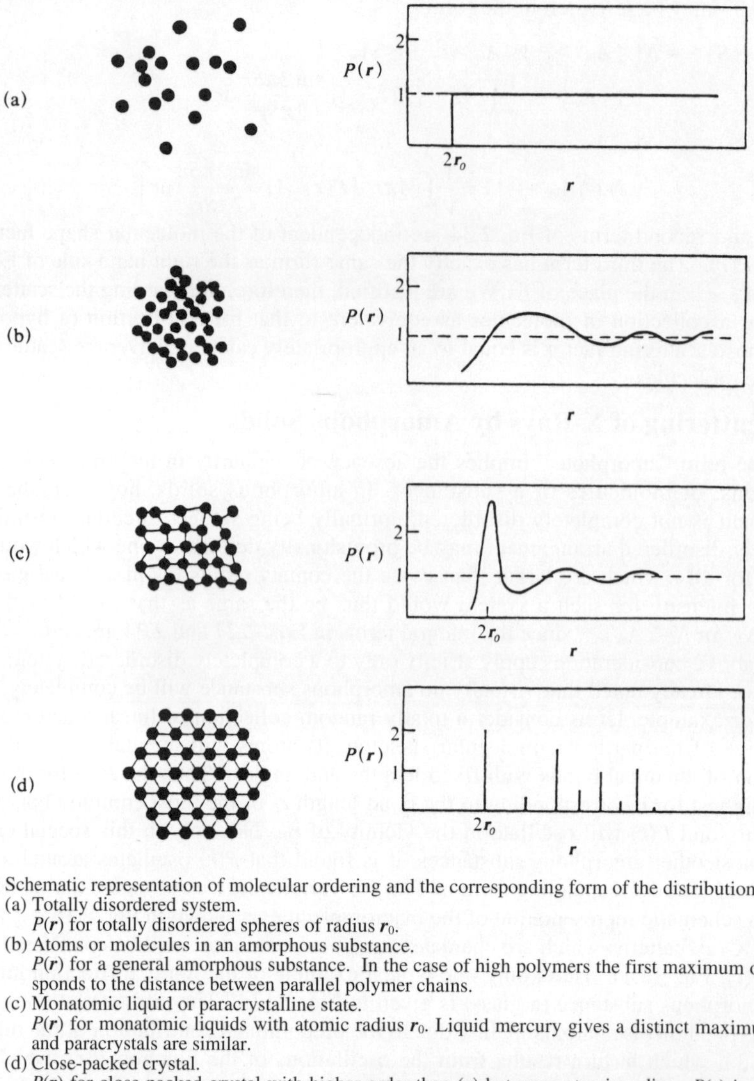

Fig. 2.16 Schematic representation of molecular ordering and the corresponding form of the distribution function.

(a) Totally disordered system.
$P(r)$ for totally disordered spheres of radius r_0.

(b) Atoms or molecules in an amorphous substance.
$P(r)$ for a general amorphous substance. In the case of high polymers the first maximum often corresponds to the distance between parallel polymer chains.

(c) Monatomic liquid or paracrystalline state.
$P(r)$ for monatomic liquids with atomic radius r_0. Liquid mercury gives a distinct maximum in $P(r)$, and paracrystals are similar.

(d) Close-packed crystal.
$P(r)$ for close-packed crystal with higher order than (c) but same atomic radius. $P(r)$ is discontinuous.

consists of single atoms, the appropriate X-ray diffraction intensities are given directly by substitution in Eq. 2.28.

$$I(S) = Nf^2\left\{1 + \int_0^{\infty} 4\pi r^2 P(r) \frac{\sin 2\pi Sr}{2\pi Sr}\,\mathrm{d}r\right\} \tag{2.28'}$$

$$I(S)/(Nf^2) - 1 = \boldsymbol{I'}(s) = \int_0^{\infty} 4\pi r^2 P(r) \frac{\sin sr}{sr}\,\mathrm{d}r$$

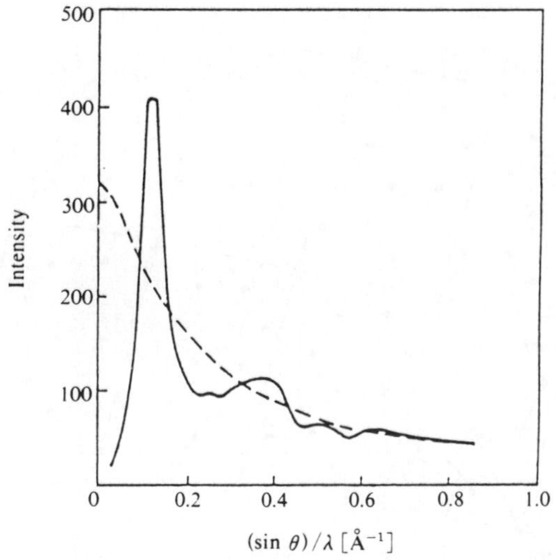

Fig. 2.17 X-ray diffraction intensity distribution for a glass.[14) [Reproduced from B. E. Warren *et al., J. Am. Ceramic Soc.*, **19**, 202, Am. Ceramic Soc. (1936)]

where

$$s = 2\pi S = (4\pi \sin\theta)/\lambda$$

Fourier transformation gives

$$4\pi r P(r) = (2/\pi)\int_0^\infty s I'(s) \sin sr ds \tag{2.35}$$

This equation is of fundamental importance in finding the radial density distribution function $P(r)$ from the X-ray diffraction intensities of amorphous substances and liquids: the function shown in Fig. 2.14(c) was calculated in this way. An example of the treatment when $P(r)$ is anisotropic is given in Section 14.3, where the case of cylindrical symmetry is discussed.

2.9 Scattering of X-Rays by Crystals

2.9.1 Amplitude and intensity of the scattered rays

In this section we shall confine ourselves to a simple explanation of X-ray diffraction with crystals as the scattering material, postponing a more detailed discussion of crystals themselves until a later section.

A crystal is a regular periodic three-dimensional arrangement of the structural units (atoms, ions, or molecules) which constitute the substance in question. Even if the periodically arranged assembly of atoms consists of complicated molecules, each molecule or group of molecules can be replaced by some fixed representative point, and these points will then form a regular arrangement having the characteristic period of the particular crystal, as shown in Fig. 2.18. Although this figure is confined to two dimensions, the concepts can readily be extended to the third dimension. Each point is referred to throughout the following discussion as if it were a single atom. This is of course legitimate even if the atom is

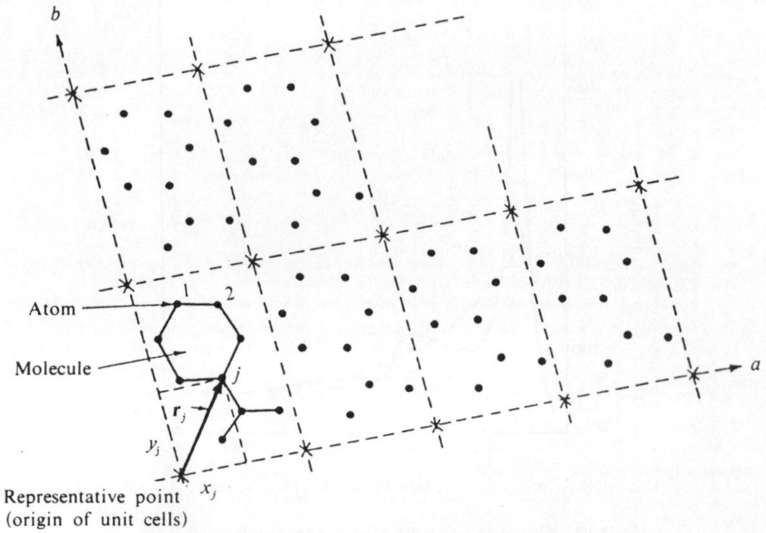

Fig. 2.18 An array of regularly repeating two-dimensional assemblies of atoms showing a lattice of representative points. Each assembly (unit cell) has composite scattering amplitude A_{cell} (see text).

ionized or forms part of a molecule: the argument still applies if the points are taken to represent whole molecules.

The broken lines in Fig. 2.18 divide the groups of atoms into identically similar assemblies, each within its own parallelogram, the parallelograms repeating periodically across the plane of the crystal. In three dimensions we visualize the atoms as within parallelepipeds which repeat along directions parallel to their three sides. A unit cell may be tentatively defined here as a parallelepiped which has the smallest regularly repeating volume element of the crystal lattice. Even this restriction on the volume, however, is not enough to define the choice of unit cell uniquely; this is usually made on the overruling considerations of crystal symmetry, which sometimes even dictate a unit cell with volume greater than the minimum possible. The position of each atom of the crystal within a unit cell of the lattice may be defined with respect to some representative point chosen as the origin. The positions of corresponding atoms in other unit cells may be found by translations through various multiples of the repeat distances along the three principal directions. This feature of crystal structure is exploited in considering the interference between X-rays scattered from a crystal.

If the set of all the atoms in one unit cell is treated as if it were a single molecule (in Fig. 2.18 the choice of "unit cell" ensures that this is actually so) and the position vectors $\boldsymbol{r}_j$ of the various atoms are taken with respect to the origin of the cell, the composite amplitude for the set is given by an equation of the same form as Eq. 2.18. In fact it is unusual for one molecule to be associated with each unit cell. If we let this amplitude be A_{cell} (not, in general A_m, which is a special case) then, omitting A_e as before, we obtain

$$A_{cell} = \sum_{\substack{unit \\ cell}} f_j \exp\{-2\pi i(\boldsymbol{S}\cdot\boldsymbol{r}_j)\} \tag{2.36}$$

Since all the atoms in the crystal can be located in terms of a three-dimensional parallel translation of the unit cell, and since the value of A_{cell} is the same for each unit cell, the scattering from the crystal as a whole is exactly the same as if there were a hypothetical atom with a scattering amplitude A_{cell} situated at each lattice point in the crystal. In Fig. 2.19,

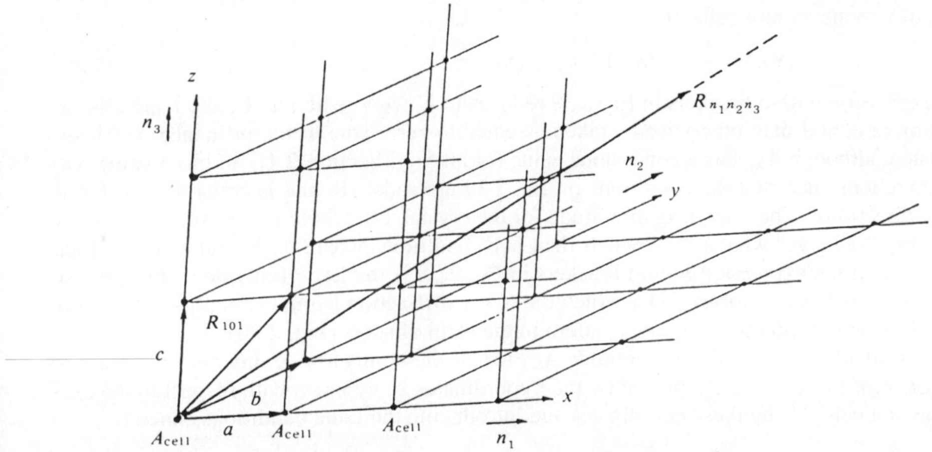

Fig. 2.19 A three-dimensional lattice of representative points, *i.e.* unit cell origins (*cf.* Fig. 2. 18).

with one lattice point as the origin, the length and direction of the sides of the unit cell are represented by the vectors $\boldsymbol{a}$, $\boldsymbol{b}$ and $\boldsymbol{c}$, in the x, y, and z directions (known as the crystallographic axes) respectively. Using these vectors, the positions of all the lattice points in the crystal can be expressed as follows.

The vector $\boldsymbol{R}_{n_1n_2n_3}$ for a lattice point displaced n_1 units in the $\boldsymbol{a}$ (or x) direction, n_2 units in the $\boldsymbol{b}$ (or y) direction, and n_3 units in the $\boldsymbol{c}$ (or z) direction is given by

$$\boldsymbol{R}_{n_1n_2n_3} = n_1\boldsymbol{a} + n_2\boldsymbol{b} + n_3\boldsymbol{c} \tag{2.37}$$

where n_1, n_2, and n_3 are integers. The amplitude A_{cr} of the scattered X-rays from the crystal as a whole is therefore found by a procedure similar to that used hitherto; summation over all values of $n_1n_2n_3$ for the product of A_{cell} with the associated phase factors $\exp\{-2\pi i(\boldsymbol{S}\cdot\boldsymbol{R}_{n_1n_2n_3})\}$. Thus, where N_1 N_2. and N_3 are the limits,

$$A_{cr} = \sum_{n_1=1}^{N_1}\sum_{n_2=1}^{N_2}\sum_{n_3=1}^{N_3} A_{cell}\exp\{-2\pi i(\boldsymbol{S}\cdot(n_1\boldsymbol{a}+n_2\boldsymbol{b}+n_3\boldsymbol{c}))\}$$

$$= A_{cell}\cdot\frac{\sin\pi N_1(\boldsymbol{S}\cdot\boldsymbol{a})}{\sin\pi(\boldsymbol{S}\cdot\boldsymbol{a})}\cdot\frac{\sin\pi N_2(\boldsymbol{S}\cdot\boldsymbol{b})}{\sin\pi(\boldsymbol{S}\cdot\boldsymbol{b})}\cdot\frac{\sin\pi N_3(\boldsymbol{S}\cdot\boldsymbol{c})}{\sin\pi(\boldsymbol{S}\cdot\boldsymbol{c})}\cdot\exp(i\alpha_0)$$

$$I_{cr} = A_{cr}A^*_{cr} = |A_{cell}|^2\left\{\frac{\sin\pi N_1(\boldsymbol{S}\cdot\boldsymbol{a})}{\sin\pi(\boldsymbol{S}\cdot\boldsymbol{a})}\cdot\frac{\sin\pi N_2(\boldsymbol{S}\cdot\boldsymbol{b})}{\sin\pi(\boldsymbol{S}\cdot\boldsymbol{b})}\cdot\frac{\sin\pi N_3(\boldsymbol{S}\cdot\boldsymbol{c})}{\sin\pi(\boldsymbol{S}\cdot\boldsymbol{c})}\right\}^2 \tag{2.38}$$

$$= |A_{cell}|^2L_1^2(N_1)L_2^2(N_2)L_3^2(N_3) \tag{2.38'}$$

where $L_1 = \dfrac{\sin\pi N_1(\boldsymbol{S}\cdot\boldsymbol{a})}{\sin\pi(\boldsymbol{S}\cdot\boldsymbol{a})}$, $L_2 = \dfrac{\sin\pi N_2(\boldsymbol{S}\cdot\boldsymbol{b})}{\sin\pi(\boldsymbol{S}\cdot\boldsymbol{b})}$, $L_3 = \dfrac{\sin\pi N_3(\boldsymbol{S}\cdot\boldsymbol{c})}{\sin\pi(\boldsymbol{S}\cdot\boldsymbol{c})}$.

The $\exp(i\alpha_0)$ factor is the phase contribution which depends upon the particular choice of origin used in the calculation of the composite amplitude A_{cell}. If the center of symmetry of the lattice can be chosen as the origin, the combination of $\pm N_1/2$, $\pm N_2/2$ and $\pm N_3/2$ ensures that this phase term disappears ($\exp(i\alpha_0)$ becomes unity).

When N_1, N_2, and N_3 are large, *i.e.* as they are in an ordinary crystal, which usually consists of very many unit cells, if

$$(\boldsymbol{S}\cdot\boldsymbol{a}) = h', \quad (\boldsymbol{S}\cdot\boldsymbol{b}) = k', \quad (\boldsymbol{S}\cdot\boldsymbol{c}) = l' \tag{2.39}$$

the expression within brackets in Eq. 2.38 reduces to $N_1 \cdot N_2 \cdot N_3$ when h', k', and l' are all integers or zero, and may otherwise be taken as equal to zero. Due to the influence of this expression, although A_{cell} has a continuous value for all $\boldsymbol{S}$ (*cf.* Section 2.4), A_{cr} has a value only for directions such that the conditions on Eq. 2.39 are satisfied, and is equal to zero for all other directions. The scattering amplitude for the crystal as a whole is therefore discontinuous. Eq. 2.39 gives what are known as the Laue conditions, where h', k', and l' are the Laue indices, and the expression within brackets in Eq. 2.38 is the Laue function. This analysis demonstrates the conditions under which the X-ray diffraction from a crystal is a discontinuous function of direction, giving a pattern in the form of spots (Fig. 2.20).

To calculate the resultant amplitude A_{cell} for all the atoms in one unit cell, the position vectors $\boldsymbol{r}_j$ of the atoms are replaced by their coordinates x_j, y_j, and z_j with respect to the axes of the unit cell. Using these coordinates and introducing the Laue conditions, since

$$\boldsymbol{r}_j = \frac{x_j\boldsymbol{a}}{a} + \frac{y_j\boldsymbol{b}}{b} + \frac{z_j\boldsymbol{c}}{c}$$

then

$$(\boldsymbol{S}\cdot\boldsymbol{r}_j) = h'\frac{x_j\boldsymbol{a}}{a} + k'\frac{y_j\boldsymbol{b}}{b} + l'\frac{z_j\boldsymbol{c}}{c} \tag{2.40}$$

Now we have already established that diffraction from the whole crystal is discontinuous, and that A_{cr} of Eq. 2.38 exists only for h', k', and l' which satisfy Eq. 2.39. Let A'_{cell} be the cell amplitude that corresponds to this condition, giving

$$A'_{cell} = \sum_j f_j \exp\{-2\pi i(h'x_j/a + k'y_j/b + l'z_j/c\} = F(h'k'l') \tag{2.41}$$

$F(h'k'l')$ is a function both of the coordinates of the atoms in the unit cell and of their scattering factors (which are specific to each atom). It is therefore literally a function of the structure of the unit cell, and is known as the structure factor. It expresses the amplitude in the diffraction direction denoted by $(h'k'l')$ and therefore plays a vital role in the determination of structures. Quantities of the type x_j/a in Eq. 2.41 are the ratios of the coordinates of the atoms to the dimensions of the unit cell. These ratios will in future be represented simply x_j, y_j, and z_j (dropping the cell dimensions from the denominator); they are therefore the coordinates of the atoms expressed as *fractions* of the dimensions of the unit cell dimensions along the three principal axes. Moreover, we shall find in the next chapter that h', k', and l', the Laue indices, are identical with the indices (hkl) of the crystal lattice planes.

Since $F(h'k'l')$ gives the amplitude of the scattered X-rays, the intensity for diffraction from the unit cell I_{cell} is

$$I_{cell}(h'k'l') = F(h'k'l')\cdot F^*(h'k'l') \tag{2.42}$$

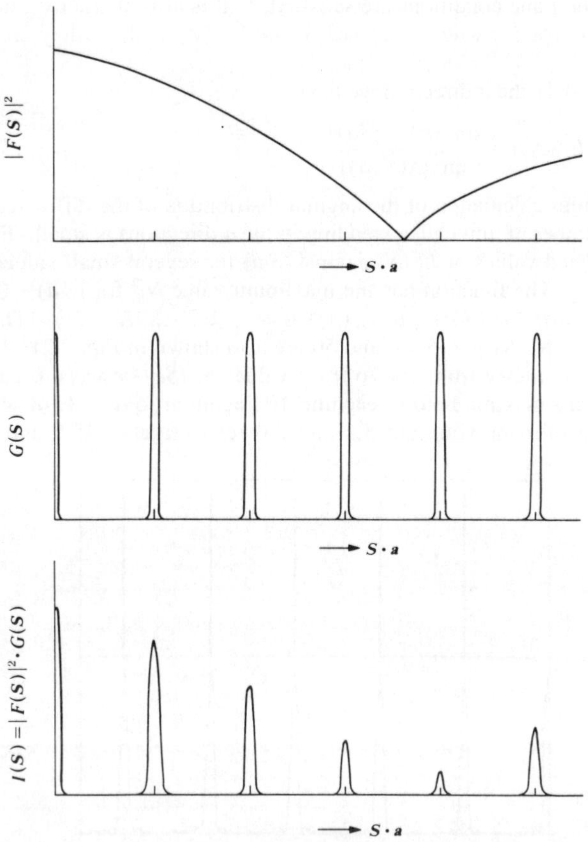

Fig. 2.20 Combination of the square of unit cell scattering amplitude (or structure factor), $|F(S)|^2$ ($=|A_{cell}|^2$ and Laue function, $G(S)$ ($=L^2(N)$) to give diffraction intensity of a crystal, $I(S)=|F(S)|^2 \cdot G(s)$ vs. ($S \cdot a$). [Reproduced with permission from R. D. B. Fraser, T. P. MacRae, *Conformation in Fibrous Proteins and Related Synthetic Polymers*, p.8, Academic Press (1973)]

2.9.2 Form of the X-ray diffraction pattern

If we write $L_1(N_1)L_2(N_2)L_3(N_3)$ for the Laue diffraction function (the expression in brackets in Eq. 2.38) we have

$$A_{cr} = A_{cell}L_1(N_1)L_2(N_2)L_3(N_3) \tag{2.43}$$

$$I_{cr} = A_{cell}A^*_{cell}L_1^2L_2^2L_3^2 = F(h'k'l')F^*(h'k'l')\,G$$

$$= |F(h'k'l')|^2G \tag{2.44}$$

where, $G = L_1^2L_2^2L_3^2$. If the crystal is sufficiently large, *i.e.* if N_1, N_2, and N_3 are at least of the order of 100, G (or L^2) has the value $N_1N_2N_3$ only over a very small solid angle (*cf.* Fig. 2.20) and the X-ray diffraction amplitude A_{cr}, as mentioned above, is zero except in those directions where the Laue conditions are satisfied.[15] It is also instructive, however, to consider the form of G (or L^2) when N_1, N_2 and N_3 are fairly small, as they are for very small crystals.

If we consider only the $\boldsymbol{a}$ direction we have

$$L_1^2(N_1) = \frac{\sin^2\{\pi N_1(\boldsymbol{S}\cdot\boldsymbol{a})\}}{\sin^2\{\pi(\boldsymbol{S}\cdot\boldsymbol{a})\}} \tag{2.45}$$

This equation enables calculation of the angular distribution of the diffracted X-ray intensity when N_1 (the number of unit cells extending in the $\boldsymbol{a}$ direction) is small. Fig. 2.21[15] gives plots of the calculated values of $L_1^2(N)$ against $(S{\cdot}\boldsymbol{a})$ for several small values of N_1, and for the limit as $N_1 \to \infty$. The function has the maximum value N_1^2 for $(S{\cdot}\boldsymbol{a}) = 0, \pm 1, \pm 2$, etc., and N_1 minima (zero) for $(\boldsymbol{S}{\cdot}\boldsymbol{a}) = p/N_1$ ($p = 0, \pm 1, \pm 2, \ldots, N_1 - 1$). There are $N_1 - 2$ small maxima; those for $N_1 = 3$, 5, 10 and 50 are also shown in Fig. 2.21. For $N_1 = 10$, the function drops continuously from 100 to 0 for values of $(\boldsymbol{Sa})$ between 0 and $1/N_1$ $(=0.1)$, with eight subsidiary maxima before reaching 100 again at $(\boldsymbol{S}{\cdot}\boldsymbol{a}) = 1$, of which the values of the second and third maxima are 5.1 and 2.0 respectively. If $\boldsymbol{S}$ and $\boldsymbol{a}$ are parallel,

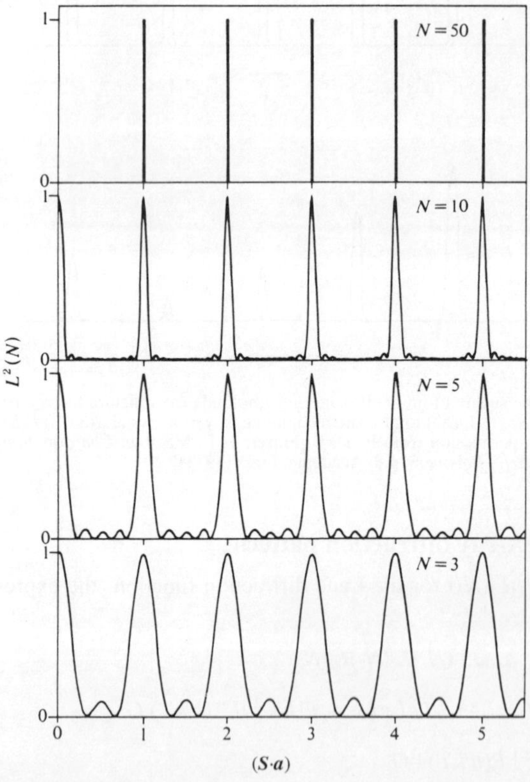

Fig. 2.21 Variation of the form $L^2(N)(=G)$ vs. $(\boldsymbol{S}{\cdot}\boldsymbol{a})$ for different values of $L^2(N)$ is normalized to unity in this diagram.[15]
[Reproduced with permission from R. D. B. Fraser, T. P. MacRae, *Conformation in Fibrous Proteins and Related Synthetic Polymers*, p.9, Academic Press (1973)]

$(\boldsymbol{S}\cdot\boldsymbol{a}) = a\cdot 2\ (\sin\theta)/\lambda$, and an edge of the major maximum is located at $\sin\theta = 0.015$ for $N_1 = 10$, with $a = 5\text{Å}$ and $\lambda = 1.5\text{Å}$. Diffraction peaks are not, therefore, very sharp, having an edge-to-edge spread of approximately $\sin^{-1} 0.03$, and we can make quantitative estimates of the broadening. Similarly for the ***b*** and ***c*** directions, the diffraction pattern becomes less clearly defined as the number of unit cells decreases. Conversely, measurement of the broadening of the X-ray diffraction pattern, by yielding information about the form of L^2, can be used to find the sizes of the diffracting crystallites. Scherrer[16)] and Jones[17)] have both developed methods based on measurement of the breadths of Laue functions. Practical applications will be discussed in Section 13.3.

2.10 Summary

This chapter has touched upon the nature of the interference that occurs when X-rays are scattered by objects, the amplitude of the resulting composite wave, and the classification of scattering bodies according to their shape. The main points concerning the amplitudes and intensities of the scattered wave will now be summarized.

1. Free electrons and electrons in the shells of atoms can give rise to coherently scattered X-rays (*i.e.* with no change in wavelength). The intensity varies with the observation direction, and is given by Eq. 2.3.

2. Every type of atom has its own characteristic number of extranuclear electrons. Their distribution throughout the space surrounding the nucleus may be taken as continuous, and the amplitude of the coherent scattering due to this atomic electron "cloud" (*i.e.* the atomic scattering amplitude *f*) can be found from the Fourier transform of the electron density $\rho(\boldsymbol{r})$ around the nucleus, see Eqs. 2.13 and 2.13'.

3. Inverse Fourier transformations can be applied to determine the angular distribution of the X-ray scattering amplitude from the electron density of the scattering substance and, conversely, to determine the electron density from the amplitude distribution, see Eqs. 2.13 and 2.14.

4. The scattering from an atom (paragraph 2) above) may be conveniently regarded as a wave of amplitude f originating from the position of the atomic nucleus. The amplitude of the scattered X-rays from an assembly of atoms (molecule, liquid, etc.) can therefore be found by appropriate summation of the scattered waves originating from all the points *j* (signifying a particular atom) having a scattering amplitude f_j. The method is that of paragraph 5. below.

5. The composite amplitude of the scattered waves from a discontinuous set of points may be found by a simple summation of the products of their individual scattering amplitudes (A_e, *f*, etc.) and the appropriate exponential phase terms for their path differences relative to a convenient origin (see Eq. 2.12).

6. If the scattering substance is a crystal, the scattering amplitude is found by combination of the scattered waves from the atoms of a single unit cell by the method of 5) above. The resultant composite amplitude is known as the structure factor $F(h'k'l')$ (see Eqs. 2.41 and 2.42).

References

1. M. von Laue, *Röntgenstrahlinterferenzen*, 2nd ed., Akadimische Verlag, Leipzig (1960).
2. R. W. James, *The Optical Principles of the Diffraction of X-Rays*, G. Bell & Sons, London (1954).
3. W. H. Zachariasen, *Theory of X-Ray Diffraction in Crystals*, John Wiley & Sons, Inc., N. Y. (1945).
4. S. Miyake, *X-Ray Diffraction* (New ed.), Asakura, Tokyo (1969) (in Japanese).

5. *X-Ray Crystallography* (I. Nitta ed.), Vol. I, Maruzen, Tokyo (1959) (in Japanese).
6. N. Kato, *Diffraction and Scattering*, Asakura, Tokyo (1978) (in Japanese).
7. *International Tables for Crystallography*, Vol. C, Mathematical, Physical and Chemical Tables, 2nd ed., (A.J.C. Wilson, E. Prince eds.) pp. 549–559, Kluwer Academic, Dordrecht (1999).
8. L. Pauling, J. Sherman, *Z. Krist.*, **81**, 24, (1932).
9. In ref. 7, pp. 572–575.
10. In ref. 7, pp. 255–257.
11. C. H. Dauben, D.H. Templeton, *Acta Cryst.*, **8**, 841 (1955).
12. *X-Ray Crystallography* (I. Nitta ed.), Vol. II, p. 423, Maruzen, Tokyo (1961) (in Japanese).
13. P. Debye, H. Menke, *Phys. Z.*, **31**, 798, Verlag von S. Hirzel (1930).
14. B. E. Warren, H. Krutter, O. Moringstar, *J. Am. Ceramic Soc.*, **19**, 202 (1936).
15. R. D.B. Fraser, T.P. MacRae, *Conformation in Fibrous Proteins and Related Synthetic Polymers*. p. 9, Academic Press, N. Y. (1973).
16. P. Scherrer, *Göttinger Nachrichten*, **2**, 98 (1918).
17. F. W. Jones, *Proc. Roy. Soc.* (London), **A166**, 16 (1938).

3. Crystal Structure

3.1 Crystal Systems and the Unit Cell

The concept of the unit cell as a regularly repeating element from which a crystal is formed by parallel displacements in three dimensions was introduced in Section 2.9. The relative positions of atoms within a unit cell are constant from cell to cell, so that any one atom in the assemblage of atoms within the cell, or indeed any fixed point in its vicinity, can act as a representative point in defining the cell's location within the crystal. The set of such points is known as the space lattice. It should be obvious, however, that although the space lattice is here defined in terms of the unit cell, the choice of unit cell is by no means unique. The lines joining the lattice points may be drawn in any number of different ways, each of which (provided it satisfies the usual condition that the volume of the repeating element should be a minimum) would satisfactorily define the shape of an appropriate unit cell (see Fig. 3.1). This two-dimensional array of atoms is the same as that of Fig. 2.18, but it is clear that in only one case will the group of atoms in the unit cell correspond to the true molecule of the crystal, although in this example each cell will contain the same number of atoms as the molecule. In fact, it is rare for only one molecule to be associated with each lattice point, so that a unit cell will usually either contain the atoms from more than one molecule or only a fraction of the atoms from a single molecule. The choice of unit celi is not, however, completely arbitrary. The lengths of the three edges of the unit cell, a, b, and c, where α, β, and γ are the angles between them as shown in Fig. 3.2, are measured in the directions of the crystallographic axes. The right choice of crystallographic axes can considerably simplify calculations and help to make apparent the inherent symmetry of the lat-

Fig. 3.1 Two-dimensional array of atoms showing various groups (molecular and other) possible within different unit cells, *cf.* Fig. 2.18.

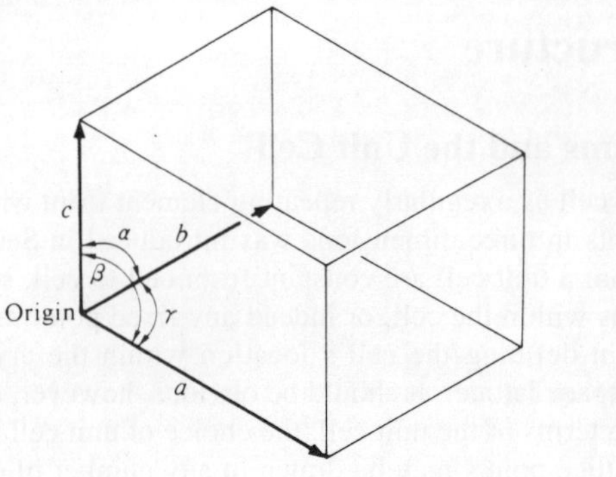

Fig. 3.2 Unit cell constructed on the origin.

tice. Crystals which belong to different symmetry groups can in some cases be referred to the same set of crystallographic axes. All crystals which can be referred to the same set of axes are said to belong to the same crystal system.[1)] Table 3.1 shows the seven crystal systems and the corresponding crystallographic axes and unit cell conditions (see also Fig. 3.2). Note that α, β, and γ are chosen to be either 90° or 120° wherever possible. We should also note, in passing, that before the advent of X-ray crystallographic techniques crystallographic axes could only be determined by inspection of the external symmetry of a crystal and its optical properties. An unambiguous identification could (and can still) be made in many cases on this basis, but the details do not concern us here.

Table 3.1 The seven crystal systems

Crystal system	Axial and angular relationships†	
Triclinic	$a \neq b \neq c$	$\alpha \neq \beta \neq \gamma$
Monoclinic	1st setting	
	$a \neq b \neq c$	$\alpha = \beta = 90° \neq \gamma$
	2nd setting	
	$a \neq b \neq c$	$\alpha = \gamma = 90° \neq \beta$
Orthorhombic	$a \neq b \neq c$	$\alpha = \beta = \gamma = 90°$
Tetragonal	$a = b \neq c$	$\alpha = \beta = \gamma = 90°$
Cubic	$a = b = c$	$\alpha = \beta = \gamma = 90°$
Rhombohedral	(Rhombohedral axes)	
(Trigonal)	$a = b = c$	$\alpha = \beta = \gamma \neq 90° (< 120°)$
	(Hexagonal axes)	
	$a = b \neq c$	$\alpha = \beta = 90°$, $\gamma = 120°$
Hexagonal	$a = b \neq c$	$\alpha = \beta = 90°$, $\gamma = 120°$

† The sign ≠ implies non-equality by reason of symmetry; accidental equality may occur.

3.2 Crystal Planes and Their Indices

3.2.1 Crystal planes

The term "crystal planes" suggests the beautiful crystal faces appearing in naturally occurring crystalline minerals such as quartz or rock salt. Natural crystals occur in many different shapes, but the crystals of a given substance are generally bounded by crystal faces which have characteristic interfacial angles to each other. In another sense, however, the term crystal plane is equally applicable to the lattice planes we have been discussing. The latter, in fact, are far more important for an understanding of X-ray diffraction than the former, which are primary concern in the fields of mineralogy or macroscopic crystallography. The set of non-visible lattice planes is directly related to the arrangements of the atoms within a crystal, as detailed in Section 3.1.

We have already seen that a simplified representation of the assembly of atoms in a crystal can be obtained by choosing representative points (lattice points) for each repetitive group of atoms. Fig. 3.3(a) shows some of the multiplicity of planes which can be drawn

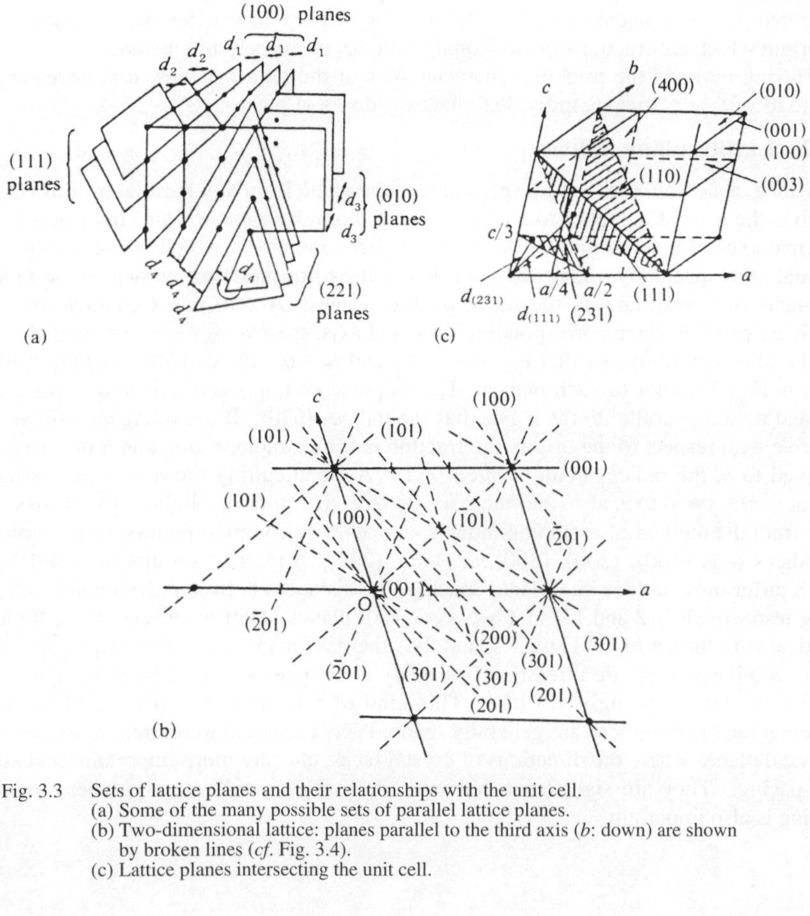

Fig. 3.3 Sets of lattice planes and their relationships with the unit cell.
(a) Some of the many possible sets of parallel lattice planes.
(b) Two-dimensional lattice: planes parallel to the third axis (*b*: down) are shown by broken lines (*cf.* Fig. 3.4).
(c) Lattice planes intersecting the unit cell.

parallel to a plane through any arbitrary group of three lattice points. It is important to note that although the number of planes may be virtually infinite, there are only a discrete number of directions in which they can run, and that any parallel group of planes will be characterized by a constant spacing. The directions and interplanar spacing of each set of lattice planes are properties of great significance for X-ray crystallographic studies.

As mentioned above, the lattice points referred to here are representative points (the origins of the unit cells) the positions of which are chosen to some extent arbitrarily. The precise locations of the lattice planes passing through them will therefore vary with the choice of lattice point. It must be clearly understood, however, that the directions of the groups of planes and the distances between them do not change. In crystal structure analysis the origins of the unit cells are generally chosen on the basis of symmetry considerations to facilitate calculations, and the positions of the lattice planes are accordingly also fixed.

When the lattice points have been chosen, since a three-dimensional diagram such as that shown in Fig. 3.3(a) is difficult to construct, a two-dimensional lattice is drawn (Fig. 3.3(b)), the third axis being taken as perpendicular (down) to the plane of the paper.

The lattice planes perpendicular to the paper (parallel to the third axis) are shown as broken lines in Fig. 3.3(b). Examination of Fig.3.3 (a), (b), and particularly (c), shows that the lattice planes adjacent to a plane through the origin will cut the axes at distances from the origin which are fractions or fractional multiples of the cell dimensions.

Having outlined the principal characteristics of the lattice planes, it is necessary to be able to identify, or "index", individual planes and sets of planes.

3.2.2 Lattice plane indices

As will be clear from Fig. 3.3(a), (b), and (c), one plane from any parallel groups (the plane which is the nearest neighbor to a plane passing through a given origin) intersects the three principal axes of a given unit cell at distances from the origin which we may write as a/h, b/k, and c/l respectively. h, k, and l, written in the form (hkl), are known as the indices of the plane: they also identify the set of parallel planes. A zero index characterizes planes which are parallel to the corresponding principal axis: the planes may be regarded as meeting the axis at infinity, so that *e.g.* $a/h \rightarrow \infty$, and $h=0$. Indies ($h01$) in Fig. 3.3(b) and ($hk0$) in Fig. 3.4 refer to such planes. Thus a plane cutting the a axis at $a/5$, the b axis at $b/3$, and running parallel to the c axis, has the indices (530). If the intercept with an axis is negative with respect to the origin, the fraction is for example, $-a/h$, and a negative sign is assigned to h, the indices being written ($\bar{h}kl$). A plane cutting the a axis on the negative side at $-a/3$, the b axis at b, and the c axis at $c/2$, therefore has indices ($\bar{3}12$). As will be clear from the method of assigning indices, indices with common factors, or multiple series of indices (*e.g.* (100), (200), (300),...; (111), (222),...) refer to groups of parallel planes which differ only in their interplanar spacing, the distances between (200) and (300) planes being respectively 1/2 and 1/3 that between (100) planes. Half of the set of (222) planes is identical with the set of (111) planes, and the other half is interleaved equidistantly between them. Similarly there are three times as many (333) planes as (111) and their interplanar spacing is correspondingly 1/3 less. This kind of relationship is illustrated in Fig. 3.5. Common factors in indices are generally ignored (*i.e.* canceled) in references to macroscopic crystal planes where the directions of crystal faces, etc., are more important than interplanar spacing. They are significant, however, in X-ray crystallography, where interplanar spacing is also important.

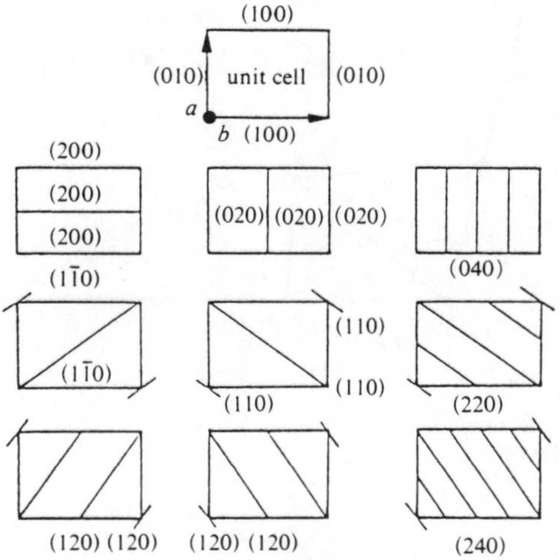

Fig. 3.4 Intersections between sets of planes parallel to the *c* axis (*c*: down) and the unit cell in the *c* plane (or *ab* plane), and their indices.

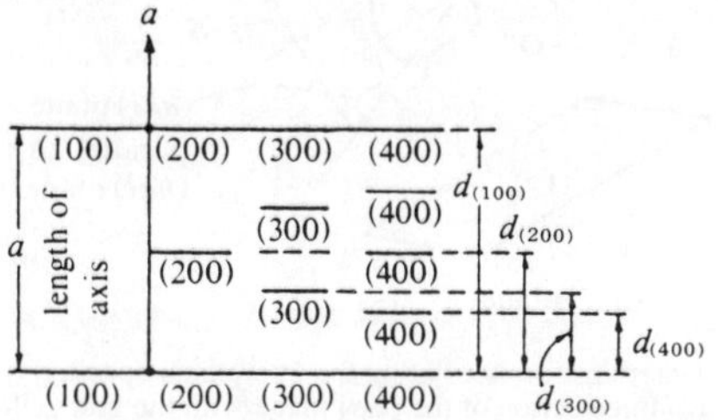

Fig. 3.5 Relationship between multiple indices and interplanar spacing(s).

3.2.3 The spacing of lattice planes and the relationship between plane indices and Laue indices

We shall first present the basic method for calculating the absolute value of the distance between crystal planes.

Let OP (Fig. 3.6(a) and (b)) be the perpendicular from the origin to the (*hkl*) plane in Fig. 3.3(c) where OP $= d_{(hkl)}$. We must consider whether the Laue conditions $(\boldsymbol{S}\cdot\boldsymbol{a}) = h'$ etc. of Eq. 2.39 have some geometrical significance with respect to these planes.

If the direction consines of OP in the diagram are $\cos\alpha$, $\cos\beta$, and $\cos\gamma$, then OP $= d_{(hkl)}$ is given by

$$d_{(hkl)} = (a/h)\cos\alpha = (b/k)\cos\beta = (c/l)\cos\gamma \tag{3.1}$$

If OP*, the direction of $\boldsymbol{S}$ for which the Laue conditions are satisfied (*i.e.* for which diffraction occurs) coincides with OP, then we have

$$(\boldsymbol{S}\cdot\boldsymbol{a}) = Sa\cos\alpha, \quad (\boldsymbol{S}\cdot\boldsymbol{b}) = Sb\cos\beta, \quad (\boldsymbol{S}\cdot\boldsymbol{c}) = Sc\cos\gamma \tag{3.2}$$

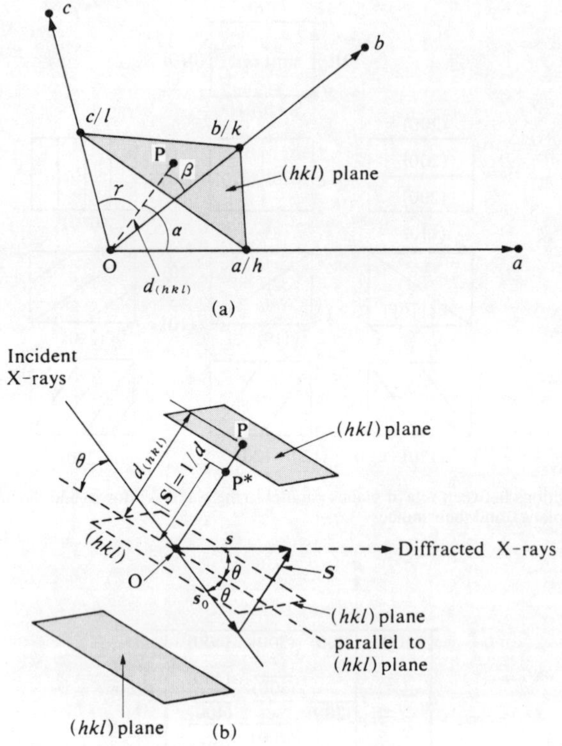

Fig. 3.6 Diagrams for calculating the interplanar spacing(s) $d_{(hkl)}$.
(a) Intersection of the (*hkl*) plane with the unit cell.
(b) Relationship between the (*hkl*) planes and the diffraction vector ***S***.

Elimination of α, β, and γ from Eqs. 3.1 and 3.2 gives

$$d_{(hkl)} = \frac{(\boldsymbol{S}\cdot\boldsymbol{a})}{Sh} = \frac{(\boldsymbol{S}\cdot\boldsymbol{b})}{\boldsymbol{S}k} = \frac{(\boldsymbol{S}\cdot\boldsymbol{c})}{Sl} \tag{3.3}$$

If $h = h'$, $k = k'$, and $l = l'$, we obtain

$$d_{(hkl)} = \frac{1}{S} \tag{3.4}$$

and conversely if $d_{(hkl)} = 1/S$ we obtain

$$h = h', k = k', l = l' \tag{3.5}$$

This shows that where the direction of ***S*** coincides with the perpendicular to a lattice plane, and its scalar magnitude is equal to the reciprocal of the spacing between similar planes, the Laue indices are also the crystallographic indices (*hkl*) of the planes associated with the diffraction, *i.e.* the planes for which Bragg diffraction occurs (*cf.* Section 4.1).

3.2.4 Coordinates of atoms, lattice points, and reciprocal lattice points and indices of crystal planes and zone axes

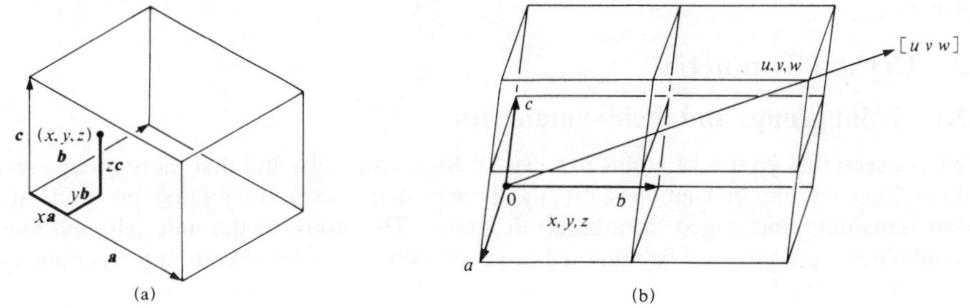

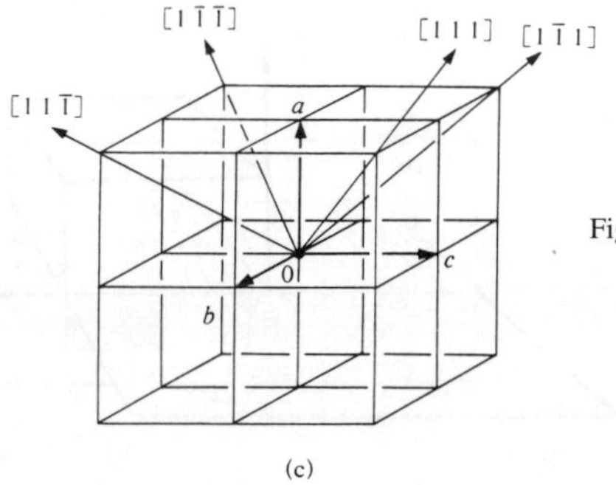

Fig. 3.7 Coordinates and indices.
a, ***b*** and ***c*** are unit cell edges, and a^*, b^* and c^* are those of the reciprocal lattice.
(a) Atomic coordinates, *x, y, z*
(b) Coordinates of a lattice point, *u, v, w*, and index of a zone axis, [*uvw*][2)]
(c) Four zone axes [111], [1$\bar{1}$1], [11$\bar{1}$], and [1$\bar{1}\bar{1}$] are equivalent in this example (cubic crystal), which are expressed as <111> as a whole[2)]

(b), (c) : [Reproduced with permission from T. Sakurai, *X-ray Crystal Structure Analysis*, p. 44, Shokabo (1967)]

Table 3.2 Coordinates and indices

Index	Description
x, y, z	Coordinates of an atom. (*x, y,* and *z* are expressed *in fraction* of cell edges, *a, b,* and *c,* respectively (Fig. 3.7(a)).
u, v, w	Coordinates of a lattice point. (*u, v,* and *w* are expressed *as multiples* of cell edges, *a, b,* and *c,* respectively (Fig. 3.7(b)).
[*uvw*]	Index of a zone axis. (Direction is from the origin toward the lattice point, *u, v, w* (Fig. 3.7(c)).
⟨*uvw*⟩	Index of equivalent zone axis. For example, in a cubic crystal, four zone axes [1 1 1], [1 $\bar{1}$ 1], [1 1 $\bar{1}$], and [1 $\bar{1}$ $\bar{1}$] are equivalent (Fig. 3.7(c), which are expressed as ⟨1 1 1⟩ as a whole.
hkl	Coordinates of a reciprocal lattice point (Fig. 4.4), or Index of a diffraction.
(*hkl*)	Index of a crystal (lattice) plane (Figs. 3.3, 3.4, and 4.4, etc.)
{*hkl*}	Index of equivalent lattice planes (*cf.* Fig. 3.5)

The indices and properties of crystal planes are discussed in the preceding section. In relation to this, here we tabulate and confirm the definitions of coordinates of atom, lattice point, and reciprocal lattice point and indices of zone axes and the crystal plane (Fig. 3.7, Table 3.2).

3.3 Crystal Symmetry[1)]

3.3.1 Point groups and their symmetries

We have seen that groups of atoms in a crystal form unit cells, and that these groups repeat with periodic parallel translations along the three principal axes, the relative positions of the atoms remaining unchanged from group to group. The atoms in the unit cell, and indeed throughout the crystal, may be regarded as an assembly of points exhibiting a certain sym-

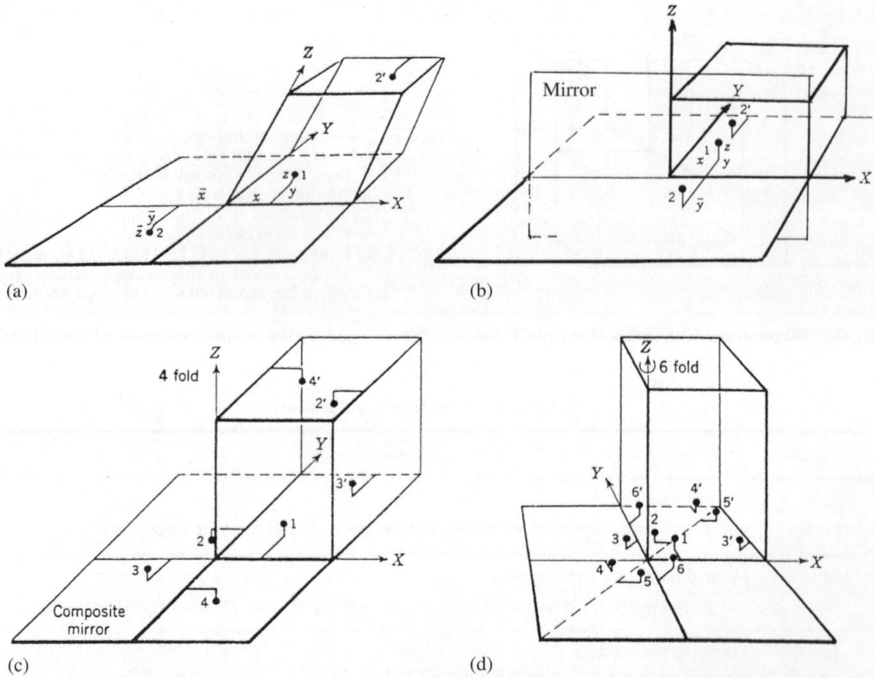

Fig. 3.8 Point-group, symmetry elements and equivalent points.[3)]
(a) $\bar{1}$(or C_i). Center of symmetry, $\bar{1}$ at the origin (oblique axis). Two equivalent points: 1(x, y, z) and 2($\bar{x}, \bar{y}, \bar{z}$).
(b) m(=$\bar{2}$, or C_s). Plane of symmetry (mirror symmertry), m. The mirror plane: ZX plane. Two equivalent points: 1(x, y, z) and 2($x, \bar{y}, z$).
(c) $\bar{4}$ (or S_4). Four-fold inversion axis, $\bar{4}$ on the Z axis. Four equivalent points: 1(x, y, z), 2($\bar{y}, x, \bar{z}$), 3($\bar{x}, \bar{y}, z$) and 4($y, \bar{x}, \bar{z}$).
(d) 6(or C_6). Six-fold axis of rotation, 6 on the Z axis. Six equivalent points: 1(x, y, z), 2($y, y\text{-}x, z$), 3($y\text{-}x, \bar{x}, z$), 4($\bar{x}, \bar{y}, z$), 5($\bar{y}, x\text{-}y, z$) and 6($x\text{-}y, x, z$).
[Reproduced with permission from Dan McLachlan, *X-Ray Crystal Structure*, pp.47, 49, 50, 57, McGraw-Hill (1957)]

metry. Illustrations of this symmetry will be given later, but it will be helpful, first, to examine Fig. 3.8(a). In considering the symmetry of an arrangement of atoms in space, the atoms are all treated as points. Real crystals, however, contain atoms of different species, and it becomes necessary to distinguish between the corresponding points. The most useful distinction to make in this case is between those points which can be mutually transposed by means of some symmetry operation (reflection, rotation, etc.), and those which cannot. The origin in this diagram is a center of symmetry, and an arrangement of atoms displaying this symmetry is shown. Atoms such as 1 and 2, and 1′ and 2′ can obviously be transposed by means of the center of symmetry: the two sets of points are known as sets of equivalent points. In a crystal, 1 and 2, and 1′ and 2′ must be atoms of the same type. Since no symmetry relationship holds between 1 and 1′, or between 2 and 2′, they are known as sets of non-equivalent points (or atoms): the positions of points 1 and 1′ are unrelated and mutually independent, so that they may (though not necessarily will they) be atoms of different types.

If the positions of points are related to an origin and a coordinate system, the points may display symmetry elements other than the center of symmetry. These include a plane of symmetry (*i.e.* mirror symmetry), an axis of symmetry (rotational symmetry), and an inversion axis, each of which may be displayed alone or in various combinations. The equivalent points in (b) have mirror symmetry, in (c) rotational inversion symmetry, and in (d) rotatory symmetry. In (b), reflection of the point 1 (xyz) in the plane of symmetry (ZX) gives 2 ($x\bar{y}z$). Fourfold rotatory inversion symmetry is illustrated in (c), the inversion axis in this case coinciding with the Z axis. This symmetry operation gives the set of equivalent points I (xyz), 2 ($\bar{y}x\bar{z}$), 3 ($\bar{x}\bar{y}z$), and 4 ($y\bar{x}\bar{z}$).

The symmetry elements found in real crystals are the center of symmetry (symbol $\bar{1}$), the plane of symmetry (symbol m), twofold, threefold, fourfold, and sixfold axes of symmetry (symbols 2, 3, 4, and 6), and threefold, fourfold, and sixfold inversion axes (symbols $\bar{3}$, $\bar{4}$, and $\bar{6}$) (*cf.* Table 3.3). The number of ways in which a set of points can be symmetrically arranged in space is surprisingly small, and it has been shown on theoretical grounds that for crystals in which all the atoms are identical there is a maximum of only 32 possible combinations from the most highly symmetrical to the completely asymmetric class. These 32 types are known as the point group symmetries, and in crystallography as the 32 crystal classes. Even of this limited number of possibilities not all have naturally occurring representatives.

As can be seen from Fig. 3.8(d), planes normal to the planes passing through equivalent points and the origin have the same symmetry as the points themselves (sixfold rotational symmetry in this example). Sets of such planes for the equivalent points of the various

Table 3.3 Symbols of lattice and symmetry elements

(a) Symbols of lattice

Symbol	Lattice type
P	Primitive lattice
A, B, C	Base-centered lattice
F	Face-centered lattice
I	Body-centered lattice
R	Rhombohedral lattice

(*Continued*)

Table 3.3 Symbols of lattice and symmetry elements (*Continued*)

(b) Symbols of symmetry elements

1) Symmetry elements without translation

Symbol	Symmetry axis	Graphical symbol	Symbol	Symmetry axis	Graphical symbol	Symbol	Symmetry axis	Graphical symbol
1	Rotation onefold	None	6	Rotation sixfold	⬢	$\bar{4}$	Inversion fourfold	◆
2	Rotation twofold	⬮ (normal to paper) → (parallel to paper)†	$\bar{1}$	Inversion onefold or Center of symmetry	○	$\bar{6}$	Inversion sixfold	⬢
			$\bar{2}$	Inversion twofold or Reflection plane	▁ (normal to paper) ┐ ┘ (parallel to paper)†			
3	Rotation threefold	▲	(*m*)	(Mirror)				
4	Rotation fourfold	◆	$\bar{3}$	Inversion threefold	△			

† If the plane is at $z=\frac{1}{4}$ this is shown by printing $\frac{1}{4}$ beside the symbol.

2) Symmetry elements with translation

Symbol	Symmetry axis or plane	Graphical symbol	Translation along the axis
2_1	Screw twofold	(normal paper)	Right-handed screw $c/2$
		⟶ (parallel to paper)	Either $a/2$ or $b/2$
		(normal to paper)	
3_1	Screw threefold	▲	$c/3$
3_2		▲	$2c/3$
4_1	Screw fourfold	◆	$c/4$
4_2		◆	$2c/4$
4_3		◆	$3c/4$
6_1	Screw sixfold	⬢	$c/6$
6_2		⬢	$2c/6$

Symbol	Symmetry axis or plane	Graphical symbol	Translation along the axis
6_3		⬢	$3c/6$
6_4		⬢	$4c/6$
6_5		⬢	$5c/6$
a, *b*	Glide plane axial	- - - - - - (normal to paper) ┐, ┐, ┐† (parallel to paper)	$a/2$ along [100] or $b/2$ along [010]; or along ⟨100⟩ (parallel to paper)
c	Glide plane axial	·········· (normal to paper)	$c/2$ along z-axis; or $(a+b+c)/2$ along [111] or rhombohedral axes (normal to paper)
n	Glide plane diagonal	—·—·— (normal to paper) ↗ † (parallel to paper)	$(a+b)/2$ or $(b+c)/2$ or $(c+a)/2$; or $(a+b+c)/2$ (tetragonal and cubic)
d	Glide plane "diamond"	—→—→— (normal to paper) $\frac{3}{8}$ ↗ $\frac{1}{8}$ ┐ (parallel to paper)	$(a\pm b)/4$ or $(b\pm c)/4$ or $(c\pm a)/4$; or $(a\pm b\pm c)/4$ (tetragonal and cubic)

† If the plane is at $z=\frac{1}{4}$ this is shown by printing $\frac{1}{4}$ beside the symbol.

[Reproduced with permission from *International Tables for Crystallography*, Vol. A (Teo Hahn ed.), pp. 7–10, IUCr. (1983)]

symmetries of the crystal correspond to the symmetries possessed by the macroscopic crystal planes, *i.e.* those of the outer faces of the crystal as found in nature. Thus the 32 point groups for the arrangement of the atoms in crystals are identical with the 32 crystal classes which refer to the possible symmetries of the crystal shapes. Conversely, since the faces of a crystal can be represented by points, the external symmetries of the crystal can be understood in terms of point group symmetries, and are directly related to the internal symmetry of the atomic arrangements in the crystal.

3.3.2 Space groups

The positions of atoms in the crystal are not determined by the point groups alone. The point groups are related to the symmetries for a collection of points around an origin in space, and in crystals they correspond to the symmetries of the collection of atoms in one unit cell around the lattice origin. The groups of atoms forming a real crystal are therefore not arranged only according to point group symmetries, but also form a lattice by means of regular translation along the directions of the three principal axes, the lattice exhibiting the symmetry inherent in its periodicity. Combination of translation with rotation gives a symmetry element known as a screw axis, while combination of translation with reflection gives yet another symmetry element called a glide plane. Combining lattic type, screw axis, and glide plane symmetries with the point group symmetries gives 230 different possible symmetries for the arrangement of points in a periodic array in space. These 230 symmetries are known as space groups. Fig. 3.9 (p. 56) shows the configurations of representative points for various types of lattice. Fig. 3.10 and 3.11 give examples of screw axis and glide plane symmetry respectively.

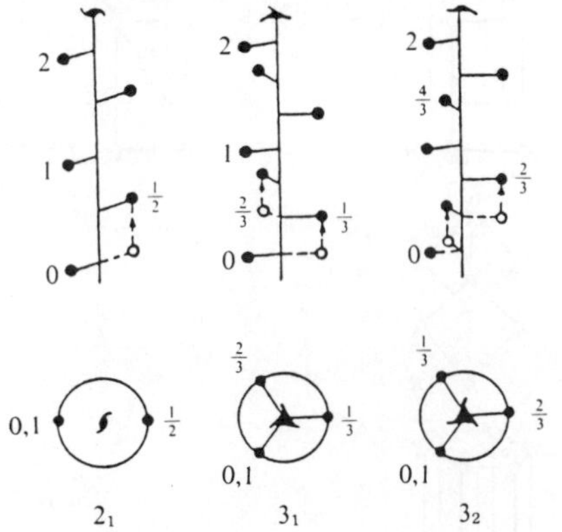

Fig. 3.10 Twofold and threefold screw axes 2_1, 3_1 and 3_3 (*cf.* text).

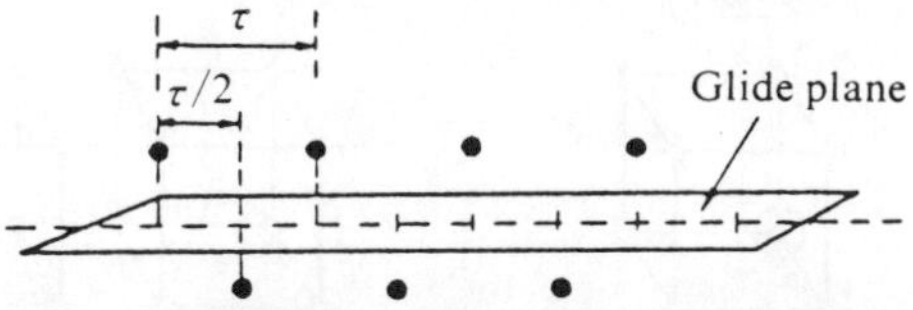

Fig. 3.11 Glide plane symmetry.
τ is the unit cell dimension in the glide direction.

Figure 3.9 illustrates the characteristic features of all 14 possible lattices (the "Bravais" lattices). Each lattice point corresponds to an atom or group of atoms. The letter P identi-

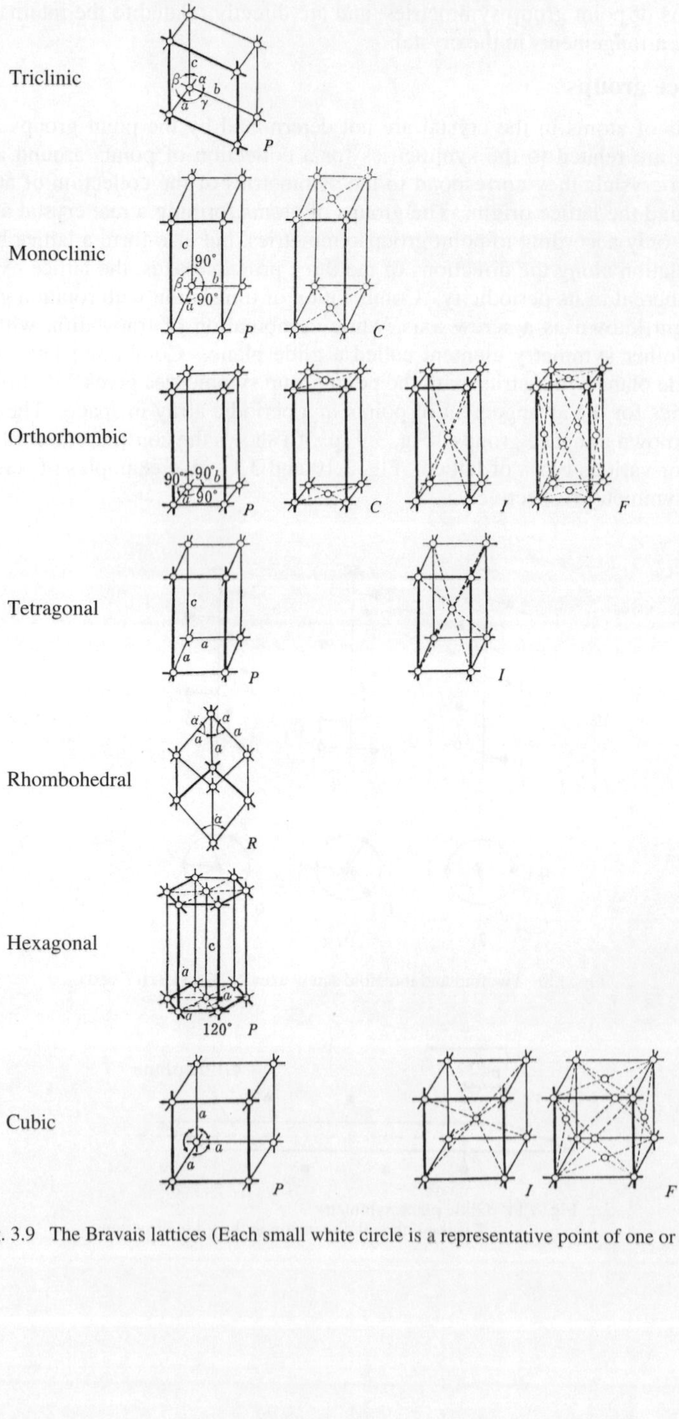

Fig. 3.9 The Bravais lattices (Each small white circle is a representative point of one or more atoms).

fies primitive lattices, *i.e.* those in which the three axes pass through each and every lattice point. *R* (used for the rhombohedral lattice) is also primitive. The other letters refer to various degrees of centering, *i.e.* the inclusion of lattice points through which axes do not pass. It might seem that the possibility of centering introduces further arbitrariness and complexity into the choice of lattice unit cell. The main requirement, however, is that the unit cell should show at least the same degree and kinds of symmetry as the whole crystal, and this resolves most ambiguities. Those which remain can be settled by reference to the X-ray diffraction pattern. Such lattices are termed multiple primitive.

Screw axes may be two-, three-, four-, or sixfold, as indicated by the numbers 2, 3, 4, or 6, with suffixes which reveal the relationship between the angle of rotation and the translation along the axis expressed as a fraction of the repeat distance. Fig. 3.10 shows a twofold screw axis two threefold screw axes, 2_1, 3_1 and 3_2. With the 2_1 screw axis a translation of one half the repeat distance is accompanied by a rotation of $360° \div 2/1$ ($= 180°$), and with the 3_1 screw axis a translation of one third has an angle of rotation $360° \div 3/1$ ($= 120°$). Note that a 3_2 axis is equivalent to a 3_1 axis with the direction of rotation reversed. Table 3.3 gives the complete list of symmetry elements.

A simple glide plane, as shown in Fig. 3.11 involves successive reflections in a plane, each accompanied by a translation equal to one half of the unit cell dimension along the *a*, *b*, or *c* axes: the letters *a*, *b*, or *c* are used appropriately to designate this kind of glide plane. It is also possible for translation to occur simultaneously along two axes (a diagonal glide plane, symbol *n*) or along all three (a diamond glide plane, symbol *d*). The symbols given in Table 3.3, some of which have been detailed above, may be used to describe the various symmetry elements present in a crystal, although an exhaustive description is not possible in the most complex cases. The details of nomenclature have little direct connection with X-ray diffraction, and moreover the point groups, space groups, equivalent points, and formular for structure factor calculations, and for electron density and Patterson functions (see Section 11.2.5), are all available in the *International Tables for Crystallography*, Vol. A. This, and the fact that to cite superfluous crystallographic information here might obscure, rather than clarify, the underlying X-ray diffraction-related phenomena, render it unnecessary to give more than a few examples.

$P2_1/c$ (No. 14), for instance, is a type found very often for crystalline organic compounds. The *P* indicates a primitive cell, and 2_1 denotes a twofold screw axis which, because it is a monoclinic crystal, must necessarily be parallel to the unique axis. If we take the *b* as the unique axis (second setting), the 2_1 must be parallel to the *b* axis, *i.e.* perpendicular to both the *a* and *c* axes. The oblique stroke means "perpendicular to," so that there is a glide plane normal to the 2_1 axis and having a translation of $c/2$.

Finally, the space group *Pnma* (No. 62) for the crystal structure of polyethylene (Section 11.2.8) is detailed in Table 3.4, which is taken from the *International Tables for Crystallography*.[1)] This table gives the symmetry elements present in the unit cell of the space group *Pnma*, as well as the positions assumed by any point with coordinates (x, y, z) in the unit cell after application of the appropriate symmetry operation. The symmetry elements are shown in the top left (the other two projection in the top right and bottom left) diagram of Table 3.4, and the positions occupied by the points as a result of symmetry operations are shown in the bottom right diagram.[10)] In these diagrams the *a* and *b* axes are presumed to lie within the plane of the paper, being drawn vertically and horizontally respectively, with the *c* axis passing through the plane of the paper. Thin solid lines represent the axes of the unit cell; thick solid lines represent sections of glide planes *n*. The symbol $\ulcorner\frac{1}{4}$ at the top left of the unit cell indicates the presence of *a*-type glide planes parallel to the

Table 3.4 Equivalent positions, extinction rules, etc. for the space group *Pnma*[1a)]

Pnma	D_{2h}^{16}	***mmm***	**Orthorhombic**
No. 62	$P2_1/n2_1/m2_1/a$		Patterson symmetry *Pmmm*

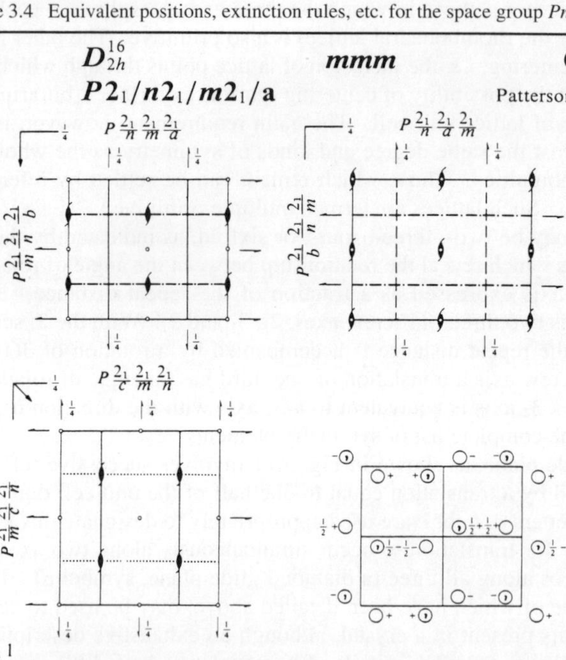

Origin at $\bar{1}$ on $1\,2_1\,1$

Asymmetric unit $0 \le x \le \frac{1}{2}$; $0 \le y \le \frac{1}{4}$; $0 \le z \le 1$

Symmetry operations

(1) 1	(2) $2(0,0,\frac{1}{2})$ $\frac{1}{4},0,z$	(3) $2(0,\frac{1}{2},0)$ $0,y,0$	(4) $2(\frac{1}{2},0,0)$ $x,\frac{1}{4},\frac{1}{4}$
(5) $\bar{1}$ 0,0,0	(6) a $x,y,\frac{1}{4}$	(7) m $x,\frac{1}{4},z$	(8) $n(0,\frac{1}{2},\frac{1}{2})$ $\frac{1}{4},y,z$

Generators selected (1); $t(1,0,0)$; $t(0,1,0)$; $t(0,0,1)$; (2); (3); (5)

Positions

Multiplicity, Wyckoff letter, Site symmetry	Coordinates				Reflection conditions
					General:
8 d 1	(1) x, y, z	(2) $\bar{x}+\frac{1}{2}, \bar{y}, z+\frac{1}{2}$	(3) $\bar{x}, y+\frac{1}{2}, \bar{z}$	(4) $x+\frac{1}{2}, \bar{y}+\frac{1}{2}, \bar{z}+\frac{1}{2}$	$0kl$: $k+l=2n$
	(5) $\bar{x}, \bar{y}, \bar{z}$	(6) $x+\frac{1}{2}, y, \bar{z}+\frac{1}{2}$	(7) $x, \bar{y}+\frac{1}{2}, z$	(8) $\bar{x}+\frac{1}{2}, y+\frac{1}{2}, z+\frac{1}{2}$	$hk0$: $h=2n$
					$h00$: $h=2n$
					$0k0$: $k=2n$
					$00l$: $l=2n$
					Special: as above, plus no extra conditions
4 c m	$x, \frac{1}{4}, z$	$\bar{x}+\frac{1}{2}, \frac{1}{4}, z+\frac{1}{2}$	$\bar{x}, \frac{3}{4}, \bar{z}$	$x+\frac{1}{2}, \frac{1}{4}, \bar{z}+\frac{1}{2}$	
4 b $\bar{1}$	$0, 0, \frac{1}{2}$	$\frac{1}{2}, 0, 0$	$0, \frac{1}{2}, \frac{1}{2}$	$\frac{1}{2}, \frac{1}{2}, 0$	hkl: $h+l, k=2n$
4 a $\bar{1}$	$0, 0, 0$	$\frac{1}{2}, 0, \frac{1}{2}$	$0, \frac{1}{2}, 0$	$\frac{1}{2}, \frac{1}{2}, \frac{1}{2}$	hkl: $h+l, k=2n$

Symmetry of special projections

Along [001] $p2gm$	Along [100] $c2mm$	Along [010] $p2gg$
$a' = \frac{1}{2}a$ $b' = b$	$a' = b$ $b' = c$	$a' = c$ $b' = a$
Origin at $0, 0, z$	Origin at $x, \frac{1}{4}, \frac{1}{4}$	Origin at $0, y, 0$

[Reproduced with permission from *International Tables for Crystallography*, Vol. A. (Teo Hahn ed.), pp. 288–289, IUCr. (1983)]

plane of the paper at heights of $z = \frac{1}{4}$ and $\frac{3}{4}$. The single arrows ↾ indicate twofold screw axes; those with no accompanying numbers indicate that the axis lies in the plane of the paper ($z = 0$), while the $\frac{1}{4}$ beside others shows that the axis is situated at a height of $\frac{1}{4}$ ($z = \frac{1}{4}$) from the $z = 0$ plane in the c direction. The unshaded small circles denote centers of symmetry, and the black ⬮ symbols are the normal indication for the 2_1 screw axes parallel to the c axis (*cf.* Fig. 3.10). It should be clearly understood that not all symmetry elements can be included in this kind of diagram. For example, there are two twofold screw axes parallel to the a axis at a height of $\frac{1}{4}$, and a further two at a height of $\frac{3}{4}$ in the unit cell, but since these coincide in a two-dimensional projection of this type, only those at $\frac{1}{4}$ are shown.

The bisectors of the unit cell are included only to facilitate identification of the positions of the points (bottom right diagram). The various positions adopted by some arbitrary point in the unit cell as the successive symmetry operations of the space group are applied to it may readily be traced from this diagram. The initial position of such a point is here shown at the top-left marked "+". With the top left-hand corner as the origin, the coordinates of this point in the a and b directions are x and y, and that in the c direction is z (positive, *i.e.* above the plane of the paper). For rotation about the center of symmetry the coordinates of this point will therefore become ($\bar{x}$, $\bar{y}$, $\bar{z}$). The positions assumed by the point as a result of each of the symmetry elements in the top-left hand diagram are shown. The symbols $\frac{1}{2}-$ and $\frac{1}{2}+$ indicate that the z coordinates are $\frac{1}{2} - z$ and $\frac{1}{2} + z$ (given as $\bar{z} + \frac{1}{2}$ and $z + \frac{1}{2}$ in the *International Tables for Crystallography Vol. A*) respectively. The points shown in the bottom right diagram are, of course, the equivalent points of the arbitrarily chosen point (x, y, z) with respect to the various symmetry operations. There are eight general equivalent points in the unit cell of this space group. If the initial choice of point lies actually upon one of the symmetry elements, *e.g.* on a center of symmetry, the corresponding symmetry operation leaves the position of the point unchanged, so that the number of equivalent points obtained when the other symmetry operations are applied is immediately reduced by a half. The locations of points with coordinates (x, y, z) chosen so that they do not lie on symmetry elements are referred to as *general positions*, while those on symmetry elements, with a consequent reduction in the number of equivalent points by $\frac{1}{2}$, $\frac{1}{4}$,... compared with the number for the most general position, are called *special positions*.

3.3.3 Relationship between the atoms present in the unit cell and the equivalent points of the space group

It is appropriate at this point to summarize the relevance of the equivalent points of the space groups to the space groups to the actual positions of atoms in the unit cell.

Equivalent points are occupied by atoms of the same species. It is possible for one set of equivalent points to be occupied by the different atoms of the same species within a single molecule, but it is more usual for them to be occupied by atoms of the same species in corresponding positions within different molecules. It follows, therefore, that in molecular crystals the number of molecules (or more generally the number of chemical units) in the unit cell (Z) is usually equal to the number of general positions. Since the number of molecules in the unit cell (Z) should be integral, the number of equivalent points in the space group of such molecular crystals should therefore be an integral multiple or sub-multiple of Z. If the molecule possesses an inherent internal symmetry, some of the atoms of the molecule may occupy certain equivalent positions related by symmetry operations, and certain kinds of atoms may occupy special positions. This has the advantage for crystal structure analysis that it may not be necessary to determine the positions of every atom in the molecule, but only of (say) one half or one third of them, since the locations of the remaining atoms can found by application of the appropriate symmetry operation or operations.

References

1. a) *International Tables for Crystallography*, Vol. A, Space-Group Symmetry, (Teo Hahn ed.), D. Reidel Pub., Dordrecht (1983).

   a´) *International Tables for Crystallography*, Brief Teaching Edition of Vol. A, Space-Group Symmetry, (Teo Hahn ed.), D. Reidel Pub., Dordrecht (1985).

   b) *International Tables for X-Ray Crystallography*, Vol. I, Kynoch Press, Birmingham (1952).
2. T. Sakurai, *X-ray crystal Structure Analysis*, Shokabo (1967) (In Japanese).
3. Dan MacLachlan, *X-Ray Crystal Structure*, pp.47, 49, 50, 57. McGraw-Hill, N.Y. (1957).

4. Detailed Interpretation of the Diffraction of X-Rays by Crystals

4.1 The Bragg Diffraction Condition

As was mentioned in Section 3.2.3, if X-rays are to be diffracted in a certain direction by a crystal, the Laue conditions must be satisfied. This occurs whenever the orientation of the crystal with respect to the direction of the incident X-rays is such that a particular lattice plane (*hkl*) is normal to **S**, and $d_{(hkl)} = 1/|\,\mathbf{S}\,|$. These relationships have been illustrated in Fig. 3.6(a) and (b). The condition that **S** should be perpendicular to (*hkl*) is equivalent to the condition that both the incident and diffracted beams subtend the same angle with respect to (*hkl*). Before diffraction can occur, however, the second condition must also be satisfied. Thus if θ, the angle that the incident beam makes with (*hkl*), is gradually altered, diffraction occurs only at those angles for which $|\,\mathbf{S}\,| = 1/d_{(hkl)}$. These values of θ are necessarily those which satisfy the Laue conditions. While the Laue conditions were derived from a rigorous analytical treatment, it is also possible, because of the geometrical disposition, to understand X-ray diffraction from crystals intuitively in terms of reflection of the incident beam by the (*hkl*) planes, as shown in Fig. 4.1. This approach also enables a restatement of the conditions upon the integers h', k', and l' in Eq. 2.39 in simpler form.

Figure 4.1 shows four of a set of parallel crystal lattice planes at which both incident and diffracted (reflected) X-ray beams subtend angles θ. The path difference between waves reflected by the first plane and those reflected by the second plane (and between each successive pair of planes) is $2d \sin\theta$. If the path differences between successive waves are equal to the wavelength of the X-rays used, or some integral multiple of it, there is constructive interference, *i.e.* all waves are in phase, and the reflected beam is very intense. Thus if $2d \sin\theta = \lambda$ an intense diffracted wave is produced in the direction indicated, the intensity decreasing abruptly with any deviation from the critical value of θ. Note that there is no path difference between waves scattered from different points on the same plane, *i.e.* there will be no difference in phase between waves scattered from point 4 in

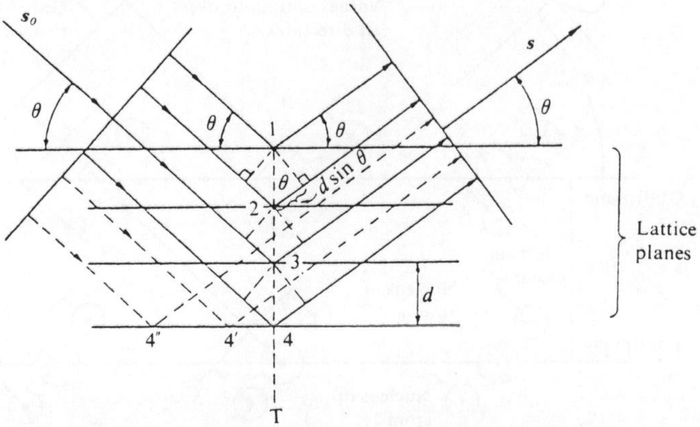

Fig. 4.1 Diffraction of X-rays considered as reflection from a set of lattice planes.

plane 4, and those from points 4′ and 4″. The difractions from the set of planes may therefore be treated as the set of reflections from the points 1, 2, 3,... on a normal T through the planes. The intuitively derived geometrical condition that $2d \sin\theta = \lambda$ can also be derived from Section 3.2.3, for since $|\boldsymbol{S}| = (2\sin\theta)/\lambda$, we have

$$1/d_{(hkl)} = |\boldsymbol{S}| = (2\sin\theta)/\lambda$$

$$2d_{(hkl)} \sin\theta = \lambda \tag{4.1}$$

Eq. 4.1 is known as the *Bragg diffraction condition* or *Bragg's law.*

Figure 4.2 shows an essentially similar situation to that of 4.1, but expressed in terms of the analytical and semi-graphical methods developed in Chapter 2. The two planes shown are the (100) planes of an imaginary crystal. There are two atomic species, and each unit cell contains three atoms. The incident wave excites many atoms in the crystal, and each atom gives rise to a scattered X-ray amplitude of f_1 or f_2 according to its species. The scattered waves from the three members of the unit cell combine to give an amplitude A_{cell} (*cf.* Eq. 2.41). The individual amplitudes A_{cell}, each of which is continuous over all directions, combine to form a diffracted X-ray of amplitude $F(100)$ (the structure factor of Eq. 2.41) and intensity $I \propto F(100)F^*(100)$, where I has a non-zero value only in directions θ such that $2d \sin\theta = \lambda$.

The scattering of X-rays by a crystal gives a diffraction pattern in the form of an array of spots, the total number of which corresponds to the number of (hkl) planes for which the Laue conditions are satisfied. The particular directions of the spots are given by specular reflection of the incident beam in the corresponding lattice planes, *i.e.* the Bragg reflection condition, although the latter gives no information about their relative intensities. In so far that it indicates diffraction directions, the Bragg condition resembles the Laue function G (or L^2) of Eq. 2.45, which is also concerned with the possible directions in which diffraction

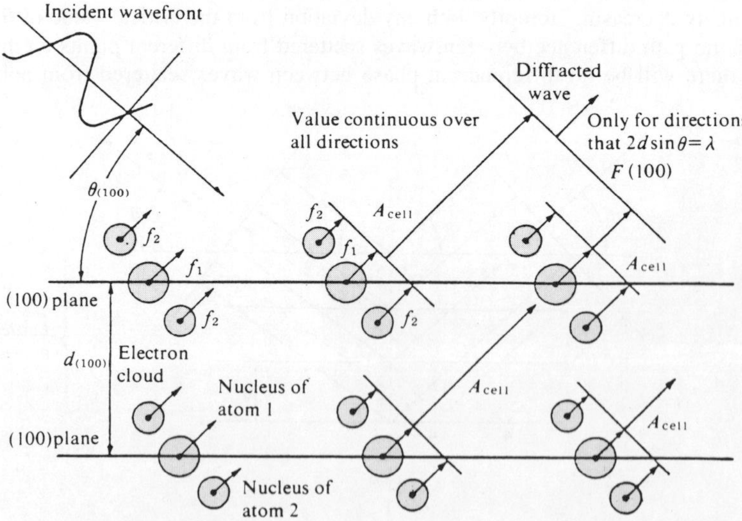

Fig. 4.2 Diffraction of X-rays in terms of the combined scattering amplitudes from atoms and unit cells.

by crystal planes can occur. The Laue function, however, is of more general significance than the Bragg condition in that it yields information on the angular distribution of the diffracted beam about the average direction (the so-called diffraction (or reflection) profile).

4.2 Lattice Structure Factors

As outlined above, the Bragg condition simply gives the average directions for which diffracted X-rays are produced by a crystal. The more general Laue conditions were detailed in Section 2.9.2. Incorporating the Laue function in a combination of Eqs. 2.42 and 2.44 gives the intensity distribution for all diffracting planes

$$I(hkl) = I_0 \frac{1}{{R_\mathrm{F}}^2} \frac{e^4}{m^2 c^4} \frac{1+\cos^2 2\theta}{2} G\, F(hkl) F^*(hkl) \tag{4.2}$$

where I_0 is the intensity of the incident X-rays (*cf.* Eq. 2.3). This expression for the intensities due to the crystal consists of the product of the Laue function G, which gives the distribution of the diffracted rays due to the lattice structure, and $|F(hkl)|^2$, the composite amplitude due to the atoms of the unit cell. Provided that the crystal is not too small, the Laue function may be assumed to take finite values only at a limited number of points, and to be zero at all others (*cf.* Fig. 2.21). The essentially continuous function F is governed by the Laue function, and it follows that the intensity, too, will have finite values at certain fixed points, the value of F at these giving the amplitude of the diffracted beam (*cf.* Fig. 2.20). If, however, the crystal is very small, or the lattice highly disordered, the Laue function is no longer a sharply defined point function, but has a certain broadening. There is consequently, as mentioned in Section 2.9.2, an inherent broadening in the diffraction spot. The factor $F(hkl)$, which depends upon the structure of the unit cell, is however the same in both cases. If $(h'k'l')$ in Eq. 2.44 is replaced by (hkl), and if x_j, y_j, and z_j are the coordinates of the jth atom in the unit cell expressed as fractions of the unit cell dimensions a, b, and c. then

$$F(hkl) = \sum_j f_j \exp\{-2\pi i(hx_j + ky_j + lz_j)\} \tag{4.3}$$

This is known as the lattice structure factor, or simply as the structure factor. This in fact governs the diffracted intensities in the case of most ordinary single crystals, for the latter are almost infinitely large in terms of their effect on the Laue function, which assumes its constant, non-zero value over very narrow angles.

4.3 Reciprocal Space and Reciprocal Lattice

Section 2.2.3 first dealt with the relationship between real space, in which the function $\rho(\boldsymbol{r})$ expresses the structure of the scattering substance, and reciprocal space, which is used in representing the angular distribution $A(\boldsymbol{S})$ of the X-ray scattering amplitude. Fig. 4.3(a) (p.64) shows the geometrical relationship between reciprocal space and the angles at which X-rays are diffracted by a crystal. If diffraction (reflection) occurs at a plane, the direction of $\boldsymbol{S}$ is normal to the plane, and, as we have already seen, its magnitude equals the reciprocal of the interplanar spacing. If the angle θ subtended at a particular plane is gradually increased by changing the direction of the incident beam or (which is physically equivalent) by rotating the crystal, reflection occurs successively from the planes with multiple indices. The reflected directions $\boldsymbol{s}_1$, $\boldsymbol{s}_2$, and $\boldsymbol{s}_3$,... have angles θ_1, θ_2, and θ_3,... as shown. In general, $\boldsymbol{s}_1$ will be associated with a particular (hkl) plane, and $\boldsymbol{s}_n$ at angle θ_n with the nth multiple plane (nh, nk, nl). The planes of Fig. 4.3(a) are all parallel to the (100) planes, so that diffraction

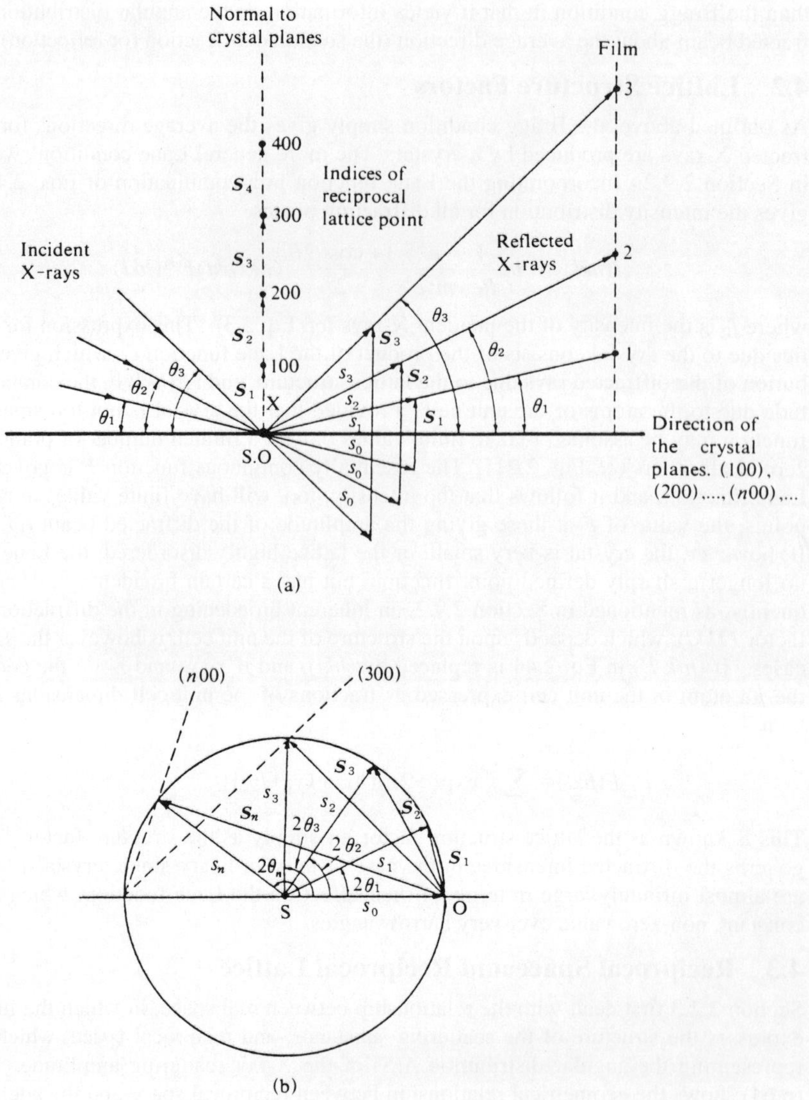

Fig. 4.3 Geometrical relationship between reciprocal space and the angles at which X-rays are diffracted.
(a) Reciprocal lattice points 100, 200,... and diffractions from the corresponding crystal lattice planes with changing angle of X-ray incidence.
(b) Diffractions of (a) redrawn as for crystal rotation, showing the diffraction sphere in reciprocal space (see text).

from the latter gives way in turn to that from the (200), (300),... planes as θ increases. It follows that $| \boldsymbol{S}_2 | = 2| \boldsymbol{S}_1 |$, and $| \boldsymbol{S}_3 | = 3| \boldsymbol{S}_1 |$, and in general $| \boldsymbol{S}_n | = n| \boldsymbol{S}_1 |$.

Consider now Fig. 4.3(b), which demonstrates the relevance of the concepts of reciprocal space to X-ray diffraction. Here the directions $\boldsymbol{s}_0$ of Fig. 4.3(a) all coincide along SO, corresponding to the case where the crystal rotates in the X-ray beam. The angle between $\boldsymbol{s}_0$ and the reflected directions $\boldsymbol{s}_1$, $\boldsymbol{s}_2$,... are those of Fig. 4.3(a), and the magnitudes of $\boldsymbol{S}_1$, $\boldsymbol{S}_2$,... are also identical. Therefore $\boldsymbol{S}_1$, $\boldsymbol{S}_2$,... meet $\boldsymbol{s}_1$, $\boldsymbol{s}_2$,... at points on the circumference of a circle with center O. The argument may be extended to all planes in the three-dimensional crystal, so that if all possible directions are to be included, the general vector $\boldsymbol{S}_n$ (with O as its origin) should meet $\boldsymbol{s}_n$ on the surface of a sphere with center S. Since $\boldsymbol{s}_0$, $\boldsymbol{s}_1$,... $\boldsymbol{s}_n$ are unit directional vectors (in wave number units), $| \boldsymbol{s}_0 |$, $| \boldsymbol{s}_1 |$,... $| \boldsymbol{s}_n |$ (and therefore the radius of the sphere) are all $1/\lambda$. The diffraction sphere, radius $1/\lambda$, is known as the Ewald sphere.

It follows from Fig. 4.3(b) that if a certain group of planes (*nh*, *nk*, *nl*) is reflecting, $\boldsymbol{S}_n$ (with its origin at O) should touch the surface of a sphere with radius $1/\lambda$ and center S; moreover, the (*nh*, *nk*, *nl*) planes should be normal to $\boldsymbol{S}_n$, with an interplanar spacing $1/| \boldsymbol{S}_n |$. Conversely, if an arc on the perpendicular from O to a particular lattice plane cutting off a length equal to the reciprocal of the interplanar spacing does not intersect it at the surface of the sphere, that plane does not give rise to diffraction. It also follows that *n* in the above discussion is not some indefinitely large integer, but is subject to strict physical limitations associated with the X-ray wavelength used and the geometrical features of the crystal lattice planes.

We have effectively restated the conditions for diffraction from a crystal (or reflection from the lattice planes) in terms of a diffraction sphere in reciprocal space rather than an incident direction $\boldsymbol{s}_0$, where the center of the diffraction sphere is at the center of diffraction, and its radius is $1/\lambda$. The reflections from lattice planes, in contrast to those of light from a mirror, are thus shown to be inherently discontinuous: planes in real space are represented by points in reciprocal space, and contact between a set of points and the surface of a sphere is necessarily discrete.

It is evident from Fig. 4.4 (a) that the set of points in reciprocal space which are at distances $1/d_{(hkl)}$ along the normal from the origin to the (*hkl*) lattice planes for all values of *h*, *k*, and *l*, forms a lattice not unlike that of the crystal lattice in real space. This is a consequence of the geometrical arrangement of the crystal lattice planes discussed in Section 3.2. There is a one-to-one correspondence between the points of the reciprocal lattice and the sets of crystal lattice planes. For example, the sets of planes (100), (200), (300),..., with interplanar spacing $d_{(100)}$, $d_{(100)}/2$, $d_{(100)}/3$,..., are represented by points similarly designated 100, 200, 300,..., at increasing distances from the origin, the reciprocals of these distances being related in exactly the same way as the interplanar spacing. The reciprocal lattice is obviously a very convenient device for representing crystal lattice planes: a set of planes can be represented by a single point, rendering unnecessary complex diagrams like Fig. 3.3. The one point represents both the inclinations of the planes to the unit cell and their interplanar spacing. It is vital that these facts should be clearly grasped, for more use will be made in this volume of Fourier theory and the concepts of reciprocal space and reciprocal lattices than of geometrical explanations based on real space relationship (like those we have employed hitherto) in interpreting diffraction effects.

The above geometrical relationship between real and reciprocal lattices requires that the reciprocal lattice axes a^*, b^*, and c^* be perpendicular to the planes containing the *b* and *c*, *c* and *a*, and *a* and *b* axes, respectively, of the real lattice (*cf.* Fig. 4.4), and that the pairs of angles α and α^*, β and β^*, and γ and γ^* between the axes should be supplementary.

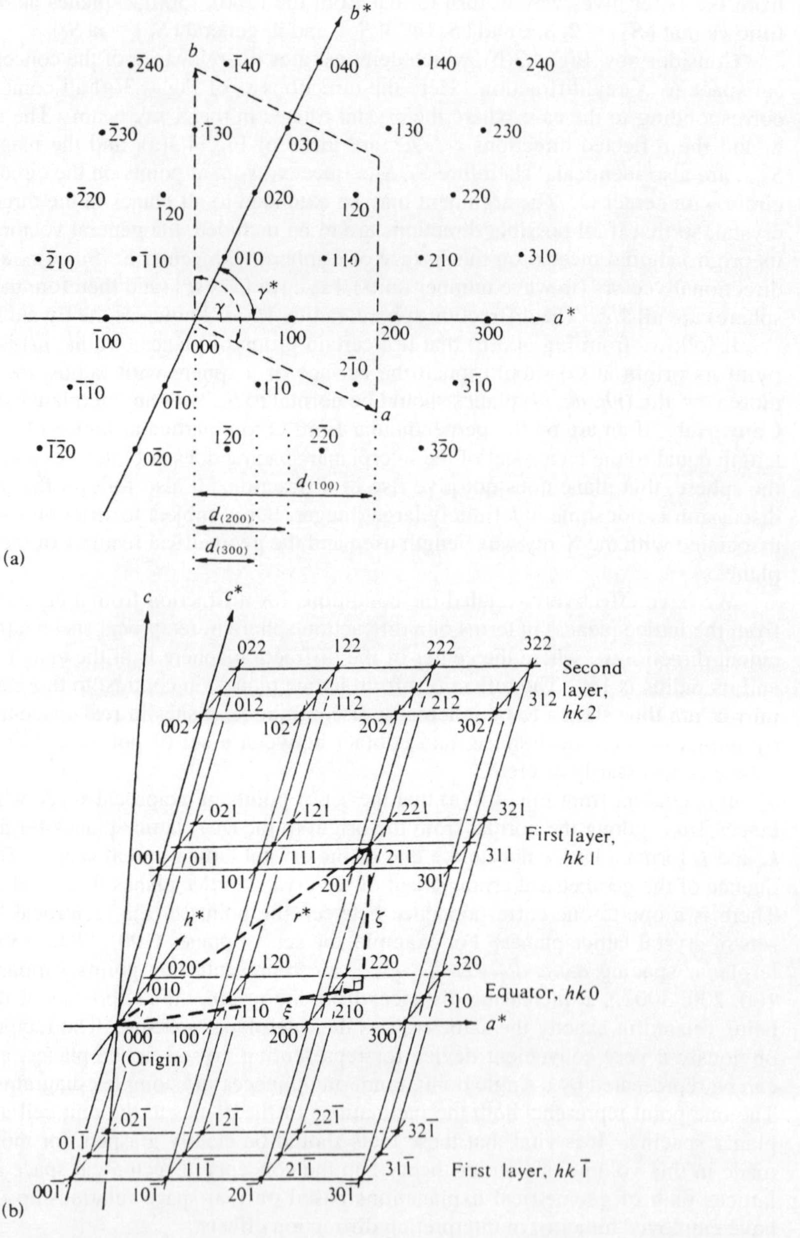

Fig. 4.4 Geometrical relationships between real and reciprocal lattices.
(a) Geometrical relationship for the lattices of a monoclinic crystal (1st setting, the c and c^* axes vertical at the origin.) (*cf.* Table 3.1). The broken lines indicate the unit cell and the dotted lines indicate ($h00$) planes.
(b) Reciprocal lattice of a triclinic crystal, and the vectors r^*, ξ, and ζ. The c and c^* axes do not coincide.

These relationships may be expressed by the following conditions upon the scalar products of the vectors representing the various axes.

$$(\boldsymbol{a}\cdot\boldsymbol{b}^*)=0, \qquad (\boldsymbol{a}\cdot\boldsymbol{c}^*)=0, \qquad (\boldsymbol{a}\cdot\boldsymbol{a}^*)=1,$$

$$(\boldsymbol{b}\cdot\boldsymbol{a}^*)=0, \qquad (\boldsymbol{b}\cdot\boldsymbol{c}^*)=0, \qquad (\boldsymbol{b}\cdot\boldsymbol{b}^*)=1,$$

$$(\boldsymbol{c}\cdot\boldsymbol{a}^*)=0, \qquad (\boldsymbol{c}\cdot\boldsymbol{b}^*)=0, \qquad (\boldsymbol{c}\cdot\boldsymbol{c}^*)=1 \tag{4.4}$$

A reciprocal lattice point that is the *h*th lattice point from the origin in the direction of the a^* axis, the *k*th in the b^* direction, and the *l*th in the c^* direction is designated *hkl*, and corresponds exactly to the (*hkl*) planes in real space. Moreover, using vector $\boldsymbol{r}^*_{(hkl)}$ for the location of this point (*cf.* Fig. 4.4(b)), we have

$$\boldsymbol{r}^*_{(hkl)}= h\boldsymbol{a}^* + k\boldsymbol{b}^* + l\boldsymbol{c}^* \tag{4.5}$$

or

$$r^{*2} = \xi^2 + \zeta^2 \qquad (cf.\ \text{Section 4.4.1}) \tag{4.6}$$

where

$$1/r^*_{(hkl)} = d_{(hkl)} \tag{4.7}$$

The relationships between parameters a, b, c, α, β, and γ of the real lattice and the corresponding parameters of the reciprocal lattice are given in Table 4 of the Appendix for the seven crystal systems.

The geometrical conditions for X-ray diffraction by a crystal in terms of reciprocal space and the reciprocal lattice are shown in Fig. 4.5. In Fig. 4.4(a) the origins of the crystal and reciprocal lattices were drawn to coincide so that the significance of indices, coordinates, angles, etc., could be more readily grasped. Fig. 4.5, however, is similar to Fig. 4.4(b): the center of the diffraction sphere S and the origin of the reciprocal lattice O are separated by a distance SO $= 1/\lambda$, where λ is the wavelength of the X-rays. The incident direction is XSO. The broken line PQ indicates the inclination of a set of planes (*hkl*) with interplanar spacing *d*. If the X-ray beam is diffracted by the PQ planes, it will pass through a point R on the surface of the diffraction sphere as shown, with an angle 2θ between the

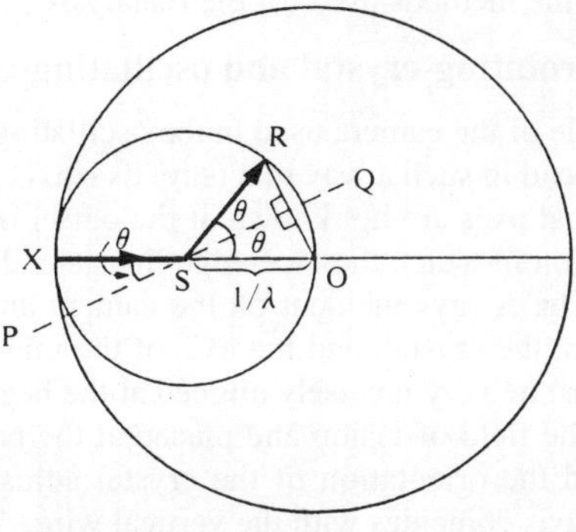

Fig. 4.5 Geometrical conditions upon the reciprocal lattice for X-ray diffraction by a crystal.

unit direction vectors SO and SR of the incident and diffracted beams. The Bragg conditions holds, which requires that $2d \sin \theta = \lambda$. Thus $2(1/\lambda) \sin \theta = 1/d$, and $\mathrm{OR} = 1/d$ (*cf.* Eq. 4.1). The point R is the reciprocal lattice point *hkl* for the (*hkl*) planes.

The possibility remains that for certain orientations of the crystal the direction of the incident X-rays will be unsuitable, with no reciprocal lattice points lying on the surface of the diffraction sphere, and therefore no reflection from the planes (*i.e.* no diffraction). In this case a change in the direction of the incident X-rays (corresponding, in reciprocal space, to rotation of the entire reciprocal lattice about its origin O, or to rotation of the diffraction sphere about O while the lattice remains fixed) will usually bring a lattice point into coincidence with the surface of the sphere, ensuring that the crystal lattice planes corresponding to this point will satisfy the Bragg condition and give rise to reflection (diffraction). This relative motion between diffraction sphere and reciprocal lattice sweeps out a spherical volume of reciprocal space with radius $2/\lambda$ (the larger circle with center at O in Fig. 4.5). All lattice points within this volume can be brought into coincidence with the sphere, and the associated planes will give rise to diffraction. When the volume, which decreases with increasing λ, is so small that it encloses no reciprocal lattice points, all directions of the incident beam alike fail to give rise to reflection. There is thus an upper limit to the wavelength of X-rays which can be used effectively for X-ray diffractional analysis.

In cases where the crystallographic axes, *i.e.* the axes of the unit cell, have already been established from the external shape of the crystal, all information concerning the directions of the incident X-ray beam, the orientation of the crystal, and the diffracted beam, can be found from the geometrical relationships of Figs. 4.5 and 4.6 by drawing the real lattice and deriving the reciprocal lattice. It is then possible to tell whether or not reflection can occur. Conversely, the indices of crystal planes causing a diffraction pattern can be readily found from the positions of the pattern with the aid of this diagram. This will be discussed in greater detail in the next section.

4.4 Wider Applications of the Reciprocal Lattice

We saw in the last section that the phenomena of X-ray diffraction (reflection) by crystals can be very readily understood from diagrams which replace X-rays by a diffraction sphere and crystal planes by reciprocal lattice points. Complex sets of crystal planes and their X-ray diffraction can be represented very simply in this way. We shall now illustrate various methods of making diffraction photographs in actual X-ray diffraction experiments and explain the broad principles of the methods used for their analysis.

4.4.1 Interpretation of rotating-crystal and oscillating-crystal photographs

Figure 4.6 shows the principle of the camera used in the oscillating-crystal method. A crystal is fixed to a goniometer head in such a way that (say) its *c* axis coincides with the axis of rotation. Though the principal axes are not known at the outset of the experiment, it is assumed at first that the direction in which the face-edge lines are clearly visible is the *c* axis, as shown in the diagram. The X-rays incident on the camera are fixed direction, and the axis of the collimator system, the crystal, and the axis of the microscope used for accurate positioning of the crystal must be very precisely aligned at the beginning of the experiment. The crystal is brought into the field of vision and placed at the point of intersection of the microscope cross-wires, and the orientation of the crystal adjusted by means of the goniometer screw so that its *c* axis coincides with the vertical wire. To facilitate the construction of the reciprocal lattice, one of the well-developed faces of the crystal is assumed to be a (100) plane, *i.e.* an *a* plane. This plane is fixed parallel to the direction of the X-rays

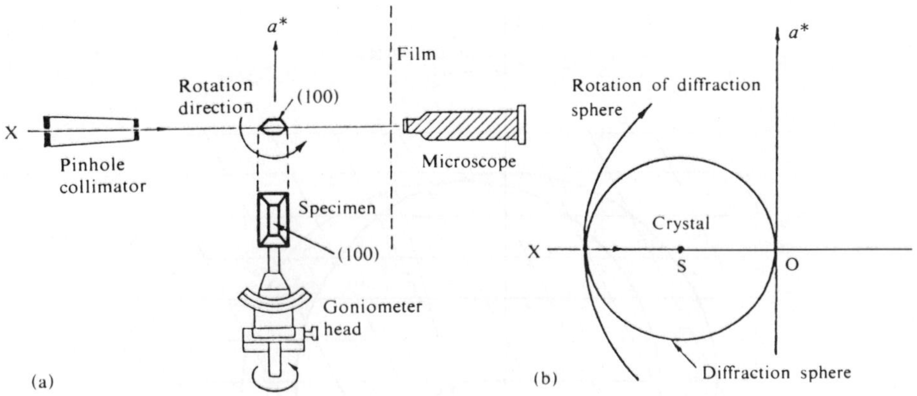

Fig. 4.6 Physical disposition of apparatus for rotation/oscillation photographs (a) and the corresponding reciprocal space diagram (b).

(Section 8.7.1 describes the practical adjustment), and the rotation is started from this position. Since this is the starting point in the experiment, the reciprocal lattice is constructed for this situation, and this juxtaposition must be noted. The construction representing the relationship between the reciprocal lattice in this situation and the incident X-rays and crystal planes is shown in Fig. 4.6(b). Reference to Fig. 4.5 should make the construction clear. In this case, however, no more than the direction of $\boldsymbol{a}^*$ (normal to the *a* planes, *i.e.* (*h*00) planes) is known. The incident X-rays are parallel to the *a* planes and perpendicular to $\boldsymbol{a}^*$, and the diffraction sphere is drawn with its radius equal to the reciprocal of the wavelength used. It will be convenient later if the diagram is drawn on graph paper, the large circle having a radius of 20 cm and the smaller a radius of 10 cm. For plotting the lattice points, the length of the vector $\xi = 2 \sin \theta$ is multiplied by 10 and expressed in centimeters. If the crystal is rotated through 15° to the left (as seen from above) and caused to oscillate between this and the original position, the crystal lattice oscillates, and so does the reciprocal lattice. The same effect would be produced if the incident direction were oscillated between the 0° position and 15° to the right, and it is convenient to think of it in this way. Rotation of the incident direction through 15° to the right causes rotation of the diameter XO of the diffraction sphere to the right with O as the center of rotation. During this rotation the surface of the sphere is brought successively into coincidence with the reciprocal lattice points. In accordance with the principles described above, therefore, diffraction (reflection) occurs at the crystal planes at the instants when the surface of the diffraction sphere passes through the corresponding reciprocal lattice points.

Diagram (a) in Fig. 4.7 shows the equatorial plane (*hk*0) (the plane containing the a^* and b^* axes) of the reciprocal lattice, the axis of rotation of the crystal in this case being the *c* axis. XO is the incident direction, S is crystal, and O is the origin of the reciprocal lattice. The crystal oscillates through 15° to the left from the starting position, in which the *a* planes are parallel to XO. The crystal appears to move from S to S′, but this is of course because our diagrams are drawn on the assumption that the reciprocal lattice remains fixed while the direction of the X-rays changes. Diagram (b) shows the appearance of the diffraction pattern obtained by this method when a film is placed behind the crystal and normal to the incident X-rays. The incident X-ray beam meets the film at its center, O′, which is also the point of intersection of the film with the projection of XO in the reciprocal lattice space. If

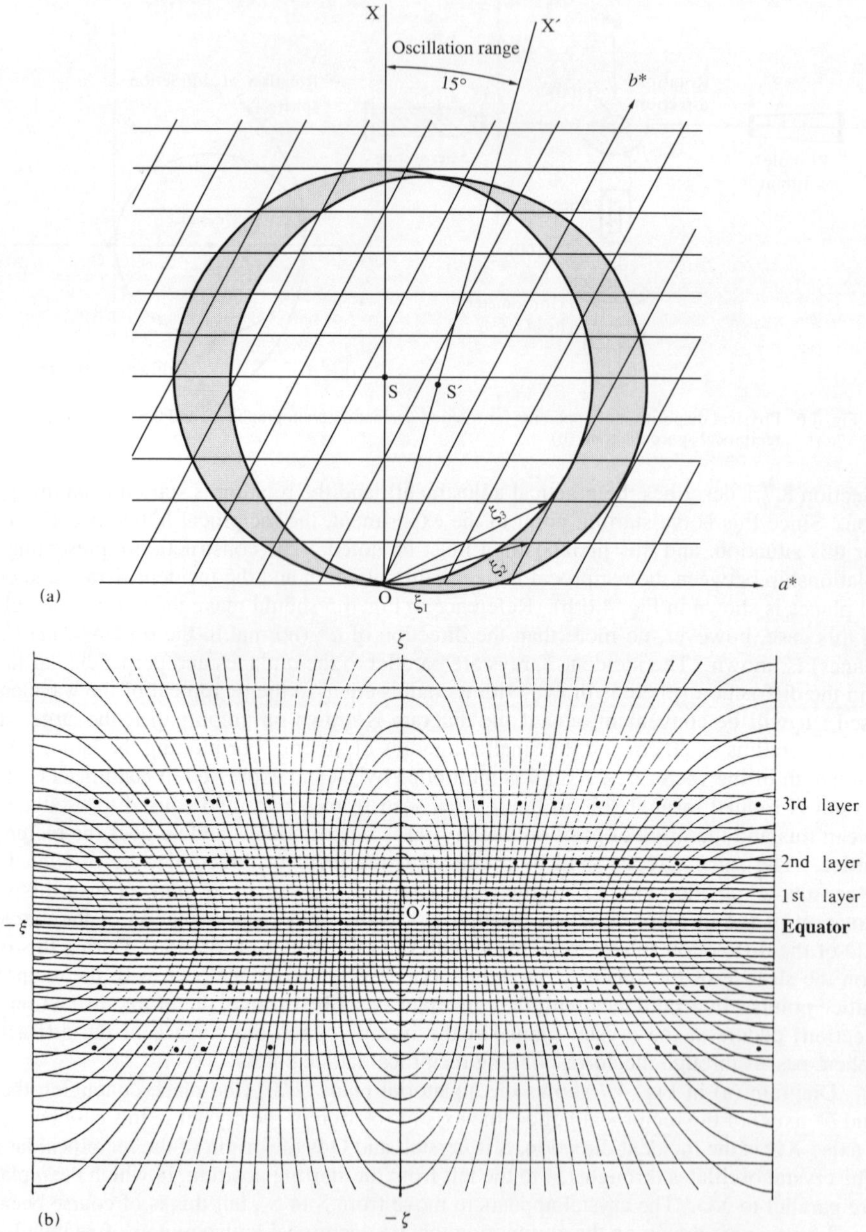

Fig. 4.7 Reciprocal lattice and oscillation photograph.
(a) Reciprocal lattice diagram for oscillation photograph of a monoclinic crystal, (equatorial level).
(b) Schematic representation of the corresponding oscillation photograph (cylindrical film, superimposed by the Bernal chart (Fig. 2 in the Appendix)).
(b) : [Reproduced from M. J. Buerger, *X-Ray Crystallography*, p.148, John Wiley & Sons, Inc. (1942)]

the film is observed from the rear of the camera, the crystal is situated at a fixed distance behind the film, with its c axis vertical. The rotation or oscillation photograph is analyzed with the aid of the geometrical relationships which follow from the orientation of the crystal and the interpretation of the diffraction pattern recorded on the film given below.

If we consider the crescent-shaped areas swept by the surface of the diffraction sphere on which X and X′ lie, the diffraction pattern on the right-hand half of the film corresponds to diffraction by the reciprocal lattice points contained in the right-hand crescent, and the pattern on the left-hand half of the film corresponds to the reciprocal lattice points in the left-hand crescent.

The horizontal row of diffractions in the middle of the pattern is known as the equator. The diffractions on this line correspond to reciprocal lattice points on the a^*b^* plane, *i.e.* on the lattice plane for which $\zeta = 0$ (*cf.* Fig. 4.4(a)), and are therefore all diffractions of planes of the type ($hk0$). The diffractions on the first rows above and below the equator consequently correspond to the next layer of reciprocal lattice points above and below the $hk0$ layer, *i.e.* to the $hk1$ and $hk\bar{1}$ layers respectively. This follows immediately from the information we have already gained concerning the reciprocal lattice, the diffraction sphere, and the direction of diffraction. The interpretation of oscillating-crystal photographs thus reduces to the analysis of the reciprocal lattice planes normal to the axis of rotation of the crystal. We can therefore determine h and k, and the analysis should then be possible simply from the vector ξ on the plane of the reciprocal lattice. The analysis and indexing of the equatorial diffractions is given below by way of illustration.

Several diffractions close to the center, where ξ is small, are chosen and labeled ξ_1, ξ_2,..., ξ_n,... The distance from the diffracted point to the center is measured, and tan 2θ found from this and specimen-to-film distance, R_F. ξ is then found from $\xi = 2 \sin \theta$ (Fig. 4.8).

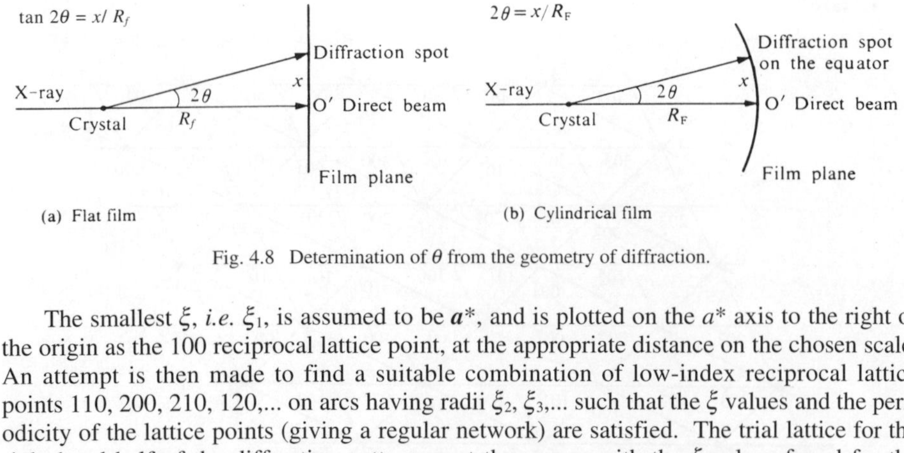

Fig. 4.8 Determination of θ from the geometry of diffraction.

The smallest ξ, *i.e.* ξ_1, is assumed to be $\boldsymbol{a}^*$, and is plotted on the a^* axis to the right of the origin as the 100 reciprocal lattice point, at the appropriate distance on the chosen scale. An attempt is then made to find a suitable combination of low-index reciprocal lattice points 110, 200, 210, 120,... on arcs having radii ξ_2, ξ_3,... such that the ξ values and the periodicity of the lattice points (giving a regular network) are satisfied. The trial lattice for the right-hand half of the diffraction pattern must then agree with the ξ values found for the left-hand crescent from the diffractions in the left-hand half of the film. The lattice can be established with the aid of a mathematical representation of the reciprocal lattice, but here we give only the simpler graphical method. It should be particularly noted that this procedure yields reciprocal lattice points, the lattice lines connecting them being selected after careful scrutiny, generally with the aim of making the reciprocal lattice as nearly rectangular as possible. It is also found that diffraction does not necessarily occur in practice from all sets of planes associated with reciprocal lattice points which fulfill the conditions we have so far been considering. There are other conditions, connected with the symmetry properties of the crystal, which may be responsible for the systematic absence of certain diffractions, *e.g.* for ($h00$) or ($0k0$) planes for which h or k are not even (*cf.* Fig. 4.9(a)), or for $h0l$ for which $h + l$ is not even (*cf.* Fig. 4.9(b) and Section 11.2.3). Other diffractions

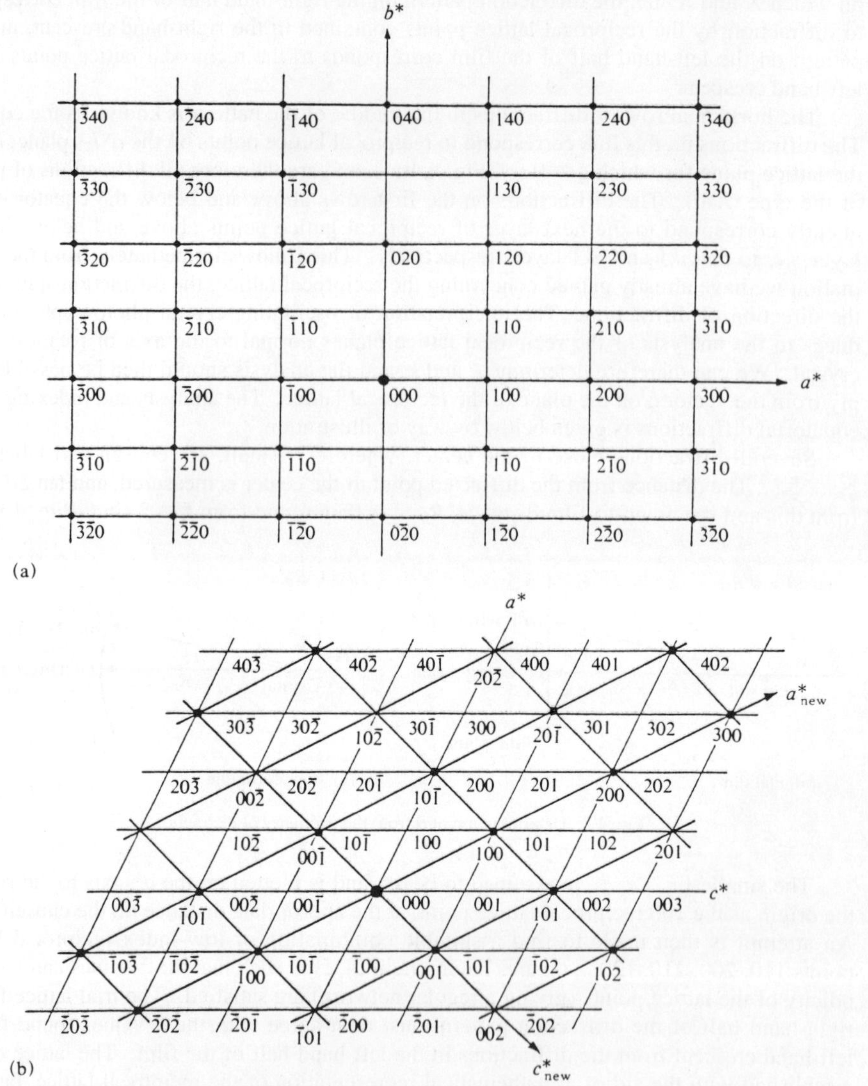

Fig. 4.9 Schematic representation of diffractions with systematic absences as they might be observed in practice (see text).
(a) The systematic absences, $0k0$: $k \neq 2n$, in the reciprocal lattice.
(b) The systematic absences, $h0l$: $h + l \neq 2n$, in the reciprocal lattice indicate that a larger reciprocal unit cell (*i.e.* a smaller real cell) can be chosen.

may be absent because their structure factors, $F(hkl)$, dictate an intensity which is too low to darken the film. The true reciprocal unit cell is found by the use of the higher order diffraction data (*cf.* Fig. 4.9(b)).

The indexing procedure for the other diffractions, *i.e.* general diffractions of the type hkl, is as follows. As was mentioned earlier, the first layer diffractions above and below the equator are the $hk1$ and $hk\bar{1}$ diffractions respectively. The indices to be determined in this case are h and k: let us refer again to Fig. 4.4(a). If the third axis, *i.e.* the c axis in this case, is normal to the a and b axes, c^* will coincide with the c axis, also being perpendicular to both a^* and b^*, and lying in the $(hk0)$ plane. In this case the ξ values of all reciprocal lattice points $hk1$ and $hk\bar{1}$ will be equal to the ξ values of the reciprocal lattice points having the same values of h and k in that equator plane. To index the points, therefore, a point whose ξ value is the same as that of an $hk0$ point is given the same indices h and k, and so becomes $hk1$ or $hk\bar{1}$. The values of x on the photograph in Fig. 4.8 will be different because of the different crystal-to-film distance, but a chart giving the different values of ξ and ζ for different positions on the film is available, and is known as a Bernal chart, see Fig. 2 in the Appendix. This chart is simply placed over the film, and the value of ξ for each diffraction spot can be read off (Fig. 4.7(b)). If this is not used, ξ must be found by measuring the

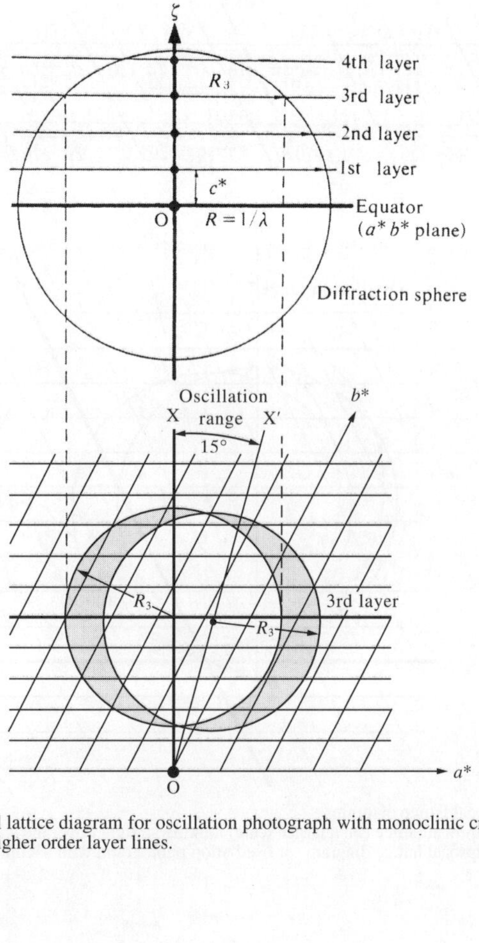

Fig. 4.10 Reciprocal lattice diagram for oscillation photograph with monoclinic crystal: first and higher order layer lines.

distance from the diffraction spot to the origin to find θ and then utilizing the equation $r^{*2} = \xi^2 + \zeta^2$ where $r^* = 2 \sin\theta$ (*cf.* Section 11.2.2). The indices h and k can be found by determining only the values for the spots on the layer lines by comparison with the corresponding spots on the equator. It must be noted that equator corresponds to the $hk0$ reciprocal lattice plane, so that it is correct to represent the diffraction sphere by an equatorial circle. The $hk1$ reciprocal lattice plane, on the other hand, is at a distance, ξ above the equator: intersection between this plane and the diffraction sphere is therefore a circle of slightly

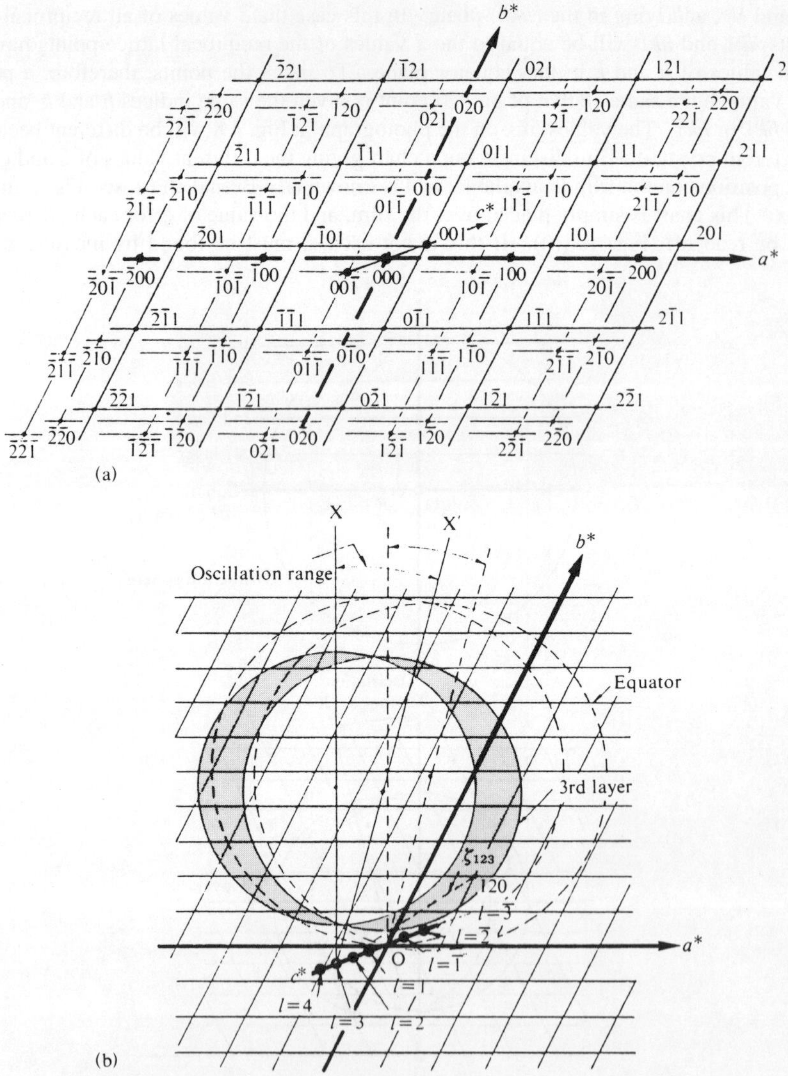

Fig. 4.11 Reciprocal lattice diagrams.
(a) Reciprocal lattice (on a larger scale) looking straight down the *c* axis.
(b) Reciprocal lattice diagram for oscillation photograph with triclinic crystal for the third layer line.

smaller radius, around the circumference of which diffraction can occur. This can be represented by drawing a full-size circle (diameter equal to that of the diffraction sphere) for the *hk*0 reciprocal lattice plane, and concentric circles of successively smaller diameter for the second and subsequent diffracting planes. The radii can be calculated from the values of ξ, and the reciprocal lattice points lying on the crescents formed by these circles correspond to planes that are capable of diffraction (*cf.* Fig. 4.10 (p. 73)).

If the third axis is not normal to a^* and b^*, as in Fig. 4.4(b), the procedure is more tedious. In this case the ξ values of the points on the layer lines are no longer the same as those of the *hk*0 points having the same values of *h* and *k*. This is because, as can be seen from Fig. 4.11, the projections of the lattice points on the equatorial plane do not coincide with the equatorial lattice points. However, utilizing the fact that all the lattice layers are identical, we can still make provisional use of this equatorial plane. The origin (the point with both *h* and *k* zero) of the first layer of the reciprocal lattice is projected on the equatorial plane, and a new lattice is drawn about the projected point as lattice origin. It will therefore be slightly offset from the previous lattice. The original origin of the equatorial plane is taken as the center of the diffraction sphere, which is then rotated. The same result is obtained if the center of rotation of the diffraction sphere is moved from the origin by exactly the same amount as the offset of the points in the first layer, the diffraction sphere then being rotated in the usual manner. By repetition of this procedure for the first layer below the equator, and for the second, third, and other higher order layers, all the diffractions *hkl* on the film can be indexed. The smaller diffraction circles forming the intersections of the lattice layers with the diffraction sphere are shown in Fig. 4.11(b).

We have now established the basis of interpreting rotation and oscillation photographs in terms of the reciprocal lattice. In most cases the normal flat film may be replaced by a cylindrical film, the axis of which coincides with the axis of rotation of the specimen. This also applies for high polymer samples. Since the film almost completely encircles the specimen in this case, diffractions can be recorded up to large diffraction angles. Moreover, instead of being curved in the case of flat film, the layer lines are parallel to the equator as in Fig. 4.7(b). The exact values of ξ and ζ for given camera conditions can be derived by referring to Section 11.2.2 and utilizing Eqs. 11.15 and 11.16.

4.4.2 Interpretation of Weissenberg photographs

The method most commonly used in practice for crystal structure analysis is the Weissenberg method. The analysis of the diffraction pattern obtained will now be described with the aid of the reciprocal lattice. The special feature of this method is that as the crystal is rotated the film is moved parallel to its cylinder axis. This movement is illustrated in Fig. 4.12 by means of the reciprocal lattice. Experimental details are given in Section 8.2.3C.

In Fig. 4.12, for simplicity, we consider only the diffractions produced by the reciprocal lattice points 300 and $3\bar{3}0$ during the rotation of the crystal. The film is cylindrical, the axis of the cylinder being normal to the reciprocal lattice plane a^*b^*. The left-hand diagrams represent the relation between reciprocal lattice and diffraction sphere, while the right-hand diagrams represent films unrolled (not to scale). Fig. 4.12(a) shows that the direct X-ray beam strikes the film at O′ when the crystal is set with its *a* plane parallel (*i.e.* the a^* axis perpendicular) to the incident X-rays. When the crystal rotates through an angle ω_1 the reciprocal lattice point 300 just touches the diffraction sphere and a diffraction spot 300 is produced on the film at a distance x_{300} from O'_1 on the direct beam trace, the film meanwhile shifting leftward a distance z_1 (Fig. 4.12(b)). Further rotation of the crystal (ro-

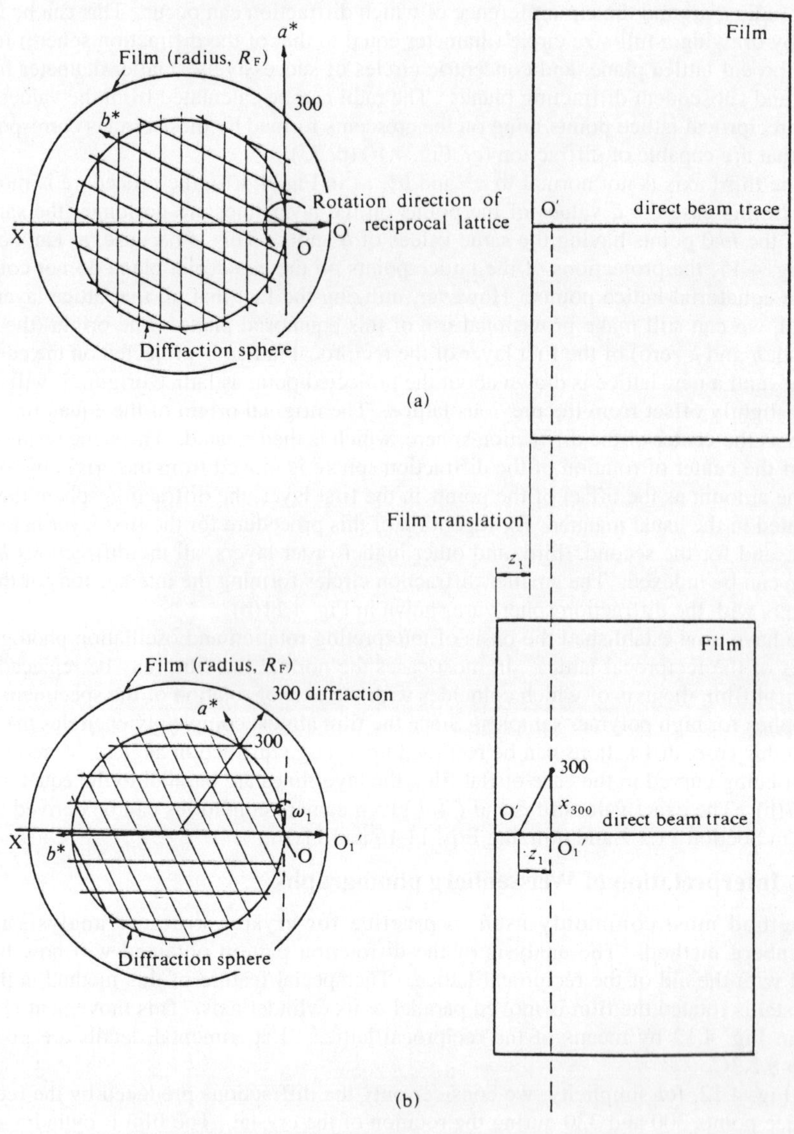

Fig. 4.12 Weissenberg projection of reciprocal lattice lines in the equatorial level (See text. *cf.* Fig. 8.22 for actual Weissenberg photographs). [Reproduced with permission from E. W. Nuffield, *X-Ray Diffraction Methods*, p. 306, John Wiley & Sons, Inc. (1966)]

tation angle ω_2) brings the $3\bar{3}0$ point into coincidence with the surface of the diffraction sphere and a diffraction spot $3\bar{3}0$ is produced at a distance $x_{3\bar{3}0}$, the distance of film translation being z_2 (Fig. 4.12(c)). The difference ω_2-ω_1 between the rotation positions at which the two diffractions occur is related to z_2-z_1 by the translation/rotation constant of the Weissenberg camera (*cf.* Section 8.2.3C), and can therefore be found after measurement of

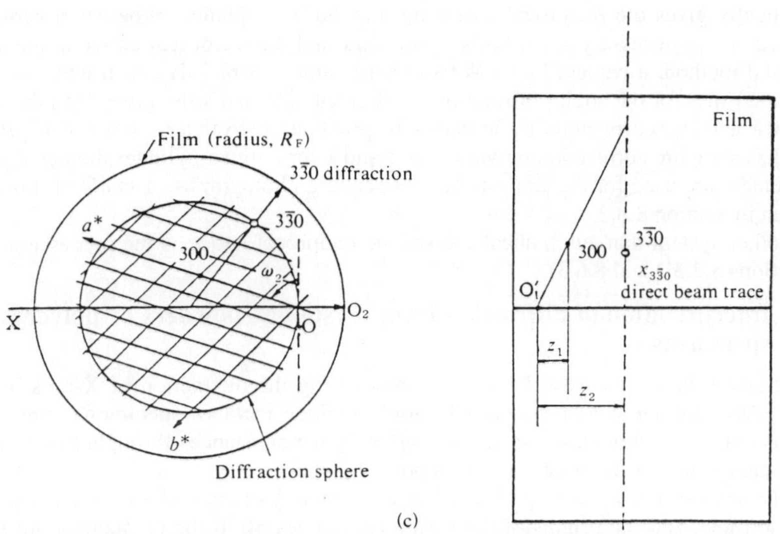

(c)

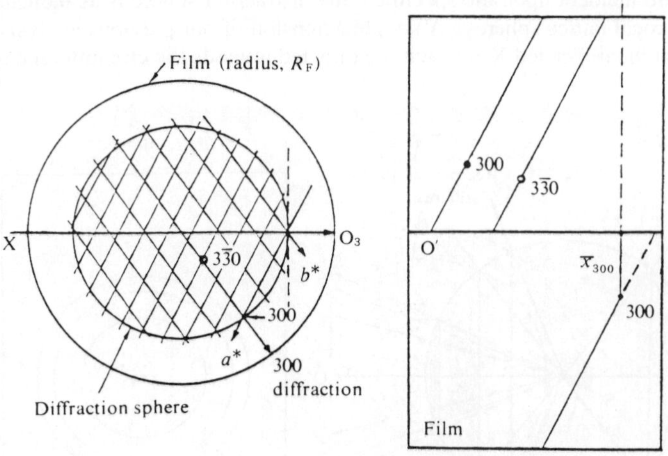

(d)

Fig. 4.12 —*Continued*

the distance of z_2-z_1 on the film. This gives the angle between the diffraction planes of the crystal. Further rotation brings the 300 point into coincidence with another part of the surface of the diffraction sphere again, and another diffraction spot 300 is produced at the lower half of the film at a distance $\bar{x}_{300}$ from the direct beam trace (Fig. 4.12(d)). A similar interpretation for all the reciprocal lattice points that diffract in the course of a 360° rotation automatically gives the reciprocal lattice net for the a^*b^* plane. Whereas a geometrical construction was necessary to index the sports and find the reciprocal lattice in the oscillating-crystal method, a camera like a Weissenberg camera, which is constructed so that the diffraction times for the angles of rotation are derived automatically, gives both the reciprocal lattice and the corresponding indices. In practice, a Wooster chart (*cf.* Fig.1 in the Appendix) showing corresponding values of ξ and x for a given cylinder diameter and film displacement are used for the analysis of Weissenberg photographs. Details of this method are given in Section 8.6.2.

Another system that automatically gives the reciprocal lattice is the precession camera (*cf.* Sections 8.2.3D and 8.6.3).

4.4.3 Interpretation of diffraction from crystalline powders or polycrystalline specimens

The reciprocal lattice can also be used in interpreting the diffraction of X-rays by finely powdered crystalline specimens (which would include metals), specimens consisting of small crystals (crystallites), or crystalline high polymer substances throughout which small crystalline regions are distributed (such as polyethylene).

If the tiny crystals in such specimens are all randomly orientated, the systems are effectively isotropic, and the origins of the reciprocal lattices of all the crystallites are brought into coincidence, all the reciprocal lattice points of a given type *hkl* should be uniformly distributed in the space around the origin at a distance $\boldsymbol{r}^*_{(hkl)}$. The reciprocal lattice points would therefore form concentric spheres about the origin as center. In Fig. 4.13, if X-rays from direction X are incident upon the specimen, the diffraction sphere is as indicated, and intersects the reciprocal lattice spheres. A simple extension of our previous analysis to this situation shows that the diffracted X-rays will be directed towards the circumferences of the

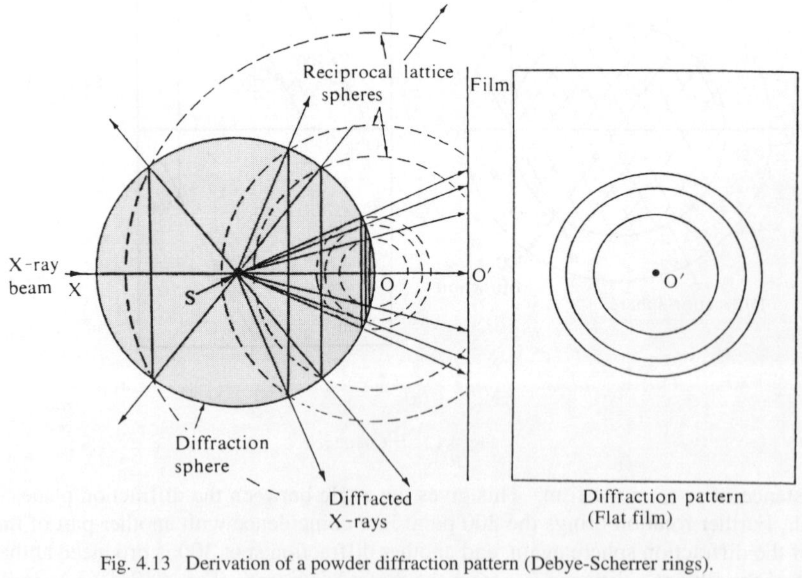

Fig. 4.13 Derivation of a powder diffraction pattern (Debye-Scherrer rings).

circles of intersection between the diffraction sphere and the reciprocal lattice spheres. If the film is flat, as shown in the diagram, the diffracted X-rays will clearly produce concentric circles with point of intersection of the incident direction and the film as center. These circles are known as Debye-Scherrer rings. It is easy to deduce how the pattern will change with the position and shape (*e.g.* cylindrical rather than flat) of the film. It will also be obvious from the above that powder diagrams of this type can yield only the lengths $|r^*|$ of the reciprocal lattice vectors, and tell us nothing about their directions. Even in oscillating-crystal photographs, where the directions of the axes and the plane corresponding to the position from which rotation commences can be found, the construction of the reciprocal lattice is generally very troublesome. It is considerably more difficult to determine the reciprocal lattice by combination of the scant data which can be obtained from powder diagrams, *i.e.* the distances to the reciprocal lattice points. In the case of highly symmetrical lattices, such as those of the cubic, tetragonal, and hexagonal systems, it is possible to assign the indices and determine the unit cell by certain graphical methods. In general, however, for lattices with low symmetry, particularly if their interplanar spacings are large, an accurate analysis is practically impossible. Consequently, although the Debye-Scherrer rings may be readily understood in terms of the reciprocal lattice, they are not used to determine lattices; the importance of the powder method lies rather in comparative and qualitative analysis.

4.4.4 Fibrous polycrystalline specimens

Fibrous substances are generally polycrystalline systems with one axis of the crystals arranged parallel to the fiber axis, as will be described in more detail in Chapter 10. Many systems of this type, in which the crystals are not entirely disordered but exhibit a certain degree of preferred orientation, are found in nature as fibers, membranes, etc. The distribution of the unit cells and their directions for the simplest case of this type, *i.e.* simple fiber structure (uniaxial orientation), is shown in Fig. 4.14(a). Since the reciprocal lattice points in such a system also have the same orientation, they are arranged in groups of concentric circles around the fiber axis (Fig. 4.14(b) and (c)). The type of diffraction pattern obtained with an incident X-ray beam normal to the fiber axis is the same as the rotation photograph of a single crystal, as shown in Fig. 4.14(d). The rotating crystal photograph is produced by diffractions that occur as the crystal rotates, and the same effect is produced by the random orientation of the crystals around the fiber axis. Since the reciprocal lattice in this case is in the form of concentric circles, intersection with the diffraction sphere always occurs, even if the direction of the incident X-rays is not varied, so that a large number of diffraction spots arise. Only part of the group of concentric circles forming the reciprocal lattice is shown in the diagram in order to illustrate how the so-called X-ray fiber patterns are produced. As an example of a real fiber diffraction pattern, Fig. 4.15[1)] shows the diffraction photograph of a fibrous DNA gel. The analysis of such patterns is precisely similar to that for the rotation photographs of single crystals. The ξ values obtained for the diffractions on the equator are used to construct the equatorial plane of the reciprocal lattice and to find the two axes a^* and b^*. The third axis c^* is found by measurement of ξ (see Section 8.6.2.A.d)). These results are used to identify the crystal system and to establish the unit cell parameters.

It should be noted that in the case of diffraction by fibers, diffraction can also arise from reciprocal lattice circles which fail actually to intersect the diffraction sphere. This is in part due to imperfect alignment within the fibers for there is generally a distribution of orientations about the mean which is not necessarily uniform, giving rise to broad bands rather than clearly defined circles in reciprocal space. Furthermore, if the crystals are small (*cf.* Section 2.9.2) or have low crystallinity (*cf.* paracrystals in Section 5.1), the reciprocal lattice points are no longer properly so called; instead, they have definite three-dimensional extension. Consequently, the reciprocal lattice of the assembly is not in the form of circles

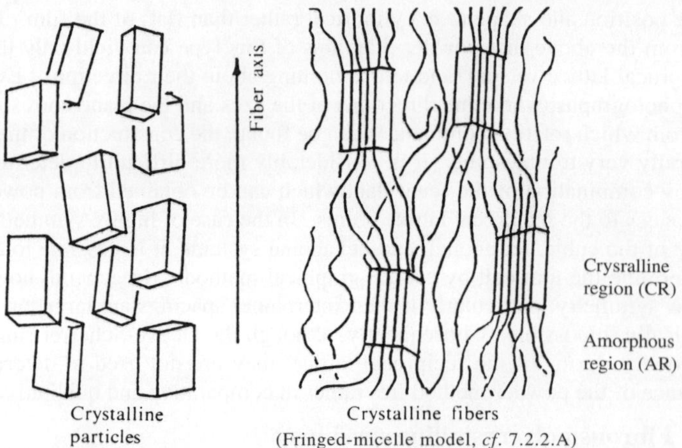

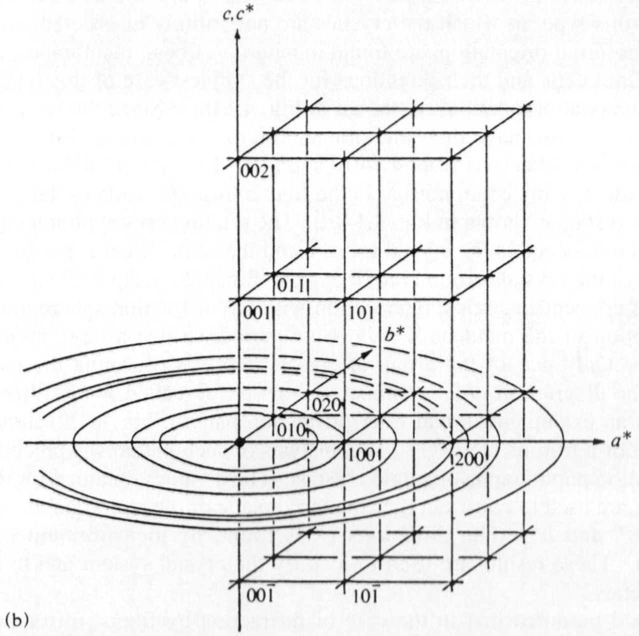

Fig. 4.14 Diffraction pattern from a uniaxially oriented fibrous specimen.
(a) Distribution of the crystalline particles or crystalline regions within the specimen.
(b) Reciprocai lattice of a crystallite, and concentric circular distribution of reciprocal lattice

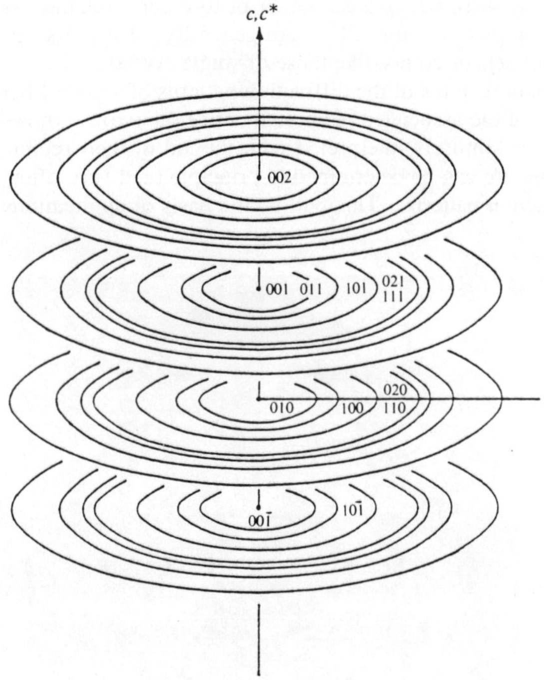

(c)

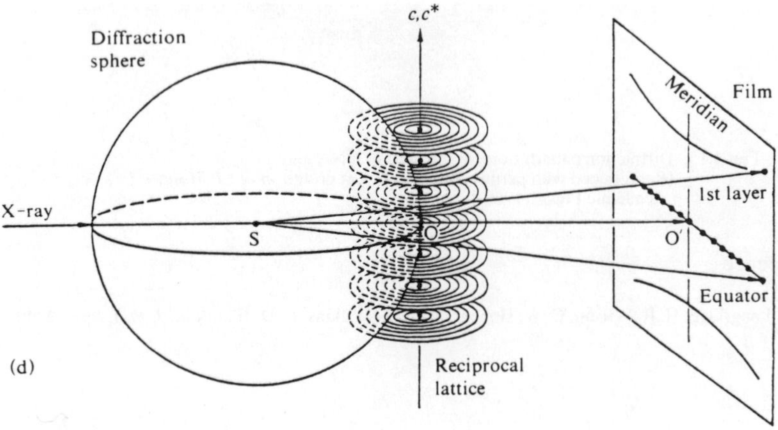

(d)

Fig. 4.14 —*Continued*
(c) Distribution of reciprocal lattice points along the fiber axis.
(d) Geometrical relationship between the distribution of reciprocal lattice points and the diffraction pattern (fiber diagram).

as shown in Fig. 4.14(d), but consists rather of toroidal (doughnut-shaped) zones. This is why the diffraction patterns from fibers are generally diffuse, like that of Fig. 4.15, rather than in the form of sets of points, like those of single crystals.

Just as the main features of the diffraction patterns of simple fiber structures can be understood in terms of the associated reciprocal lattice (above), so those of other special types of assembly can be similarly interpreted with the aid of their reciprocal lattices, and conversely, their structure can be determined in principle (and very often in practice) by analysis of their diffraction patterns. This will be the basis of applications discussed in Chapter 10.

Fig. 4.15 Diffraction pattern from an oriented L*i* DNA gel.[1)]
[Reproduced with permission from R. Langridge *et al.*, *J. Mol. Biol.*, **2**, 30, Academic Press (1960)]

Reference

1. R. Langridge, H. R. Wilson, C. W. Hooper, M. H. F. Wilkins, L. D. Hamilton, *J. Mol. Biol.*, **2**, 30 (1960).

5. Diffraction of X-Rays by Imperfect Crystals and Paracrystals

5.1 Ideal Crystals and Imperfect Crystals

The term "crystalline" as we have so far used it in this volume has usually referred to ideal crystals with perfectly regular arrangements of atoms. High polymer solids are often referred to as crystalline, but, unlike ionic crystals or metals, the internal atomic arrangement is by no means as simple as this might imply. Although direct observation of the internal arrangement is impossible, reliable inferences from physical and chemical evidence lead us to believe that high polymers consist of various regions, each with a characteristic degree of internal order, ranging continuously from something close to the ideal crystalline state to the completely amorphous state.[†1] Chain polymers, for example, can exhibit a very wide variety of arrangements, with varying degrees of disorder. Even in one and the same highly crystalline region of such a substance the arrangement of the atoms and molecules is not always perfectly regular, and even in amorphous regions there are often areas in which the molecules are roughly parallel to one another, *i.e.* in which the arrangement is quasi-crystalline.[†2] This complexity of polymer structure means that the distinction between a substance which forms "ideal" crystals and that which is merely "crystalline" must be clearly drawn when referring to polymers. Cotton, silk, and nylon are "crystalline" high polymers. However, the crystalline regions of these substances are not "crystals" in the strict sense, and are more generally referred to as "paracrystals". The X-ray diffraction patterns of these materials give further evidence of a variety of atomic arrangements, in that they contain elements which range from those more characteristic of a crystal with a perfect lattice to those resembling the diffuse halos given by amorphous substances. Three examples of such patterns are given in Fig. 5.1. The order of decreasing crystallinity is from left to right.

There is another state which is regarded as crystalline, but in which the crystals are soft and plastic like wax. Substances in this state are known as plastic crystals. While the positions of the molecules in the crystal remain unchanged, the molecules as a whole, or their side chains, rotate, weakening the intermolecular interactions to give these crystals plasticity and other characteristic properties. Liquid crystals, either nematic, smectic, or cholesteric, are formed mostly by rod-like molecules. At elevated temperatures before melting, a regular arrangement of molecules in a crystalline state is disturbed by thermal motions of their end, side chain groups, or others, changing the crystal into a liquid crystalline mesophase (thermophilic liquid crystal). In lyophilic liquid crystals, this kind of disorder is caused by water or solvent molecules penetrated into the crystalline lattice, instead of thermal agitations. Their unique structures and properties are known well. Discotic mesophase mostly formed by disc-shaped molecules is also one of the mesophases between crystal and liquid (or solution) phases.

---

†1 We have to write here as "*non-crystalline* solid state" instead of amorphous state. Originally, "*amorphous*" means a-morphous (non morphological shape, non-crystalline.). "Amorphous" also means that the state in question belongs to a non-equilibrium state. However, the word "amorphous" is customarily and widely used in the field of high polymer science to mean non-crystalline. Accordingly, in this volume the word "amorphous" is used instead of "non-crystalline."

†2 Not to be confused with the "quasicrystal," which was first discovered in rapidly solidified Al-Mn alloys.[1] The quasicrystals examined hitherto show icosahedral symmetry, or eight-, ten-, or twelve-fold symmetry. The discovery of quasicrystals has stimulated considerable research activity mainly because it appears to violate the classical rules of crystallography.

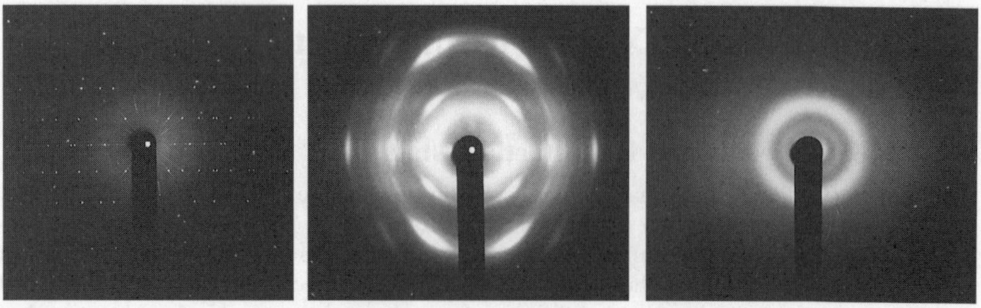

(a) (b) (c)

Fig. 5.1 Examples of X-ray diffraction diagrams of a single crystal and high polymer substances of varying crystallinity (cylindrical film).
(a) Dimethyltin bis(dithiocarbamate) (Oscillation photograph, single crystal); (b) Polyvinylidene-fluoride (Fiber diagram, draw direction vertical); (c) Curdlan (or β-(1 → 3)-D-glucan).

Figure 5.2 is an attempt to illustrate these states using representative points for the unit assemblages of atoms or molecules. In the case of a high polymer the representative point corresponds to the repeating unit, *e.g.* -CH_2-CH_2- for polyethylene. Diagram (a) in Fig. 5.2 shows the amorphous state, in which the arrangement is completely disordered; (d), at the opposite extreme, shows the ideal crystalline state, in which the atoms, ions or molecules have a completely regular arrangement in three dimensions. Crystals of substances having identical molecules of fixed shape can exhibit a very high degree of regularity; these are the so-called ideal crystals. To fall within this category, distortion of the arrangement within the crystal due to all operative factors should be such that the displacement of the lattice points from the ideal positions, expressed as a statistical average over the whole crystal, is

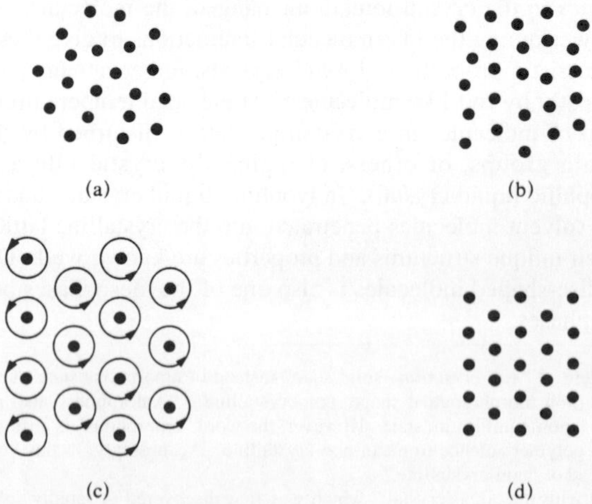

Fig. 5.2 Schematic representation of atomic and molecular assemblages.
(a) Amorphous substance; (b) Paracrystal; (c) Plastic crystal (circular arrow indicates the rotation of each molecule around its center of gravity); (d) Crystal.

(say) less than 1% of the repeat distance (*cf.* Fig. 15.30). The crystalline state which is "perfect" within this 1% limit is more likely to be observed in substances of low molecular weight, where incipient distortion is generally corrected under the equilibration of the interparticle forces within the solid.

The situation is different, however, with high polymers. It is particularly improbable that the molecular arrangement in the fibrous solids formed by linear chain polymers should be that of an ideal crystal as shown in Fig. 5.2(d). The crystal structures of various high polymers beautifully drawn in many textbooks are all ideal structures, and may be far from the real structure. In long chain macromolecules and very large three-dimensional ions such as are found in glasses, all atoms may not have time to assume the arrangement with the highest degree of crystallinity before they are "frozen" into position by the rapid cooling of the melt. Again, there may be thermodynamic factors which make the state with residual distortion inherently stable. This kind of imperfection is represented by Fig. 5.2(b). Diagram (c) in Fig. 5.2 represents a plastic crystal.

A classification of crystal imperfections (lattice distortions) follows.

5.1.1 Lattice distortions of the first kind

A. Thermal vibration

The lattice points constitute an ideal crystal, but there is continuous thermal vibration about the lattice points. The ideal crystalline lattice is, however, maintained by the equilibrium positions of the atoms (Fig. 5.3(b)).

B. Frozen structure

The lattice points are displaced away from their theoretical positions, the displacements being small compared with the interatomic distances, but the average lattice in the crystal is

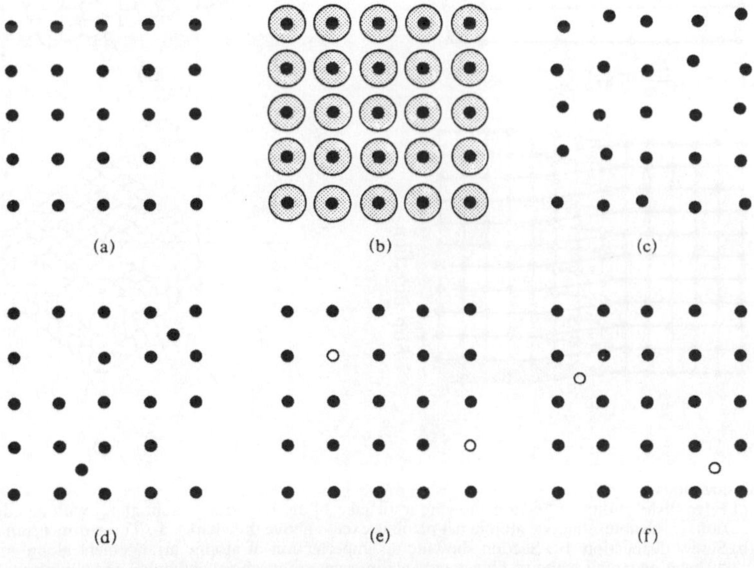

Fig. 5.3 Schematic representation of distortions of the first kind.
(a) Ideal two-dimensional crystal lattice, (b) Thermal motion. (c) Frozen thermal motion, (d) Defect lattice (vacancies and interstitials), (e) Mixed crystal or solid solution (substitution type), (f) Mixed crystal or solid solution (interstitial type).

ideally maintained. This imperfection corresponds to a thermally vibrating crystal lattice, the structure of which has been frozen at a certain instant (Fig. 5.3(c)).

C. Vacancies and interstitials

Most of the lattice points are correctly positioned, but a small number of atoms are missing from the lattice points, or a small number of extra atoms are inserted into the lattice but not on regular lattice sites (Fig. 5.3(d)).

D. Mixed crystals or solid solutions

The atoms, atomic groups, ions, or molecules forming the crystal are not all of the same type, but are a substituted mixture of different components which, on average, form an ideal lattice (Fig. 5.3(e) and (f)).

E. Dislocations

Edge and screw dislocations (Fig. 5.4)[2] are typical one-dimensional defect of the atomic (or molecular) arrangement in crystals. In the top half of the Fig. 5.4 a_1), a projection of a distorted structure looks like as if an extra atomic net plane is inserted perpendicular to the paper like a knife edge (edge dislocation). By shear the slip runs to left (or right) along with the broken line between the upper and lower structures, and the dislocation moves (*cf.* Fig. 5.5(f)). Fig. 5.4 b_1) and b_2) show that the mistake of the atomic arrangement is helical

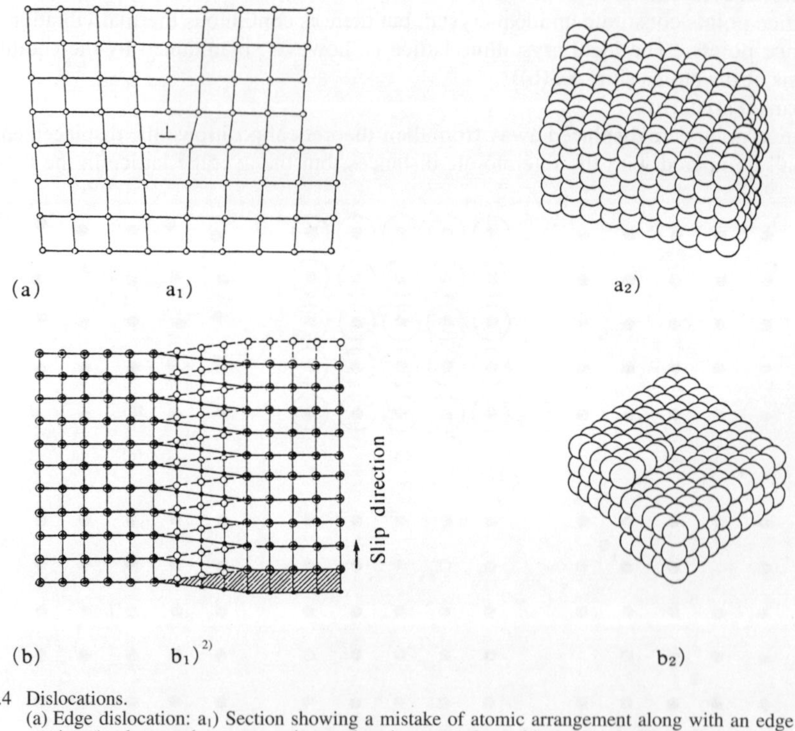

Fig. 5.4 Dislocations.
(a) Edge dislocation: a_1) Section showing a mistake of atomic arrangement along with an edge dislocation; (⊥ denotes that one atomic net plane is excess above this mark), a_2) Three-dimensional model
(b) Screw dislocation: b_1) Section showing an imperfection of atomic arrangement along with a screw dislocation (small white and black circles represent atoms above and below the slip plane, respectively); b_2) Three-dimensional model.
b_1): [Reproduced from W. T. Read, Jr, *Dislocations in Crystals*, p. 17, McGraw-Hill (1953)]
a_2), b_2): [Reproduced with permission from S. Koda, *Introduction to Metal Physics* (Revised ed.), pp. 134, 135, Corona Pub. (1973)]

along the central, vertical line (screw dislocation). In this dislocation, the slip direction of atoms is parallel to dislocation line (*cf.* Fig. 7.11(b_2)).

Basic defects in crystalline regions of high polymer substances are also schematically shown in Fig. 5.5. In a regular arrangement of high polymer chain molecules, there is a defect where the number of repeating units of a chain is more (or less) than the other near-by polymer chains (Fig. 5.5(a)). Fig. 5.5(b) shows an interchange, but not a crossing, of two adjacent chains. A defect caused by a pair of back-foldings of chain molecules is illustrated in Fig. 5.5(c). Back-folding(s) or end(s) of polymer chains also make a defect (Fig. 5.5(d) and (e)). Fig. 5.5(f) may be considered formed from (d) or (e) by the successive movement of adjacent polymer chains to fill vacancies, which may be considered as a movement of an edge dislocation.

The above imperfections of crystals introduce fluctuations into the distances between corresponding atoms throughout the substance, but preserve a long-range order which is distributed only to about the same extent as the short-range order.

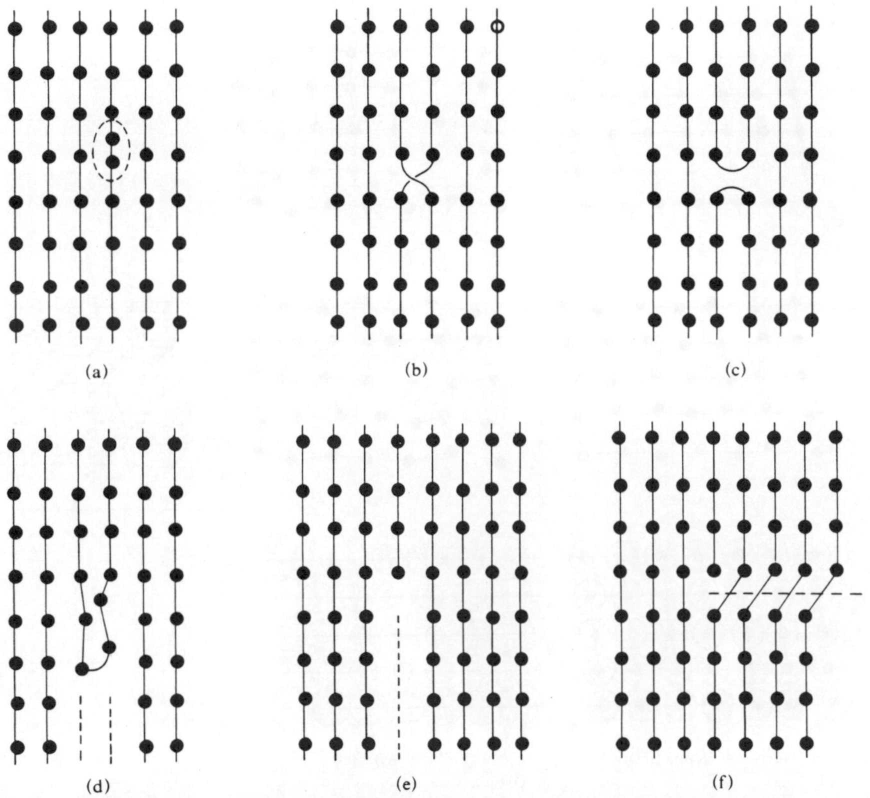

Fig. 5.5 Schematic illustration of basic defects in high polymer crystalline regions (*cf.* Fig. 7.18). (Small black circles represent repeating units in high polymer chain molecules).
(a) Mistake of the number of repeating units, (b) Spatial interchange of adjacent polymer chains, (c) Backfoldings of polymer chains, (d) Backfolding of polymer chain and vacancy, (e) End of polymer chain and vacancy, (f) Kinking of polymer chain to fill vacancy followed by moves of vacancy.

5.1.2 Lattice distortions of the second kind

Whereas the average positions of all the representative points in distortions of the first kind correspond to the lattice points of an ideal crystal, there is a further kind of distortion in which not even the statistical averages of the positions of the representative points form an ideal lattice, the distortion being so great that they deviate significantly from the ideal lattice positions. Three-dimensional periodicity resembling that of an ideal crystal only persists over short ranges; over long ranges there is a permanent disorder resembling that of the amorphous state. These are known as distortions of the second kind, and it is this type of imperfection which Hosemann introduced as paracrystalline.[3,4] In one sense this classification applies to all types of structure intermediates between the ideal crystal and amorphous states. It also, however, covers those mixtures of crystalline and amorphous regions with continuous mutual gradation which are formed by chain molecules, so that the entire non-homogeneous structure may properly be described as paracrystalline.

These different types of distortion are illustrated in Fig. 5.6,[5] in which, again, only the representative points are shown. Fig. 5.6 a_1) shows an ideal oblique crystal lattice. The lat-

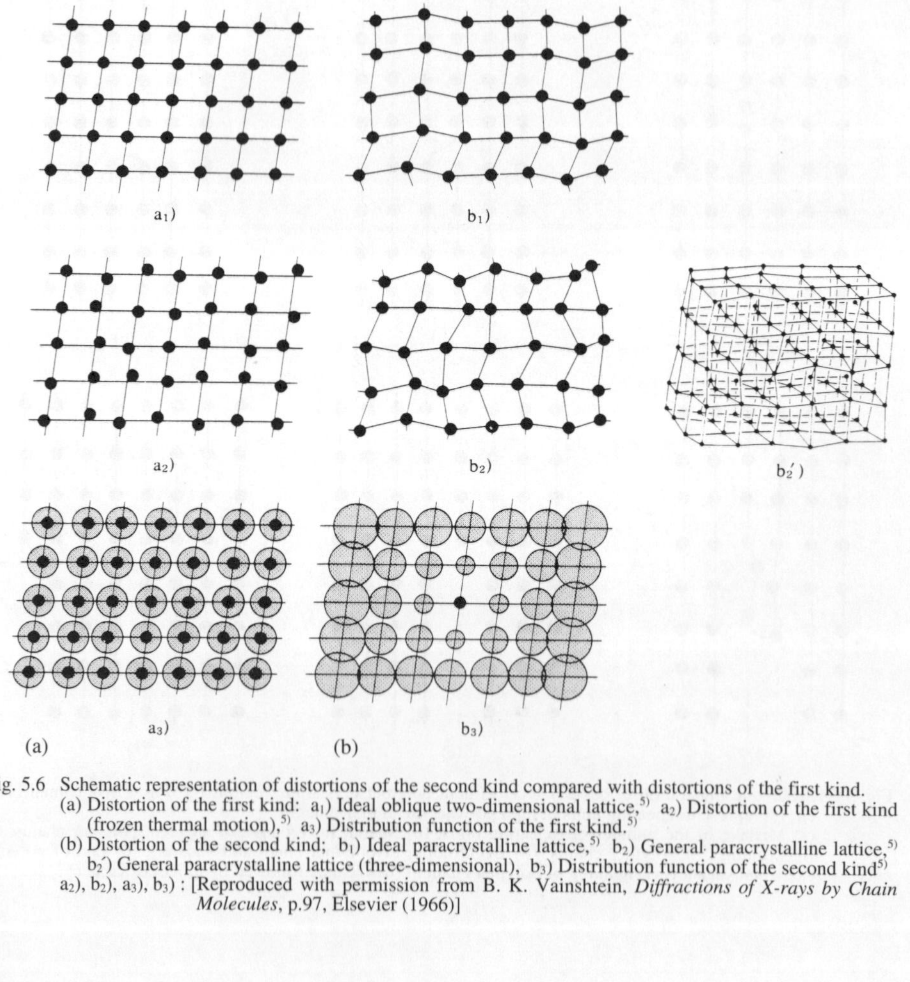

Fig. 5.6 Schematic representation of distortions of the second kind compared with distortions of the first kind.
(a) Distortion of the first kind: a_1) Ideal oblique two-dimensional lattice,[5] a_2) Distortion of the first kind (frozen thermal motion),[5] a_3) Distribution function of the first kind.[5]
(b) Distortion of the second kind; b_1) Ideal paracrystalline lattice,[5] b_2) General paracrystalline lattice,[5] b_2') General paracrystalline lattice (three-dimensional), b_3) Distribution function of the second kind[5]
a_2), b_2), a_3), b_3) : [Reproduced with permission from B. K. Vainshtein, *Diffractions of X-rays by Chain Molecules*, p.97, Elsevier (1966)]

tice in b_1) still appears to possess the regularity of a crystal lattice. All unit cells are parallelpipeds (parallelograms in this two-dimensional representation), so the vectors in any sequence, *e.g.* left to right, bottom to top, are always parallel and equal in continuous part (an ideal paracrystalline lattice: lattice distortion of the second kind). Fig. 5.6 a_2) again illustrates the frozen structure of thermal vibration in an ideal crystal; the positions of the atoms fixed at an instant are denoted by the small black circles. In b_2) the lattice is still discernible with the aid of the lines joining the points, but if the lines were removed, the arrangement would be practically indistinguishable from the amorphous state (a general paracrystalline lattice). Fig. 5.6 b_2') depicts a three-dimensional, general paracrystalline lattice. In the paracrystalline lattice we assume that we can clearly trace the three main directions or three axes corresponding to those of the crystal, that each of these directions is a continuation of paracrystalline unit cell vector $\boldsymbol{a}$, $\boldsymbol{b}$, or $\boldsymbol{c}$, the values and directions of which differ slightly from cell to cell, and that these directions do not fold back and never cross spatially. Fig. 5.6 a_3) and b_3) respectively show distribution function for the lattice distortions of the first and the second kinds.

This concludes our brief description of the paracrystalline state, but since the analytical methods we have so far developed are inadequate to explain X-ray diffraction by such substances, it is necessary to introduce Fourier transform theory.

5.2 Fourier Transform Theory of X-Ray Diffraction

5.2.1 Fourier transform theorem

Our discussion of X-ray diffraction theory has so far been confined mainly to simple methods for calculating the amplitudes of X-rays diffracted from substances with various atomic structures. These are the most general ways of treating X-ray diffraction phenomena, although there is another straightforward way (applicable only to diffractions from crystals) which finds a place in most works on the subject. A far more refined mathematical analysis is needed, however, to explain the diffraction of X-rays by more complex structures, such as the paracrystalline substances dealt with in this chapter. A simple exposition of the fundamentals of Fourier transform theory as it applies to X-ray diffraction follows. This should equip the reader to both interpret and apply X-ray diffraction from paracrystals and other substances.

The relationship between the structure of a substance (in terms of its electron density distribution) and the X-ray scattering amplitude was shown in Eqs. 2.13 and 2.14. Eq. 2.13 shows how the amplitude $A(\boldsymbol{S})$ of the scattered X-rays may be derived from the structure of the substance $\rho(\boldsymbol{r})$ by phase-dependent integration of the contributions from all elements of the system:

$$A(\boldsymbol{S}) = \int_0^\infty \rho(\boldsymbol{r}) \exp\{-2\pi i(\boldsymbol{S}\cdot\boldsymbol{r})\} \mathrm{d}v_r \tag{5.1}$$

According to the Fourier integral theorem, a function $F(\boldsymbol{S})$ in Fourier space is equal to the integral over real space of a function $\rho(\boldsymbol{r})$ in real space, *i.e.*

$$F(\boldsymbol{S}) = \int_0^\infty \rho(\boldsymbol{r}) \exp\{-2\pi i(\boldsymbol{S}\cdot\boldsymbol{r})\} \mathrm{d}v_r \tag{5.2}$$

Thus if $\rho(\boldsymbol{r})$ in Eq. 5.2 is a function representing the electron density of the atoms in a substance, it can be seen from the form of this equation that $F(\boldsymbol{S})$ corresponds exactly to the X-ray scattering amplitude $A(\boldsymbol{S})$ for the substance. From the inverse Fourier transform theorem (Eq. 2.14), we may put

$$\rho(\boldsymbol{r}) = \int_0^\infty A(\boldsymbol{S})\exp\{2\pi i(\boldsymbol{S}\cdot\boldsymbol{r})\}\mathrm{d}v_s \tag{5.3}$$

The physical meaning of this is that the structure of a given substance is equivalent to the Fourier transform of its X-ray scattering amplitude. Using this theorem of Fourier transforms, we can take the mathematical analysis of X-ray diffraction a stage further.

As we saw earlier, the intensity of the scattered X-rays, in units of the scattering intensity of an electron, is simply

$$I(\boldsymbol{S}) = A(\boldsymbol{S})A^*(\boldsymbol{S}) = |A(\boldsymbol{S})|^2 \tag{5.4}$$

Let us now examine the Fourier transform of this intensity *I*.

$$\begin{aligned} Q(\boldsymbol{r}) &= \int_0^\infty I(\boldsymbol{S})\exp\{2\pi i(\boldsymbol{S}\cdot\boldsymbol{r})\}\mathrm{d}v_s \\ &= \int_0^\infty A(\boldsymbol{S})A^*(\boldsymbol{S})\exp\{2\pi i(\boldsymbol{S}\cdot\boldsymbol{r})\}\mathrm{d}v_s \end{aligned} \tag{5.5}$$

Eq. 5.5 is simply derived using Eqs. 5.1 and 5.3. Another useful theorem will now be introduced.

For given functions $\rho(\boldsymbol{r})$ and $\sigma(\boldsymbol{r})$, the function $\rho(\boldsymbol{r}) * \sigma(\boldsymbol{r})$ defined by

$$\int_0^\infty \rho(\boldsymbol{y})\sigma(\boldsymbol{r}-\boldsymbol{q})\mathrm{d}v_q = \rho(\boldsymbol{r}) * \sigma(\boldsymbol{r}) (= \widehat{\rho(\boldsymbol{r})\sigma(\boldsymbol{r})}) \tag{5.6}$$

is known as the convolution of $\rho(\boldsymbol{r})$ and $\sigma(\boldsymbol{r})$, and the corresponding operation is known as convolution or folding, denoted by $*$ (or $\frown$) in the place of a multiplication sign. A two-dimensional illustration of this mathematical operation is given in Fig. 5.7.

If $A(\boldsymbol{S})$ and $S(\boldsymbol{S})$ are Fourier transforms of $\rho(\boldsymbol{r})$ and $\sigma(\boldsymbol{r})$ respectively, the Fourier transform of the product of $\rho(\boldsymbol{r})$ and $\sigma(\boldsymbol{r})$ is

$$\int_0^\infty \rho(\boldsymbol{r})\sigma(\boldsymbol{r})\exp\{-2\pi i(\boldsymbol{S}\cdot\boldsymbol{r})\}\mathrm{d}v_r = \int A(\boldsymbol{S}')S(\boldsymbol{S}-\boldsymbol{S}')\mathrm{d}v_{s'} = A(\boldsymbol{S}) * S(\boldsymbol{S}) \tag{5.7}$$

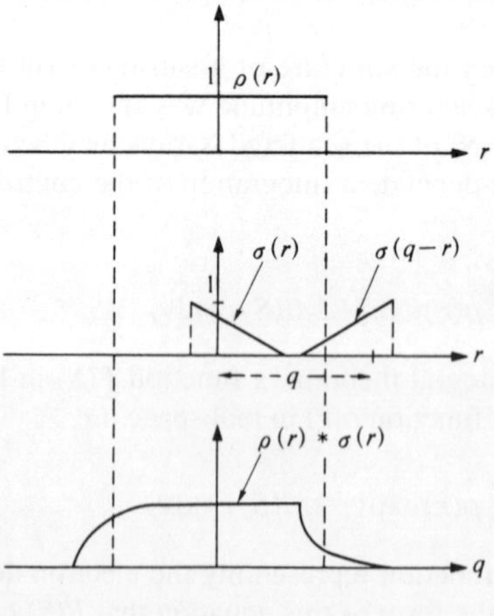

Fig. 5.7 The convolution operation on two functions $\rho(\boldsymbol{r})$ and $\sigma(\boldsymbol{r})$.

Thus the Fourier transform of the product of the functions is the convolution of their respective Fourier transforms $A(S)$ and $S(S)$. The inverse theorem is therefore

$$\int_0^\infty A(\boldsymbol{S}) * S(\boldsymbol{S}) \exp\{2\pi i(\boldsymbol{S}\cdot\boldsymbol{r})\} \mathrm{d}v_s = \rho(\boldsymbol{r})\cdot\sigma(\boldsymbol{r}) \tag{5.8}$$

i.e. the Fourier transform of a convolution is equal to the product of the Fourier transforms of the respective functions.

Let us now return to Eq. 5.5. Since $Q(\boldsymbol{r})$ is the Fourier transform of the product of two functions $A(\boldsymbol{S})$ and $A^*(\boldsymbol{S})$, application of the theorem of Eq. 5.7 to Eq. 5.5 gives

$$Q(\boldsymbol{r}) = \int \rho(\boldsymbol{q})\rho(\boldsymbol{r}+\boldsymbol{q})\mathrm{d}v_q = \rho(\boldsymbol{r}) * \rho(\bar{\boldsymbol{r}}) \tag{5.9}$$

$Q(\boldsymbol{r})$ is the self convolution of $\rho(\boldsymbol{r})$. In physical terms, when the vector $\boldsymbol{r}$ between a certain two points in the real space of the scattering substance is subjected to parallel displacements into every position in all space, the total of the products of the densities $\rho(\boldsymbol{r})$ at the two ends of the vector is $Q(\boldsymbol{r})$. Thus if there are atoms at two points in space separated by a vector corresponding in magnitude and direction to $\boldsymbol{r}$, the value of $Q(\boldsymbol{r})$ will be large. Otherwise the value of $Q(\boldsymbol{r})$ for this $\boldsymbol{r}$ will be zero. $Q(\boldsymbol{r})$ therefore seems to be geometrically of the same nature as the density distribution functions of Eqs. 2.26, 2.33 and 6.74. This correspondence will be discussed later. If $\rho(\boldsymbol{r})$ is the electron density of an infinitely large crystal, $Q(\boldsymbol{r})$ is known as the Patterson function (usually written $P(\boldsymbol{r})$). The value of $Q(\boldsymbol{r})$ is large if there are atoms at both ends of the vector $\boldsymbol{r}$, and this important function assists materially in structural analysis by determining the vectors between the atoms in the crystal. Thus,

$$\begin{aligned}\rho(\boldsymbol{r}) * \rho(\bar{\boldsymbol{r}}) &= \int I(\boldsymbol{S})\exp\{2\pi i(\boldsymbol{S}\cdot\boldsymbol{r})\}\mathrm{d}v_s \\ &= P(\boldsymbol{r}) \text{ (the Patterson function in the case of crystals, cf. Eq. 11.41, Section 11.2.5)} \\ &= Q(\boldsymbol{r}) \text{ (general case)}\end{aligned} \tag{5.10}$$

These functions are obtained by Fourier transformation of the intensity distribution of the diffracted X-rays. The inverse transform is

$$I(\boldsymbol{S}) = \int Q(\boldsymbol{r})\exp\{-2\pi i(\boldsymbol{S}\cdot\boldsymbol{r})\}\mathrm{d}v_r \tag{5.11}$$

so that if the self convolution $Q(\boldsymbol{r})$ of the electron densities in a substance is known, the X-ray diffraction intensity is given by the Fourier transform of $Q(\boldsymbol{r})$. This is therefore an important equation.†

We shall now use the above Fourier transform and convolution theorems to derive a number of factors involved in X-ray analyses. (Fourier transformation of helical molecules is discussed in Section 11.4.)

---

† We have so far put forward Eq. 5.1 as the most convenient starting point for deriving X-ray intensities from the wave amplitude. What we fmd from experimental measurements are, however, not the wave amplitudes, but the intensity, and it is impossible to isolate the effect of phase from the amplitudes. Therefore, when we are seeking to determine structures on the basis of diffraction patterns it is safer to use Eq. 5.11.

5.2.2 Shape factor for the scattering body

The integrations in the equations beginning with Eq. 5.1 should in principle be carried out over all real space or reciprocal space. In X-ray diffraction experiments, however, even if the specimen is large enough to enable an approximation to this condition, only the scattering from a limited region is in fact observed. The following procedure minimizes the attendant difficulties.

The electron density distribution $\rho(\boldsymbol{r})$ in a particle of matter scattering X-rays, or in the case of a large specimen in the entire volume v within the X-ray beam, is expressed in terms of the electron density distribution $\rho_\infty(\boldsymbol{r})$ for the unbounded body as follows.

$$\rho(\boldsymbol{r}) = \rho_\infty(\boldsymbol{r}) \cdot \sigma(\boldsymbol{r}) \tag{5.12}$$

In this equation, the function $\sigma(\boldsymbol{r})$, which is known as the shape function, always takes the value unity when $\boldsymbol{r}$ lies within v, and is zero for all $\boldsymbol{r}$ outside v. On substitution of Eq. 5.12 in Eq. 5.1, by the convolution theorem, convolution of the Fourier transform $S(\boldsymbol{S})$ of $\sigma(\boldsymbol{r})$ and the fourier transform $A_\infty(\boldsymbol{S})$ of $\rho_\infty(\boldsymbol{r})$ gives the following equation.

$$A(\boldsymbol{S}) = A_\infty(\boldsymbol{S}) * S(\boldsymbol{S}) \tag{5.13}$$

If the scattering matter is represented as an infinite periodic arrangement similar to the lattice points of a crystal, then since $A_\infty(\boldsymbol{S})$ is the Fourier transform of the spatial distribution, it, too, is a periodic function with very narrow peaks, like the Laue function L in Eqs. 2.43 and 2.45. This is shown diagrammatically in Fig. 5.15(a) and (d). The convolution of $A_\infty(\boldsymbol{S})$ with $S(\boldsymbol{S})$ thus brings $S = 0$ in $S(\boldsymbol{S})$ to the positions of all the peaks (including the peaks with $S = 0$) of $A_\infty(\boldsymbol{S})$, as in Fig. 5.15(g). The width of the X-ray beam is usually large enough to ensure that the breadth of $S(\boldsymbol{S})$ is very small, and so there is seldom any practical necessity to take Eq. 5.13 into account when analyzing the structures of single crystals. Under these conditions $|S(\boldsymbol{S})|^2$ produces an effect only at very small values of 2θ, the so-called "form scattering" of the sample. On the other hand, in cases where the crystals are very small, as in solid high polymers and metals, the volumes v of the crystals themselves that are responsible for the interference effects are small, irrespective of the width of the X-ray beam, and $S(\boldsymbol{S})$ is found to have an appreciable broadening. This is seen in the broadening of each diffraction spot, and around the central spot formed by the incident beam (the small-angle scattering).

Where the scattering body is an assembly of particles, if $\rho(\boldsymbol{r})$ is a function having a fixed average density ρ_1 only inside the particles and a density ρ_0 outside the particles (which corresponds to the case of a single particle in space in small-angle scattering calculations), the Fourier transform $A_\infty(\boldsymbol{S})$ of ρ_0 over all space again has a very small peak like a point function $\delta(\boldsymbol{r})$ but now only at $\boldsymbol{S} = 0$. Its convolution with $S(\boldsymbol{S})$ is thus $S(\boldsymbol{S})$ displaced to the origin on $\boldsymbol{S} = 0$. $|S(\boldsymbol{S})|^2$ is therefore in itself a measure of the intensity distribution of the central small-angle scattering, which only depends on the shape (form) of the sample and, hence, is called form scattering (See text accompanying Eq. 6.1). To grasp the general application of this concept, let us consider the calculation of the diffraction intensity for a system of particles having an average electron density ρ_1 in a medium (*e.g.* in a solution) having an average density ρ_0. If we consider the density ρ_0 as being continuous over all space throughout the system, within which particles of density $\rho_1 - \rho_0$ float and if $\sigma_1(\boldsymbol{r}) = 1$ inside all particles and zero outside, the density distribution of the system is given by

$\rho_0 + (\rho_1 - \rho_0)\sigma(\boldsymbol{r})$. From the Fourier transform, the amplitude and intensity of the small-angle scattering are

$$A(\boldsymbol{S}) = \{\rho_0\delta(\boldsymbol{S}) + (\rho_1 - \rho_0)S_1(\boldsymbol{S})\} * S(\boldsymbol{S}) \tag{5.14}$$

The first term, like the Dirac point function $\delta(\boldsymbol{r})$, is usually unobservable since it occurs at the small angle 2θ, and it is called extremely form scattering:

$$I_0(\boldsymbol{S}) = \rho^2 \mid S(\boldsymbol{S}) \mid^2 \tag{5.14a}$$

The second term of Eq. 5.14 remains observable however. Assuming that the particles are small compared with the shape of the whole sample, one obtains:

$$I(\boldsymbol{S}) = I_0(\boldsymbol{S}) + (\rho_1 - \rho_0)^2 \mid S_1(\boldsymbol{S}) \mid^2 \tag{5.15}$$

It is evident that this intensity is proportional to $\Delta\rho^2$, where $\Delta\rho = \rho_1 - \rho_0$; we have thus verified the statement, made at the end of Section 6.2.2, concerning the influence of $\Delta\rho$ on the scattering intensity.

We shall now deal briefly with the self convolution of $\sigma_1(\boldsymbol{r})$, *i.e.*

$$\sigma_1(\boldsymbol{r}) * \sigma_1(\boldsymbol{r}) = Q_s(\boldsymbol{r}) \tag{5.16}$$

This function has a central peak at $\boldsymbol{r} = 0$ consisting of the self-convolution of the shape function σ_k of the N particles in the samples. If the particles have a Gaussian distribution, the particle distribution function $P(\boldsymbol{r})$ (see Section 2.5) practically $= 1$ for $r > 2R$ and, hence

$$Q_s(\mathrm{r}) = \sum_{k=1}^{N} \sigma_k(\mathrm{r}) * \sigma_k(\overline{\mathrm{r}}) = N < \sigma_k(\mathrm{r}) * \sigma_k(\overline{\mathrm{r}}) > \tag{5.16a}$$

By Fourier transformation Q_s, like $Q(\boldsymbol{r})$, again gives a diffraction intensity function that depends only on the number average self-convolution of the shape function $\sigma_k(\boldsymbol{r})$ of the particles. Applying Eq. 5.8 we obtain

$$I_s(\boldsymbol{S}) = \int_0^\infty Q_s(\boldsymbol{r})\exp\{2\pi i(\boldsymbol{S}\cdot\boldsymbol{r})\}\mathrm{d}v_r$$

$$I(\boldsymbol{S}) = I_0(\boldsymbol{S}) + \Delta\rho^2 N < S_k(\boldsymbol{S}) S_k * (\boldsymbol{S}) > = I_0(\boldsymbol{S}) + \Delta\rho^2 N < \mid S_k(\boldsymbol{S}) \mid^2 > \tag{5.17}$$

From Eq. 5.13

$$A(\boldsymbol{S}) = A_\infty(\boldsymbol{S}) * S(\boldsymbol{S}), \qquad A(\boldsymbol{S})A^*(\boldsymbol{S}) = \mid A_\infty(\boldsymbol{S}) \mid^2 * \mid S(\boldsymbol{S}) \mid^2 = I(\boldsymbol{S}) \tag{5.18}$$

It is evident that Eq. 5.18 contains terms which correspond to the intensity $I_s(\boldsymbol{S})$ and the form scattering ($\mid S(\boldsymbol{S}) \mid^2$). The two together give the distribution of the small-angle scattering which corresponds to the shape of the particle and the shape of the whole sample, which is generally unobservable.

Finally, we shall explain the geometrical significance of $Q_s(\boldsymbol{r})$ utilizing Fig. 5.8 (*cf.* Eq. 5.9)

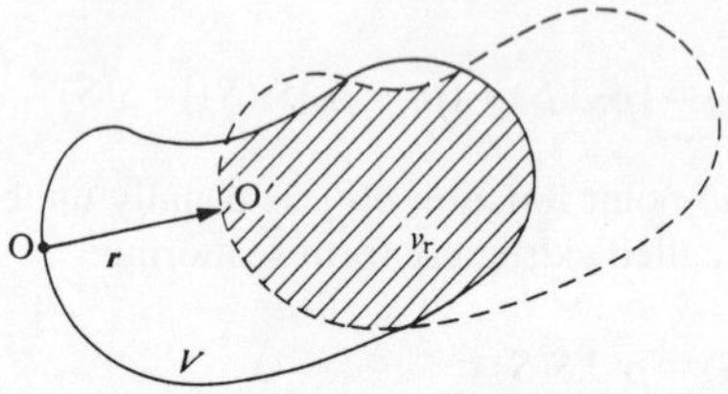

Fig. 5.8 Particle of volume V undergoing displacement $\boldsymbol{r}$ (see text for meaning of $Q_s(\boldsymbol{r})$ in this context).

Figure 5.8 shows the shape of a particle (continuous line) whose volume is V. If the particle as a whole undergoes a parallel displacement in the direction of an arbitrary vector $\boldsymbol{r}$ from a fixed point O in the particle to bring O into coincidence with the other end of the vector $\boldsymbol{r}$, its new position will be as shown by the broken line. The shape bounded by the broken line is called a ghost of the original particle (the scattering body). The value of the self convolution $Q_s(\boldsymbol{r})$ of the shape function of the particle for a given $\boldsymbol{r}$ is the sum of only those terms for which the product of the ends of $\boldsymbol{r}$ is $1 \times 1 = 1$ for parallel displacement of $\boldsymbol{r}$ to all possible positions in the original particle. If the volume common to the original particle and the ghost is v_r, as shown in the diagram, and if the ratio of this part to the total volume is $V(\boldsymbol{r})$,

$$Q_s(\boldsymbol{r}) = v_r = VV(\boldsymbol{r}) \tag{5.19}$$

This is a three-dimensional analogue of the situation in Fig. 5.7.

5.2.3 Scattering factor of atoms undergoing thermal vibrations in a crystal

As was shown in Eq. 2.15, the scattering factor of an atom at rest in a crystal lattice is given by the Fourier transform of its electron density $\rho_0(\boldsymbol{r})$ in the theoretical position. This is the atomic scattering factor given by Table 3 of the Appendix. In reality, however, the atoms in a crystal are not at rest, but undergo thermal vibrations about their respective "rest" positions. The displacements of the various atoms from their mean positions at a given instant must therefore be taken into account in calculating the structure factor $F(hkl)$.

To simplify the calculation it is assumed that each atom vibrates with isotropic harmonic motion, and that if the displacements from their rest positions of all the crystallographically equivalent atoms at a given moment are the same as that of any single vibrating atom, the electron densities of the individual atoms would be expressed by the average of a Gaussian distribution function having a mean square displacement $\overline{u}^2$. Thus if the function $\rho_0(\boldsymbol{r})$ is averaged for all positions of $\boldsymbol{r}$ by a function $\rho_1(\boldsymbol{r})$, the effective electron density $\rho_{at}(\boldsymbol{r})$ is the convolution $\rho_0 * \rho_1$. Taking ρ_1 as a Gaussian function, therefore, we find

$$\rho_{at}(\boldsymbol{r}) = \rho_0(\boldsymbol{r}) * \rho_1(\boldsymbol{r}), \qquad \rho_1(\boldsymbol{r}) = (2\pi\overline{u}^2)^{-3/2} \exp(-r^2/2\overline{u}^2) \tag{5.20}$$

The effective atomic scattering factor f_{at} is therefore the Fourier transform of $\rho_{at}(\boldsymbol{r})$, and application of Eq. 5.8 gives

$$f_{at}(S)=\int \rho_0(r) * \rho_1(r)dv_r = f(S)\exp(-2\pi^2\overline{u}^2S^2)$$
$$= f(\boldsymbol{S})\exp\{-8\pi^2\overline{u}^2(\sin^2\theta)/\lambda^2\}$$
$$D = \exp\{-8\pi^2\overline{u}^2(\sin^2\theta)/\lambda^2\} = \exp\{-B(\sin^2\theta)/\lambda^2\} \tag{5.21}$$

In the above expressions, f, is the Fourier transform of ρ_0, *i.e.* the scattering factor of the atom at rest. D is known as the Debye factor and B is called the temperature factor. In many cases the thermal motions of atoms are not isotropic, and assuring the thermal motion of atoms to be ellipsoidal the anisotropic thermal parameters are used as the better approximation. These are important factors in X-ray analyses, and their practical use will be shown later (Eqs. 11.34 and 11.36 (a, b)).

5.2.4 Optical experiments on Fourier transforms

The Fourier transformations in Eqs. 5.2, 5.6, 5.9, and 5.11, and their inverse transformations, can only be performed numerically or analytically for a given function if the form is already known. If the form is not given, Eq. 5.2 cannot be calculated mathematically. However, Eq. 5.1 (Eq. 2.13), which is of the same form as Eq. 5.2, shows that the amplitude of the resultant diffracted wave is given combination of the phase-dependent contributions from every element of the structure. Therefore, even if the mathematical form of the electron density function, $\rho(\boldsymbol{r})$, of a substance is unknown, it may still be possible to construct a model structure which will embody suspected characteristic features of the substance. Comparison of the diffraction pattern from such a model structure with that from the substance itself can provide a means of evaluating the tentative structure. It is impossible to construct prepared structures of the same order of magnitude as atoms themselves; the scale is usually some tens of thousands of times greater. The model may consist of a two-dimensional array of black dots or holes marked on film or glass, and the scale multiplication is such that visible light can be used to produce the diffraction effects. The Fourier transform in this sense is effectively performed by so-called optical inversion. Eq. 5.1 is, of course, equally valid for all forms of wave motion, including X-rays, visible light, and electron beams. Diffraction experiments with atomic models using visible light (*e.g.* the sodium D line doublet) are often useful for checking the validity of structures of great complexity, such as those involved in the detailed investigation of fiber paracrystals. Examples where such light diffraction patterns have been employed are given in Fig. 5.9,[3)] and 5.13.[4)] The reader is referred to the various experiments which Hosemann has performed in this connection.[3)]

The experimental arrangements used and examples of their application are shown in Fig. 5.9. The upper diagrams of Fig. 5.9(a) and (b) show optical systems for Fourier transforms and convolutions (including self convolutions), respectively. Fig. 5.9(a) includes the model of the lattice function $\rho(r)$ to be transformed, and its transform $A(\boldsymbol{S})$ is recorded photographically as the intensity $|A(\boldsymbol{S})|^2$, giving the Fraunhofer diffraction pattern shown. In Fig. 5.9(b), if the second model differs from the first, the convolution of the two is obtained. If, however, both models are the same, which is the case illustrated, the self convolution of the model is obtained. This is the means by which $\rho(\boldsymbol{r}) * \rho(\overline{\boldsymbol{r}})$, *i.e.* $Q(\boldsymbol{r})$, can be determined by visible light diffraction.

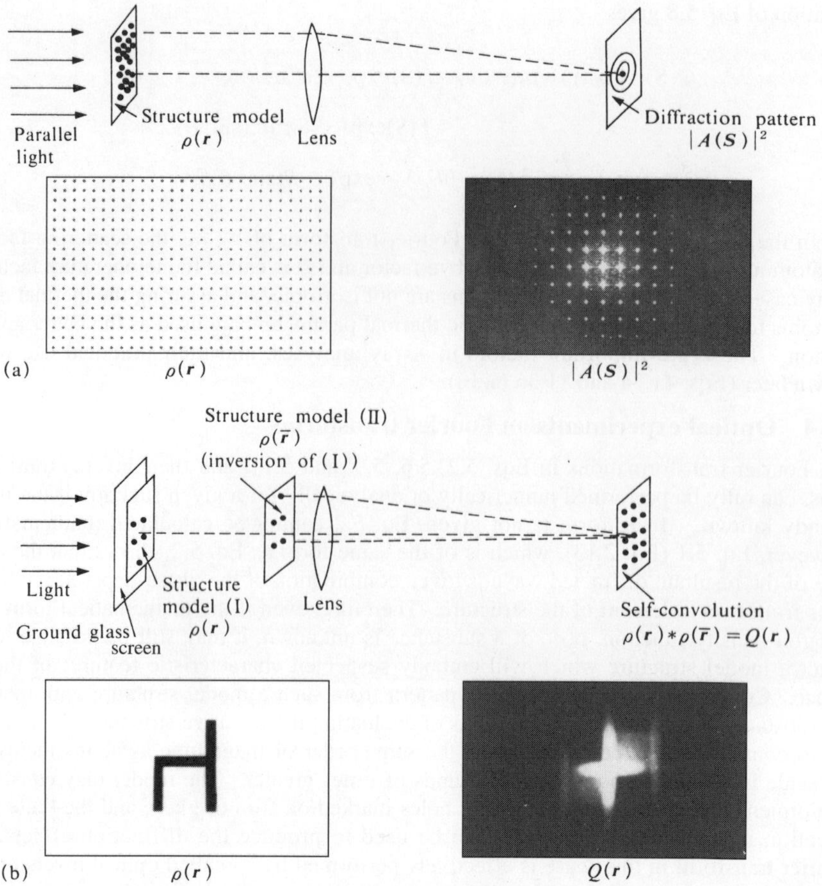

Fig. 5.9 Optical methods for Fourier transformations and convolutions.[3,4)]
(a) Fourier transformation. (b) Convolution (and self-convolution).
[Reproduced with permission from R. Hosemann, *Polymer*, **3**, 349, IPC Business Press (1962)]

5.3 Diffraction of X-Rays by Paracrystals†

For diffraction from a crystal, since the unit cells are arranged in a regular manner in accordance with Eq. 2.37, the phases of their scattering amplitude A_{cell} can be combined as shown in Eq. 2.38. For paracrystals, on the other hand, the positions of the lattice points can be expressed only as a statistical average, as was mentioned earlier. It is therefore necessary to find some adequate means of representing the positions of the lattice points, just

† after Hosemann[3,4)]

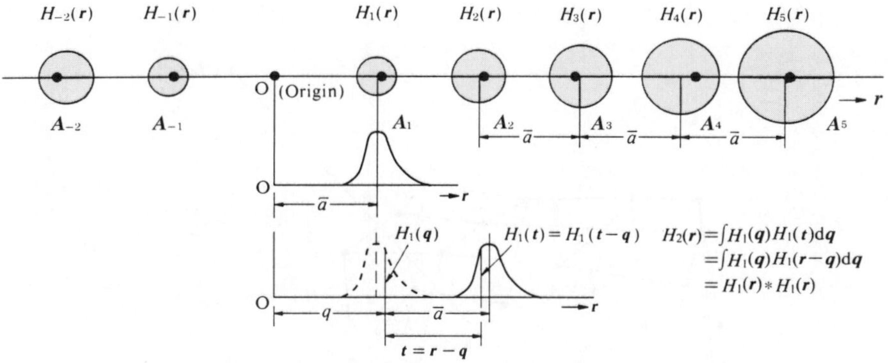

Fig. 5.10 Diagram showing the distribution of lattice points in a linear paracrystal.

as Eq. 2.37 does for crystals, and of combining their phases to find the diffraction due to the paracrystalline lattice. We first consider the paracrystal as a one-dimensional system having significant extension only in the $\boldsymbol{a}$ direction (Fig. 5.10).

5.3.1 Statistical representation of paracrystalline lattice points and the derivation of their function $Q(\boldsymbol{r})$

Let the lattice points in the $\boldsymbol{a}$ direction be A_1, A_2, etc., where A_1 is the nearest neighbor to some arbitrary origin, O. If the probability that A_1 be located by a vector $\boldsymbol{q}$ from the origin is $H_1(\boldsymbol{q})$, then the probability that A_2 is located by an independent vector $\boldsymbol{t}$ from the end of $\boldsymbol{q}$ will be $H_1(\boldsymbol{q}) \cdot H_1(\boldsymbol{t})$. The total probability, $H_2(\boldsymbol{r})$, that A_2 (the next nearest neighbor) will lie at the end of a single direct vector $\boldsymbol{r} = \boldsymbol{q} + \boldsymbol{t}$ from the origin is not, however, the product of two individual probabilities $H_1(\boldsymbol{q})$ and $H_1(\boldsymbol{t})$ if we assume that there is no statistical correlation between the individual $\boldsymbol{q}$ and $\boldsymbol{t}$ vectors. Since we are interested in the probability for a certain $\boldsymbol{r}$, whatever the value of $\boldsymbol{q}$ we have to integrate overall $\boldsymbol{q}$ for a fixed $\boldsymbol{r} = \boldsymbol{q} + \boldsymbol{t}$. Hence $\boldsymbol{t} = \boldsymbol{r} - \boldsymbol{q}$ and we finally obtain

$$H_2(\boldsymbol{r}) = \int H_1(\boldsymbol{q})H_1(\boldsymbol{r}-\boldsymbol{q})\mathrm{d}q = H_1(\boldsymbol{r}) * H_1(\boldsymbol{r}) \tag{5.22}$$

which is a convolution of $H_1(\boldsymbol{r})$ with $H_1(\boldsymbol{r})$ (*cf.* Eq. 5.6) and, in general, the probability fo the nth lattice point is

$$H_n = H_1 * H_1 * \cdots * H_1 \text{ (the } (n-1) \text{ fold convolution of } H_1) \tag{5.23}$$

$H_n(\boldsymbol{r})$ is normalized as unity, and $H_{-n}(\boldsymbol{r}) = H_n(-\boldsymbol{r})$, with $\int \boldsymbol{r} H_n(\boldsymbol{r})\mathrm{d}r = n\bar{a}$.

Let us now consider the extension of this treatment to three dimensions. In Fig. 5.11 the three points A_1, A_2, and A_3 are located by vectors $\boldsymbol{a}_1$, $\boldsymbol{a}_2$, and $\boldsymbol{a}_3$, corresponding to directions $\boldsymbol{a}$, $\boldsymbol{b}$, and $\boldsymbol{c}$ in an ideal lattice. The "nearest neighbor" probabilities in the three directions (corresponding to H_1 along $\boldsymbol{a}$ in the preceding discussion) may be written H_{100}, H_{010}, and H_{001} respectively, where $H_{pqr}(\boldsymbol{r})$ denotes the probability that the vector $\boldsymbol{r}$ from the origin will locate the point which is the pth in the $\boldsymbol{a}_1$ direction, the qth in the $\boldsymbol{a}_2$ direction, and the rth in the $\boldsymbol{a}_3$ direction. This probability can be expressed in terms of the $(p-1)$ fold convolution of H_{100}, the $(q-1)$ fold convolution of H_{010}, and the $(r-1)$ fold convolution of H_{001},

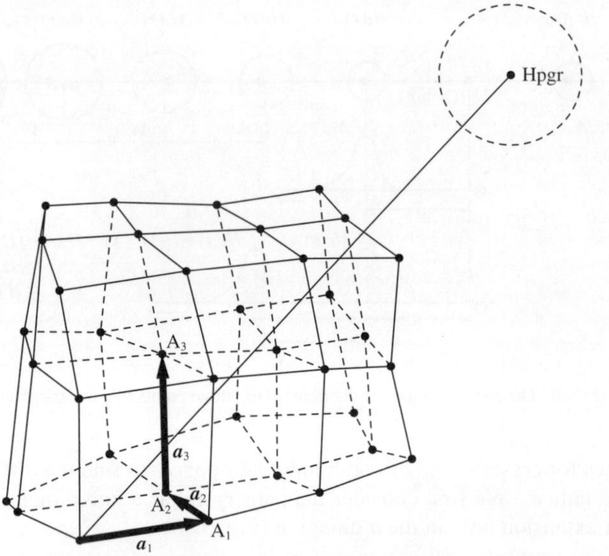

Fig. 5.11 Paracrystalline lattice in three dimensions (*cf.* Fig. 5.6(b_2')).

with an extra normalizing term if the probability at the origin is taken as unity. Thus

$$H_{pqr} = H_{000} * \underbrace{H_{100} * H_{100} \cdots * H_{100}}_{(p-1)\text{fold}} * \underbrace{H_{010} * \cdots * H_{010}}_{(q-1)\text{fold}} * \underbrace{H_{001} * \cdots * H_{001}}_{(r-1)\text{fold}} \tag{5.24}$$

H_{000} is a point function having a definite value only at the origin. Eq. 5.24 expresses the probability that a lattice point (pqr) will exist at a point displaced by the appropriate number of multiples (p, q, and r) of the average lengths $\overline{a}_1$, $\overline{a}_2$, and $\overline{a}_3$ of the sides of the cell from an origin at some arbitrary lattice point.

A lattice point (pqr) can be reached by several independent paths, depending upon the directions taken from point to point along the route through the paracrystal. The H_{pqr} of Eq. 5.24 is therefore but one element in the total probability, which must be summed for all possible paths from an arbitrary lattice-point origin to the point (pqr) located by vector $\boldsymbol{r}$, *i.e.*

$$\sum_p \sum_q \sum_r H_{pqr} = \sum_{p=-\infty}^{\infty} H_{p00} * \sum_{q=-\infty}^{\infty} H_{0q0} * \sum_{r=-\infty}^{\infty} H_{00r} = z(\boldsymbol{r})_{pqr} \tag{5.25}$$

5.3.2 Lattice factor and diffraction intensity for a paracrystal

Having derived a statistical distribution, $z(\boldsymbol{r})$, for the distance between lattice points in a paracrystalline lattice, we proceed to use it to calculate the diffraction intensity.

Let the Fourier transform of $H_k(\boldsymbol{r})$ be $\mathcal{F}_k(\boldsymbol{S})$, where H_k replaces H_{pqr} in the preceding analysis for (pqr) equal to (100), (010), or (001). We shall now calculate the scattering intensity for the X-ray scattering from all the lattice points of a paracrystal, taking the scattering factor A_{cell} of the unit cells as unity and considering only the intensity due to the lattice structure.

The quantity $z(\boldsymbol{r})$ is just the probability function $P(\boldsymbol{r})/v$ of Section 2.5, no longer spherically symmetric but with paracrystalline lattice character. It may be regarded as the probability distribution for the presence of lattice points of the same unit density at both ends of a given vector $\boldsymbol{r}$. This is simply the self convolution of a density distribution $\rho(\boldsymbol{r})$, *i.e.* $Q(\boldsymbol{r})$, (*cf.* Eq. 5.9) having point function $\delta(\boldsymbol{r})$ only at positions occupied by lattice points. The X-ray diffraction intensity due to the paracrystalline lattice structure itself is therefore given directly by the Fourier transform of $z(\boldsymbol{r})$, according to Eq. 5.11. Let the Fourier transform of $z(\boldsymbol{r})$ for the k direction ($k=(100)$, (010), or (001)) be $K_k(\boldsymbol{S})$. If M is the number of unit cells along each axis in the paracrystal, then$\mathcal{F}$

$$K_k(\boldsymbol{S})=\sum_{p=0}^{\infty}(\mathcal{F}_k{}^{p}+\mathcal{F}_k{}^{*p})-1=\lim_{M\to\infty}2\,\mathrm{Re}\left(\frac{1-\mathcal{F}_k{}^{M+1}}{1-\mathcal{F}_k}\right)-1$$

$$=\lim_{M\to\infty}\mathrm{Re}\left(\frac{2-2\mathcal{F}_k{}^{M+1}-1+\mathcal{F}_k}{1-\mathcal{F}_k}\right)=\mathrm{Re}\left(\frac{1+\mathcal{F}_k}{1-\mathcal{F}_k}\right)+\lim_{M\to\infty}\mathrm{Re}\left(\frac{-2\mathcal{F}_k{}^{M+1}}{1-\mathcal{F}_k}\right) \quad (5.26)$$

$\mathcal{F}_k{}^*$ is the Fourier transform of H_k for points on the negative side of the origin, and is the complex conjugate function of that for points on the positive side, $\mathcal{F}_k$. Re denotes the real part. If we let the second term in Eq. 5.26 be $K(0)$, then

$$K(0)=-2\lim_{M\to\infty}\frac{\mathrm{Re}(1-\mathcal{F}_k{}^*)\cdot\mathcal{F}_k{}^{M+1}}{|1-\mathcal{F}_k|^2} \quad (5.27)$$

The complex $\mathcal{F}_k$ may be represented by modulus and phase, thus

$$\mathcal{F}_k(\boldsymbol{S})=|\,\mathcal{F}_k(\boldsymbol{S})\,|\exp\{-2\pi i(\bar{\boldsymbol{a}}_k\cdot\boldsymbol{S})\},$$

$$\mathcal{F}_k{}^*(\boldsymbol{S})=|\,\mathcal{F}_k(\boldsymbol{S})\,|\exp\{2\pi i(\bar{\boldsymbol{a}}_k\cdot\boldsymbol{S})\},\ \text{and}$$

$$\bar{\boldsymbol{a}}_k=\int rH_k(r)dv \quad (5.28)$$

if $H_k(r)$ has a center of symmetry.
We obtain

$$K(0)=\lim_{M\to\infty}|\mathcal{F}_k|^M\frac{\sin\pi(2M+1)(\bar{\boldsymbol{a}}_k\cdot\boldsymbol{S})}{\sin\pi(\bar{\boldsymbol{a}}_k\cdot\boldsymbol{S})} \quad (5.29)$$

$$K_k(\boldsymbol{S})-K(0)=\mathrm{Re}\frac{1+\mathcal{F}_k}{1-\mathcal{F}_k} \quad (5.30)$$

The form of the second term of $K(0)$ in Eq. 5.29 is evidently similar to the form of the Laue function (Eq. 2.38). The presence of the $|\,\mathcal{F}_k\,|^M$ factor (with $|\,\mathcal{F}_k\,|<1$), however, ensures that $K(0)$ approximates to a peak function having significant values only for a small range of $\boldsymbol{S}$ centered about $\boldsymbol{S}=0$. Eq. 5.30, from which the central peak of $K(\boldsymbol{S})$ has been subtracted, gives the most important part of the diffuse pattern which characterizes second and higher order diffraction from paracrystals. In fact, when M is not infinitely large, but is a finite, small number, we find a "broadening" similar to that for the Laue function in Section 2.9.2. This will be discussed further at a later stage. From Eq. 5.30,

$$K_k(\boldsymbol{S})-K(0)=\frac{1-|\mathcal{F}_k|^2}{1+|\mathcal{F}_k|^2-2\,|\,\mathcal{F}_k|\cos 2\pi(\bar{\boldsymbol{a}}_k\cdot\boldsymbol{S})} \quad (5.31)$$

Since the principal term on the left-hand side is $K_k(\boldsymbol{S})$, we obtain from Eq. 5.31

$$K_k(\boldsymbol{S})(\max) = \frac{1+|\mathcal{F}_k|}{1-|\mathcal{F}_k|} \quad (\text{if } \bar{\boldsymbol{a}}_k \cdot \boldsymbol{b} = h_k) \tag{5.32}$$

$$K_k(\boldsymbol{S})(\min) = \frac{1+|\mathcal{F}_k|}{1-|\mathcal{F}_k|} \quad (\text{if } \bar{\boldsymbol{a}}_k \cdot \boldsymbol{b} = h_k + \frac{1}{2}) \tag{5.33}$$

where h_k is an integer. These are the basic equations giving the maximum and minimum values of the intensity of the diffuse diffraction pattern from a paracrystal. They are a generalization of the Laue conditions, Eq. 2.39. Gathering together the expressions for the three dimensions, one can prove that the result is

$$Z(\boldsymbol{S}) = K_1 K_2 K_3 = \frac{1}{v}\delta(\boldsymbol{S}-0) + \prod_{k=1}^{3} \mathrm{Re}\frac{1+|\mathcal{F}_k|}{1-|\mathcal{F}_k|} \tag{5.34}$$

$\delta(\boldsymbol{S}-0)$ represents the term $K(0)$, which is a three-dimensional point function at $\boldsymbol{S}=0$, the suffixes 1, 2, and 3 refer to the three directions, and v is the average volume of the unit cell. Eq. 5.34 is the most important equation for the X-ray intensity for a substance having a paracrystalline structure. $Z(\boldsymbol{S})$ is called the paracrystalline lattice factor, and it corresponds to the Laue function for the ideal crystal. The diffraction maxima occur at the same positions as the diffraction spots for an ideal crystal, whereas the diffuseness of the diffraction peaks increases with the index of the diffraction (reflection) due to the contribution of the second term. The broadening also increases with increasing lattice distortion, and finally, for a given index, the intensity peaks merge indistinguishably with the background or with the peaks of the next higher index. This is schematically shown in Fig. 5.12. The right figures show, from top to bottom, square of the average unit cell scattering amplitude of an average net plane (say the (h00) plane), $|<A_{\text{cell}}(\boldsymbol{S})>|^2$, paracrystalline lattice factor, $Z(\boldsymbol{S})$, and intensity, $I(\boldsymbol{S}) = |<A_{\text{cell}}(\boldsymbol{S})>|^2 \cdot Z(\boldsymbol{S})$. In the left those for an ideal crystal (Fig. 2.20) are shown again for comparison. This topic will be covered in a later section (*cf.* 13.6.3).

It should be noted that although several important equations were developed in terms of three dimensions, the treatment is basically one-dimensional in that it presumes no relationship between the dimensions. In the general case certain problems arise. Thus it was assumed at the beginning of this section that the "original ideal lattice" could be discerned. In the case of completely disordered amorphous substances, however, it is meaningless to speak of a vector in the $\boldsymbol{a}$, $\boldsymbol{b}$, or $\boldsymbol{c}$ directions, because the directions are indistinguishable in isotropic amorphous substances. The applicability of this method therefore depends upon the degree of distortion from the ideal lattice for a given single direction in the specimen, *e.g.* the direction of the molecular chains or a specified direction in the plane normal to the direction of the chains. A measure of the degree of distortion can be obtained by forming a weighted average of the probabilities H_k of finding displaced lattice points in the various directions. This average is taken over the whole of the specimen but, since H is very diffuse and small in value for the highest degree of distortion, it should be clear that the average will not be greatly affected by contributions from amorphous regions. Fig. 5.13[4)] (p.102) gives a comparison between the theoretical value of H and experimental values. Fig. 5.13(a) shows H_{00} H_{10} and H_{01} for a paracrystalline lattice, and Fig. 5.13(b) shows the lattice point model constructed on this basis. Fig. 5.13(c) illustrates H for various lattice points as found by two-dimensional synthesis of the H values calculated from the model. Fig. 5.13(d) shows the optical diffraction pattern obtained in an actual experiment with this model; this diffraction pattern corresponds to the Fourier transform of the H of Fig. 5.13(b).

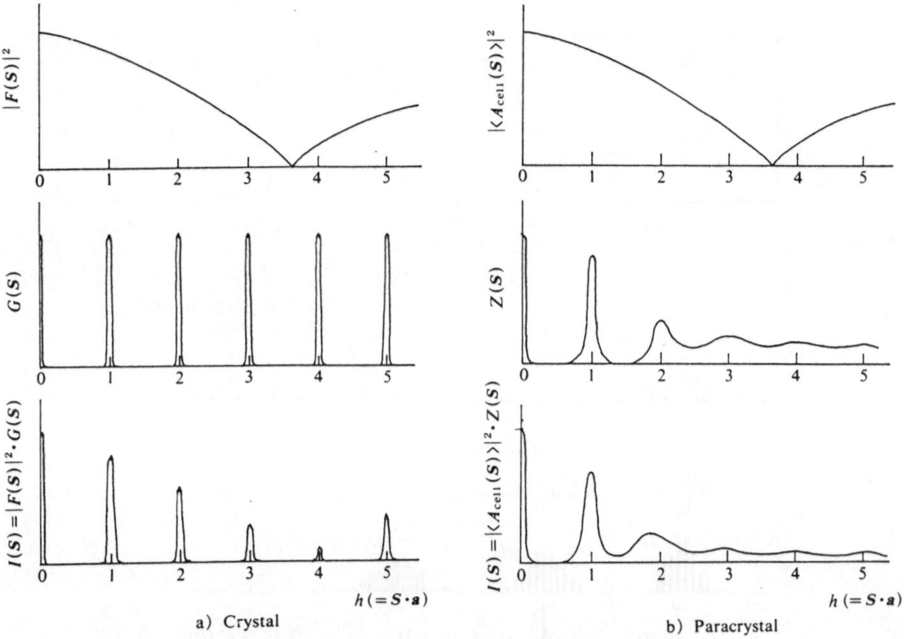

Fig. 5.12 A schematic comparison of diffraction intensities by crystal and paracrystal.
a) Crystal: (from top to bottom) Square of structure factor, $|F(S)|^2$; Laue function, $G(S)$; Intensity, $I(S)=|F(S)|^2\cdot G(S)$ [Reproduced with permission from R. D. B. Fraser, T. P. Mac Rae, *Conformations in Fibrous Proteins and Related Synthetic Polymers*, p.8, Academic Press (1973)]
b) Paracrystal: (from top to bottom) Square of average unit cell scattering amplitude, $|\langle A_{cell}(S)\rangle|^2$ (for simplicity, magnitude of $|\langle A_{cell}(S)\rangle|^2$ is taken the same as $|F(S)|^2$); Paracrystalline lattice factor $Z(S)$; Intensity, $I(S)=|\langle A_{cell}(S)\rangle|^2\cdot Z(S)$.

We have so far referred to H only as the probability of the presence of a lattice point, and no reference has been made to its actual form. In many paracrystalline substances, bearing in mind the physical causes of distortion in the crystal, it is often legitimate to choose a Gaussian-type function for the H that expresses the distribution. As shown in Fig. 5.14 (p.103),[4] therefore, if we take the $H_k(\boldsymbol{r})$ for the three principal directions of the orthogonally paracrystalline lattice cell ($k=1$, 2, or 3) as a three-dimensional Gaussian function, it can be expressed in terms of a matrix T_{jk} consisting of nine elements of mean-square displacements from the regular lattice, $\Delta^2\boldsymbol{r}_{jk}$, in groups of three for each principal axis. The Fourier transform of H, when it takes the form of a one-dimensional Gaussian function is $\mathcal{F}_k(\boldsymbol{S})$, which is already given in Eq. 5.28, and in the equation

$$\mathcal{F}_k(\boldsymbol{S})=\exp\{-2\pi^2(\boldsymbol{S}\cdot T_k\cdot\boldsymbol{S})\} \tag{5.35}$$

$$(\boldsymbol{S}\cdot T_k\cdot\boldsymbol{S})=\sum_{kj}T_{jk}S_jS_k \tag{5.36}$$

$$T_{jk}=\begin{pmatrix}\Delta^2 r_{11} & \Delta^2 r_{12} & \Delta^2 r_{13}\\ \Delta^2 r_{21} & \Delta^2 r_{22} & \Delta^2 r_{23}\\ \Delta^2 r_{31} & \Delta^2 r_{32} & \Delta^2 r_{33}\end{pmatrix},\quad (\Delta^2 r_{jk}=\Delta^2 r_{kj}), \tag{5.37}$$

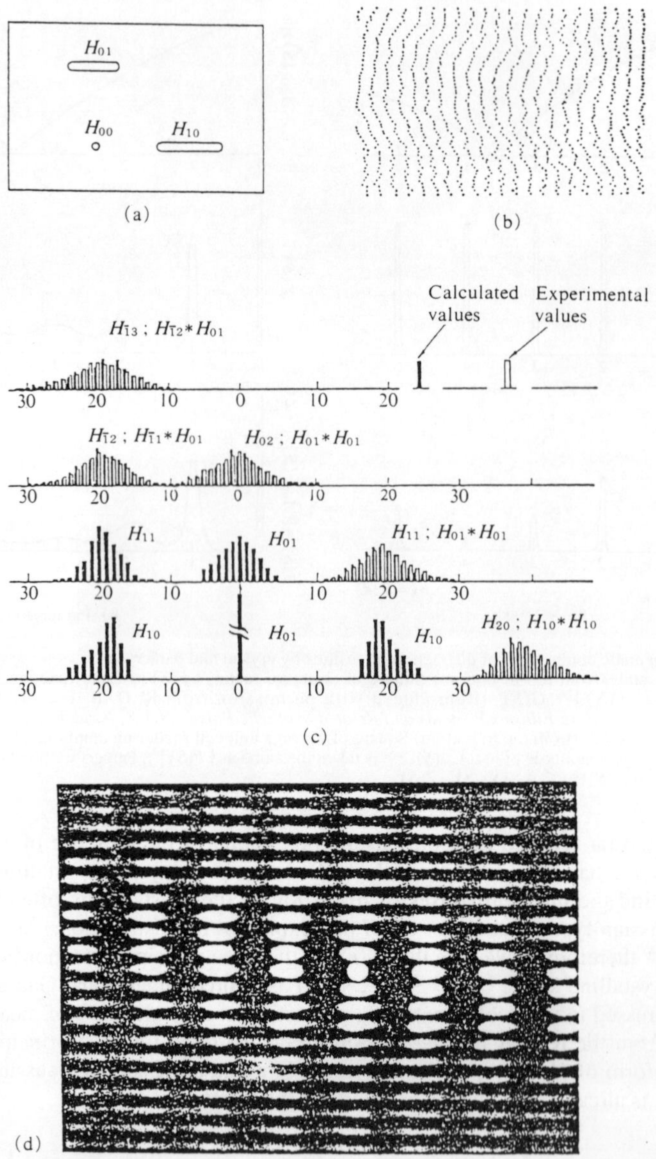

Fig. 5.13 A comparison between theoretical and experimental values of H.[4)]
(a) H_{00}, H_{01}, and H_{10} for a paracrystalline lattice; (b) The paracrystalline lattice model corresponding to (a); (c) Convolution values found from the diffraction pattern (d) and those calculated directly from the model; (d) Optical (Fraunhofer) diffraction pattern derived from this model (b).
[Reproduced with permission from R. Hosemann, S. N. Bagchi, *Direct Analysis of Diffraction by Matter*, pp.143, 144, 146, North-Holland Pub. (1962)]

(see Fig. 5.14). It should be noted that Eq. 5.35 is of exactly the same form as the Debye factor, D of Eq. 5.21.

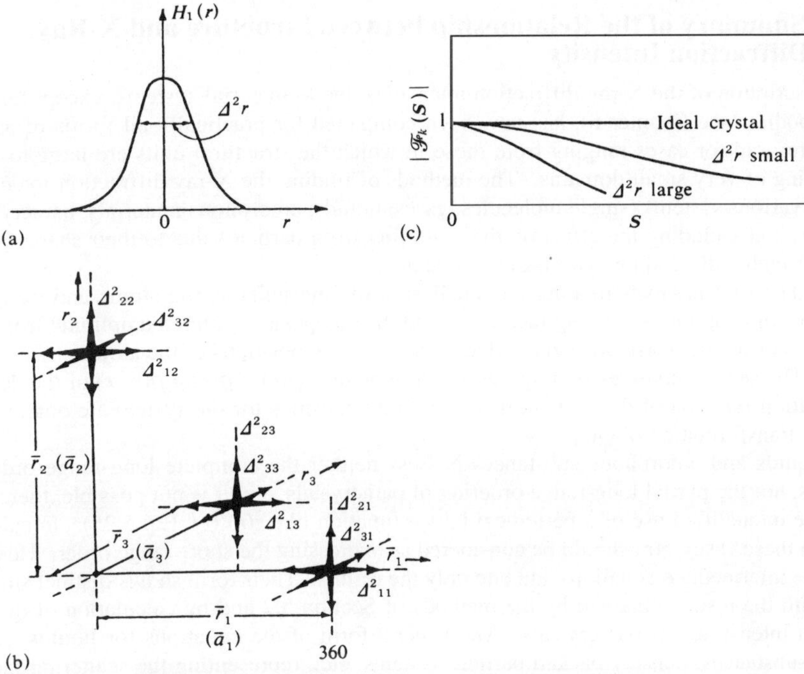

Fig. 5.14 Delineation of H, and its effect on $|\mathcal{F}_k|$ (see text, and also *cf.* Fig. 11.8 for Z).[4)]
(a) One-dimensional $H_1(r)$ (Gaussian, *cf.* Fig. 5.10 for the general case, unsymmetrical distribution);
(b) Tensor representation of the corresponding paracrystalline lattice distortion in the directions 1, 2, and 3 by means of the nine symmetrical matrix elements $\Delta_{11}{}^2$, $\Delta_{22}{}^2$, etc.; (c) $|\mathcal{F}_k(\boldsymbol{S})|$ vs. $\boldsymbol{S}$ for various degrees of distortion.
[Reproduced with permission from R. Hosemann, *Polymer*, **3**, 349, IPC Business Press (1962)]

This concludes the fundamental analysis of X-ray diffraction intensities for paracrystalline structures, and we can combine the results to obtain an equation for the X-ray diffraction intensity found in practice. A paracrystal differs from a normal crystal in that the unit cells are not all identical, so there is no unique A_{cell} which expresses the composite amplitude due to all the atoms in the unit cell. It is therefore necessary to use an averaged $< A_{\text{cell}} >$ over all unit cells. Taking the system as a dense array of cells having different structure factors, and applying the reasoning which gave Eqs. 2.30, 2.33, and 2.34, we can find the intensity with the aid of the $Z(\boldsymbol{S})$ obtained by the statistical method of this chapter for the distortion of the lattice period, thus:

$$I(\boldsymbol{S})=\overline{N}(<A_{cell}{}^2>-\mathrm{D}_{\mathrm{I}}^2<A_{cell}>^2)+\frac{1}{v}<A_{\text{cell}}>^2<\mathrm{D}_{\mathrm{I}}^2>^2(Z(\boldsymbol{S})*|S(\boldsymbol{S})|^2) \tag{5.38}$$

$\mathrm{D}_{\mathrm{I}}^2$ is the distortion factor of the first kind, and is of the same form as Eq. 5.21. $\overline{N}$ is the total number of unit cells, and v is the average volume of one unit cell. $|S(\boldsymbol{S})|^2$ is the shape factor of the domain of the paracrystalline lattice given in Eq. 5.17, or may be regarded as the contribution due to the shape of the coherent boundaries with shape factor $|S(\mathbf{S})|^2$ in a polyparacrystalline material (see Eq. 5.17).

5.4 Summary of the Relationship between Structure and X-Ray Diffraction Intensity

The description of the X-ray diffraction intensities due to material systems, except for very small bodies (see Chapter 6), has now been completed for practically all forms of atomic structures and for cases ranging from those in which the structural units are large to those consisting of very small domains. The methods of finding the X-ray diffraction intensities for the various systems (single molecules, gases, liquids, amorphous materials, paracrystals, crystals, and including the effect on the scattering from particles due to their shape factor) can be roughly divided into two basic procedures.

1) The first involves treating the smallest scattering units as the atoms and using the corresponding atomic scattering factor f to find the composite scattered amplitude from calculations using the phase differences due to the differing positions of the atoms.

2) The second involves finding the self-convolution $Q(\mathbf{r})=\rho(\mathbf{r}) * \rho(-\mathbf{r})$ of the density distribution function of the system; the scattering intensities for the system are obtained by Fourier transformation of $Q(\mathbf{r})$.

Liquids and amorphous substances possess neither the complete long-range order of crystals, nor the partial long-range ordering of paracrystals, and it is not possible, therefore, to make unqualified use of a periodical lattice function like $z(\mathbf{r})$ (*cf.* Eq. 5.25). In calculations in these cases, $z(\mathbf{r})$ should be considered as expressing the short-range order. Here we omit the intermediate equations and cite only the results. Their form shows distinct similarities with the results obtained by the methods of Section 5.2 and by calculation of the diffraction intensities for paracrystals. The general form of the equations for liquids, amorphous substances, densely packed particle systems, etc., representing the scattering factors of the atoms or molecules of the liquid or amorphous substance or the scattering factors of the particles as $A(\mathbf{S})$, is as follows:

$$I(\mathbf{S})=M[<|A^2(\mathbf{S})|>|S(\mathbf{S})|^2+|<A(\mathbf{S})>|^2\Phi(\mathbf{S}) * |S(\mathbf{S})|^2] \tag{5.39}$$

where

$$\Phi(\mathbf{S})=1+\frac{1}{v_1}\int\{P(\rho)-1\}\exp\{2\pi i(\Sigma\cdot\rho)\}\mathrm{d}v \tag{5.40}$$

For the individual case, the reader is referred to earlier sections (*cf.* Sections 2.5 − 2.8).

For crystals

$$I(\mathbf{S})=\frac{1}{v}|F(\mathbf{S})|^2\, G(\mathbf{S}) * |S(\mathbf{S})|^2 \tag{5.41}$$

For crystals containing distortions of the first kind we have

$$I(\mathbf{S})=|F(\mathbf{S})|^2\,[N(1-\mathrm{D}_1^2)+\frac{1}{v}\cdot\mathrm{D}_1^2 G(\mathbf{S}) * |S(\mathbf{S})|^2] \quad \text{(Thermal motion)} \; (5.42)$$

$$I(\mathbf{S})=N[<|F(\mathbf{S})|^2>-|<F(\mathbf{S})>|^2]+\frac{1}{v}|<F(\mathbf{S})>^2 G(\mathbf{S}) * |S(\mathbf{S})|^2$$

(Mixed crystals) (5.43)

For paracrystals containing distortions of both the first and second kinds (distortion factor D_t) we have

$$I(S) = N(<|A_{cell}|^2> - D_t^2|<A_{cell}>|^2) + \frac{1}{v}|<A_{cell}>|^2 D_t^2 Z(S) * |S(S)|^2 \quad (5.44)$$

In the above equations, $G(S) = L_1^2(N_1)L_2^2(N_2)L_3^2(N_3)$ is the Laue function, v_1 is the volume occupied by one particle (atom or molecule), v is the volume of a unit cell, M the total number of particles, and N the number of unit cells. Since the shape of the crystalline or paracrystalline single lattice is taken into consideration by the shape factor $|S(S)|^2$, the numbers N_1, N_2, and N_3 in the Laue function (*cf.* Eqs. 2.38 and 2.45) are infinitely large, and hence $G(S)$ has point-like peaks in the reciprocal lattice points.

1) For amorphous substances, the integral containing $P(r)$ is the Fourier transform of the statistical distribution of the distances of the scattering points, while for paracrystals and crystals, $Z(S)$ and $G(S)$ are the Fourier transforms of the statistical distribution of the distances of the lattice points. The similarity between Eqs. 5.43 and 5.44 is particularly obvious.

2) When the different phases of a substance (crystalline or paracrystalline, amorphous or liquid) are finely divided into very small regions, or in the case of fine particles, the small-angle scattering of X-rays to be discussed in Chapter 6 is quite appreciable because of the $|S(S)|^2$ factor (see Eq. 5.17). In the case of crystals or paracrystals, small-angle scattering is observed not only for $S = 0$ (*i.e.* in the direction of the incident beam), but at each diffraction spot, owing to the convolution of the shape factor $|S(S)|^2$ of the lattices with $G(S)$ or $Z(S)$. For amorphous substances, liquid or solution it is observed only at the center spot.

3) When the crystals are small, broadening of $G(S)$ occurs in accordance with Fig. 2.16, and it should be understood that the resultant broadening of the diffraction spots will be further increased by convolution with the factor $|S(S)|^2$. The latter contribution, of course, is at its most intense in the direction of the incident beam.

Reference to Fig. 5.15 will clarify the above conclusions. It also illustrates the method in 2) above for calculating X-ray diffraction intensities.

Since direct Fourier transformation for a crystal of finite size (represented in one dimension in Fig. 5.15(c)) is impossible (*cf.* Section 5.2.2), the lattice function of the crystal with finite size is obtained by considering an infinitely large crystal (a) conjunction with a shape function $\sigma(r)$, (b). The self-convolution $Q(r)$ (e) of the lattice function (c) is the product of the two self-convolutions z and $Q_s(r)$ (in Fig. 5.15(e). The diffraction intensity is then found by Fourier transformation of $Q(r)$ (*i.e.* of $z(r) \cdot Q_s(r)$). The result is the convolution (in (g)) of the individual Fourier transforms shown in diagram (f). The intensity distribution (g) can also be derived as the broadening of the Laue function due to the fact that in the case we are considering the crystals are finite with small numbers of unit cells, but here we have obtained it as the broadening of the shape function $|S(S)|^2$. The diagram shows that the small-angle scattering is not restricted to the center, but is associated with all the diffraction spots. In Fig. 5.15 $\mathcal{F}$ denotes a Fourier transform.

A similar analysis for paracrystals in Fig. 5.16 gives the intensity distribution shown in diagram Fig. 5.16(g). Although we are considering a paracrystal with infinite extension, the broadening of $z(r)$ increases with increase in the order of diffraction due to the distortion of the lattice, and the diffraction intensity also broadens out with increasing order number, as shown in Figs. 5.12 and 13.16. For finite crystals, therefore, this broadening will be further increased by convolution with the broadening of $|S(S)|^2$ (*cf.* Section 13.6).

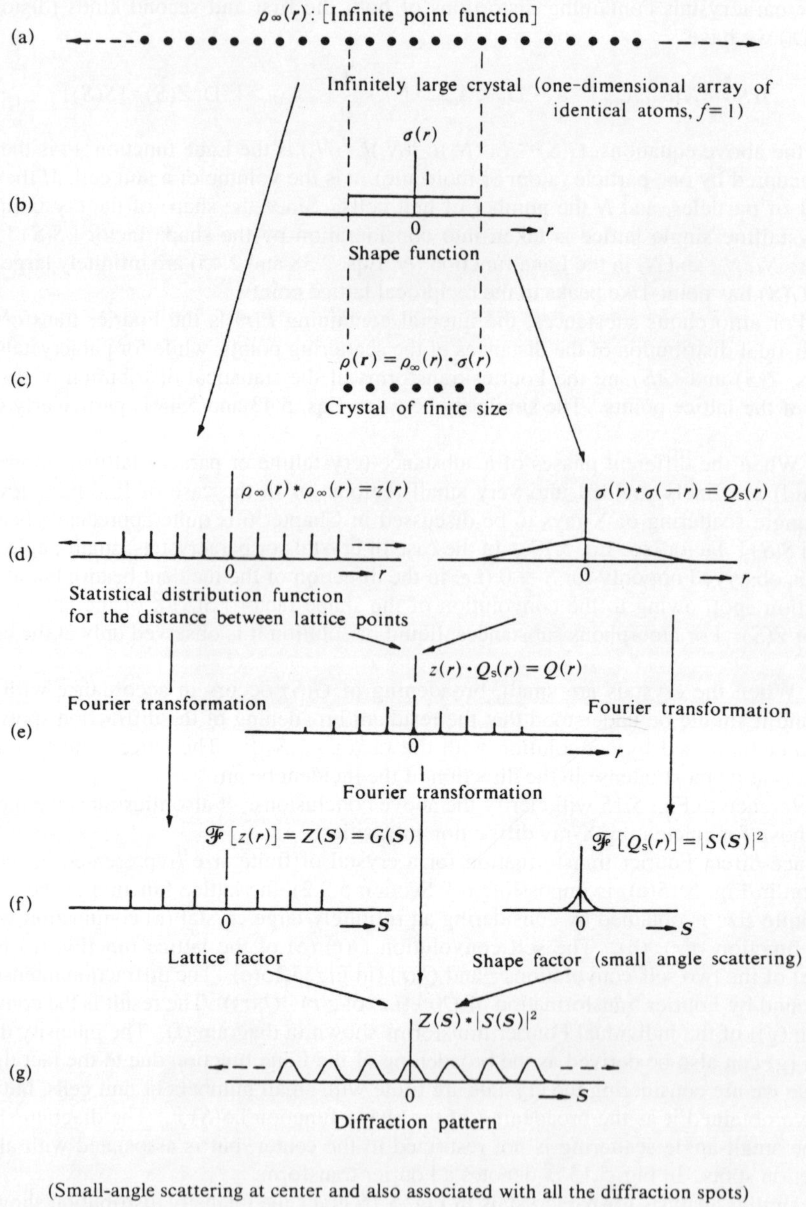

Fig. 5.15 Schematic representation of the steps involved in calculating the diffracted X-ray intensities (bounded crystals). Atomic scattering factors, f's are taken as unity. $\mathscr{F}$ denotes a Fourier transform.

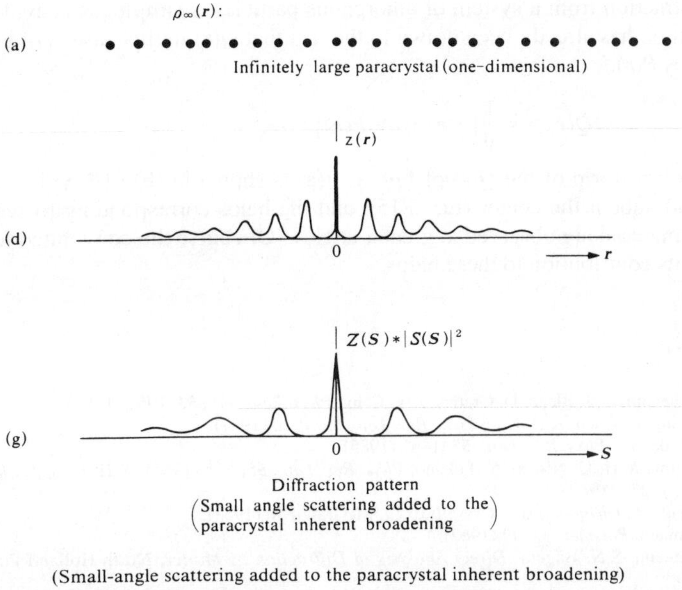

Fig. 5.16 Schematic representation of the steps in calculating X-ray intensities (bounded paracrystals). Only those steps corresponding to (a), (d), and (g) in Fig. 5.15 are shown. Atomic scattering factors are taken as unity.

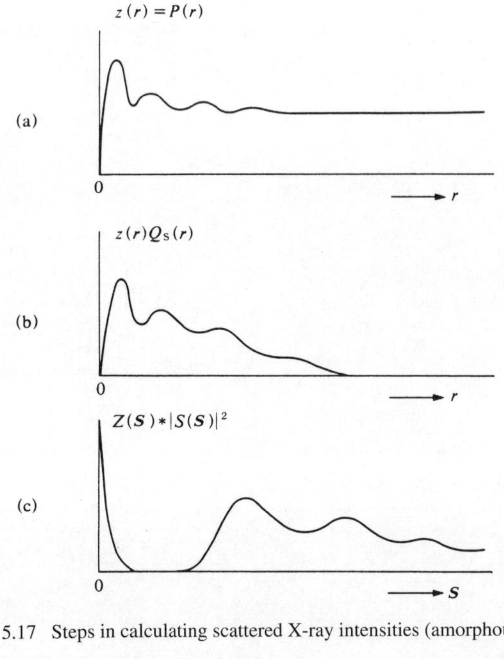

Fig. 5.17 Steps in calculating scattered X-ray intensities (amorphous substances).

The diffraction from a system of amorphous particles is simple, as is evident from Fig. 5.17. Attention has already been drawn to the fact that $Q(\boldsymbol{r})$ in this case is of essentially the same form as $P(\boldsymbol{r})$.

$$<Q(\boldsymbol{r})> = \frac{1}{v}[\delta(\boldsymbol{r}-0) + P(\boldsymbol{r})] \tag{5.45}$$

The Fourier transform of the $Q(\boldsymbol{r})$ of Fig. 5.17(a) is shown in (b). $|S(\boldsymbol{S})|^2$ is concentrated predominantly about the center (Eq. 5.15), and the halos corresponding to scattering from amorphous material are observed at greater angles. Of course the convoluted form of $|S(\boldsymbol{S})|^2$ also makes its contribution to these halos.

References

1. a) D. Schechtman, I. Blech, D. Gratias, J.W. Cahn, *Phys. Rev. Lett.*, **53**, 1951 (1984).
   b) N. Wang, H. Chen, K.H. Kuo, *Phys. Rev. Lett.*, **59**, 1010 (1987).
   c) L. Bendersky, *Phys. Rev. Lett.*, **55**, 1461 (1985).
   d) T. Ishimasa, H.-U. Nissen, Y. Fukano, *Phys, Rev. Lett.*, **55**, 511 (1985); K.H. Kuo, *J. Electron Microsc. Tech.*, **7**, 277 (1987).
2. W.T. Read, Jr., *Dislocations in Crystals*, p.17, McGraw-Hill (1953).
3. R. Hosemann, *Polymer*, **3**, 349 (1962).
4. R. Hosemann, S.N. Bagchi, *Direct Analysis of Diffraction by Matter*, North-Holland Pub., Amsterdam (1962).
5. B.K. Vainshtein, *Diffraction of X-rays by Chain Molecules*, p.97, Elsevier, Amsterdam (1966).

6. Scattering of X-Rays by Very Small Bodies

6.1 Small-angle Diffuse Scattering

When monochromatic X-rays are passed through very small particles, normally of the order of 10 to 2,000 Å or through a body containing zones of non-uniform density of about this size, a diffuse scattering pattern generally results within a very small range of angles around the incident direction (Fig. 6.1). Since the scattering is dependent upon the geometrical structure of the minute inhomogeneous zones, it is possible to establish the size, shape, state of aggregation, etc. of the small particles by analysis of the scattering pattern. This scattering differs in principle from diffraction, but is sometimes accompanied by diffraction due to normal Bragg diffraction (Section 4.1) and characteristic long repeat distances which can occur at very small angles. Since it is frequently impossible to distinguish unambiguously between these two contributions the term "small-angle scattering" is generally applied, without distinction, to them both.

In this section we attempt no more than an elementary explanation of the principle, analytical methods, and applications of small-angle scattering in the strict sense. If considerations of space did not preclude it, it would still be out of place in what is intended as a primer for experimental workers: the difficult theory and considerable overlap with other X-ray experimental methods have led the authors to refer the reader, instead, to the literature cited and to suitable textbooks for details.

Small-angle scattering is produced by substances containing small zones of non-uniform density which may be linear, planar, or particulate. The phenomenon has a remarkably wide application to the measurement of small particles, particularly in comparison with other physical methods. Thus, the breadths of the Debye-Scherrer rings are used in well-known method for the investigation of crystalline powders, and measurement methods based on viscosity, light scattering, osmotic pressure, surface tension, centrifugation, and electron microscopy have all been widely used for colloids and high polymers. However, these methods are all to some extent limited by the necessity to prepare the sample: X-ray small-angle scattering, on the other hand, has the very important advantages that the sample may be liquid, solid, crystalline, amorphous, or a mixture of these, and may take the form

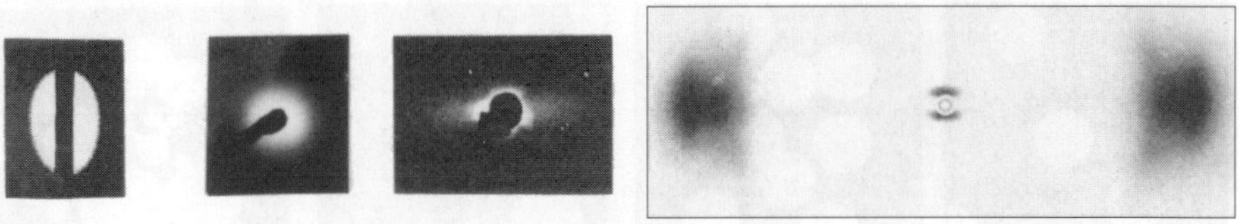

(a) (b) b₁) b₂) b₃)

Fig. 6.1 X-ray small-angle scattering patterns. (a) With slit system, carbon black (*cf.* Fig. 15.1). (b) With pin-hole system. b_1) Polyethylene (undrawn), isotropic, central diffuse scattering. b_2) Curdlan(or β-(1→3)-D-Glucan), (draw direction vertical), diffuse scattering along the equator. b_3) Polyurethane, (draw direction vertical), merridional small-angle scattering and wide-angle diffractions on the equator. [Reproduced from K. Hess, H. Kiessig, cited by R. Hosemann, in "*Zur Struktur und Materie der Festkörper*", p.137, Deutsch. Mineral. Ges. (1951)]

of minute inclusions or voids. It also requires no special preparation of the sample, and is essentially non-destructive. Nevertheless, the method is not a simple panacea for all problems of small particle analysis: as with so many physical measurements, the legitimacy of an interpretation can only be established by checking very carefully the connection between theory and results (and the purpose for which they were obtained).

6.2 Small-angle Scattering Theory

The principles of X-ray small-angle scattering were established as early as 1930, and by about 1938, Guinier,[1)] Kratky,[2)] and Hosemann[3)] had developed a quantitative treatment and shown how this scattering could be used. Further progress has been made since then as a result of various investigations into the problem of mutual interference based on differences between aggregations of particles, as well as into analytical methods and types of experimental equipment.

6.2.1 X-ray scattering by a substance of any structure

As summarized in Section 5.4, Eqs. 5.39 to 5.43 clearly show that, the central diffuse small-angle scattering is observed regardless the cohesive state of the substance, *i.e.* liquid, amorphous, crystalline, or paracrystalline, and in the case of crystalline or paracrystalline state the small-angle scattering is observed associated with each diffraction spot.

The other important conclusion obtained at the same time is that the intensity of the small-angle scattering depends upon the magnitude of $|\langle A \rangle|^2$ (or $|F|^2$ for crystal or $\langle |F| \rangle^2$ for paracrystal). This means that voids in substance can also give small-angle scattering as well as particles (Fig. 6.2).[4)] Particle-like density heterogeneities in substances

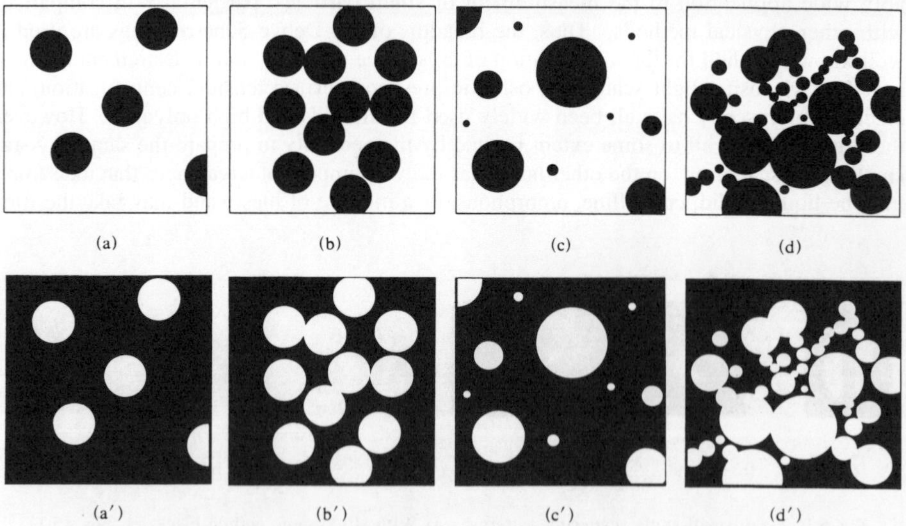

Fig. 6.2 Random assemblies of spherical particles.[4)]
(a) Dilute system of identical particles; (b) Densely packed system of identical particles; (c) Dilute system of non-identical particles; (d) Densely packed system of non-identical particles.
(a´), (b´), (c´) and (d´) Complimentary system of (a), (b), (c) and (d), respectively.
[Reproduced with permission from *X-Ray Crystallography*, (I. Nitta ed.), Vol. II, p.519, Maruzen (1961)]

such as precipitation in alloys (Guinier-Preston zone) and inclusion in the matrix lattice of grain with a different lattice but with the same composition (Fig. 6.3[5)]), also can give central diffuse small-angle scattering. Their intensities depend upon the electron density difference between void and matrix and that between the precipitation and matrix. This fact leads to the development of the "contrast variation method" (Section 6.2.2D).

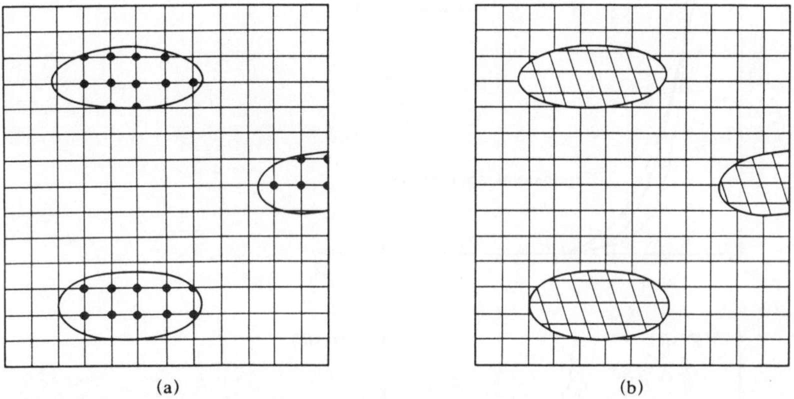

Fig. 6.3 Inhomogeneities in substance.[5)]
(a) Precipitation of foreign atoms; clustered without lattice deformation.
(b) Inclusion in the matrix lattice of a grain with a different lattice but with the same composition.
[Reproduced from A. Guinier, G. Fournet, *Small-Angle Scattering of X-Rays*, p.201, John Wiley & Sons, Inc. (1955)]

6.2.2 Small-angle scattering from systems of dilutely dispersed particles (or voids)[6)]

The small-angle scattering intensity by one particle with uniform density, ρ_∞ can be written as (*cf.* Eq. 5.14a)

$$I(S) = I_e \rho_\infty \mid S(S) \mid^2 \tag{6.1}$$

where the shape factor

$$S(S) = \int_{particle} \exp(2\pi i(\boldsymbol{S}\cdot\boldsymbol{r}))\,\mathrm{d}v_r \tag{6.2}$$

and then

$$I(S) = I_e \rho_\infty^2 \mid S(S) \mid^2 = I_e \rho_\infty^2\, V^2\, \Psi(S) = I_e n^2\, \Psi(S) \tag{6.3}$$

where $\mid S(S) \mid^2 = V^2\,\Psi(S)$, $\Psi(S)$ is the scattering function of the particle and $n(= V\rho_\infty)$ is the number of electrons in a particle, respectively.

If the shape of the particle is known as, for example, parallelepiped, sphere, ellipsoid of revolution, cylinder, etc. based on the theoretical equation for the three-dimensional particle shape, the scattering function can be obtained analytically or at least by numerical calculation. Some examples of the scattering functions are given below.

A. Globular particles

a) A globular particle fixed in a space.

1) A spherical particle of uniform density with radius *R*. This is the simplest case, and

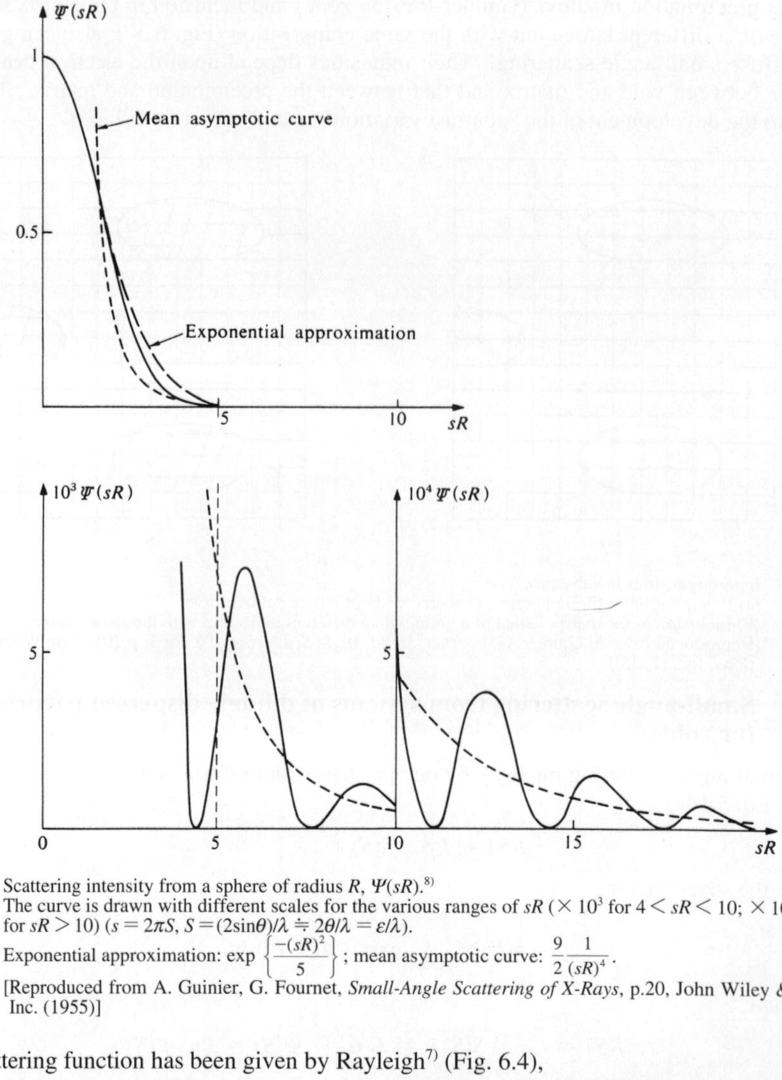

Fig. 6.4 Scattering intensity from a sphere of radius R, $\Psi(sR)$.[8)]
The curve is drawn with different scales for the various ranges of sR ($\times 10^3$ for $4 < sR < 10$; $\times 10^4$ for $sR > 10$) ($s = 2\pi S$, $S = (2\sin\theta)/\lambda \fallingdotseq 2\theta/\lambda = \varepsilon/\lambda$).
Exponential approximation: $\exp\left\{\frac{-(sR)^2}{5}\right\}$; mean asymptotic curve: $\frac{9}{2}\frac{1}{(sR)^4}$.
[Reproduced from A. Guinier, G. Fournet, *Small-Angle Scattering of X-Rays*, p.20, John Wiley & Sons, Inc. (1955)]

the scattering function has been given by Rayleigh[7)] (Fig. 6.4),

$$\Psi(S) = \Phi^2(2\pi SR)$$
$$= \left[3\frac{\sin(2\pi SR) - (2\pi SR)\cos(2\pi SR)}{(2\pi SR)^3}\right]^2 = \frac{9\pi}{2}\left[\frac{J_{3/2}(2\pi SR)}{(2\pi SR)^{3/2}}\right]^2 \qquad (6.4)$$

2) A rectangular parallelepiped with three edge lengths of a, b, and c.[6)] We assume that the a, b, and c edges of the parallelepiped are parallel to the x, y, and z axes, respectively (Fig. 6.5), and that the y axis is parallel to, and the x and z axes perpendicular to the incident X-ray beam, and that three components of S, the ξ, η and ζ axes are parallel to the x, y, and z axes, respectively.

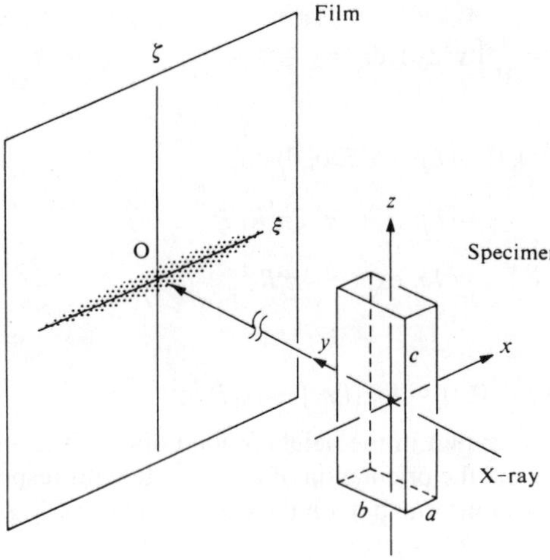

Fig. 6.5 Small-angle scattering from a fixed parallelepiped.

Then, the scattering function is given

$$\Psi(S) = \Psi(\xi, \eta, \zeta) = \frac{\sin^2(\pi\xi a)}{(\pi\xi a)^2} \cdot \frac{\sin^2(\pi\eta b)}{(\pi\eta b)^2} \cdot \frac{\sin^2(\pi\zeta c)}{(\pi\zeta c)^2} \tag{6.5}$$

This equation indicates that the diffuse small-angle scattering appears along the ξ, η and ζ directions with the broadening of $1/a$, $1/b$, and $1/c$, respectively. As the X-ray incidence is parallel to the y (or b) axis the scattering diagram observed corresponds to the $\xi\zeta$ section of $\Psi(S)$.

If the particle is a crystal, this small-angle scattering associates with all the reciprocal lattice points: the resultant is the Laue diffraction function (Eq. 2.38 and Fig. 5.12).

3) A globular particle with any shape at a fixed position (Guinier's approximation).[6]

In general, the shape of the particle is unknown or is so complicated that it is difficult to express its shape by analytical equation in order to obtain the scattering function. In such a case, the only information obtainable about the magnitude of the particle is the radius of gyration, R_g of the particle with respect to its center of gravity.

Taking the y axis again parallel to and the x and z axes perpendicular to the incident X-ray beam, and the origin at the center of gravity of the particle, the scattering observed on the xz plane is, since $\eta = \zeta = 0$,

$$S(\xi, 0, 0) = \int \exp(2\pi i \xi x) \mathrm{d}x\mathrm{d}y\mathrm{d}z \tag{6.6}$$

As the scattering angle is small and therefore ξ is small, the exponential function in Eq. 6.6 can be expanded by a series,

$$S(\xi, 0, 0) = \int (1 + 2\pi i \xi x - 2\pi^2 \xi^2 x^2 + \ldots) \mathrm{d}x\mathrm{d}y\mathrm{d}z \tag{6.7}$$

The second term in Eq. 6.7 is zero by integration since the origin is taken at the center of gravity of the particle. If we write

$$R_x{}^2 = \frac{1}{V}\int x^2 \mathrm{d}x\mathrm{d}y\mathrm{d}z \tag{6.8}$$

the scattering intensity is

$$\begin{aligned} I(\xi, 0, 0) &= I_e\rho^2 \mid S(\xi, 0, 0) \mid^2 \\ &= I_e\rho^2\, V^2(1 - 2\pi^2 R_x{}^2\xi^2 + \ldots)^2 \\ &= I_e n^2 \exp(-4\pi^2 R_x{}^2 \xi^2) \end{aligned} \tag{6.9}$$

and

$$\log I(\xi, 0, 0) = \log[I_e n^2] - 4\pi^2 R_x{}^2 \xi^2 \tag{6.10}$$

The tangent of the log I *vs.* ξ^2 plot in the neighborhood of $\xi = 0$ gives the value of R_x and hence if the particle shape and the orientation of the particle with respect to the x and z axes is known the tangent gives an information on the particle size. (*cf.* Fig. 6.7 and also Table 5 in the Appendix).

b) Globular particles with random orientation.

1) Spherical particles with radius R. The scattering function $\Psi(S)$ shows no change from that given by Eq. 6.4, since the sphere shows no directional difference in shape, and the small-angle scattering intensity of a spherical particle is:[7)]

$$< I(S) > = I(S) = I_e n^2\ \Psi(S). \tag{6.11}$$

$$\Psi(S) = \left[3\frac{\sin(2\pi SR) - (2\pi SR)\cos(2\pi SR)}{(2\pi SR)^3} \right]^2$$

If the specimen is a dilute system consist of M identical particles the intensity given in the above equation must be multiplied by M.

$$< I(S) > = I_e\, M n^2 \Psi(S) \tag{6.11a}$$

The scattering function $\Psi(2\pi SR)$ *vs.* $2\pi SR$ or $\Psi(sR)$ *vs.* sR (where, $s = 2\pi S$ and $S \doteqdot 2\theta/\lambda = \varepsilon/\lambda$ (for very small angles)) is shown in Fig. 6.4,[8)] which is compared with its exponential approximations (*cf.* Fig. 15.5).

2) Ellipsoids of revolution with axial lengths of $2a$, $2a$, and $2wa$ (axial ratio $w = b/a$).[1)]

$$<I(S)> = I_e M n^2 \int_0^{\pi/2} \psi(2\pi S a\sqrt{(\cos^2\theta + w^2 \sin^2\theta)}\ )\cos\theta d\theta \tag{6.12}$$

The scattering curves for ellipsoids of revolution with various axial ratios w are shown in Fig. 6.6.[9)] The log $I(S)$ *vs.* log S (or log $I(\varepsilon)$ *vs.* log ε, etc.) plot of the scattering curve (log-log plot) is often used to obtain a rough information on the overall shape of the particle (curve-fitting method *cf.* Section 15.2.2)

3) Globular particles with any shape. If the orientation of particles is not fixed, Eq. 6.7 must be averaged for all orientations of particles. The summation of the $R_x{}^2$ and its analogs along the y and z axes gives the square of R_g, the radius of gyration with respect to the center of gravity (Guinier's approximation).

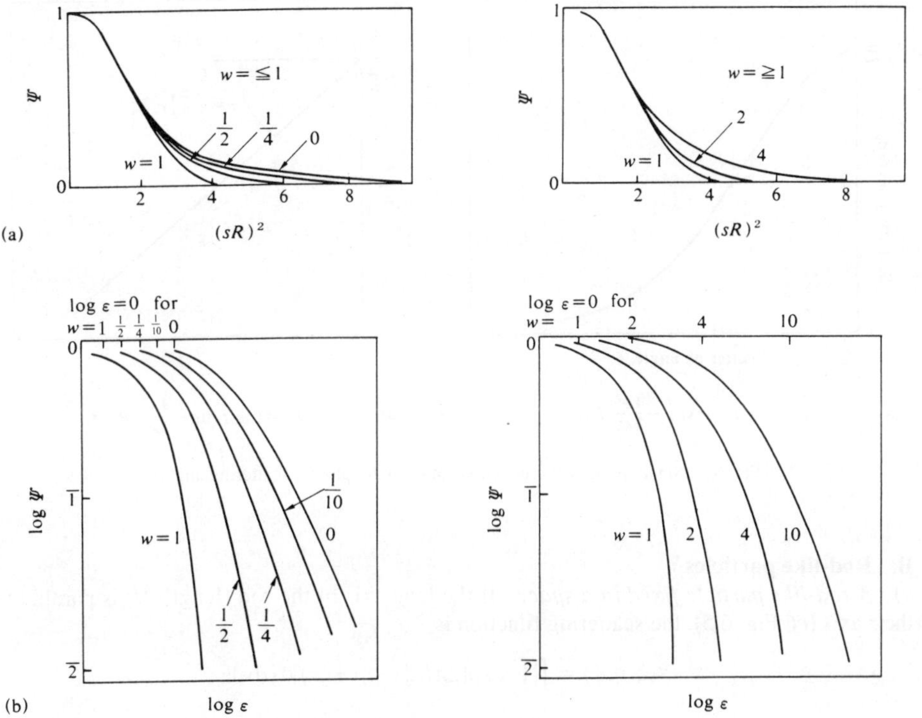

Fig. 6.6 Theoretical scattering curves for ellipsoids of revolution with various axial ratios w.[9] (a) Scattering function Ψ vs. $(sR)^2$, (b) Log-log plot derived from (a) where $sR = 2\pi SR = 2\pi \varepsilon R/\lambda$. [Reproduced from A. Guinier, *Ann. Phys. (Paris)*, **12**, 161, Masson et Cie. (1939)]

$$R_x{}^2 + R_y{}^2 + R_z{}^2 = \frac{1}{V}\int (x^2+y^2+z^2)\mathrm{d}x\mathrm{d}y\mathrm{d}z = R_g{}^2 \tag{6.13}$$

For random orientation all the averages of $R_x{}^2$, $R_y{}^2$, and $R_z{}^2$ are equal, and $< R_x{}^2 >$ can be written equal to $R_g{}^2/3$.

$$< \Psi(S) > = \exp[-4\pi^2 R_g{}^2 S^2/3] \tag{6.14}$$

$$< I(S) > = I_e\, Mn^2 \exp[-4\pi^2 R_g{}^2 S^2/3] \tag{6.15}$$

$$\log < I(S) > = \log(I_e\, Mn^2) - 4\pi^2 R_g{}^2 S^2/3 \tag{6.16}$$

$$= \log[I(0)] - 4\pi^2 R_g{}^2 S^2/3 \tag{6.16a}$$

If we plot values of log $I(S)$ against S^2 (experimentally, log $I(\varepsilon)$ *vs.* ε^2; $\varepsilon = 2\theta$: scattering angle) we can obtain a straight line in a small S (or ε) region[1] (Guinier plot, Fig. 6.7), and we can easily determine the value of R_g from its slope, α $(= -4\pi^2 R_g{}^2/3\lambda^2)$ (Section 15.2). If the shape of the particles is known, information about the particle size will be obtained.

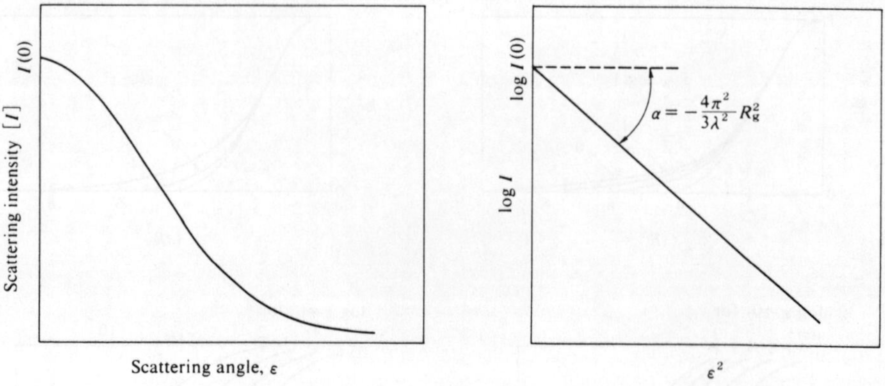

(a) $I = I(0)\exp\left(-\frac{4\pi^2}{3\lambda^2}R_g^2\varepsilon^2\right)$ (b) $\log I = \log I(0) - \frac{4\pi^2}{3\lambda^2}R_g^2\varepsilon^2$

Fig. 6.7 Small-angle scattering curve and Guinier plot (ε: scattering angle).

B. Rod-like particles [6)]

a) A rod-like particle fixed in a space. If the long axis of the rod (length H) is parallel to the z axis (*cf.* Fig. 6.5), the scattering function is

$$S(\xi,\eta,\zeta) = \iiint_{\text{section}\,-\frac{H}{2}}^{\frac{H}{2}} \exp[2\pi i(\xi x+\eta y+\zeta z)]\mathrm{d}x\mathrm{d}y\mathrm{d}z$$

$$= \frac{\sin(\pi H\zeta)}{\pi L\xi}\iint \exp[2\pi i(\xi x+\eta y)]\mathrm{d}x\mathrm{d}y \tag{6.17}$$

For the scattering observed on the $\xi\zeta$ plane, $\eta = 0$, and by the approximation for the small-angle Eq. 6.17 can be written as

$$S(\xi,0,\zeta) = H\frac{\sin(\pi H\zeta)}{\pi H\zeta}\iint \exp[2\pi i\xi x]\mathrm{d}x\mathrm{d}y$$

$$= HA\frac{\sin(\pi H\zeta)}{\pi H\zeta}\exp[-2\pi^2\xi^2 R_{xq}{}^2] \tag{6.18}$$

and then

$$I(\xi,\zeta) = I_e M n^2 \frac{\sin^2(\pi H\zeta)}{(\pi H\zeta)^2}\exp[-4\pi^2\xi^2 R_{xq}{}^2] \tag{6.19}$$

where, A is the transverse sectional area of the rod, $V = HA$, and

$$R_{xq}{}^2 = \frac{1}{A}\iint x^2\mathrm{d}x\mathrm{d}z \tag{6.20}$$

If H is large (*i.e.* the rod is long), the small-angle scattering intensity concentrates in the neighborhood of the ζ axis to give a streak along the ξ axis, of which the integral breadth is $1/H$.

b) Identical rod-like particles, of which all the long axes are parallel to the z axis but their transverse sections are randomly oriented.

Average R_{xq}^2 is

$$< R_{xq}^2 > = R_{gs}^2/2 \tag{6.21}$$

where, R_{gs} = is the radius of gyration of the section perpendicular to the long axis of the rod, and

$$R_s^2 = \frac{1}{A}\int (x^2+y^2)\mathrm{d}x\mathrm{d}y \tag{6.22}$$

and the small-angle scattering intensity on the ζ axis is

$$I(S) = I_e\, Mn^2 \exp[-2\pi^2 S^2 R_{gs}^2] \tag{6.23}$$

$$\log I(S) = \log[I_e\, Mn^2] - 2\pi^2 S^2 R_{gs}^2 \tag{6.24}$$

The Guinier plot of the equatorial small-angle scattering gives the radius of gyration of the section, R_{gs} of the rod-like particle (*cf.* Fig. 6.8(a) and Section 15.2.2).

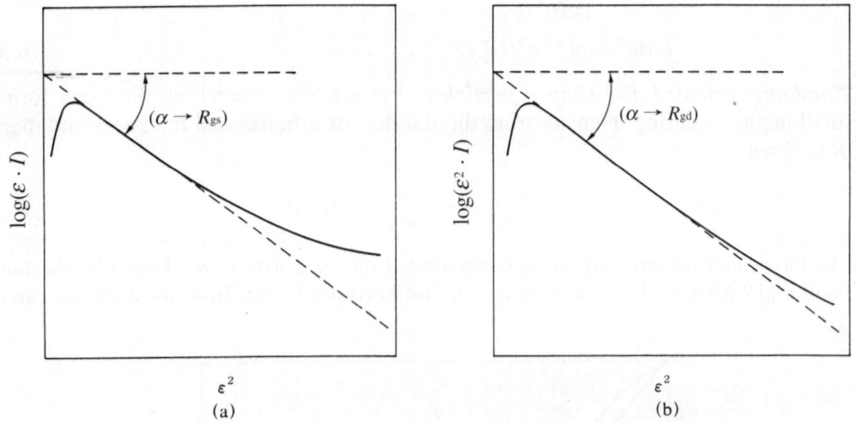

Fig. 6.8 Guinier plot of $\log(\varepsilon \cdot I)(= \log I_{BT})$ *vs.* ε^2 for R_{gs} and $\log(\varepsilon^2 \cdot I)$ *vs.* ε^2 for R_{gd}.

c) Randomly oriented rod-like particles (1/H ≪ S ≪ 1/R, H: length, R: radius of the section).

1) General case.

The scattering intensity concentrated on the $\xi\eta$ plane (Fig. 6.5) is dispersed for all directions and the intensity given by Eq. 6.23 is weakened by the factor $1/(2SH)$

$$<I(S)> = I_e Mn^2 \frac{1}{2SH}\exp[-2\pi^2 R_{gs}^2 S^2] \tag{6.25}$$

and

$$\log[S < I(S) >] = \text{Const.} - 2\pi^2 R_{gs}^2 S^2 \tag{6.26}$$

As special cases, the scattering intensities from uniform cylindrical particles and from rods of infinitesimal transverse dimensions are given in the following formulae.

2) Cylinders of revolution of diameter $2R$ and length H[10)]

$$<I(S)> = I_e M n^2 \int_0^{\pi/2} \frac{\sin^2(\pi SH\cos\theta)}{(\pi SH\cos\theta)^2} \cdot \frac{4J_1{}^2(2\pi SR\sin\theta)}{(2\pi SR\sin\theta)^2} \sin\theta \mathrm{d}\theta \tag{6.27}$$

3) Fibers of extremely small radius and length H[10–12)]

$$<I(S)> = I_e M n^2 \left[\frac{Si(2\pi SH)}{\pi SH} - \frac{\sin^2(\pi SH)}{(\pi SH)^2}\right] \tag{6.28}$$

where

$$Si(x) = \int_0^x \frac{\sin t}{t} \mathrm{d}t.$$

C. Disc-shaped particle ($1/H \gg S \gg 1/R$)[6)]

The scattering intensity concentrate in the direction perpendicular to the disc plane.

a) A disc-shaped particle fixed in a space.

If the disc plane is parallel to the x axis, assuming that $\xi = 0$ in Eq. 6.19

$$I(\zeta) = I_e M n^2 \frac{\sin^2(\pi H\zeta)}{(\pi H\zeta)^2} \tag{6.29}$$

$$= I_e M n^2 \exp[-\pi^2 H^2 \zeta^2/3] \tag{6.30}$$

b) Randomly oriented disc-shaped particles. For random orientation, the exact form of the small-angle scattering intensity from the flat disc of infinitesimal thickness and diameter $2R$ is given[13)]

$$<I(S)> = I_e M n^2 \frac{2}{(2\pi SR)^2}\left[1 - \frac{1}{2\pi SR} J_1(2\pi SR)\right] \tag{6.31}$$

That is, the scattering intensity to be concentrated on the ξ axis is weakened by the factor $1/S^2$, and $\log[S^2 I(S)]$ *vs.* S^2 (or $\varepsilon^2 I(\varepsilon)$ *vs.* ε^2) plot becomes linear, from its slope we can de-

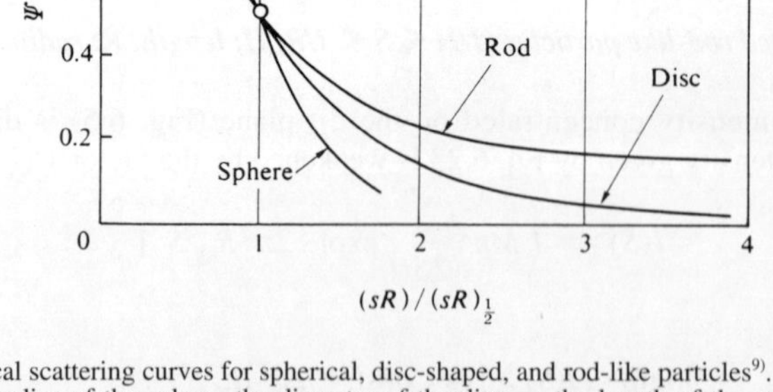

Fig. 6.9 Theoretical scattering curves for spherical, disc-shaped, and rod-like particles[9)].
R is the radius of the sphere, the diameter of the disc, or the length of the rod. The abscissa is transformed as $(sR)/(sR)_{1/2}$ so that all three curves pass through the same point, which is defined as the point where the scattering function has exactly half its maximum value. ($sR = 2\pi SR = 2\pi\varepsilon R/\lambda$)
[Reproduced from O. Kratky, G. Porod, *J. Colloid Sci.*, **4**, 35, Academic Press (1949)]

termine the thickness of the disc H (Fig. 6.8(b) and *cf.* Section 15.2.2B).

In Fig. 6.9 the scattering curves for spherical, rod-like and disc-shaped particles are compared.[9]

If we assume the overall shape of the particle to be a hollow sphere, cylinder, hollow cylinder, rod or other, and if we can calculate the scattering functions numerically, we can apply the curve-fitting method (log-log plot) to simulate the most suitable shape and size of the particle.

D. Particles suspended in a medium

If the particles are dispersed within a suspending medium and the particles are impermeable, the number of electrons n in Eq. 6.11 and other equations in the preceding sections must be replaced by the product of the difference $\Delta\rho = \rho_1 - \rho_2$ between the electron densities in the two phases and the volume of the particles V_1, *i.e.* by the effective electron excess or deficiency $\Delta\rho\, V_1$ in the particles.

$$< I(S) > = I_e\, M(\Delta\rho\, V_1)^2\, \Psi(S) \qquad (6.32)$$

This means that increasing the electron density outside V_1 is equivalent to decreasing the scattering intensity uniformly within the particle. In the early times Barton and Brill[14] confirmed this fact experimentally for the small-angle scattering from carbon black suspended in aqueous solutions of inorganic salt by changing the concentration of salt.

a) Contrast variation method.[15–18] To obtain more information on the structure of inhomogeneous particles the contrast variation method is often used, especially in the field of neutron small-angle scattering. This method uses solvents of different electron densities.

The contrast $< \Delta\rho >$ is the mean difference in scattering density between the particle ρ_1 and solvent ρ_2.

$$< \Delta\rho > = \bar{\rho}_1 - \bar{\rho}_2 \qquad (6.33)$$

where $\bar{\rho}_1$ and $\bar{\rho}_2$ are averages of ρ_1 and ρ_2 over the volume V_1, respectively.

If we assume that within the volume V_1 of the particle the electron density remains unchanged while the surrounding solvent has a variable homogeneous electron density ρ^2, the difference in electron density is then assumed to be the sum of the two terms. One describes the effect of excluding solvent from the region occupied by the particle; this term is directly affected by the contrast between the particle and solvent. The other term covers any variation within the particle (Fig. 6.10).

$$< \Delta\rho(r) > = < \Delta\rho > \sigma_1(r) + \rho_3(r) \qquad (6.34)$$

$\sigma_1(r) = 1$ inside the particle and zero outside. Thus, $\int \sigma_1(r)\mathrm{d}r = V_1$. The ρ_3 is the deviation from the mean difference electron density $< \rho >$: it arises from any internal structure within the particle. It further holds that

$$\int_{V_1} \rho_3(r)\mathrm{d}r = 0 \qquad (6.35)$$

The scattering amplitude of the particle is calculated as

$$A(S) = < \rho > S_1(S) + A_3(S) \qquad (6.36)$$

where $S_1(S)$ and $A_3(S)$ are the Fourier transforms of $\sigma_1(r)$ and $\rho_3(r)$, respectively. The scattering intensity is

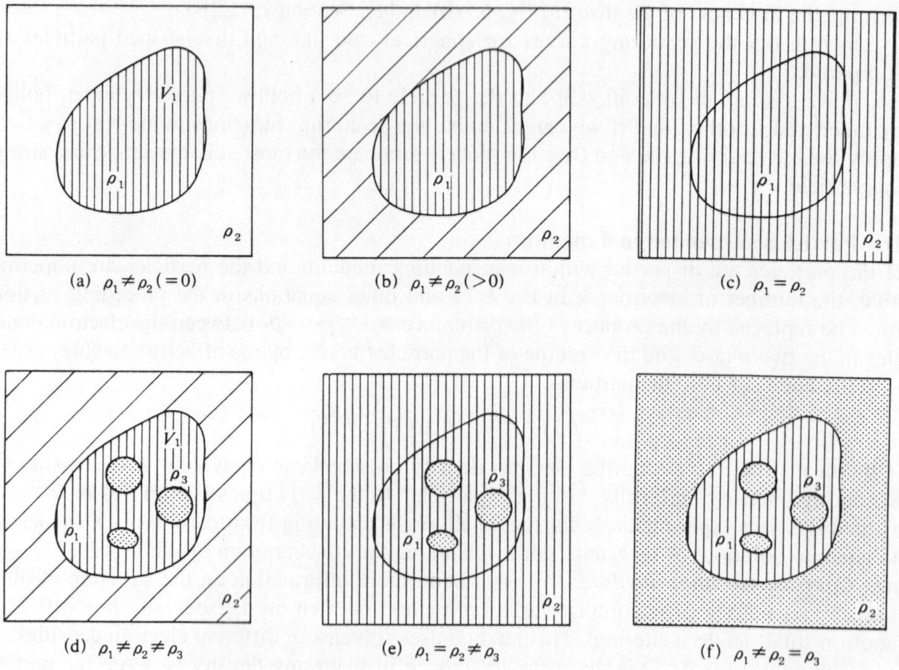

Fig. 6.10 Effect of the contrast.
V_1 denotes the volume of a particle; ρ_1, ρ_2 and ρ_3 are the densities of a particle, of the dispersion medium and of internal inhomogeneity in the particle, respectively,
(a) $\rho_1 \neq \rho_2 (=0)$, particle with uniform density in vacuum; (b) $\rho_1 \neq \rho_2 (>0)$, particle with uniform density in a medium. A similar but less intense small-angle scattering(a) is observed; (c) $\rho_1 = \rho_2$, particle has the same density as the medium. No small-angle scattering is observed; (d) $\rho_1 \neq \rho_2 \neq \rho_3$, particle which has internal inhomogeneity in a medium; (e) $\rho_1 = \rho_2 \neq \rho_3$. Only internal inhomogeneities with density ρ_3 give small-angle scattering; (f) $\rho_1 \neq \rho_2 = \rho_3$. Only inhomogeneity ρ_1 gives small-angle scattering.

$$I(S) = A(S)A*(S)$$

$$= <\Delta\rho>^2 | S_1(S) |^2 + <\Delta\rho> | S_1(S)A_3(S) | + | A_3(S) |^2 \tag{6.37}$$

In Eq. 6.37, $| S_1(S) |^2$ is the "shape factor" of the volume V_1 and can be determined by extrapolation because $A_3(S)$ vanishes at $S=0$. The square root of the scattering intensity at zero angle is linear in $<\Delta\rho>$; and $[I(0)]^{1/2} = <\Delta\rho> | S_1(0) |$. $| A_3(S) |^2$ originates from $\rho_3(r)$ and is measurable if $<\Delta\rho> = 0$. $| A_3(S) |^2 = 0$ according to Eq. 6.35. The cross term $| S_1(S)A_3(S) |$ establishes the correlation between $| S_1(S) |^2$ and $| A_3(S) |^2$ by the following general inequality

$$| S_1(S)A_3(S) | \leqq \frac{1}{2}\sqrt{| S_1(S) |^2 \cdot | A_3(S) |^2} \tag{6.38}$$

Introduction of Eq. 6.34 into the definition of the radius of gyration R_g yields

$$R_g^2 = R_3^2 + \alpha\big/_{<\Delta\rho>} - \beta\big/_{<\Delta\rho>^2} \tag{6.39}$$

where

$$R_3{}^2 = \frac{1}{V_1}\sigma_1(r)r^2\mathrm{d}r \tag{6.40}$$

$$\alpha = \frac{1}{V_1}\rho_3 r^2\mathrm{d}r \tag{6.41}$$

$$\beta = \frac{1}{V_1}\rho_3(r)\rho_3(r')\cdot r\cdot r'\mathrm{d}r\mathrm{d}r' \tag{6.42}$$

R_3 represents the radius of gyration of a homogeneous particle and does not depend on $<\Delta\rho>$. α is the second moment of the electron density fluctuations. If the sign of α is positive the positive contributions dominate. β is an asymmetry parameter according to Eq. 6.40. The $R_g{}^2$ *vs.* $1/<\Delta\rho>$ is a straight line if $\beta = 0$, and the slope of this line is α and the value of $R_g{}^2$ at $1/<\Delta\rho> = 0$ is $R_s{}^2$.

The contrast variation method for X-ray small-angle scattering is performed by variations in the concentration of sugar or other low molecular weight solutes. Difficulties may arise from preferential interactions between solvent components and the macromolecule under investigation.

E. S^{-4} Rule of scattering intensity

As shown earlier the small-angle scattering intensity of a system of spherical particles of radius R and density ρ is given by Eq. 6.11, which can be written as follows as

$$n^2 = (\rho V)^2 = (4/3\cdot\pi R^3\rho)^2.$$

$$I(S) = \frac{\rho^2}{8\pi^3}\left[\frac{4\pi R^2}{S^4} + \frac{1}{\pi S^6} - \frac{4R}{S^5}\sin(4\pi SR) - \frac{4R^2}{S^4} - \frac{1}{\pi S^6}\cos(4\pi SR)\right] \tag{6.43}$$

For sufficiently large S, the second and higher terms in the bracket can be neglected and

$$I(S) = \frac{\rho^2}{8\pi^3}\cdot\frac{4R^2}{S^4} \tag{6.44}$$

This means that theoretically the tail end (or fmal slope) of the small-angle scattering curve should conform to the asymptotic course for S^{-4}.[19]

For the two-phase system the asymptotic value of $S^4I(S)$ is rational to the square of density difference and the area of the interface.

$$\lim_{S\to\infty} S^4 I(S) = \frac{1}{8\pi^2}(\rho_1 - \rho_2)\cdot(\text{Area of the interface}) \tag{6.45}$$

6.2.3 Correlation function and distance distribution function[20,21]

A. Correlation function

a) Dilute particle system. The particles are randomly oriented and far from each other so that all the interparticle interference may be neglected. The scattering intensity is given

$$<I(S)> = \rho^2 V\int_0^\infty \gamma_0(r)\frac{\sin Sr}{Sr}4\pi r^2\mathrm{d}r \tag{6.46}$$

where $\gamma_0(r)$, the correlation function[22,23] (characteristic function[19] or distance probability function) of the particle, represents the probability that a point at a distance r in an arbitrary direction from a given point in the particle will itself also be in the particle (See Fig. 5.8).

The properties of $\gamma_0(r)$ are: from the definition $\gamma_0(0) = 1$ and γ_0 decreases as r increases. It becomes zero when r is greater than the maximum particle diameter.

$$\int_0^\infty 4\pi r^2 \gamma_0(r) \mathrm{d}r = V = \frac{4\pi}{\rho^2} \int_0^\infty S^2 I(S) \mathrm{d}S, \tag{6.47}$$

and

$$\int_0^\infty \gamma_0(r) \mathrm{d}r = \frac{<l>}{2} = \frac{\pi}{\rho^2 V} \int_0^\infty S<I(S)> \mathrm{d}S \tag{6.48}$$

$<l>$ is the average of the diameters drawn through every point within the particle in all directions.

$$\left. \frac{\mathrm{d}\gamma_0}{\mathrm{d}r} \right|_{r\to 0} = \frac{O}{4V} \tag{6.49}$$

where O is the surface area of the particle. This relation is a consequence of the fact $S^4 < I(S) >$ must tend toward a constant limiting value

$$\lim S^4 <I(S)> = \frac{\rho^2 O}{8\pi^3} \tag{6.50}$$

as S increases.

By Fourier inversion the correlation function γ_0 is obtained from the small-angle scattering intensity measured.

$$\gamma_0(r) = \frac{2}{\rho^2 V r} \int_0^\infty S<I(S)> \sin 2\pi r S \mathrm{d}S \tag{6.51}$$

b) General case. The specimen composed of matter of uniform electron density ρ filling a fraction χ_1 of the total volume V_0 is

$$I(S) = V_0 \rho^2 \chi_1 \chi_2 \int_0^\infty \gamma(r) \frac{\sin 2\pi S r}{2\pi S r} 4\pi r^2 \mathrm{d}r \tag{6.52}$$

where $\chi_2 = 1 - \chi_1$, $\gamma(r)$ is the correlation function of the heterogeneous system, and for a system composed of identical particles $\gamma(r)$ becomes $\gamma_0(r)$, the correlation function of the particle. In the general case $\gamma(r)$ is defined in the following way: Let $Y(r)$ be the probability that a point a distance r from another point occupied by matter is also occupied by matter.

Then

$$Y(r) = \chi_1 + \chi_2 \gamma(r)$$

By Fourier inversion $\gamma(r)$ may be obtained from the experimental data.

The limit of the product $S^4 I(S)$ is still given by $(\rho^2/8\pi^3)O_0$. (O_0: total surface area).

B. Distance distribution function[24)]

a) Globular particles. Using the correlation function the distance distribution function of the particle is defined as

$$p(r) = \gamma_0(r) \cdot r^2. \tag{6.53}$$

$4\pi p(r)$ represents the number of lines with lengths r, which are found in the combination of any volume element j with any other volume element k.

Inhomogeneous particles may have regions with positive and negative contributions to the $p(r)$.

1) Sphere. The distance distribution function of a sphere is given analytically.[20]

$$p(r) = \text{const.} \cdot x^2(2 - 3x + x^3) \tag{6.54}$$

where $x = r/2R$ and R is the radius of the sphere. The maximum of the $p(r)$ is near $r = R(x = 0.5)$ (Fig. 6.11). Any deviation from spherical symmetry shifts the maximum to smaller r values if R is kept constant.

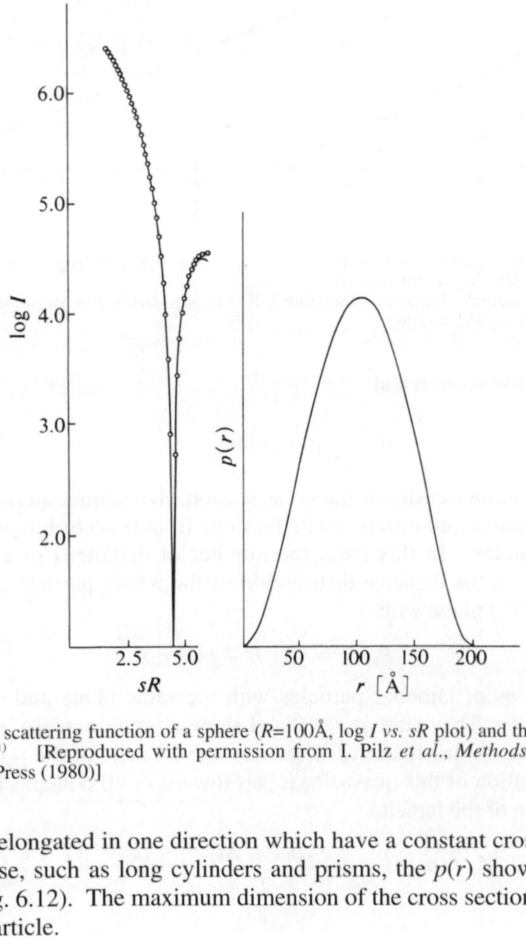

Fig. 6.11 Theoretical scattering function of a sphere (R=100Å, log I *vs.* sR plot) and theoretical distance distribution, $p(r)$.[21] [Reproduced with permission from I. Pilz *et al.*, *Methods in Enzymology*, **61**, 168, Academic Press (1980)]

2) Particles elongated in one direction which have a constant cross section of arbitrary shape. In this case, such as long cylinders and prisms, the $p(r)$ show a linear decrease at large r values (Fig. 6.12). The maximum dimension of the cross section d_{max} is much smaller than that of the particle.

$$2R/d_{max} \geqq 2.5 \tag{6.55}$$

The slope of the linear region is given by

$$\tan\alpha = -\frac{dp}{dr} = \frac{A^2 < \rho_c >^2}{2\pi} \tag{6.56}$$

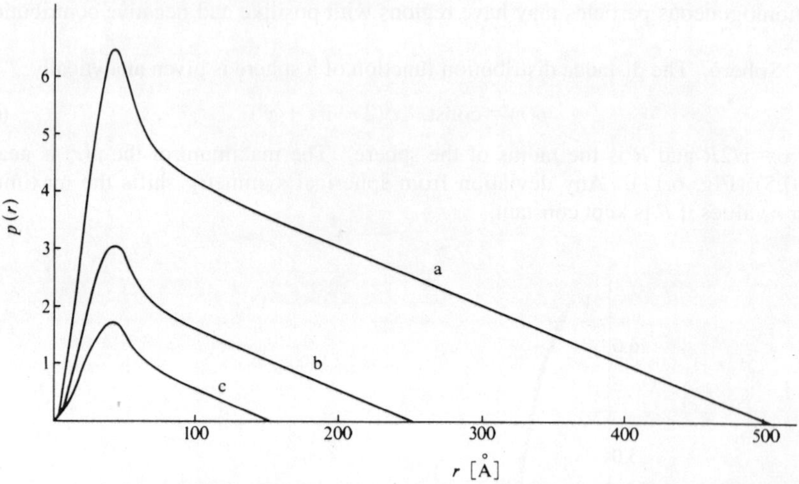

Fig. 6.12 Distance distributions from homogeneous prisms with edge lengths of (a) 50:50:500, (b) 50:50:250, (c) 50:50:150.[21)]
[Reproduced with permission from I. Pilz *et al.*, *Methods in Enzymology*, **61**, 169, Academic Press (1980)]

A is the area of the cross section and

$$\rho_c = \frac{1}{A}\int_A \rho_c(x)\mathrm{d}a \tag{6.57}$$

where $\rho_c(x)$ is the electron density in the cross section. Difference in the area of the cross section results in a pronounced difference in the slope from α according to Eq. 6.56.

3) Lamellar particles. In this case, the number of distances in a plane is equal to $2\pi r\gamma_0(r)$. In analogy to the distance distribution of the whole particle $p(r)$, we define the distance distribution of a plane with

$$P_L(r) = \gamma_0(r)r = p(r)/r \tag{6.58}$$

The $p_L(r)$ functions of lamellar particles with the same plane and different thickness t are given in Fig. 6.13. They start at $r = 0$ and show a rapid increase, and at a point near $r = t$ begin to decrease almost linearly. The limiting value A_L, of the $p_L(r)$ function resulting from the extrapolation of this quasi-linear part toward $r = 0$ contains information on the area of the basal plane of the lamella.

$$A_L = p_L(r)\Big|_{r\to 0} = \frac{<\rho_t>^{-2}At^2}{2} = \frac{<\rho_t>^{-2}Vt}{2} \tag{6.59}$$

where

$$<\rho_t> = \frac{1}{t}\int_t \rho(x)\mathrm{d}x$$

The extrapolation to $r = 0$ is more accurate the larger the ratio of $2R/t$.

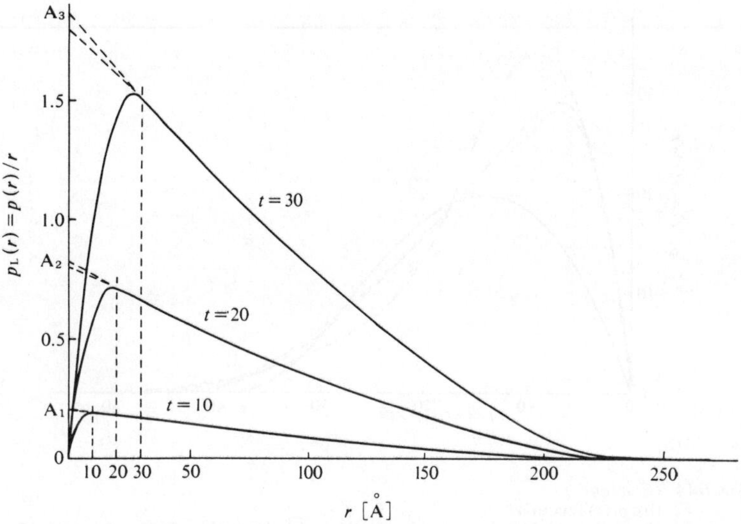

Fig. 6.13 Functions of lamellar particles with the same plane (100 × 100) and varying thickness *t* (10, 20, 30). The transition points are signalized by the vertical dashed lines.[21)]
[Reproduced with permission from I. Pilz *et al.*, *Methods in Enzymology*, **61**, 170, Academic Press (1980)]

Figure 6.14 gives a comparison of $p(r)$ and $p_L(r)$ between a sphere, a prolate ellipsoid and an oblate ellipsoid having the same number of excess electrons and radius of gyration.

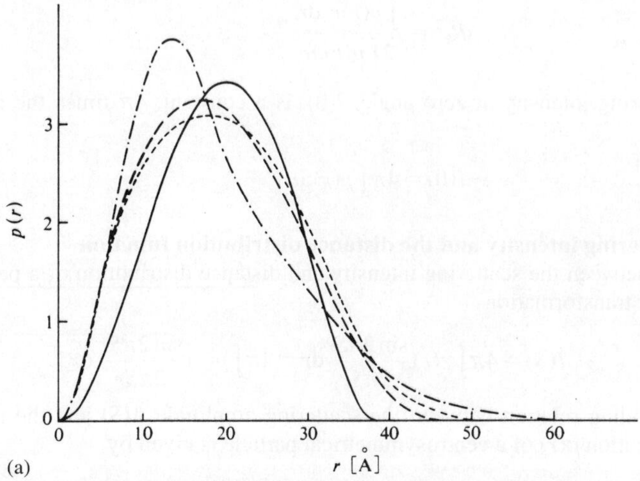

Fig. 6.14 Comparison of the distance distribution functions of a sphere (——), a prolate ellipsoid of revolution 1 : 1 : 3 (–·–·–), an oblate ellipsoid of revolution 1 : 1 : 0.2 (----) and a flat prism 1 : 1 : 0.23 (---) with the same radius of gyration. (a): $p(r)$ function.

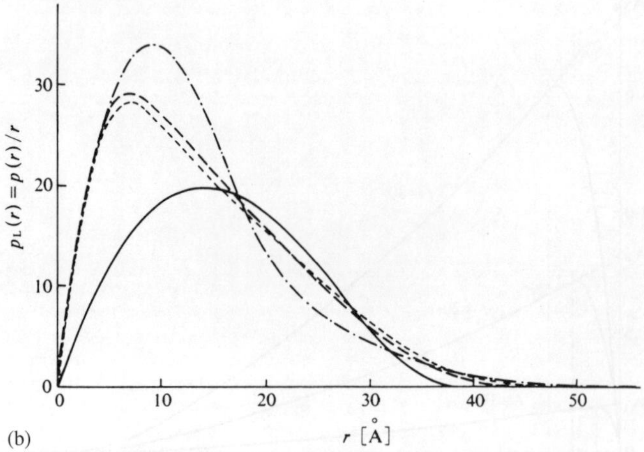

Fig. 6.14 (*Continued*)
(b); $p_L(r)$ function.[21)]
[Reproduced with permission from I. Pilz *et al.*, *Methods in Enzymology*, **61**, 171, Academic Press (1980)]

C. Distance distribution function and radius of gyration and zero angle intensity

The radius of gyration of the particle, R_g, mentioned earlier is defined by the distance distribution function, $p(r)$,

$$R_g{}^2 = \frac{\int p(r)r^2 \mathrm{d}r}{2\int p(r)\mathrm{d}r} \tag{6.60}$$

and the scattering intensity at zero angle, $I(0)$, is a constant: 4π times the integral of the $p(r)$:

$$I(0) = 4\pi \int p(r)\mathrm{d}r \tag{6.61}$$

D. The scattering intensity and the distance distribution function

The relation between the scattering intensity and distance distribution of a particle is given by the Fourier transformation

$$I(S) = 4\pi \int \gamma(r) \frac{\sin 2\pi Sr}{2\pi Sr} \mathrm{d}r = 4\pi \int p(r) \frac{\sin 2\pi Sr}{2\pi Sr} \mathrm{d}r \tag{6.62}$$

The corresponding relation between the scattering amplitude $A(S)$ and the radial electron density distribution $\rho(r)$ of a centrosymmetrical particle is given by

$$A(S) = 4\pi \int \rho(r) r^2 \frac{\sin 2\pi Sr}{2\pi Sr} \mathrm{d}r \tag{6.63}$$

Inverse transformation gives

$$\gamma(r) = 4\pi \int I(S) S^2 \frac{\sin 2\pi S r}{2\pi S r} \mathrm{d}S \tag{6.64}$$

or

$$p(r) = 2\int I(S) \cdot Sr \cdot \sin(2\pi Sr) \mathrm{d}S \tag{6.65}$$

and from Eq. 6.63

$$\rho(r) = 4\pi \int A(S) \cdot S^2 \frac{\sin 2\pi S r}{2\pi S r} \mathrm{d}S \tag{6.66}$$

6.2.4 Polydispersed system of particles with uniform shape

When the shape is uniform but there is a distribution of particle diameters we may take $M(R_g)$ as the mass distribution function of particles between R_g and $R_g + \mathrm{d}R_g$, and the scattering intensity will be given by

$$I = K_0 \int_0^\infty M(R_g) R_g^{\,3} \exp(-s^2 R_g^{\,3}/3) \mathrm{d}R_g \tag{6.67}$$

If this particle size distribution is approximately divided into N rectangular distributions and the ratio of the weight of the particles in the jth component to the total particle weight is $W(R_{gj})$, then

$$I = K_0 \sum_{j=1}^{N} W(R_{gj}) R_{gj}^{\,3} \exp(-s^2 R_{gj}^{\,3}/3)$$

$$= K_0 \sum_{j=1}^{N} W(R_{gj}) R_{gj}^{\,3} \exp\{-(4\pi^2/3\lambda^2)\varepsilon^2 R_{gj}^{\,2}\} \tag{6.68}$$

In practice, a tangent drawn to the lower end of the log I *vs.* ε^2 curve (*cf.* Fig. 13.22) gives R_{g1}, the minimum radius of gyration. The radius of gyration R_{g2} of the second component is found by subtracting the intensity corresponding to this tangent from the original curve and drawing a tangent to the curve produced in this way. Successive repetitions of this operation give R_{g1}, R_{g2}, R_{g3},... The weight ratios $W(R_{gj})$ of the size fractions can then be found from the ratios of the intercepts of the tangents on the ordinate $K_j = K_0 W(R_{gj}) R_{gj}^{\,3}$, the Jellinek-Solomon-Fankuchen method.[25]

The subject of errors due to the slit system and their elimination is covered in Section 15.1.2. Correction in the case of the Guinier plot is, however, very simple. For measurements made with a slit that is so narrow and long that it can be considered as approximating a slit of infinite height and zero width, calculation of I_H from Eq. 6.68 and 15.2 gives

$$I_H = K_0 \sum_{j=1}^{N} W_j(R_{gj}) R_{gj}^{\,3} \int_0^\infty \exp\{(-4\pi^2/3\lambda^2)(\varepsilon^2 + t^2/r^2) R_{gj}^{\,2}\} \mathrm{d}t$$

$$= K_0' \sum_{j=1}^{N} W_j(R_{gj}) R_{gj}^{\,2} \exp\{(-4\pi^2/3\lambda^2)\varepsilon^2 R_{gj}^{\,2}\} \tag{6.69}$$

Accordingly, since the gradient of the Guinier plot for each component is not affected by the slit height, R_{gj}, may be calculated directly from Eq. 15.5. Since the value of the intercept is $K_j' = K_0' W_j(R_{gj}) R_{gj}^{\,2}$, as shown by Eq. 6.69 if the weight ratio is calculated using $W_j R_{gj}^{\,2}$ in place of $W_j R_{gj}^{\,3}$, the correct result is obtained without correction for the effect of errors caused by the slit.[26]

A very convenient formula is given by Hosemann,[27] if $M(R_g)$ is a Maxwellian,

$$M(R_g)=\left(\frac{R_g}{R}\right)^n \exp\left\{-\left(\frac{R_g}{R}\right)^2\right\}\frac{2}{R\Gamma\left(\frac{n+1}{2}\right)} \tag{6.70}$$

$\Gamma\left(\frac{n+1}{2}\right)\cong\left(\frac{n-1}{2}\right)!$ is the Γ-function. The weight averaged particle radius $<R_g>$ and the polydispersity g_R are given by

$$<R_g>= R\Gamma\left(\frac{n+3}{2}\right)\Big/\Gamma\left(\frac{n+1}{2}\right)\cong R(n+1)/(2n+3)^{\frac{1}{2}} \tag{6.71}$$

$$g_R=\frac{(<R_g{}^2>-<R_g>^2)^{\frac{1}{2}}}{<R_g>}=\left[\frac{\Gamma\left(\frac{n+5}{2}\right)\Gamma\left(\frac{n+1}{2}\right)}{\left\{\Gamma\left(\frac{n+3}{2}\right)\right\}^2}\right]^{\frac{1}{2}}\cong\{2(n+1)\}^{\frac{1}{2}} \tag{6.72}$$

The intensity is given by

$$I = I_e\, Mn^2[1+(sR_g)^2/3]^{-(n+4)/2} \tag{6.73}$$

For small g_R, hence $n > 100$ it reduces to the Guinier equation (Eq. 15.4).

6.2.5 Small-angle scattering from systems of densely packed particles

If particles producing small-angle scattering are sufficiently densely packed we must consider not only the interference contributed by individual particles, but also the mutual interference between neighboring particles, just as the mutual interference between neighboring atoms had to be taken into account in extending the treatment of the X-ray diffraction by monatomic gases to the case of liquids. Since the scattering amplitude for one particle is $(\Psi)^{1/2}$, the system may be treated as a densely packed assembly of hypothetical atoms having scattering factors $(\Psi)^{1/2}$, and the composite scattering amplitude may be found by the method of Section 2.5. The scattering theory for a monatomic liquid cannot, however, be applied without modification. The details of the treatment will not be given here, but small particles differ from atoms in that their size and shape are not constant. Moreover, the regular cohesive forces which exist between atoms (Van der Waals forces, etc.) lead to fairly regular statistical distributions of neighboring atoms, whereas the cohesive forces between small particles differ with their nature and external shape, so that the statistical distributions of such systems are rather complex. Thus there are very few cases in which the theory of X-ray scattering for liquids is strictly applicable, and special care is generally necessary in the interpretation of small-angle scattering from densely packed particles.

The following is a brief description of the main points of scattering theories for densely packed systems of particles with certain specific shapes.

A. Spheres of uniform size

In the scattering Eq. 2.28, derived initially by Zernicke and Prins[28] and by Debye and Menke[29] for monatomic liquids, replacement of the atomic scattering factor f by the number of electrons n in one article and the scattering function, $\Psi(s)$, for a sphere (see again Table 5 of the Appendix) gives

$$I(S)=I_e Mn^2\Upsilon(s)^2\left\{1+\int_0^\infty 4\pi r^2[\rho(r)-\rho_0]\frac{\sin sr}{sr}\,\mathrm{d}r\right\} \tag{6.74}$$

ρ_0 is the average particle density in the system and $\rho(r)$ is the particle density in the immediate environment of a particle. It should be noted that Eq. 6.74 is of the same form as Eq. 2.28, with the scattering intensity Ψ for a single spherical particle in place of the square of atomic scattering factor, f^2.

If the radial particle density distribution $P(r)$ is zero for $0 < r < R$ and constant for $r > R$ (the case with Fig. 2.16(a)), the value for intensity given by Eq. 6.74 is dependent upon the particle density, since particle density $\rho(r) = P(r)/v$. The effect is such that interference increases with increasing particle density, but approximates more closely to single particle scattering as the density decreases. This is shown in Fig. 6.15. The dependence does not change in character even if the scattering body is rod-shaped or lamellar rather than spherical.

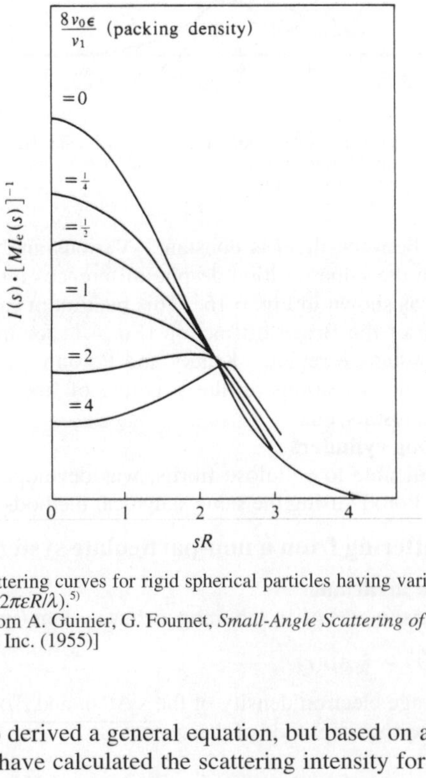

Fig. 6.15 Theoretical scattering curves for rigid spherical particles having various packing densities ($sR = 2\pi S R = 2\pi\varepsilon R/\lambda$).[5]
[Reproduced from A. Guinier, G. Fournet, *Small-Angle Scattering of X-Rays*, p.50, John Wiley & Sons, Inc. (1955)]

Fournet[30] has also derived a general equation, but based on a rather different approach. Lund and Vineyard[31] have calculated the scattering intensity for clusters of several spherical particles.

B. Spheres and ellipsoids of revolution of different sizes

Fournet[12] calculated the scattering intensity for dense assemblies of spheres having two different radii, and Kratky and Porod[9] and Hosemann[27] calculated intensities for dense assemblies of spheres whose radii have a Maxwellian distribution.

Roess and Shull[32] calculated the scattering by an ensemble of ellipsoids of revolution.

C. Dense accumulations of lamellae

Kratky and Porod[9] calculated the intensity for this case as a model of the structure of the crystalline regions in natural and synthetic fibers. In this case neither the thickness of the

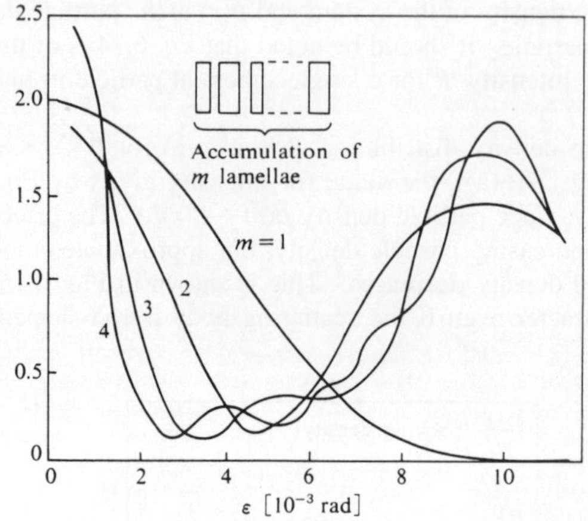

Fig. 6.16 Influence of m on the scattering curve for lamellar packing.[9) [Reproduced from O. Kratky, G. Porod, *J. Colloid Sci.*, **4**, 35, Academic Press (1949)]

lamellae nor the distance between them is constant. Various analyses all give scattering curves with a maximum in the same position despite differences in the statistical treatment of the thickness variation, as shown in Fig. 6.16.[9)] This maximum can be interpreted as corresponding to the position of the Bragg diffraction (Eq. 4.1) for the average distance between the layers in the crystalline region. Kratky and Porod extended the theory further and carried out numerous investigations on the structure of fibers. Section 13.6.3 gives more details of techniques, results, etc.

D. Parallel bundles of long cylinders

This theory, which is applicable to cellulose fibrils, was developed, and the calculations performed, by Kratky and Porod[9)] using the same statistical methods as in **C** above.

6.2.6 Small-angle scattering from a non-particulate system

A. Scattering power and invariant

If we write the electron density $\rho(r)$ of an inhomogeneous system (volume V) as

$$\rho(r) = <\rho(r)> + \Delta\rho(r), \tag{6.75}$$

where $<\rho(r)>$ is the average electron density of the system and $\Delta\rho(r)$ is fluctuation of the electron density. The scattering amplitude of the system is

$$S(S) = \int <\rho(r)> \exp[2\pi i(\boldsymbol{S}\cdot\boldsymbol{r})]\mathrm{d}v_r + \int \Delta\rho(r)\exp[2\pi i(\boldsymbol{S}\cdot\boldsymbol{r})]\mathrm{d}v_r \tag{6.76}$$

In an ordinary case, the first term in the right side can be neglected in the very neighborhood of $Sr = 0$, and if $r = r_2 - r_1$

$$S(S) = \int \Delta\rho(r_1)\exp[-2\pi i(\boldsymbol{S}\cdot\boldsymbol{r_1})]\mathrm{d}v_{r1} \int \Delta\rho(r_2)\exp[2\pi i(\boldsymbol{S}\cdot\boldsymbol{r_2})]\mathrm{d}v_{r2}$$

$$= \iint_{(r=r_2-r_1)} \Delta\rho(r_1+r)\exp[2\pi i(\boldsymbol{S}\cdot\boldsymbol{r})\mathrm{d}v_{r1}]\mathrm{d}v_r \tag{6.77}$$

Let

$$Q(r) = \int \Delta\rho(r_1)\Delta\rho(r_1+r)\mathrm{d}v_{r1} \tag{6.78}$$

then

$$I(S) = \int Q(r)\exp[2\pi i(\boldsymbol{S}\cdot\boldsymbol{r})]\mathrm{d}v_r \tag{6.79}$$

or inversely

$$Q(r) = \int I(S)\exp[-2\pi i(\boldsymbol{S}\cdot\boldsymbol{r})]\mathrm{d}v_s \tag{6.80}$$

If $r = 0$ in Eqs. 6.78 and 6.80

$$\frac{1}{V}Q(0) = \frac{1}{V}\int(\Delta\rho(r_j))^2\mathrm{d}v_j = <(\Delta\rho(r))^2> = \frac{1}{V}\int I(S)\mathrm{d}v_s \tag{6.81}$$

Eq. 6.81 shows that the mean square deviation of the electron density $<(\Delta\rho(r))^2>$ is equal to the total scattering intensity per unit volume. This may be called the "scattering power"of the system.

For isotropic scattering the right side of Eq. 6.81 may be written as

$$Q(0) = 4\pi\int_0^\infty S^2 I(S)\mathrm{d}S \tag{6.82}$$

This type of integral scattering intensity has been termed the "invariant".[19]

The volume of the particle can be determined, using the Q and $I(0)$, the scattering intensity at zero angle, as

$$V = I(0)/Q \tag{6.83}$$

For the two-phase system, in which ρ_1 and χ_1 are the electron density and the volume fraction of the disperse phase (or solute), respectively, and ρ_2 and χ_2 ($= 1 - \chi_1$) those of dispersant (solvent), respectively.

$$(\rho_1 - \rho_2)^2\chi_1\chi_2 = (\Delta\rho)^2\chi_1\chi_2 = <(\Delta\rho)^2> \tag{6.84}$$

Substituting this into Eq. 6.81

$$(\rho_1 - \rho_2)^2\chi_1\chi_2 = Q(0)/V \tag{6.85}$$

This can be applied to the crystallinity (volume) measurement of the system consisting of crystalline and amorphous regions by means of X-ray small-angle scattering.

B. Correlation function and specific internal surface

The correlation function[22,23] (or characteristic function[19–21]), $\gamma(r)$ mentioned before (Section 6.2.3A) can be written as follows using the Q and $Q(0)$.

$$\gamma(r) = \frac{Q(r)}{Q(0)} = \frac{\int I(S)\exp[2\pi i(\boldsymbol{S}\cdot\boldsymbol{r})]\mathrm{d}v_s}{\int I(S)\mathrm{d}v_s} \tag{6.86}$$

In a randomly distributed two-phase system, for electron densities ρ_1 and ρ_2, such as holes in a solid or particles interspersed with voids, it has been shown[23] that

$$\gamma(r) = \exp[-r/\langle l_c \rangle], \tag{6.87}$$

and

$$\frac{O}{V} = \frac{4\chi_1\chi_2}{\langle l_c \rangle} \tag{6.88}$$

The term O/V is the specific internal surface defined in terms of the overall volume of the system, V. l_c is the correlation length (or distance) or reduced inhomogeneity length,[19,33] a measure of the size of inhomogeneities, which gives the integral breadth of $\gamma(r)$. If we define the transversal, or inhomogeneity, lengths l_1 and l_2 as given below,

$$\langle l_1 \rangle = \frac{4\chi_1}{O/V} \quad \text{and} \quad \langle l_2 \rangle = \frac{4\chi_2}{O/V} \tag{6.89}$$

by comparison with Eq. 6.44 it can be written as

$$\frac{1}{\langle l_c \rangle} = \frac{1}{\langle l_1 \rangle} + \frac{1}{\langle l_2 \rangle} \tag{6.90}$$

According to Kratky and coworkers, the $\langle l_1 \rangle$ and $\langle l_2 \rangle$ can be visualized as follows.[34] If we imagine the system to be pierced in all directions at random by rays (Fig. 6.17), the mean length of all the line segments intercepted. by the disperse phase is $\langle l_1 \rangle$ whereas the mean length of all the segments intercepted by the dispersant is $\langle l_2 \rangle$.

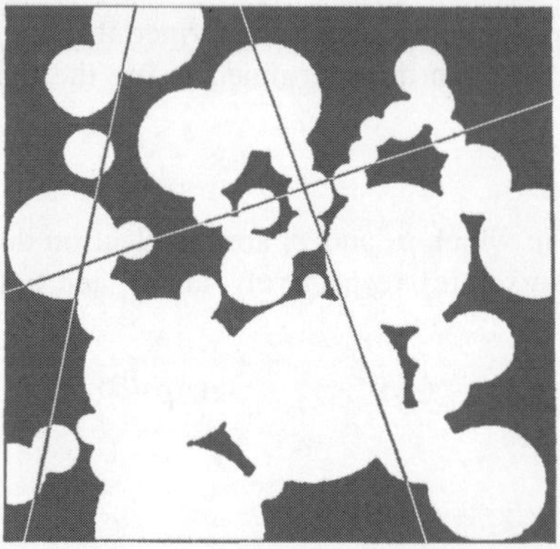

Fig. 6.17 Transversal or inhomogeneity length.[43] [Reproduced with permission from O. Kratky, *Angew Chem.*, **72**, 478, Verlag Chemie (1960)]

The theory of small-angle scattering has been presented in outline. For further details the reader is referred to other books on the subject.[5,10a),16,18,35–43)] Experimental methods and applications will be described in later chapters.

References

1. A. Guinier, *Ann. Phys. (Paris)*, **12**, 161 (1939).
2. O. Kratky, *Naturwissenschaften*, **26**, 94 (1938).
3. R. Hosemann, Z. *Physik*, **113**, 759 (1938); *ibid.*, **114**, 133 (1939).
4. M. Kakudo, in: *X-Ray Crystallography* (I. Nitta ed.), Vol. II, p. 519, Maruzen (1961).
5. A. Guinier, G. Fournet, *Small-Angle Scattering of X-Rays*, p. 201, John Wiley & Sons, Inc., N.Y. (1955).
6. T. Seto, in: *Structure and Physical Properties of High Polymers*, Vol. II. pp. 307–319 Baifukan, Tokyo (1973).
7. Lord Rayleigh, *Proc. Roy. Soc., (London)*, **A90**, 219 (1914).
8. In ref. 5, p. 20.
9. O. Kratky, G. Porod, *J. Colloid. Sci.*, **4**, 35 (1949).
10. a) In ref. 9, p. 20.
    b) *International Tables for X-Ray Crystallography*, Vol. III, pp. 324–326, Kynoch Press, Birmingham (1968).
11. D. L. Dexter, *Phys. Rev.*, **90**, 1007, (Serial approximation) (1953).
12. G. Fournet, *Doctoral Thesis*, Paris Univ., series A. No. 3256, 2384 (1950).
13. *International Tables for X-Ray Crystallography*, Vol. III, 327, Kynoch Press, Birmingham (1968).
14. H.M. Barton, R. Brill, *J. Appl. Phys.*, **21**, 783 (1950).
15. a) H. Stuhrmann, Z. *Phys. (Frankfurt am Main)* [*N.S.*], **72** (1970).
    b) H. Stuhrmann, *J. Appl. Cryst.*, **7**, 173 (1974).
    c) K. Ibel, H. Stuhrmann, *J. Mol. Biol.*, **93**, 255 (1975).
    d) H. Stuhrmann, A. Miller. *J. Appl. Cryst.*, **11**, 325 (1978).
16. I. Pilz, O. Glatter, O. Kratky, *Methods in Enzymology*, **61**, 180, Academic Press, N.Y. (1980).
17. H. Stuhrmann, in: *Small-angle X-ray Scattering*, (O. Glatter, O.Kratky eds.) p.197, Academic Press, N.Y. (1982).
18. C.R. Cantor, P.R. Schimmel, *Biophysical Chemistry*, Part II, Techniques for the Study of Biological Structure and Function, p. 831, W.H. Freeman, San Francisco (1980).
19. G. Porod, *Kolloid-Z.*, **124**, 83 (1951); *ibid.*, **125**, 51 (1952).
20. G. Porod, *Acta Phys. Austriaca*, **3**, 255 (1948).
21. a) In ref. 16, pp. 152–167.
    b) O. Glatter, O. Kratky eds., *Small Angle X-ray Scattering*, pp.167–196, Academic Press, London (1982).
22. P. Debye, A.M. Bueche, *J. Appl. Phys.*, **20**, 518 (1949).
23. P. Debye, H.R. Anderson, H. Brumberger, *J. Appl. Phys.*, **28**, 679 (1957).
24. In ref. 16, pp. 167–180.
25. M.H. Jellinek, E. Solomon, I. Frankuchen, *Ind. Eng. Chem.*, **18**. 172 (1946).
26. M. Kakudo, N. Kasai, M. Kimura, Y. Kubota, T. Watase, *J. Chem. Soc. Jpn., Pure Chem. Sect.* (in Japanese), **78**, 821 (1957); N. Kasai, M. Kakudo, T. Watase, *Technol. Rept. Osaka Univ.*, **8**, 433 (1958).
27. R. Hosemann, *Kolloid-Z.*, **117**, 13 (1950).
28. F. Zernike, J.A. Prins, Z. *Physik*, **41**, 184 (1927).
29. P. Debye, H. Menke, *Ergeb. Tech. Röntgenkunde*, **2**, 1 (1931).
30. G. Fournet, *Acta. Cryst.*, **4**, 293 (1950).
31. L.H. Lund, G.H. Vineyard, *J. Appl. Phys.*, **20**, 593 (1949).
32. L.C. Roess, C.G. Shull, *J. Appl. Phys.*, **18**, 308 (1947).
33. L. Kahovec, G. Porod, H. Ruck, *Kolloid-Z*, **133**, 16 (1953).
34. O. Kratky, *Pure and Appl. Chem.*, **12**, 483 (1966).
35. R. Hosemann, S.N. Bagchi, *Direct Analysis of Diffraction by Matter*, North-Holland, Amsterdam (1962).
36. V. Luzzati, *Small-angle X-Ray Scattering on an Absolute Scale*, (H.H. Pattee, V.E. Coslett eds.) and A. Engström, *X-Ray Optics and X-Ray Microanalysis*, Academic Press, N.Y. (1963).
37. W.O. Statton: *Small-angle X-Ray Studies of Polymers in Newer Methods of Polymer Characterization, Handbook of X-Rays* (B. Kaelbe ed.), Interscience, N.Y. (1963).
38. *Small-angle Scattering* (H. Brumberger ed.), Gordon & Breach, Reading (1967).
39. L.E. Alexander, *X-Ray Diffraction Methods in Polymer Science*, Wiley-Interscience, N.Y. (1969).
40. J. Schelten, R.W. Hendricks, *J. Appl. Cryst.*, **11**, 297 (1978).
41. H.B. Stuhrmann, A. Miller, *J. Appl. Cryst.*, **11**, 325 (1978).
42. O. Glatter, X-ray techniques and, Neutron techniques in: Small-angle techniques, in: *International Tables for Crystallography* Vol. C, (A.J.C. Wilson ed.), Kluwer Academic Pub., Dordrecht (1992).
43. O. Kratky, *Angew. Chem.*, **72**, 478 (1960).

7. Structure of High Polymeric Substances

As a basis for the discussion of the diffraction of X-rays by high polymers, we have discussed the nature of X-rays themselves and the scattering of X-rays by various states of matter. The theory of X-ray diffraction as so far presented is, in general, applicable to the particular subject of this volume namely high polymers. High polymers, however, unlike low molecular weight substances in general, and ionic and metallic crystals, consist of huge molecules, and the shapes and detailed structures assumed by the molecules or their aggregates in liquid and solid states are very complex. These substances consequently form a unique class of structural systems. Since X-ray diffraction phenomena vary with structure, the structural uniqueness of high polymers should give rise to X-ray diffraction patterns which exhibit important characteristics not found with other substances. Before proceeding to a discussion of the practical applications of X-ray analysis to the special features of high polymers, we must first in this chapter make a brief general survey of the peculiarities of high polymer structure.

The word "structure" applies to a substance at many different levels: in the normal crystalline substance, in so far as it affects X-ray diffraction, the first problem to be considered is that of the structure at the atomic level ("fine structure"). This involves the steric arrangement of the atoms in the molecules from which the substance is formed. This, in turn, leads to an examination of the bonding relationships between atoms and the mutual orientations and contacts between the molecules. This kind of structural information gives valuable insights into the nature (*i.e.* the physical and mechanical properties) of the substance. The situation is not so simple in the case of high polymers.

In general, as will be discussed later on, crystalline high polymer substances can be thought of as consisting of two phases, crystalline and amorphous. This concept is important for "fine texture" (see brief remarks below) and has a strong effect on the physical and mechanical properties of high polymers.

As this stage it becomes necessary to consider "crystallinity," that elusive variable which indicates the proportion of the crystalline to the amorphous regions in the polymer. Imperfections in crystalline regions (*cf.* Chapter 5) and some ordered arrangement of molecules in amorphous regions break down the clear distinction between the crystalline and amorphous regions. Differences in experimental technique can thus give rise to a different estimate of crystallinity (*cf.* Section 14.2.3), so that it is difficult to define crystallinity unambiguously. A second factor in high polymer fine texture is that of "preferred orientation of the crystalline region." This refers to the "type of orientation" and also to the "degree of orientation," an average direction—or distribution of directions—of (say) a certain axis of the unit cells with respect to some fixed direction of the polymer substance. in that type to preferred orientation. A third factor is the average "size of the crystalline regions" in such high polymer substances. In some favorable cases the average "shape of crystalline regions" may also be determined. "Lattice imperfections in crystalline regions" affect the average size estimated of the crystalline regions and must be separated before determining the average size. The fourth factor is the detailed "structure of amorphous regions", and the fifth is the "arrangement of crystalline and amorphous regions throughout the polymer substance" (higher order structure). Table 7.1 summarizes various factors of solid high polymer structure.

Table 7.1 Structure of solid high polymers

Classification	Itemization	Regions
Fine structure	Molecular structure Crystal structure Distortion of the atomic arrangement (crystalline / paracrystalline / amorphous)	Crystalline
Fine texture	Preferred orientation of crystalline regions Type of orientation (simple fiber structure, helical fiber structure, ring fiber structure, etc.) Degree of orientation	
	Size and shape of and / or lattice imperfection in crystalline region	
	Crystallinity	
	Structure of amorphous regions (or amorphous state) Short range order Chain orientation	Amorphous
Higher order structure	Arrangement of crystalline and amorphous regions and their interconnection with each other	Crystalline and Amorphous

7.1 Structure of High Polymer Chains in the Liquid State and in Solution

It is practically impossible to determine accurately the shapes of high polymer molecules in a liquid. However, if the molecule is globular of the like in shape or if the molecule has a random coil structure, some information on the structure of the molecule can be obtained by means of X-ray small-angle scattering (Chapter 15) or light scattering of its dilute solution.

To determine the detailed structure of a molecule, it is usual first to assume the molecular shape and size from the arrangement of atoms found by crystal structure analysis. In the case of linear high polymers, however, the general form of the molecules is often established without the need for a complete crystal structure analysis. For instance, the molecules of a linear high polymer may be built up from a constant monomer unit, and their stereoregularity may be predictable from the feature of the polymerization reaction and from the fiber period found by X-ray analysis. Their three-dimensional structures may be investigated with the aid of models, and further information may be obtainable by infrared absorption, especially by nuclear magnetic resonance (NMR),[1] or by other physical methods.

7.1.1 Configuration and conformation

The structure of high polymer chain molecules is determined by two factors, *i.e.* configuration and conformation.

A. Configuration

In vinyl polymers consisting of $-CH_2-CH\mathbf{R}-$ monomer units, for instance, head-to-tail $(-CH_2-CH\mathbf{R}-CH_2-CH\mathbf{R}-)_n$ and head-to-head $(-CH_2-CH\mathbf{R}-CH\mathbf{R}-CH_2-)_n$ isomers may be obtained during the polymerization process. The head-to-tail isomers again have isotactic, syndiotactic, and atactic isomers due to the stereoregularity, *i.e.* due to the regular or irregular appearance of the **R** substituents along the main chain (Fig. 7.1(a)). In the polymers of diene compound, $\overset{1}{C}H_2=\overset{2}{C}H-\overset{3}{C}\mathbf{R}=\overset{4}{C}H_2$, there exist isomers due to 1,4-*cis*-, 1,4-*trans*-, 1,2-syndiotactic, and 1,2-isotactic addition (Fig. 7.1(b)).[2a-2d]

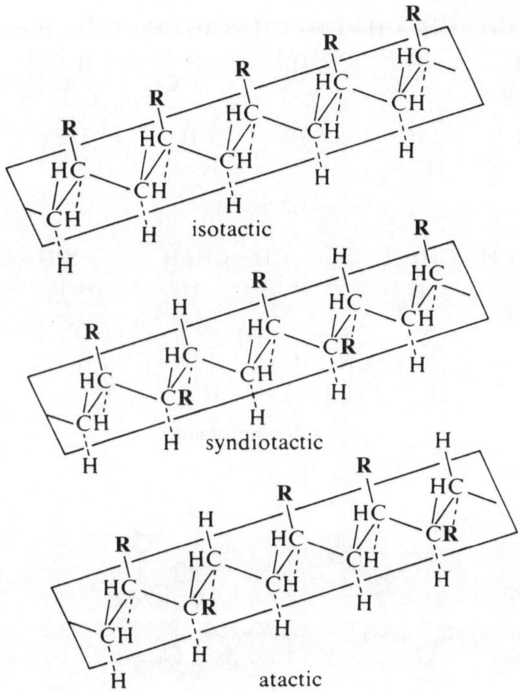

(a)

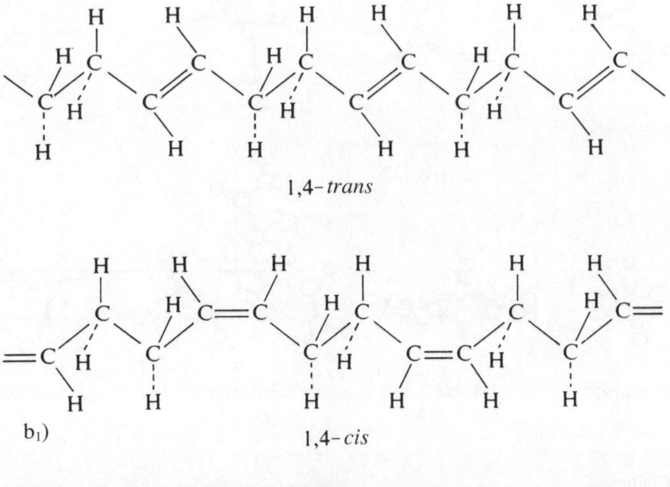

(b) b$_1$)

Fig. 7.1 Configuration of polymer chains.
(a) Stereoregularity of vinyl polymers head-to-tail isomers.
b$_1$) Configurational isomers of polybutadiene.
(*Continued on p.138.*)

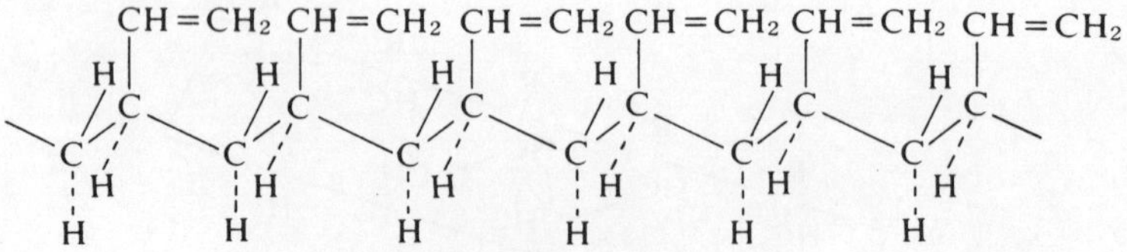

1,2-isotactic

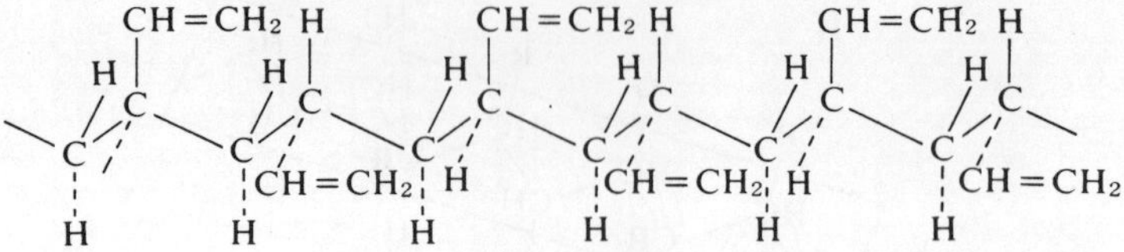

1,2-syndiotactic

b$_1$)—*Continued from p.137*

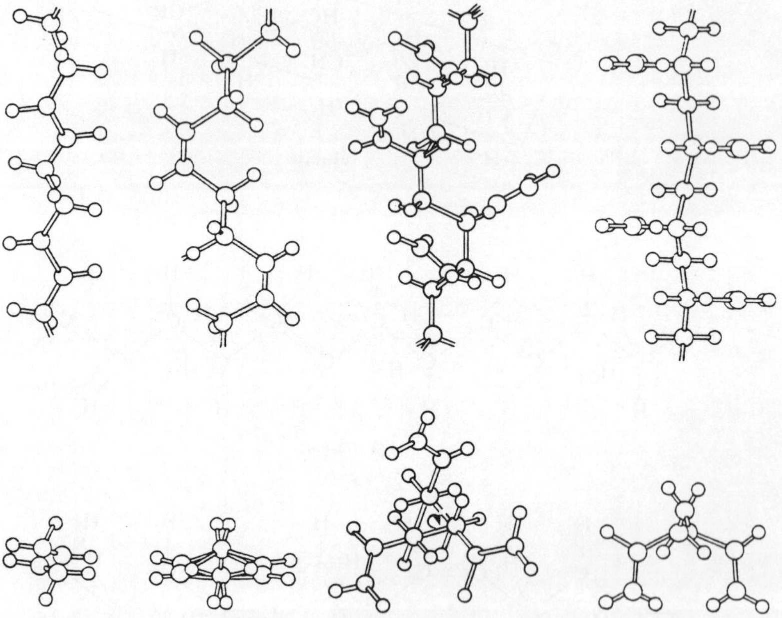

1,4-*trans*[2a)] $c = 4.9$Å | 1.4-*cis*[2b)] $c = 8.60$Å | 1,2-isotactic[2c)] $c = 6.50$Å | 1,2-sybdiotactic[2d)] $c = 5.14$Å

b$_2$)

Fig. 7.1 — *Continued.*
b$_2$): Molecular structure of polybutadiene.
[Reproduced with permission from G. Natta, *Makromol. Chem.*, **35**, 112, Hüthig & Wepf Verlag (1960)]

B. Conformation

Basic structures of high polymer chains are determined by three states, *trans*(T), *gauche*(G),

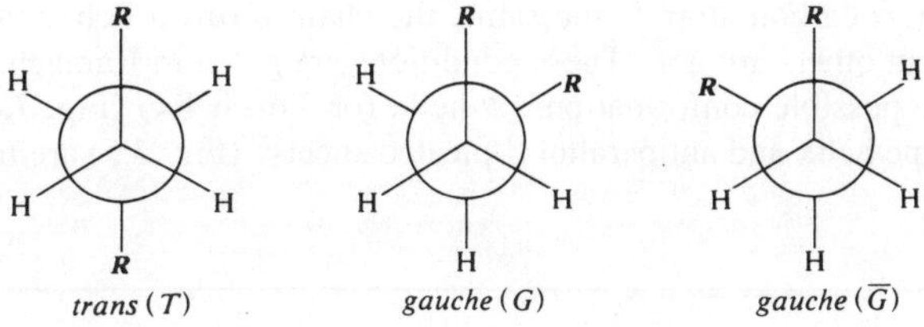

a₁)

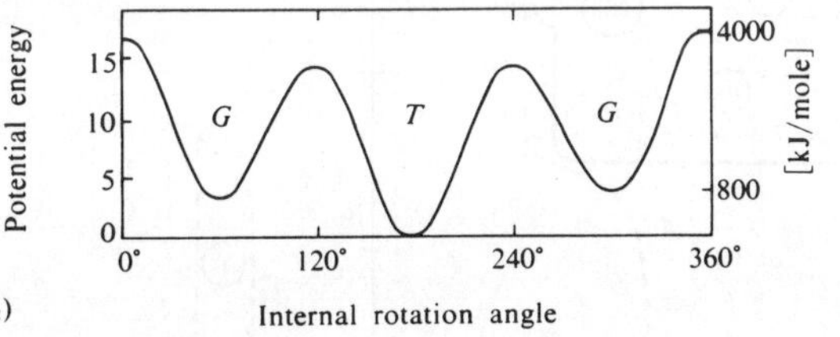

(a) a₂)

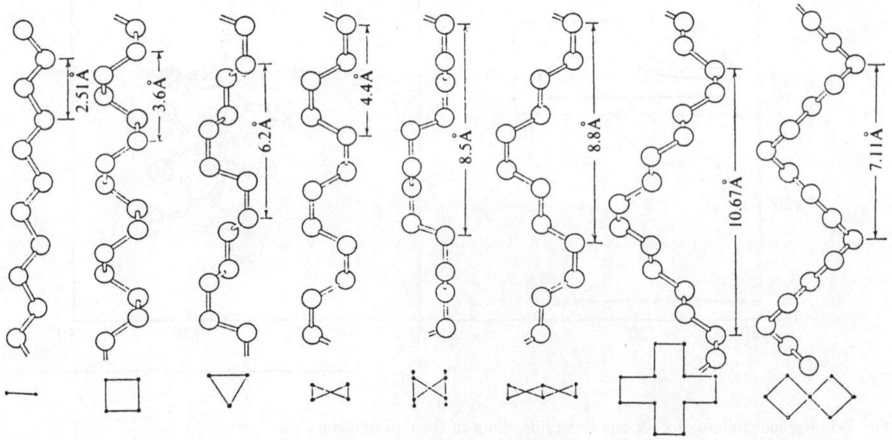

(b) T_2[4a)] G_4[4b)] $(TG)_3$[4c)] $TGT\overline{G}_3$[4d)] $(TG)_2(T\overline{G})_2$[4e)] $T_3GT_3\overline{G}$[4f)] $(T_2G)_4$[4g)] $(T_2G_2)_2$[4h)]

Fig. 7.2 Conformation of polymer chains.[4)]
(a) a₁) Conformations along the C – C single bond in 1,2-dihalogenoethane; *trans* and two *gauche*.
a₂) Internal rotation angle along the central C – C single bond of *n*-butane *vs.* potential energy.
(b) Conformations of polyethylene chains determened (except $(TG)_2(T\overline{G})_2$), among which the left 6 conformations, from T_2 to $T_3GT_3\overline{G}$, were proposed by Bunn.[3b)] H. Tadokoro: Right 2)
[Reproduced with permission from H. Tadokoro, *Structure of Crystalline Polymers*, p.18, Kagaku Dojin, Kyoto (1976)]

and the other *gauche*($\bar{G}$)[†1] caused by the internal rotation along the C − C single bond of the main chain (Fig. 7.2).[4a)-4h)]

The conformation of a polypeptide chain is determined by two factors φ and ψ: the twist about the Cα − N bond axis, φ, and that about the Cα − C axis, ψ, since the two amide planes of amino acid residues are joined by the tetrahedral bond of α carbon atom. If the twist at every α carbon atom is the same, the chain forms a helix. Some foldings are easily achieved but others are not. These conditions are given in Ramachandran plot[5)] (Fig. 7.3). Among the possible conformations, α-helix (or 3.6_{13} helix) (Fig. 7.4),[6)] 3_{10} helix,[†2] 2_7 ribbon, polyproline helix and antiparallel β-pleated sheets[5)] (Fig. 7.5) are important types.

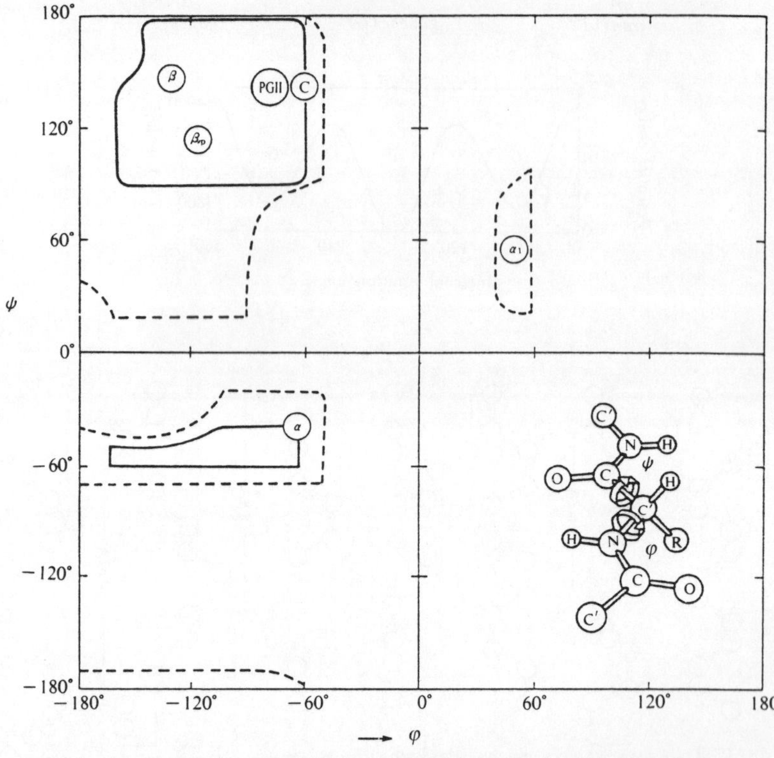

Fig. 7.3 Ramachandran plot for a polypeptide chain of fixed dimensions.[5)]
φ: The twist about the C_α − N bond axis; φ: The twist about the C_α − C bond axis.
α: Right-handed α-helix α_L: left-handed α-helix; β: Antiparallel-chain β pleated sheet; β_p: Parallel chain β pleated sheet; C: Collagen-like coiled coil; PGII: Poly(glycine) II type helical structure. [Reproduced with permission from G. N. Ramachandram, V. Sasisekaran, *Adv. Protein Chem.*, **23**, 283, Academic Press (1968)]

---

†1 Bunn first used A; B and C for *T*; *G* and $\bar{G}$.[3b)] Mizushima and Shimanouchi adopted *T*; *G* and *G*′.[3c)] In the field of high polymer solution, *t*, g^+ and g^- are being used.[3d)]

†2 The 3/1 helix is customarily expressed as 3_{10} helix in polypeptides. Three amino residues make one turn and hydrogen bonds complete rings of 10 atoms in the helix (Fig. 7.4).[6)]

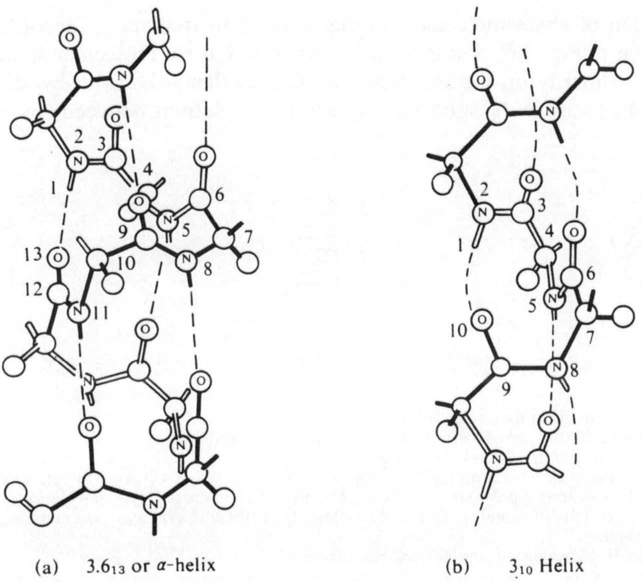

Fig. 7.4 Helical chain.[6)]
(a) Right-handed $3.6_{1.3}$ of α-helix (α_R); hydrogen bonds complete 13-atom rings.
(b) Right-handed 3_{10} helix; hydrogen bonds complete l0-atom rings.
[Reproduced with permission from T. L. Blundell, L. N. Jhonson, *Protein Crystallography*, p.32, Academic Press (1976)]

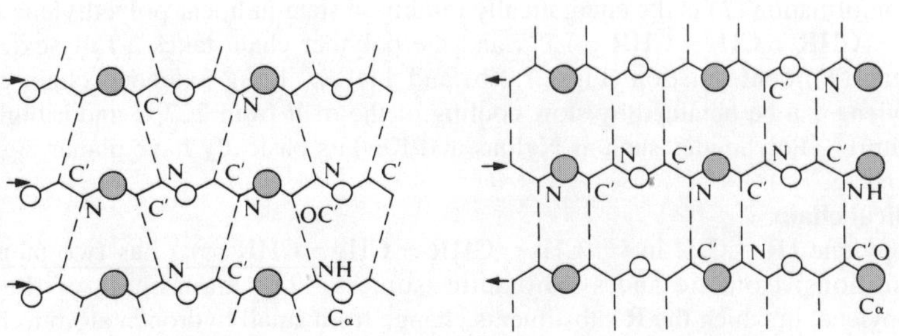

Fig. 7.5 β Pleated sheets.
○ and ● show that the side chains are below and above the paper, respectively.

7.1.2 Classification of chain molecules

The classification of chain molecules on the basis of their shape in accordance with this scheme is shown in Fig. 7.6. Let us now consider how these molecular shapes arise. The shape depends primarily on the individual molecules themselves. New stable structures may, however, be found in the solid state, owing to interactions between the molecules.

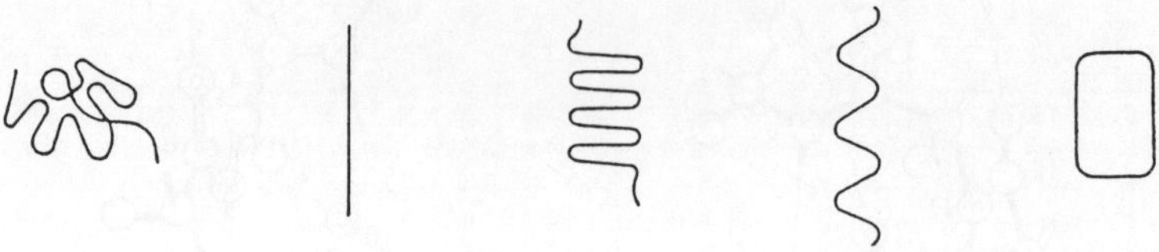

(a) (b) (c) (d) (e)

Fig. 7.6 Classification of chain molecules by shape.
(a) Random coil: disordered form in liquid and amorphous states.
(b) Straight chain: fully extended backbone.
(c) Folded: linear or helical molecule folded with period of about 100Å. This group includes lamellar crystals of synthetic polymers (*cf.* Figs. 7.11 and 7.12) and cross-β type polypeptides (*cf.* Fig. 7.5).
(d) Helical: α-helix of proteins and helical molecules of isotactic synthetic polymers: single, double, and triple helices.
(e) Cyclic: ring-condensed peptides and siloxanes.

A. Random coil

Chain molecules in solution, for instance, have many possible, energetically similar conformations, and they have no fixed structure. They are like a mobile coil constantly changing due to rotation caused by thermal motion, intermolecular forces, and / or interaction with the solvent (*cf.* Fig. 15.24).

B. Straight chain

Trans conformation (T) is the energetically minimum state in linear polyethylene ($\mathbf{R} = \mathrm{H}$ in $(-\mathrm{CH_2}-\mathrm{CH}\mathbf{R}-\mathrm{CH_2}-\mathrm{CH}\mathbf{R}-)_n$)[4b], and the polymer chain takes a fully-extended planar zigzag (T_2) conformation (Figs. 7.2(b) and 11.12). Long extended chain crystals of polyethylene can be obtained by slow cooling of the melt from 227°C under high pressure (5,000 atm).[8] Polyamides such as Nylons and Kevlars basically have planar zigzag structures. (*cf.* Fig. 7.5).

C. Helical chain

Polypropylene ($\mathbf{R} = \mathrm{CH_3}$ in $(-\mathrm{CH_2}-\mathrm{CH}\mathbf{R}-\mathrm{CH_2}-\mathrm{CH}\mathbf{R}-)_n$) has two stereoregular configurations: isotactic and syndiotactic isomers. The main chain of the isotactic polypropylene, in which the **R** substituents change from small hydrogen atom to bulky $\mathrm{CH_3}$ group, takes $(TG)_3$ conformation (Fig. 7.2(b)) because of the intramolecular repulsion between adjacent **R** ... **R** substituents along the main chain. The molecular shape changed to take a 3/1 helix (Fig. 7.7(a_1)).[9] On the other hand, syndiotactic polypropyrene takes a T_2G_2 (or $T_2\bar{G}_2$) conformation (Figs. 7.2(b) and 7.7(a_2)).[4b] Isotactic polystyrene molecule ($\mathbf{R} = \mathrm{C_6H_5}$) also has a similar 3/1 helical shape[4c] as isotactic polypropyrene ($(TG)_3$, Fig. 7.7(b_1)). However, isotactic poly-*o*-methylstyrene molecule ($\mathbf{R} = \mathrm{C_6H_4(CH_3\text{-}}o)$) and isotactic poly-*m*-methylstyrene ($\mathbf{R} = \mathrm{C_6H_4(CH_3\text{-}}m)$) have a 4/1 and a 11/3 helical structures, respectively (Fig. 7.7(b_2) and (b_3)).[9] Isotactic poly(methyl methacrylate) (PMMA) has a double helical structure consisting of two 10/1 helical PMMA molecules bound by van der Waals attraction.[10]

α-Helix is one of the most important conformations found as a basic unit in many fibrous and globular proteins (Fig. 7.4(a)). The keratins are mostly α-helix. Myosin, epidermin, fibrinogen, etc. are also α-helical. Whereas nucleic acids have a double helix structure[11] (Fig. 7.8 (p.145)) and collagen is a triple-helix.[12]

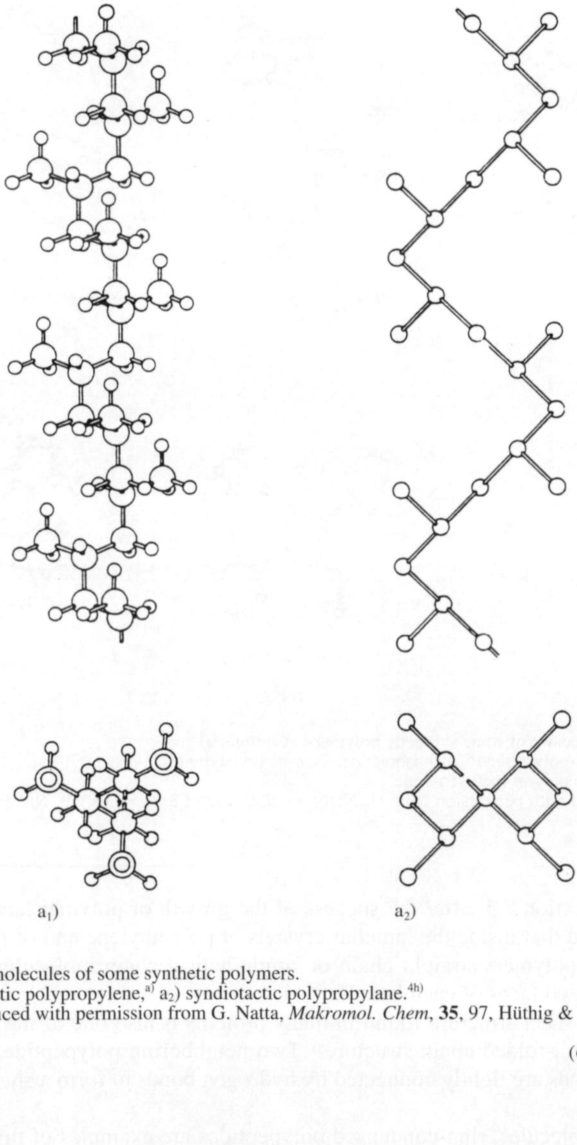

Fig. 7.7 Helical molecules of some synthetic polymers.
a_1) isotactic polypropylene,[a)] a_2) syndiotactic polypropylane.[4h)]
[Reproduced with permission from G. Natta, *Makromol. Chem*, **35**, 97, Hüthig & Wepf Verlag (1960)]

(*Continued on p.144*)

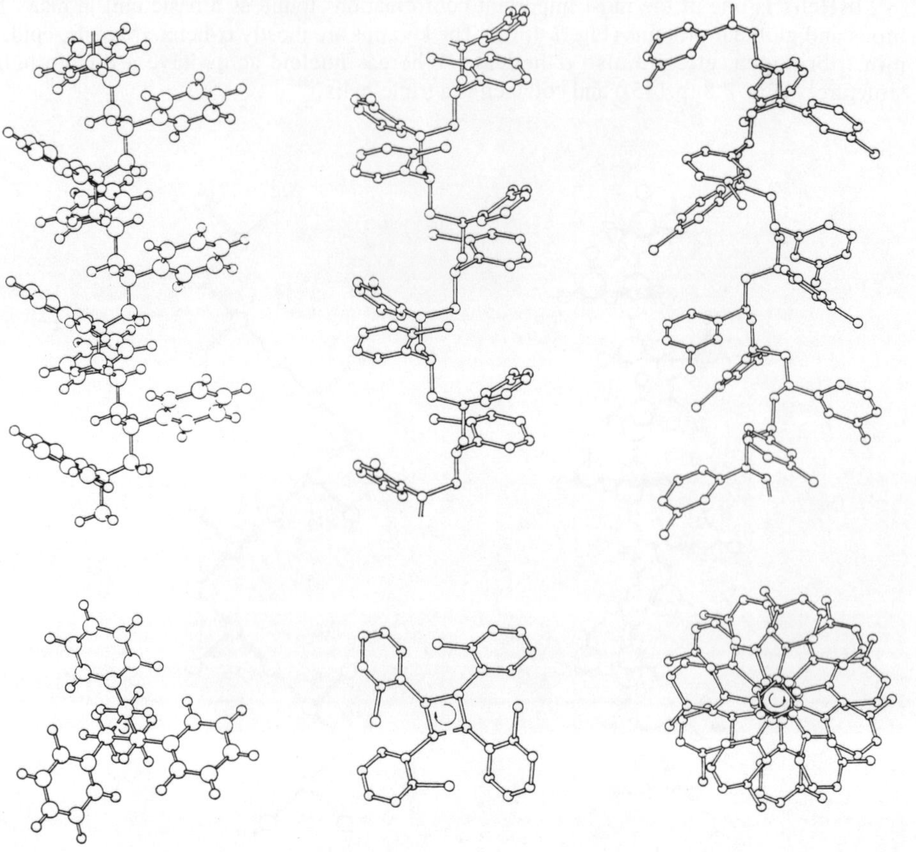

b1) b2) b3)

Flg. 7.7 —Helical molecules of some synthetic polymers. *(Continued)*
b_1): isotactic polystyrene[4c], b_2): isotactic poly-*o*-methylstyrene,[a)]
b_3); isotactic poly-*m*-methylstyrene[a)].
[Reproduced with permission from G. Natta, *et al.*, *Nuovo Cimento*, **15**, 69, Nicola Zanichelli Editore (1960)]

D. Folded chain

As described in Section 7.3, after the success of the growth of polymer lamellar crystals it has been concluded that inside the lamellar crystals of polyethylene and of many other synthetic and natural polymers straight chain or single helical chain molecules fold back and forth between the two faces of each lamella.

The β-pleated sheet structure found in many proteins is also one of the most important basic units to have a folded chain structure. Two neighboring polypeptide chains running in opposite directions are tightly connected by hydrogen bonds to form a sheet (Fig. 7.5).

E. Cyclic chain

Cyclic siloxane molecules, ring-condensed polypeptides are examples of this type of structure.

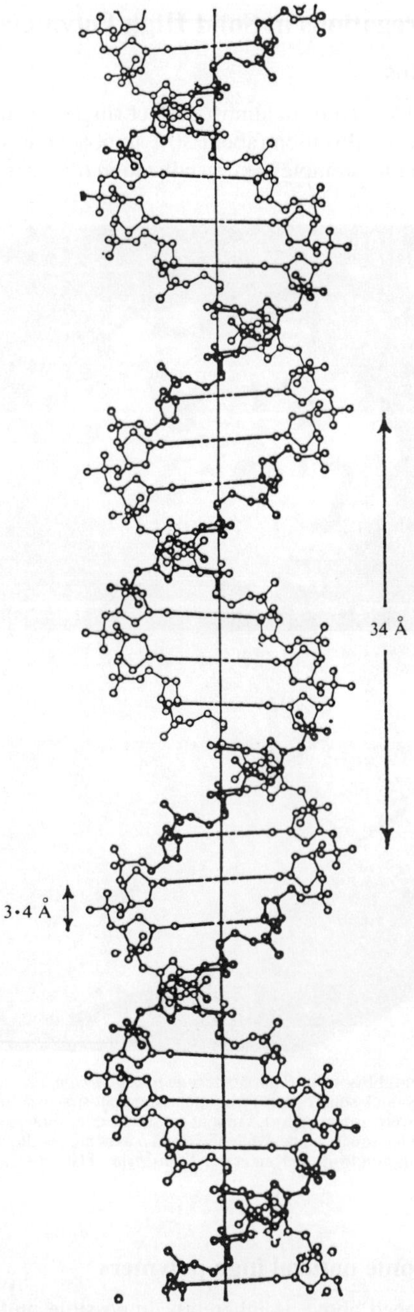

Fig. 7.8 Double helix of nucleic acid (B-type).[11] [Reproduced with permission from H. R. Wilson, *Diffraction of X-Rays by Proteins, Nucleic Acid and Viruses*, p.103, Edward Arnold Pub. (1966)]

7.2 Molecular Aggregations in Solid High Polymers

7.2.1 Globular proteins

Globular proteins and the like can form almost perfect single crystals (Fig. 7.9), which contain water (or solvent) of crystallization (about 30% to 60%) and other molecules added to the protein solution in order to promote the crystallization (*cf.* Section 12.2).

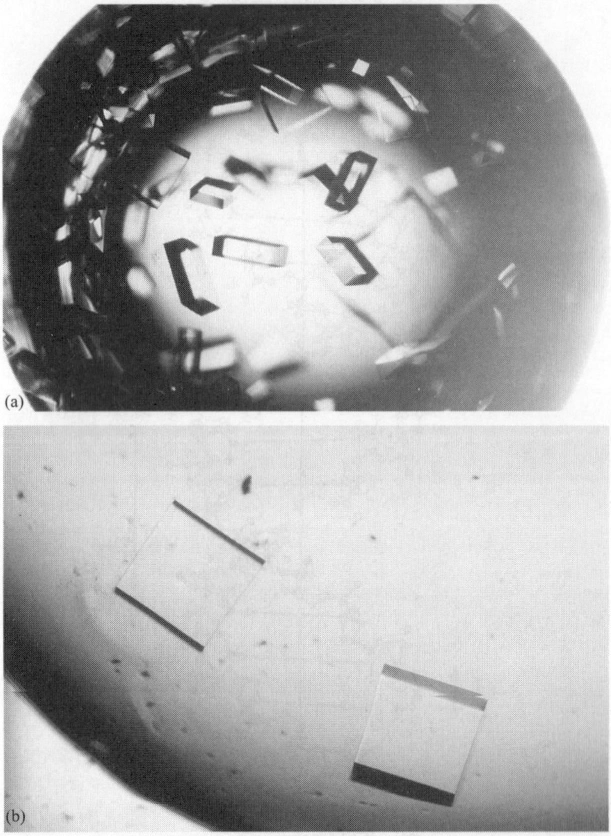

Fig. 7.9 Crystals of a flavoprotein, FP_{390} from *Photobacterium phosphoreum.*[13)]
(a) Crystals appeared as thick square plates from potassium phosphate solutions.
(b) The largest crystals were grown up to 1.5 mm in the square edge dimension. The crystallographic *c* axis lies perpendicular to the square plane, while the *a* and *b* axes are parallel to the edges of the square.
[Reproduced with permission from A. Kita *et al.*, *J. Biochem.*, **110**, 749, Japanese Biochem. Soc. (1991)]

7.2.2 Synthetic and some natural high polymers

The crystallization mentioned above is inherently impossible with ordinary synthetic and some natural high polymers; the non-uniformity of the molecules makes impossible.

The structure of these crystalline high polymers is usually described by reference to the crystalline and amorphous components. Essentially two models have been proposed hitherto.

A. Fringed-micelle model[14)]

Chain molecules are stacked in a regular fashion in some regions and so give rise to a crystal structure (crystalline region or crystalline part) (*cf.* Fig. 11.12). The portions of molecules between crystalline regions, where molecules wander from one crystalline region to the next constitute amorphous regions (or amorphous parts) (Fig. 7.10).[15)] There may exist an intermediate region (paracrystalline region) between crystalline and amorphous regions (see Fig. 7.17).

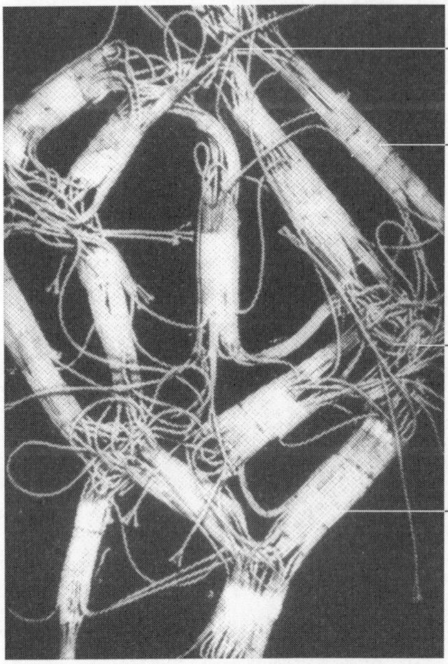

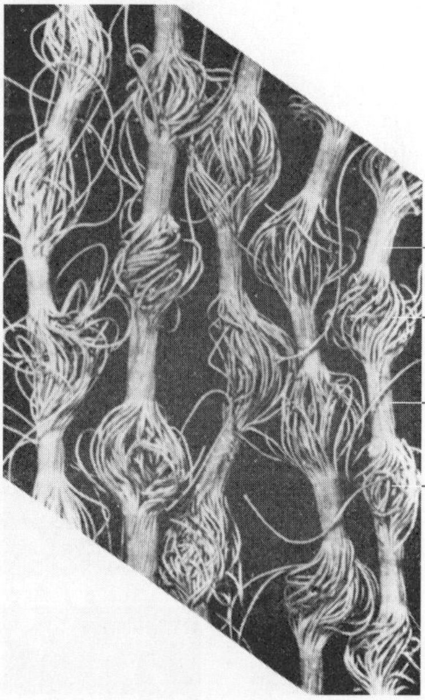

Fig. 7.10 Fringed-micelle model of a polymer.
CR: cryotalline region, AR: amorphous region.
(a) random orientation (unstretched)[15a)]; (b) uniaxial orientation (stretched).[15b)]
(a): [Reproduced with permission from M. Takayanagi, *Buturi*, **16**, 19, Phys. Soc. Jpn. (1961)]
(b): [Reproduced with permission from R. Hosemann, *Polymer*, **3**, 380, Butterworths (1962)]

B. Folded-chain model

In 1957–1958, Till,[16)] Keller,[17)] Fischer,[18)] and Kobayashi[19)] discovered that linear polyethylene molecules form lamellar crystals (so-called polymer single crystals) when crystallized from dilute solution under conditions approximating equilibrium (Fig. 7.11).[20)] Based on the results of electron microscopic and electron diffraction studies, it was concluded that polyethylene molecules in the lamellar crystal assume a folded-chain structure.[21)] Outer, minor portions of the lamellar crystal where chain molecules fold back are considered to be an amorphous region, whereas major portions where chains stack regularly as crystalline re-

(a)

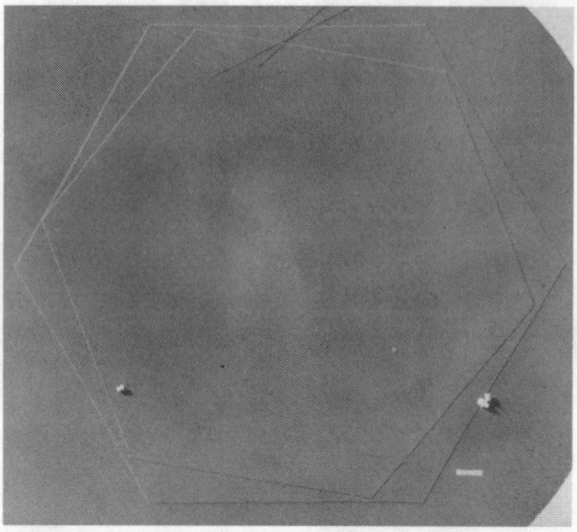

(b₁)

(b₂)

Fig. 7.11 Lamellar crystals (*cf.* Fig. 7.12).[20]
(a) Polyethylene: the *a* and *b* axes of the unit cell coincide with long and short diagonals of rhombohedron respectively. The *c* axis vertical to the plane of lemella.
(b) Polyoxymethylene: (b_1) two thin lamellae, (b_2) spiral growth.
[Reproduced with permission from N. Kasai, K. Kobayashi, *Structure and Properties of High Polymers*, pp.101. 102. 105, Maruzen (1963)]

gion (Fig. 7.12). This discovery led to the wide acceptance of the folded-chain model instead of the fringed-micelle model for the structure of many synthetic and natural crystalline polymers.

However, modifications such as short chain loops and long chain loops, and others were later introduced in the uniform chain folding assumed at the initial stage. Then the switch-board model,[22] in which some (or many) molecular chains go through from a lamella to the upper and / or lower lamellae, closer to the fringed-micelle model provided a much better explanation for the physical, mechanical, and rheological properties observed in the polymer specimen.

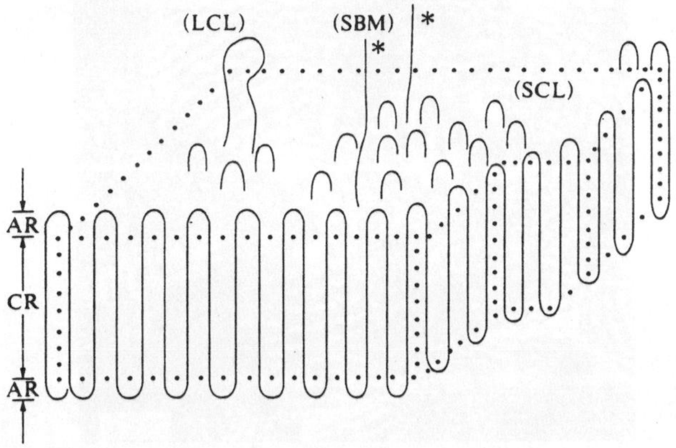

Fig. 7.12 Folded-chain model (lamellar crystal) of a polymer.
CR: Crystalline region, AR: Amorphous region (looped portion), SCL: Short chain loop, LCL: Long chain loop, SBM: Switch board model part. *Unlooped chains to adjacent lamella(e).

C. Some special cases of crystallization of polymers

As a special case the urea adduct of poly(ethylene oxide) gives prismatic, single crystal-like crystals, the prism faces of which appear to be very similar to those of single crystals of small molecules but both ends of the prism do not have any well-defined shapes (Fig. 7.13).[23]

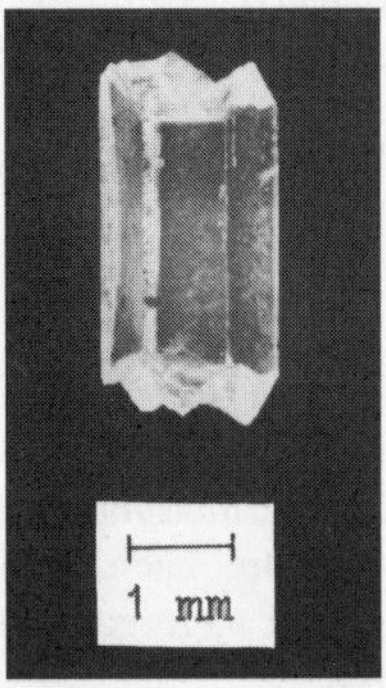

Fig. 7.13 Single crystal-like crystal of poly(ethylene oxide)-urea complex. Obtained by cocrystallization from a solution containing poly(ethylene oxide) and urea.[21] [Reproduced with permission from H. Tadokoro, *et al.*, *J. Polym. Sci.*, **B2**, 366, John Wiley & Sons, Inc. (1964)]

Spherulites (Fig. 7.14)[20] are birefringent, usually spherical, structures (for X-ray studies see Section 10.2.2) observable in polymers by optical (or electron) microscopy during the crystallization process through various stages of crystal growth. Formation of spherulites is recognized to be the most characteristic mode of crystallization from the melt.

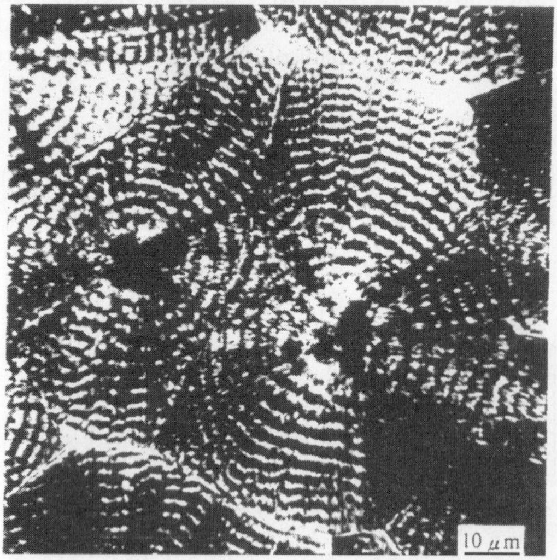

Fig. 7.14 Spherulites of polyethylene.[20]
[Reproduced with permission from N. Kasai, K. Kobayashi, *Structure and Properties of High Polymers*, p.110, Maruzen (1963)]

Opal consists of spherical particles of silicon dioxide (SiO_2) with fairly uniform size and shape. Synthetic opals, made of either silicate or polystyrene, are commercially available. Inside the spherical particle the silicate or the polystyrene does not form a crystal. It is amorphous by X-ray examination. However, these spherical particles pile up regularly to form a three-dimensional lattice. The size of these uniform amorphous spheres are of the order of the wavelengths of visible light. Therefore, it is supposed that an X-ray amorphous but optically crystalline lattice of opal gives Bragg reflections of visible light, which gives the fire and play of color in the opalescence of reflected and transmitted light.[24]

7.3 Structure of the Amorphous State and of Amorphous Regions in Solid High Polymers

7.3.1 Random-coil model

Figure 7.15(a) shows a random coil model for the structure of the amorphous state in linear high polymers consisting of chemically bonded long chains (*cf.* Fig. 15.24). This model was presented by Flory[25] based on his thermodynamical theory on high polymers. All long chain molecules have random coil structure; each of the chain molecules has locally no ordered structure and none of the molecular aggregates possess well-defined arrangement between molecules.

However, as was mentioned in Section 2.8, the spatial arrangement of the atoms in matter which is identified as being in an amorphous state generally exhibits at least some

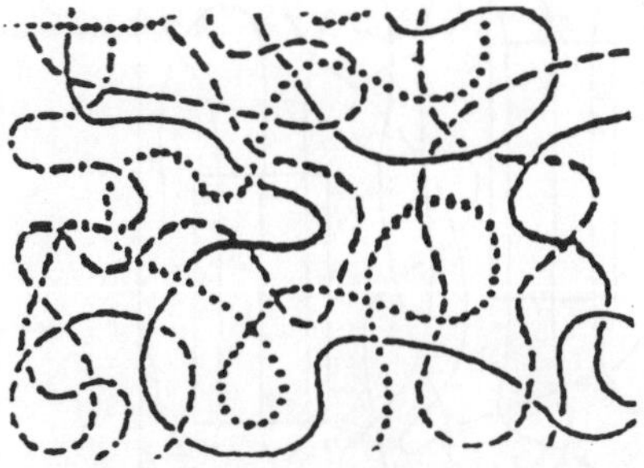

(a) Random-coil model (Flory).[25]
Solid line, dotted line, broken line, etc. are used in order to make individual molecular chain clear.

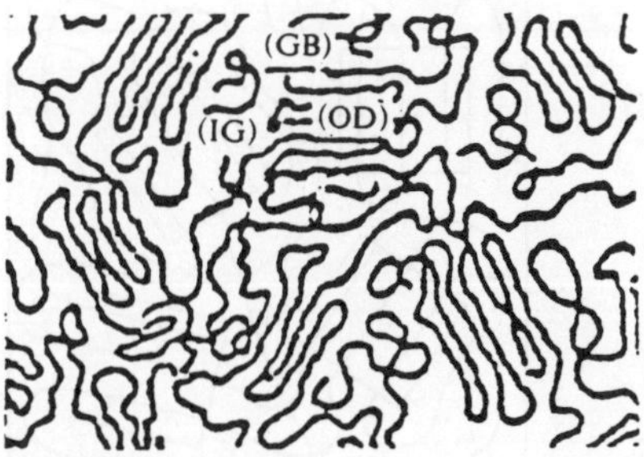

(b) Folded-chain-fringed-micellar-grain model (Yeh).[26]
OD: ordered region, GB: grain boundary, IG: inter-grain.

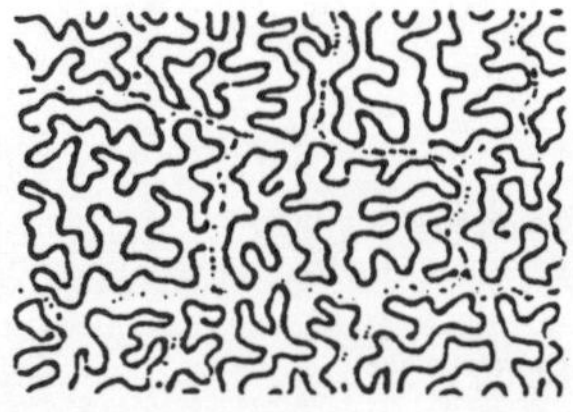

(c) Collapsed-ball model (Vollmert).[27]

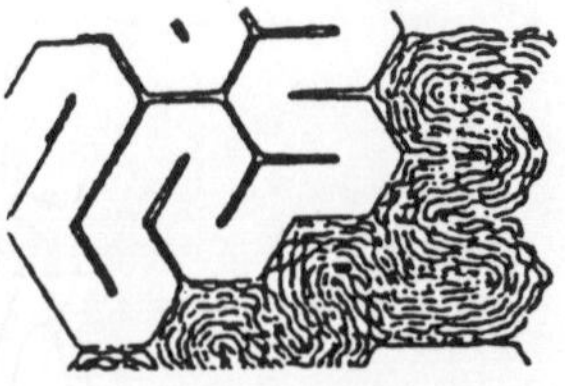

(d) Meander model (Pechhold).[28]

Fig. 7.15 Models of amorphous structures.
(a) Random-coil model (Flory); (b) Folded-chain-fringed-micellar-rain model (Yeh); (c) Collapsed-ball model (Vollmart); (d) Meander model (Pechhold).
(b): [Reproduced with permission from G. S. Y. Yeh, *J. Macromol. Sci. Phys.*, **86** 465, Marcel Dekken, (1972)]
(c): [Reproduced with permission from B. Vollmert, H. Stutz, *Angew. Makromol. Chem.*, **3**, 188, Hüthig & Wepf Verlag (1968)]
(d): [Reproduced with permission from W. Pechhold, S. Blasenbrey, *Kolloid-Z. Z. Polymere.* **241**, 955, Dr. Dietrich Steikopff Verlag (1971)]

short range order. The experimental results on amorphous materials suggest that the aggregates of these chain molecules have some ordered structures and arrangements.

7.3.2 Folded-chain-fringed-micellar-grain model

Among the alternate models proposed so far for the structure of amorphous state (Fig. 7.15(b)-(d)),[26-28] the folded-chain-fringed-micellar-grain model (Fig. 7.15(b)), presented by Yeh[26] based on the results of electron microscopic and other studies, is presently considered the most plausible for the structure of amorphous state.

7.4 Fine Texture in Solid High Polymers

From the above account of the special features of crystalline and amorphous structures based on polymer molecular structure, it will be readily appreciated that actual structures of solid high polymers can be rather complicated. To close this brief chapter, the complexities of polymer structure will be systematized (*cf.* Table 7.1) and some characteristic types illustrated.

Figure 7.16 shows two model structures of crystalline linear polymer with different fine

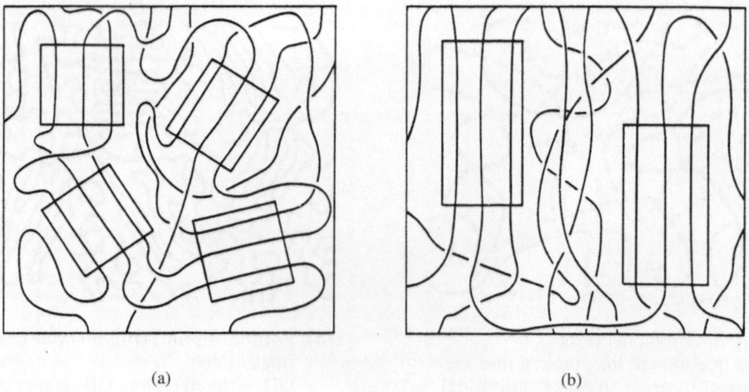

Fig. 7.16 Two model structures (fringed micelle model) of crystalline linear polymer with equal crystallinity but different size of crystalline region (rectangles represent crystalline regions).[29)]
[Reproduced with permission from G. Bodor, *Structural Investigation of Polymers*, p.264, Akademiai Kiado (1991)]

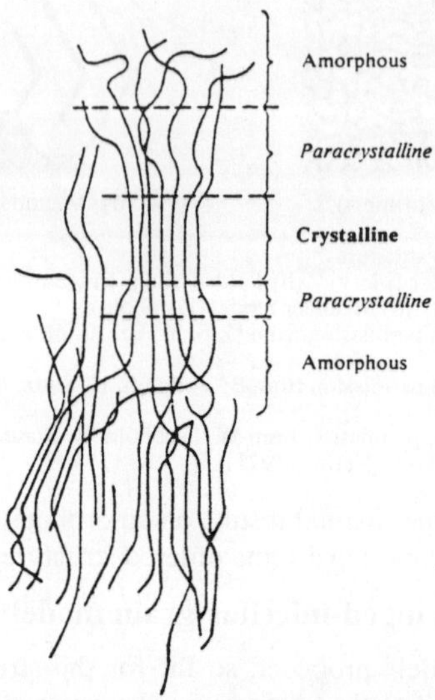

Fig. 7.17 Paracrystalline regions between crystalline and amorphous regions (fringed micelle model).

texture.[29)] They are well-known fringed micelle-type model of structures. In Fig. 7.16(a) small regions of crystalline order (crystalline region, crystalline part, crystallite) are formed by sections of molecules: the polymer consists of crystalline and amorphous regions. For simplicity, in these models paracrystalline regions transitional between crystalline and amorphous regions (Fig. 7.17) are not considered. In this model 1/4 of the structure is crystalline region. The crystalline regions are small, the section of which are square, and these regions are spread in an amorphous matrix with random orientation. The other model is given in Fig. 7.16(b). Crystalline regions are two times larger than those in (a), their sections are rectangular, and in an amorphous matrix they dispersed so that the chain axes (say the c axes) are vertical (presumably the structure has a uniaxial orientation). However, both structures have equal crystallinity of 25%. Naturally since there is a density difference between the crystalline and amorphous regions in the two models, both structures may give central diffuse small-angle scattering with slightly different feature with each other. Fig.

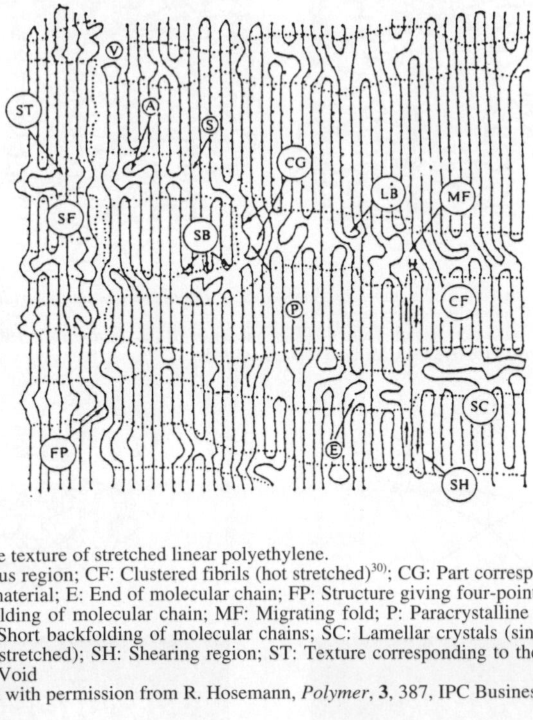

Fig. 7.18 Model of fine texture of stretched linear polyethylene.
A: Amorphous region; CF: Clustered fibrils (hot stretched)[30]; CG: Part corresponding to crystal growth in the bulk material; E: End of molecular chain; FP: Structure giving four-point diffraction pattern; LB: Long backfolding of molecular chain; MF: Migrating fold; P: Paracrystalline layer lattice; S: Straight chains; SB: Short backfolding of molecular chains; SC: Lamellar crystals (single crystals); SF: Single fibrils (cold stretched); SH: Shearing region; ST: Texture corresponding to the Statton model (*cf.* Fig. 7.10(b)); V: Void
[Reproduced with permission from R. Hosemann, *Polymer*, **3**, 387, IPC Business Press (1962)]

7.18 is a model of fine texture of stretched linear polyethylene.[30] In this figure many structural elements of the fine texture are shown.

In Figs. 7.19 – 7.21 are shown mostly models for the structure of some biologically interesting substances, such as photoreaction center in a membrane of photosynthetic bacteria,[31] riboflavin synthase,[32] and muscle.[33]

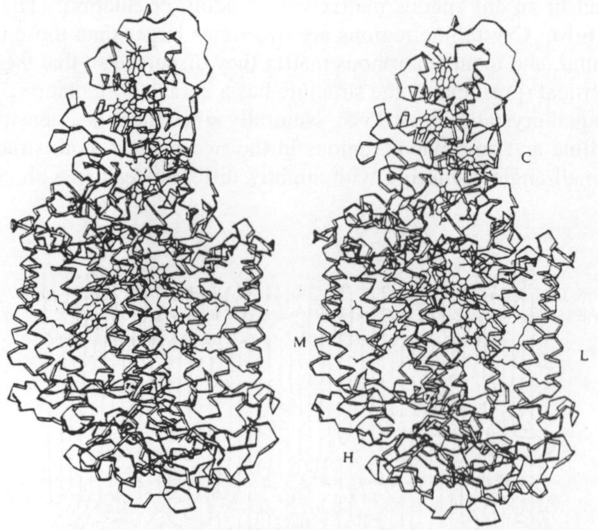

Fig. 7.19 Stereoscopic drawing of the photoreaction center of *Rhodopseudomonas viridis*. H, M, L, and C show H, M, L, and C subunits, respectively.[31]
[Reproduced with permission from J. Deisenhofer *et al.*, *J. Mol. Biol.*, **180**, 385, Academic Press (1984), *Nature*, **318**, 618, MacMillan (Journals)(1985)]

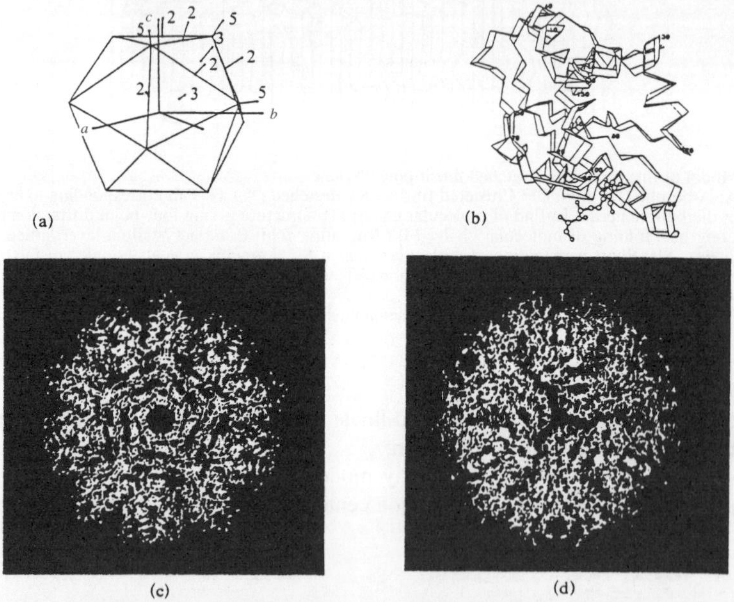

Fig. 7.20 Riboflavin synthase of snake $\alpha_3\beta_{20}$. Three α-subunits (*M*r. 23,500) are included in an eicosahedrally symmetrical 60-mer of β-subunits (*M*r. 16,200, 154 amino residues).[32]
(a) Eicosaheadral symmetry of a 60-mer with crystallographic symmetry axes; (b) Structure of a β-subunits; (c) Structure of a 60-mer of β-subunits; Projected along a five-fold axis of an eicosahedral structure. (Three α-subunits are not located.); (d) As above. Projected along a three-fold axis.
[Reproduced with permission from R. Ladenstein *et al.*, *J. Mol. Biol.*, **203**, 1045, Academic Press (1988)]

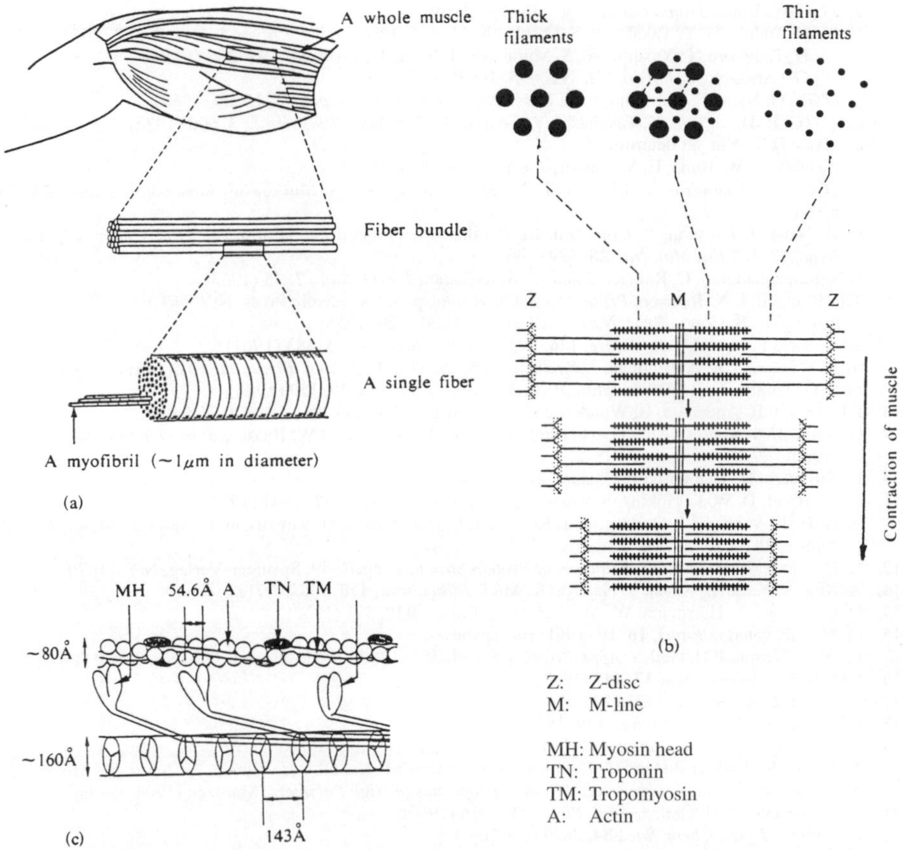

Fig. 7.21 A skeletal muscle and actin-myosin interaction.[33]
(a) Structure of a skeletal muscle. Each myofibril is made up of thousands of identical repeating units called sarcomeres, which are the smallest functional units of the muscle; (b) Structure of a sarcomere; (c) Actin-myosin interaction.
[Reproduced with permission from K. Wakabayashi, Y, Amemiya, *Handbook of Synchrotron Radiation*, (S. Ebashi *et al.* eds.), **4**, pp.601, 602, 606, Elsevier (1991)]

References

1. For example,
   a) F.A. Bovey, *Nuclear Magnetic Resonance Spectroscopy*, Academic Press, N.Y. (1969).
   b) E. Klesper, G. Sielaff, *High Resolution Magnetic Resonance Spectroscopy* (D.O. Hummel ed.), Polymer Spectroscopy, Verlag Chemie, Weinheim, (1974).
   c) K. Würthrich, *NMR of Proteins and Nucleic Acids*, John Wiley & Sons, Inc. (1986).
2. a) G. Natta, P. Corradini, *Angew. Chem.*, **68**, 615 (1956).
   b) S. Iwayanagi, I. Sakurai, T. Seto, *J. Macromol. Sci. Phys.*, **2**, 163 (1968).
   c) G. Natta, P. Corradini, *J. Polym. Sci.*, **20**, 251 (1956).
   d) G. Natta, *Angew. Chem.*, **68**, 393 (1956).
3. a) H. Tadokoro, *J. Polymer Sct., part C*, **15**, 1 (1966).
   b) C.W. Bunn, *Proc. Roy. Soc.*, **Al80**, 67 (1942).
   c) S. Mizushima, T. Shimanouchi, *J. Am. Chem. Soc.*, **86**, 3521 (1964).
   d) I. Sakurada. K. Kaji, *J. Polymer Sci., Part C*, **31**, 57 (1970).

4. a) T_2: C. W. Bunn, *Trans. Farady Soc.*, **35**, 482 (1939).
   b) G_4: H. Tadokoro, Y. Takahashi, S. Otsuka, K. Mori, F. Imaizumi, *J. Polym. Sci., B*, **3**, 697 (1965).
      H. Tadokoro, T. Yasumnoto, S. Murahashi, I. Nitta, *J. Polym. Sci.*, **44**, 266 (1960).
      G. Carazzolo, *J. Polym. Sci.*, **A1**, 695 (1963).
   c) $(TG)_3$: G. Natta, P. Corradini, I. W. Bassi, *Nuovo Cimento.* (*Suuppl. 1*), **15**, 69 (1960).
   d) $TGT\bar{G}$: R. Hasegawa, Y. Takahashi, Y. Chatani, H. Tadokoro, *Polymer J.*, **3**, 600 (1972).
   e) $(TG)_2(T\bar{G})_2$: Not yet determined.
   f) $T_3GT_3\bar{G}$: C. W. Bunn, E. V. Garner, *J. Chem. Soc.*, 654 (1942).
   g) $(T_2G)_4$: H. Tadokoro, Y. Chatani, T. Yoshihara, S. Tahara, S. Murahashi, *Makromol. Chem.*, **73**, 109 (1954).
   h) G. Natta, I, Pasquon, P. Corradini, M. Peraldo, M. Pergeraro, A. Zambelli, *Atti. Accad. Naz. Lincei, Rend, Cl. Sci. Fis. Mat. Nat.* **28**, 539 (1960).
5. G.N. Ramachandran. C. Ramacrishnan, V. Sasisekaran, *J. Mol. Biol.*, **7**, 95 (1963).
6. T.L. Blundell, L.N. Johnson, *Protein Crystallography*, p. 32, Academic Press, N.Y. (1976).
7. L. Pauling, R. B. Corey, *Proc. Natl. Acad. Sci. U. S.*, **37**, 729 (1955).
   a) A. Rich, F. H. C. Crick, *Nature*, **176**, 593 (1955); *J. Mol Biol.*, **3**, 483 (1961).
   b) R. S. Bear, O. E. A. Bolduan, *J. Polym. Sci.*, **5**, 159 (1950); *J. Appl. Phys.*, **22**, 191 (1951).
   c) S. G. Tomlin, C. R. Worthington, *Proc. Roy. Soc.*, (*London*), **17**, 159 (1955).
8. P.H. Geil, F.R. Anderson, B. Wunderlich, T. Arakawa, *J. Polymer Sci.*, **A2**. 3707 (1964).
9. G. Natta, *Makromol. Chem.*, **35**, 94 (1960); G. Natta, P. Corradini, I.W. Bassi, *Nuovo cimento*, (*Suppl. 1*) **15**, 69 (1960).
10. H. Kusanagi, H. Tadokoro, Y. Chatani, *Macromolecules*, **9**, 531 (1976).
11. a) S. Arnott, D. W. L. Hukins, *Biochem. Biophys. Res. Commun.*, **47**, 1504 (1972).
    b) A. H.-J. Wang, G. J. Quigley, F. J. Kolpak, J. L. Crawford, J. H. van Boom, G. van cler Marel, A. Rich, *Nature*, **282**, 680 (1974).
12. G. E. Schulz, R. H. Shirmer, *Principle of Protein Structure*, pp.76-89, Springer-Verlag, N.Y. (1979).
13. A. Kita, N. Kasai, S. Kasai. T. Nakaya, K. Miki, *J. Biochem.*, **110**, 748 (1991).
14. O. Grengross, K. Herrmann, W. Abitz, *Z. Phys. Chem.*, **B10**, 371 (1930).
15. a) M. Takayanagi, *Buturi*, **16**, 19 (1961) (in Japanese).
    b) W.O. Statton, P.H. Geil, *J. Appl. Polymer Sci.*, **4**, 357 (1960); R. Hosemann, *Polymer*, **3**, 380 (1962).
16. P.H. Till, *J. Polymer Sci.*, **17**, 447 (1957).
17. A. Keller, *Phil. Mag.*, **2**, 1171 (1957).
18. E.W. Fischer, *Z. Naturforschg.*, **129**, 753 (1957).
19. K. Kobayashi, *IVth Intemational Congress of Electronmicroscopy* (1958).
20. N. Kasai, K. Kobayashi, in: (I. Uematsu, S. Okamoto M. Kakudo, H. Kawai, N. Saito, A. Miyake, M. Yamamoto eds.), *The Structure and Physical Properties of High Polymers*, Maruzen (1963) (in Japanese).
21. D.H. Renecker, P.H. Geil, *J. Appl. Phys.*, **31**, 1916 (1960).
22. P. S. Flory, *J. Am. Chem. Soc.*, **84**, 2859 (1962).
23. H. Tadokoro, T. Yoshihara, Y. Chatani, S. Murahashi, *J. Polym. Sci., B.*, **2**, 366 (1964).
24. P.J. Darrah, A.J. Gaskin, J.V. Sanders, *Scientific American*, **234**, 84 (1976).
25. P.J. Flory, *Principle of Polymer Chemistry*, Cornell Univ., Ithaca. N.Y. (1953).
26. G.S.Y. Yeh, *J. Macromol. Sci.-phys.*, **B6**, 465 (1972).
27. B. Vollmert, H. Stutz, *Angew. Makromol. Chem.*, **3**, 182 (1968); *ibid.*, **29**, 71 (1971).
28. W. Pechhold, S. Blasenbrey, *Kolloid-Z. Z. Polymer*, **241**, 955 (1971).
29. G. Bodor, *Structural Investigation of Polymers*, p.264, Akadémiai Kiadó, Budapest (1991).
30. R. Hosemann. *Polymer*, **3**, 349 (1962).
31. J. Deisenhofer, O. Epp, K. Miki, R. Huber, H. Michael, *Nature*, **318**, 618 (1985); *J. Mol. Biol.* **180**, 385 (1984).
32. R. Ladenstein, M. Schneider, R. Huber, H, B. Bartunik, K. Wilson, K. Schott, A. Backer, *J. Mol. Biol.*, **203**, 1045 (1988).
33. a) J.M. Squire, *The Structural Basis of Muscular Contraction*. pp. 266, 271, 318, Prenum Press, N.Y. (1981).
    b) H.E. Huxley, *Proc. Roy. Soc.* London. **B 178**, 131 (1971).

Part II

Experimental

8. Experimental Methods

8.1 Preliminary Considerations

Before commencing an experimental investigation, careful attention must be given to the choice of apparatus and procedure: the resultant X-ray diffraction pattern should be that which most readily yields the required information about the specimen. The right decision on the type of diffraction pattern to be obtained obviously depends upon a knowledge of the kind of information educible from each particular type of pattern. A similar knowledge of the types and applications of available apparatus is necessary for a suitable choice of experimental equipment (X-ray camera, diffractometer, etc.), X-ray parameters (wavelength, intensity, monochromatization, etc.), and the establishment of experimental procedures (method of recording diffraction intensities, etc.). Knowledge of the state of the specimen (*e.g.* its dimensions, chemical and physical stability, toxicity, etc.), information on its pretreatment, and any information already obtained by other methods, all provide essential data for planning the investigation.

When X-rays are incident upon an object in one direction, the diffracted rays are distributed among all directions in space centered upon the object. If the orientation of the object with respect to the incident X-rays is changed, a different diffraction pattern results depending on its internal structure, and the intensity of each diffracted ray also varies. The object of all X-ray diffraction investigations is to find one or more of the following three essential variables, which therefore form the three basic criteria in X-ray diffraction studies.

1. The features of the diffraction pattern (whether sharp or broad spots, lines, or arcs, or broad halos)

2. The directions (in 2θ) in which the scattered X-rays are diffracted (which necessarily involves measuring the orientations of the specimen and the observation point with respect to the incident X-rays)

3. The intensity of the diffraction in the various directions (in the case of a continuous pattern, the intensity distribution within the pattern)

The importance of these evaluations, and the precision with which they must be made, will depend upon the objective of the analysis. Even during the course of an investigation, the ultimate usefulness of a particular X-ray diffraction pattern may come into question, and it may be that the need to verify some tentative conclusion (or a requirement for more detailed information) will dictate the subsequent course of the investigation.

8.2 X-Ray Equipment

As mentioned earlier, X-rays obtained from X-ray generators or from synchrotron radiations are used in the experiments.

8.2.1 X-ray generators

X-ray generators can be divided roughly into (A) equipment using sealed tubes and (B) equipment using demountable tubes. Sealed tubes are employed in ordinary investigations. On the other hand, demountable tubes can incorporate rotating anodes (or anticathodes) which will support a high current density, and can therefore give greater X-ray intensities. Both are designed for convenience in use, and various safety devices are incorporated, so that even the beginner should experience no trouble if they are used in accordance with instructions.

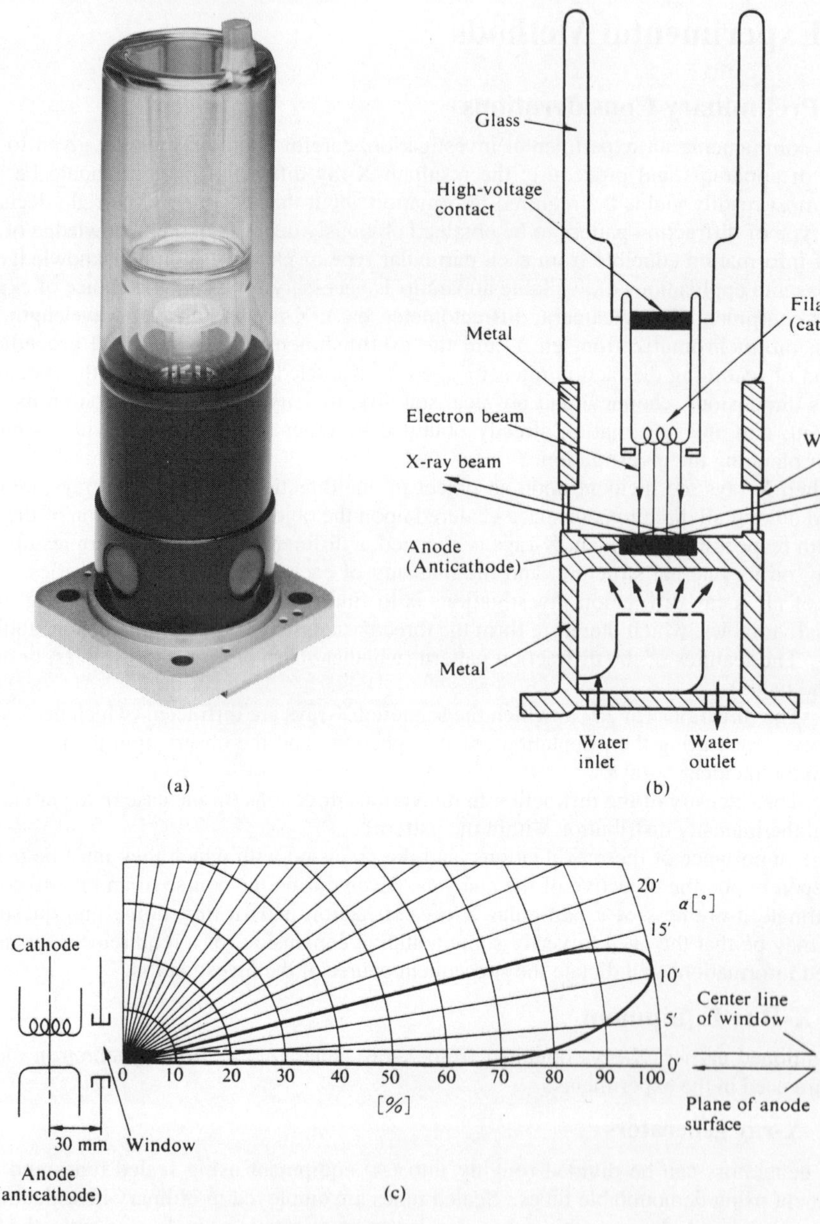

Fig. 8.1 Sealed X-ray tube.
(a) Commercial sealed X-ray tube (Philips); (b) Diagram; (c) X-ray intensity distribution.

A. Sealed X-ray tubes

The X-rays generally emerge through four windows; a point focus is seen in one direction and a line focus in the perpendicular direction. The direction of maximum X-ray intensity varies from tube to tube, but is inclined at an angle of about 6–10° to the surface of the anode (*cf.* Fig. 8.1). The maximum operating current for this type of tube depends upon the anode material.

B. Rotating anodes X-ray tubes

To increase the intensity of the X-rays, the anodes of these tubes are made to rotate, so that cooling is improved and greater currents are possible. At present, tubes which permit currents at 200–300 mÅ are widely used (Fig. 8.2), and those up to 1.5 A (90 kVA) are commercially available.

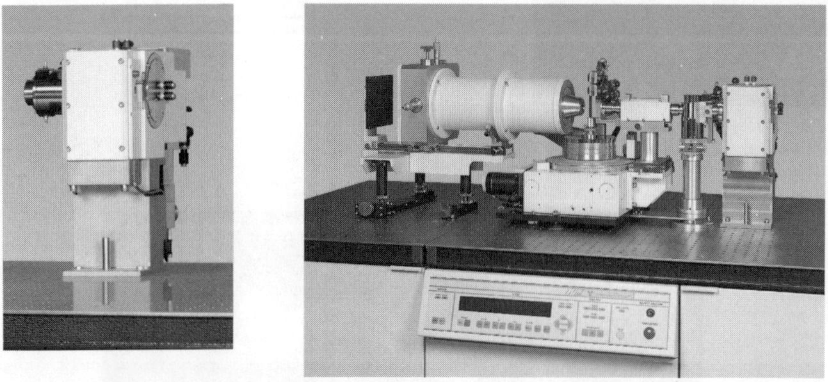

(a) (b)

Fig. 8.2 Rotating anode X-ray generator. (a) X-ray generator (Rigaku, UltraX 18). (b) UltraX 18 (Right) installed with small- and wide-angle diffraction apparatus with Imaging plate (IP) (Left) (Rigaku, SWXD).

Figure 8.3 (p.162) shows an X-ray generator with a self-rotating anode combined with a high-frequency power supply, which bears no direct or indirect anode drive system. This unit has the advantages of generating X-rays under a smooth high voltage and cutting off vibration or belt friction noise as well as minimizing the window-to-instrument distance.

C. Micro-focus X-ray tubes

If the width of the electron focus on the anode is less than about 0.1 mm, the effectiveness of the anode cooling system is improved, and the permissible current density is greatly increased. It therefore becomes easy to obtain a sharp X-ray beam of high intensity, which is very effective for the production of intense diffracted rays from small crystals or from small areas of the specimen. The electron beam of a micro-focus X-ray tube is focused with electrostatic or electromagnetic lenses. Tubes with cored electromagnetic lenses are smaller than those with electrostatic lenses, and give a stable focus. Very high brilliancy is obtained with the rotating anode micro-focus tube.

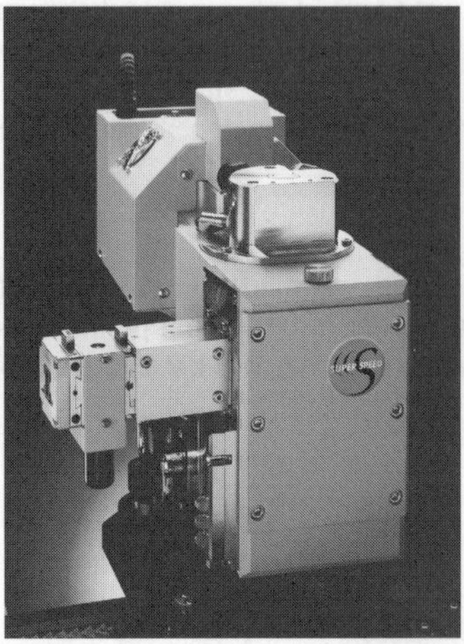

(a)

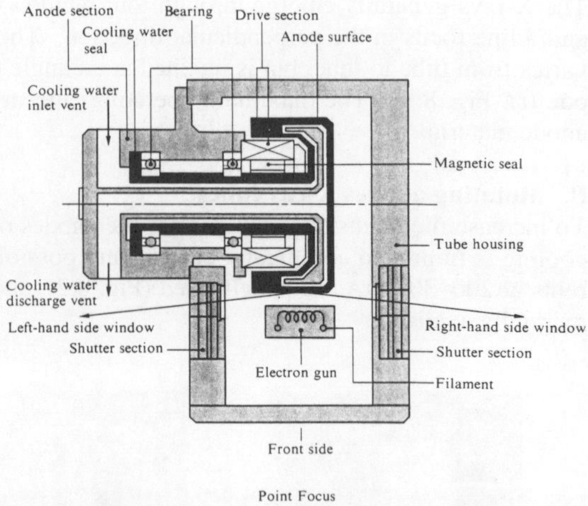

(b)

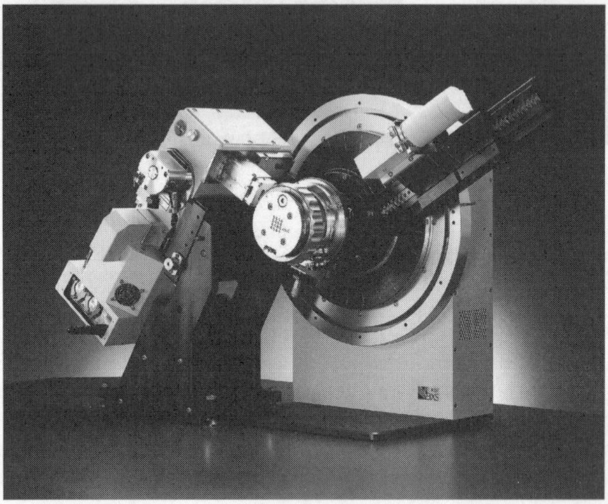

(c)

Fig. 8.3 Rotating anode X-ray generator (Turbo X-ray Source (TXS), Bruker AXS).
(a) Outlook of D8 Super Speed Solutions Turbo X-ray Source (D8 SSS TXS).
(b) Elevation view of TXS (Only point focus is shown).
(c) D8 SSS TXS (Left) installed with Multilayer Optics (Göbel mirror) and D8 Goniometer (Right) (Bruker AXS).

8.2.2 X-ray detectors

A. Point detectors

a) Scintillation counters and proportional counters. Both counter tubes are widely used for the collection of diffraction intensity data by point by point scanning. Structures and characteristics of these counter tubes and a GM counter tube are compared briefly in Table 8.1[1)] (*cf.* Fig. 8.5).

Table 8.1 Comparison of counter tubes[1)]

	GM counter tubes	Proportional counter tubes	Scintillation counter tubes
Cross section†	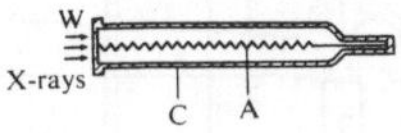	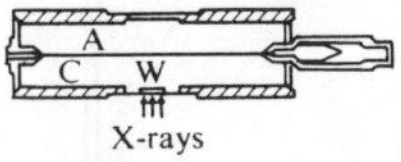	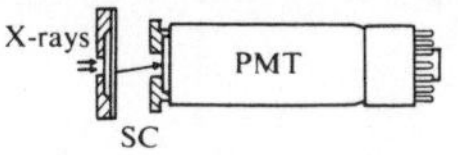
Energy analysis	Not possible	Excellent	Good
Pulse-height analyzer	No	Yes	Yes
Sensitivity to short wavelength X-rays	Poor	Poor	Good
Dead time	50-300 μs	about 0.2 μs	about 0.2 μs
Operating voltage	about 1,300 V	about 1,850 V	about 1,000 V
Other remarks		Incident X-ray quanta remaining after excitation of fluorescence in the gas give rise to low energy (escape) peaks (*cf.* Fig.8.33(b))	Noise level higher than in proportional counters
	Since the sensitivity near the center is low, this area is avoided.		

† Key—A: anode; C: cathode; W: window; SC: scintillation crystal; PMT: photomultiplier tube.
Figs: [Reproduced with permission from IUCr, *International Tables for X-Ray, Crystallography*, Vol. **III**, p.145, Kynoch Press (1962)]

b) Solid-state detectors (SSD) (or semiconductor detectors).[2)] The detector consists of two conducting electrodes and the region between these electrodes is filled with a semiconductor crystal. A voltage difference is applied across the electrodes, producing an electric field in the semiconductor. An X-ray entering the semiconductor is totally absorbed and loses its energy by producing free charge carriers, *i.e.* electron-positive hole pairs, in the semiconductor. The carriers, the number of which is proportional to the X-ray energy, move under the influence of the electric field until they are collected at the electrodes or trapped internally in the crystal. The signal is amplified and shaped to produce a pulse whose amplitude is proportional to the energy lost by the X-ray.

The structure of the detector most commonly used is shown in Fig. 8.4.[2)] A trench defines the sensitive volume of the detector and also serves to protect the highly sensitive junction between the i and p type semiconductor regions from contamination (Fig. 8.4(a)). For optimum resolution the detector capacitance must be low; a typical high performance detector has an area of 25 mm^2 and a thickness of 3 mm, which keeps the capacitance be-

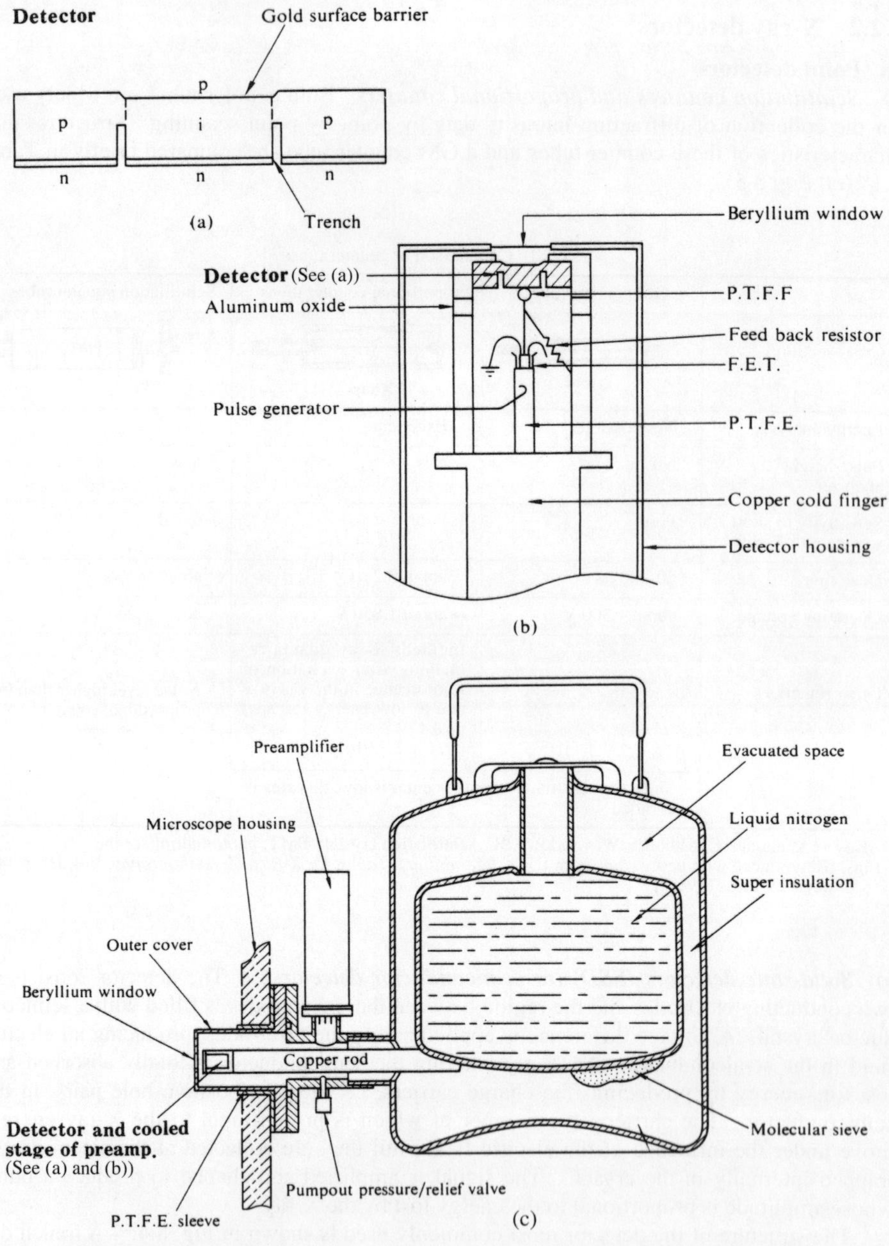

Fig. 8.4 Solid state detector (SSD).[2)]
(a) Configuration of a semiconductor detector; (b) X-ray detector mount; (c) Detector-cryostat assembly.
[Reproduced with permission from A.F.H. Meggleton, *Nucl. Instr. Methods*, **101**, 114, North-Holland Pub. (1972)]

low 1.5 pF. The detector entrance window is formed by a gold surface barrier having thickness equivalent to 0.2 μm of silicon or 1 μm of germanium. In order to stabilize the compensation by lithium in germanium or in silicon, in operation the detector must be cooled to liquid nitrogen temperature (77 K) to minimize electrical noise due to thermally excited carriers in the bulk material. The diode is required to operate at liquid nitrogen temperature with a collection field of up to 2,000 V/cm and a leakage current less than 10 pA. As the intrinsic surface is extremely sensitive to atmospheric contamination, efficient cryogenic equipment is required to contain the device. Currently cooling by liquid nitrogen or cooling by Peltier effect is employed. Resolution between scintillation, proportional and semiconductor detectors for Ag*K* radiations is compared in Fig. 8.5.

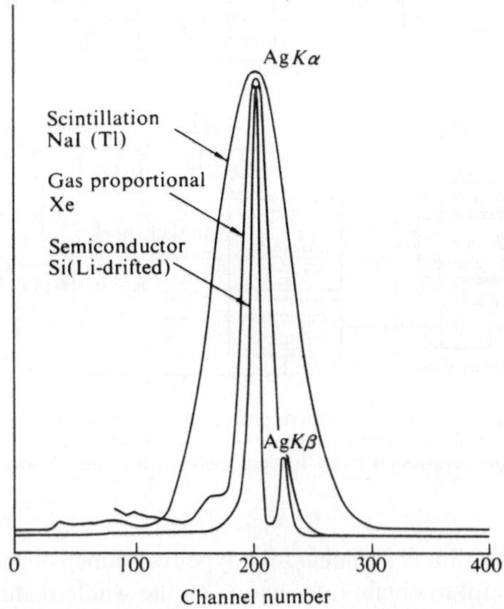

Fig. 8.5 Comparison of resolution between scintillation, proportional and semiconductor Si(Li) detectors for Ag*K* radiations.[2] [Reproduced with permission from A.F.H. Meggleton, *Nucl. Instr. Methods*, **101**, 114, North-Holland Pub. (1972)]

B. Line detectors

Position sensitive proportional counter (PSPC). In this proportional counter, the position where an electron avalanche, caused by an incident X-ray photon, hits the positive electrode is read out electronically by delay-line,[3a)] resistive wire[3b)], charge division[3c)] or wire-free-wire[3d)] read-out technique.

In the delay-line method, due to the high electrical potential of the anode wire, the electrons are collected on the anode wire and generate an electrical charge, which is transferred to the capacitively coupled delay-line. This charge is carried away by two pulses moving in opposite directions and they arrive at different times at the end of the delay-line. This time difference is measured, which corresponds to the location of incident photon (Fig. 8.6). Both linear and curved PSPC (Fig.8.17) are available.

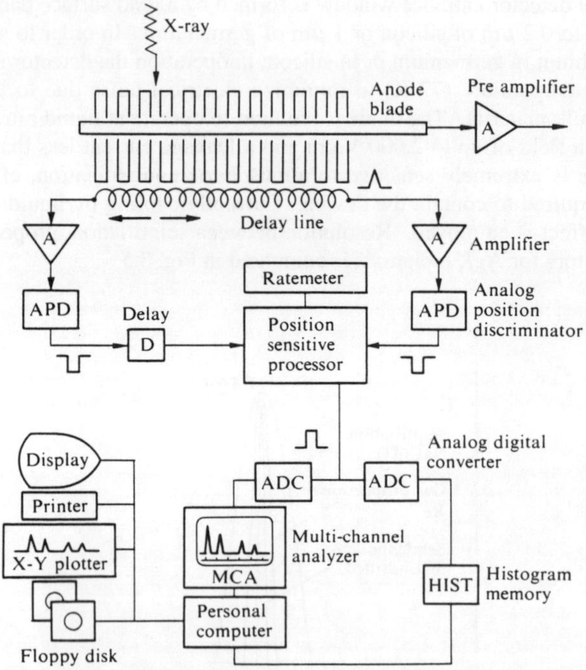

Fig. 8.6 Scheme for delay-line position sensitive proportional counter and intensity data transfar system with multi-channel analyzer and computer-aided measurement and control.[3a)]
[Reproduced with permission from T. Yamanaka, *et al.*, *Adv. X-Ray Anal.*, **35**, 417, Plenum Press (1992)]

C. Area detectors

a) X-ray film. X-ray film is an integration type two-dimensional position-sensitive detector, which is very useful to obtain information on the whole feature of the diffraction. The spatial resolution is very high. X-ray film is cheap, and the diffraction diagram obtained can be kept semi-permanently. However, it has some drawbacks in that careful chemical processing such as development, fixing, washing, and drying is necessary before obtaining the X-ray diffraction photograph, densitometry is necessary to obtain intensity data digitally, the S/N ratio is not so large due to the fogging, and the dynamic range of the non-linear energy/sensitivity characteristic curve is narrow compared to the counter tubes and imaging plate.

Reports have been published on the characteristics of X-ray film produced in various countries.[4)] However, very few are commercially available at present.

Coating both sides of the base by emulsion practically doubles the blackening for incident X-rays normal to the film. In some cases, however, the X-rays strike the film obliquely so that the diffraction pattern is slightly displaced on the side away from the incident beam. In such cases, the emulsion on the backside must be masked by an appropriate tape during development or the backside emulsion must be removed after fixing by an aqueous solution of sodium hydroxide (*ca.* 6N in concentration) for accurate measurement.

b) Storage phosphor detector. A storage phosphor detector, called an ***Imaging plate (IP)*** is an erasable new two-dimensional detector to supersede X-ray film. It was originally developed by Fuji Film Co. Ltd., Japan, mainly for diagnostic radiography.[5] The IP is approximately 0.5 mm in thickness and is composed of a flexible plastic plate coated with fine photostimulable phosphor crystals, $BaFBr : Eu^{2+}$, typically 150 μm in thickness, combined with an organic binder. The photostimulable phosphor is capable of storing a fraction of the absorbed incident energy from irradiation with X-rays (or ultraviolet light, electrons or protons). When later stimulated by visible or infrared radiation, it emits photo-stimulated luminescence (PSL), the intensity of which is proportional to the radiation energy absorbed. As shown in Fig. 8.7, the X-ray image stored in IP is read out by a scanning He-Ne laser beam ($\lambda = 632.8$ nm) which releases the PSL radiation ($\lambda \sim 390$ nm). Scanning is

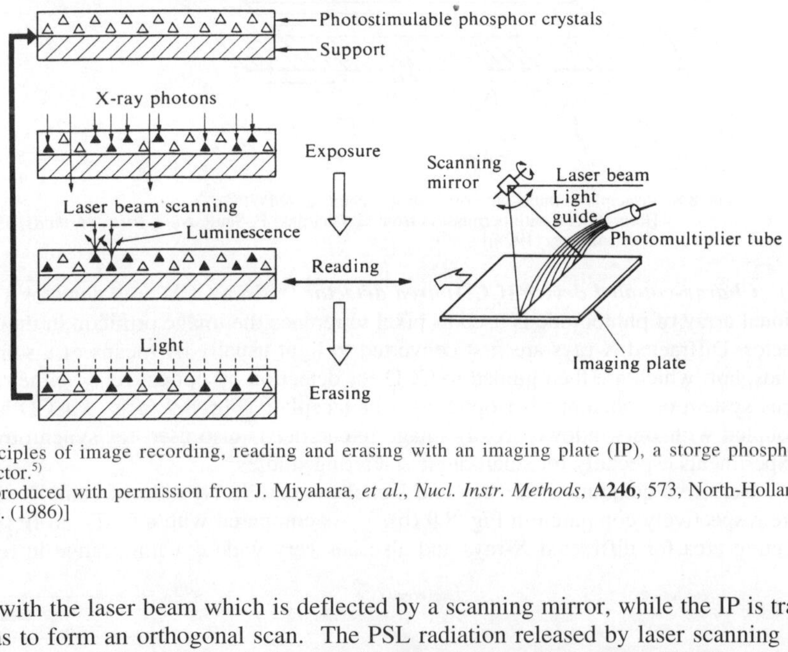

Fig. 8.7 Principles of image recording, reading and erasing with an imaging plate (IP), a storge phosphor detector.[5]
[Reproduced with permission from J. Miyahara, *et al.*, *Nucl. Instr. Methods*, **A246**, 573, North-Holland Pub. (1986)]

performed with the laser beam which is deflected by a scanning mirror, while the IP is traversed so as to form an orthogonal scan. The PSL radiation released by laser scanning is collected through a light guide and channeled into a photomultiplier tube (PMT), which converts the PSL radiation into electrical signals. The output signals form histogram slices of the incident X-ray intensity in the IP. In ordinary cases, the residual X-ray image in the IP can be erased simply by exposing it to a large dose of visible light.

In order to obtain an X-ray diffraction diagram, the digital signals are reconverted into analog signals that modulate the intensity of another He-Ne laser beam. The modulated beam is then used to scan a photographic film to imprint the image on it. The film is finally developed to display the image. Many IP detector systems for X-rays are commercially available.

c) Multi-wire proportional counter (or multi-wire proportional chamber) (MWPC). The MWPC is simply a chamber containing a large number of anode wires, each acting independently as a position-sensitive proportional counter. The MWPC usually consists of a parallel array of wires in a plane which is separated from the plane of anode wires by a few millimeters. The MWPC used for X-ray experiments normally has two planes of cathode wires, one on either side of the anode plane (Fig. 8.8), and the delay-line read-out technique is ordinarily applied.[6)]

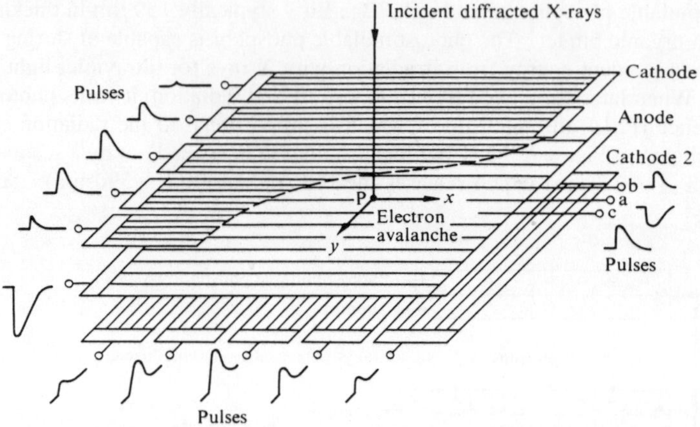

Fig. 8.8 Principle of multi-wire proportional chamber (MWPC).[6)]
[Reproduced with permission from G. Charpak, F. Sauli, *Nucl. Instr. Methods*, **152**, 185, North-Holland Pub. (1978)]

d) Charge-coupled device (CCD) area detector.[7a)] In a CCD area detector a two-dimensional array of photodiode is used as pixel to replace the image orthicon in the TV area detector. Diffracted X-rays are first converted to light usually by means of a scintillator or a phosphor, which are then guided to CCD for detection by optical relay, either high-quality lens system or coherent fiber-optic taper or faceplate (Fig. 8.9 (a)). A CCD area detector coupled with Be-windowed X-ray image intencifier is also used for synchrotron radiation experiments especially for small-angle scattering studies.

Detective quantum efficiency and signal-to-noise (S/N) ratio of some X-ray detectors are respectively compared in Fig. 8.9 (b).[7c)] As compared with a CCD, an IP has large detecting area for diffracted X-rays and also has very wide dynamic range in recording the

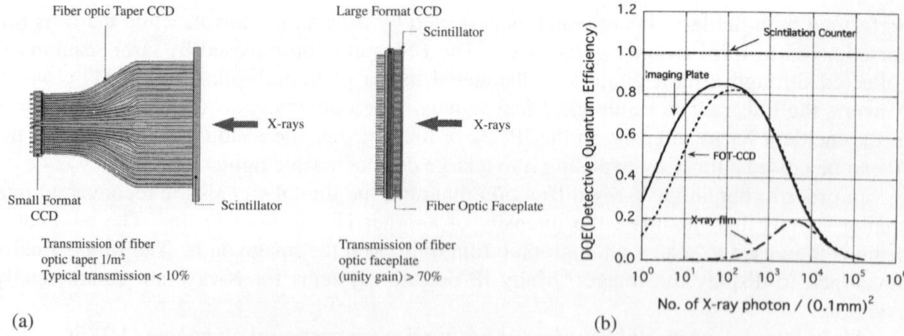

Fig. 8.9 Charge-coupled device (CCD) area detector.
(a) Fiber-optic taper (FOT) CCD (Left) and direct imaging CCD (Right) (Bruker AXS).
(b) Comparison of detective quantum effieiency of X-ray detectors[7c)]
[Reproduced with permission from A. Amemiya, K. Ito, *J. Cryst. Soc. Jpn.*, **45**, 171, Cryst. Soc. Jpn. (2003)]

diffraction intensity. However, the read-out time of X-ray image stored in CCD detector is much faster than that in IP. The CCD detector is therefore often used for synchrotron radiation experiments for alignment of protein crystal, native or heavy atom derivative, and also for qualification examination of a crystal within a short time.

*e) **Television area detector***. In short this is a scintillation screen with television-type read-out.[7b] Direct and indirect type detector systems have been developed: the former utilizes an X-ray sensitive image orthicon to detect X-ray diffractions whereas the latter uses an ordinary image orthicon to detect scintillations on a screen caused by X-ray diffraction. This TV detector also has higher sensitivity and wider dynamic range.

8.2.3 X-ray cameras

A. Laue cameras

This is the simplest type of camera, with both specimen and film stationary. A flat film (or IP) is used, and diffraction patterns can be recorded by the transmission method, the back-reflection method, or side-reflection method. The specimen-to-film distance depends on the greatest angle at which diffraction is to be recorded, but is normally about 3–5 cm.

Powder diagrams and fiber photographs are commonly made using characteristic X-rays. Methods using this simple camera have the advantage that the specimen can be readily heated, cooled, stretched, or compressed. They suffer from the disadvantages, however, that the range of measuring angles is small, the layer lines are hyperbolic, and specimen-to-film distance and the shapes of the spots vary from spot to spot. It is therefore recommended that cylindrical film be used in quantitative studies.

With regard to alignment of the camera, the fiber axis need not be perpendicular to the X-ray beam, and specific high-order diffraction patterns can be obtained by inclining the axis of the specimen in the vertical plane containing the X-ray beam (Tilt photograph). Since the front of the film is completely exposed in this type of camera, care is necessary to avoid fogging by X-rays scattered from the sides.

Goppel camera.[8] With this camera (Fig. 8.10) the diffraction patterns of a polymer specimen and the inner lines of the patterns of a standard sample are recorded simultaneously on the flat film or IP. Any variation in the intensity of primary beam can be ignored and those

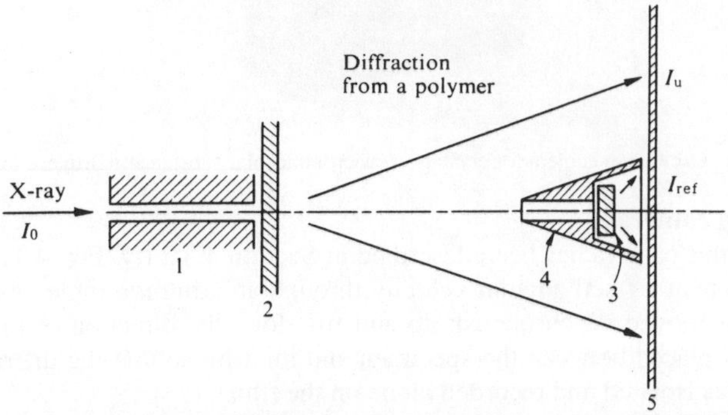

Fig. 8.10 Diagram of Goppel camera for simultaneously recording the diffraction patterns of polymers and the inner lines of the pattern of a standard sample.[8] 1, Collimator; 2, Polymer specimen; 3, Reference specimen; 4, Goppel reference chamber for the reference specimen; 5, Film.
[Reproduced from J. M. Goppel, *Appl. Sci. Res.*, **A1**, 8. Academic Press (1949)]

due to thickness change of polymer specimen can be calibrated. This camera of equivalent diffractometer technique[9] is useful for X-ray crystallinity measurement.

B. Rotating- (oscillating-) crystal cameras

The crystal is rotated about one axis (*cf.* Section 4.1), at a rate of one cycle per 3–6 min, either through one complete revolution (360°) in one direction or with an oscillatory motion through 5–10°. In the latter case the crystal may be oscillated by a reversing synchronous motor (or an eccentric cam). The diameter of the film is usually 57.296 mm, which ensures that 1 mm on the film corresponds to a scattering angle of 2°.

The reciprocal lattice of a uniaxially oriented specimen consisting of an aggregation of crystallites gives concentric circles (*cf.* Fig. 4.12), and the photograph obtained *without* rotation of the specimen is the same as if the specimen, with its crystal lattice and reciprocal lattice, had turned through 360°. The arrangement is basically the same as in the Laue method, except that a cylindrical film is always used. In the case of a doubly oriented specimen, however, if the direction of a second axis in the specimen is known, the reciprocal lattice is drawn as described in Section 4.4.1. The diffraction patterns can then be recorded separately by oscillation in such a way that diffractions are produced separately from specific regions of the reciprocal lattice. However, the Weissenberg method is more convenient for this purpose.

Figure 8.11 shows an automated oscillation camera for the intensity measurement of crystals of macromolecular substances.[10] This camera has eight film cassettes and a maximum of eight different oscillation diagrams can be taken.

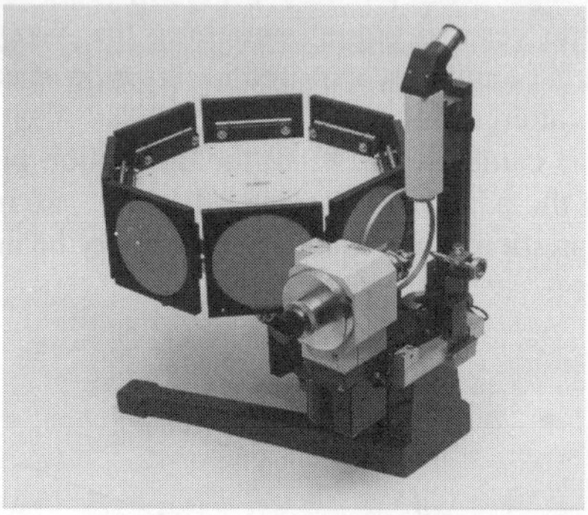

Fig. 8.11 Oscillation camera for crystals of macromolecular substances (Bruker-Nonius).

C. Weissenberg cameras

The principle of this camera has been described in Section 4.4.2 (*cf.* Fig. 4.12). The crystal is made to oscillate at a fixed angular velocity through an arbitrary angle, and the cylindrical film (or IP) is moved synchronously to and fro along the direction of the axis of rotation. A screen is placed between the specimen and the film so that the diffractions from a certain layer line is isolated and recorded alone on the film.

A diameter of 57.296 mm and a gearing constant of 2 °/mm is the standard for conventional Weissenberg cameras. If the film-cassette carriage is disconnected from the crystal drive the camera can be used for oscillating-crystal or rotating-crystal experiments. In this case the screen is replaced by a beam stop.

a) ***Equi-inclination method*** (*cf.* Section 8.6.2 and Fig. 8.42). It is possible to measure spots close to the origin of the reciprocal lattice on the layer lines. Moreover, comparison of the diffraction intensities is very easy since the specimen-to-film distance and the angle of incidence on the film are constant for the spots on a given layer line with an single film. Because of the use of a screen, background scattering is reduced to one-tenth that of the oscillating crystal method and one-hundredth that of the rotating crystal method, so weak spots can be measured more easily.

In the case of highly oriented polymer specimens (*e.g.* those which are biaxially or doubly oriented), this is the most convenient method of separating the diffraction patterns, as well as of indexing diffractions, measuring intensities, and accurately locating the axis (see Fig. 8.47).

If the unit cell is very large as in protein crystallography, a specially designed Weissenberg camera with large diameter utilizing IP may be used in combination with syn-

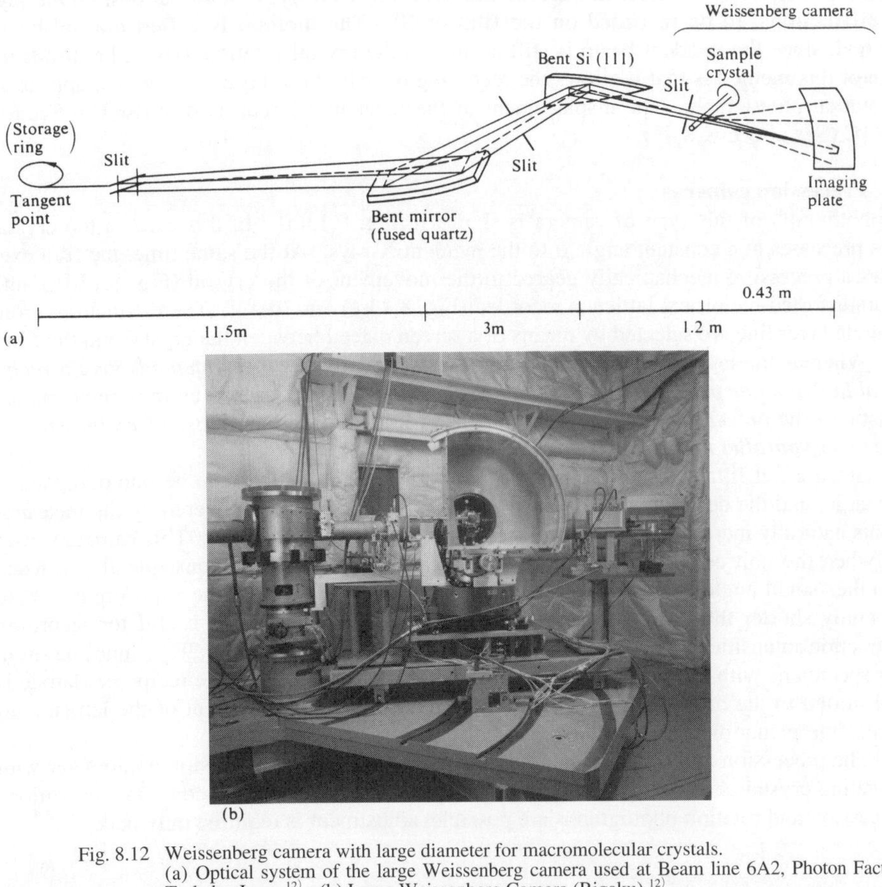

Fig. 8.12 Weissenberg camera with large diameter for macromolecular crystals.
(a) Optical system of the large Weissenberg camera used at Beam line 6A2, Photon Factory, Tsukuba, Japan.[12] (b) Large Weissenberg Camera (Rigaku).[12]
[Reproduced with permission form Y. Satow, *et al.*, *Rev. Sci. Instr.*, **60**, 2394. Am. Inst. Phys. (1989)]

chrotron radiation.[12] Fig. 8.12 shows a picture of this type of Weissenberg camera system: with normal beam method, controlling the ratio of IP translation (0.005–4 mm/s) and specimen rotation speed (0.002–2 °/s), the best data collection conditions, such that there is no overlap of the diffraction spots on the IP and as many diffractions as possible are recorded on a single sheet of IP can be chosen. The radius of the film cassette can be set to either 143.25, 286.5, 430.0 or 57.296 mm. In principle, the camera requires no screen, but the use of a screen with multi layer-line sets (radius of 132.5 mm) improves the S/N ratio by a factor of at least 2 or 3. One of the weak points of this camera is that when the longest axis of the crystal is approximately parallel to the incident beam, diffraction spots will have a tendency to overlap each other with horizontally adjacent spots. This drawback can be avoided if another movement of the cassette along the 2θ direction perpendicular to the base line is added.

b) Normal beam method.[11] The experimental conditions are similar to those for the equator, except that the layer line screen has to be shifted to prevent all but desired *n*th layer diffractions, to be recorded on the film or IP. The method is called normal-beam method, since the incident beam is still normal to the crystal rotation axis. The disadvantage of this method is that it allows the recording of only those layer lines which appear on a rotation photograph. The displacement of the layer line screen is discussed in Section 8.7.1C (See also Fig. 8.47).

D. Precession cameras

A photograph of this type of camera is shown in Fig. 8.13(a). In this camera the crystal axis precesses at a constant angle μ to the incident X-rays. At the same time, the film executes a precession mechanically geared to the movement of the crystal (Fig. 8.13(b)), and an undistorted reciprocal lattice is recorded (Fig. 8.13(c) and (d)).[13] The diffractions from a single layer line are selected by means of a screen placed between the crystal and the film.

Whereas the layer lines in the *rotating-crystal* and *Weissenberg methods* give a *reciprocal lattice plane* ***perpendicular*** *to the axis of rotation* of the crystal, an important characteristic of the *precession method* is that the layer lines selected are those of *reciprocal lattice planes* ***parallel*** *to the axis of rotation.*

Since a flat film is used, the part of the reciprocal lattice that can be photographed is not large, and the design is usually such that $\mu \leqq 30°$. The region covered by the measurements naturally increases as the wavelength of the X-rays is decreased. This camera is useful when the unit cell is large and the reciprocal lattice points are consequently crowded into the "small angle" part of the film, as with proteins. Moreover, the exposure times are generally shorter than for Weissenberg photographs. The method is useful for recording diffraction intensities, but the multiple film technique (*cf.* Section 8.6.2D) cannot be used. For specimens with diffuse patterns, such as polymers, the fact that the reciprocal lattice is undistorted in the diffraction photograph permits accurate measurement of the lattice constants, interplanar distances, and degree of orientation.

The precession and rotation chamber, the combination of a precession goniometer with a rotating crystal chamber, is also commercially available. Under exactly equal conditions precession and rotation photographs are possible, adjustment is required only once.

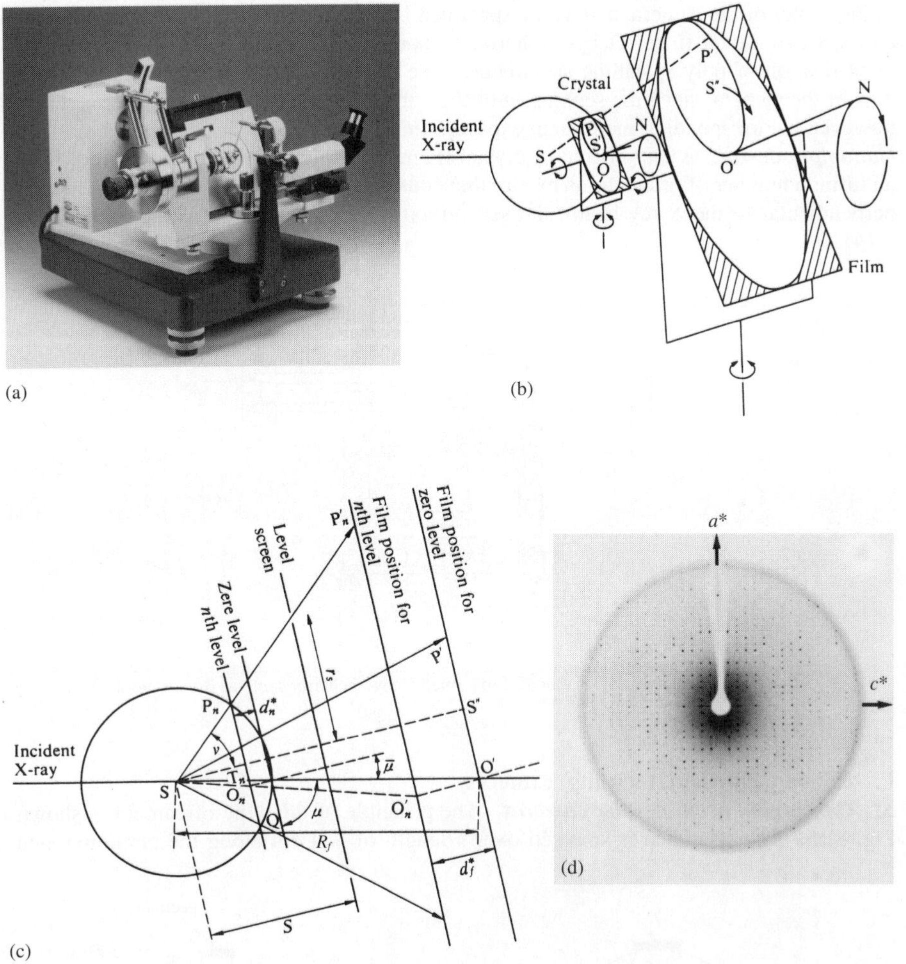

Fig. 8.13 Precession camera.
(a) Commercial precession camera (Bruker-Nonius); (b) Movement mechanism: The crystal and the film are moved in such a way that each can rotate about two perpendicular axes. A pantograph ensures that the single reciprocal lattice plane (representing the crystal in the diagram) remains parallel with the film; [13a)] (c) Geometry for recording the *n*th reciprocal lattice point. To achieve same magnification for *n*th and zero levels the film must be moved a distance $d_f^* = R_f d_n^*$ nearer to the crystal (*cf.* Section 8.7.2B); [13a)] (d) *h0l* Precession photpgraph of Pseudoazurin from *Methylobacterium extorquens* AM1.[13b)]
(b), (c): [Reproduced with permission from M. J. Buerger, *The Photography of the Reciprocal Lattice*. pp.4,5, Am. Cryst. Assoc. (1994)]
(d): [Reproduced with permission from T. Inoue, *Doctral Thesis*, p.20, Osaka Univ. (1994)]

E. Debye-Scherrer cameras (powder cameras)

In this camera (*cf.* Section 4.4.4) the sample, often in powder form, is molded into a small rod or packed into a glass capillary tube. The resulting polycrystalline specimen is placed in the center of the camera, and as the specimen is rotated, the diffraction pattern is recorded on a cylindrical film. Debye-Scherrer cameras are generally used for the qualitative identification of polycrystalline specimens. The accuracy naturally increases with the radius of the camera, since the separation of the various diffraction lines improves. There is, however, a corresponding increase in exposure time.

Gandolphi device. A tiny crystal or crystal fragment, which cannot be powdered, is given an infinite number of orientations by simultaneous rotation about two axes. The first axis is perpendicular to the X-ray beam; the second axis is inclined at 45° to the first axis (Fig. 8.14).

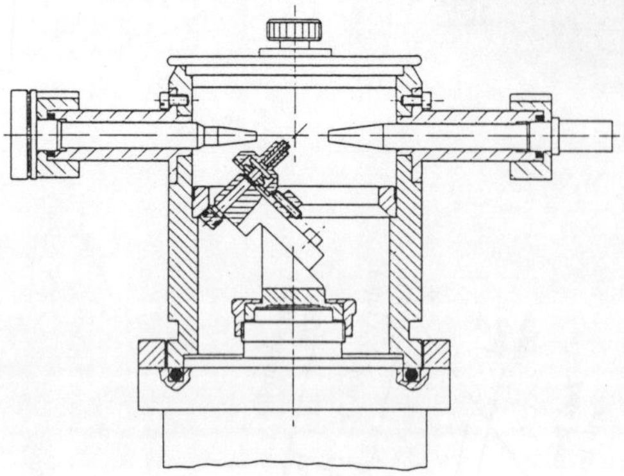

Fig. 8.14 Gandolphi device mounted in a small Debye-Scherrer camera (Bruker-Nonius).

F. Convergent beam (focusing) cameras

a) Convergent beam powder cameras. The principle of this type of camera is shown in Fig. 8.15. The specimen is smeared over a length of 3–4 cm along the circumference in

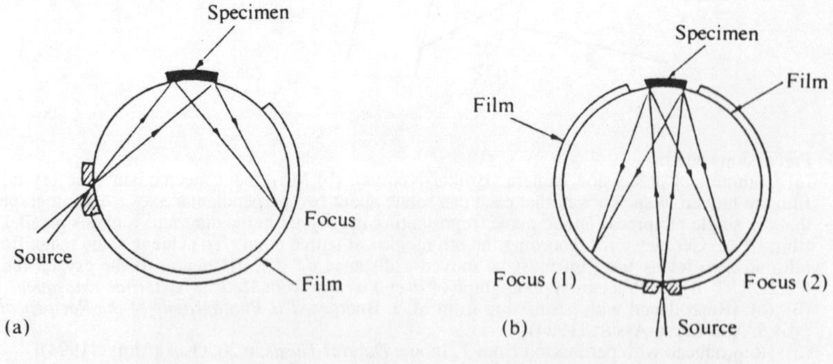

Fig. 8.15 Principle of convergent beam cameras. (a) Asymmetrical type; (b) Symmetrical type. [Reproduced with permission from *International Tables for X-Ray Crystallography*, (C.H. MacGillavry, G. D. Rieck eds.) Vol. III, p.85, IUCr. (1968)]

both the symmetrical (b) and the asymmetrical (a) types. Since the diffracted X-rays converge in accordance with the Bragg condition, the exposures are short. However, the method has the disadvantage that the low-order diffractions are weak because of the shallow angle at which they strike the film. Polymer films can be curved and fixed directly to the specimen mount.

b) Guinier cameras. In this type of camera a curved-crystal monochromator is used to cause the X-rays to converge. As shown in Fig. 8.16, both transmission types and reflection types are available. In the normal Guinier camera, the convergent monochromatic X-rays pass through the specimen, and the converging diffracted X-rays are recorded. By this method, all wavelengths except the K_α doublet are removed, and the background is good. Moreover, the quantity of specimen is comparatively large and the diffracted rays are made to converge. Consequently the exposures are not too long despite the use of the monochromator. Since the incident X-rays are separated from the diffracted X-rays by an angle of 4θ at the center of the camera, the dispersion is twice that of a Debye-Scherrer camera of the same diameter. This method is therefore more suitable for qualitative identifications.

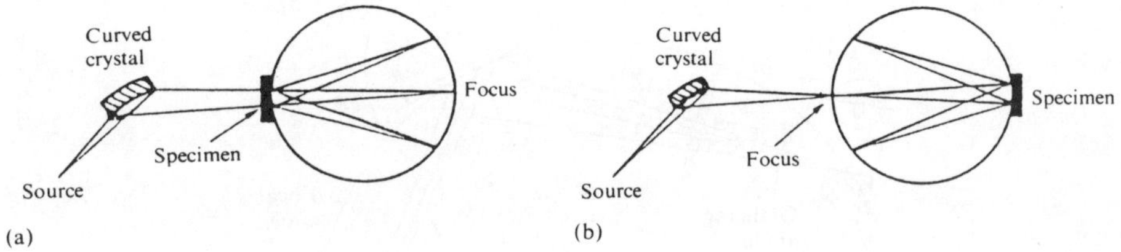

Fig. 8.16 Principle of Guinier cameras.
(a) Transmission type (Bragg angles < 45°); (b) Reflection type (Bragg angles > 45°).

G. Microbeam cameras

This type of camera is suitable for the examination of very small crystals or particles, or for the investigation of the structure of texture of very small regions of polymers. Clear diagrams of weak diffractions from small crystals or small areas can be obtained with short exposures by the use of an intense X-ray beam 10–100 μm wide. The apparatus is simple, and instead of using commercially obtainable equipment, the experienced investigator could well design and construct equipment to suit the purpose of his experiments.

8.2.4 X-ray diffractometers

Instead of recording diffraction pattern on a film, X-ray diffractometers measure the intensity distribution, either one-dimensionally or tw-dimensionally by means of a counter tube or an area detector.

A. Powder diffractometers

a) Scanning type diffractometers. The optical system of a goniometer, which, like the majority actually in use are based on the Bragg-Brentano technique, has its counter tube set to scan point by point, in a horizontal plane, is shown in Fig. 8.17. Using a "line-focus" X-ray source, the Soller slits, S_1 and S_2, limit the vertical divergence of the incident and diffracted beams to (say) 2°, so that the error in defining the equatorial direction is small. The

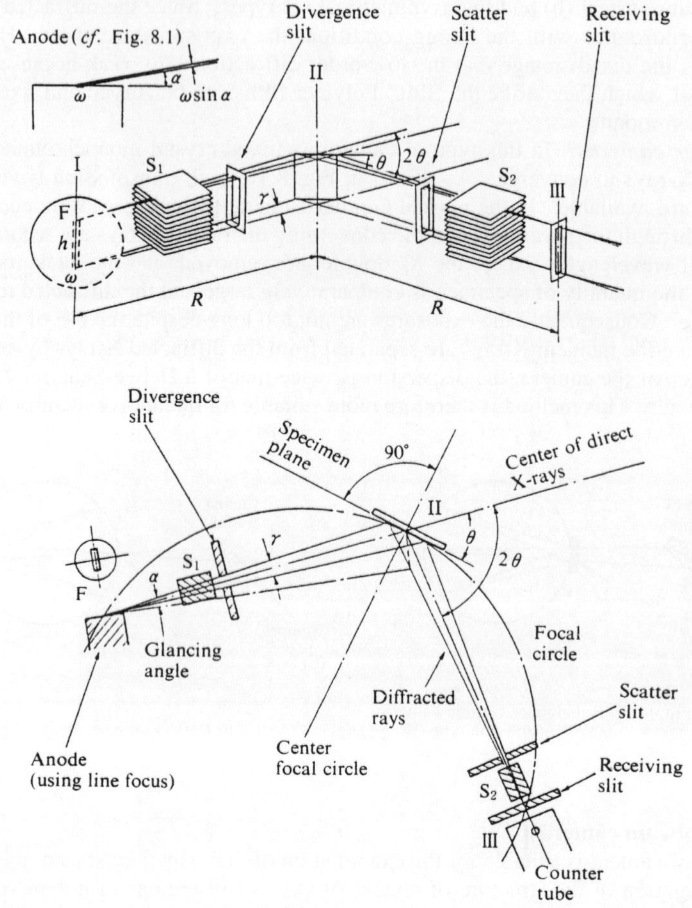

Fig. 8.17 X-ray powder diffractometer (scanning type) : Principle of finding the diffracted intensity at Bragg angle θ (equatorial).
F: Focus (width ω, height h); α: Glancing angle; S_1, S_2: Soller slits; I: Center of focus; II: Center of goniometer; III: Center of receiving slit; γ: Horizontal divergence angle; R: Goniometer radius.
[Reproduced from L. Alexander, H. Klug, *X-ray Diffraction Procedures for Polycrystalline and Amorphous Materials*, p.241, John Wiley & Sons, Inc. (1954)]

X-rays, which are also limited horizontally by the divergence slit, are diffracted by the specimen, and the rays having a diffraction angle of 2θ are focused on the receiving slit and enter the counter tube. Vertical goniometers are also available and, although the movement of force exerted by the counter tube affects the spring loading of the worm gear during the scanning, they have the advantage that powder or liquid specimen is less likely to be lost.

Scintillation counters or proportional counters give reduced background counts and great accuracy in intensity measurements. The intensities may be determined by 1) the strip-chart recording method using a rate mater or 2) the fixed-count method using a scaler and timer. Method 1) is suitable for measurement of the diffraction profile, but it is necessary to use a suitable combination of scanning speed and time constant. Method 2) is useful for accurate measurement, since the random error of the measurements is constant, but the time required is excessive for angles at which the intensity is very low.

Possible ancillary equipment for the investigation of high polymers includes monochromators, facilities for heating and cooling the specimen, and holders for liquid specimens. Special equipment may also be provided for investigation of stretched or swollen specimens. Holders for fibrous specimens and equipment for the determination of pole figures may be useful for determining the orientation of crystallites, and computer-controlled measuring devices are also available for the construction of pole figures. The blind region caused by the specimen holder is eliminated by a combination of the reflection and transmission methods.

A Guinier diffractometer is also commercially available (*cf.* Section 8.2.3F(b)).

b) Position-sensitive type diffractometers. A curved position-sensitive proportional counter with a large radius, installed horizontally on the 2θ arm of the usual powder diffractometer, has made possible the simultaneous collection of diffracted X-rays over 120° in 2θ (Fig. 8.18). Only a few seconds are needed to obtain a whole powder diffraction pattern because of the extremely high counting efficiency due to the streamer mode. This type of diffractometer has advantages for the time-resolved observations of solid reactions and structure transitions (Fig. 8.18 (c)) and other.

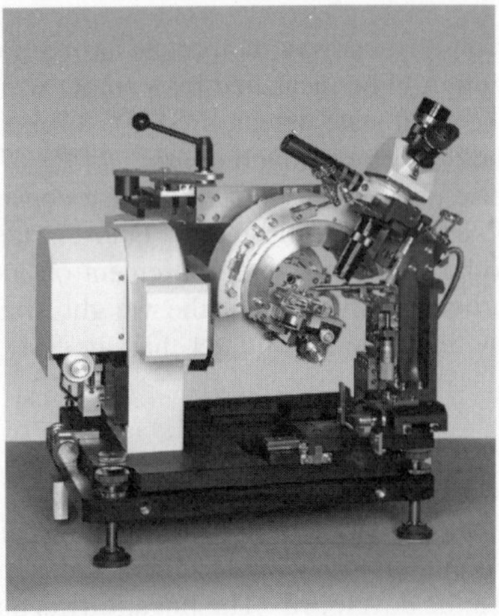

(a)

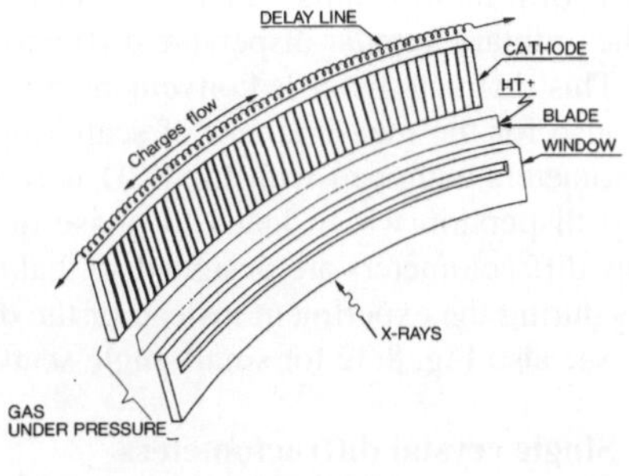

(b)

Fig. 8.18 X-ray powder diffractometer with curved position-sensitive proportional counter (PSPC).
(a) Diffractometer for the time-resolved diffraction study (Rigaku, PSPC-MDG).
(b) Illustration of a curved PSPC (*cf.* Fig. 8.6).

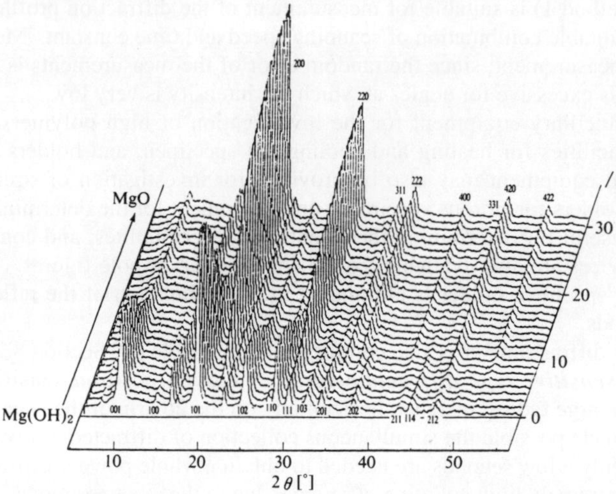

Fig. 8.18 —*Continued*
(c) An example of the time-resolved diffraction pattern of the dehydration of $Mg(OH)_2$ taken every 30 s. Every tenth pattern is presented in the figure.[16)]
[Reproduced with permission from T. Yamanaka, *et al.*, *Adv. X-Ray Anal.*, **35**, 421, Plenum Press (1992)]

c) Energy dispersive diffractometers. Diffractions by various reciprocal lattice points due to several characteristic X-rays from a X-ray tube can be measured by a single scan by means of a multi-channel analyzer connected to the solid state detector (SSD). Compared to the ordinary angular dispersive diffractometer the S/N ratio is much improved.[14a)]

This diffractometer is convenient for the precise determination of the lattice constants and also for the measurement of scattering by non-crystalline substances. Four-circle diffractometers equipped with the SSD is reported to be good for the measurement of anomalous dispersion which leads to phase determination.[14b)] Because of the weight of SSD many diffractometers are designed so that the X-ray tube, which is lighter than the SSD, rotates during the experiment instead of the detector.

See also Fig. 8.32 for small-angle scattering.

B. Single crystal diffractometers

a) Four-circle diffractometers. For the accurate determination of crystal structures, accurate measurement of the positions and intensities is important. Counter diffractometers of the point by point measurement type were the first to be developed. The linear diffractometers were the first commercially available, and in some laboratories efforts have been made to produce a scanning equi-inclination Weissenberg counter diffractometer. However, all of them were replaced by four-circle single crystal diffractometers after several years.

The crystal orientation is varied by rotation about three axes (φ, χ, and ω or φ, κ, and ω in Fig. 8.19), and the direction of the diffracted beam is brought to coincide with the equatorial direction, so that the measurements are always carried out on the equator. The blind regions due to each crystal axis are very small, especially in the latter κ-geometry diffrac-

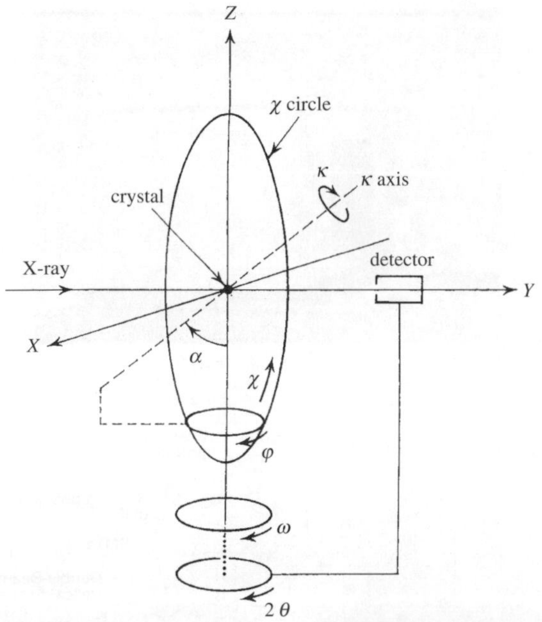

(a)

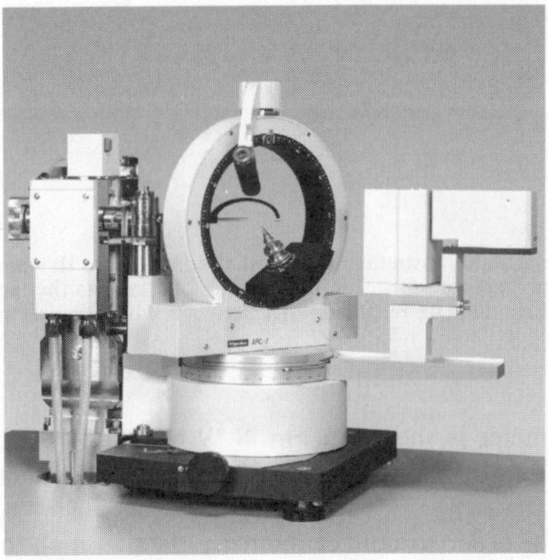

(b)

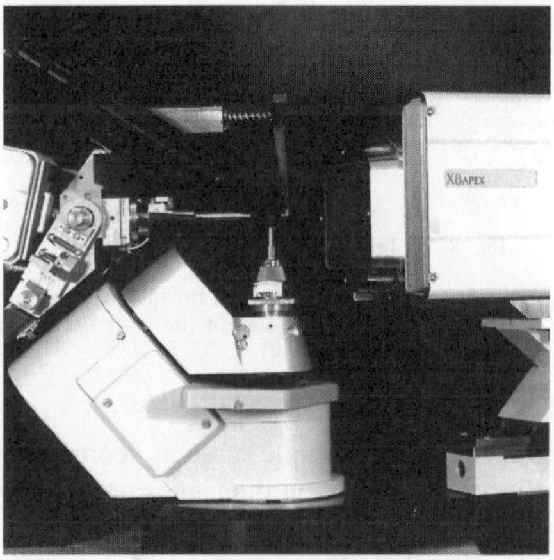

(c)

Fig. 8.19 Four-circle single crystal diffractometer.
(a) Geometry of crystal rotation about the four axes (2θ, φ, χ, ω,) and (2θ, φ, κ, ω). 2θ is the diffraction angle. The dotted line of the 2θ-axis indicates that the movement around the 2θ-axis is independent from those around the other three axes. *χ-Geometry*: χ-circle rotates along the Z-axis by rotation of the ω-axis. The φ-axis and hence the crystal on the φ-axis moves on the χ-circle by rotation of the χ-circle. κ-Geometry: The φ-axis makes precession movement around the κ-axis by rotation of the κ-axis. (b) χ-Geometry four-circle goniometer (Rigaku, AFC-7R + scintillation counter) χ or $1/4 \cdot \chi$ circle with CCD is commercially available. (c) κ-Geometry four-circle goniometer (Bruker AXS, X8APEX (X8 + CCD). (a): [Reproduced with permission from Chem. Soc. Jpn, ed. *Experiments in Chemistry*, 4th ed., Vol. 10, *Diffraction,* p.58, Maruzen (1992)]

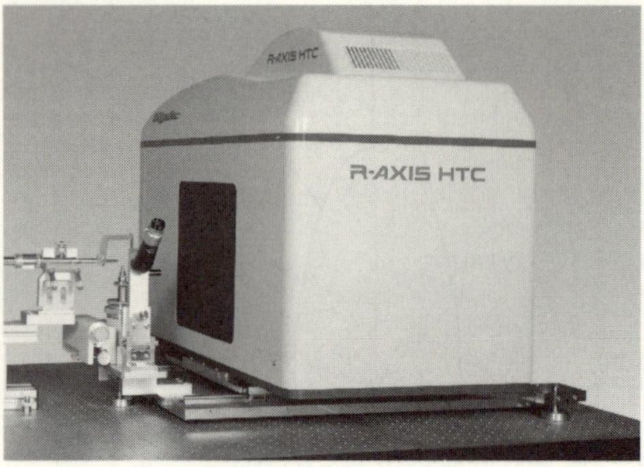

(a)

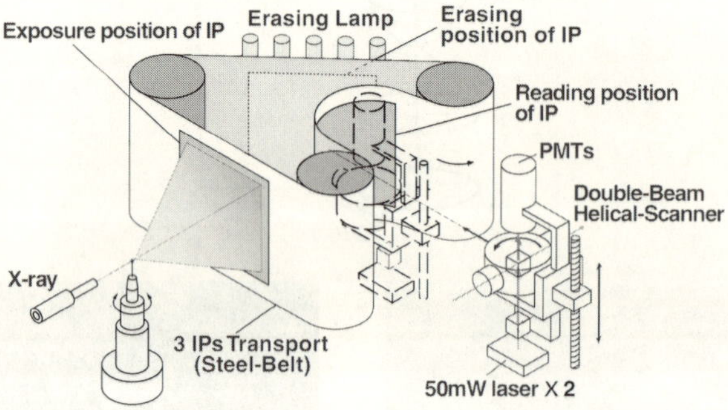

(b)

Fig. 8.20 Imaging Plate (IP) diffractometer.
(a) Outlook (Rigaku, R-AXIS HTC). (b) Schematic illustration of optical system. Three IP's are continuously used for Intensity data collection, exposure for recording, reading out the data (by the usual X-Y scan of the laser beam), and erasing the data recorded. [Reproduced with permission from Rigaku]

tometer. An example of each geometry goniometer is shown in Fig. 8.19(b) and 8.19(c). Since the diffraction intensities are measured point by point, it takes a long time to collect all the intensity data desired.

b) Area-sensitive detector diffractometers. In order to collect intensity data in a short time, especially for data collection in protein crystallography, single crystal diffractometers using the IP (Figs. 8.19, 8.20 and 8.21), multi-wire proportional counter or CCD detector are commercially available. This kind of diffractometer is also very useful for precise intensity measurement of fiber diagrams of high polymer specimens.

The adequate use of IP, which is flexible and can be used in a cylindrical shape, and its read-out and erasing system accelerated the development of various types of IP diffractometers with original or modified camera geometries, enabling rapid and precise intensity measurement and even time-resolved study of the fast structural changes in crystals or other crystalline materials.

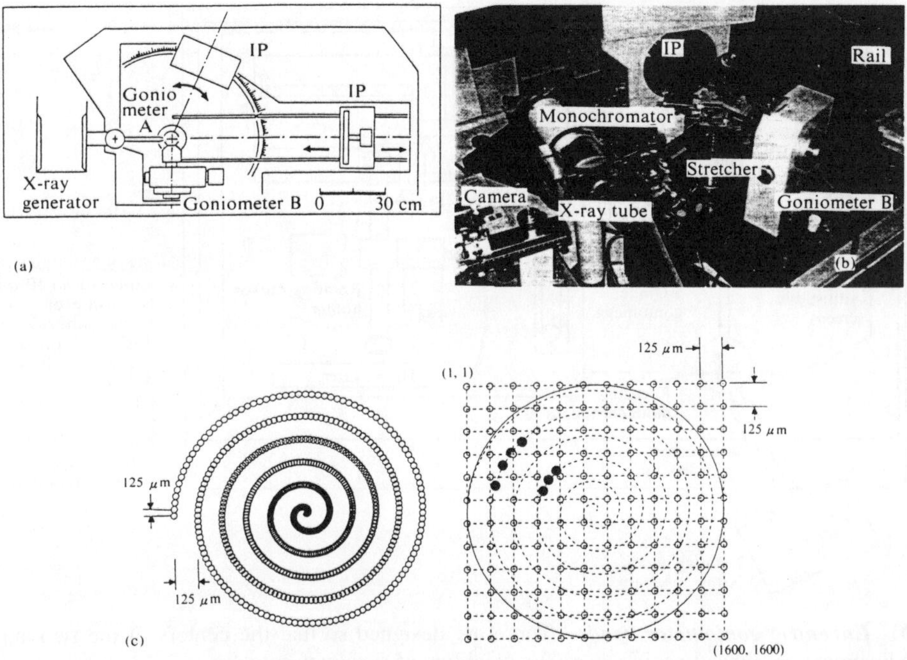

Fig. 8.21 Imaging Plate diffractometer for polymers (Bruker AXS).
(a) Plan view. (b) Optical system. Specimen stretcher may be replaced by usual goniometer head for single crystal studies. (c) Intensity data recorded on the IP are read out by the spiral scan of the laser beam, the coordinates are converted to the orthogonal coordinates (pixel).

An example of the IP diffractometer of the Weissenberg type is that of Sakabe mentioned above and the other is that of Iwasaki.[15] The latter consists of three parts: a Weissenberg chamber with adjustable screens (Fig. 8.22), an IP read-out mechanism and a partition board between these two parts. The Weissenberg chamber is composed of an X-ray optics, a three-axes goniometer, an adjustable multi-layer line screen system and two cylindrical IP's. The IP read-out mechanism is composed of a rotary optics for the He-Ne laser light and photostimulated luminescence and a halogen lamp for erasing the IP. The X-ray optics consists of a self rotating anode X-ray tube (Mo, Cu, or Ag anode), a focusing monochromator of graphite or LiF, a collimator and other optical elements such as a beam tunnel, a shutter, attenuators, slits and a beam stop.

C. Goniometer heads

These are devices for attaching the specimen to the mount of the camera or diffractometer in such a way that the orientation can be adjusted as desired (Fig. 8.23). Close manufacturing tolerances and stability are necessary to ensure precision in the orientation of the crystal. Goniometer heads can be broadly divided into two types.

a) Cross-type goniometer heads. These are the normal type, permitting adequate parallel movement for adjustment of the orientation of the specimen.

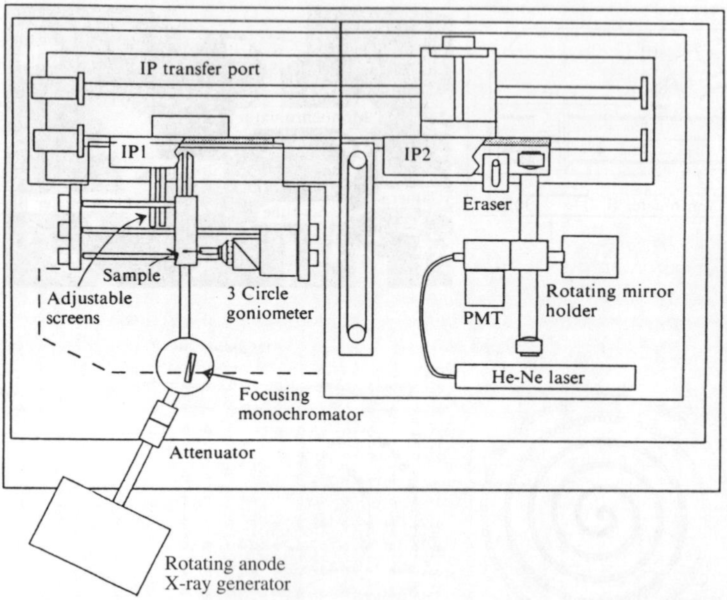

Fig. 8.22 Block diagram of an IP diffractometer of Weissenberg type with adjustable screen (Bruker AXS).[15)] [Reproduced with permission from N. Kamiya, *et al.*, *J. Appl. Cryst.*, **28**, 746, IUCr. (1995)]

b) Eucentric goniometer heads. These are designed so that the centers of the two axis adjustment arcs exactly coincide on the axial line of specimen rotation.

The fitting for attaching the goniometer head should be the same for cameras and specimen mounts of diffraction equipment to enable the specimen and the head to be transferred as a unit form one piece of equipment to another without alteration of the setting. The international standard is gradually gaining ground.

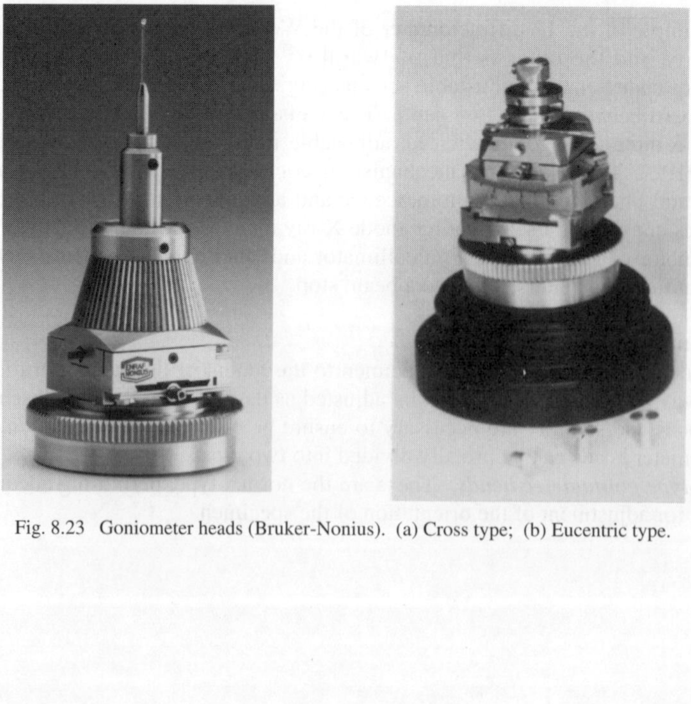

Fig. 8.23 Goniometer heads (Bruker-Nonius). (a) Cross type; (b) Eucentric type.

8.2.5 X-ray small-angle scattering cameras

This equipment is used for the measurement of diffraction intensities at small angles ($2\theta \leqq 5^\circ$) and/or sometimes at middle angles between the small and wide angles around the incident X-ray beam, normally by means of the transmission method.[17] Modifications of the cameras mentioned below equipped with IP, CCD, MWPC or other area detector systems can be easily used as small-angle diffractometers.

A. Collimating cameras

Fine parallel X-ray beam is produced by using an optical system consisting of small pinholes or narrow slits as apertures in this camera. The optical system is simple and easy to handle, but in order to obtain a small-angle scattering diagram with good resolution both strong monochromatic X-ray source and long exposure time are required. The geometry of the slits camera has been discussed by Bear and Bolduan.[18]

Figure 8.24(a) is a sketch of a simple small- and medium-angle scattering camera designed by Statton.[19] Specimen-to-film distance can be changed stepwise to suit the purpose of the experiment. In the other type of apparatus size and shape of aperture, collimator length, and specimen-to-film distance can be adjusted. The optical system can be placed under vacuum. This kind of apparatus is fitted with equipment allowing heating and swelling of the specimen, and is therefore very versatile. By the addition of a simple flat

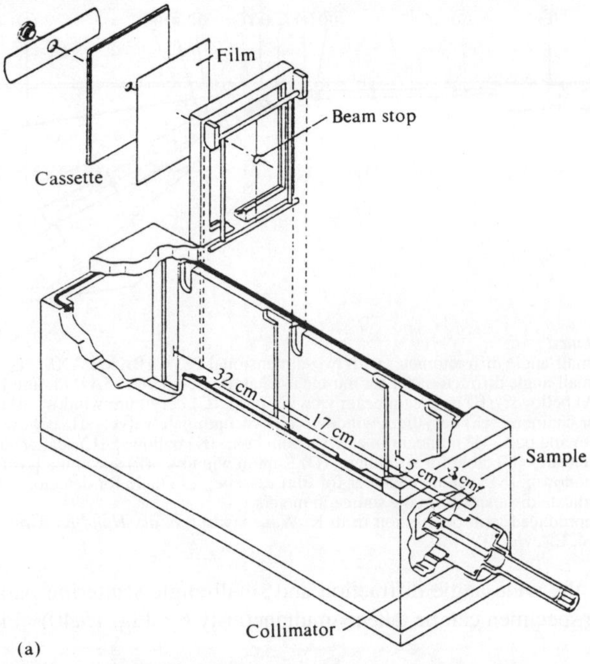

Fig. 8.24 X-ray small-angle scattering apparata.
(a) Small-angle scattering camera (W.O. Statton)[19].
[Reproduced with permission from L. E. Alexander, *X-Ray Diffraction Method in Polymer Science*, p.103, John Wiley & Sons, Inc. (1969)]

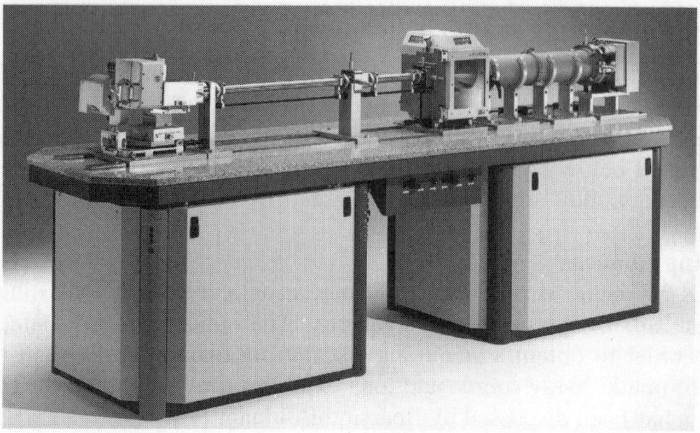

(b)

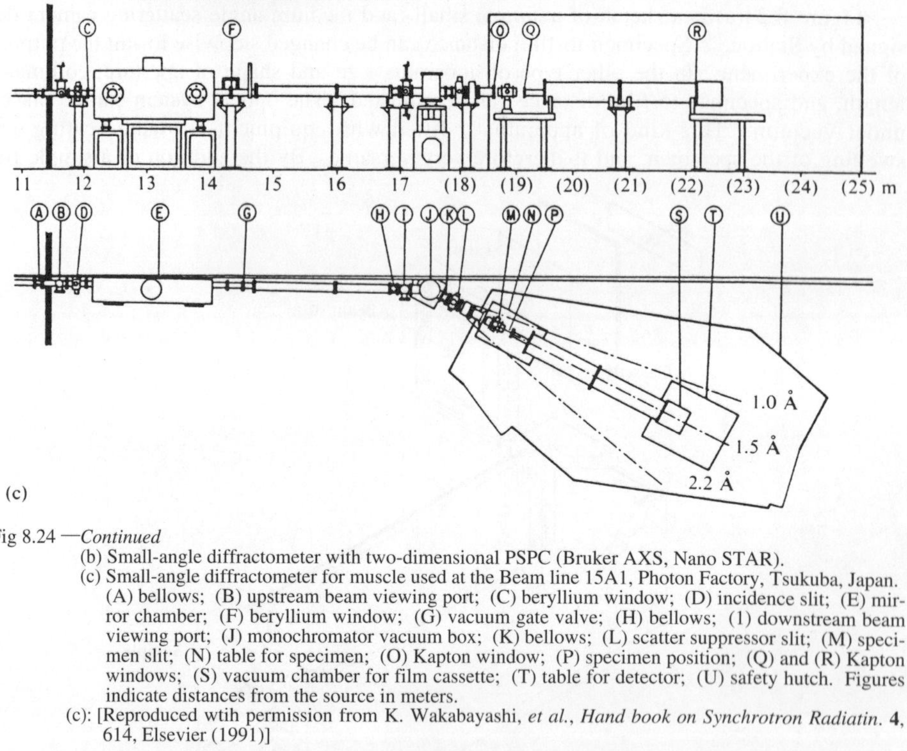

(c)

Fig 8.24 —*Continued*

(b) Small-angle diffractometer with two-dimensional PSPC (Bruker AXS, Nano STAR).

(c) Small-angle diffractometer for muscle used at the Beam line 15A1, Photon Factory, Tsukuba, Japan. (A) bellows; (B) upstream beam viewing port; (C) beryllium window; (D) incidence slit; (E) mirror chamber; (F) beryllium window; (G) vacuum gate valve; (H) bellows; (1) downstream beam viewing port; (J) monochromator vacuum box; (K) bellows; (L) scatter suppressor slit; (M) specimen slit; (N) table for specimen; (O) Kapton window; (P) specimen position; (Q) and (R) Kapton windows; (S) vacuum chamber for film cassette; (T) table for detector; (U) safety hutch. Figures indicate distances from the source in meters.

(c): [Reproduced wtih permission from K. Wakabayashi, *et al.*, *Hand book on Synchrotron Radiatin.* **4**, 614, Elsevier (1991)]

film cassette, the wide-angle diffraction and small-angle scattering patterns from the same portion of the specimen can be taken simultaneously (*cf.* Fig. 15.40). Fig. 8.24(b) shows an apparatus.

Gigantic long path small-angle cameras[20,21] are used at some laboratories. Kratky made a cone camera,[22] to overcome both the problems of difficult slit-length collimation correction in the weak tails of scattering curves obtained in long-slit geometry and the problems of the weak scattering in point geometry.

B. Block collimating cameras

Block slits, made of stainless steel or glass the surfaces of which are cut and finished with high precision, are used in the optical system of this camera (Fig. 8.25). The *small-angle scattering intensity at* ***one side*** *of the direct beam* is measured free from the parasitic scattering.

Figure 8.25 illustrates the Kratky U-slit camera.[23] A diaphragm D_3 lies across the gap between the two ends of the U-shaped body. The incident X-rays passing through the slit S cross the space between the edge E_1 of the U and the diaphragm D_3 to strike the sample, where they produce small-angle scattering. The top surface D_2, the bottom surface D_3, and the top surface of beam stop are located in the same plane, so that parasitic small-angle scattering from the diaphragm is avoided. The small-angle scattering is recorded on only one side of the slit; a very fine linear X-ray beam is necessary. Measurements are possible at scattering angles down to the order of 16″ (0.7×10^{-4} rad, corresponding to a Bragg spacing of about 2×10^4 Å) with a properly adjusted instrument having a high resolving power. An X-ray tube with a line focus of uniform intensity distribution is preferable as the source of X-rays. This type of camera (or diffractometer) is particularly recommended for small-angle scattering measurements on high polymer solutions.

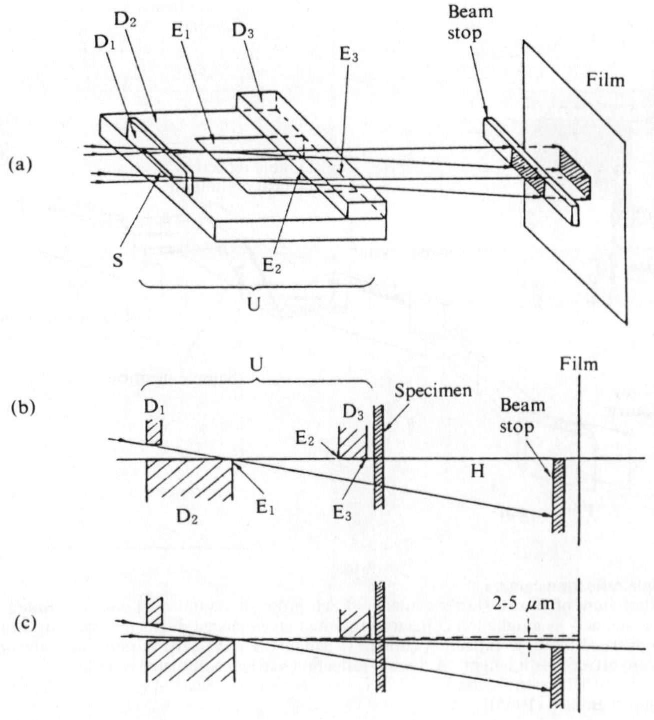

Fig. 8.25 Block collimating camera (Kratky U-slit camera).[23]
(a) General view; (b) Correct positioning of the diaphragm D_3. The lower edge of D_3 and the upper surface of D_2 are in the same plane; (c) A means of avoiding parasitic small-angle scattering. The lower edge of D_3 is chamfered.
[Reproduced with permission from H. Brunberger ed., *Small-Angle X-Ray Scattering*, pp.78, 79, Gordon & Breach (1965)]

C. Multiple reflection cameras (Bonse-Hart cameras)

Figure 8.26 shows the multiple reflection camera developed by Bonse and Hart.[24] This instrument offers very high angular resolution, 9 seconds of arc, with intense highly monochromatic radiation. With a pair of grooved crystals in place of the usual pair of plane crystals in a double crystal diffractometer, the angular resolution is obtained by using multiple (alternately *h* and *h*) reflections from nearly perfect crystals of germanium or silicon and at the same time the troublesome background of the rocking curve (the tails of the single crystal reflection curve) is eliminated. By using two double crystal arrangements with mutually orthogonal axes the effective slit height can be rendered negligible (Fig. 8.26(b)).

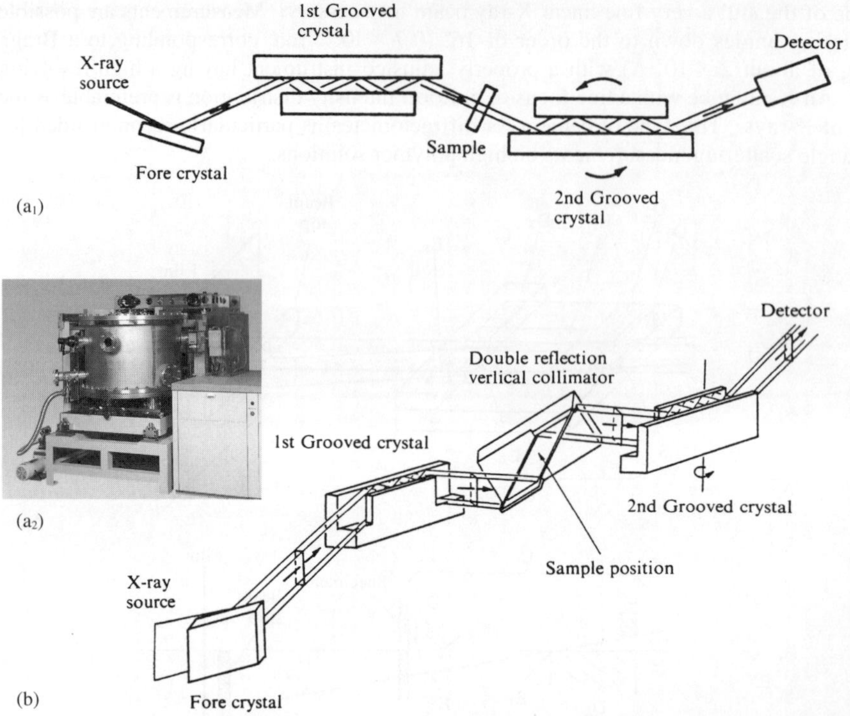

Fig. 8.26 Multiple reflection camera.[24]
(a_1) Plan view of Bonse-Hart's camera—Each grooved crystal contains five Bragg reflections. The fore crystal acts as a radiation collector providing an enhanced diffracted intensity from the source. (a_2) A commercial multiple reflection camera (Rigaku, USAXS). (b) A multiple-reflection diffractometer with zero effective slit height. A double-reflection vertical collimator is added.
[Reproduced with permission from U. H. Brumberget ed., *Small-Angle X-Ray Scattering*, pp.123,124, Gordon & Breach (1965)]

D. Focusing cameras

In the optical system of this type of camera a curved total reflection mirror, curved crystal monochromators (Tables 8.2 and 8.3) or a combination of curved total reflection mirror and curved monochromator are applied in order to focus the small-angle scattering on the surface of the detector. With this kind of camera strong small-angle scattering can be observed with better resolution than the collimating small-angle scattering cameras. However, the optical system of these cameras is much more complicated than that of the collimating cameras.

a) Curved total reflection mirror cameras (Franks cameras).[25] The point focusing camera using two curved total reflection mirrors with mutually orthogonal arrangement (Fig. 8.27(a)) or the line-focusing camera using a curved total reflection mirror is used for the studies on supra-structure of viruses, membranes, and muscles or studies on fiber structures of biological or synthetic high polymer substances. Small-angle resolution of this camera is about 1,000 Å. Fig. 8.27(b) shows a structure and the optical system of the latter type Franks camera. The curved mirror is located on a midpoint, c between the center of the line focus, F_1 in a X-ray tube and the center of the detector surface, F_2. In Fig. 8.27(b) $l = m$, where l and m are the distances between F_1 and c and c and F_2, respectively.

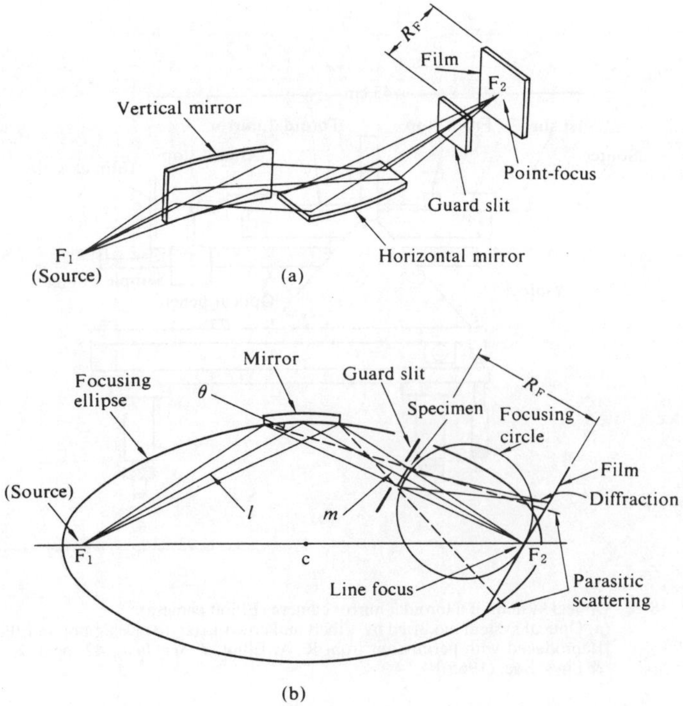

Fig. 8.27 Optical system of a curved total reflection camera (Franks camera).[25]
(a) Point-focusing camera; (b) Line-focusing camera
[Reproduced from A. Franks, *Proc. Phys. Soc.*, **68**, 1054, Phys. Soc. (1955)]

b) Toroidal mirror cameras (Elliott cameras).[26] Only a toroidal mirror is necessary to obtain point-focusing X-rays. This camera is useful for structural studies on fibrous macromolecules and membranes by using middle-angle and wide-angle scatterings of the specimen. The optical system and the structure of this camera is given in Fig. 8.28.

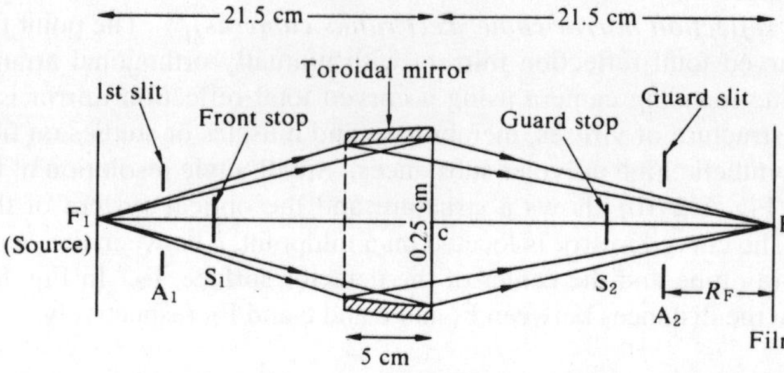

(a)

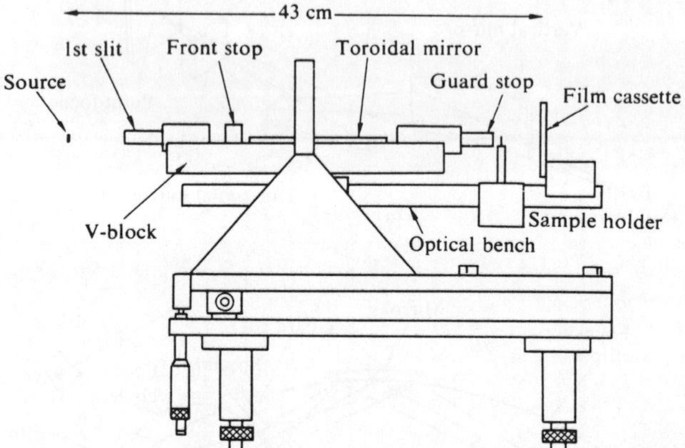

(b)

Fig. 8.28 Optical system of a toroidal mirror camera (Elliott camera).[26]
(a) Optical system modified by Vibert and coworkers; (b) Design of an Elliott camera [Reproduced with permission from R. A. Elliot, *J. Sci. Inst.*, **42**, pp.312, 314, Inst. Phys. & Phys. Soc. (1965)]

c) Cameras with curved total reflection mirror and curved monochromator. Point-focusing X-rays are obtained by using an optical system in which a curved total reflection mirror (Franks type mirror) is used in combination with a curved crystal monochromator (Fig. 8.29).[27]

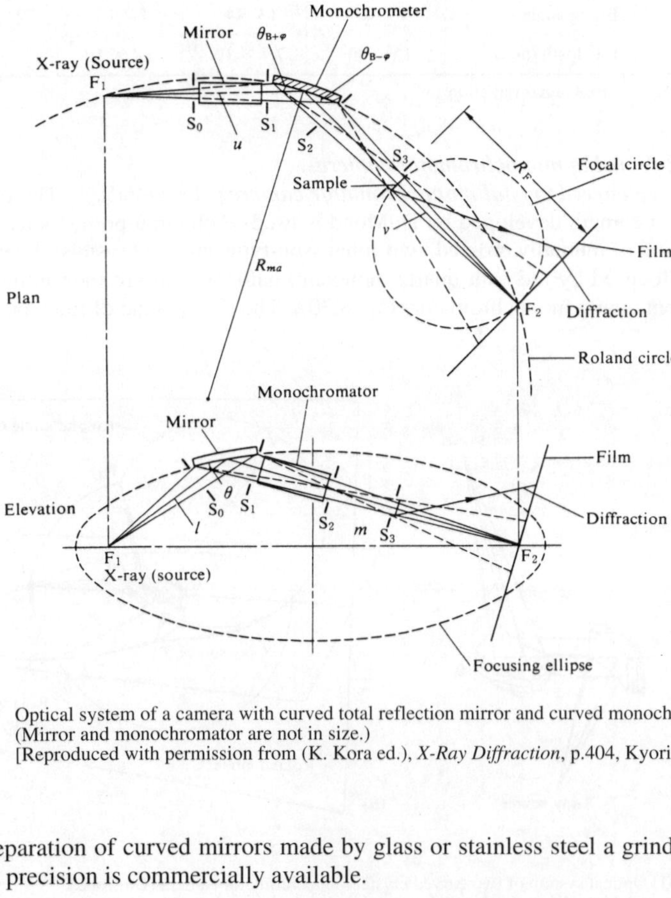

Fig. 8.29 Optical system of a camera with curved total reflection mirror and curved monochromator.[27] (Mirror and monochromator are not in size.)
[Reproduced with permission from (K. Kora ed.), *X-Ray Diffraction*, p.404, Kyoritsu (1988)]

For the preparation of curved mirrors made by glass or stainless steel a grinding apparatus with high precision is commercially available.

Table 8.2 Characteristics of total reflection mirrors (for the Cu$K\alpha$ radiation ($\lambda = 1.54$ Å))

	Glass	Nickel-deposited	Gold-deposited
Critical angle (θ_c)	11′–14′	23′	32′
Refractivity ($R(\theta_c)$)	0.73	0.65	0.45
$(R(\theta_c)(\mathrm{Cu}K\alpha))/(R((\theta_c)(\mathrm{Cu}K\alpha))$ at θ_c of $K\alpha$	0.048	0.021	0.13

[Reproduced with permission from J. Witz, *Acta Cryst.*, **A25**, 34, IUCr. (1969)]

Table 8.3 Crystals used for curved crystal monochromator

	Quartz	Ge	Si	LiF
Crystal system	hexagonal	cubic	cubic	cubic
Crystal face	$(10\bar{1})$	(111)	(111)	(111)
Bragg angle	13° 20′	13° 44′	14° 13′	22° 29′
Bandpath ($\omega_{cryst}{}^{\dagger}$)	1.6×10^{-5}	7.1×10^{-5}	3.4×10^{-5}	

† for Cu $K\alpha$ radiation.

d) Point-focusing monochromator cameras.

1) Two curved crystal monochromator cameras (DuMond).[28] The optical system of this type of camera developed by DuMond is used to obtain a point-focused X-ray using a combination of mutually crossed, two Johansson-type, curved crystals: divergent X-rays are first line-focused by the bent quartz monochromator, which are then point-focused by the second bent quartz monochromator (Fig. 8.30). The (310) plane of the quartz is used.

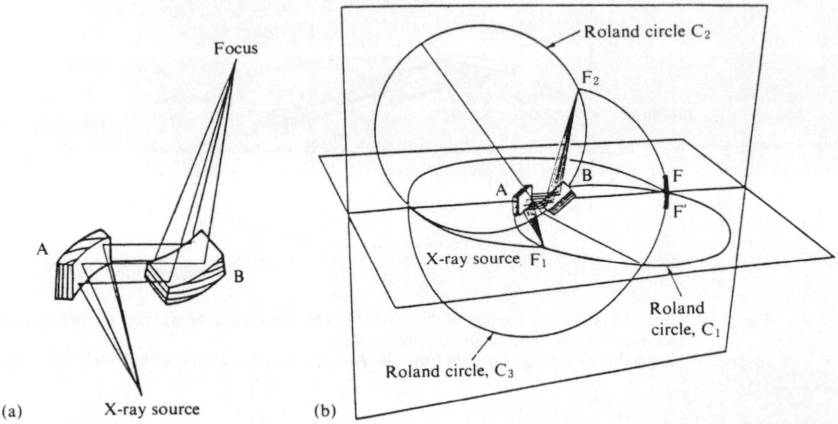

Fig. 8.30 Optical system of two curved crystal monochromator camera (DuMond).[28] (a) Mutually crossed, two Johausson-type curved crystals; (b) Optical system for point-focusing [Reproduced from J. W. M. DuMond, L. Sheufil, W. E. Danielson, *Rev. Sci. Instr.*, **21**, 188, Am. Inst. Phys. (1950)]

2) Double-bent crystal monochromator cameras (Beereman).[29] Beereman devised a method bringing the divergent X-rays to a point focus with a crystal having double curvatures at right angles to each other (Fig. 8.31(b)). The double-bent point-focusing LiF monochromator gives about 10 times stronger intensity, but the intensity distribution in the cross-section was reported not uniform. However, an improved technique applied to the preparation of this type of monochromator succeeded is giving uniform point-focused X-ray beam.

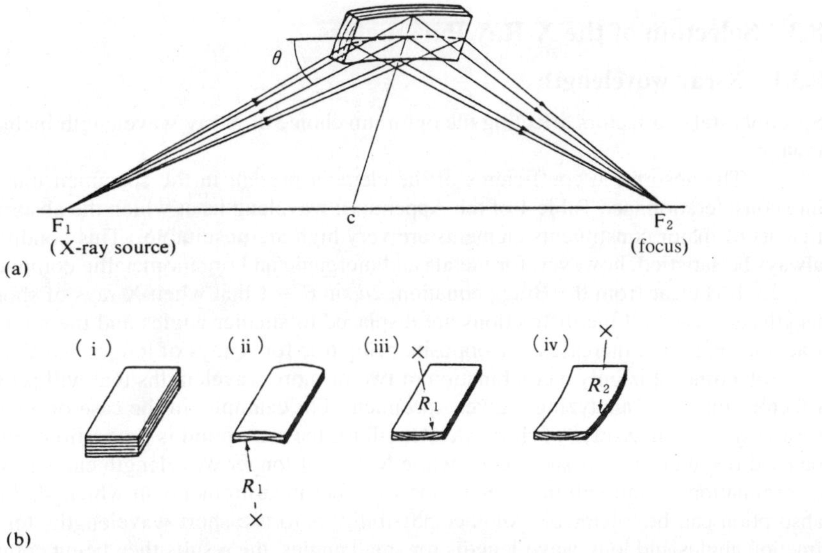

Fig. 8.31 Double-bent crystal monochromator camera (Beereman).[29,30]
(a) Optical system (θ: Bragg angle).; (b) Preparation of point-focusing double-bent monochromator (X, focus of curvature, R_1, R_2: Radii of curvature).
[Reproduced from D. W. Berreman, *et al.*, *Rev. Sci. Instr.*, **25**, 1219, Am. Inst. Phys. (1954)]

E. Energy-dispersive X-ray small-angle scattering system[31)]

An example of this type of system used at NINA in Daresbury Laboratory is shown in Fig. 8.32.

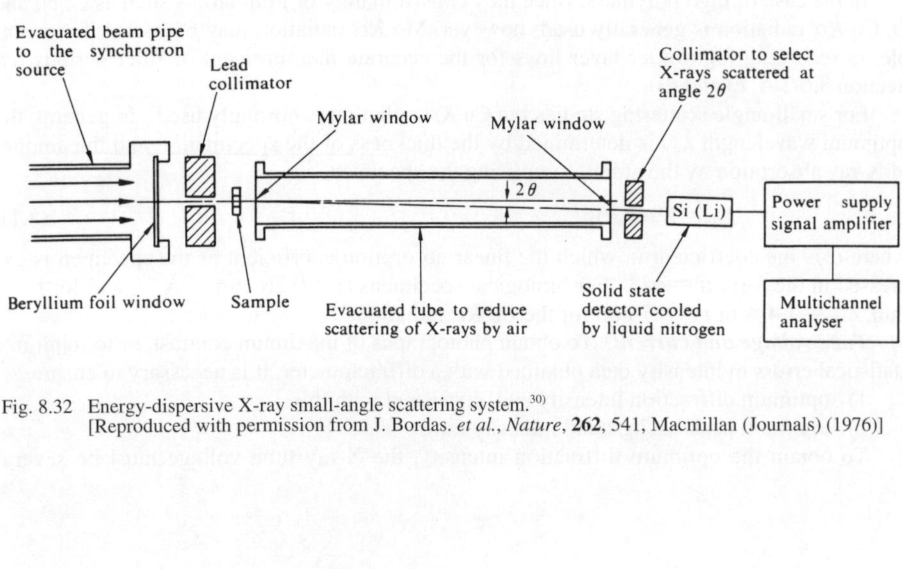

Fig. 8.32 Energy-dispersive X-ray small-angle scattering system.[30)]
[Reproduced with permission from J. Bordas. *et al.*, *Nature*, **262**, 541, Macmillan (Journals) (1976)]

8.3 Selection of the X-Ray Parameters

8.3.1 X-ray wavelength

Specimen-related factors affecting the optimum choice of X-ray wavelength include the following.

1. The absorption coefficients of the element present in the specimen must be taken into consideration (see Table 1 of the Appendix); wavelengths at which the absorption coefficients of major constituents elements are very high are unsuitable. This condition cannot always be satisfied, however, for metals and inorganic and organometallic compounds.

2. It is clear from the Bragg equation, $2d \sin \theta = \lambda$ that when X-rays of shorter wavelengths are used, all the diffractions are displaced to smaller angles and the number of diffractions measured increases, the opposite being true for X-rays of longer wavelength.

Sometimes it is only a combination of two or more wavelengths that will provide a satisfactory means of analyzing a given specimen. For example, in the case of a strongly absorbing specimen, X-rays of short wavelength for the absorption is proportionately less may be used for intensity measurements, while X-rays of longer wavelength can be used for the determination of unit cell parameters and for other measurements in which slightly greater absorption can be tolerated. Another possibility is to use short wavelengths for large diffraction angles and long wave-lengths for small angles, the results then being combined.

A. X-ray tubes

a) Anode materials. Cr, Fe, Co, Cu, Mo, Ni, and Ag anode (anticathode) tubes are commercially available. Among these, Cu and Mo tubes are widely used, the most common being Cu. With rays of longer wavelength, such as Cr $K\alpha$ and Fe $K\alpha$, absorption and scattering by air are significant. In the case of Mo $K\alpha$ radiation, the characteristic K line may be superimposed on the maximum of the continuous X-rays, thus rendering the measurement of intensity more difficult. Moreover, the low sensitivity of the film to this radiation entails long exposures. Another disadvantage with Mo $K\alpha$ and radiations of shorter wavelength is the danger of stray emission of X-rays from points other than the slit aperture of the X-ray tube.

In the case of high polymers, since they consist mainly of light atoms such as C, N, and O, Cu $K\alpha$ radiation is generally used; however, Mo $K\alpha$ radiation may be useful, for example, in recording high order layer lines for the accurate measurement of fiber periods (*cf.* Section 8.6.2 A, c) and d)).

For small-angle scattering studies the Cu $K\alpha$ radiation is normally used. In general, the optimum wavelength λ_{opt} is determined by the thickness of the specimen, t, and the amount of X-ray absorption by the atoms comprising the specimen.[32)]

$$\lambda_{\mathrm{opt}} = [2/(3\alpha t)]^{1/3} \tag{8.1}$$

where α is the coefficient in which the linear absorption coefficient of the specimen is expressed in the form $\mu = \alpha\lambda^3$. For biological specimens $\alpha = 0.26\ \mathrm{mm}^{-1}\ \mathrm{Å}^{-3}$, and for $t = 1$ mm, $\lambda_{\mathrm{opt}} \doteqdot 1.4$ Å or $t = 0.7$ mm for the Cu $K\alpha$ radiation.[33)]

b) Tube voltage and current. To obtain photographs of maximum contrast, or to minimize statistical errors in intensity data obtained with a diffractometer, it is necessary to ensure

1) optimum diffraction intensity and, consistent with this,
2) optimum peak/background intensity ratio.

To obtain the optimum diffraction intensity, the X-ray tube voltage must be several times as large as the excitation voltage of the characteristic radiation (25–60 kV, *cf.* Table 1.2). As shown by Eq. 1.2, the X-ray intensity increases with increasing tube voltage, but a limit is imposed by the fact that the intensity of the continuous X-rays also increases (*cf.* Section 8.2.1). In order to satisfy condition 2), background scattering must be reduced to a minimum (*cf.* Section 8.3.3 below). In order to minimize the exposure time reasonably the X-ray tube must be operated with an economically maximum current.

B. Synchrotron radiation

By using the monochromator the most desirable wavelength may be selected. It is easy to select the suitable wavelengths for anomalous dispersion studies or others.

8.3.2 Production of monochromatic X-rays[34]

A. β-Filters

a) Elimination of $K\beta$ radiation by filtration. Table 1.3 shows the filters and their thickness for commonly used X-rays. The filter is normally a metal foil. Where a foil cannot be made, filters can be obtained by compressing the element to form a disc, spraying a solution on filter paper, or dispersing a powder on a polymer film. Even the use of a $K\beta$ filter, of course, does not ensure the complete elimination of $K\beta$ radiation, which will play a part in diffraction at very high intensities. A considerable fraction of the continuous X-radiation will also pass the filter.

b) Use of $K\beta$ filters in conjunction with pulse-height analyzers. In measurements using scintillation or proportional counters, the X-ray can be rendered more nearly monochromatic by the use of a pulse-height analyzer to eliminate the pulses due to continuous X-rays passed by the $K\beta$ filter (Fig. 8.33).[35]

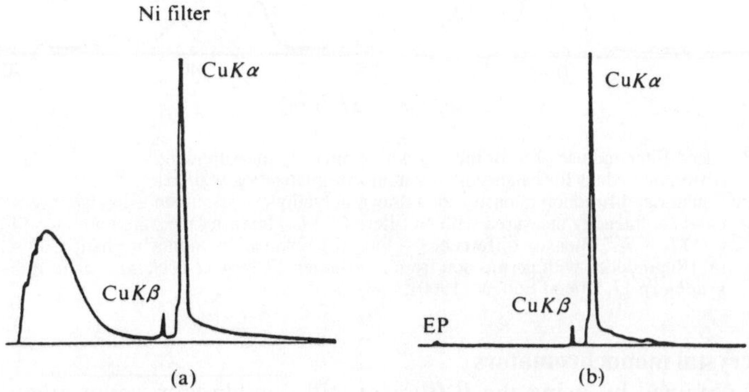

Fig. 8.33 Effect of pulse-height analyzer.[35]
(a) Residual continuous radiation with use of $K\beta$ filter alone. (b) With addition of pulse-height analyzer (EP is the escape peak registered by the proportional counter; *cf.* Table 8.1). [Reproduced with permission from *International Tables for X-Ray Crystallography*, (C.H. McGillavry, G.D. Rieck eds.) Vol. **III**. p.151, IUCr. (1962)]

c) Balanced filters.[36] This method is mainly used in measurements with counters (Fig. 8.34). The standard balanced filters are listed in *International Tables for X-Ray Crystallography,* Vol. III.[37] Using a respective β-filter of the balanced pair *the measurements must be carried out two times under equal conditions.* An example of a diffraction intensity measurement with balanced filters is shown in Fig. 8.34

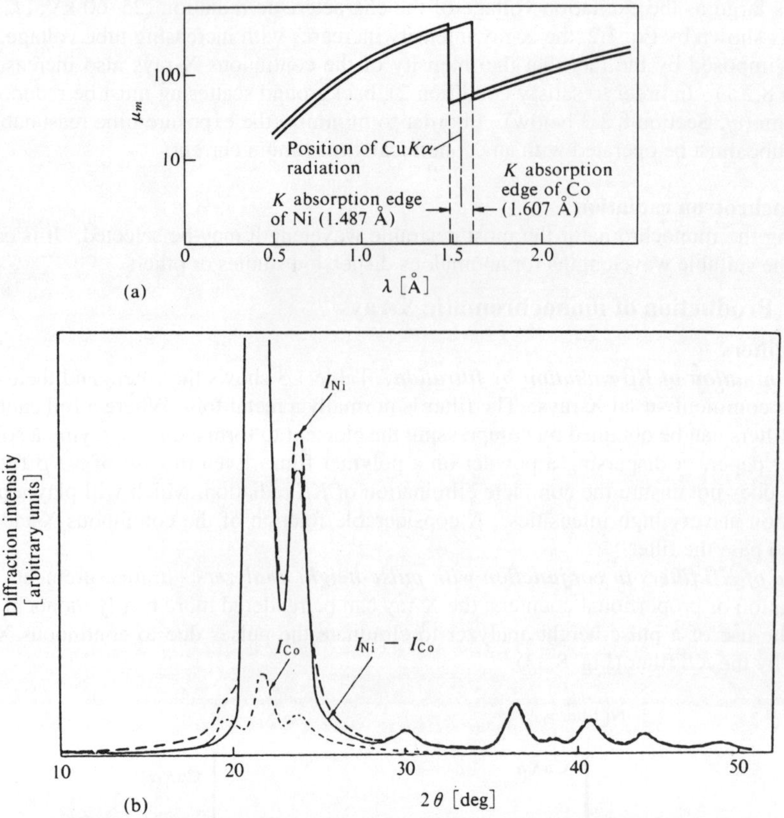

Fig. 8.34 Balanced-filter and use of it for the diffraction intensity measurement.
(a) Absorption edges for balanced-filter monochromatization (Cu$K\alpha$).
(b) Equatorial diffraction intensity for a drawn polyethylene specimen using balanced-filters.
(----) I_{Ni}: Intensity measured with Ni filter; (....) I_{Co}: Intensity measured with Co filter;
(—) $I_{Ni} - I_{Co}$: Intensity difference (= Intensity obtained by monochromatic Cu$K\alpha$ radiation).
(a):[Reproduced with permission from A. Guinier, *Théorie et Technique de la Radiocristallographie*, p.17, Dunod Editeur (1960)]

B. Single crystal monochromators

The X-rays obtained by using the β-filter(s) still contain some continuous radiation. Almost completely monochromatic X-rays can be obtained if use is made of rays diffracted by a crystal. The crystals which find use as monochromators include graphite, LiF, pentaerythritol, NaCl, MgO, calcite, and quartz. LiF is one of the most efficient crystals for Cu $K\alpha$ radiation, its reflective power amounting to some 30%. Graphite monochromator is used in most experiments. For the selection and monochromatization of X-rays from synchrotron radiation a silicon monochromator is currently used.

a) Flat crystal monochromators. These make use of the fact that intensity can be increased by the use of a crystal cut in such a way that the reflection is asymmetrical with respect to the surface of the crystal (Fankuchen cut, see Fig. 8.35(a_2). A transmission-type monochromator is also applicable (Fig. 8.35(b_2)).

b) Curved crystal monochromators. The use of a suitably curved crystal surface ensures not only that the X-rays from the tube or from the synchrotron radiation can be monochromatized, but also brought to a line (Figs. 8.15, 8.35(b_1), b_2)) or point (Figs. 8.29, 8.30, 8.31) focus after a certain distance, allowing very efficient measurement. Even when a monochromator is used, harmonics ($\lambda/2$, $\lambda/3$,...) may be present at high tube voltages, but these are generally of negligible intensity.

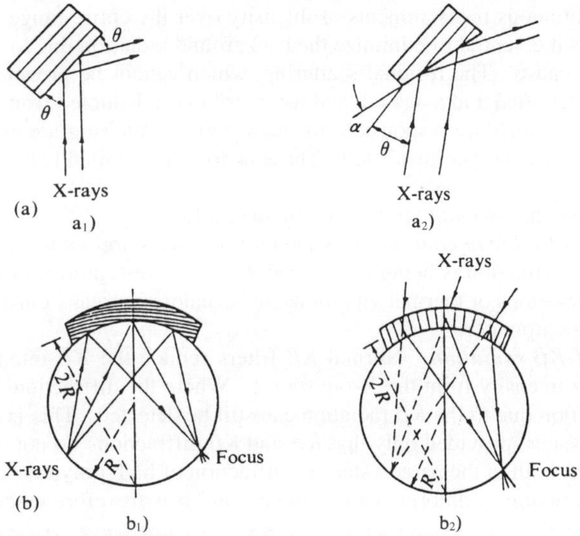

Fig. 8.35 Single crystal monochromators (θ: Bragg angle).
(a) Flat crystal monochromators
a_1) Flat crystal; a_2) Fankuchen-cut crystal (cut angle α)
(b) Curved-crystal monochromators
b_1) Johanson type; DuMond-Kirkpatrick type; b_2) DuMond-Kirkpatrick type
R: Radius of focal circle; $2R$: radius of curvature of the crystal planes. The curved-crystal surface is cut to fit the focal circle. In some cases an uncut crystal is used, but this does not give a perfect point focus.
[Reproduced with permission from *International Tables for X-Ray Crystallography*, (C. H. McGillavry, G. D. Rieck eds.), Vol. **III**, p. 81, IUCr. (1962)]

C. Multilayer optics*

Development of multilayer deposition technology has made possible to produce a thin film coating consisting of alternating layers of high atomic number material (*e.g.* W, Mo or Ni) and low atomic number material (*e.g.* Si, C or B_4C).

Multilayer diffracts X-rays in a fashion analogous to the Bragg reflection by the atomic planes of a crystal. For Cu$K\alpha$ radiation the reflectivity is 70 % and the half width of the rocking curve is 0.055° (*cf.* LiF (200): 18 %, 0.05° or pyro-graphite (002): 39%, 0.53°).

Multilayer can be bent along a geometrical curve (*e.g.* elliptical or parabolic), which made possible to produce either focusing or collimating multilayer optics. These optics

* Max-Flux Optics, Marketed by all major X-ray instrument manufacturers under a variety of names such as Gutman Optics, Arndt Optic, Prefix Optics, Göbel Mirrors.
Ref. Catalogue issued by Osmic Inc., or J. Harada, *J. Cryst. Soc. Jpn*, **45**, 306 (2003).

transform a divergent polychromatic X-ray beam into either a focused or parallel *monochromatic* X-ray beam.

High quality optics require precise *d*-spacing profiles, whether uniform or graded, on curved or flat optical surfaces. Side-by-side connected confocul optics solves problems inherent in both traditional and crosse-coupled optical scheme.

8.3.3 Elimination of unwanted scattered X-rays

The X-ray diffraction analyses of high polymers, unlike those of single-crystal structures, often require continuous measurements of intensity over the entire range of diffraction angles. This makes it essential to minimize the background radiation due to unwanted scattering whatever its cause. The residual scattering, which cannot be prevented despite every effort, must be identified and assessed, and its contribution deducted from the measured results. The factors which are responsible for background scattering are associated with the X-ray source, or with the specimen itself. These factors are detailed below.

A. Use of X-rays that are not strictly monochromatic

a) Halos or streaks due to continuous X-rays. The effects are not too serious where only sharp crystalline diffraction is being considered, but for investigations of diffuse scattering due to crystal distortions or thermal vibrations, these halos or streaks must be eliminated by the use of a monochromator.

b) Presence of $K\beta$ radiation. Normal $K\beta$ filters reduce the $K\beta$ intensity to less than 1/ 100 of the $K\alpha$ intensity from the X-ray source. Where the diffractions are very intense, however, diffraction due to the $K\beta$ radiation can still be detected. This is of no importance in the case of crystals, provided only that $K\alpha$ and $K\beta$ diffractions do not overlap. It is significant, however, both in the paracrystalline diffraction of high polymers and in the continuous diffuse diffraction of amorphous specimens, and it is therefore essential to minimize the $K\beta$ radiation.

B. Scattering due to contamination of the target or window of the X-ray tube

During the life of an X-ray tube, its output radiation contains increasing proportions of the characteristic radiation from Fe, Ni, Co, W, among others, with a corresponding reduction in the intensity of the original characteristic radiation of the tube. These elements evolve from the filament and its stem, etc. Since the presence of such extraneous radiation interferes with the recognition of desired diffractions and with accurate intensity measurements, the spectrum of the X-rays from the tube must be checked from time to time.

"Clean" X-rays can be obtained again from rotating-anode tubes by dismantling and polishing the anode and renewing the windows.

C. Scattering by the slits, screens, and other parts of the apparatus and by air

The diffraction pattern as recorded can be deleteriously affected by scattering between the window of the tube and the camera and by scattering of X-rays from adjacent windows in front of the camera. This radiation reaches the film through chinks in the camera. Although careful attention is given to this point in the manufacture of most cameras, care is still necessary in certain methods which are particularly vulnerable to the fogging caused by this type of unwanted scattering. Scattering due to air itself can be reduced to a tolerably low level by a beam stop of suitable shape and size placed close to the specimen, and by placing the camera under vacuum or by filling it with hydrogen or helium.

D. Scattering by materials used to seal or cover the specimen and by adhesives and mounting materials

Scattering may be caused by materials used to seal or cover specimens that are unstable in air, or by adhesives or mounting substances. A certain minimum amount of these substances is essential to the preparation of a specimen, and to that extent the scattering is unavoidable. The absence–or if present, the positions–of diffuse halos or Debye-Scherrer rings produced in this way must be ascertained beforehand, and any contributions subsequently deducted from the results.[38)] Diffraction due to NaC1 or other inorganic salts mixed with the material and, in low temperature experiments, due to the presence of ice adhering to the specimen or the perimeter of the slit, may also contribute to unwanted scattering.

E. Fluorescent X-rays from various elements in the specimen

This can be reduced by selection of X-rays of suitable wavelength with respect to the absorption coefficient of the specimen. Fluorescent X-rays from the specimen can sometimes also be blocked by a suitable filter. In the investigation of a specimen containing iron with Cu $K\alpha$ radiation, for example, an aluminum foil covering the film is used in addition to the $K\beta$ filter. This is because aluminum has a higher absorption coefficient for Fe $K\alpha$ than for Cu $K\beta$ radiation. Background scattering is also caused by recoil electrons and photoelectrons resulting from absorption by the specimen.

F. Natural radiation counts

Cosmic rays and radiation due to radioactive substances must be taken into account, particularly in measurements with GM counter tubes. With proportional counters, background of this type is reduced to about 1/10 of that found with GM counters.

G. Noise due to nearby electrical equipment

This applies only to measurements with counter tubes. The pulse due to the opening or closing of a switch can have a considerable effect. Measurements taken with a rate meter are not too badly affected, but care is necessary when a scaler is used.

8.4 The Specimen

8.4.1 Preparation of the specimen

An adequate knowledge of the properties of the specimen is obviously essential for the formulation of satisfactory experimental procedures. Desirable information would include the name of the substance and its chemical or structural formula (which affect its chemical stability, X-ray absorption, fluorescence, etc.). Physical constants including the melting point, boiling point, and various phase transition points are relevant. They affect the magnitude of the thermal vibrations at the measuring temperature, recrystallization, molding, and the stability of the specimen. The solubility of the substance, also, naturally has a marked effect on the conditions for recrystallization, molding, etc. Decisions on the desirability of sealing the specimen in a capillary, the changing of the specimen during the investigation, and the necessity of performing the investigation at low temperatures will obviously depend upon any tendency to sublime, and upon hygroscopicity, etc. Handling of the specimen must also be affected by the degree of toxicity.

These are therefore the most pressing reasons for a thorough examination of the substance prior to the investigation. The methods of preparing the specimen once this preliminary information has been obtained will now be described.

A. Single crystal specimens

Experience has shown that the most critical property governing the choice of single crystal specimens for analysis is the linear absorption coefficient μ (*cf.* Section 1.3.2A) of substance for the wavelength of X-rays used. For a cylindrical specimen, the optimum radius R is such that $\mu R = 1.0$.[39] This may also be taken as a suitable criterion for needle- or prism-like specimens with cross sections which are not circular. As a rough measure for an organic crystal, the best is that with approximate dimensions of $0.2 \times 0.2 \times 0.2$ mm and a needle crystal with approximate sectional dimensions of 0.2×0.2 mm. The wealth of information available on single crystal specimens in other publications renders any further discussion here unnecessary.[40]

If the crystal is unstable in air it should be properly coated by collodion or other film or sealed in a thin-walled glass capillary tube.

For macromolecular single crystal specimens see Chapter 12.

B. Polycrystalline specimens

In view of the very wide variety of polycrystalline specimens, rather than attempting to discuss those methods which are suitable for all such specimens under this heading,[41] the various methods which apply generally to high polymer specimens are entered under the appropriate headings below.

C. Solid high polymer specimens

a) ***Bulk specimens***. As a result of chemical or mechanical treatment during the preparation of the sample, the fine structure and fine texture of the surface often differ from those of the interior. Therefore the specimen should be taken from the interior of the sample, avoiding the superficial regions.

For measurements by the reflection method with an X-ray powder diffractometer, a flat specimen with a large enough surface to receive the whole of the X-ray beam from the vertical collimator slit should be used. If the thickness of the specimen in this case is $3/\mu$ cm or more (typically 2.4–3.0 mm or more for high polymers), not only does correction for absorption become unnecessary, but the intensity of the diffracted X-rays is increased.

b) ***Fiber specimens***. Since the diameter is usually very small, many filaments are held in a parallel bundle by means of holders such as those shown in Fig. 8.36. If twisting is present, the filaments are held so as to minimize it.

Single filaments, say 0.4–0.5 mm thick or less, may be examined individually, although thicknesses of this order are by no means ideal, for special purposes such as very precise measurement of preferred orientation or studies with microcameras.

Figure 8.36 shows some examples of specimen holders. The provision of notches at the points where the specimen is grasped, as in Fig. 8.36(b), has the advantage that distortion of the bundle of filaments is avoided. The device in (c) allows the specimen to be placed under slight tension to eliminate any slack, and also permits the sample to be stretched to some extent should this be advisable. A simple arrangement consisting of two scraps of X-ray film or card with their centers cut out and with the two ends of the specimen fixed between them (d) may also be improvised.

c) ***Film specimens***. The intensity of diffracted X-rays from a single thickness of film is too low unless the film is at least 0.5 mm thick. Exposures can be shortened by the use of several layers of film to give a total thickness of about 0.8–1.0 mm. If the layers of specimen are cut from one large film, they must be cut from areas that are not likely to possess special characteristics as a result of the method of manufacture or subsequent treatment.

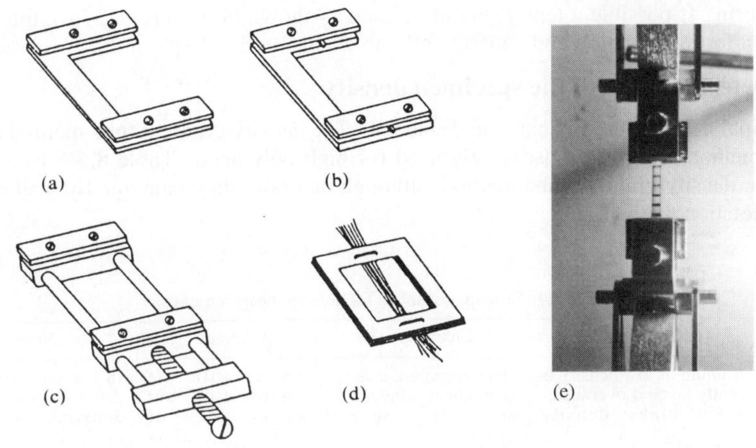

Fig. 8.36 Fiber-specimen holders.

The cutout layers should be placed together in such a way that their directions with respect to the original film are the same (Fig. 8.37). The layers are cemented together at both ends or fixed in a specimen holder. Specimens cut as shown in Fig. 8.37(a) are too long for X-ray measurements in the *Z* direction, and specimens cut as shown in (b) should be used for this purpose.

In drawn specimens, simple stretching may give local differences in elongation and the cross section of the film may no longer be uniform (Fig. 8.37(c)). A fairly wide piece of film is therefore drawn, and the specimen is cut from the central part, where the thickness is

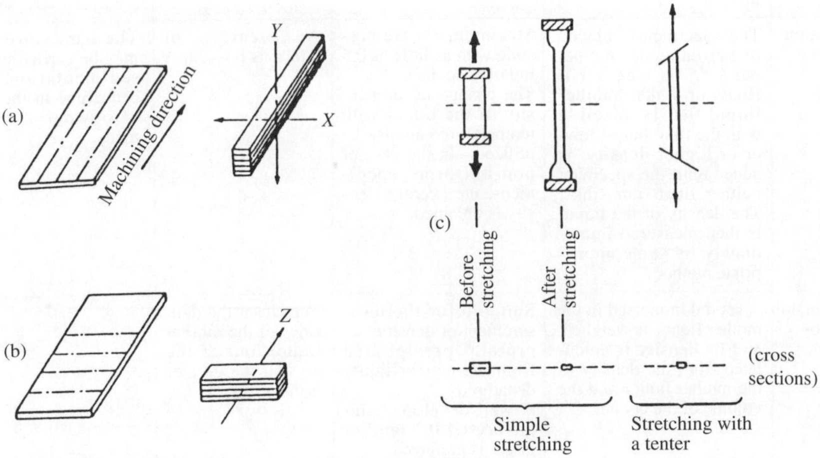

Fig. 8.37 Specimen preparation from film samples.
(a), (b) Strips cut from the sample are laid on top of one another to give the required thickness, with directions in which the strips were machined all parallel.
X: Direction normal to the machining and parallel to the film plane; *Y*: Normal to both the direction of machining and the film plane; *Z*: Machining direction.
[Reproduced from W. O. Statton, G. M. Goddard, *J. Appl. Phys.*, **28**, 1112. Am. Inst. Phys. (1957)]
(c) Stretching a film sample.

more uniform. If possible a tenter should be used as shown in the Fig. 8.37(c); this should ensure a uniform elongation and consequently no change in the shape of the cross section.

8.4.2 Determination of the specimen density

The principal methods are indicated in Table 8.4. The density gradient tube method and the flotation method are those most widely used for high polymers. Table 8.5[43] lists liquids used in the density gradient tube method, although naturally they can equally well be used with the flotation method.

Table 8.4 Principal methods for density measurement

Method	Principle	Characteristics	Accuracy	Notes
Density gradient tube[42a]	A liquid of low density is gently layered over a liquid of higher density with which it is readily miscible, and a density gradient is produced by diffusion. The specimen is introduced into the density gradient tube, and the position at which it comes to rest is noted. Its density can then be found from the density gradient. The density gradient is calibrated by introduction of floats or droplets of an immiscible liquid of known density into the tube.	Once prepared, a density gradient tube can be used for several months. Measurement can be carried out any temperature between the freezing point and the boiling point of the liquids by placing the density gradient tube in a thermostatically controlled bath at the desired temperature. In the case of a non-uniform specimen. the density distribution can be found.	Varies with the liquids used, but an accuracy of $\pm$ 0.002 g/ml is generally quite feasible.	In the case of powdered specimens. the density gradient method may be carried out in a centrifuge (though the gradient is practically unchanged, correction is necessary). Has the disadvantage that there is no satisfactory technique for removing the specimen used.
Flotation method	The specimen is placed in a given liquid, and depending on whether it floats or sinks, another liquid that is miscible with the first, but of lower or higher density, is added until the specimen neither floats nor sinks. The density of the liquid is then measured immediately by some appropriate method.	Measurements are possible with as little as 0.5 mg of specimen. The change in the density of the liquid with temperature can also be utilized. In the case of non-uniform specimens, the average density is obtained.	An accuracy of 0.02% is possible.	The measurement may be carried out by sedimentation in a centrifuge in the case of powdered specimens.
Immersion micro-balance[42b]	A crystal immersed in the mother liquor is weighed, and its density is calculated from the density of the mother liquor and the volume of the crystal.	Suitable for the measurement of densities of protein precipitated from the mother liquor denatured. A well developed single crystal 0.1 mm^3 or larger is preferred.	The closer the densities of the mother liquor and of the crystal, the greater is the accuracy.	
Volume-nometry	The difference in the gas pressures in a vessel when the specimen is present and when it is absent is measured.	A gas that is inert towards the specimen is used.	Down to 0.1%	This is the only method available if the specimen dissolves in all suitable liquids.

Table 8.5 Liquids for use in the density gradient tube method† [43)]

Lipophilic liquids			Aqueous solutions (with max. density of concentrated solutions)	
Liquid	Approximate density at 25°C [g/ml]	$\alpha^{\dagger}$	Solute	Approximate density at 25°C [g/ml]
Iso-octane	0.69	− 0.8	Water	0.99823 (20°C)
Kerosene	0.79	− 0.5, − 0.8	Sodium chloride	1.20
Toluene	0.867		Potassium tartrate	1.40
m-Xylene	0.86	− 0.85	Potassium iodide	1.63
Benzene	0.8799		Ferric sulphate	1.80
	(10°C)		Zinc bromide	2.00
Chlorobenzene	1.10	− 1.1	Zinc iodide	2.39
Bromobenzene	1.49	− 1.3	Thallium formate	3.5
Carbon tetrachloride	1.60	− 1.9	Toulette solution	4.3
Methyl iodide	2.28	− 2.7	(prepared by adding	
Methylene bromide	2.496		aqueous thallium	
Bromoform	2.89	− 0.9	malonate solution to	
s-Tetrabromomethane	2.96	− 2.2	aqueous thallium	
Methylene iodide	3.32	− 2.6	formate solution)	

† The density at t°C is found from the density at 25°C with the aid of the formula $d_t = d_{25} + 10^{-3} \cdot \alpha(t - 25)$.
[Reproduced with permission from *IUCr, International Tables for X-Ray Crystallography*, (C. H. McGillavry, G. D. Rieck eds.), Vol. **III**, p.19, IUCr. (1962)]

8.5 Diffraction Studies for Identification Purposes

8.5.1 Qualitative identification using polycrystal diffraction data (unoriented X-ray diagrams)

Qualitative identification using polycrystalline diffraction depends ultimately upon measurement of the precise locations and intensities of the diffraction lines. The interplanar spacings corresponding to the diffracted X-rays are found first, and the intensities are then measured. Either a Debye or a Guinier camera with a large radius is suitable for the photographic method. Where considerations of speed and accuracy of measurement of diffraction lines are paramount, X-ray powder diffractometers especially those equiped with curved PSPC give excellent results. (*cf.* 8.2.4.A) Graphite monochromatized or Ni-filtered Cu $K\alpha$ radiation is used. (*cf.* Section 8.3.1).

The following special points concerning preparation of the specimen should be noted (*cf.* Section 8.5.2). It is essential that the specimen be crushed to form uniform particles no larger than, say, 10 micrometers. If the particles are plate- or rod-like in form, external force in inserting the particles into the sample holder, etc. may produce a preferred orientation of the individual particles: care must therefore be taken to avoid the undue application of force in the preparation of the specimen. Mixing a small amount of isotropic material (like MgO) usually reduces the orientation effect, and if an internal standard is desirable, the proper choice of standard can serve both purposes.[44)]

8.5.2 Treatment of the results

The positions of the diffraction lines on a photographic film can be located on a scale divided into tenths of a millimeter. A simple comparator is generally used. A higher degree of accuracy is possible if a reference substance with precisely known interplanar spacings is mixed with the specimen as an internal standard.

When the diffractions are very intense, the residual $K\beta$ radiation due to incomplete filtering produces weak diffractions with smaller diffraction angles. Care is also necessary in interpreting results at large diffraction angles, for separation between the $K\alpha_1$ and $K\alpha_2$ diffractions can occur.

The Bragg equation $2d \sin \theta = \lambda$ is used to fmd the interplanar spacing from the measured angle θ. Useful tables[45,46)] are available for the ready determination of d. The relative intensity, I/I_1 (usually expressed as a percentage of the highest intensity observed in the diffraction pattern I_1) is measured.

When a powder diffractometer is used, the angular position of the diffraction line is read off with the aid of an angular scale marked on the recording chart. Where accuracy is crucial, the diffraction diagram of a reference specimen is used for calibration. If the specimen has high crystallinity, the intensities are found by simply subtracting the height of the background from the heights of the diffraction peaks on the chart.

If the reader is using the latest model of powder diffractometer and if the powder data file, JCPDS-PDF (See Section 9.2.2) is loaded on the computer attached to the diffractometer, measurement of the diffraction leads directly to the identification of the specimen.

High polymers generally give a few broad diffraction lines, on which is superimposed the diffuse scattering due to the amorphous regions. They are not, therefore, in this respect, suitable for identification purposes. If the degree of orientation is fairly high, however, the whole of the X-ray diffraction pattern of the specimen is compared with that of the reference substance, and then the latter should, ideally, be as nearly identical to the specimen as possible in its mode and degree of orientation and its crystallinity. There is also the virtual impossibility of positively identifying polycrystalline proteins by the powder method, which means that the identification should be made on the basis of single-crystal diffraction patterns whenever possible.

8.6 Diffraction Studies for Crystal Structure Analysis

8.6.1 General remarks

Basically both the photographic and diffractometer methods, using IP, CCD, MWPC or else, involve the following two steps.

A. Determination of the unit cell parameters and the space group
In order to make these determinations it is necessary to locate precisely the positions of a large number of diffraction spots, and to obtain a semiquantitative estimate of their intensities.

B. Collection of intensity data
The principal object is the accurate measurement of the intensities of the diffraction spots, and it is necessary only to verify the indexing to the spots by determining their locations.

C. Apparatus to be used
a) Use of four-circle single crystal diffractometer. For single crystals of small molecules, it is recommended that an oscillation photograph be taken or, better, that the crystal chosen be examined to check whether it is good and stable for the diffraction studies. However, in general, this step is apt to be skipped. For the crystal mounted on a goniometer head attached to the goniometer of the diffractometer, rocking curves (profiles) of some diffractions are usually examined instead before the collection of intensity data (*cf.* Fig. 8.18).

If the crystal looks good and stable, collection of intensity data starts on the diffractometer, followed by background correction, *Lp* correction, indexing and estimation of measuring weight from the counting statistics for each diffraction, automatically leading to the establishment of the structure by using an appropriate program stored in the computer attached to the diffractometer.

b) Photographic method. The apparatus normally used is a Weissenberg camera, which can provide all the data necessary for the structure analysis. Precession cameras are also very effective in combination with Weissenberg cameras, and are widely used on their own for specimens with large unit cells, such as proteins. The use of single-crystal diffractometers has been increasing recently, and the accuracy of the measurements is being further improved by the use of computer-controlled diffractometers.

The X-rays are usually monochromatized by means of filters in the case of single-crystal specimens, since the diffraction spots are distinct. Cu $K\alpha$ radiation is employed most frequently, though Mo $K\alpha$ radiation is often used with precession cameras because of the narrow range of angles which is the most that can be photographed with these cameras. For further information affecting the choice of wavelength see Section 8.3.1.

For polymers, the cameras mentioned above in connection with single crystals may be used for highly oriented specimens. Measurements on specimens with cylindrically symmetrical (*e.g.* uniaxial or spiral) orientation are generally carried out with rotating-crystal cameras. In the case of fibrous polymers, highly oriented specimens should, if possible, be used for structure analyses. When specimen is in the form of very fine fibers, a microbeam camera used in combination with a microfocus X-ray tube gives good results.

We have already stressed (*cf.* Section 8.3.2) the desirability of monochromatic X-rays for use with high polymers in view of the diffuse amorphous scattering that always accompanies the crystalline diffraction, not to mention the broadening of the diffraction pattern due to lattice distortions,

8.6.2 Weissenberg photographs

A. Oscillation (or rotation) photographs

These are taken for single-crystal and fiber specimens in order to align a fixed crystal axis accurately with the axis of rotation of the camera. The film cassette of the Weissenberg camera is fixed (diameter of the film: say 57. 296 mm), and the specimen is made to oscillate through an angle of $\pm 5 \sim 10^\circ$ (for crystals of small molecules).

a) Adjustment of the axis. Since the axis of rotation is unlikely to coincide directly with the principal axis of the crystal at first, a photograph with distorted equator and layer lines similar to that shown in Fig. 8.38 will be obtained. Fig. 8.38(a) shows the result when the principal axis is inclined out of the plane of the paper, while (b) shows that for downward inclination of the crystal axis (the axis of rotation being horizontal). In practice, the result is more likely to be a combination of these two examples, as shown in Fig. 8.38(c). The misalignment of the principal axis is corrected by means of the two arcs of the goniometer head. Different workers use different methods of adjustment, but the following method is recommended.

For ease of adjustment of the axis, the oscillation photograph is taken with oscillation about a position in which the face of one of the two arcs of the goniometer head is practically parallel to the incident X-rays. On the far side of the film from the incident X-rays, and with the film spread out so that the axis of rotation of the crystal is on the right, the deviations a and b of the equator (zero layer line) from the normal position at distances of 45 mm ($2\theta = 90^\circ$) from the center (*cf.* Fig. 8.38(c)) are measured in millimeters. If the displacement components due to inclination out of the plane of the film and to downward inclination are A and B respectively, then

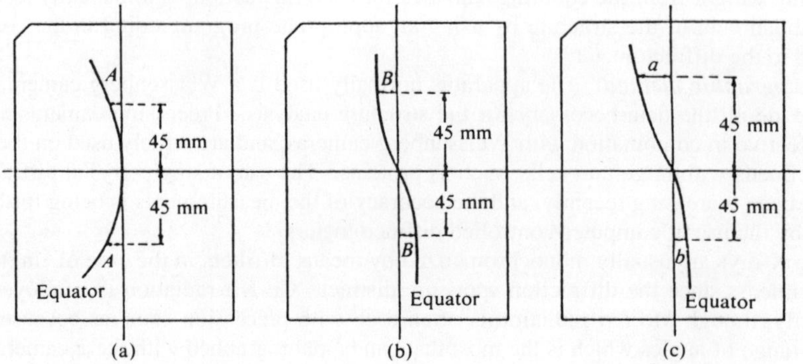

Fig. 8.38 Oscillation photographs showing the effect of misalignment of the principal axis. (a) Principal crystal axis inclined out of the plane of the paper. (b) Principal axis inclined downward. (c) Principal axis inclined both outward and down ward. The axis of rotation is horizontal, the cylindrical film vertical. X-rays rise out of the plane of the paper. Diameter of the film: 57.296 mm.

$$a = A + B, \qquad b = A - B$$

The angular adjustments of the arcs required to correct the inclination of the crystal axis are therefore

$$2A = a + b \quad \text{(in the horizontal plane)}$$

$$2B = a - b \quad \text{(in the vertical plane)}$$

The required adjustments of the arcs are numerically equal to $2A$ and $2B$ respectively.

This procedure is repeated until the inclination of the principal axis has been corrected. The crystal is then rotated through about 90°, and another oscillation photograph is made to verify that the principal axis of the crystal coincides exactly with the axis of rotation. Once one axis of the crystal has been accurately aligned, the identity period (axial length) for the axis that coincides with the axis of rotation can be found from oscillation- or rotation-crystal photographs (*cf.* Subsection A.*d*) below). We can then proceed directly to take a Weissenberg photograph of the equator.

b) Orientation of high polymers (*cf.* Section 4.4.4). When the high polymer specimen is oriented, the axis of rotation of the camera should either coincide with or be parallel to an orientation axis. However, a precise adjustment of the orientation cannot be made as it was in the case of single crystals. An alternative approach, not dealt with in this book, which gives satisfactory alignment, is to examine the specimen under a microscope between crossed nicols.

c) Higher order layer lines of high polymer specimens. In normal oscillation or rotation cameras, the height of the film cassette is sometimes insufficient to record the high-order layer lines. In such cases the Weissenberg camera has the advantage that the film cassette can be displaced to record the high-order layer lines on one side. The values of the identity periods (fiber periods) obtained from the lower-order layer lines must be used to calculate whether the required higher-order layer lines are capable of being recorded. For example, polyethylene has a fiber period of 2.534 Å. It is clear, therefore, that measurements beyond the first layer lines are inherently impossible with Cu $K\alpha$ radiation. In such cases, X-rays of shorter wavelength, *e.g.* Mo $K\alpha$, may be used. For the method of indexing *cf.* Section 4.4.1.

d) Measurement of identity period along the axis of rotation. The layer line separation z_n on the film from the equator (the zero layer line) to the nth layer line is measured, and the identity period can then be calculated from the film radius R_F and the wavelength λ of the X-rays using the following equations.

$$I \sin \mu_n = n\lambda,$$

$$\tan \mu_n = \frac{z_n}{R_F} \tag{8.2}$$

I is the identity period of the crystal parallel to the axis of rotation (usually the fiber period in the case of oriented high polymers) and μ_n is the angle of elevation of the nth layer line (see Fig. 8.39). The layer line separation z_n is measured as follows. The film is spread out flat and distance between corresponding layer lines on either side of the equator is measured. The measurements are taken from center to center of corresponding diffraction spots, although in some cases the distances from peak to peak and from trough to trough may be measured and averaged. In oscillating-crystal photographs in which the layer lines are situated asymmetrically with respect to the equatorial line, straight lines are drawn on the film through the layer lines, and the distances between these lines are measured. When a flat film has been used, an approximate value can be obtained with the aid of a Bernal chart (*cf.* Appendix, Fig. 2).

The following should be noted.

1) Precise alignment of the crystal is essential.
2) Layer lines of low intensity may be overlooked if exposures are short.
3) The value of the axial length found by this method should be regarded as tentative, for use only in the early stages of the analysis, and not to be used as the final value of the

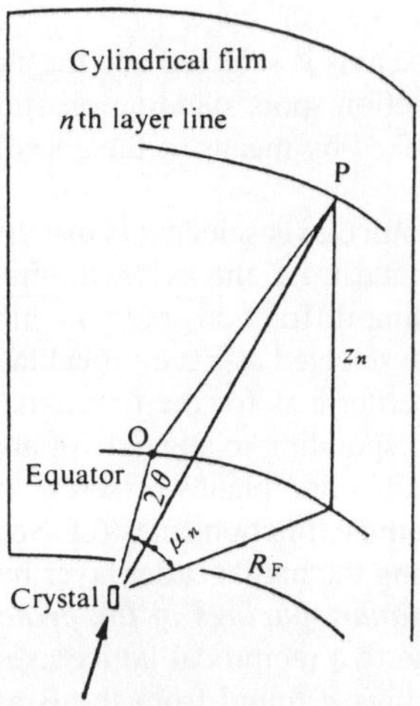

Fig. 8.39 Geometry for measurements of identity period along the axis of rotation (cylindrical film).

lattice constant. It will undoubtedly need correction in the light of accurate measurement of the corresponding unit cell parameter obtained, for example, from the higher order diffractions of an equatorial Weissenberg photograph (*cf.* Subsection B.*d*) below).

4) Tables are available for finding the identity period I from the layer-line separation z for Cu, Ni, Mo, and other X-rays, and for camera radii of 28.65, 30 and 45 mm.[47)]

B. Equatorial Weissenberg photographs

a) Screen test.

1) The layer line screen is used to select only the equatorial line, and the oscillation photograph of this line is recorded.

2) The position of the film cassette is altered, and at the same time the crystal is rotated through about 90°, and the oscillation photograph is again recorded.

3) It is ascertained whether the equator has been correctly selected by the screen or not.

b) The Weissenberg photograph. The Weissenberg photograph is made by means of simultaneous oscillation of the crystal and displacement of the film cassette. The crystal is made to oscillate through an angle of about 200°, and the length of the exposure is usually a matter of hours. Examples of Weissenberg photographs are shown in Fig. 8.40. If the various elements in the diffraction pattern shown by the Weissenberg photograph are unduly elongated, or if the intensity distribution is uneven, or if there are clusters of small diffraction spots, it can be assumed that twinning or other faults are present in the crystal. A new Weissenberg photograph should then be made with another crystal.

c) Indexing.

1) A print is made from the Weissenberg photographic film, or the positions of the various, diffraction spots are marked on tracing paper. The position of the incident X-rays on the center line of the film is also marked.

2) A Wooster chart (see Appendix, Fig. 1) is placed on the resulting diffraction distribution diagram. The chart is positioned so that the bottom edge passes through a row of diffraction spots lying on one reciprocal lattice axis, *i.e.* lying in a straight line.

3) A curve is drawn joining diffraction spots standing on similar curves in the Wooster chart.

4) Another reciprocal lattice axis is selected and the sloping edge of the Wooster chart is placed along this axis. Diffraction spots standing on similar curves are again joined up, and the diffraction spots are indexed by means of the coordinates of the intersections of the two curves (*cf.* Fig. 8.40(c)).

NOTE: (1) The first reciprocal lattice axis selected is usually that in which the distances between the colinear points are smallest, i.e. the axis with the largest interplanar spacings. If the intensities of the corresponding diffraction spots on either side of a given row of spots are symmetrical, this row may be selected as a reciprocal lattice axis.

(2) Following the same criteria as for the first axis, the second axis is taken as one about 45 mm from the first (corresponding to an angle of about 90° between the axes).

(3) If the unit cell contains glide planes or screw axes, only diffractions with even indices will appear according to the extinction rules (cf. Section 11.2.3), and the correct indices are determined after indexing the higher order layer lines.

d) Measurement of the interplanar spacings of the principal planes. The Bragg angles θ_n for the diffraction spots on the two reciprocal lattice axes chosen for indexing are determined, and the interplanar spacings d found from the Bragg equation $2d \sin\theta = \lambda$. The Bragg angle itself is found from the distance $2\chi_n$ between corresponding diffraction spots

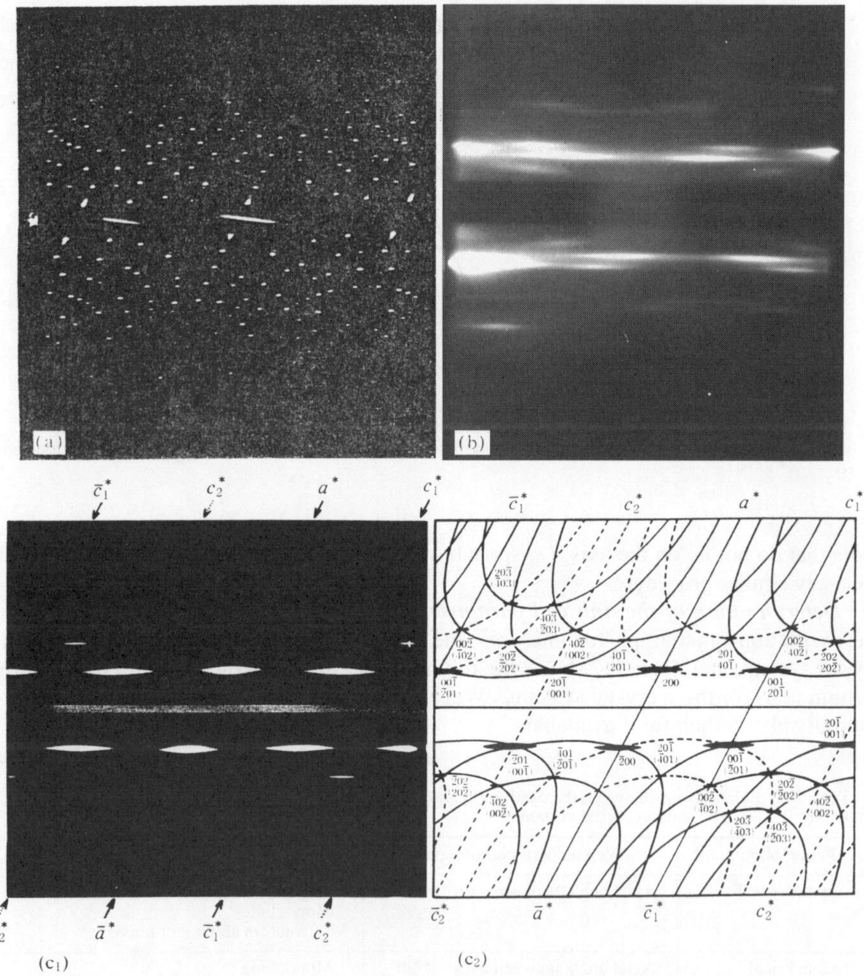

Fig. 8.40 Weissenberg photographs.
(a) 2-Aminoethanesulphonic acid (taurine). Equatorial photograph for the *c* axis.
(b) Doubly oriented Nylon 6. Equatorial photograph for the fiber axis (*b* axis).
(c_1) Doubly oriented Nylon 12. Equatorial photograph for the fiber axis (*b* axis).
(c_2) Indexing of (c_1) (*hkl*, indices in parentheses are the diffractions of the second component indexed by the a^* and $c_1{}^*$ axes). [Reproduced with permission from T. Ishikawa, *et al.*, *J. Chem. Soc. Jpn.*, 105, Chem. Soc. Jpn. (1973)]

on either side of the center line of the film (Fig. 8.41) with the aid of the equation $\theta_n = \chi_n/(2R_F)$. Since the value of *d* obtained is generally the interplanar spacing for arbitrary indices, *e.g.* for $h00$ in the case of the a^* axis, multiplication of this value by *h* gives the interplanar spacing for the (100) planes, *i.e.* $1/a^*$. The interplanar spacings obtained are corrected by photographing the diffraction lines of a reference substance (*e.g.* aluminum) on the same film, using the higher order diffractions. The interplanar spacings for diffraction

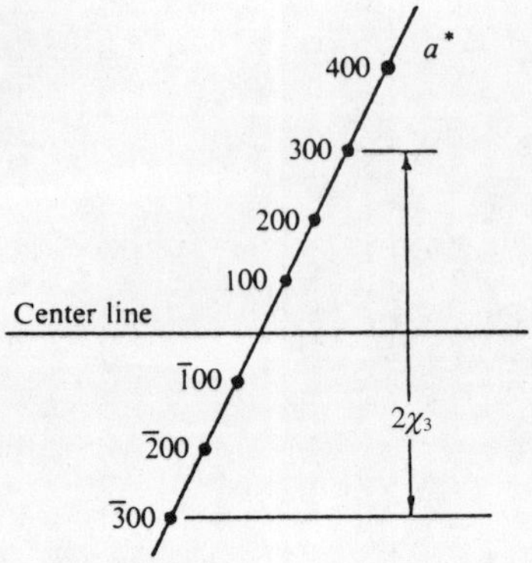

Fig. 8.41 Measurement of parameters on a Weissenberg photograph.
The diagram illustrates the measurement of $2\chi_3$ (in the general case $2\chi_n$) for the 300 diffractions on the a^* axis.

spots that do not lie on the axis, *i.e.* spots having the general indices *hkl*, are measured by a precisely similar procedure.

e) Determination of the unit cell parameters. Table 8.6 shows the relationship between the oscillation photograph for one axis, the equatorial Weissenberg photograph, and the crystal system, and Table 8.7 shows the minimum set of photographs required to determine the unit cells for these crystal systems. Weissenberg photographs for other crystal axes will naturally also be helpful if available.

Table 8.6 Relationships between the oscillation and the equatorial Weissenberg photograph for one crystal axis, and the crystal system

Oscillation photograph	Weissenberg photograph	Crystal system
Asymmetrical[†1]	Axial angle between axes[†2] $\neq 90°$ $= 90°$	Triclinic Monoclinic[†3] (rotation about *c* or *a* axis)
Symmetrical[†1]	Axial angle between axes[†2] $\neq 90°$ $= 90°$	Monoclinic[†3] (rotation about *b* axis) Orthorhombic or system of higher symmetry

[†1] *Asymmetrical* denotes that the *n*th layer line and $(-n)$th layer line do not agree, while *symmetrical* means that corresponding diffraction spots in both layer lines agree in position and naturally in intensity.
[†2] The angle found from the distance between the two reciprocal lattice axes on the direct beam trace of the Weissenberg photograph.
[†3] The *b* axis is the unique axis (2nd setting of the monoclinic system, see Table 3.1).

Table 8.7 Minimum photographs required for unit cell determination in different crystal systems

Orthorhombic system	Oscillation and Weissenberg photograph for one axis
Monoclinic system	Oscillation and Weissenberg photograph for *b* axis (unique axis)
Triclinic system	Oscillation and Weissenberg photographs for two axes

If the minimum number of photographs given in Table 8.7 for the determination of the unit cell cannot be made, *e.g.* in the case of an acicular crystal of the triclinic system having a pronounced cleavage parallel to the needle axis, or if the axis of rotation of a crystal of the monoclinic system (2nd setting) is the *a* axis or *c* axis and it is difficult to mount the crystal for rotation about an axis other than the needle axis, the following alternatives are available.

1) The *n*th layer line Weissenberg photograph for this axis is made by the equi-inclination method,[48] and the equatorial Weissenberg photograph is made by the anti-equi-inclination method, these two photographs being utilized.[49]

2) The zero level precession photograph is made for the reciprocal lattice plane containing this axis.

The second of these two methods is the more accurate. The data obtained from the above measurements are used to calculate the lattice parameters *a*, *b*, *c*, α, β, and γ utilizing Table 5 of the Appendix.

f) Number of chemical formula units per unit cell.

1) The equation used is

$$Z = \frac{VD_m}{M\, m_H} \tag{8.3}$$

D_m: measured density of the crystal. Methods of measuring the density were detailed in Section 8.4.2. In the case of crystalline high polymers, the density of the polymer including amorphous regions is provisionally used.

V: volume of the unit cell. *V* is found from the lengths and angles of the unit cell as given in Table 5 of the Appendix for the various crystal systems. Thus it is generally necessary to know *a*, *b*, *c*, α, β, and γ, or at least the reciprocal lattice of the basal plane and the cell length along the axis of rotation.

M: molecular weight of the chemical formula unit. Where the unit cell is made up of parts of huge molecules, as in the case of high polymer crystals, the sum of the atomic weights of the repeat unit is used.

m_H: mass of one hydrogen atom (1.66×10^{-24} g).

NOTE: (1) The value found for *Z* should be a whole number within the limits of experimental error. Thus, if the value differs significantly from a whole number, it is probable that the unit cell size used was in error, or that the value assumed for the density was incorrect, and these must therefore be checked. If they are found to be correct, the chemical formula of the specimen must be checked. In the case of high polymers, since the specimen is generally not 100% crystalline, the value for *Z* will be 5 ~ 10% lower than the true value, dependent upon the actual degree of crystallinity.

(2) The density of the crystal is given by the following equation, which is derived from Eq. 8.3:

$$D_m = \frac{1.660ZM}{V} \tag{8.4}$$

The value calculated in this way for the density of the crystalline regions will often be 5 ~ 10% higher than the experimental value, *e.g.* when the specimen is a high polymer typically containing amorphous regions having lower densities than the crystalline regions.

(3) Molecular weight can be found from the size of the unit cell and the density by employing knowledge of the space group to which the crystal belongs:

$$M = \frac{VD_m}{1.660Z} \tag{8.5}$$

***Example*: *2,2,3,3,5,5,6,6-octamethyl-1,4-dioxa-2,3,5,6-tetrasilacyclohexane*,** $[(CH_3)_4Si_2O]_2$

Experimental values :[50] $a = 7.67$ Å, $b = 6.64$ Å, $c = 17.39$ Å, $\beta = 111.0°$, $D_m = 1.04$ gcm^{-3} (or Mg m^{-3}) (20°C)

Calculation: from Eq. 8.3

$$Z = \frac{VD_m}{Mm_H} = \frac{7.67 \times 6.64 \times 17.39 \times 0.9336 \times 1.04}{264.6 \times 1.660} = 1.96 \cong 2$$

NOTE: The crystal in this example was originally thought to be tetramethyldisilanediol $HO(CH_3)_2SiSi(CH_3)_2OH$ (I). However, this gave a value for Z of 3.45, which deviates unacceptably (−13.7%) from the next highest whole number, $Z = 4$. This difference was significantly greater than the errors in the measurements of unit cell parameters and of the density. Since the crystals had an extremely high vapor pressure and readily sublimed, it was thought possible that the molecule contained the siloxane linkage −Si−O−Si−. The calculations were therefore performed for the linear and cyclic dimers (II) and (III) resulting from condensation of the hydroxyl groups in (I), and the results were used to calculate the molecular weight (Table 8.8). The experimental errors for (II) and (III) were both found to be small, and it was therefore difficult to decide which was correct. However, the error was slightly smaller for (III), and on re-examination of the infrared spectra and element analyses, it was concluded that the substance was in fact (III).[50]

Table 8.8 A comparison between the calculated and experimental molecular weights for possible structural formulae

	Structural formula	Molecular wt.(calc.)	Z	Molecular wt.(exptl.)	Error
(I)	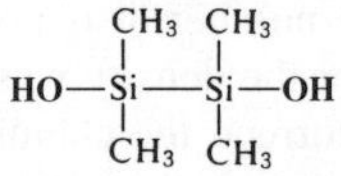	150.3	3.45 ≅ 4	130	−13.7%
(II)	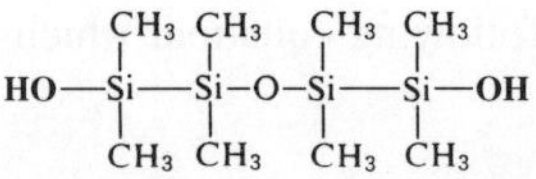	282.6	1.83 ≅ 2	259	−8.5%
(III)	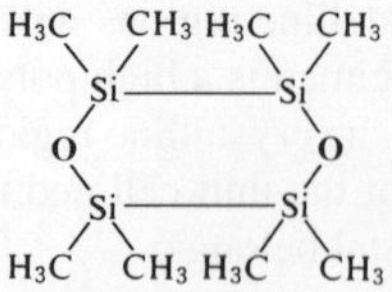	264.6	1.96 ≅ 2	259	−2.0%

C. Layer line Weissenberg photographs

Layer line Weissenberg photographs can be made by the equi-inclination method,[48] the normal beam method,[51] or the flat cone method.[52]

a) Equi-inclination method. The equi-inclination method is in most common use, and the procedure is detailed below. The principle is shown in Fig. 8.42.[53] The axis of rotation of the crystal is inclined at an angle ν_n to the incident X-rays, and the Weissenberg photograph of the layer line in question is recorded when the screen is displaced at the correct location. Fig. 8.42(c) shows the case where the axis of rotation of the crystal coincides with the reciprocal lattice axis, *i.e.* in the case of crystals of orthorhombic or higher symmetry and monoclinic crystals where the axis of rotation is the *b* axis (unique axis). Indexing is exactly the same as for the equatorial diffraction, except that diffractions for rows of reciprocal lattice points having the same d^* move farther away from the center line with increasing order number. Moreover, the 0*k*0 spots for the various layer lines always satisfy the diffraction conditions and form a line on the center line.

Figure 8.42(d) shows the situation for the triclinic system or for the monoclinic system with rotation about the *a* or *c* axis. Since the reciprocal lattice axis does not coincide with the axis of rotation (though a principal crystal axis naturally does), the reciprocal lattice points on the reciprocal lattice axis are increasingly displaced from the axis of rotation with increasing order number. Consequently, *e.g.* for rotation about the *c* axis, the 0*kl* and *h*0*l* rows (both *l*'s being the same, and for monoclinic crystals only the 0*kl* row) curve away from the normal, so that the 00*l* diffraction may be found at a point outside the center line of the *l*th layer line.

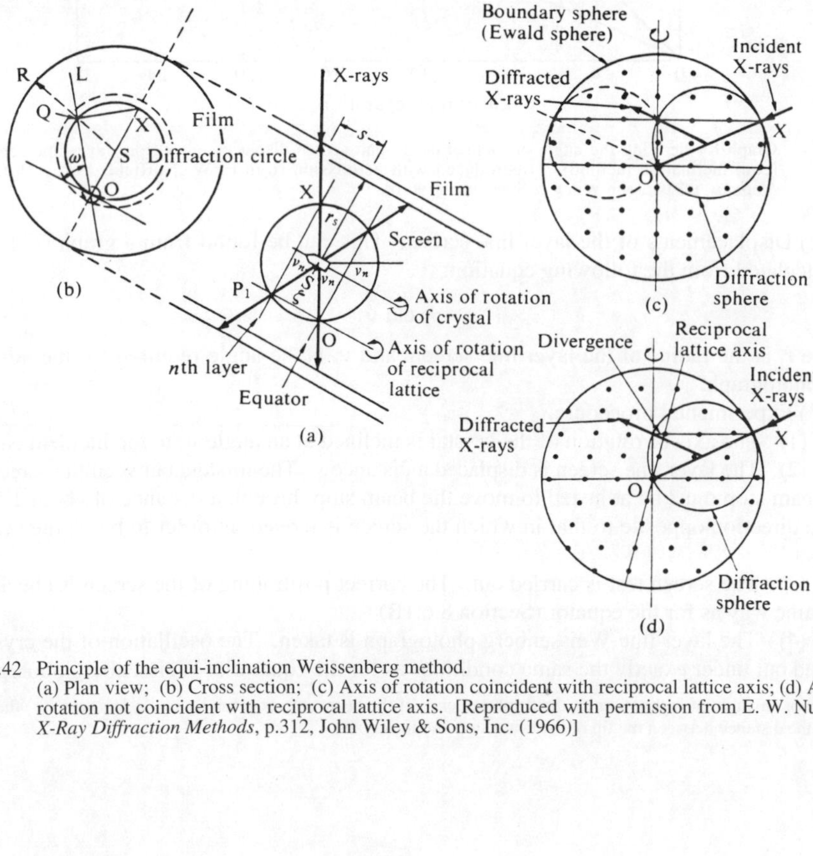

Fig. 8.42 Principle of the equi-inclination Weissenberg method.
(a) Plan view; (b) Cross section; (c) Axis of rotation coincident with reciprocal lattice axis; (d) Axis of rotation not coincident with reciprocal lattice axis. [Reproduced with permission from E. W. Nuffield, *X-Ray Diffraction Methods*, p.312, John Wiley & Sons, Inc. (1966)]

1) Fixing the angle of inclination ν_n. If I is the axial length of the crystal in the direction of the axis of rotation as found from the oscillation photograph, the angle of inclination required for the nth layer line Weissenberg photograph is the angle ν_n that satisfies the following Bragg condition:

$$2\,I \sin \nu_n = n\lambda$$

ν_n may be calculated from the above formula, but it is usually found in practice from a graph (Fig. 8.43).[54)]

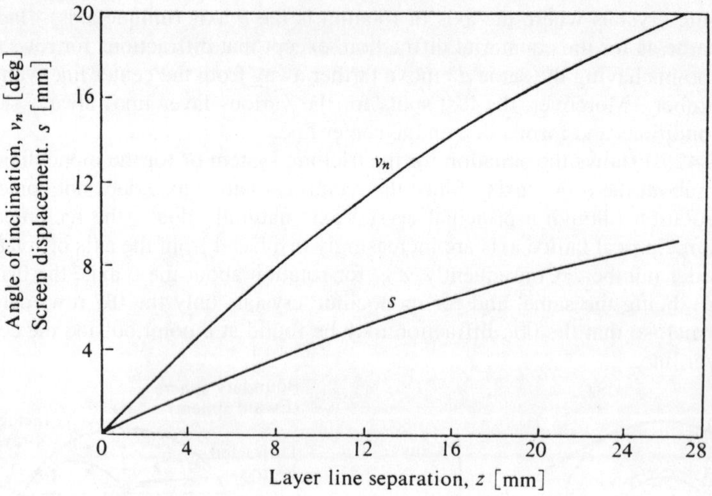

Fig. 8.43 Graph for deriving the angle of inclination, ν_n and screen displacement s from layer-line separation (equi-inclination method). [Reproduced with permission from E. W. Nuffield, *X-Ray Diffraction Methods*, p.314, John Wiley & Sons, Inc. (1966)]

2) Displacements of the layer line screen. This can be found from a graph (Fig. 8.43) or calculated from the following equation:

$$s = r_s \tan \nu_n \tag{8.6}$$

where r_s is the radius of the layer line screen, and ν_n is the angle required for the nth layer line photograph.

3) Experimental procedure.

(1) The axis of rotation of the crystal is inclined at an angle ν_n to the incident X-rays.

(2) The layer line screen is displaced a distance s. The linkage between the screen and the beam stop must be adjusted to move the beam stop through a distance of about $1.5 \times s^{\dagger}$ in the direction opposite to that in which the screen is moved, in order to block the primary X-rays.

(3) The screen test is carried out. The correct positioning of the screen is checked in the same way as for the equator (Section 8.6.1B)

(4) The layer line Weissenberg photograph is taken. The oscillation of the crystal is carried out under exactly the same conditions as for the equatorial Weissenberg photograph.

†1 For an angle of inclination ν_n, the displacement t of the beam stop is given by $t = (r_s + u) \tan \nu_n$, where u is the distance between the tip of the beam stop and the crystal.

Unless the film cassette is moved through a distance equal to the displacement of the screen and in the same direction, one side of the Weissenberg photograph will not be recorded on the film. Since the angle of inclination of the crystal with respect to the incident X-rays becomes more acute with increasing order number, the absorption increases, and the exposures accordingly become longer. Crystal samples which give layer line photographs that are, unlike those of the equatorial photograph, split up as a result of twinning or other faults, should be replaced, and the experiment repeated from the beginning with the new crystal.

(5) Indexing. Where the crystal is of orthorhombic or higher symmetry, or monoclinic with rotation about the *b* (unique) axis, the diffractions can be indexed in exactly the same way as for the equator (*cf.* Fig. 8.42(c)). Care is necessary, on the other hand, in the case of monoclinic crystals with rotation about the *a* or *c* axis (b^* axis curved) and of triclinic crystals (both reciprocal lattice axes curved) (*cf.* Fig. 8.42(d)).

b) Normal beam method.

Figure 8.44 illustrates the theoretical basis for the reconstruction of the reciprocal lattice.[51]

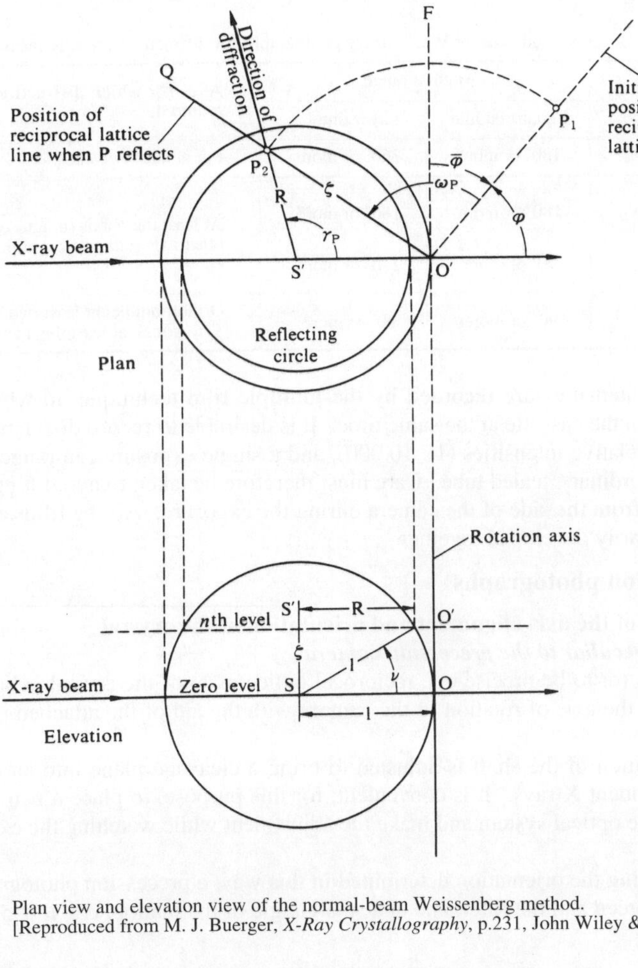

Fig. 8.44 Plan view and elevation view of the normal-beam Weissenberg method.
[Reproduced from M. J. Buerger, *X-Ray Crystallography*, p.231, John Wiley & Sons, Inc. (1942)]

The screen displacement s_n for the nth layer line can be calculated from

$$s_n = r_s \tan \mu_n \tag{8.7}$$

where r_s is the radius of screen and μ_n the angle of elevation of the nth layer line ($\mu_n = z_n/R_F$; z_n is the layer line separation on the film and R_F the film radius).

D. Photographs for intensity measurements

To obtain the three-dimensional diffraction intensity data required for crystal structure analysis, data must be accumulated from as many layer lines as possible in Weissenberg photographs for at least two axes. The oscillation-angle ranges and other required conditions are summarized in Table 8.9. In practice, the oscillation-angle range should preferably be considerably above the minimum cited in Table 8.9. The film cassette can only cover an angular range of about 220°, so photographs that must cover a range of 270° or 360° (or more) are made in two parts. Moreover, the angular ranges given in Table 8.9 are those for cases where only one side of the center line of the Weissenberg photograph is used.

Table 8.9 Angular ranges and axes for Weissenberg photographs for diffraction intensity measurements

Crystal system	Angular range		Axes for which diffraction data are required.
	Equatorial line	Layer line	
Triclinic	180° or more	360° or more	For at least two axes
Monoclinic (*b* axis (unique axis))	180° or more	180° or more	At least the *b* axis (unique axis) and one other axis (either the *a* or the *c* axis)
Monoclinic (*a* or *c* axis)	90° or more	270° or more	
Orthorhombic (tetragonal, cubic)	90° or more	90° or more	Orthorhombic: at least two axes Tetragonal: at least the *a* axis

Diffraction intensities are recorded by the multiple-film technique, in which several films are placed in the cassette at the same time. It is desirable to record diffractions having a wide range of relative intensities (1 : 10,000), and a single exposure can range from 24 h to a week for an ordinary sealed tube. Care must therefore be taken to avoid fogging due to X-rays scattered from the side of the camera during the exposure. All the films exposed together are necessarily developed together.

8.6.3 Precession photographs

A. Adjustment of the axis alignment and orientation of the crystal

a) Procedures peculiar to the precession camera.

1) One axis (or to be precise one reciprocal lattice axis) of the crystal is brought into coincidence with the axis of rotation of the camera with the aid of the attached magnifying glass or telescope.

2) The position of the shaft is adjusted to bring a cleavage plane into an orientation normal to the incident X-rays. It is convenient, for this purpose, to place a pair of crossed nicol prisms in the optical system and make the adjustment while watching the extinction of the crystal.

3) Maintaining the orientation determined in this way, a precession photograph is taken, without the screen and the filter, and at a small angle of inclination, *e.g.* $\mu = 5°$ or $10°$.

4) Further orientation adjustment is made subsequent to analysis of the precession photograph. Without the screen, the higher order level lines are recorded at the same time as the zero level; however, these can be distinguished, for the continuous X-rays give radial streaks with the position of the incident X-rays at the center for the zero level, whereas for the higher order levels they give loops. In this case, if the camera is fitted with a circular (not annular) screen of radius $r_s = 5$ mm, corresponding to $\mu = 10°$, the zero level alone will be recorded. The radial streaks due to the continuous X-rays generally have the form shown in Fig. 8.45. The adjustment of the crystal orientation starts with the inclination ΔV of the vertical direction of the crystal. The angle of inclination ΔV of the streak with respect to the spindle axis is measured on the photograph (Fig. 8.45(a)), and the inclination of the vertical arc of the goniometer head is adjusted. Another precession photograph is made, and the inclination ΔH of the horizontal arc and the stagger ΔD of the spindle direction are adjusted with the aid of the following relationship:

$$R_f \Delta H = x_{Rt} - x_{Lt}$$

$$R_f \Delta D = y_{Up} - y_{Dn} \tag{8.8}$$

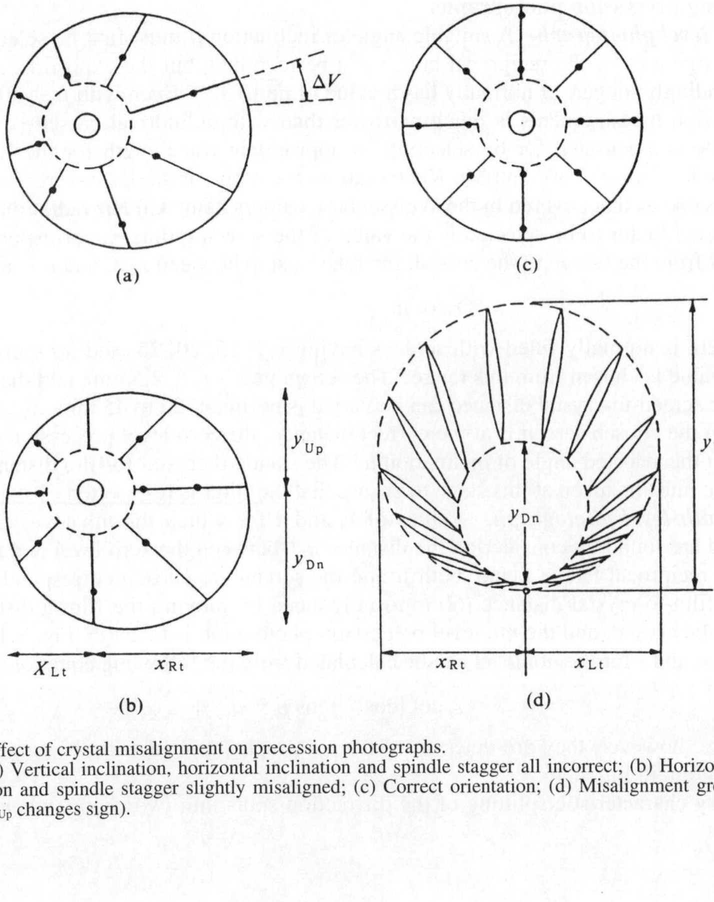

Fig. 8.45 Effect of crystal misalignment on precession photographs.
(a) Vertical inclination, horizontal inclination and spindle stagger all incorrect; (b) Horizontal inclination and spindle stagger slightly misaligned; (c) Correct orientation; (d) Misalignment greater than μ (y_{Up} changes sign).

If the two arcs of the goniometer head are not approximately vertical and horizontal, the adjustment should strictly be made by dividing ΔV and ΔH into two components each.

If the crystal is seriously misaligned, the zero level will become a collection of loops instead of streaks, *e.g.* as shown in Fig. 8.45(d). Since the lengths of the various loops approximately $4R_f \sin \mu$, the area due to the zero layer is found from the precession photograph taken without a screen, and the orientation is adjusted by the above method.

b) Adjustment of the alignment by using a Weissenberg camera. A certain crystal axis, *e.g.* the *c* axis, is aligned with the aid of a Weissenberg camera. The goniometer head and the crystal are then transferred as a unit to the precession camera. In crystals for which the *c* axis does not coincide with the c^* axis of the reciprocal lattice, the arcs of the goniometer head are shifted to bring the c^* into coincidence with the axis of rotation. The orientation of the c^* axis with respect to the *c* axis is determined beforehand from the Weissenberg photograph. The a^* or b^* axis is then positioned normal to the incident X-rays. This is a simple matter, for the orientation of the a^* or b^* axis can readily be found from Weissenberg photographs. Accurate adjustment of the crystal orientation on the basis of a precession photograph taken without screen or filter, and with $\mu = 5^\circ$ or 10° (as described under *a)* above) is possible as soon as these adjustments have been made. Preliminary alignment of the axis can also be made on the basis of Laue photographs.

B. Taking precession photographs

a) Zero level photograph. A suitable angle of inclination μ must first be selected. If μ is large, a large area of the reciprocal lattice can be recorded, but the exposures required are correspondingly longer. μ normally has a value of up to 30°. Even with $\mu = 30^\circ$, the range of diffraction measurements is much narrower than with cylindrical cassette cameras, but this can be compensated for by selecting an appropriate wavelength for the X-rays used. For example, with $\mu = 30^\circ$ and Mo $K\alpha$ radiation, the range of angles covered is almost exactly the same as that covered in the Weissenberg camera using Cu $K\alpha$ radiation.

The next factor to be selected is the value of the screen radius r_s, corresponding to the distance *s* from the screen to the crystal; the relationship between r_s, *s*, and μ is as follows:

$$s = r_s \cot \mu \tag{8.9}$$

The camera is normally fitted with screens having $r_s = 15$, 20, 25, and 30 mm. The most suitable value is chosen from this range. The screen gap is $r_s \pm 2.5$ mm, and the range over which the screen-to-crystal distance can be varied is normally 20 to 45 mm.

When the chosen screen is at the correct distance, the zero level precession photograph is taken at the selected angle of inclination μ. The standard crystal-to-film distance, R_f is 60 mm. Care must be taken at this stage to ensure that the filter is re-inserted.

b) The nth level photograph. Values of r_s and *s* for which the *n*th level can be photographed are found by considering the distance d_n^* between the zero level and the *n*th level on the reciprocal lattice plane; both μ and the screen are chosen correspondingly. The standard film-to-crystal distance (60 mm) is reduced by moving the film a distance $R_f d_n^*$ closer to the crystal, and the *n*th level precession photograph is taken(*cf.* Fig. 8.17(c)). The values of r_s and *s* for the *n*th level can be calculated from the following equation:

$$s = r_s \cot [\cos^{-1}(\cos \mu - d_n^*)] \tag{8.10}$$

In practice, however, they are determined from a graph supplied with the camera or from the nomogram in Fig. 8.46.[55]

A very characteristic splitting of the diffraction spots into two occurs when $R_f d_n^*$ (the

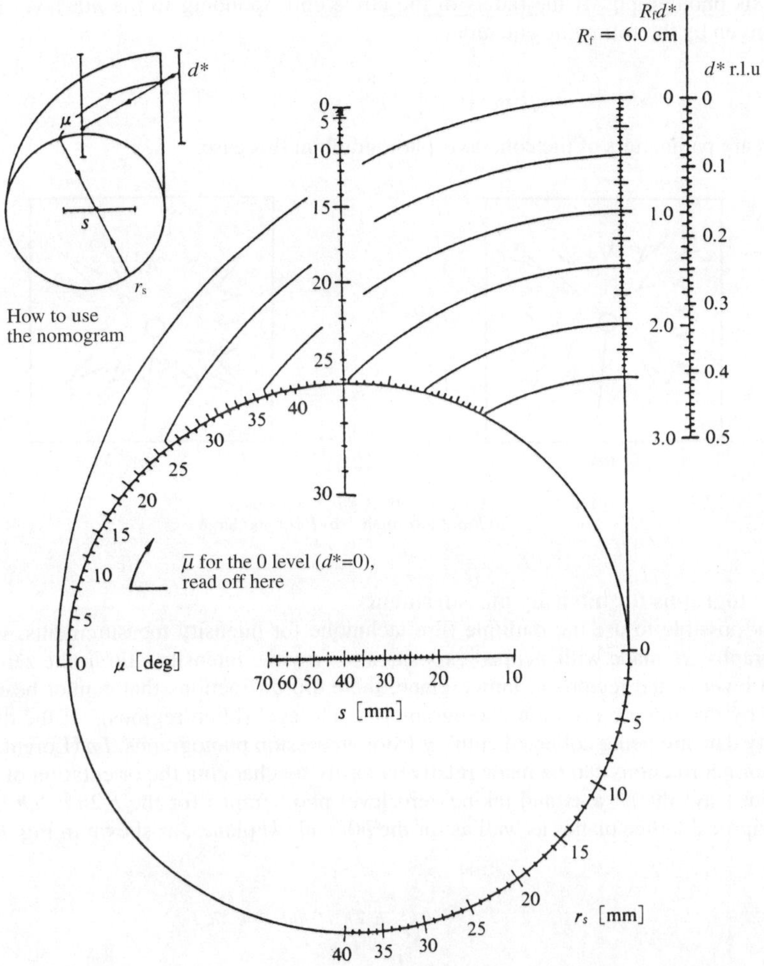

Fig. 8.46 Adams-Evans nomogram.
Values of the screen radius, r_s and the screen-to-crystal distance, s are obtained by the nomogram for making the nth level precession photogrtaph. (r. l. u.: reciprocal lattice unit)
[Reproduced with permission from E. W. Nuffield, *X-Ray Diffraction Methods*, p.262, John Wiley & Sons, Inc. (1966)]

crystal-to-film distance for the nth level) is incorrectly adjusted. This results from a difference between the positions at which diffraction occurs when the reciprocal lattice point enters the small circle of Fig. 8.17 and when it leaves the circle during the precessional motion. If $R_f d_n^*$ is too small, needle-1ike streaks diverging outwardly appear as shown in Fig. 8.47(a); if it is too large, the streaks diverge inwardly as shown in Fig. 8.47(b).

If d^* is unknown, it may be determined by wrapping a film in light-proof paper and inserting it in the screen-holding frame before taking a precession photograph. The diffractions from the reciprocal lattice levels are recorded on concentric rings, the zero level being nearest the center, followed by the first and subsequent levels in order. This is known as a cone-axis photograph. If the radius of the circle corresponding to the nth level is r_n, then d_n^* is given by the following equation:

$$d_n^* = \cos\mu - s / \{(r_n^2 + s^2)^{\frac{1}{2}}\} \tag{8.11}$$

μ and s are parameters of the cone-axis photograph in this case.

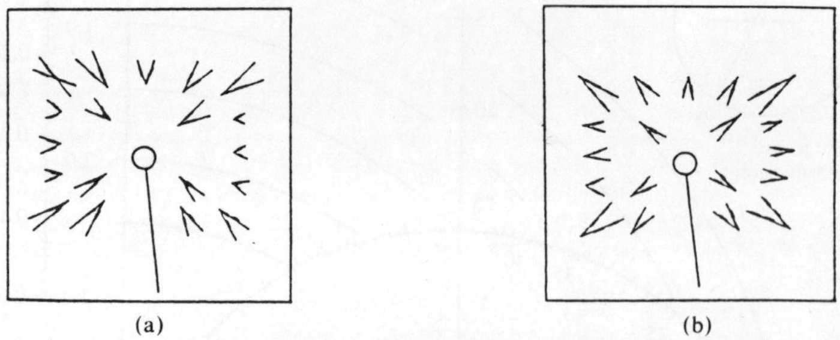

Fig. 8.47 Effect of incorrect $R_f d_n^*$ adjustment.
(a) $R_f d_n^*$ too small; (b) $R_f d_n^*$ too large.

C. Photographs for intensity measurements

It is not possible to use the multiple film technique for intensity measurements, so several photographs are made with stepped exposures to vary the intensity. From the zero level to the nth level of the reciprocal lattice plane, there are diffractions that cannot be measured simply by making a precession photograph of each level (blind regions). If the diffraction intensity data are being collected entirely from precession photographs, Lp (Lorentz and polarization) corrections can be made relatively easily by changing the orientation of the crystal about (say) the c^* axis and taking zero level photographs for the $h,2h,l$; h,h,l; $3h,2h,l$; etc. reciprocal lattice planes as well as for the $h0l$ and $0kl$ planes, as shown in Fig. 8.48.

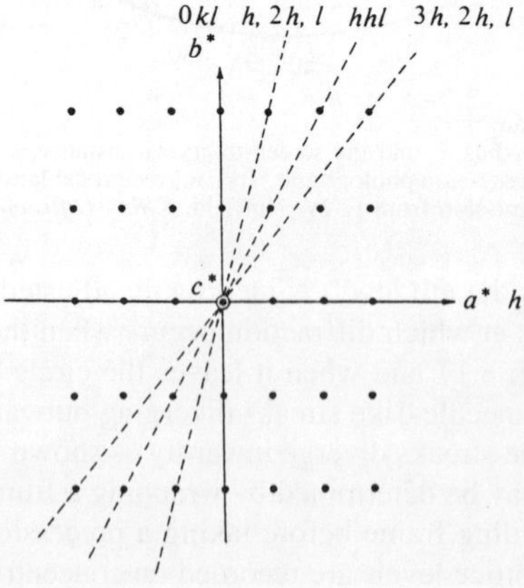

Fig. 8.48 Making zero-level precession photographs to derive intensity data for easy Lp correction (see text).

D. Analysis of precession photographs

a) Indexing. Since the reciprocal lattice is reproduced without distortion, no explanation is necessary.

b) Determination of the unit cell parameters. The translation distance t^* of the diffraction spots in the direction of the principal axis of the reciprocal lattice may be measured on the film. It is, however, more accurate to measure the distance d^* between the rows of diffraction spots and to find t^* with the aid of the axial φ^*, which is measured at the same time (*cf.* Fig. 8.49):

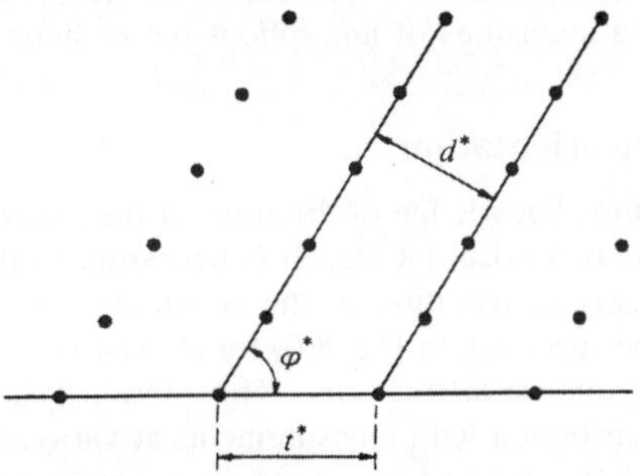

Fig. 8.49 Measurement of interplanar spacing along the principal reciprocal lattice axis.

$$t^* = d^*/\cos(90^\circ - \varphi) = d^*/\sin \varphi^* \tag{8.12}$$

The interplanar spacing d along the principal axis is calculated from t^* and the specimen-to-film distance R_f by using the following relationship:

$$d = R_f\lambda/t^* = \lambda/d^* \tag{8.13}$$

The axial angle can be determined by measurement of the angle of intersection φ^* of two principal axes of the reciprocal lattice on the film with an accurate protractor.

With the exception of crystals of the triclinic system, the unit cell dimensions can be readily found from the interplanar spacing d^* (the distance between rows of reciprocal lattice points) along the principal axis of the reciprocal lattice with the aid of the following relationships:

$$a = \lambda/d^*_{100}, \quad b = \lambda/d^*_{010}, \quad c = \lambda/d^*_{001} \tag{8.14}$$

For triclinic crystals, the reciprocal lattice is determined by the values of a^*, b^*, and c^* along the axes of the reciprocal lattice and the inter-axial angles, and the dimensions of the unit cell are found from the equations given in Table 6 of the Appendix.

8.7 Diffraction Studies for Analysis of Fine Textures

8.7.1 Measurement of crystallinity

The normal method for the determination of crystallinity is based on the theoretical presupposition that the specimen exhibits no preferred orientation. The photographic method is therefore used first to verify that the specimen is unoriented; this entails scrutiny of the entire X-ray diffraction diagram. If the diagram shows a preferred orientation, the specimen may be reduced to an unoriented (randomly oriented) state before measurements are made.

For example, the specimen may be chopped as finely as possible and the pieces aggregated in a completely random manner. Alternatively, the specimen may be retained in its original form if it is rotated by using a Gandolfi (Fig. 8.14) or other devices during measurements with the diffractometer.

The continuous intensity distribution in the diffraction direction (2θ) is measured. Since the intensity values found in this way are essential to the analysis, the measurements must be carried out over the widest possible range of angles, and must be of adequate accuracy. Adequate data can normally be obtained with a rotation camera. However, a powder diffractometer or area detector diffractometer is preferable for accurate intensity measurements. The need to measure the intensities of amorphous halos dictates the use of monochromatic X-rays.

8.7.2 Analysis of crystallite orientation

The analysis of orientation requires knowledge of the form of the entire diffraction pattern, *i.e.* the spatial distribution of the reciprocal lattice. It is necessary to determine accurately the continuous azimuthal intensity distribution in the ψ direction (*cf.* Fig. 10.16; corresponding to the χ direction of the specimen in Fig. 8.6) for at least two sets of crystal planes (reciprocal lattice points) giving intense diffractions. This obviously involves determining a three-dimensional intensity distribution with measurements at various angles of incidence φ.

If it is necessary only to check qualitatively for the presence of orientation or the determine the type of orientation, a Laue camera or a simple oscillation or rotation camera may be used.

The investigation of sheet-like specimens often requires diffraction diagrams from at least three directions (one normal to the plane of the sheet and two at right angles to each other in a plane parallel to the sheet) (*cf.* Figs. 10.1, 10.2 and 15.36). Fiber specimens, when it is thought that the orientation will be cylindrically symmetrical, require only the use of two directions (parallel to the fiber axis and in a direction normal to the fiber axis). The systematic classification of types of orientation is discussed in Section 10.2.2).

The most accurate quantitative measurements of the degree of orientation are obtained with X-ray diffractometers fitted with fiber-specimen mount or apparatus for readily obtaining pole figures. A single-crystal X-ray diffractometer is the most suitable for this purpose. Weissenberg, Sauter, and other goniometer cameras are also generally suitable for quantitative measurements. Since the measurements are concerned with the crystal diffraction pattern, satisfactory results can be obtained with filtered X-rays, provided that the filtering effectively removes the $K\beta$ radiation and eliminates superposition of $K\beta$ diffractions.

8.7.3 Measurement of the size and shape of and/or lattice distortion in crystallites

This involves accurate determination of the three-dimensional form of the reciprocal lattice, *i.e.* of the diffraction pattern. Thus, although a Laue camera may be used, a rotation camera that allows movement of the specimen is preferable. High resolution is desirable, because the accuracy with which the crystallite size can be determined depends upon the accuracy of measurement of the diffraction line breadths.

High polymer diffractions are generally diffuse, with low intensities, because the crystallites are small and the lattice imperfect. The diffractions consequently merge into the background at higher order numbers. It is therefore most important to minimize extraneous background scattering. Moreover, since it is necessary to measure the diffraction profile, the X-rays should be monochromatic if at all possible.

Any broadening of the diffraction lines induced by the geometry of the optical system (shape and size of the specimen, errors in the collimating system, etc.) must be minimized. The extent of this broadening due to experimental conditions can readily be determined. The diffractions for a standard specimen of adequate crystal size, and for the specimen with crystallite size as yet undetermined, are measured simultaneously or separately under identical conditions. The broadening of the diffractions due to the experimental conditions is estimated from a comparison of the results, and the inherent broadening for the specimen in question can then be determined. Descriptions of this procedure are given in Chapter 13; it is also described in many other books.[57)]

When two or more broad diffraction lines overlap it is necessary to resolve them and determine their individual shapes.

The relevant information theory is usually applied by using an electronic computer. Line profiles of the components of an incompletely resolved diffraction are first assumed to belong to a single type such as the Gauss, Lorentz (Cauchy), or some other distribution, *e.g.* a Voigt, pseudo Voigt, or Pearson VII function.[58)] The line profile of the diffraction is then calculated for various positions, heights, and half widths of the component diffractions. The method of least-squares indicates the best fit between observed and calculated profiles of the incompletely resolved diffraction. Some methods for the numerical calculations which give a rapid convergence of the least-square residuals have been reported.[59.60)] On the other hand, though the precision is a little inferior to the above method, analogue computers are also convenient for this purpose. A commercial curve analyzer is also available.

For the conventional graphic resolution we may apply the method of DuMond-Kirkpatrick,[61)] Rachinger,[62)] or Pease.[63)] The DuMond-Kirkpatrick method is given below.

If the profiles of the two diffraction lines are similar, the profile $F(x)$ of the double line can be expressed as $F(x)=f(x)+kf(x-\delta)$, where $f(x)$ and $kf(x-\delta)$ are the profiles of two component diffraction lines separated by a distance δ (Fig. 8.50). We find that

$$f(x)=F(x)-kF(x-\delta)+k^2F(x-2\delta)-k^3F(x-3\delta)+\cdots \tag{8.15}$$

In these expressions, k is the ratio of the heights of the two component lines.

If k and δ are known, as in the resolution of the $K\alpha_1/K\alpha_2$ doublet or of overlapping lines in the diffraction pattern of a crystal the structure of which is already known, a set of ordinates spaced at intervals of δ are read off for the profile obtained in the experiment.

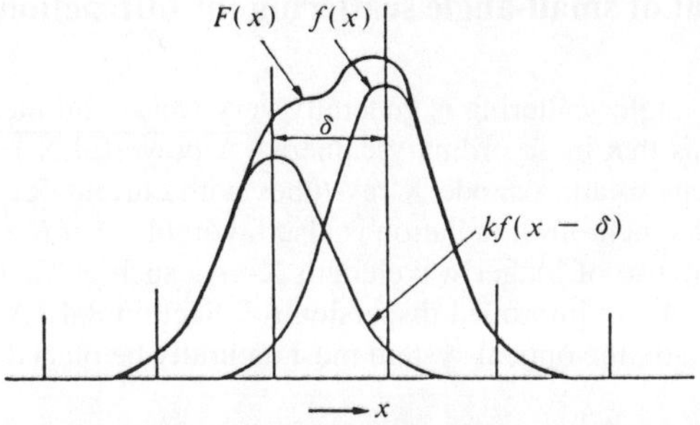

Fig. 8.50 Resolution of overlapping diffraction lines with similar profiles.
[Reproduced from J. W. M. DuMond *et al.*, *Phys. Rev.* **37**, 136, Am. Inst. Phys. (1931)]

These ordinates are substituted in the right-hand side of Eq. 8.15 to find $f(x)$. If either or both k and δ are unknown, or if the profiles of the component peaks are not similar, two suitable component peaks must be found by trial and error.

8.7.4 Measurement of diffuse halos due to amorphous solids and liquids

In contrast with the study of diffractions from crystals, it is necessary to record continuously and accurately the intensity distribution of a diffraction pattern extending over a wide angular range. As in the measurement of the degree of crystallinity, therefore, background scattering must be reduced to a minimum. Since most of the diffraction patterns are isotropic and simple, the measurements can be made with a Debye-Scherrer camera, but an X-ray diffractometer meets the requirement for accuracy more satisfactorily. Oscillation or rotation cameras, Weissenberg cameras, or single crystal diffractometers may naturally also be used. A monochromator must be employed to obtain monochromatic X-rays. Since the scattering measurements should in principle be carried out up to large values of $(\sin\theta)/\lambda$, radiation of shorter wavelength than, Mo $K\alpha$ such as Ag $K\alpha$ or W $K\alpha$ is used.

Except in the special case of diffuse halos exhibiting orientation, the intensity may be recorded by scanning the film with a photometer in any radial direction. A large area of the film must be kept unexposed by masking with a lead plate to allow assessment of the base density for comparison with an intensity scale. The specimen is mounted as in the powder method, and diffraction or scattering due to any supporting substance used must be taken into account (*cf.* Section 8.4.3).

8.7.5 Analysis of distorted crystalline diffraction

A monochromator should preferably be used to ensure a monochromatic X-ray beam. Study of the diffraction must be particularly comprehensive because the overall intensity throughout reciprocal space is to be recorded, and must include both crystalline and amorphous elements. Suitable apparatus includes rotation, Weissenberg, or precession cameras and automatic single crystal diffractometers. A complete analysis would require a three-dimensional intensity representation of the three-dimensional reciprocal lattice over all space. In the case of a fibrous high polymer, however, it is enough if the ξ and ζ projections of the reciprocal lattice (*i.e.* all the diffraction intensities in planes containing the fiber axis and the equator) are accurately recorded. This may be achieved directly in a precession photograph with the fiber axis as the axis of rotation.

8.7.6 Measurement of small-angle scattering (or diffraction)

A. X-rays

The intensity of small-angle scattering is generally very weak, and the path of the X-rays is several times as long as that in an ordinary camera. A powerful X-ray source is therefore desirable, and fine-focus rotating anode X-ray tubes with currents of 100 mA or more are often used. The use of synchrotron radiation is also favorable. Cu $K\alpha$ radiation is normally employed, although the use of longer wavelength X-rays such as Cr $K\alpha$ or Al $K\alpha$ may be considered in order to obtain improved dispersion (*cf.* Section 8.4.1A). With Cr $K\alpha$ or X-rays of longer wavelength, the optical system must naturally be placed under vacuum.

B. Optical system

The accuracy of the analysis increases with decreasing slit width and with increasing slit distance and specimen-to-film distance. On the other hand, the small-angle scattering intensity simultaneously decreases. The experimental conditions must therefore be chosen to give the shortest exposures consistent with the degree of accuracy appropriate to the nature of the specimen and the purpose of the measurements. This is discussed in some detail in other books,[64.65] and will not be dealt with here.

8.7.7 Special experimental methods

It is sometimes necessary to study the crystal structures and fine textures of polymeric specimens at high and low temperatures in connection with the investigation of the temperature dependence of their properties. The methods used for high- and low-temperature studies of low molecular weight single crystals and powder specimens are applicable in such cases, and the reader is referred to accounts of these techniques.[66–68] Whereas experiments at a fixed temperature (whether high or low) are fairly straightforward, studies in which the temperature is continuously varied generally present considerable difficulties.

Stein and his co-workers have developed so-called "rheo-optics" which permits dynamic X-ray diffraction measurements, *i.e.* the measurement of the change in the diffraction pattern during tensile stress and deformation measurements.[69] Sakurada *et al.*[70] have determined crystal elasticity from the change in the diffraction pattern.

References

1. a) *International Tables for X-Ray Crystallography,* Vol. III, p. 145, Kynoch Press, Birmingham (1962).
   b) *International Tables for Crystallography*, Vol. C, 2nd ed. (A.J.C. Wilson, E. Prince eds.), Kluwer Academic, Dordrecht (1992).
2. A.F.H. Meggleton, *Nucl. Instr. Methods*, **101**, 114 (1972).
3. a) C. Cork, R. Fehr, R. Hamlin, W. Vernon, N.H. Xuong, V. Prez-Mendez, *J. Appl. Cryst.*, **7**, 319 (1974).
   b) J. Borkowski, M.K. Kopp, *Rev. Sci. Instrum.*, **39**, 1515 (1968); *IEEE Trans. Nucl. Sci.*, **17**, 340 (1970); *Rev, Sci. Instrum.*, **46**, 951 (1975).
   c) J. Alberi, J. Fischer, V. Radeka, *IEEE Trans. Nucl. Sci.*, **NS-22**, 255 (1975).
   d) C. Parkman, Ż. Hajduk, A. Jeavons, N. Ford, B. Lindbeng, *Proc. 2nd. ISPRA Nuclear Electronics Symposium, EUR5370*, **139**, 237 (1975).
4. a) H. Morimoto, R. Uyeda, *Acta Cryst.*, **16**, 1107 (1963).
   b) *J. Appl. Cryst.*, **14**, 256 (1981).
5. a) M. Sonoda, M. Takano, M. Miyahara, H. Kato, *Radiology*, **148**, 833 (1983).
   b) Y. Amemiya, J. Miyahara, *Nature*, **336**, 89 (1988).
   c) J. Miyahara, K. Takahashi, Y. Amemiya, N. Kamiya, Y. Satow, *Nuclear Instr. Methods*, **A246**, 572 (1986).
6. U.W. Arndt, A.R. Faruqi, in: *The Rotation Method in Crystallography* (U.W. Arndt, A.J. Wanacott eds.) pp. 219–243, North-Holland Pub., Amsterdam (1977).
7. a) N.M. Allinson, *J. Synchrotron Rad.*, **1**, 54–82 (1984).
   b) In ref. 1a), p.627.
   c) Y. Amemiya, K. Ito, *J. Cryst. Soc. Jpn.*, **45**, 171 (2003).
8. J.M. Goppel, *Appl. Sci. Res.*, **A1**, 3 (1949); J.J. Arlman, *Appl. Sci. Res.*, **A1**, 347 (1949); J.M. Goppel, J.J. Arlman, *Appl. Sci. Res.*, **Al**, 462 (1949).
9. L.E. Alexander, S. Ohlberg, G.R. Taylor, *J. Appl. Phys.*, **26**, 1068 (1955).
10. U.W. Arndt, J.N. Champness, R.P. Pizackerley. A.J. Wanacott, *J. Appl. Cryst.*, **6**, 457.
11. For example, M.J. Buerger, *X-Ray Crystallography*, pp. 230–251, John Wiley & Sons. Inc., N.Y. (1942).
12. a) N. Sakabe, *J. Appl. Cryst.*, **16**, 542 (1983).
    b) N. Sakabe, *Phys. Res.*, **A303**, 448 (1991).
13. a) M.J. Buerger, *The Photography of the Reciprocal Lattice*, pp. 4–5, American Society for X-Ray and Electron Diffraction (1944).
    b) T. Inoue, Doctoral Thesis p.20, Osaka Univ.(1994).
14. a) T. Fukamachi, Y. Nakano, O. Shimomura, *J. Appl. Cryst.*, **11**, 688 (1978).
    b) T. Sakamaki, S. Hosoya, T. Fukamachi, *J. Appl. Cryst.*, **13**, 433 (1980); *idem.*, *Acta Cryst.*, **A36**, 183 (1980).
15. a) H. Iwasaki, *Proc. Taiwan–Japan Joint Seminar on Crystallography*, Tokyo, p.145 (1990).

b) H. Iwasaki, *Proc. Second R.O.C. –Japan Joint Seminar on Crystallography*, Taipei, p.241 (1992).
c) N. Kamiya, H. Iwasaki, *J. Appl. Cryst.*, **28**, 745 (1995).
16. T. Yamanaka, S. Kawasaki, T. Shibata, *Adv. X-Ray Anal.*, **35**, 415 (1992).
17. a) A. Guinier, G. Fournet, *Small-Angle Scattering of X-Rays*, John Wiley & Sons, Inc., N.Y. (1955).
b) J. Schelten, R.W. Hendricks, *J. Appl, Cryst.*, **11**, 297(1978).
18. R.S. Bear, O.E.A. Bolduan. *J. Appl. Phys.*, **20**, 983 (1949).
19. W.O. Statton, in: L. Alexander, *X-Ray Diffraction Method in Polymer Science*, p.103, John Wiley & Sons, Inc., N.Y. (1969).
20. R.W. Hendricks, *J. Appl. Cryst.*, **11**, 15 (1978).
21. C.J. Borkowski, M.K. Kopp, *IEEE Trans. Nucl. Sci.*, **NS19**, 161 (1972).
22. O. Kratky, H. Stabinger, E. Wrentschur, P. Zipper, *Acta Phys. Austriaca*, **44**, 173 (1976).
23. a) O. Kratky, *Z. Elektrochem.*, **58**. 49 (1954).
b) O. Kratky, *Kolloid Z.*, **144**, 110 (1955).
c) O. Kratky, A. Sekora, *Monatsh. Chem.*, **85**. 660 (1954).
d) O. Kratky *Z. Elektrochem.*, **62**. 66 (1958).
e) O. Kratky, Z. Skala, *Z. Elektrochem.*, **62**. 73 (1958).
f) O. Kratky, in: *Progress in Biophysics*, Vol.13, pp. 139–141, Pergamon, N.Y. (1963).
g) O. Kratky, in: *Small-Angle X-Ray Scattering* (H. Brumberger ed.) pp. 75–88, Gordon and Breach, N.Y. (1967).
24. a) U. Bonse, M. Hart, *Appl. Phys. Letters*, **6**, 155 (1965).
b) U. Bonse, M. Hart. *Z. Physik*, **189**. 151 (1966).
c) U. Bonse, M. Hart, in: *Small-Angle X-Ray Scattering* (H. Brumberger ed.) pp. 121–130, Gordon & Breach, N.Y. (1967).
25. a) A. Franks, *Proc. Roy. Soc. London*, **B68**, 1054 (1955).
b) A. Franks, *J. Appl. Phys.*, **9**, 349 (1958).
c) A. Franks, P.R. Breakwell, *J. Appl. Crys.*, **7**, 122 (1974).
26. a) G.F. Elliott, C.R. Worthington, *J. Ultrastruc. Res.*, **9**, 166 (1963).
b) G.F. Elliott, *J. Ultrastruc. Res.*, **9**, 171 (1963).
c) G.F. Elliott, *J. Sci. Instrum.*, **42**. 312 (1965).
27. H.E. Huxley, W. Brown, *J. Mol. Biol.*, **30**, 383 (1967).
28. J.W. M. DuMond, B. Henke, *Phys. Rev.*, **89**, 1300 (1953).
29. In ref. 1a), p. 84.
30. D.W. Beereman, *Rev. Sci. Instr.*, **26**. 1048 (1955).
31. J. Bordas, I.H. Munro, A.M. Grazer, *Nature* (*London*), **262**, 541 (1976).
32. G. Rosenbaum, K.C. Holmes, in: *Synchrotron Radiation Research*, (H. Winick, S. Doniak eds.), p. 530, Plenum Press, N.Y. (1980).
33. J. Kirtz, D. Sayre, in: *Synchrotron Radiation Research*, (H. Winick, S. Doniak eds.), p. 280, Plenum Press, N.Y. (1980).
34. *Methods of Obtaining Monochromatic X-Rays and Neutrons*, (F.H. Herbstein ed.), Biblio. 3, International Union of Crystallography, Commission on Crystallographic Apparatus (1967).
35. In ref. 1a), p. 151.
36. R.A. Young, *Z. Krist.*, **118**, 233 (1963).
37. In ref. 1a), p. 79.
38. In ref. 1a), pp. 30–34.
39. In ref. 1a), pp. 291–312.
40. For example, G.H. Stout, L.H. Jensen, *X-Ray Structure Determination*, pp. 62–73, Macmillan, Oxford (1968).
41. In ref. 1a), p.19.
42. B.W. Low, F.M. Richards, *J. Am. Chem. Soc.*, **74**, 1660 (1952): *idem.*, *ibid.*, **76**, 2552 (1954).
43. In ref. 1a), pp. 21–29.
44. *Handbook of X-Rays in Research and Analysis*, (E. Kaelbe ed.), pp. 9–32, McGraw-Hill, N.Y. (1967).
45. *Tables for Conversion of X-Ray Diffraction Angles to Inter-planar Spacing*, U.S. Government Printing Office (1950).
46. W. Parrish, M. Mack, *Data for X-Ray Analysis*, Vol. 1 (Cu), Vol. 2 (Mo, Co, W), Vol. 3 (Fe, Cr), Philips Technical Laboratory (1963).
47. For example, a) M.J. Buerger, *X-Ray Crystallography*, John Wiley & Sons, Inc., N.Y. (1942).
b) G.H. Stout, L.H. Jensen, *X-Ray Structure Determination*, Macmillan, Oxford (1968).
48. In ref. 1a), pp. 102–121.
49. In ref. 47a), pp. 252–295.
50. T. Takano, N. Kasai, M. Kakudo, *Bull. Chem. Soc. Jpn.*, **36**, 585 (1963).
51. In ref. 47), p. 230–251.

52. In ref. 47), p. 301–304.
53. E.W. Nuffield, *X-Ray Diffraction Methods*, p. 312, John Wiley & Sons, Inc., N.Y. (1966).
54. In ref. 48), pp. 294–295.
55. Ref. 4, Vol. II, p. 194 (1959).
56. For example, W.O. Statton, *J. Appl. Polym. Sci.*, **7**, 803 (1963).
57. H.P. Klug, L.E. Alexander, *X-Ray Diffraction Procedures for Polycrystalline and Amorphous Materials*, pp. 494-511, John Wiley & Sons, Inc., N.Y. (1954).
58. R. Shimizu, K. Murata, G. Shinoda, *Technol. Repts, Osaka Univ.*, **16**, 664, 691 (1965).
59. R.H. Hobine, L. Pinsonneault, *Rev. Sci. Instr.*, **34**, 1445 (1963).
60. H. Stone, *J. Opt. Soc. Am.*, **52**, 998 (1952).
61. J.W.M. DuMond, H.A. Kirkpatrick, *Phys. Rev.*, **37**, 136 (1931).
62. W.A. Rachinger, *J. Sci. Instr.*, **25**, 254 (1948).
63. R.S. Pease, *J. Sci. Instr.*, **25**, 353 (1948).
64. A. Guinier, G. Fournet, *Small-Angle Scattering of X-Rays*, pp. 83-125, John Wiley & Sons, Inc., N.Y. (1955).
65. L.E. Alexander, *X-Ray Diffraction Method in Polymer Science*, John Wiley & Sons, Inc. (1969).
66. H.J. Goldschmidt ed., *High-Temperature X-Ray Diffraction Techniques*, Bibliog., 1, International Union of Crystallography (IUCr), Commission on Crystallographic Apparatus.
67. a) B. Post, ed., *Low-Temperature X-Ray Diffraction*, Bibliog., 2, IUCr, Commission on Crystallographic Apparatus (1966).
    b) R. Rundman, *Low-Temperature, X-Ray Diffraction*, Apparatus and Techniques, Plenum Press (1966).
68. M.J. Buerger, *The Precession Methods in X-Ray Crystallography*, pp. 247-260, John Wiley & Sons. Inc., N.Y. (1964).
69. T. Kawaguchi, T. Itoh, H, Kawai, D.A. Keedy, R.S. Stein, *J. Polym. Sci.*, **82**, 1075 (1964).
70. L. Sakurada, T. Nukushina, T. Itoh, *J. Polym. Sci.*, **57**, 651 (1962).

Part III

Analytical

As we have already often had cause to note, the diffraction patterns of high polymers generally consist of a rather broad crystalline diffraction pattern which is superimposed upon strong background scattering that includes the effect of amorphous diffraction. In view of these special characteristics of high polymers, Chapters 9 to 12 and part of Chapter 13 deal mainly with methods of analysis based on only the crystalline diffraction patterns of high polymers, whereas, Chapters 13 and 14 proceed to describe methods applicable to the complete diffraction intensity distribution (including the background). Chapter 15 treats the analysis of small-angle scattering from high polymers. By dealing with these topics separately, and with the aid of practical examples, we hope to make it easier for the reader to understand the methods used for the analysis of the X-ray diffraction patterns of solid high polymers and the structural features of the diffracting substances. The diagram below gives a systematic scheme listing the various peculiarities which may be exhibited by X-ray diffraction patterns from high polymers and correlating them with the structural feature or features from which they may arise.

Correlation between X-ray fiber diagram and structural elements of diffracting substances

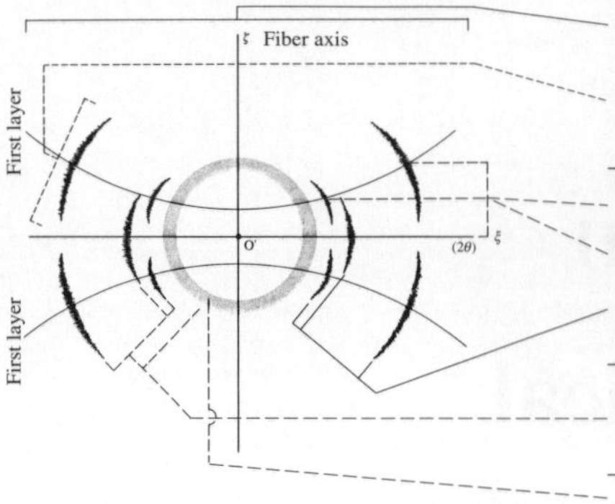

Wide angle diffraction pattern (Flat plate, schematic)	Characteristics of diffraction pattern	Possible structual features of specimen
	Special feature of whole diffraction pattern	Type of crystallites orientation
	Azimuthal intensity distribution, $I(\psi)$	Degree of orientation
	Position of layer lines, ζ_n Position of diffractions, ξ_j, ζ_k	Fiber period } Unit cell
	Intensity of independent diffractions, $I(hkl)$	Crystal and molecular structure
	Line broadening, $I(2\theta)$	Size and shape and / or Imperfection of crystallites
	Sharp diffraction(s) and diffuse halo(s) (of *UNORIENTED* pattern)	Crystallinity
	Diffuse halo(s)	Amorphous structure

θ: Bragg angle, ψ: angle from the equator (ξ axis) measured on the Debye-Scherrer ring.

Correlation between X-ray small-angle scattering and structural elements of scattering substances

Small-angle scattering pattern (schematic)	Characteristic of scattering pattern	Method of Analysis	Possible structual features of specimen
a) Central diffuse scattering 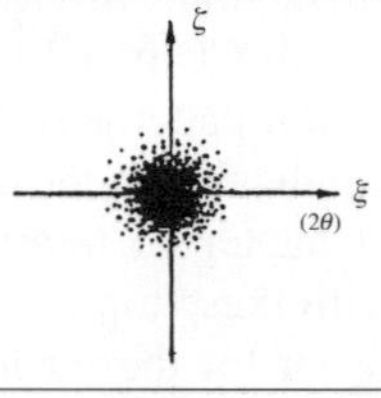	Isotropic (or circular), Anisotropic (elliptical, streak, or..)	$\log I$ *vs.* s^2 (Guinier plot) $\log I$ *vs.* $\log s$	Radius of gyration of particle (or void), R_g Size and shape of particle (or void)
	$\left[I(s) \text{ vs. } s,\ s = (4\pi\sin\theta)/\lambda = 2\pi S/\lambda \fallingdotseq 4\pi\theta/\lambda \fallingdotseq 2\pi\varepsilon/\lambda \right]$	(Contrast variation method)	Particle or void? or further detailed structure
		Loose chain molecule $s^2{\cdot}I$ *vs.* s, s^*(knick)	Persistence length
	(Fanning, along the equator)		Preferred orientation
b) Long period pattern 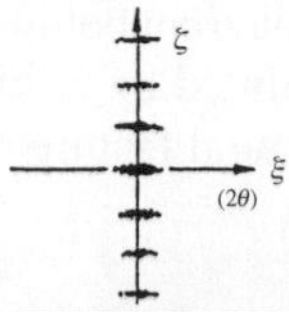	Along the meridian	$L{\cdot}\varepsilon = n\lambda$ ($D_L{\cdot}\varepsilon = n\lambda$)	Long period(s)
	First order only,		L
	or Several orders		$L_1, L_2, .., L_n \rightarrow Z$
	Tilted (Fanning, along the meridian)		
c) Small-angle diffraction 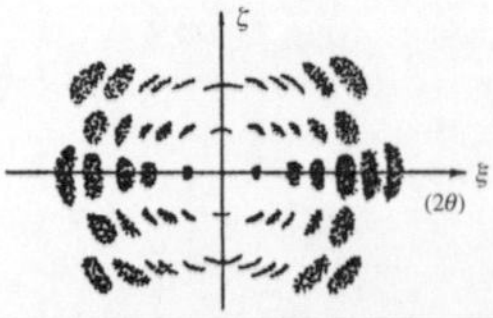	Fiber diagrams at small-angle region	May be analysed in the similar way to the wide angle fiber diagram. (Much more careful and precise measurement is required)	Fiber structure at much larger dimensions (If the central diffuse scattering is observed at very small-angles, this should be separated and may be analysed in the way similar to that in a))

ε: scattering angle (= 2θ), D_L replaces L for the long period with circular and elliptical scattering. Z: Paracrystalline lattice factor

9. Identification of Crystals by X-Ray Diffraction

9.1 Principles of Identification

The structure of a crystal is essentially a three-dimensional periodic repetition of a basic structural element (the unit cell), as was shown in Section 3.2.1. The various groups of crystal planes may be identified in terms of a particular unit cell by means of indices (*hkl*) which refer to their intercepts with the axes of the unit cell. According to Eq. 4.1, the angle of diffraction 2θ for X-rays diffracted by a set of crystal planes depends upon the interplanar spacings d. This implies that if two different crystalline substances happen to have unit cells of the same size and shape, all the spots in the X-ray diffraction patterns of the two crystals will appear in exactly the same positions. Even if the size and shape of the unit cells are identical, however, the nature and numbers of the atoms in the cells and their spatial arrangement must be different because the substances are different. From Eq. 4.3, therefore, the diffracted intensities for corresponding crystal planes should differ. Thus, even if there is a strong superficial resemblance between the diffraction patterns of two different substances, it will always be possible to distinguish between them; there is a one-to-one correspondence between crystal and diffraction pattern. In other words, under identical conditions, crystals of the same substance will always give identical diffraction patterns (in both position and intensity), and identity of diffraction patterns is definitive evidence for identity between the diffracting crystals.

It is clear from the above that differences in the positions of the diffraction lines or spots (*i.e.* in the diffraction angles) are indicative of differences in the unit cells. Differences in the relative intensities of the various diffraction lines or spots reflect differences in the internal structures of the unit cells. It follows, therefore, as we have seen in Section 8.5, that any meaningful comparison of diffraction patterns for qualitative identifications depends upon effective standardization of all factors that affect the locations or intensities of the diffraction pattern elements.

9.2 Identification by the Powder Method

If it is possible to hazard an informed guess at the nature of the unknown substance, a simple and positive identification is often possible by comparison of its X-ray diffraction pattern with that of the substance it is thought to be.

The diffraction data of the unknown substance are generally compared with standard diffraction data. This presupposes the availability of standard X-ray diffraction data for the widest range of substances possible. The most comprehensive compilation of this type, both in quality and quantity, is the JCPDS[†1] file (former ASTM[†2] file), which is supplemented annually. There are others suitable for identifications purposes, though none on the same scale.

9.2.1 The JCPDS system

The JCPDS Powder Diffraction File (JCPDS-PDF),[1)] containing over 46,000 data on organic and inorganic compounds, is available on cards and microfiche and in book form.[1)] It should be noted, however, that from Set 37 no data on cards have been released. Electronic database on CD-ROM and magnetic tape convenient for quick search is also available.

†1 JCPDS: Joint Committee for Powder Diffraction Standards.
†2 ASTM: American Society for Testing and Materials.

9.2.2 Locating a JCPDS card

A. Search by computer

Rapid identification is possible by electronic computer with the aid of the JCPDS database. This database can be installed in the computer attached to most of the latest versions of the X-ray powder diffractometer, and identification can be made immediately after measurement of the diffraction patterns.

Among the search program systems, Johnson/Vand,[2] PDSM,[3] and SANDMAN[4] are well known. The Jonson/Vand program system is the earliest one, and is offered free by the JCPDS to those who purchase the JCPDS-PDF in magnetic tape: the source program is open to the public.

B. Search by hand

Indexing card(s) for a given substance can be found manually with aids such as the JCPDS Index by Hanawalt method[5] or other indexing methods such as Fink Inorganic Index and Mathews Coordinate Index (Termatrex Index).

C. Procedure for identification

The procedure to identify the unknown substances is divided into the two cases discussed below.

a) Single phase. If the unknown specimen consists of a single substance, and

1) if the diffraction data observed can be located in the JCPDS-PDF, searching by hand may not be so difficult. However,

2) if the diffraction data observed cannot be located in the JCPDS-PDF, but crystal data are obtainable elsewhere, or

3) if the unknown specimen is a new phase, determination of unit cell parameters followed by indexing of all the diffraction patterns observed is required. Graphical method by Hull,[6] Bijurström,[7] Harrington,[8] or the methods of Hesse[9] and Lipson[10] may be used for a crystal belonging to a system with higher symmetry. For any crystal system, the use of Ito's method and related articles[11–16] is recommended. If the crystal data determined, searching will be carried out in a way similar to the identification of single crystals described below (Section 9.3).

b) Two or more phases. If the unknown specimen consists of two or more component substances, and

1) if the diffraction data of all the component substances can be located in the JCPDS-PDF, hand search may take longer, and computer search is much better.

2) If the diffraction data of one component is not found but those of the other components are found in the JCPDS-PDF, subtract the contributions of all the components identified by the JCPDS-PDF from the diffraction patterns observed, then try to identify the unknown component using residual diffraction patterns.

3) If the diffraction data of more than two components are not found, a very rare and difficult case, try using the last method 2) described above.

In practice, the identification will be carried out in the following order:

[a) Single phase -1)] → [b) Two or more phases -1] → [a)-2)] → [a)-3)] → [b)-2)] → [b)-3)].

9.3 Identification by the Single Crystal Method

Databases of structural information on crystals, such as their lattice constants, space group, measured and calculated densities, and coordinates of the atoms in the unit cell, as well as optical and morphological data may by used for this purpose. All these databases are updated yearly.

9.3.1 Computer databases

A. Cambridge structural database system

Among the database systems, the Cambridge Structural Database System (CSDS), compiled and offered by the Cambridge Crystallographic Database Center, U.K, is excellent. The CSDS system consists of two major components: the Cambridge Structural Database (CSD) and Software for Search, Retrieval, Analysis and Display of CSD contents.

The CSD is mainly a database of small organic and organometallic molecules. The CSD continues to grow in number of entries: presently the entries are more than a quarter of million, 251,515 (30 Oct. 2001) and they are expected to increase to half a million in 2010. A total of 227,181 (90.3%) entries is with $R^{\dagger} < 0.100$, which includes 202,848 (80.7%) with $R < 0.075$ [125,112] (49.7%) with $R < 0.050$ and 22,346 (8.9%) with $R < 0.030$].[17)]

The software system is available in two forms: Basic software system and Graphics software system.

B. Protein data bank

The Protein Data Bank (PDB) is the single worldwide archive of primary structural data of biological macromolecules, which contains a total of 18,770 molecules (17,233 protein, peptides and viruses, 839 protein/nucleic acid compleses, 684 nucleic acids and 14 carbohydrates) and also a total of 3,352 NMR experimental entries (2 July 2002).[18)]

C. Database for inorganic compounds and minerals and for metals and alloys

For inorganic compounds and minerals, the Inorganic Crystal Structure Database, FIZ, Germany (41,629 entries (1996)) is available[19a)], and CRYSMET:CITI/NRC, Canada[19b)] (> 4,000 entries) can be used for the identification of metals and alloys.

9.3.2 Others

Besides the databases mentioned above, the Crystal Structure Determinative Tables[20a)] may be of some help for identification purposes. These are also distributed by magnetic tape as NBS Crystal Data File,[20b)] and an indexing program system, NBS*LATTICE, was released by NBS in 1986.

Strukturberichte (Vols. 1–7) succeeded by Structure Reports (Vols. 8–49 and the latest publications) and Molecular Structure and Dimensions also can be referred to for this purpose.

9.4 Identification of High Polymers

This is not necessarily confined to identification of the substance from the crystalline region by the usual means described above, but can extend to complete identification starting with analysis on the basis of each one of the elements of fine texture discussed in Section 7.4.

† for R, see Eqs. 11.38 and 11.39.

However, it is more usual simply to identify the substance from the X-ray diffractions of the crystalline region. The reader is also referred to Section 9.5.

9.4.1 Identification by unoriented X-ray patterns

There is basically no difference between this and the identification of low molecular weight substances by their powder diffraction data. In the case of high polymers, however, the diffraction lines are broadened owing to the imperfections of the crystal, and the number of lines is small. These factors, together with the superposition of the diffuse halo due to amorphous regions, make identification very difficult. Except where the diffraction pattern is very distinctive, therefore, unoriented X-ray diagrams are not often used for identification; oriented specimens are used as far as possible.

9.4.2 Identification by oriented X-ray patterns

This is the same, in principle, as identification by single-crystal X-ray diffraction. Whereas the powder method is a one-dimensional analysis making use only of the interplanar spacing d, this method involves a two-dimensional analysis of the reciprocal lattice using ξ and ζ. It should therefore be capable of higher precision than identification attempted on the basis of unoriented specimens and patterns.

Crystal data of high polymers are given by Miller.[21)]

9.5 X-Ray Diffraction Patterns of Copolymers and Polymer Blends

9.5.1 X-ray diffraction patterns of copolymers

The propensity of a copolymer to crystallize depends upon whether the monomer units A and B are three-dimensionally very similar in shape and size (as in styrene/*o*-fluorostyrene copolymers, *cf.* Table 14.2) or only moderately similar (as in ethylene/carbon monoxide copolymers, *cf.* Table 9.2 below), and upon any differences between the lateral dimensions of the molecular chains. There is a corresponding variety in the possible X-ray diffraction patterns displayed by copolymers, as detailed below.

1. The crystalline diffraction pattern of the homopolymer of just one component may be very distinct (though the lattice constants and intensities will not necessarily be those it would have displayed in isolation).

2. The diffraction patterns of homopolymers of A and B may be superimposed (in this case, too, the lattice constants and intensities may be modified).

3. A new diffraction pattern, distinct from the homopolymer of either component, may appear.

4. The crystallinity may disappear, whereupon the pattern will consist entirely of amorphous diffraction halos.

These are idealized categories, and the diffraction pattern is affected in practice by other factors including the following.

1) The proportions of monomer units A and B.

2) The mutual affinities of the monomer units of A and B (this may be thought of as their mutual solubilities).

3) The relative sizes of monomer units A and B.

4) Any difference in the crystallinities of the homopolymers of A and B.

5) The incidence of blocks of either A or B.

In view of the complexity of the above effects, the analysis of the X-ray diffraction patterns of copolymers can be very tedious, and a positive identification of crystals is often much more difficult than when the identities of solid solutions of low molecular weight are being investigated. Some examples of data derived from the X-ray diffraction patterns of copolymers are given below, together with some typical patterns.

Table 9.1 shows the unit cells derived at different monomer ratios for copolymers of two monomer units whose sizes are very similar, but whose homopolymers differ in the sizes of the side chains, the numbers of repeat units per turn of the helices, and the unit cells.[22] Table 9.2[23] lists the unit cell parameters of a polyketone (ethylene/carbon monoxide copolymer).

Table 9.1 3-Methyl-1-butene/4-methyl-1-pentene copolymers[22]

3-Methyl-1-butene [%]	4-Methyl-1-pentene [%]	m.p. [°C]	No. of units perturn	Unit cell cross section	c [Å]
100	0	300	4/1	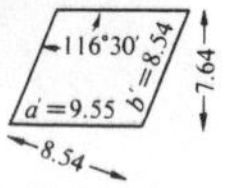	6.85
75	25	256		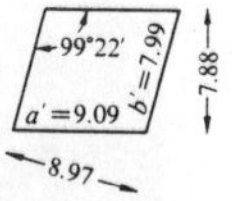	6.85
50	50	218		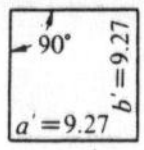	6.85
25	75	208			13.85
13	87	225			
0	100	240	3/1	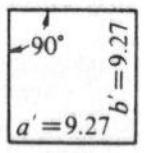	13.85

[Reproduced from F. P. Redding, E. R. Walter, *J. Polym. Sci.*, **37**, 555, John Wiley & Sons, Inc. (1959)]

Table 9.2 Ethylene/carbon monoxide copolymers[23]

Ethylene/CO ratio	a [Å]	b [Å]	c [Å](fiber axis)
∞ (polyethylene)	7.36	4.29	2.534
3.5/1	7.53	4.91	2.54
2.2/1	7.73	4.85	2.54
1.3/1	7.86	4.81	2.54
1.1/1			7.57
1/1	7.97	4.76	7.57

[Reproduced with permission from Y. Chatani *et al.*, *J. Polym. Sci.*, **62**, S27, John Wiley & Sons, Inc. (1962)]

Figure 9.1[24] shows the diffraction patterns of some vinylidene chloride/vinyl chloride copolymers. The detailed crystal structure of copolymers presents many difficult problems, but some interesting articles on the X-ray diffraction patterns of various copolymers have been published. Examples of such articles include a paper by Beevers and White[25] on copolymers of acrylonitrile and styrene and an article by Ishibashi.[26]

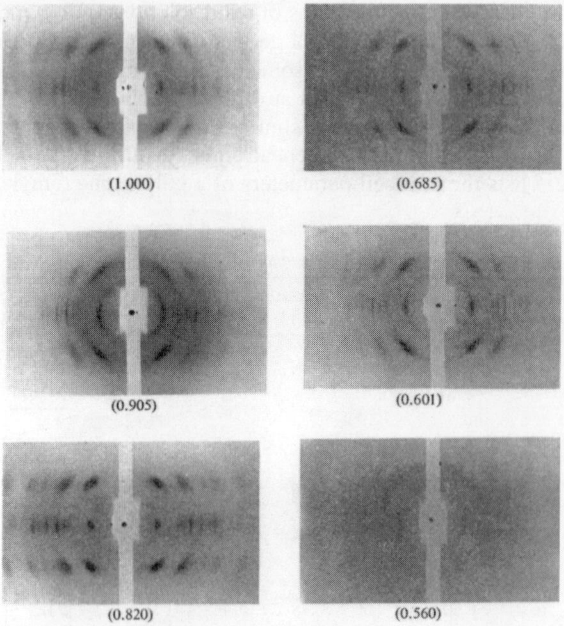

(a)

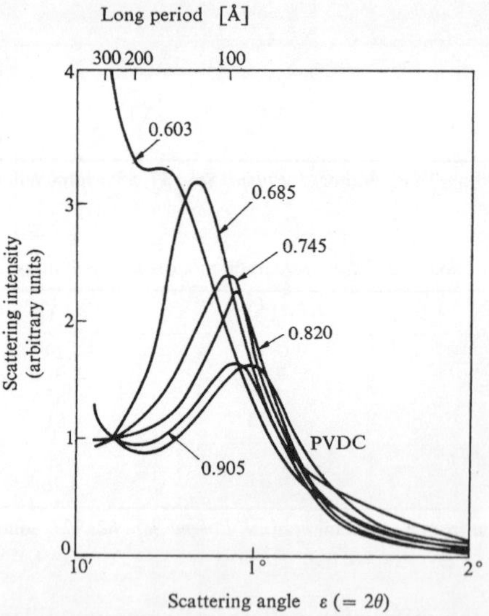

(b)

Fig. 9.1 X-ray diffraction from vinylidene chloride (VDC) /vinyl chloride (VC) copolymers.[24)]

(a) Wide-angle diffraction patterns—The figures refer to mole fractions of VDC. From fractions 1.000 to 0.603 the patterns change little. Crystallinity and orientation suddenly collapse at mole fraction 0.560.

(b) Small-angle scattering intensity curves—The specimens were cold drawn then heat-treated. Long period diffraction is visible down to mole fraction 0.603, but only central diffuse scattering is found at 0.560.

[Reproduced with permission from K. Okuda, *J. Polymer Sci.*, **A2**, 1749, John Wiley & Sons, Inc. (1964)]

9.5.2 X-ray diffraction patterns of polymer blends

Identification seems to be often rather simpler for blends than for copolymers. Thus the ratio of the components should be proportional to the intensities of the crystalline diffraction

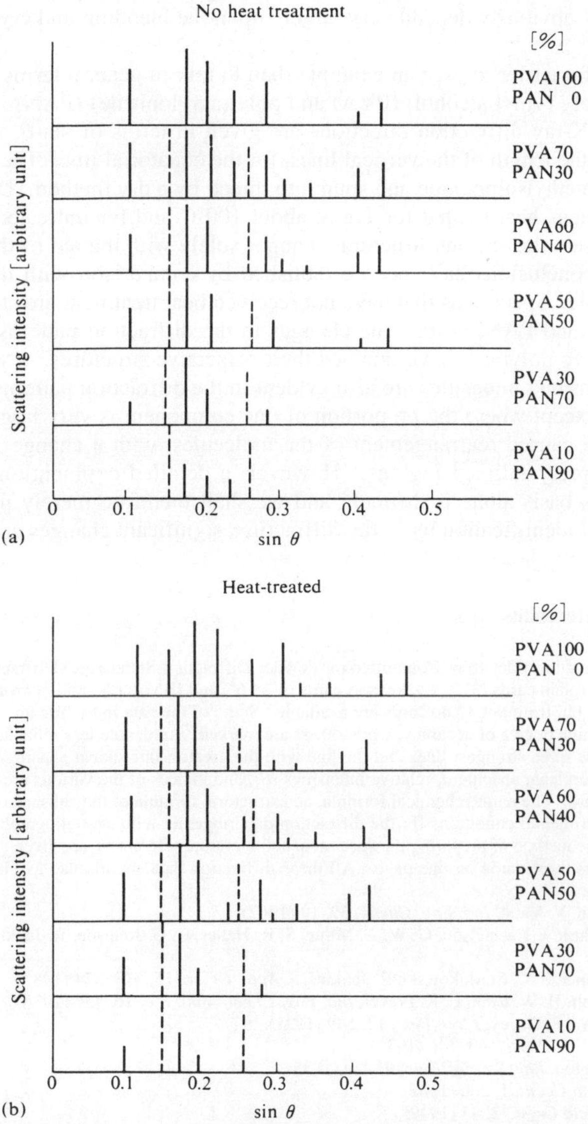

Fig. 9.2 Equatorial X-ray diffractions from PVA-PAN blended fibers.

patterns of the components. The crystal growth of the two components may, however, be affected in some cases by the degree of mutual dispersion of the component polymers, and by the interactions between polymers. If the mixing is such as to bring the molecular chains of the two components into fairly intimate contact, the eventual structure may be entirely different from any of its components. The degree to which this is observed in a particular case will obviously depend very largely upon the blending and crystallization conditions.

Since it is far easier to give an example than to talk in general terms here, we cite data for blends of poly (vinyl alcohol) (PVA) and poly (acrylonitrile) (PAN) in various ratios in Fig. 9.2. The X-ray diffraction directions are given in terms of sin θ, and the intensities represented by the length of the vertical lines, for the equatorial line. The components were dissolved in dimethylsulphoxide and spun into thread by a dry method. Diagrams are given both for specimens heat-treated for 1 h at about 100°C and for untreated specimens. It is difficult to explain the various structural changes solely with the aid of these diagrams, but the following conclusions have been established by comparison with the diffraction diagrams as a whole. Specimens that have not received heat treatment are still recognizable as blends of PVA and PAN, despite the changes in the diffraction patterns. It therefore appears that the two polymers have retained their respective structures. Some changes in the relative positions and intensities are also evident in the diffraction patterns of the heat-treated specimens except where the proportion of one component is very large. The heat treatment may have caused rearrangement of the molecules, with a change in the structure at least near the recrystallized regions. However, a detailed explanation of the structural changes on this basis alone is difficult, and we shall therefore merely note that, from the point of view of identification by X-ray diffraction, significant changes are observable.

Notes and References

1. Powder Diffraction File. Joint Committee on Powder Diffraction Standards. Diffraction data are available in card form (plain cards (3″ × 5″), keysort cards (4″ × 6″) and IBM cards) and microfiche and in book form up to Set 36, but from Set 37 no cards are available. Some of the data in the file are marked ★ to indicate a particularly high degree of accuracy, while others are marked ○ to denote less reliable data. At top left each card gives the three strongest lines and the line with the greatest interplanar spacing, together with intensities. The interplanar spacing d, relative intensities I/I_1, and indices of the various arcs are given in the right half of the card. The name, chemical formula, and structural formula of the substance are given at top right, and the experimental conditions for the diffraction data, together with crystallographic data and optical information, the method of preparing the specimen, and literature references, are given at lower left. The book form contains three cards on one page. All these diffraction data set are also available on CD-ROM and magnetic tape.

2. C. G. Johnson, V. Vand, *Ind. Eng. Chem.*, **59**, 19 (1967).
3. R. G. Marquardt, I. Katsnelson, G. W. A. Milne, S. R. Heller, C. G. Johnson, R. Jenkins, *J. Appl. Cryst.*, **12**, 629 (1979).
4. W. N. Schreiner, S. C. Surdukowski, R. Jenkins, *J. Appl. Cryst.*, **15**, 513, 524 (1982).
5. J. D. Hanawalt. H. W. Linn, L. K. Frevel, *Ind. Eng. Chem. Anal. Ed.*, **10**, 457 (1936).
6. A. W. Hull, W. P. Davey, *Phys. Rev.*, **17**, 549 (1921).
7. T. Bijurström, *Z. Physik*, **69**, 346 (1931).
8. R. A. Harrington, *Rev. Sci. Instrum.*, **9**, 429 (1938).
9. R. Hesse, *Acta Cryst.*, **1**, 200 (1948).
10. H. Lipson, *Acta Cryst.*, **2**, 43 (1949).
11. T. Ito, *Nature*, **164**, 755 (1949).
12. T. Ito, *X-Ray Studies on Polymorphism*, pp. 87–228, Maruzen, Tokyo (1950).
13. D. Louër, M. Louër, *J. Appl. Cryst.*, **5**, 271 (1972).
14. P. E. Werner, *Z. Krist.*, **120**, 75 (1964).

15. J. W. Wisser, *J. Appl. Cryst.*, **2**, 89 (1969).
16. P. M. De Wolf, *J. Appl. Cryst.*, **1**, 108 (1968).
17. a) F. H. Allen, *Acta Cryst.*, **B58**, 380–388 (2002).
    b) http://www.ccdc.cam.ac.uk/
18. a) H. M. Berman, T. Battistuz, *et al.*, *Acta Cryst.*, **D58**, 899–907 (2002).
    b) H. M. Berman, J. D. Westbrook *et al.*, *Nucleic Acids Res.*, **28**, 235–242 (2000).
    c) http://www.pdb.org/
19. J. D. H. Donnay, G. Donnay, E. G. Cox, O. Kennard, M. V. King, *Crystal Data Determinative Tables*, 3rd ed., American Crystallographic Association. Vol. 1 (organic compounds) and Vol. 2 (inorganic) issued in 1973 contain over 24,000 crystal data reported before 1966. Vol. 3 (organic) and Vol. 4 (inorganic) published in 1978 include about 22,000 data from 1967–69. These are stored on magnetic tape as the NBS Crystal Data File.[20)] NBS*LATTICE is an indexing program system released in 1986.
    a) http://www.fiz-karlsruhe.de/products/iscad.html
    b) http://www.nrc.ca/programs/toth/crystnet.html
20. V. L. Himes, A. D. Mighell, *NBS Technical Note*, 1214 (1985).
21. R. L. Miller, Crystallographic Data for Various Polymers, in:
    a) L. E. Alexander, *X-Ray Diffraction Methods in Polymer Science*, pp. 473–523, John Wiley & Sons, Inc. N.Y. (1973).
    b) Polymer Handbook, (J. Branup, E. H. Immergut eds.) Wiley-Interscience, N.Y. (1969).
22. F. P. Redding, E. R. Walter, *J. Polymer Sci.*, **37**, 555 (1959).
23. Y. Chatani, T. Takizawa, S. Murahashi, *J. Polymer Sci.*, **62**, S27 (1962).
24. K. Okuda, *J. Polymer Sci.*, **A2**, 1749 (1964).
25. R. B. Beevers, E. F. T. White, *J. Polymer Sci.*, **B2**, 793 (1964).
26. M. Ishibashi, *J. Polymer Sci.*, **B2**, 781 (1964).

10. Analysis of Crystallite Orientation

10.1 Crystallite Orientation and the X-Ray Diffraction Diagram

Figure 10.1 shows a schematic illustration of the structural changes of a high polymer specimen by drawing, according to the fringed-micelle model (*cf.* Section 7.2.2.A), and the corresponding X-ray diffraction diagrams obtained from three different directions, mutually orthogonal two directions perpendicular to the drawing direction and the direction parallel

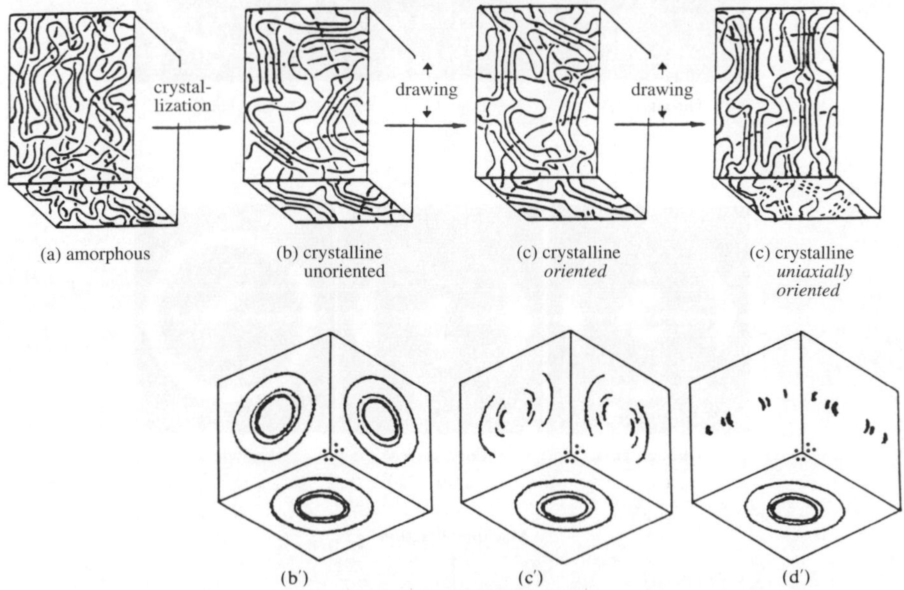

Fig. 10.1 Schematic illustration of structural changes by crystallization and by drawing for a linear high polymer specimen (according to the fringed-micelle model) and the corresponding X-ray diffraction diagrams.

to the drawing direction. An example of X-ray diffraction diagrams obtained for a drawn Nylon 12 specimen from the three directions is given in Fig. 10.2.[1] By drawing or by other mechanical treatments, the central diffuse small-angle scattering taken from the directions mentioned above also show different diagrams corresponding to the different wide-angle diffraction diagrams. An example of a doubly oriented poly(ethylene terephthalate) specimen is given in Fig. 15.36. Fig. 10.3 depicts a schematic illustration of structural change in the metal wire by drawing from a random orientation to a uniaxial orientation. By drawing the randomly oriented $< 111 >$ axis of each cubic unit cell has become parallel to the draw direction.[2] This chapter deals with the preferred orientation of crystallites in high polymer substances. We can easily apply this method to determine the preferred orientation of other substances included in high polymer specimens, such as the preferred orientation of platelet crystals of $Mg(OH)_2$ in a rolled rubber sheet. The preferred orientation of molecular chains in the amorphous state will be discussed briefly in Section 14.5.

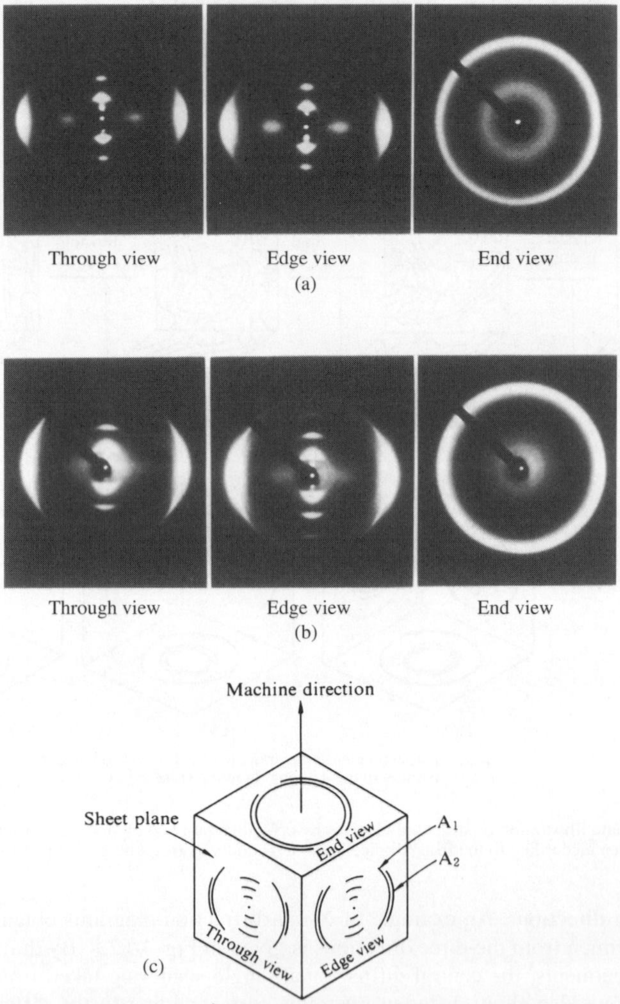

Fig. 10.2 X-ray diffraction diagrams of a drawn Nylon 12 specimen taken from three different directions.[1)]
(a) drawn out uniaxially to 3.0 times from its original length at 100°C (A_1 and A_2 not split).
(b) as above at 160°C (A_1 and A_2 split).
(c) schematic representation of (b).
[Reproduced with permission from T. Ishikawa *et al.*, *J. Chem. Soc. Jpn., Chem. Industrial Chem.*, p. 103, Chem. Soc. Jpn. (1978)]

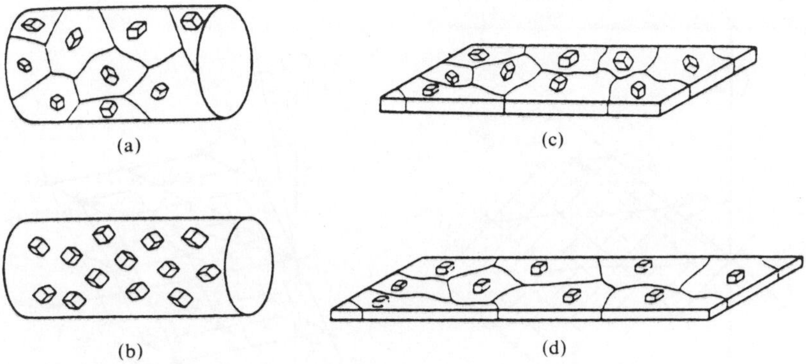

Fig. 10.3 Schematic illustration of preferred orientation in metal wire by drawing.[2)]
(a) Random orientation in a cylindrical specimen.
(b) $<111>$ Axis of each crystallite (represented by a cubic unit cell) is parallel to the draw direction (horizontal direction).
(c) Random orientation in a plate specimen.
(d) (100) plane of each crystallite is parallel to the plane of the specimen.
[Reproduced with permission from T. Imura, *X-Ray Diffraction* (K. Kohra ed.), p. 600, Kyoritsu Pub. (1980)]

10.1.1 General survey

In a well-oriented specimen, the crystallites or crystalline regions[†] are generally aligned in such a way that, say, all the *c* axes lie in the same direction, while the directions of the other two axes are uniformly distributed around the *c* axis direction. Moreover all the directions of the *c* axes are uniformly inclined at a constant angle φ to a reference axis (*e.g.* draw direction), OZ in the specimen (*cf.* Fig. 10.4, p. 242).

Let a point O in the specimen be the origin, and let the origins of all the crystallites be transposed to this point. The reciprocal lattice points R of an arbitrary set of planes (hkl) in the crystallites (where OR is inclined at an angle σ to the c axis, at a distance from the origin $OR = r^* = 1/d_{(hkl)}$) trace out a circle (the reciprocal lattice point circle) with the *c* axes passing through its center, as shown in Fig. 10.4(b). Since the *c* axes of the crystallites are uniformly distributed around the fixed direction OZ in the specimen at an angle φ to OZ, the small circle traced by the reciprocal lattice points R generates a band (the reciprocal lattice point distribution band) by revolution about OZ as shown in Fig. 10.4(b). There is a corresponding band for the ($\overline{hkl}$) planes, and the two together form a symmetrical arrangement along the Z axis (Fig. 10.4(c)).

If X-rays are incident upon the specimen from (say) the $\overline{Y}$O direction, the reciprocal lattice points on the intersection of the distribution band with the diffraction sphere, *i.e.* on the arc R_1R_2 in Fig. 10.5, give rise to diffraction. A diffraction arc P_1P_2, corresponding to the arc R_1R_2, will now appear on a flat film placed normal to the incident X-rays. Due to the symmetrical arrangement of the reciprocal lattice distribution bands along the X and Z axes, the result will be four diffraction arcs arranged symmetrically with respect to the ξ and ζ axes on the film.

In Fig. 10.4(b), the distribution of the reciprocal lattice point R around the *c* axis is constant, but as this axis itself revolves, the density distribution on the reciprocal lattice point

---

† "Crystallite" will be used throughout this chapter to indicate "crystallite", "crystalline region" or "crystalline part."

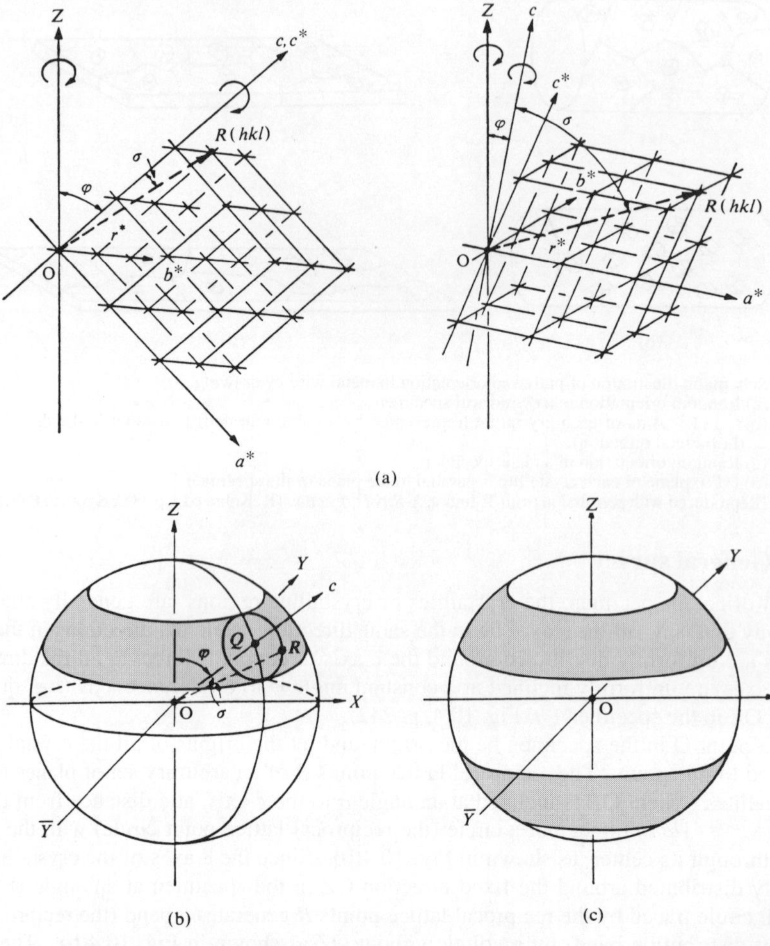

Fig. 10.4 Crystallite orientation, the reciprocal lattice, and the reciprocal lattice point distribution bands.
(a) Reciprocal lattice orientation corresponding to the orientation of the crystallites. Left: orthorhombic; right: triclinic system.
(b) Distribution band for an arbitrary reciprocal lattice point R.
(c) Symmetrical arrangement of the distribution bands about the Z axis.

distribution band is at its densest along the upper and lower edges of the band, and is least along the path of the center Q. The intensity distribution along the diffraction arc P_1P_2 is accordingly greatest at its ends and least in the center (Fig. 10.5(b)). The two ends of the diffraction arc P_1P_2 are located by the following relationships (*cf.* Fig. 10.5(a)):

$$\cos\delta_1 = \cos(\varphi - \sigma)/\cos\theta, \qquad \cos\delta_2 = \cos(\varphi + \sigma)/\cos\theta \tag{10.1}$$

The form of the reciprocal lattice point distribution bands, and hence also of the diffraction patterns, varies with the relationship between φ and σ. The four arcs may be joined in pairs to give two arcs intersecting either the ξ or the ζ axis, or all four may even

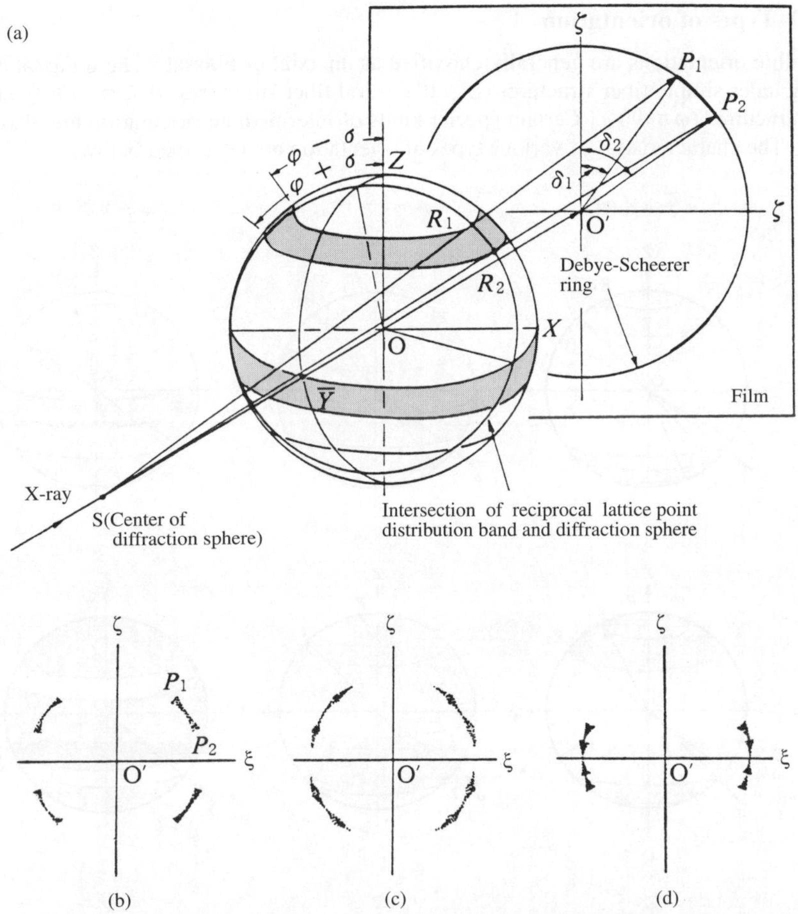

Fig. 10.5 Spiral Orientation.
(a) Geometrical relationship between lattice point distribution and the diffraction pattern.
(b) Form of the diffraction pattern that would be given by an ideally oriented sample.
(c) General form of the diffraction pattern as actually observed.
(d) Form of the diffraction pattern that would be given by an ideally oriented sample where $(\varphi + \sigma) > 90°$ (see Fig. 10.6), *i.e.* where the distribution bands overlap.

be joined to form a complete circle. The principal variants of the forms of the reciprocal lattice point distributions and of the diffraction patterns are shown in Figs. 10.5, 10.6, 10.7, and 10.9. Diffraction patterns for the ideally oriented state are shown, together with sketches of those actually observed. In practice, the orientations of the crystallites in a specimen never correspond to a single fixed value of φ. Some crystallites with orientations at different angles will always be present, the number of crystallites inclined at a given angle decreasing as the deviation $\Delta\varphi$ of this angle from φ increases (see Fig. 10.7(b) and(d)). Examples of the kinds of diffraction patterns obtained under these circumstances are also given in Figs. 10.5, 10.7, and 10.9.

10.1.2 Types of orientation

Crystallite orientations are generally classified as uniaxial or biaxial. The uniaxial orientation includes simple fiber structures ($\varphi = 0°$), spiral fiber structures ($0 < \varphi < 90°$), and ring fiber structures ($\varphi = 90°$). Certain special kinds of intermediate orientation are also recognized. The characteristics of various types of orientation are described below.

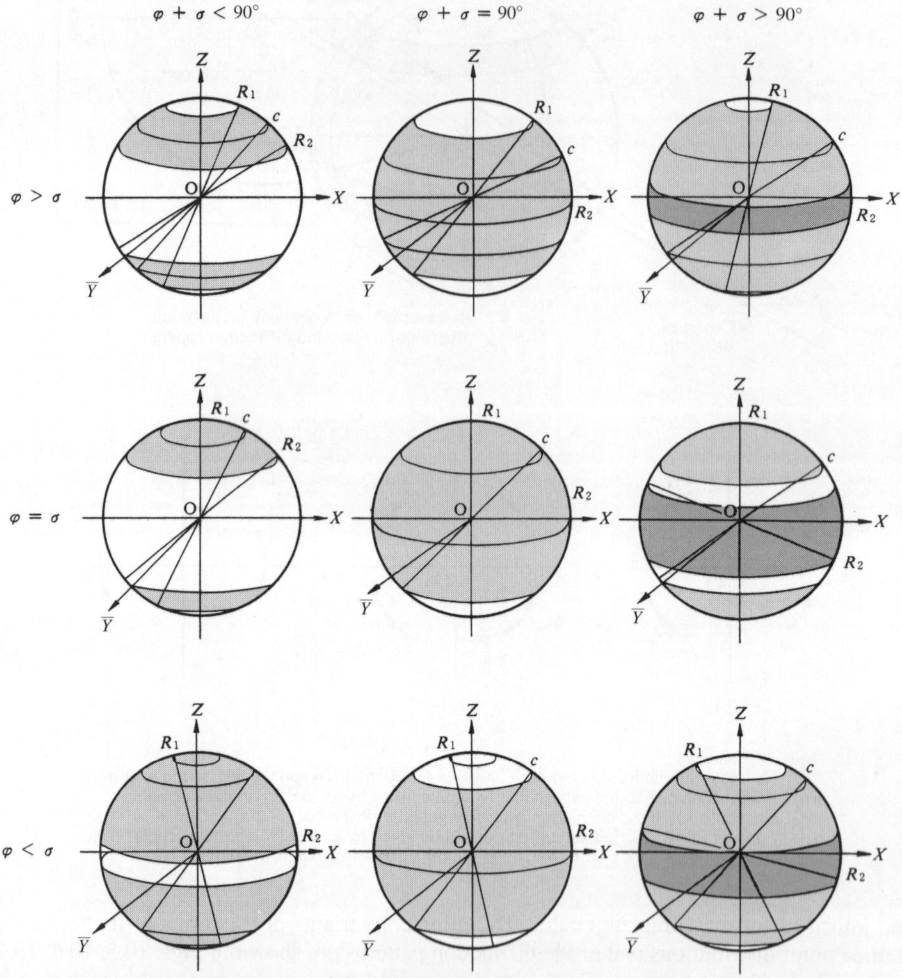

Fig. 10.6 Various forms of reciprocal lattice point distributions for spiral orientation. As in Figs. 10.4 and 10.5, φ is the helix angle (*i.e.* the inclination of the c axis to the fiber axis) and σ is the angle between the reciprocal lattice vector and the c axis.

A. Uniaxial orientation (simple fiber structure)

Here ($\varphi = 0°$, *i.e.* the c axis coincides with the Z axis (Fig. 10.7). The reciprocal lattice is cylindrically symmetrical with respect to the fiber axis (*cf.* Fig. 4.14(c)), and the X-ray dia-

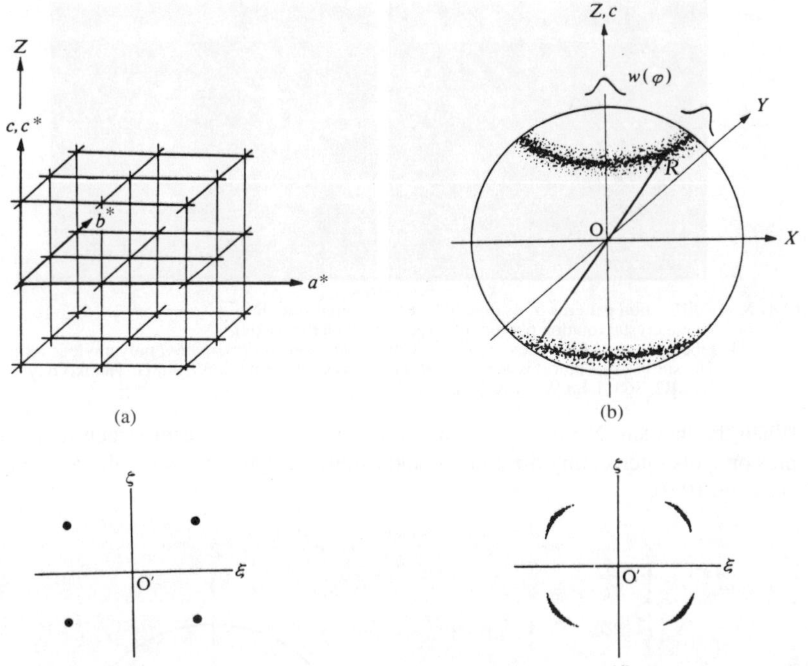

Fig. 10.7 Uniaxial orientation, the reciprocal lattice point distribution, and the resultant diffraction pattern (*cf.* Figs. 4.14, 4.15).
(a) Orientation with respect to the fiber axis Z (ideal case, $\varphi = 0$).
(b) Distribution for a reciprocal lattice point R, showing the effect of small deviations, $w(\varphi)$.
(c) Diffraction from an ideal specimen.
(d) Diffraction from an actual specimen with imperfectly aligned crystallites (*cf.* (b)).

gram obtained with the incident X-rays normal to the fiber axis is essentially the same as a complete rotation photograph for a single crystal (*cf.* Section 4.4.4 and Fig. 10.8). Since $\varphi = 0°$, we find for the general reciprocal lattice point:

$$\delta = \delta_1 = \delta_2, \quad \therefore \quad \cos\delta = \cos\sigma/\cos\theta \tag{10.2}$$

and a four-point diagram is obtained instead of a four-arc diagram. If the incident X-rays are parallel to the fiber axis, on the other hand, the diffraction pattern made for this direction is as if the specimen were unoriented, forming a complete circle (*i.e.* a Debye-Scherrer ring see Fig. 10.1(c′)and (d′)).

B. Spiral orientation (spiral fiber structure)

Spiral fiber structure is the most general case, $0° < \varphi < 90°$, and all that appears in connection with Fig. 10.4 to 10.6 (above) applies to this category of structures.

C. Ring orientation (ring fiber structure)

Here $\varphi = 90°$, and Eq. 10.1 gives, therefore,

$$\cos\delta_1 = \sin\sigma/\cos\theta = -\cos\delta_2 \quad \therefore \quad \delta_2 = 180° - \delta_1 \tag{10.3}$$

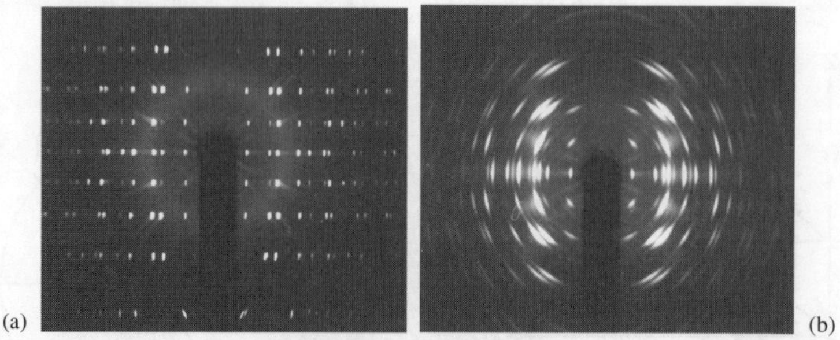

Fig. 10.8 X-ray diffraction patterns of a polyethylene oxide-urea complex.
(a) Single crystal rotation photograph (rotation about the *c* axis).
(b) Fiber diagram of complex prepared by immersing *uniaxially oriented* polyethylene oxide in urea. The similarity with (a) is evident. [Reproduced with permission from H. Tadokoro *et al.*, *J. Polym. Sci.*, **B2**, 363, John Wiley & Sons, Inc. (1964)]

When the incident X-rays are normal to the *X* axis, this type of structure always gives two arcs or spots intersecting the equator and extending symmetrically above and below the equator (Fig. 10.9).

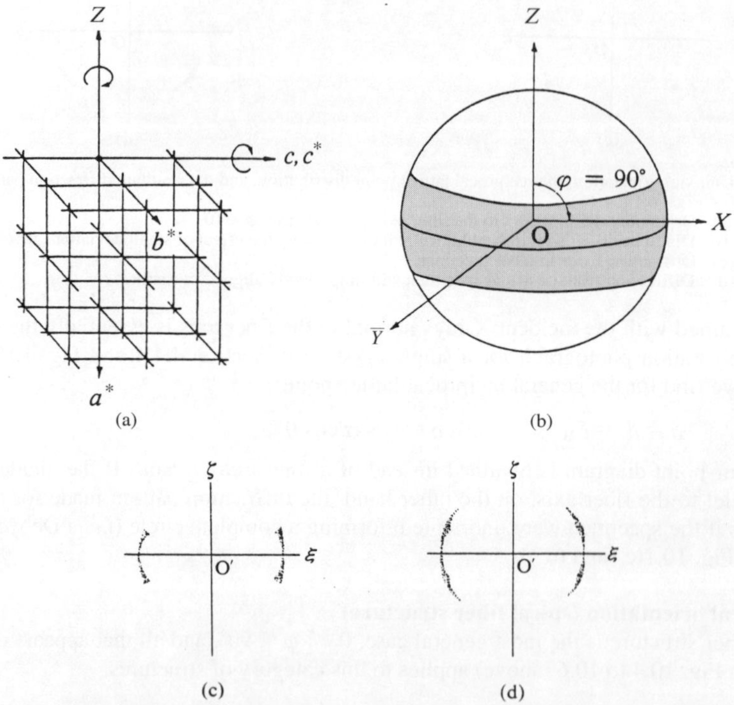

Fig. 10.9 Ring orientation, the reciprocal lattice point distribution, and the resultant diffraction pattern.
(a) Orientation with respect to the fibre axis *Z* (ideal case, $\varphi = 90°$).
(b) Distribution for a reciprocal lattice point.
(c) Diffraction from an ideal specimen.
(d) Diffraction from an actual specimen.

If the incident X-rays are in the Z direction, however, the diffraction sphere and the reciprocal lattice point distribution band are parallel. Thus, when the diffraction sphere intersects the distribution band, a circle of uniform intensity is produced. The diffraction diagram is therefore of the same form as that given by an unoriented specimen. To distinguish between an unoriented structure and a ring-oriented structure in such cases, the specimen is inclined (*cf.* Section 10.1.3). The diffraction diagram of an unoriented specimen will be unchanged, whereas a change to two arcs intersecting the equator indicates ring orientation.

D. Tilted orientation

In tilted orientation, the crystallite distribution corresponds to a cylindrically symmetrical distribution of the tilted c axis, *with no rotation* about the c axis. The reciprocal lattice is shown in Fig. 10.10. It is evident that the equatorial line and the layer lines are generally not well developed, and vary with the angle of inclination. The diffraction is observed around the position in which it should appear as a layer line in the case of uniaxial orientation. This corresponds to a rotation photograph for a single crystal with inaccurate alignment of the crystal (a tilted single crystal), and where the axis of rotation corresponds to the fiber axis.

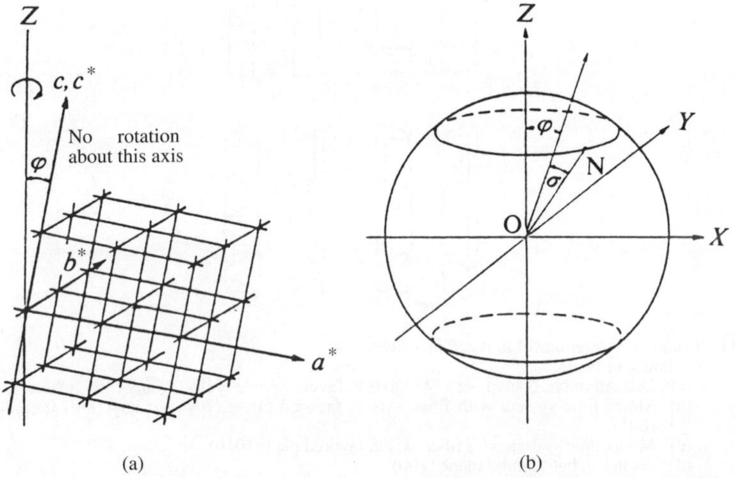

Fig. 10.10 Tilted orientation and the corresponding reciprocal lattice point distribution.
(a) Orientation with respect to the fiber axis Z: $0^\circ < \varphi < 90^\circ$, but this orientation is not to be confused with spiral orientation; there is no rotation about the c axis.
(b) Distribution for a reciprocal lattice point R. Compare the uniaxial orientation in Fig. 10.7.

E. Biaxial orientation

This type of orientation is the most restrictive one. Like a single crystal of low molecular weight compound mounted on a goniometer head, a crystal axis is oriented parallel to the Z axis (reference axis) and one of the other two axes is also oriented parallel to a reference plane (say) the film plane (for further details see Section 10.3.2.). This type of oriented structure cannot be further ordered except by decreasing the number of crystallites by joining them into larger ones. The fact that a Weissenberg or precession photograph is very similar to that of a single crystal enables ready identification of this type of orientation.

The density distributions of the various reciprocal lattice points can be determined from the intensity distribution of diffractions on a Weissenberg or precession photograph.

F. Double orientation

In this structure, a crystal axis is oriented parallel to the Z axis and a particular crystal plane to lie in the reference plane, the film plane. In many circumstances, there may be two or more different orientations of the unit cell in the specimen; four different orientations are possible for crystals of the monoclinic and triclinic systems (see Fig. 10.11).[3] This type of orientation gives photographs which are like the oscillation photographs of single crystals or twinned crystals (see Fig. 10.8).

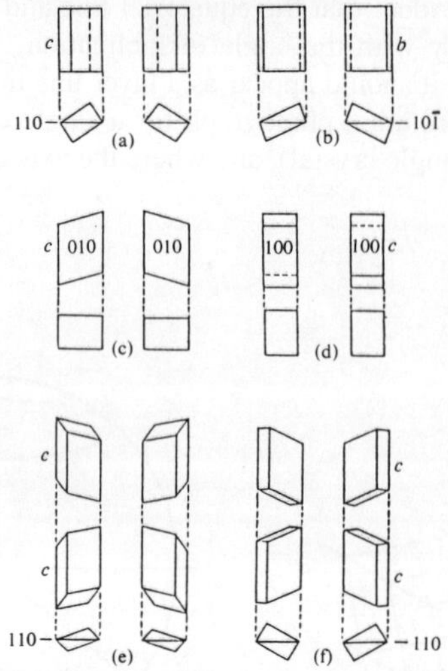

Fig. 10.11 Twinned orientations of unit cell in specimen.[3]
Examples of twins.
(a) Orthorhombic system with fiber axis c, favored plane (110) (or any $(hk0)$ where $h, k \neq 0$).
(b) Monoclinic system with fiber axis b, favored plane (101) (or any $(h0l)$ including (100) and (001)).
(c) Monoclinic system with fiber axis c, favored plane (010).
(d) As in (c), but favored plane (100).
Examples of quadruplets.
(e) Triclinic system with fiber axis c, favored plane (110) (or any $(hk0)$, including (100) and (010)).
(f) Monoclinic system with fiber axis c, favored plane (110) (or any $(hk0)$ where $h, k \neq 0$).
[Reproduced from C. W. Bunn, *Chemical Crystallography*, p. 179, The Clarendon Press (1945)]

10.1.3 Interpretation of inclined X-ray diagrams

These are the diffraction diagrams obtained when the incident X-rays are inclined at an oblique angle with respect to the Z axis. The locus of the reciprocal lattice point distribution is unchanged. However, since the diffraction sphere is inclined as shown in Fig. 10.12(a) (angle of inclination $= \mu$), the top two of the four arcs are situated closer together, while the bottom two are farther apart.

If the angular displacements of the ends of the upper and lower arcs from the ζ axis are δ_1, δ_2, δ_3, and δ_4, we obtain relationships of the type (*cf.* Fig. 10.12 (b))

$$\left.\begin{aligned}\cos\delta_1 &= \{\cos(\varphi-\sigma)+\sin\mu\sin\theta\}/\cos\mu\cos\theta\\ \cos\delta_2 &= \{\cos(\varphi+\sigma)+\sin\mu\sin\theta\}/\cos\mu\cos\theta\end{aligned}\right\} \quad (10.4)$$

$$\left.\begin{aligned}\cos\delta_3 &= \{-\cos(\varphi+\sigma)+\sin\mu\sin\theta\}/\cos\mu\cos\theta\\ \cos\delta_4 &= \{-\cos(\varphi-\sigma)+\sin\mu\sin\theta\}/\cos\mu\cos\theta\end{aligned}\right\} \quad (10.5)$$

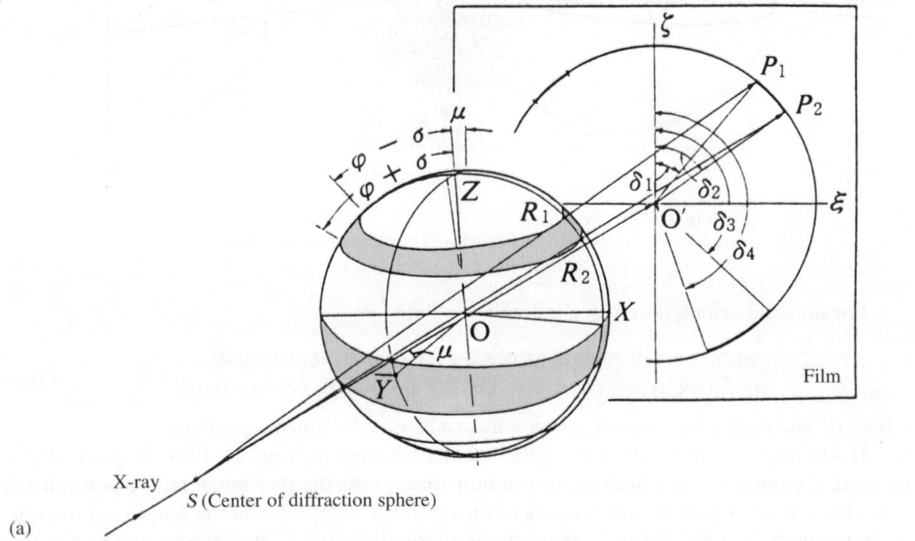

Fig. 10.12 The effect of oblique X-ray incidence upon the diffraction pattern.
(a) Geometry of the reciprocal lattice point distribution and the diffraction pattern.
(b) Variation of the diffraction pattern with inclination angle, μ of X-ray incidence. The arrows indicate movement of the diffraction elements.

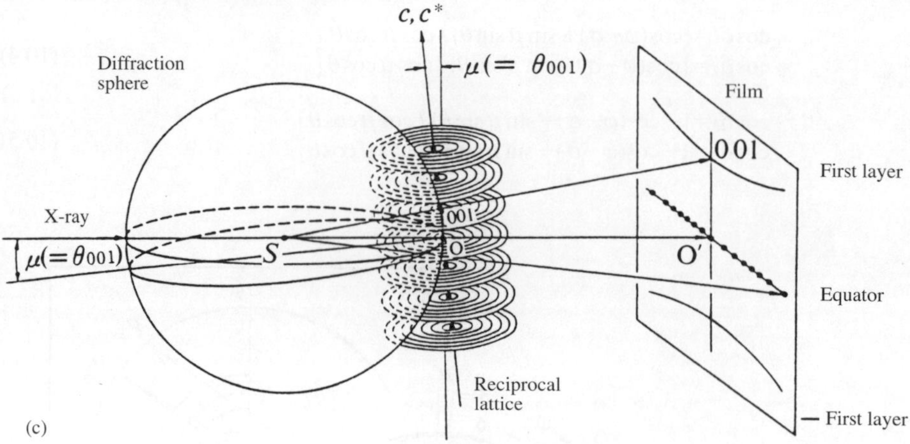

Fig. 10.12 — *Continued*
(c) Ideal case for uniaxial orientation (*cf.* Fig. 4.14(d))

For uniaxial orientation, since $\varphi = 0$, we obtain

$$\left.\begin{aligned} \cos\delta_U = \cos\delta_1 = \cos\delta_2 = \{\cos\sigma + \sin\mu\sin\theta) / \cos\mu\cos\theta \\ \cos\delta_D = \cos\delta_3 = \cos\delta_4 = \{-\cos\sigma + \sin\mu\sin\theta) / \cos\mu\cos\theta \end{aligned}\right\} \qquad (10.6)$$

where δ_U and δ_D are the upward and downward displacements respectively.

If the angle of inclination μ of the incident X-rays in Fig. 10.12(a) is gradually increased, the positions at which the diffraction sphere cuts the reciprocal lattice point distribution band in the upper half of the sphere move closer together, and the lengths of the intersections increase, until the two intersections merge into one. The changes in the positions of the diffraction arcs on either side of the meridian with μ are shown in Fig. 10.12(c).

Inclined X-ray diagrams are very useful for measurements of the 001 diffractions, which do not appear with a normal incident beam. However, for the reasons given above, non-meridional diffractions which appear near the 00*l* diffractions merge to give a single-arc diffraction pattern on the meridian (see also Section 10.3.1A).

10.2 Analysis of the Type of Crystallite Orientation

10.2.1 Establishing the presence or absence of orientation

This is established by inspection of the X-ray diffraction photographs of the specimen. It follows from the relationship between the form of the reciprocal lattice point distribution band and the diffraction pattern (established in Figs. 10.4 to 10.9) that diffractions in the form of points or arcs, *i.e.* rings with non-uniform intensity distributions, indicate the presence of orientation. The presence of orientation may also be assumed if, even when Debye-Scherrer rings of uniform intensity are obtained, certain of the rather intense diffractions that would be expected from an unoriented specimen are missing; there is a strong possibility that uniaxial or ring orientation is responsible in this case (*cf.* Section 10.1). If a single X-ray diffraction photograph does not indicate the presence of orientation, the shapes of the diffraction patterns obtained for various incident directions of the X-rays are examined.

The specimen may safely be assumed to unoriented in the majority of cases if this still fails to produce evidence of orientation. It is advisable, however, to check for the presence of spherulites under (say) a polarizing microscope, particularly for specimens in which spherulites readily develop. It is then finally possible to say whether the uniform intensity distribution of the X-ray diagram is due to averaging of the orientations in all directions caused by the presence of many spherulites small in size relative to the X-ray beam, or whether the specimen is in fact quite unoriented. Section 14.4 deals with the case where oriented diffraction may be masked by amorphous halos.

10.2.2 Identification of the type of orientation

A. Utilizing special features of the X-ray diffraction diagram

Table 10.1 presents various criteria for the determination of the type of orientation in comparatively highly oriented specimens. Using a schedule of this nature, the possible types of orientation may be deduced from the features of the diffraction diagram (*cf.* Figs. 10.4 to 10.9). The next stage is to index the diffraction spots (or arcs), and establish the type of orientation from the reciprocal lattice.

Table 10.1 Criteria for determination of the type of crystallite orientation

1st pattern; X-rays are normal to the specimen axis†	**2nd pattern;** X-rays still normal, but in different direction	**3rd pattern;** X-rays are oblique to the specimen axis	Possible types of orientation
(1) *Layer lines clearly developed*			
Resembles single crystal oscillation/rotation photograph; period in axis direction ≈unit cell edge length →	Pattern *differs*, but still resembles oscillation/rotation photograph	→	Biaxial (Double)
	Pattern unchanged	→	Uniaxial
(2) *Layer lines not very clearly developed* Consists of many arcs	Pattern unchanged	→	Tilted or Spiral
(3) *Layer lines unrecognizable*			
Arcs, intensity concentrated at equator →	Pattern unchanged →	Rings, each with uniform intensity	→ Ring
or			
Rings, each with uniform intensity →	Pattern unchanged →	Arcs, intensity concentrated at equator	→ Ring
		Pattern unchanged →	Unoriented

† Taken as the direction of rolling or drawing.

If the degree of orientation is low, it may not be possible to determine the type of orientation, even though the presence of some kind of orientation may be evident. In such cases the degree of orientation must be increased by suitable treatment such as drawing or rolling to enable identification of the type.

a) Determination of the helix angle φ for specimens with spiral orientation. The helix angle φ for specimens with spiral orientation, *i.e.* the inclination of the c axis with respect to the fiber axis, is found from the azimuthal angle δ as follows:

1) δ_e is found from the $hk0$ diffractions (the equatorial diffractions). Since $\sigma = 90^\circ$ for the equatorial ($hk0$) planes, Eq. 10.1 gives

$$\cos\delta_1 = \sin\sigma/\cos\theta = -\cos\delta_2 \quad \therefore \quad \delta_2 = -\delta_1$$

The complementary angle δ_e of δ_1 satisfies the following equation

$$\sin\varphi = \sin\delta_e \cos\theta \tag{10.7}$$

Thus (1) $\cos\theta \cong 1$, and $\delta_e \cong \varphi$ when the diffraction angle is small, but (2) δ_e is generally slightly greater than φ.

2) δ_m is found from the $00l$ diffractions (the meridional diffractions). If the a and b axes of the crystallites are both perpendicular to the c axis, $\sigma = 0$ for the $(00l)$ planes, and Eq. 10.1 gives

$$\cos\delta_1 = \cos\varphi/\cos\theta = \cos\delta_2 \quad \therefore \quad \delta_1 = \delta_2$$

Whereas the diffractions from the other crystal planes are all in the form of arcs, those from the $(00l)$ planes are points, and give a four-point diagram.

It we let $\delta_1 = \delta_m$, we obtain

$$\cos\delta_m \cos\theta = \cos\varphi \tag{10.8}$$

Thus (1) $\cos\theta \cong 1$, and $\delta_m \cong \varphi$ when the diffraction angle is small, but (2) δ_m is generally slightly smaller than φ.

3) The average of the measured values of δ_m and δ_d gives a reasonably close approximation to the helix angle φ.

4) If only one of δ_m and δ_d can be found, this may be taken as a measure for φ.

b) Determination of the azimuthal angle δ.

1) When a flat film is used, the azimuthal angle δ can be obtained by direct measurement of the angle between the meridian and the point of maximum intensity of the diffraction arcs (or points) on the film.

2) There are two cases when a cylindrical film is used; (1) the fiber axis may be parallel to the cylinder axis, and (2) the axis may be perpendicular. In 1), with the film unrolled (see Fig. 10.13(a)), the value of Z for the point of maximum intensity is measured, and δ found from Eq. 10.9.

$$\cos\delta = \sin\nu/\sin 2\theta, \qquad \tan\nu = z/R_F \tag{10.9}$$

where R_F is the radius of the cylindrical film camera. In the case of 2) the procedure is exactly the same (see Fig. 10.13(b)), with z measured and δ found from Eq. 10.10.

$$\sin\delta = \sin\nu/\sin 2\theta, \qquad \tan\nu = z/R_F \tag{10.10}$$

c) Determination of the fiber axis. The fiber axis may be found from the "layer-line conditions." Thus, in a rotation photograph with the fiber axis $[uvw]$ as the rotation axis, if a diffraction having the indices hkl appears in the nth layer line, the following relationship holds.[4)]

$$hu + kv + lw = n \tag{10.11}$$

The fiber axis can therefore be easily found from the layer line in which the hkl diffraction appears, with the aid of Eq. 10.11.

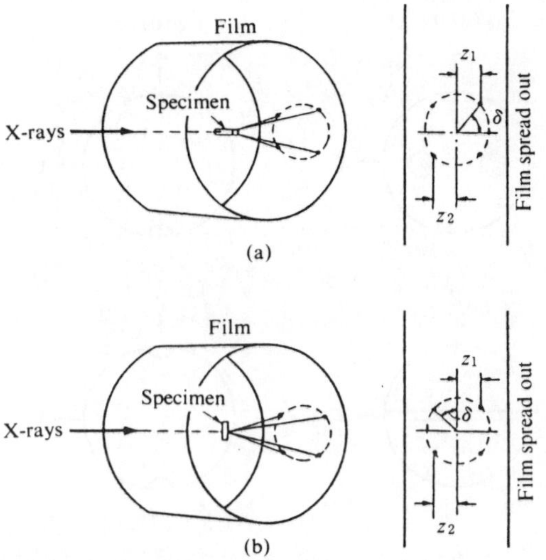

Fig. 10.13 Measurement of the azimuthal angle δ in the case of a cylindrical film.
(a) Fiber axis parallel to the axis of rotation of the camera.
(b) Fiber axis perpendicular to the axis of rotation of the camera Film as unrolled.
[Reproduced with permission from IUCr., *International Tables for X-Ray Crystallography*, (C.H. Mac Gravry, G.D. Rieck eds.), Vol. III, p. 295, IUCr. (1968)]

B. Utilizing the pole figure[5)]

Pole figures are stereographic projections showing the density of crystallographic poles of certain planes as a function of orientation,[6)] and so provide a good method of representing orientation. A pole is the point of intersection of the normal to a crystal plane with the surface of a sphere having the crystal at its center. If the radius of the sphere is $r^* = 1/d_{(hkl)}$, the pole coincides with the reciprocal lattice point of this plane, and the density of the poles on the surface of the sphere is a faithful representation of the reciprocal lattice point density distribution.

If the crystallite orientation is completely random, the poles will be scattered all over the stereographic projection.† If orientation is present, on the other hand, the poles will tend to be concentrated in certain areas within the projection, while the remaining areas will be completely unpopulated.

Figure 10.14 shows pole figures for the (00*l*) planes and those for (*hk*0) planes containing the fiber axis (*c* axis) for various types of orientation. These can be used for comparison with experimental pole figures to determine the type of orientation.

The degree of orientation can also be found from the pole density. The fiber axis can be determined graphically by using a pole figure.[5)]

C. Utilizing small-angle scattering

This heading is included for completeness here, but a full description accompanied by actual examples is deferred until Section 15.4.2.

† For a completely unoriented specimen, the number of poles in a given area on the surface of the projection sphere with the specimen as its center will be constant. Since the stereographic projections is a planar projection of the spherical surface, however, areas are not faithfully reproduced, and the apparent pole distribution even in the pole figure of an unoriented specimen will have a lower concentration near the center.

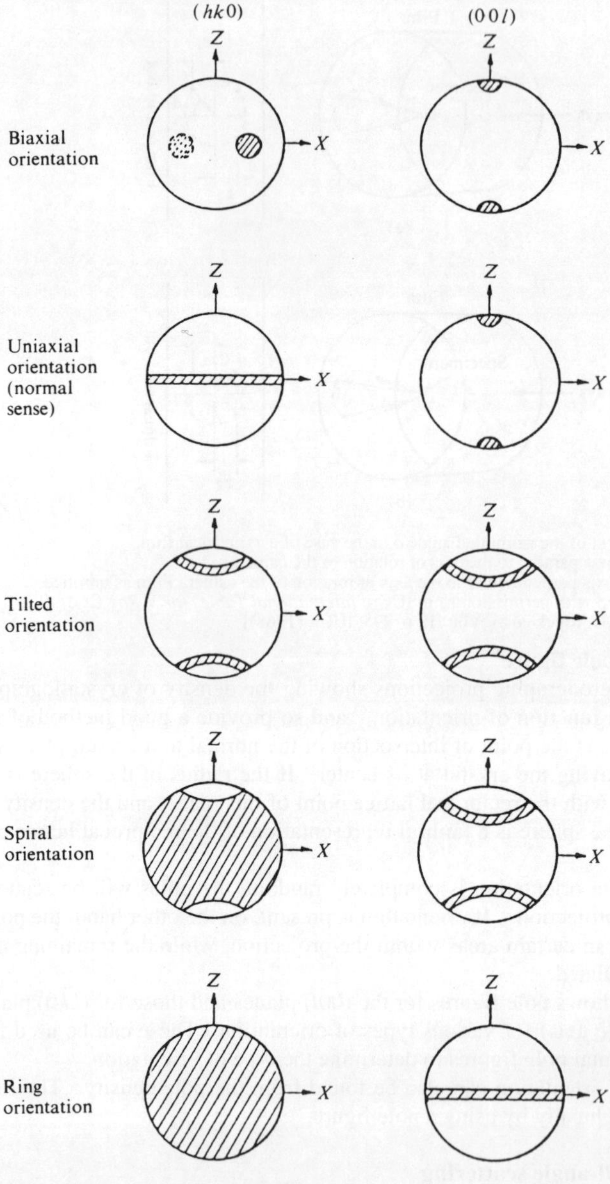

Fig. 10.14 Pole figures for various types of orientation.
The *Z* direction is taken as parallel to that of drawing or rolling (milling), and in the latter the *X* direction is parallel to the rolled surface and normal to the rolling direction.
[Reproduced with permission from C.J. Heffelfinger, R.L. Burton, *J. Polym. Sci.*, **47**, 290, John Wiley & Sons, Inc. (1960)]

D. Determination of the type of orientation

The variation of the type of orientation in the necking portions of cold-drawn polyethylene is described in Section 15.4.1 in connection with the results of small-angle scattering analyses. In the present section we shall discuss the analysis of the type of orientation in the interior of spherulites and of the orientation of extruded polyethylene film.

a) Radial variation of the orientation in spherulites. As was mentioned earlier, if the X-ray beam is very large in proportion to the size of the spherulites in the specimen, the orientations in the various regions inside the spherulites average out, and a completely unoriented X-ray diagram may be obtained. If, on the other hand, we use microbeam X-ray diffraction (*cf.* Section 8.2) with a sufficiently small beam in proportion to the size of the spherulites, the X-ray diagram will give evidence of orientation.

Figure 10.15[7] shows a large polyethylene spherulite measuring 0.5 ~ 1 mm. Extinction rings are observed at intervals of about 50 micrometers. As shown in the diagram, the continuous radial variation of the orientation was traced by making X-ray diffraction patterns at intervals of 13 micrometers along a radius, using a microbeam having dimensions of about 2 × 1.3 micrometers. The variations of the 110 and 200 diffractions are shown in Fig. 10.15(b).

The results were analyzed in terms of the reciprocal lattice as follows. Fig. 10.15(c) shows the limited distribution of the reciprocal lattice points for part of the reciprocal lattice of polyethylene. The intersections of the reciprocal lattice points with the diffraction sphere are therefore arcs, and the diffractions are accordingly also arcs rather than spots. Fig. 10.15(d) shows the projection of the reciprocal lattice onto the a^*c^* plane; I, II, III,..., VII identify the successive directions of the Incident X-rays. The shapes of the intersec-

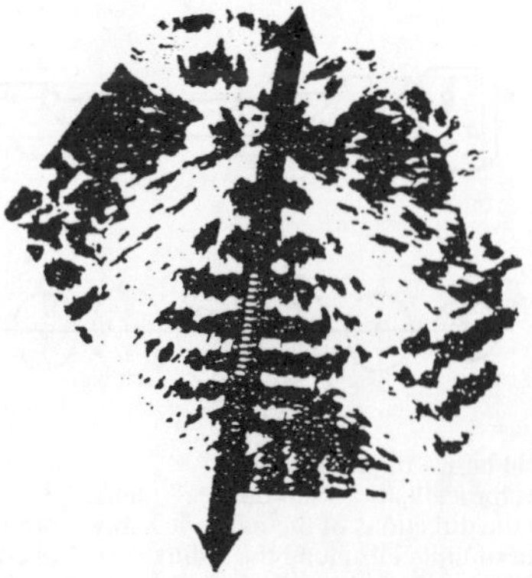

(a)

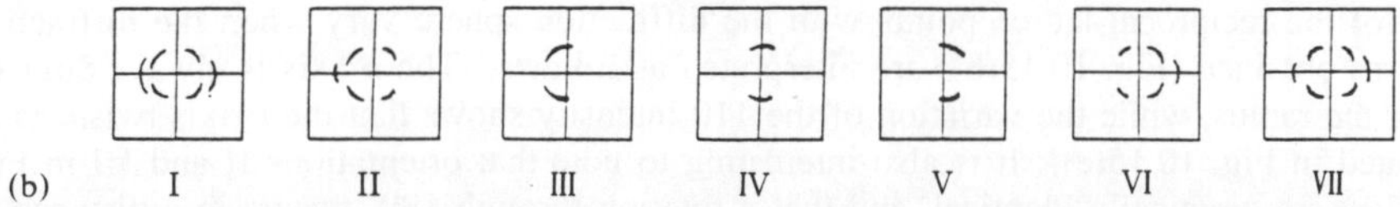

Fig. 10.15 Investigation of the orientation in polyethylene spherulites by microbeam X-ray diffraction.[7]
(a) Photomicrograph of the specimen taken in polarized light. The points marked along the superimposed arrow show the positions at which diffractions were obtained.
(b) Sketches of the 110 and 200 diffractions obtained at successive positions.

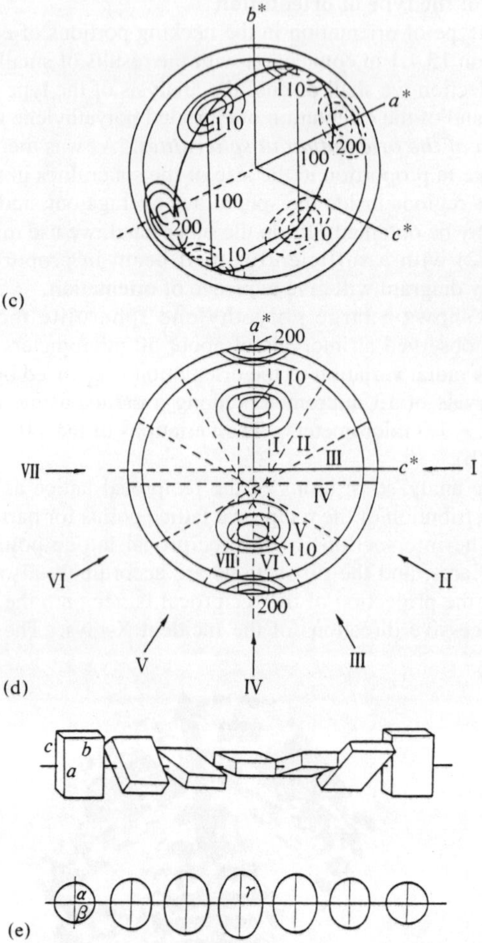

Fig. 10.15 —*Continued*
(c) Part of the reciprocal lattice of polyethylene.
(d) Projection of the reciprocal lattice onto the a^*c^* plane.
I, II,...,VII indicate the directions of the incident X-rays corresponding to the diffractions in (b).
(e) Helical arrangement of unit cells along the radius of the spherulite (upper diagram).
Corresponding rotation of the index ellipsoid with refractive indices α, β, and γ (lower diagram).
[Reproduced with permission from Y. Fujiwara, *J. Appl. Polym. Sci.*, **4**, 11, John Wiley & Sons, Inc. (1960)]

tions of the reciprocal lattice points with the diffraction sphere vary when the diffraction patterns obtained (Fig. 10.15(b)) are interpreted as follows. The *b* axis is always directed along the radius, while the variation of the 110 intensity shows that the *c* axis twists as illustrated in Fig. 10.15(e). It is also interesting to note that orientations II and VI in Fig. 10.15(b) are practically identical, and that a rotation through 180° occurs in a distance of $4 \times 13 = 52$ micrometers. The half-period of this twisting of the *c* axis thus coincides with the spacing of the rings due to double refraction, and a dark ring is observed whenever the *c* axis is perpendicular to the basal plane.

b) Pole figures of extruded polyethylene film. Figure 10.16 shows the pole figures obtained for the 200, 020, and 002 diffractions from extruded polyethylene (Marlex 50) tubular film.[8)]

It is clear from the (200) pole figures that at a blow ratio of 1.7 : 1, the *a* axis is directed along the extrusion direction, but with maximum concentrations at 45° and at 135° to the plane of the film. This result shows that the *a* axes in the extruded material are not parallel to the extrusion direction, as had been thought, but are oriented at an angle of 45° to this direction. As the blow ratio is increased the *a* axes naturally become parallel to the extrusion direction in accordance with the earlier view.

Examination of the (020) pole figures shows that the *b* axes are distributed in a plane perpendicular to the extrusion direction, and that the degree of orientation is much higher than that of the *a* axes. This is also something that had not previously been established conclusively, since the 020 intensity is much lower than the 200 intensity. Moreover, the direction of orientation of the *b* axes shows practically no change with increasing blow ratio.

The (002) pole figures show that at a blow ratio of 1.7 : 1, the orientation of the *c* axis

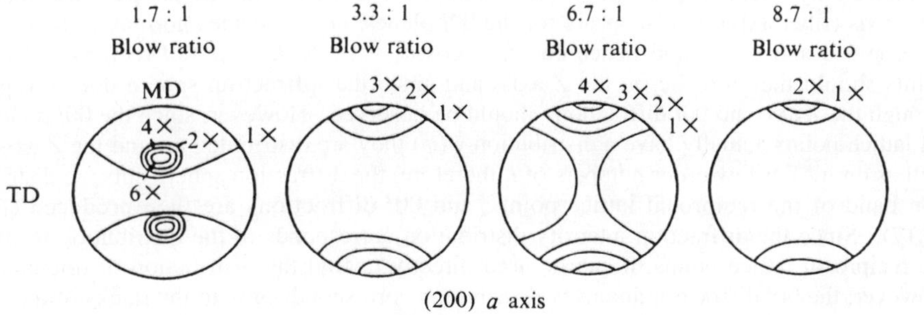

Fig. 10.16 Pole figures of extruded polyethylene film.[8)]
MD: Machine (rolling) direction.
TD: Transverse direction (parallel to rolled surface and normal to rolling direction)
[Reproduced with permission from P.H. Lindenmeyer, S. Lustig, *J. Appl. Polym. Sci.*, **9**, 235, John Wiley & Sons, Inc. (1965)]

is rather complex, but corresponding to the orientation of the *a* axis, the orientation distribution has maxima at angles of 45° and 135° to the plane of the film.

10.3 Determination of the Degree of Orientation

Specimens in which the crystallites exhibit the various types of orientation discussed hitherto can differ widely in the degree of the orientation. The reciprocal lattice point distribution bands are obviously affected by any spread in the orientations, and there is a corresponding effect on the X-ray diffraction patterns. The latter generally have an intensity distribution taking the form of arcs on Debye rings, but may (depending on the degree of orientation) vary from sharp points to uniform rings. The distribution of the reciprocal lattice points for a given crystal plane, *i.e.* the degree of orientation, can therefore be inferred from the intensity distribution of the corresponding diffraction on the Debye ring.

10.3.1 Criteria of the degree of orientation

A. Using meridional diffractions 00*l*

In a uniaxially oriented specimen, if the *a* and *b* axes of the crystallites are perpendicular to the *c* axis (fiber axis), δ_l disappears for the 00*l* planes, *i.e.* the planes normal to the *c* axis, since $\varphi = 0$ and $\sigma = 0$ and hence $\cos\delta_m = \cos\delta_l = 1/\cos\theta$. The 00*l* reciprocal lattice points should therefore lie on the *Z* axis, and since the diffraction sphere does not pass through the *Z* axis, no 00*l* diffractions should be observed. However, since the 00*l* reciprocal lattice points actually have a distribution $w(\varphi)$ they are distributed around the *Z* axis as well as on it. For low order numbers of *l*, therefore, the diffraction sphere cuts the distribution band of the reciprocal lattice points, and 00*l* diffractions are then produced (Fig. 10.17). Since the diffraction intensity distribution corresponds to the distribution $w(\varphi)$ of the reciprocal lattice points, it can be used directly to find the distribution of orientation. However, the 00*l* diffraction intensity distribution corresponds only to the flanks of the reciprocal lattice point distribution. In practice, therefore,

a) It is necessary to know the 00*l* intensity distribution from an inclination photograph for which the specimen was inclined at an angle of $\mu = \theta_{00l}$ (where θ_{00l} is the Bragg angle for the 00*l* diffractions) to the incident X-rays in order to be able to place the distribution band

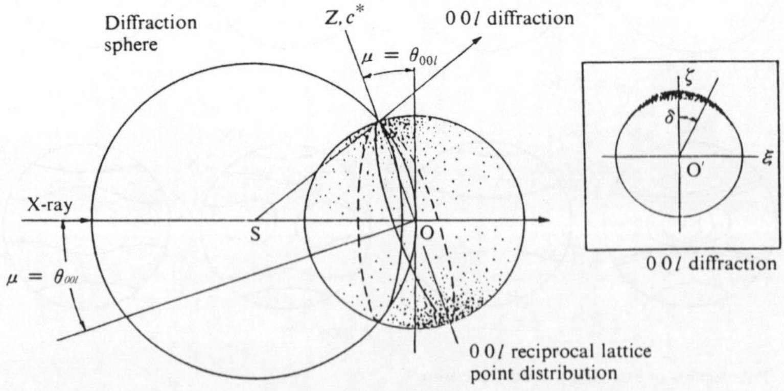

Fig. 10.17 Measurement of the degree of orientation using meridional diffractions. Measurement of the meridional diffractions is possible when the specimen is inclined appropriately (see text).

of $00l$ reciprocal lattice points exactly on the diffraction sphere. Since this intensity distribution is along a small circle of the diffraction sphere, the following equation is used to convert it into a distribution along the great circle, and so to find the correct orientation distribution.

$$\cos\varphi = \cos\delta\cos\theta_{00l} \tag{10.12}$$

φ is the angle from the OZ axis, and δ is the angle from the meridian measured on the Debye ring (*cf.* Fig. 10.17).

b) Alternatively, the $00l$ diffraction intensity distribution can be found without the need for the correction (Eq. 10.12) required in *a)* above, by taking precession photographs of reciprocal lattice planes containing the c^* axis or from Weissenberg photographs for axes perpendicular to the c axis.[†]

B. A practical measure of parallelism[9)]

If the degree of orientation is low, it is difficult to obtain pure meridian $00l$ diffraction intensity measurements. In such cases a practical measure of parallelism Π is found as follows from the intensity distribution of diffraction spots on the equator, and is used as a criterion of the degree of orientation. This quantity is defined by the equation

$$\Pi = \frac{180^\circ - H^\circ}{180^\circ} \times 100 \tag{10.13}$$

where H° is the half width of the intensity distribution on the Debye ring of the most intense diffraction on the equator. Π has a value of 0 if the specimen is completely unoriented, while if the crystallites are all arranged perfectly parallel to one another it is equal to 100.

10.3.2 Determination of the mean of the crystallite orientation distribution (orientation coefficient)

This method determines not the orientation distribution, but its mean value. The value obtained in this way should agree with the mean of the orientation distribution function described in Section 10.3.3, but the work and calculations are naturally less onerous than those involved in first determining the orientation distribution function and then its mean.

A. Uniaxial orientation[10)]

If the Z axis is the fiber axis (Fig. 10.18) and the angles made by the a, b, and c axes of the crystallite with the Z axis are φ_a, φ_b, and φ_c, respectively, the orientation coefficients, F_a, F_b, and F_c[††] representing the degrees of the three axes with respect to the Z axis are defined by the following equations.

$$\left.\begin{aligned} F_a &= (3\langle\cos^2\varphi_a\rangle - 1)/2 \\ F_b &= (3\langle\cos^2\varphi_b\rangle - 1)/2 \\ F_c &= (3\langle\cos^2\varphi_c\rangle - 1)/2 \end{aligned}\right\} \tag{10.14}$$

† The $00l$ diffraction is easily confused with the nearby $10l$ and $01l$ diffractions. These generally have small values of δ_1 and appear close to the meridian; they sometimes merge to give a single arc cutting the meridian. As the degree of orientation of the specimen increases, however, the intensity of the $00l$ diffraction spot decreases, whereas those of the $10l$ and $01l$ diffractions do not decrease but sometimes separate on either side of the meridian.

†† These F's are not to be confused with those (structure factors) in Sections 2.9, 4.2, 11.2, 11.4, etc.

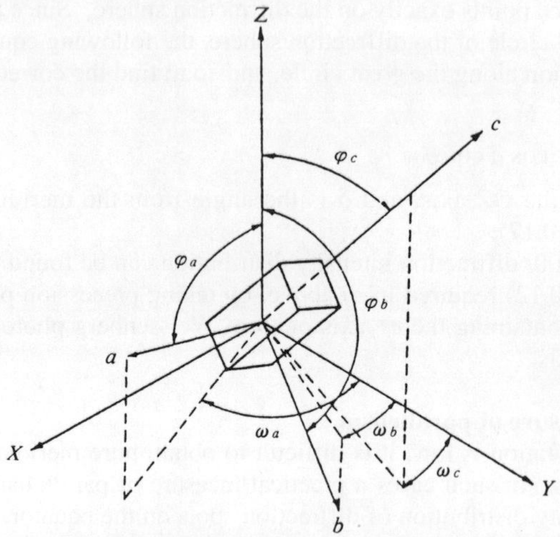

Fig. 10.18 Parameters used in the description of uniaxial crystallite orientation. [Reproduced from R.S. Stein, *J. Polym. Sci.*, **31**, 327, John Wiley & Sons, Inc. (1958)]

The values of $<\cos^2 \varphi_a>$, $<\cos^2 \varphi_b>$, and $<\cos^2 \varphi_c>$ required for the determination of the orientation coefficients are found as follows.

If we consider, say, the orientation coefficient for the a axis, we have $\cos \varphi_a = \cos \theta_{h00} \sin\psi_{h00}$, with the mean given by

$$<\cos^2 \varphi_a> = \cos^2 \theta_{h00} <\sin^2 \psi_{h00}>$$

$<\sin^2 \psi_{h00}>$ is in turn given by

$$<\sin^2 \psi_{h00}> = \frac{\int_0^{\pi/2} I(\psi_{h00}) \sin^2 \psi_{h00} \cos\psi_{h00} \mathrm{d}\psi_{h00}}{\int_0^{\pi/2} I(\psi_{h00}) \cos\psi_{h00} \mathrm{d}\psi_{h00}}$$

Hence

$$<\cos^2 \varphi_a> = \frac{\int_0^{\pi/2} I(\psi_{h00}) \sin^2 \psi_{h00} \cos\psi_{h00} \cos^2\theta_{h00} \mathrm{d}\psi_{h00}}{\int_0^{\pi/2} I(\psi_{h00}) \cos\psi_{h00} \mathrm{d}\psi_{h00}} \tag{10.15a}$$

Similarly, for the b and c axes

$$<\cos^2 \varphi_b> = \frac{\int_0^{\pi/2} I(\psi_{0k0}) \sin^2 \psi_{0k0} \cos\psi_{0k0} \cos^2\theta_{0k0} \mathrm{d}\psi_{0k0}}{\int_0^{\pi/2} I(\psi_{0k0}) \cos\psi_{0k0} \mathrm{d}\psi_{0k0}} \tag{10.15b}$$

$$<\cos^2 \varphi_c> = \frac{\int_0^{\pi/2} I(\psi_{00l}) \sin^2 \psi_{00l} \cos\psi_{00l} \cos^2\theta_{00l} \mathrm{d}\psi_{00l}}{\int_0^{\pi/2} I(\psi_{00l}) \cos\psi_{00l} \mathrm{d}\psi_{00l}} \tag{10.15c}$$

$I(\psi_{h00})$, $I(\psi_{0k0})$, and $I(\psi_{00l})$ are the intensity distributions of the $h00$, $0k0$, and $00l$ diffractions on the Debye-Scherrer rings, and θ_{h00}, θ_{0k0}, and θ_{00l} are the measured values of the Bragg angle for the $h00$, $0k0$, and $00l$ diffractions. ψ $(=90^\circ - \delta)$ is the angle from the equator, measured on the Debye-Scherrer ring (Fig. 10.19).

In the special case where the a, b, and c axes of the crystallites are orthogonal, then

$$\cos^2\varphi_a + \cos^2\varphi_b + \cos^2\varphi_c = 1 \tag{10.16}$$

$$F_a + F_b + F_c = 0 \tag{10.17}$$

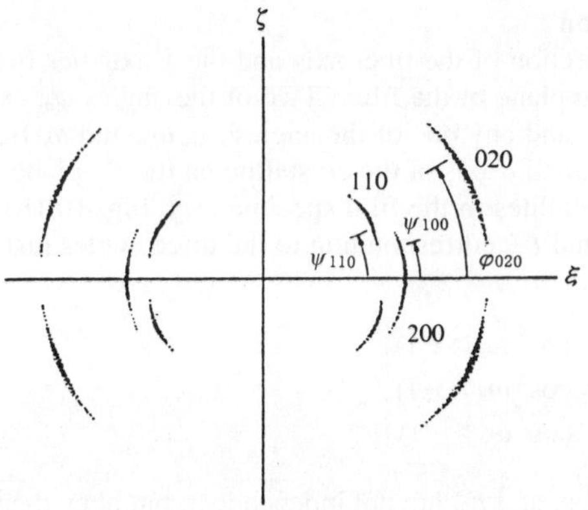

Fig. 10.19 Sketch of an X-ray diffraction pattern from oriented polyethylene. The a axes are oriented perpendicular to the drawing direction. The degree of orientation is indicated by the intensity distributions along the diffraction arcs (*i.e.* the variation with φ).

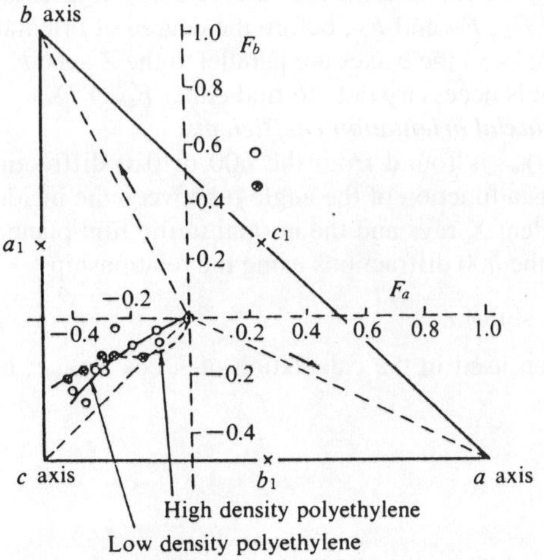

Fig. 10.20 Orientation diagram for polyethylene showing the effects of cold drawing.[11] The orthogonal dotted lines are the axes of F_a and F_b and the origin corresponds to the unoriented state. Circles indicate low density polyethylene, and crosses indicate high density polyethylene. [Experimental data supplied by R.S. Stein in private communication (1958)]

so that we need only determine the values of any two of the orientation coefficients.

Orientation coefficients of polyethylene.[10] Polyethylene belongs to the orthorhombic system (see Section 11.2.8), so Eq. 10.16 and 10.17 are normally valid. For spiral orientation as represented schematically in Fig. 10.19, F_a and F_b can be found by measuring the 200 and 020 intensity distributions $I(\psi_{200})$ and $I(\psi_{020})$ and calculating $<\cos^2 \varphi_a>$ and $<\cos^2 \varphi_b>$ from Eq. 10.15a and 15b. Fig. 10.20[11] shows how the orientation coefficients found in this way depend upon the conditions of cold drawing.

B. Biaxial orientation[12]

If the Z axis is the direction of the fiber axis and the Y axis lies in the plane of the film, the X axis is normal to the plane of the film. Two of the angles φ_a, φ_b, and φ_c used in the case of uniaxial orientation and any one of the angles ω_a, ω_b, and ω_c between the Y axis and the projections of the a, b, and c axis of the crystallite on the XY plane are required to define the orientation of the crystallites in the film specimen (*cf.* Fig. 10.18). The biaxial orientation coefficients F_{2a}, F_{2b}, and F_{2c} corresponding to the three angles just defined are given by the following equations.

$$\left.\begin{aligned} F_{2a} &= (3<\cos^2 \omega_a>-1) \\ F_{2b} &= (3<\cos^2 \omega_b>-1) \\ F_{2c} &= (3<\cos^2 \omega_c>-1) \end{aligned}\right\} \tag{10.18}$$

The coefficients F_{2a}, F_{2b}, and F_{2c} are not independent, but obey the following relationships:

$$\left.\begin{aligned} \sin^2 \varphi_a \cos^2 \omega_a + \sin^2 \varphi_b \cos^2 \omega_b + \sin^2 \varphi_c \cos^2 \omega_c = 1 \\ \sin^2 \varphi_a \sin^2 \omega_a + \sin^2 \varphi_b \sin^2 \omega_b + \sin^2 \varphi_c \sin^2 \omega_c = 1 \end{aligned}\right\} \tag{10.19}$$

Thus if F_a and F_b are found as in the case of uniaxial orientation, there is a further requirement for two of F_{2a}, F_{2b},and F_{2c}, before the degree of orientation can be expressed. In the special case where (say) the c axes are parallel to the Z axis, $F_c = 1$ and $F_a = F_b = -1/2$, and since $F_{2b} = F_{2a}$, it is necessary only to find either F_{2b} or F_{2a}.

a) Calculation of biaxial orientation coefficients.

$I(\varphi_a, \mu)_{h00}$ or $I(\varphi_b, \mu)_{0k0}$ is found from the h00 or 0k0 diffraction intensity $I(\psi_{h00}, \mu)$ or $I(\psi_{0k0}, \mu)$ measured as a function of the angle μ between the incident X-rays and the X axis, *i.e.* between the incident X-rays and the normal to the film plane. For example, $I(\varphi_a, \mu)_{h00}$ is easily found from the h00 diffractions using the relationship

$$\cos \varphi_a = \cos \theta_{h00} \sin \psi_{h00}$$

This $I(\varphi_a, \mu)_{h00}$ is then used in the calculation of $<\cos^2 \omega_{a0}>$, and hence to find F_{2a} (see Fig. 10.21).

$$\omega_a = \omega_{a0} + \mu$$

$$<\cos^2 \omega_{a0}> =$$

$$\frac{\int_0^{\pi/2} I(\varphi_a, \mu)_{h00}[(\sin^2 \theta_{h00})/(\sin^2 \varphi_a) + \{1-(2\sin^2 \theta_{h00})/(\sin^2 \varphi_a)\}\cos^2 \mu]d\mu}{\int_0^{\pi/2} I(\varphi_a, \mu)_{h00}\, d\mu} \tag{10.20a}$$

Similarly, F_{2b} is calculated with the aid of $< \cos^2 \omega_{b0} >$ from the $0k0$ diffraction

$$< \cos^2 \omega_{b0} > = \frac{\int_0^{\pi/2} I(\varphi_b, \mu)_{0k0}[(\sin^2 \theta_{0k0}) / (\sin^2 \varphi_b) + \{1-(2\sin^2 \theta_{0k0}) / (\sin^2 \varphi_b)\} \cos^2 \mu] d\mu}{\int_0^{\pi/2} I(\varphi_b, \mu)_{0k0} d\mu} \tag{10.20b}$$

If the uniaxial orientation coefficients F_a and F_b are calculated as described in the previous section, we now have all the orientation coefficients F_a, F_b, F_{2a}, and F_{2b} generally required to express the degree of orientation in cases of biaxial orientation.

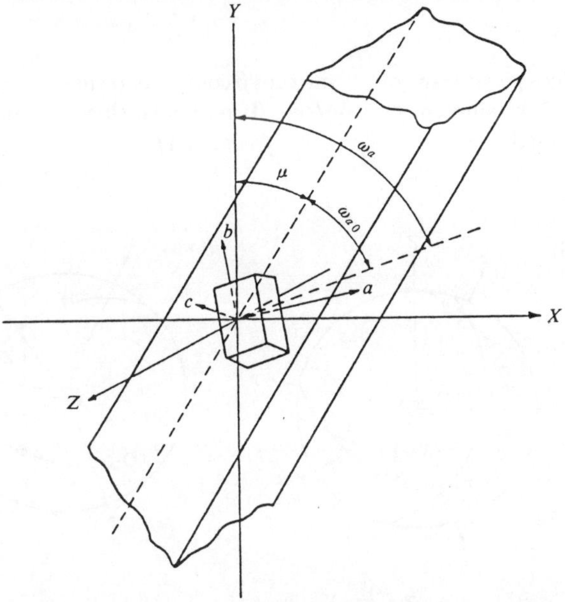

Fig. 10.21 Parameters used in the description of a tilted film.
[Reproduced from R.S. Stein, *J. Polym. Sci.*, **31**, 338, John Wiley & Sons, Inc. (1958)]

10.3.3 Analysis of the crystallite orientation distribution (orientation distribution functions)[13-15]

Crystallite orientation distributions are most suitably expressed in terms of crystallite orientation distribution functions. The calculations to determine these functions are rather laborious, and require high speed electronic computing facilities. The method has the compensating advantage, however, that the orientation distribution functions for planes giving very weak diffractions can be calculated from experimental results for many other planes.

A. Biaxial orientation distribution functions

The specimen is referred to the coordinate system O-*XYZ*, where the *Z* axis lies along the drawing direction, and the *YZ* plane is the plane of the specimen. The crystallites in the specimen have the coordinate system O-*xyz*, where the *x* direction is that of the molecular chains. The orientation of the crystallites in the specimen is expressed by the Euler angles α, β, and γ (Fig. 10.22(a)). The reciprocal lattice vector $\boldsymbol{r}^*_j$, normal to a given crystal plane in a crystallite, can therefore be defined in terms of the angular parameters Θ_j and Φ_j in the coordinate system O-*xyz* of the crystallite (Fig. 10.22(b)) and in terms of χ_j and η_j in the coordinate system O-*XYZ* of the specimen (Fig. 10.22(c)). The two sets of angles (Θ_j, Φ_j) and (χ_j, η_j) are related by the formula

$$\begin{pmatrix} \sin\chi_j & \cos\eta_j \\ \sin\chi_j & \sin\eta_j \\ & \cos\chi_j \end{pmatrix} = T^{-1}(\alpha,\beta,\gamma) \begin{pmatrix} \sin\Theta_j & \cos\Phi_j \\ \sin\Theta_j & \sin\Phi_j \\ & \cos\Theta_j \end{pmatrix} \tag{10.21}$$

for changing from the O-*xyz* system to the O-*XYZ* system. In this formula,

$$T(\alpha,\beta,\gamma)=$$

$$\begin{pmatrix} \cos\alpha\cos\beta\cos\gamma - \sin\alpha\sin\gamma & \sin\alpha\cos\beta\cos\gamma + \cos\alpha\sin\gamma & -\sin\beta\cos\gamma \\ -\cos\alpha\cos\beta\sin\gamma - \sin\alpha\cos\gamma & -\sin\alpha\cos\beta\sin\gamma + \cos\alpha\sin\gamma & \sin\beta\sin\gamma \\ \cos\alpha\sin\beta & \sin\alpha\sin\beta & \cos\gamma \end{pmatrix} \tag{10.22}$$

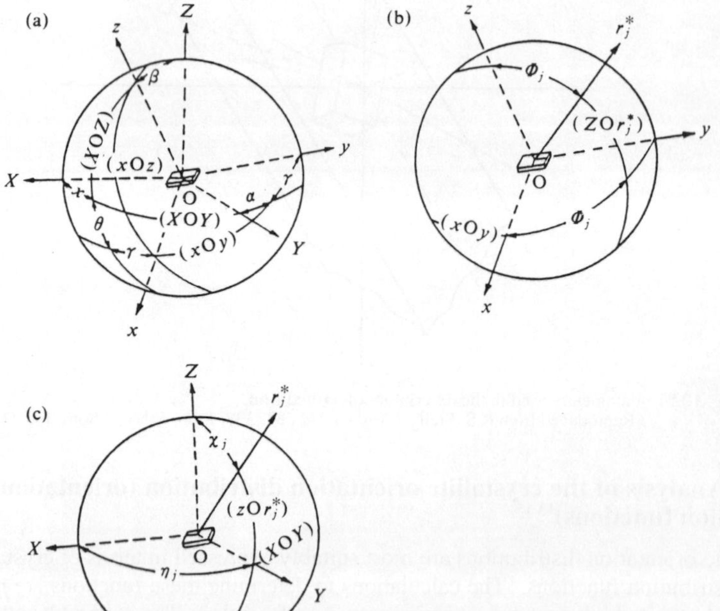

Fig. 10.22 Relationships between the coordinate systems and the reciprocal lattice vector $\boldsymbol{r}^*_j$.
(a) Specimen coordinate system, O-*XYZ*, and crystallite system, O-*xyz*, where α, β, γ are the Euler angles.
(b) Crystallite coordinate system and the arbitrary reciprocal lattice vector.
(c) Specimen coordinate system and reciprocal lattice vector $\boldsymbol{r}^*_j$.
[Reproduced with permission from R.-J. Roe, W. R. Krigbaum, *J. Chem. Phys.*, **40**, 2608, Am. Inst. Phys. (1964)]

With $\xi = \cos\beta$ and $\zeta_j = \cos\chi_j$, the crystallite orientation distribution function $w(\xi, \alpha, \gamma)$ may be expressed in terms of the experimental orientation distribution function $q_j(\zeta_j, \eta_j)$ for the reciprocal lattice vector $\boldsymbol{r}_j$ by means of the expanded forms of both $w(\xi, \alpha, \gamma)$ and $q_j(\zeta_j, \eta_j)$ which are obtained from the associated Legendre polynomials.[16] The expressions derived are of the following forms.

$$w(\xi,\alpha,\gamma)=\sum_{l=0}^{\infty}\sum_{m=-l}^{l}\sum_{n=-l}^{l} W_{lmn} Z_{lmn}(\xi)\exp(-im\alpha)\exp(-in\gamma) \tag{10.23}$$

$$q_j(\zeta_j,\eta_j)=\sum_{l=0}^{\infty}\sum_{m=-l}^{\infty} Q_{lm}{}^{j} P_l^{m}(\zeta_j)\exp(-im\eta_j) \tag{10.24}$$

$$Q_{lm}{}^{j}=2\pi\{2/(2l+1)\}^{1/2}\sum_{n=-l}^{l} W_{lmn} P_l^{n}(\Xi_j)\exp(-in\Phi_j) \tag{10.25}$$

$$\Xi = \cos\Theta$$

$P_l^m(\zeta)$ and $P_l^n(\Xi)$ are both normalized associated Legendre functions, while Z_{lmn} is a generalized associated Legendre function.[16] $w(\xi, \alpha, \gamma)$ is of course normalized.

The orientation distribution function $q_j(\zeta_j, \eta_j)$ for the reciprocal lattice vector $\boldsymbol{r}^*$ is found from the measured intensities $I(\zeta_j, \eta_j)$ for various independent lattice planes.

$$q_j(\zeta_j,\eta_j)=I(\zeta_j,\eta_j)\Big/\int_0^{2\pi}\int_{-1}^{1} I(\zeta_j,\eta_j)\,\mathrm{d}\zeta_j d\eta_j \tag{10.26}$$

$Q_{lm}{}^{j}$ is then calculated from the following equation.

$$Q_{lm}{}^{j}=\frac{1}{2\pi}\int_0^{2\pi}\int_{-1}^{1} q_j(\zeta_j,\eta_j)P_l^{m}(\zeta_j)\exp(im\eta_j)\,\mathrm{d}\zeta_j d\eta_j \tag{10.27}$$

Having found the value of a given $Q_{lm}{}^{j}$ from Eq. 10.27, we find W_{lmn} ($n = -1,...,0,...,1$) for certain fixed values of l, m, and j with the aid of Eq. 10.25 by solving $(2l + 1)$ simultaneous first-order equations in the complex plane. If the number of $q_j(\zeta_j, \eta_j)$ obtained from experimental values is N, then W_{lmn} is found for at least $(N - 1)/2$ values of l. Finally, $w(\xi, \alpha, \gamma)$ can be calculated from Eq. 10.23.

It is convenient in practical calculations to express W_{lmn} and $Q_{lm}{}^{j}$ as

$$W_{lmn} = A_{lmn} + iB_{lmn} \tag{10.28}$$

$$Q_{lm}{}^{j} = a_{lm}{}^{j} + b_{lm}{}^{j} \tag{10.29}$$

Then

$$q_j(\zeta_j,\eta_j)=\sum_{l=0}^{\infty}\sum_{m=-l}^{l} P_l^{m}(\zeta_j)[a_{lm}{}^{j}\cos m\eta_j+b_{lm}{}^{j}\sin m\eta_j] \tag{10.30}$$

$$w(\xi,a,\gamma)=\sum_{l=0}^{\infty}\sum_{m=-l}^{l}\sum_{n=-l}^{l} Z_{lm}(\xi)[A_{lmn}\cos(m\alpha+n\gamma)+B_{lmn}\sin(m\alpha+n\gamma)] \tag{10.31}$$

$$\left.\begin{aligned} a_{lm}{}^j &= 2\pi\left\{\frac{2}{2l+1}\right\}^{1/2} \sum_{n=-l}^{l} P_l^n(\Xi_j)[A_{lmn}\cos n\Phi_j - B_{lmn}\sin n\Phi_j] \\ b_{lm}{}^j &= 2\pi\left\{\frac{2}{2l+1}\right\}^{1/2} \sum_{n=-l}^{l} P_l^n(\Xi_j)[A_{lmn}\sin n\Phi_j - B_{lmn}\cos n\Phi_j] \end{aligned}\right\} \tag{10.32}$$

Moreover, from the symmetry of Z_{lmn} and P_l^m, we obtain

$$\left.\begin{aligned} Q_{lm} &= (-1)^m Q_{l\bar{m}}{}^* \\ a_{lm} &= (-1)^m a_{l\bar{m}} \\ b_{lm} &= (-1)^{m+1} b_{l\bar{m}} \end{aligned}\right\} \tag{10.33}$$

$$\left.\begin{aligned} W_{lmn} &= (-1)^{m+n} W_{l\bar{m}\bar{n}}{}^* \\ A_{lm} &= (-1)^{m+n} A_{l\bar{m}\bar{n}} \\ B_{lm} &= (-1)^{m+n+1} B_{l\bar{m}\bar{n}} \end{aligned}\right\} \tag{10.34}$$

If the statistical distribution of the crystallites in the specimen has a certain symmetry, and if the crystallites themselves possess crystallographic symmetry elements, many of the W_{lmn} become zero, as shown in Tables 10.2 and 10.3,[15] with corresponding simplification of the calculation. The coordinate systems O-xyz and O-XYZ should therefore be chosen in such a way as to give the greatest possible number of symmetry elements.

When the crystallite coordinate system is changed to O-$x'y'z'$ by rotation of the O-xyz coordinate system through the Euler angles α', β', and γ' in order to manifest the highest degree of symmetry possible, the equation for the change of coordinates is

Table 10.2 Symmetry properties of Q_{lm} and W_{lmn} due to statistical symmetry of crystallite distribution

Statistical symmetry element	$q(\zeta, \eta)$	Q_{lm}	W_{lmn}
Mirror plane normal to x axis	$= q(\zeta, \pi - \eta)$	$= Q_{l\bar{m}}$	$= W_{l\bar{m}n} = (-1)^{m+n} W^*_{lm\bar{n}}$
Mirror plane normal to y axis	$= q(\zeta, -\eta)$	$= (-1)^m Q_{l\bar{m}}$	$(-1)^m W_{l\bar{m}n} = (-1)^n W^*_{lm\bar{n}}$
Mirror plane normal to z axis	$= q(-\zeta, \eta)$	$\begin{cases} \neq 0\ (m \text{ even}) \\ = 0\ (m \text{ odd}) \end{cases}$	$\begin{cases} \neq 0\ (m \text{ even}) \\ = 0\ (m \text{ odd}) \end{cases}$
Three mirror planes normal to x, y, and z axes respectively		$\begin{cases} = Q_{l\bar{m}}\ (m \text{ even}) \\ = 0\ (m \text{ odd}) \end{cases}$	$\begin{cases} = W_{l\bar{m}n} = (-1)^n W^*_{lm\bar{n}}\ (m \text{ even}) \\ = 0\ (m \text{ odd}) \end{cases}$
Cylindrical symmetry about z axis	$= q(\zeta, 0)$	$\begin{cases} \neq 0, m = 0 \\ = 0, m \neq 0 \end{cases}$	$\begin{cases} \neq 0, m = 0 \\ = 0, m \neq 0 \end{cases}$

[Reproduced with permission from R.-J. Roe, *J. Appl. Phys.*, **36**, 2024, Am. Inst. Phys. (1965)]

Table 10.3 Symmetry properties of W_{lmn} due to crystallographic symmetry of crystallites

Crystallographic symmetry element	Equivalent vectors of Θ and Φ	W_{lmn}
Mirror plane normal to X axis	$(\Theta, \pi - \Phi)$	$= (-1)^n W_{lm\bar{n}} = (-1)^m W^*_{l\bar{m}n}$
Mirror plane normal to Y axis	$(\Theta, -\Phi)$	$= W_{lm\bar{n}} (-1)^{m+n} W^*_{l\bar{m}n}$
Mirror plane normal to Z axis	$(\pi - \Theta, \Phi)$	$\begin{cases} \neq 0\ (n \text{ even}) \\ = 0\ (n \text{ odd}) \end{cases}$
r-fold rotational symmetry with respect to Z axis	$(\Theta, \Phi + 2\pi j/r)$ $(j = 1, 2, ..., r-1)$	$\begin{cases} \neq 0\ (\text{if } n \text{ is a multiple of } r) \\ = 0\ (\text{other cases}) \end{cases}$

[Reproduced with permission from R.-J. Roe, *J. Appl. Phys.*, **36**, 2024, Am. Inst. Phys. (1965)]

$$\begin{pmatrix} \sin\Theta_j & \cos\Phi_j \\ \sin\Theta_j & \sin\Phi_j \\ \cos\Theta_j & \end{pmatrix} = T^{-1}(\alpha',\beta',\gamma') \begin{pmatrix} \sin\Theta_j' & \cos\Phi_j' \\ \sin\Theta_j' & \sin\Phi_j' \\ \cos\Theta_j' & \end{pmatrix} \tag{10.35}$$

where Θ_j' and Φ_j' are the polar and azimuthal angles respectively of the reciprocal lattice vector $\boldsymbol{r}^*_j$ in the new coordinate system.

B. Uniaxial orientation distribution functions

For uniaxial orientation, $w(\xi, \alpha, \gamma)$ and qj(ζ_j, η_j) no longer contain the variables α and η_j; the orientation distribution functions and their interrelationships therefore reduce to the following.

$$w(\xi,\gamma) = \sum_{l=0}^{\infty} A_{l0} P_l^0(\xi) + \sum_{l=1}^{\infty}\sum_{m=1}^{l} [A_{lm} P_l^m(\xi)\cos m\gamma + B_{lm} P_l^m(\xi)\sin m\gamma)] \tag{10.36}$$

$$Q_l^j = 2\pi\left\{\frac{2}{2l+1}\right\}^{1/2} \times\{A_{l0}P_l^0(\Xi_j) + 2\sum_{m=1}^{l}(-1)^m[A_{lm}P_l^m(\Xi_j)\cos m\Phi_j + B_{lm}P_l^m(\Xi_j)\sin m\Phi_j]\} \tag{10.37}$$

Q_l^j is found from the experimental $q_j(\zeta_j)$ as follows.

$$Q_l^j = \int_{-1}^{1} q_j(\zeta_j) P_l(\zeta_j) d\zeta_j \tag{10.38}$$

$$q_j(\zeta_j) = I_j(\chi) / \int_0^{\pi} I_j(\chi)\sin\chi d\chi \tag{10.39}$$

C. Procedure for the calculation of $w(\xi, \gamma)$

1) The parameters Θ and Φ are calculated for the various diffractions from the size of the unit cell.

2) $I_j(\chi)$ is measured for the *j*th diffractions, and after correction for absorption, polarization, and Lorentz factors, $q_j(\zeta_j)$ is found from Eq. 10.39.

3) Eq. 10.38 is used to find the Q_l^j for $l = 0, 2, 4,...$ for the various diffractions. The expanded form of $q_j(\zeta_j)$ obtained with the aid of Q_l^j, *i.e.*

$$q_j(\zeta_j) = \sum_{l=0}^{\infty} Q_l^j P_l(\zeta_j) \tag{10.40}$$

is tested for convergence by means of Eq. 10.41, at the same time determining the number of terms of the associated Legendre polynomial required for convergence within the expected error, *i.e.* the minimum number of diffractions required to find the orientation distribution function with the expected accuracy.

$$\sigma_q^2 = \int_{-1}^{1} [q_j(\zeta_j)]^2 d\zeta_j - \sum_{l=0}^{\infty} (Q_l^j)^2 \tag{10.41}$$

4) The coefficients A_{lm} and B_{lm} required for the calculation of $w(\xi, \gamma)$ are found by applying Eq. 10.37 to each of the Q_l^j obtained and solving the simultaneous equations.

5) $w(\xi, \gamma)$ is calculated.

6) Where the orientation distribution function of a given crystal plane cannot be determined, *e.g.* because the diffraction intensity is too low, or because there is an unacceptable

amount of overlapping with other diffractions, the coefficients A_{lm} and B_{lm} used in the calculation of the orientation distribution functions of the crystal planes from other independent diffractions are used in Eq. 10.37 to calculate the required Q_l. If $q_j(\zeta_j)$ is then calculated from Eq. 10.40, $w(\xi, \gamma)$ can be calculated by the same procedure as for the other crystal planes.

7) When several superimposed diffractions are measured as one, the measured intensity must be broken down into the contributions due to the various components using the "weights" of the reciprocal lattice vectors $\boldsymbol{r}_j^*$. These weights C_{jp} are derived from the following equation.

$$C_{jp} = (F_{jp})^2 / \sum (F_{jp})^2 \tag{10.42}$$

F_{jp} is the structure factor of the reciprocal lattice vector $\boldsymbol{r}_{jp}^*$ measured as the reciprocal lattice vector $\boldsymbol{r}_j^*$ resulting from the superposition of other diffractions with very similar Bragg angles. We obtain

$$q_j(\zeta_j) = \sum_p C_{jp} q_{jp}(\zeta_j) \tag{10.43}$$

$$Q_l^j = \sum_p C_{jp} Q_l^{jp} \tag{10.44}$$

so that Eq. 10.37 is replaced by

$$Q_l^j = 2\pi \left\{\frac{1}{2l+1}\right\}^{1/2} \{A_{l0} \sum_p C_{jp} P_l^0(\Xi_j) + 2\sum_{m=1}^{l} (-1)^m [A_{lm} \sum_p C_{jp} P_l^m(\Xi_j) \cos m\Phi_{jp}$$
$$+ B_{lm} \sum_p C_{jp} P_l^m(\Xi_j) \sin m\Phi_{jp}]\} \tag{10.45}$$

where $\Xi_{jp} = \cos\Theta_{jp}$.

Orientation distribution functions of drawn polyethylene. [14] Since poly-ethylene crystals are orthorhombic (*cf.* Section 11.2.7), the coordinate systems are chosen so that the x, y, and z axes coincide with the a, b, and c axes, respectively, of the unit cell. Owing to the presence of a mirror plane normal to the z axis, all the A_{lm} disappear from $w(\xi, \gamma)$ except those for which l and m are both even. Eqs. 10.36 and 10.45 thus become

$$w(\xi,\gamma) = \sum_{\substack{l=0 \\ (even)}}^{\infty} A_{l0} P_l^0(\xi) + 2 \sum_{\substack{l=2 \\ (even,}}^{\infty} \sum_{\substack{m=2 \\ even)}}^{l} A_{lm} P_l^m(\xi) \cos m\gamma \tag{10.46}$$

$$Q_l^j = 2\pi\{2/(2l+1)\}^{1/2}\{A_{l0} \sum_p C_{jp} P_l^0(\Xi_{jp}) + 2 \sum_{\substack{m=2 \\ (even)}}^{l} A_{lm} \sum_p C_{jp} P_l^m(\Xi_j) \cos m\Phi_{jp}\} \tag{10.47}$$

The specimen in this case a sheet of polyethylene (Alathon-10) that had received 10 Mrad of radiation and was drawn (draw ratio 5.58 : 1) to induce uniaxial orientation. The intensity distributions $I(\chi)$ of 12 diffractions were measured, and the $q(\zeta)$ were calculated. Fig. 10.23 shows the $q(\zeta)$ for the four diffractions with the smallest angles. The circles are the experimental values, and the solid curves indicate the values found from the expanded approximation obtained from Eq. 10.40 with $l = 16$. The dashes give curves obtained by expanding the intensity distribution curves of the 12 diffractions (isolated from the 23 overlapping diffractions of Table 10.4) into a series and applying corrections so as to minimize the differences from all the experimental curves. The dotted curve shown in the 200 dia-

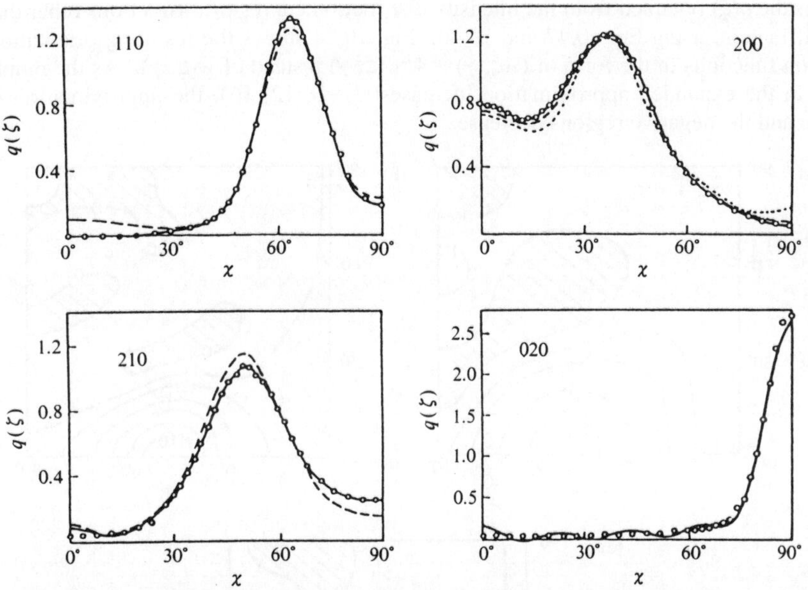

Fig. 10.23 Reciprocal lattice point distribution functions $q(\zeta)$ of the first four diffractions for uniaxially oriented polyethylene (see text).[14)]
[Reproduced with permission from R.-J. Roe, W. R. Krigbaum, *J. Chem. Phys.*, **41**, 737, Am. Inst. Phys. (1964)]

Table 10.4 Identification of the diffractions, etc. for Fig. 10.23[14)]

j	2θ, diffraction angle[deg]	ρ_j[†]	hkl	C_{jp}
1	21.62	1.0	110	
2	24.02	1.0	200	
3	30.15	1.0	210	
4	36.38	1.5	020	
5	39.79	0.7	011	0.81
			310	0.19
6	40.85	0.7	310	0.55
			011	0.30
			111	0.15
7	41.69	0.7	111	0.70
			310	0.19
			201	0.11
8	43.07	0.7	201	0.71
			220	0.15
			111	0.14
9	47.01	0.5	211	
10	55.00	0.5	311	0.88
			130	0.12
11	57.32	0.5	130	0.66
			221	0.34
12	61.92	0.5	401	0.60
			230	0.36
			420	0.04

† Weight due to superposition of diffractions.
[Reproduced with permission from W.R. Krigbaum, R.-J. Roe., *J. Chem. Phys.*, **41**, 737, Am. Inst. Phys. (1964)]

gram is the $q(\zeta)$ obtained from the intensity distribution curves of diffractions other than the 200 diffraction, using Eq. 10.47 and 10.40. Fig. 10.24 shows the resulting orientation distribution functions in the form of $G(\xi, \gamma) = 4\pi w(\xi, \gamma)$ instead of $w(\xi, \gamma)$. As the number of terms in the expanded approximation increases ($l = 6$, 12, 16), the approximation is improved and the negative regions decrease.

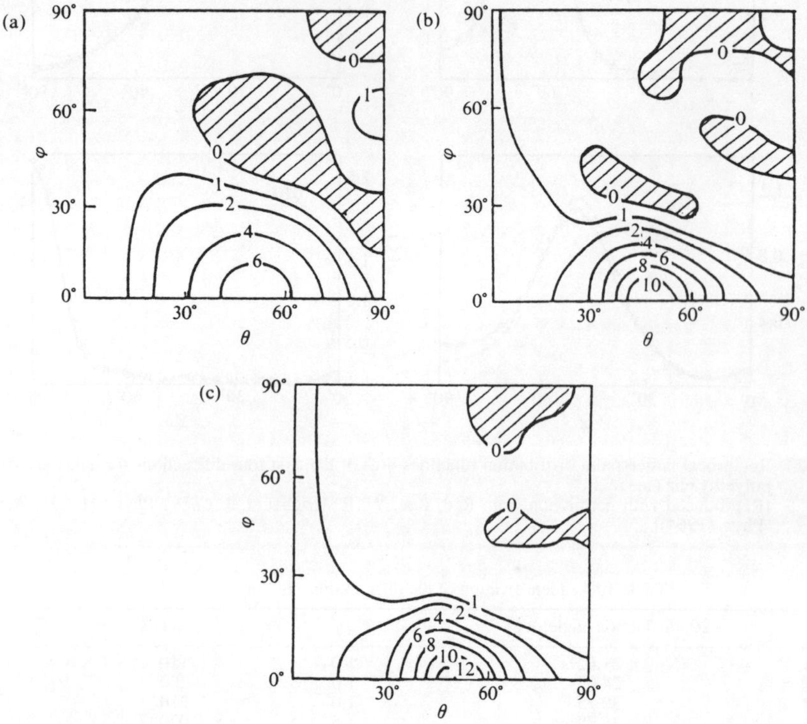

Fig. 10.24 Orientation distribution functions $G(\xi, \gamma) = 4w(\xi, \gamma)$.[14)]
The number of terms used in the expanded approximation for (a), (b), and (c) were $l = 6$, 12, and 16, respectively. As the number of terms is reduced, the negative (shaded) area due to the approximation error increases.
[Reproduced with permission from W.R. Krigbaum, R.-J. Roe., *J. Chem. Phys.*, **41**, 737, Am. Inst. Phys. (1964)]

10.4 Preferred Orientation of Two-dimensional Lattices

In the previous sections the orientation of three-dimensional lattices has been discussed. The diffraction intensity of a two-dimensional lattice is represented in reciprocal space by a periodic array of parallel rod-1ike intensity distributions. The asymmetric interference lines produced by a randomly oriented two-dimensional lattice were discussed by Laue[17)] and later by many other authors. The effect of non-random distribution of the two-dimensional lattices has been treated by Guentert and Cviekevich,[18)] and has been generalized by Ruland and Tompa.[19)] Ruland and coworkers also interpreted the effect of preferred orientation on (hk) interferences as shown by electron diffraction of carbon fibers.[20)] Orientation of polymer chains (one-dimensional cases) will be dealt with in Section 14.5 (refer also 14.4 and 14.6).

References

1. T. Ishikawa, A. Sugihara, S. Nagai, N. Kasai, *J. Chem. Soc. Jpn.*, p.102, Chem. Soc. Jpn. (1978) (in Japanese).
2. T. Imura, in: *X-Ray Diffraction* (K. Kohra ed.), p. 600, Kyoritsu, Tokyo (1988) (in Japanese).
3. C.W. Bunn, *Chemical Crystallography*, 2nd ed., p. 192, Oxford University Press, London (1961).
4. R.W. James, *Optical Principles of the Diffraction of X-Rays*, p. 606, Bell & Sons, London (1964).
5. *International Tables for X-Ray Crystallography*, Vol. III. p. 295. Kynoch Press. Birmingham (1962).
6. B.D. Cullity, *Elements of X-Ray Diffraction*, sect. 2–11, Addison-Wesley, Reading (1956).
7. Y. Fujiwara, *J. Appl. Polymer Sci.*, **4**, 10 (1960).
8. P.H. Lindenmeyer, S. Lustig, *J. Appl. Polymer Sci.*, **9**, 227 (1965).
9. Y. Go, T. Kubo, *J. Chem. Soc. Jpn.*, *Ind. Chem. Sect.*, **39**, 929, Chem. Soc. Jpn. (1939) (in Japanese).
10. R.S. Stein, *J. Polymer Sci.*, **31**, 327 (1958).
11. R.S. Stein, *ONR Technical Report.*, Nonr. 2151 (00), NR 356–78 (1957).
12. In ref. 10, p. 335.
13. R.-J. Roe, W.R. Krigbaum, *J. Chem. Phys.*, **40**, 2608 (1964).
14. W.R. Krigbaum, R.-J. Roe, *J. Chem. Phys.*, **41**, 737 (1964).
15. R.-J. Roe, *J. Appl. Phys.*, **36**. 2024 (1965).
16. For example, H. Margenau, G.M. Murphy, *The Mathematics of Physics and Chemistry*, p. 272, van Nostrand, N.Y. (1943).
17. M. von Laue, *Z. Krisl.*, **82**, 127 (1932).
18. O.J. Guentert, S. Cviekevich, *Carbon*, **1**, 309 (1964).
19. W. Ruland, H. Tompa, *Acta Cryst.*, **A24**, 93 (1968).
20. A. Fourdeux, R. Peret, W. Ruland, *J. Appl. Cryst.*, **1**, 252 (1968).

11. Crystal Structure Analysis of High Polymers

It may not be immediately apparent why the methods used for the structure analysis of single crystals of low molecular weight substances should so often prove inadequate (unless considerably modified) in the case of polymers. This, however, is to ignore the many differences between X-ray diffraction from high polymer crystals and that from single crystals. The structure analysis of high polymers is particularly complicated by the imperfect crystallite orientation and the smaller total number of diffractions. On the other hand, the fact that the molecular structure of a given high polymer can often be al least partially inferred from its chemical composition compensates to some extent for the necessity to base structure analyses on data which is limited in scope and precision.

11.1 Use of Unoriented Diffraction Patterns

In an unoriented polycrystalline specimen it is inevitable that certain information is lost as a result of the random orientation of the crystallites. A further, and in practice more serious, loss of information is a result of the overlap of independent diffraction peaks in the powder pattern. However, latest developments in the powder diffraction method have gained new importance in the field of materials science for crystal structure determination and characterization of polycrystalline specimens.

The methods of analysis for powder diffraction patterns are roughly divided in two[1]: the Rietveld method[2] and the pattern decomposition method.[3] The former involves crystal structure determination using the whole diffraction pattern to refine the structural parameters, and in the latter the observed diffraction pattern is decomposed into individual Bragg diffraction components before analysis regardless of the structural model.

11.1.1 Rietveld method[2]

In this method, the model structure assumed for the crystalline specimen is refined by a least-squares procedure fitting a calculated powder diffraction pattern to the observed whole diffraction pattern.

The quantity minimized in the least-squares refinement is

$$\Delta J = \Sigma w_i [J_{\text{obs},i} - J_{\text{calc},i}(Q_i,..)]^2 \tag{11.1}$$

where $J_{\text{obs},\,i}$ is the diffraction intensity as measured, on which no correction has been made. w_i is the weight.

The calculated diffraction intensity for the model structure $J_{\text{calc},\,i}\,(Q_i,...)$ is

$$J_{\text{calc},\,i}(Q_i,..) = I(Q_i)A(Q_i)c\;\Sigma\, j_k\, E_k\, P_k Lp(Q_k)\mid F_k\mid^2 G(Q_i - Q_k) + B(Q_j) \tag{11.2}$$

where $I(Q_j)$ is intensity distribution in the incident beam, $A(Q_j)$ absorption factor[3] and $B(Q_i)$ the background intensity, c the scale factor,[4] k order of diffractions, j_k multiplicity of a Bragg diffraction, E_k and P_k correction factors for the extinction[5] and preferred orientation[6,7] respectively, $Lp(Q_k)$ Lorentz and polarization factor and F_k structure factor. Q_k is the position of the kth peak, and Q is given by the following equation,

$$\begin{aligned} Q &= 2\pi/d = (4\pi \sin\theta)/\lambda = 2\pi S/\lambda \\ &= 2\pi(h^2a^{*2} + k^2b^{*2} + l^2c^{*2} + 2klb^*c^* \cos\alpha^* \\ &\qquad + 2lhc^*a^* \cos\beta^* + 2hka^*b^* \cos\gamma^*)^{1/2} \end{aligned} \tag{11 .3}$$

where d is the interplanar spacing and a^*, b^*, c^*, α^*, β^*, and γ^* are reciprocal unit cell parameters.

The structure factor F_k is given by

$$F_k = \Sigma o_j f_j D_j \exp[2\pi i(hx_j + ky_j + lz_j)] \tag{11.4}$$

where x_j, y_j and z_j are fractional coordinates of the jth atom in the unit cell, o_j occupancy and f_j atomic scattering factor. D_j, thermal parameter is expressed in the form either isotropic

$$D_j = \exp[-B_j(\sin\theta/\lambda)] \tag{11.5}$$

or anisotropic

$$D_j = \exp[-(h^2\beta_{11j} + k^2\beta_{22j} + l^2\beta_{33j} + 2hk\beta_{12j} + 2hl\beta_{13j} + 2kl\beta_{23j})]. \tag{11.6}$$

$G(Q_i - Q_k) \equiv G(\Delta Q)$ is the profile-shape function to approximate the diffraction profile. Among the profile-shape functions applied hitherto,[8] the Voigt, pseudo-Voigt and Pearson VII functions, especially the latter two, are mostly used for the analysis at present. For the powder diagrams obtained by the energy dispersive method different profile-shape functions suitable for the measurement should be used.

***Voigt function*:**

$$V(x) = I(0)\frac{\beta}{\beta_C\beta_G}\int C^V(z)G^V(x-z)\mathrm{d}z \tag{11.7}$$

with

$$C^V(x) = \{1 + \pi^2x^2/\beta_C^2\}^{-1} = \{1 + 4x^2/(2w_o)^2)\}^{-1}$$

$$G^V(x) = \exp\{-\pi^2x^2/\beta_G^2\} = \exp\{-(4\ ln\ 2)x^2/(2w_G)^2)\}, \tag{11.8}$$

where C^V and G^V denote the Cauchy and Gaussian components of the Voigt function with integral breadth β_C and β_G respectively, and full width at half maximum $2w_C$ and $2w_G$, respectively. $I(0)$ is the maximal intensity value at $x = 0$ and β is the integral breadth of $V(x)$.

***pseudo-Voigt function*:**

$$pV(x) = I(0)\{C^{pV}(x) + G^{pV}(x)\} \tag{11.9}$$

with

$$C^{pV}(x) = \eta\{1 + \pi^2x^2/(\beta_{C'})\}^{-1}$$

$$G^{pV}(x) = (1-\eta)\exp\{-\pi x^2/(\beta_{G'})^2\}, \qquad 0 \leqq \eta \leqq 1, \tag{11.10}$$

where $\beta_{C'}$ and $\beta_{G'}$, denote the integral breadths of the Cauchy and Gaussian components, and η and $1-\eta$ prescribe the fractions of Cauchy and Gaussian components included.

***Pearson VII function*:**

$$PVII(x) = I(0)\{(1 + x^2/ma^2)\}^{-m}, \qquad 1 \leqq m \leqq \infty, \tag{11.11}$$

with

$$a = \beta\Gamma(m)/[(m\pi)^{1/2}\Gamma(m-1/2)], \tag{11.12}$$

where $m = 1$ or $m = \infty$ yields a Cauchy or Gaussian function, respectively. The symbol Γ denotes the gamma function.

A flow chart for the structure determination is shown in Fig. 11.1. The powder diffrac-

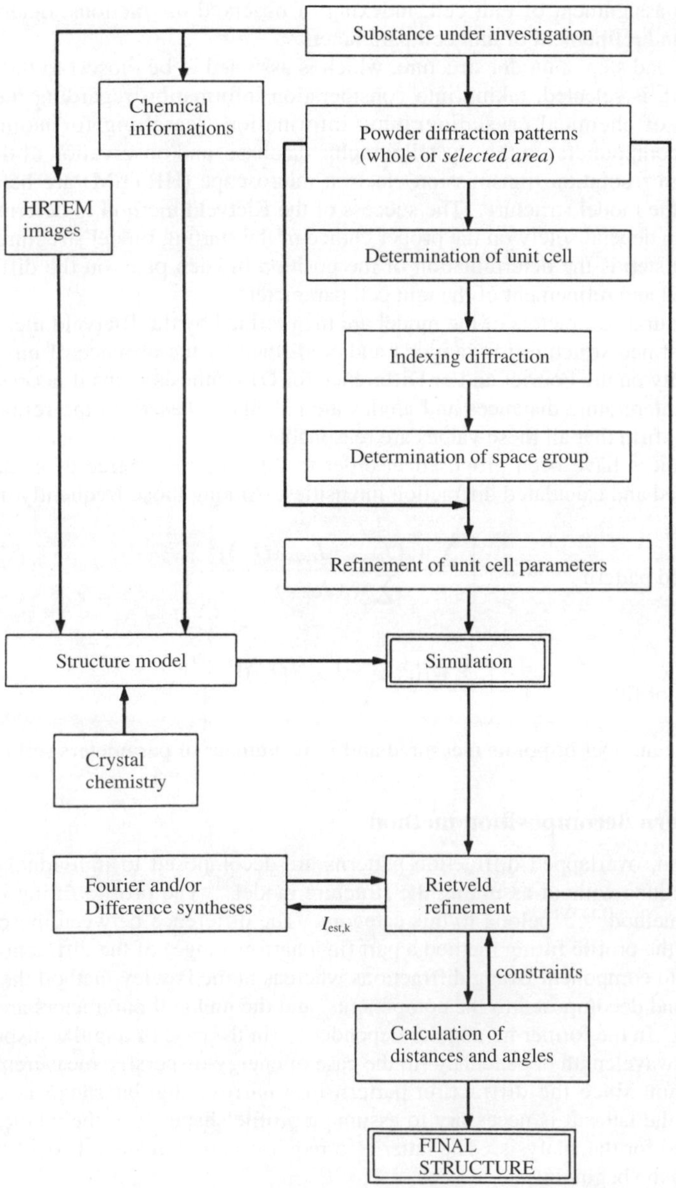

Fig. 11.1 Flow chart of the structure determination by Rietveld method.
HRTEM: high resolution transmission electron microscope, $I_{est,k}$: integrated intensity estimated from the result of the Rietveld method.
[Reproduced with permission from F. Izumi, *J. Cryst. Soc. Jpn.*, **34**, 82, Cryst. Soc. Jpn. (1992)]

tion pattern of the specimen is first obtained by precise intensity measurement. This is followed by the assignment of unit cell, indexing of observed diffractions, determination of space group and refinement of unit cell parameters.

In the second step, a model structure, which is assumed to be closest to the structure to be determined, is selected, taking into consideration information regarding the specimen. Examination of chemical crystallographic information, searching for isomorphous or isostructural compound(s) in the crystallographic database, and observation of the specimen under the high resolution transmission electron microscope (HRTEM) are helpful for the screening of the model structure. The success of the Rietveld method to determine a structure appears to depend solely on the proper choice of the starting model structure.

The third step is the determination of the position of each peak on the diffraction diagram observed and refinement of the unit cell parameters.

The structural parameters of the model are then refined by the Rietveld method (*cf.* Fig. 11.1). The refined structure is examined and confirmed by the absence of unwanted extra electron density on the Fourier and/or Difference (or D) synthesis maps if necessary.

Finally, interatomic distances and angles are calculated based on the refined structure in order to confirm that all these values are reasonable.

Many indices have been proposed in order to estimate the degree of coincidence between observed and calculated diffraction intensities. Among those frequently used are:

R-weighted pattern
$$R_{\mathrm{wp}} = \left[\frac{\sum w_j \{J_{\mathrm{obs},i} - J_{\mathrm{calc},i}(Q_{i,..})\}^2}{\sum w_i (J_{\mathrm{obs},i})^2} \right]^{1/2} \tag{11.13}$$

and

Goodness of fit
$$x = \left[\frac{\sum w_j \{J_{\mathrm{obs},i} - J_{\mathrm{calc},i}(Q_{i,..})\}^2}{N-P} \right]^{1/2} \tag{11.14}$$

where N is the number of points measured and P the number of parameters refined.

11.1.2 Pattern decomposition method

In this method, overlapped diffraction patterns are decomposed to individual component Bragg diffractions without assuming the structure model.[8] The profile fitting method[†,9–11] and Pawley method[†,12,13] belong to this category. The difference between these two methods is that in the profile fitting method a part (in a narrow range) of the diffraction pattern is decomposed to component Bragg diffractions whereas in the Pawley method the whole pattern is used and decomposed to the components, and the unit cell parameters are refined simultaneously. In the former no angular dependency (in the case of angular-dispersive measurement) or wavelength dependency (in the case of energy-dispersive measurement) is taken into account since the diffraction pattern in a narrow angular range is considered. However, in the latter it is necessary to assume a profile shape since the whole diffraction pattern is used for the analysis. The latter also requires approximate values of the unit cell parameters in the beginning.

The quantity minimized in order to optimize the structural parameters in both methods is again ΔJ (Eq. 11.1). Standard deviation of the structural parameters are obtained in a

† H. Toraya proposed a "position unconstrained pattern decomposition method" for the pattern-fitting method and a "position constrained pattern decomposition method" for the Pawley method, respectively, in *The Rietveld Method*, (R.A. Young ed.), Oxford Univ. Press, Oxford, 1992.

similar way as in the single crystal structure analysis. However, it has been pointed out that values of standard deviation are much lower than those obtained for the single crystal.[14)]

The use of convolution function in the pattern fitting method derives physically meaningful information from the profile shape (See Section 13.5).

11.1.3 Extension to fibrous materials

An extension of the Rietveld method from the one-dimensional case (powders) described above to a two-dimensional case (fibers), application to a stretched isotactic polypropylene, has been reported by Immirzi and Innelli.[15)] It is reported that in spite of the structural complexity, low resolution, and diffraction overlap, well-distinguishable fits have been obtained by considering two structural models, $P2_1/c$ and $C2/c$, which are closely related to each other. The only assumptions made are that of known bond lengths and 3_1 symmetry of the molecular chain. The results obtained are the C–C–C chain angles of 116.9° and 112.4°, chain torsion angle of 178° and 59° and methyl-to-chain angles of 108.20°.

The other example is an extended application of the pattern fitting method to the structure determination of one-dimensionally oriented fibrous crystal of tetrakis-(dimethyl-ammonium)hexamolybdate(VI) dihydrate, $[(CH_3)_2NH_2]_4Mo_6O_{20} \cdot 2H_2O$.[16)] This complex crystallizes in the form of a bundle composed of a number of fibrous crystals with their *c* axes parallel to the needle axis. The pattern fitting technique was applied by Toraya, Marumo and Yamase to obtain structure factors of overlapping diffractions.

11.2 Structure Analyses Using Uniaxially Oriented Diffraction Patterns

This method is probably the one in most common use for the structure analysis of high polymers. The steps in the analytical procedure are outlined in Fig. 11.2, and described in more detail in the following sections. See also Fig. 11.16, which details the general procedure for the crystal structure analysis of ordinary single crystals.

11.2.1 Determination of fiber period

The first step in the structure analysis of a high polymer is usually to determine the fiber period from the oriented X-ray diagram. The method was described in Section 8.6.2A. A certain amount of information about the molecular structure and the crystal structure can often be obtained from the structural formula unit and the fiber period, particularly in the case of chain polymers.

A. Fiber period and models of the molecular chain

1) The fiber period for the fully extended state of the high polymer chain is calculated from the chemical structure of the repeat unit of the polymer chain, using the standard interatomic distances given in the *International Tables for Crystallography,* Vol. C[17)] for various types of bonds. Information about planarity, partial structures, etc., of any groups contained in the molecular chain may be obtained by reference to the Cambridge Structural Database[18)] which give the molecular structures of related low molecular weight compounds and to sources of crystallographic polymer data.[19)]

2) The measured value of the fiber period is compared with the value calculated from the model of the molecular chain. If the values agree within the limits of experimental error, the proposed model often proves ultimately to have been correct. When the calculated period is slightly greater than the experimental value, this often indicates a twisted or loosely folded conformation of the molecular chains, and the degree of twisting or folding can be estimated from the discrepancy between the two values. Another possibility is that the

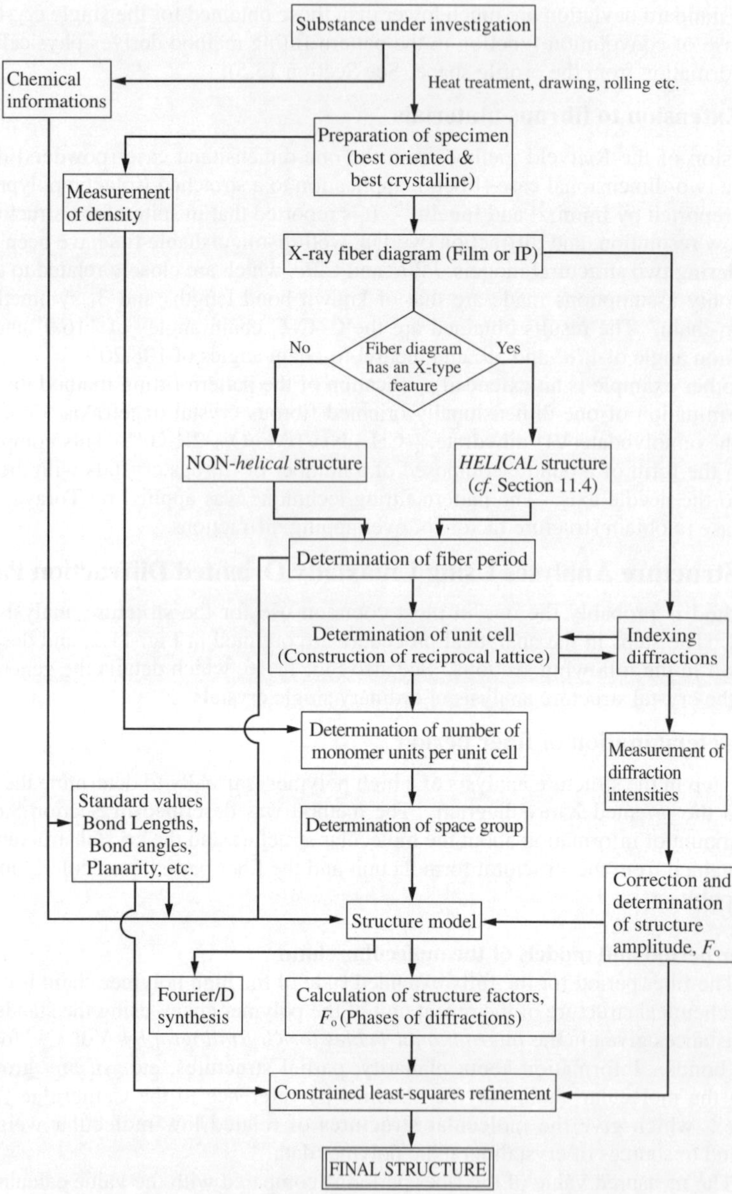

Fig. 11.2 Flow chart of the structure determination of well-oriented high polymers.

chains are slightly inclined with respect to the principal axis of the unit cell. A large discrepancy, particularly if the experimental value is much smaller than the calculated value, often points to a helical conformation of the chains. Methods of ascertaining the presence of a helical structure will be discussed in a later section (*cf.* Section 11.4.3).

3) ***Repeat period (fiber period) of poly (vinyl alcohol)* (– CH_2CH (OH) –)$_n$.** The repeat period for the carbon atoms in poly(vinyl alcohol) is 2.53 Å (Fig. 11.3(a)), while the repeat period for the oxygen atoms in the head-to-tail structure is 5.1 Å similar to the poly(vinyl chloride) (Fig. 11.3(b)). On the other hand, the observed value of the fiber period (along the *b* axis) is 2.53 Å, *i.e.* 1/2 of the calculated fiber period. Since the molecular chains are in the extended state, it must be concluded that some kind of disorder (Fig. 11.3(c)) exists in the arrangement of the oxygen atoms in the direction of the *b* axis.[20]

NOTE :

(1) In the above comparison it is also necessary to take into account not only the effects of internal rotation but also the restrictions imposed by the van der Waals radii on the shapes that the molecular chain can assume.

(2) The possibility of intermolecular hydrogen bonding must always be examined in view of the strong effect this phenomenon has on the shape of the molecular chains.

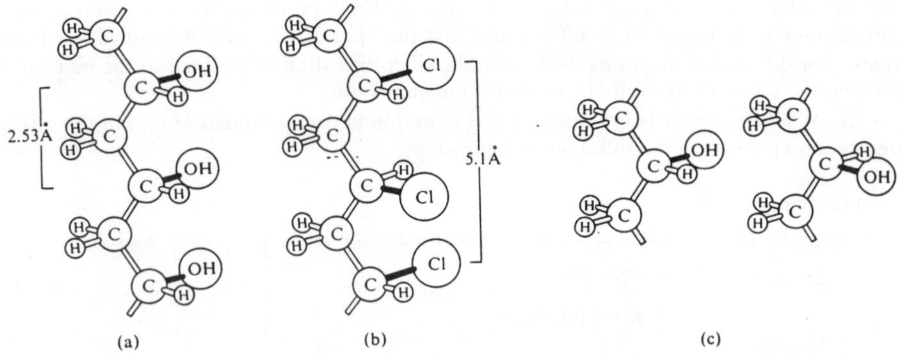

Fig. 11.3 Repeat period (fiber period) of poly(vinyl alcohol).[20]
(a) Repeat period expected for a head-to-tail structure similar to that of poly(vinyl chloride) is 5.1 Å but the period observed is 2.53 Å, which corresponds the repeat distance of the C...C atoms along the polymer chain.
(b) Repeat period observed for the head-to-tail structure of poly(vinyl chloride) is 5.1 Å.
(c) Disorder of the hydroxyl groups along the molecular chain must be considered to explain the unusually short period of 2.53 Å.
[Reproduced from *X-Ray Crystallography*, (I. Nitta ed.), Vol. I, p. 550, Maruzen (1959)]

B. Where the fiber periods determined from diffractions of different order show discrepancies

In some cases, values of the fiber period determined from diffractions of different order show considerable discrepancies. This is presumably due to broad principal maxima of the Laue function, $G(S)$ (or paracrystalline lattice factor, $Z(S)$) caused by very small crystallite size and, at the same time, to rapid decrease or increase of structure factor in the neighborhood of the principal maxima (See Section 13.2.2).

C. Where the fiber period is indeterminate

Polyacrylonitrile[21] and poly(vinyl trifluoroacetate)[21] give clear diffractions on the equator, whereas only diffuse scattering is found in the layer-line region. The approximate fiber period as found from the region of maximum intensity of the scattering in the layer-line region is not always consistent with the expected value (Table 11.1). Specimens of this type cannot be regarded as three-dimensional crystals, but must be treated as two-dimensional crystals consisting of rod-like molecules or as paracrystalline structures (*cf.* Sections 13.6.1 and 14.8).

Table 11.1 Non-equatorial diffractions for poly(vinyl trifluoroacetate)[20]

n	1	2	3	4
Intensity	Very faint	Strong	Faint (on meridian)	Faint (on meridian)
c/n[Å]	9.5	4.9	2.9	2.4
c/n[Å]	9.5	9.8	8.8	6.7

11.2.2 Indexing diffractions and determining unit cell parameters

A. Reciprocal lattice coordinates ξ and ζ from rotation or fiber photographs

1) In most cases the film is placed on the Bernal chart for the appropriate film radius and the values of ξ and ζ are read off directly. Diffraction diagrams made with a micro-camera may be enlarged to a suitable size and the values read with the aid of the Bernal chart. For diffraction diagrams made with the fiber axis slightly inclined to the incident X-rays, correction must naturally be made for the inclination.

2) Where accuracy is important, ξ and ζ are found the coordinates (x, z) of the diffraction spot on the film in the following relationships.

(1) Cylindrical film:

$$\left.\begin{aligned} &\xi=\{1+1/[1+(z/R_f)^2]-2\cos(z/R_f)/[1+(z/R_f)^2]^{1/2}\}^{1/2} \\ &\zeta=(z/R_f)/[1+(z/R_f)^2]^{1/2} \\ &\qquad R_f=\text{film radius.} \end{aligned}\right\} \tag{11.15}$$

(2) Flat film:

$$\left.\begin{aligned} &\xi=\left\{1+\frac{1+(x/R_f)^2-2[1+(x/R_f)^2+(z/R_f)^2]}{1+(x/R_f)^2+(z/R_f)^2}\right\}^{1/2} \\ &\zeta=(z/R_f)/[1+(z/R_f)^2]^{1/2} \\ &\qquad R_f=\text{specimen - to - film distance} \end{aligned}\right\} \tag{11.16}$$

B. Indexing the equatorial diffractions: subsequent unit cell determination

The procedure is identical with that for indexing rotation photographs of low molecular weight specimens. For details, see Section 4.4. If the equatorial diffractions are readily amenable to this method of indexing, the unit cell is assumed to be monoclinic, with the c axis (unique axis) along the axis of rotation (fiber axis); tentative axial lengths (a' and b') and a tentative axial angle (γ') are found from the equation

$$\xi^2\,(hk0)=h^2\,(\lambda/a')^2+k^2\,(\lambda/b')^2+2h(\lambda/a')k(\lambda/b')\cos\gamma' \tag{11.17}$$

The volume of the unit cell, the number of formula (monomer) units in the unit cell, and the crystal density can then be calculated exactly as described in Section 8.6.2B before the true unit cell is known. Subsequent to indexing the layer line diffractions the actual unit cell parameters are found and the values refined by the method of least-squares. The coordinate system should be taken right-handed.

11.2.3 Determination of the space group

A. Space groups and structure factors

As mentioned in Section 4.3, X-ray diffractions from a crystal can only occur (if at all) in directions defined by the reciprocal lattice points, and with intensities which are governed by a structure factor of the form

$$F(hkl) = \sum_{j}^{N} f_j \exp\{2\pi i(hx_j + ky_j + lz_j)\} \tag{11.18}$$

where f_j is the atomic scattering factor of the jth atom in the unit cell, and x_j, y_j, and z_j are the fractional coordinates of this atom in the unit cell. The summation is carried out over all the atoms in the unit cell. The atoms in the unit cell are not independent, but generally fall into groups within which the atoms are related by the symmetry of the crystal. Since comparatively simple relationships hold between the coordinates of the atoms in these groups, insertion of these relationships into Eq. 11.18 simplifies the equation and reduces the number of terms in the summation. This is illustrated below with the aid of some examples.

a) Center of symmetry present. In a cell with center of symmetry at the origin if there is an atom at (x_j, y_j, z_j), there must also be one at ($\overline{x}_j$, $\overline{y}_j$, $\overline{z}_j$). Collecting the terms in pairs, therefore, and continuing to write $\overline{x}_j$ in place of $-x_j$ for convenience, we obtain

$$\begin{aligned} F_j(hkl) &= f_j\ [\exp\{2\pi i(hx_j + ky_j + lz_j)\} + \exp\{2\pi i(h\overline{x}_j + k\overline{y}_j + l\overline{z}_j)\}] \\ &= f_j\ [\exp\{2\pi i(hx_j + ky_j + lz_j)\} + \exp\{-2\pi i(hx_j + ky_j + lz_j)\}] \end{aligned} \tag{11.19}$$

Since

$$2\cos\alpha = \exp(i\alpha) + \exp(-i\alpha) \tag{11.20}$$

Eq. 11.19 can be written in the form

$$F_j(hkl) = 2f_j\ [\cos 2\pi(hx_j + ky_j + lz_j)] \tag{11.21}$$

Hence

$$F(hkl) = 2\sum_{j=1}^{N/2} f_j \cos 2\pi(hx_j + ky_j + lz_j)\} \tag{11.22}$$

Thus the imaginary part has been eliminated, and the structure factor is twice the contribution due to one of two independent halves of the atoms in the unit cell.

b) Twofold screw axis present. If a twofold screw axis 2_1, passing through the point where $x = 0$ and $z = 1/4$, lies along the b axis, the atoms at (x_j, y_j, z_j) and ($-x_j$, $1/2 + y_j$, $1/2 - z_j$) are connected by the symmetry element. Therefore

$$\begin{aligned} F_j(hkl) &= f_j\ [\exp\{2\pi i(hx_j + ky_j + lz_j)\} + \exp\{2\pi i(-hx_j + k(1/2 + y_j) + l(1/2 - z_j)\} \\ &= f_j\ [\exp\{2\pi i(k/4 + ky_j + 1/4)\} \\ &\quad \times (\exp\{2\pi i(hx_j - k/4 + lz_j - 1/4)\} + \exp\{2\pi i(-hx_j + k/4 - lz_j + 1/4])] \\ &= f_j\ [\exp\{2\pi i(k/4 + ky_j + 1/4)\} \times 2\cos 2\pi(hx_j + lz_j - k/4 - 1/4) \end{aligned} \tag{11.23}$$

For $0k0$ diffractions, we get

$$F(0k0) = f_j\ [\exp\{2\pi i(k/4 + ky_j]\} \times 2\cos(k\pi/2) \tag{11.24}$$

For k odd, $\cos(k\pi/2) = 0$, hence

$$F(0k0) = 0 \qquad \text{if } k = 2n + 1. \tag{11.25}$$

If the lattice contains a twofold screw axis in the b direction, all $0k0$ diffractions with $k = 2n + 1$ are systematically absent (*axial extinction rule*).

c) c Glide plane present. Assuming that in space group $P2_1/c$, the pair of equivalent points concerning the c-glide plane is (x_j, y_j, z_j) and $(x_j, 1/2 - y_j, 1/2 + z_j)$. Then we get

$$F_j(hkl) = f_j\ [\exp\{2\pi i(hx_j + ky_j + lz_j)\} + \exp\{2\pi i(hx_j + k(1/2 - y_j) + l(1/2 + z_j)\}]$$
$$= f_j\ [\exp(2\pi i(hx_j + lz_j) \times \{\exp(2\pi iky_j) + \exp(2\pi ik(1/2 - y_j))\exp(i\pi l)\}] \tag{11.26}$$

For hkl diffractions with $k = 0$, we obtain

$$F_j(h0l) = f_j\ [\exp\{2\pi i(hx_j + lz_j)\}(1 + \exp(i\pi l))] \tag{11.27}$$

For l odd, $\exp(i\pi l) = -1$, hence

$$F_j(h0l) = 0 \qquad \text{if } l = 2n + 1 \tag{11.28}$$

If a lattice contains a glide plane perpendicular to the b axis with glide component $c/2$, all $h0l$ diffractions with $l = 2n + 1$ are systematically absent (*zonal extinction rule*).

d) Lattice is C-centered. In a C-centered lattice, if there is an atom at (x_j, y_j, z_j)there must also be one at $(x_j + 1/2, y_j + 1/2, z_j)$.

Then we have

$$F_j(hkl) = f_j\ [\exp\{2\pi i(hx_j + ky_j + lz_j)\} + \exp\{2\pi i(h(x_j + 1/2) + k(y_j + 1/2) + lz_j)\}]$$
$$= f_j\ [\exp\{2\pi i(hx_j + ky_j + lz_j)\}\{1 + \exp\{i\pi(h + k)\}] \tag{11.29}$$

With $h + k$ odd, $\exp\{i\pi(h + k)\} = -1$, hence

$$F_j(hkl) = 0, \qquad \text{if } h + k = 2n + 1. \tag{11.30}$$

That is: if a lattice is C-centered, all the hkl diffractions with $h + k = 2n + 1$ are systematically absent (*general extinction rule*).

The above examples 2) to 4) are known as an extinction rule (a systematic absence) of diffractions, and because it arises from the crystal symmetry, the space group of the crystal may be identifiable from the characteristic, systematic absence of diffractions. Table 11.2 shows the diffractions that disappear for various lattice types and symmetry elements.

B. Determination of the space group

Space group determination starts with the assignment of the Laue class and the determination of the unit cell geometry. Since the X-ray diffraction from a single crystal does not give the crystal symmetry but gives the intensity symmetry, Laue class is determined from the intensity symmetry observed (Table 11.3). The Laue class determines the crystal system.

Table 11.2 Extinction rules†

(a) Dependent on lattice types

Space lattice	General *hkl* diffractions
P	All present
A	Only those with $k+l=2n$ present
B	$l+h$
C	$h+k$
I	$h+k+l$
F	$h+k$, $h+l$, $k+l$ or h, k, and l all even or all odd
R (Rhombohedral axes)	All present
R (Hexagonal axes)	Only either $-h+k+l=3n$ (Obverse setting) or $h-k+l=3n$ (Reverse setting)
$R \rightarrow H$ (Hexagonally-centered lattice)	$h-k=3n$

(b) Dependent on glide planes (valid only where independent of the lattice types extinction rules)

Position of glide plane	normal to *a*	normal to *b*	normal to *c*	$\{1\bar{1}0\}$	$\{110\}$
diffractions observed	$0kl$	$h0l$	$hk0$	hhl	$h\bar{h}l$
a glide plane	—	$h=2n$	$h=2n$	—	—
b glide plane	$k=2n$	—	$k=2n$	—	—
c glide plane	$l=2n$	$l=2n$	—	$l=2n$	$l=2n$
n glide plane	$k+l=2n$	$h+l=2n$	$h+k=2n$	—	—
d glide plane	$k+l=4n, k=2n$	$h+l=4n, h=2n$	$h+k=4n, h=2n$	$2h+1=4n$	—
Mirror plane *m*	All present	All present	All present		

(c) Dependent on screw axes (valid only where independent of the lattice types and glide plane extinction rules)

Direction of screw axis	Parallel to *a*	Parallel to *b*	parallel to *c*
Diffractions observed	$h00$	$0k0$	$00l$
$2_1, 4_2, 6_3$	$h=2n$	$k=2n$	$l=2n$
$3_1, 3_2, 6_2, 6_4$	—	—	$l=3n$
$4_1, 4_3$	$h=4n$	$k=4n$	$l=4n$
$6_1, 6_5$	—	—	$l=6n$
2, 3, 4, 6	All present	All present	All present

† See *International Tables for Crystallography*, Volume A, Space Group Symmetry (T. Hahn ed.), D. Reidel Pub., Dordrecht, 1985.

The indices of all the diffraction spots (or arcs) for the crystal are first tabulated. Regularities are then sought in the indices of observed diffractions. This is because sometimes a very weak diffractions may be observed due to the experimental conditions, obscuring the regularity of the indices of completely missing diffractions. The symmetry is determined from the *general*, *zonal*, and *axial extinction rules* respectively corresponding to the information on lattice type, glide plane, and screw axis, *in that order*. At this stage it is advantageous to use specially compiled space-group sorting tables in the International Tables mentioned above or others.[22] Finally, the space group(s) is found by consulting the *International Tables for Crystallography*, Vol. A*

* The older edition, *International Tables for X-Ray Crystallography*, Vol. I, can also be used. In some cases W. H. Baur, E. Tillmanns, *Acta Cryst.*, **B 42**, 95 (1986) may be helpful for space group determination.

Table 11.3 Laue classes and intensity symmetry

Crystal system	Laue class	Included crystal classes	Intensity symmetry		Asymmetric unit of limiting sphere
Triclinic	$\bar{1}$	1; $\bar{1}$	$I(hkl)=I(\bar{h}\bar{k}\bar{l})$	$[\bar{1}]^{\dagger1}$	1/2 sphere
Monoclinic (*b* unique)	2/*m*	2; *m*; 2/*m*	$I(hkl)=I(\bar{h}\bar{k}\bar{l})$ + $I(hkl)=I(h\bar{k}l)$	$[\bar{1}]^{\dagger1}$	1/4 sphere
Orthorhombic	*mmm*	222; *mm*2; *mmm*	$I(hkl)=I(\bar{h}\underline{\bar{k}}\bar{l})$ $I(hkl)=I(h\bar{k}l)$ + $I(hkl)=I(\bar{h}kl)=I(hk\bar{l})$	$[2/m]^{\dagger1}$	1/8 sphere
Tetragonal	4/*m*	4; $\bar{4}$; 4/*m*	$I(hkl)=I(\bar{h}\bar{k}\bar{l})$ + $I(hkl)=I(hk\bar{l})=I(\bar{k}hl)$	$[1]^{\dagger1}$	1/8 sphere or less[†2]
	4/*mmm*	422; 4*mm*; 42*m*; 4/*mm*	$I(hkl)=I(\bar{h}\bar{k}\bar{l})$, $I(hkl)=I(hkl)=I(\bar{k}hl)$ + $I(hkl)=I(khl)$	$[4/m]^{\dagger1}$	(as above)
Rhombohedral (Trigonal)	3	3; $\bar{3}$	$I(hkl)=I(\bar{h}\bar{k}\bar{l})$ + $I(hkl)=I(k, \bar{h}+\bar{k}, l)=I(\bar{h}+\bar{k}, h, l)$	$[1]^{\dagger1}$	(as above)
	$\bar{3}m$	32; 3*m*; $\bar{3}m$	$I(hkl)=I(\bar{h}\bar{k}\bar{l})$, $I(hkl)=I(k, \bar{h}+\bar{k}, l)=I(\bar{h}+\bar{k}, h, l)$ + $I(hkl)=I(khl)$	$[3]^{\dagger1}$	(as above)
Hexagonal	6/*m*	6; $\bar{6}$; 6/*m*	$I(hkl)=I(\bar{h}\bar{k}\bar{l})$, $I(hkl)=I(k, \bar{h}+\bar{k}, l)=I(\bar{h}+\bar{k}, h, l)$ + $I(hkl)=I(hk\bar{l})$	$[3]^{\dagger1}$	(as above)
	6/*mmm*	622; 6*mmm*; 6*m*2 6/*mmm*	$I(hkl)=I(\bar{h}\bar{k}\bar{l})$, $I(hkl)=I(k, \underline{h}+\bar{k}, l)=I(\bar{h}+\bar{k}, h, l)$, $I(hkl)=I(hkl)$ + $I(hkl)=I(khl)$	$[6/m]^{\dagger1}$	(as above)
Cubic	*m*3	23; *m*3	$I(hkl)=I(\bar{h}\bar{k}\underline{\bar{l}})$, $I(hkl)=I(hk\bar{l})=I(\bar{k}hl)$ + $I(hkl)=I(klh)$	$[4/m]^{\dagger1}$	(as above)
	*m*3*m*	432; $\bar{4}3m$; *m*3*m*	$I(hkl)=I(\underline{h}\bar{k}\bar{l})$, $I(hkl)=I(\bar{h}kl)=I(kh\bar{l})$ $I(hkl)=I(klh)$ + $I(hkl)=I(khl)$	$[m3]^{\dagger1}$	(as above)

†1 The intensity symmetry may be expressed briefly as that of the Laue class given in parentheses plus extra symmetry.

†2 See *International Tables for Crystallography*, Vol. A, for structure factor expressions.

[Reproduced with permission from *Experiments in Chemistry*, 4th ed. (Chem. Soc. Jpn. ed.), Vol. 10, Diffraction, p. 41, Chem. Soc. Jpn. (1992)]

C. Notes on the determination of space groups of high polymer crystals

A very similar procedure to the above (for single crystals of low molecular weight substances) may be used. There is, however, the added complication that diffraction diagrams can often be obtained only for uniaxially oriented specimens of high polymers, and it is often impossible to verify the extinction rules because of difficulty of indexing due to overlapping of reciprocal lattice points, inability to observe the high order diffractions, etc. In such cases, one solution is to select a possible space group and carry out the analysis, assuming initially that the symmetry is high. If this fails, other space groups are tried in turn in an attempt to find the correct one. However, this is a rather trial and error method, and it is better to try to obtain as many independent $|F(hkl)|^2$ as possible, *e.g.* by preparing a biaxially drawn specimen, making inclined fiber photographs or Weissenberg photographs, or using X-rays of shorter wavelength, such as Mo radiation, to obtain data for higher order diffractions.

11.2.4 Structure analysis

A. Measurement of diffraction intensities and determination of observed structure factors $|F_o|$

Measurement of the diffraction intensity normally requires the total intensity for the various reciprocal lattice points intersecting the circumference of the diffraction sphere, *i.e.* the *integrated intensity*. From Eq. 4.2, taking the specimen-to-film distance R_f as equal to $1/\lambda$ (the radius of the diffraction sphere), we can generally write

$$I_{\text{int}} = I_o \frac{1}{R_f^2} \frac{e^4}{m^2c^4} \lambda^3 \frac{\Delta v}{v_o^3} AjL \frac{1+\cos^2 2\theta}{2} |F(hkl)|^2 D^2 \qquad (11.31)$$

where Δv is the volume of the crystal irradiated by X-rays and v_o volume of a unit cell. A and j are absorption and multiplicity factors, respectively. L in this case is the Lorentz factor: the rate at which the reciprocal lattice points pass through the circumference of the diffraction sphere varies with the diffraction angle, for which the factor makes due correction. D is the thermal parameter (see Eqs. 11.5 and 11.6). The Eq. 11.31 assumes that the diffracted X-rays are unaffected by the special absorption effect known as extinction.

If X-ray film in used to record the diffraction, and where the degree of orientation is high as in single-crystal rotation photographs, the intensities of the various diffractions are measured by comparison with a standard intensity scale. If the multiple-film technique is used for diffraction intensity measurements, for successive films, the intensities decrease by the film factor which depends on the absorbance of the sheet of film used, and which can be found from a comparison of intensity values for diffractions having the same indices.

If the degree of orientation is low, so that the various diffractions are drawn out into arcs, it is necessary to find the integral of the intensity distribution found on the various arcs (*cf.* Fig. 11.4). Where several diffractions with different indices overlap on the film, all those that can be resolved from the diffraction intensity curve are so resolved (see Fig. 11.5 and *cf.* Section 8.3.3 for a method of resolving overlapping diffractions). Where this is impossible, one solution is to take the average of the intensity reading over the number of overlapping diffractions, and to use this as a provisional value for each of the diffractions. The distribution is then adjusted on the basis of the calculated structure factor obtained as the analysis progresses.

B. Corrections to be applied to the observed diffraction intensity

a) Distance correction. A correction is naturally necessary when the distances of the diffractions from the specimen differ (*cf.* Eq. 4.2 or 11.31).

b) Absorption correction. Accurate absorption coefficients can be calculated and used when the specimen is a sphere,[23] a cylinder,[23] or a right-angled parallelepiped.[24] It is, however, difficult to determine the absorption factor for general specimens of complex shape. It is therefore usual to avoid the problem of an absorption correction by using speci-

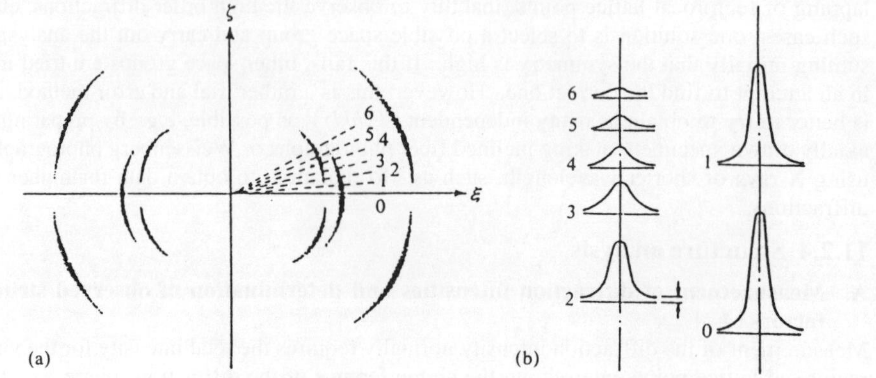

Fig. 11.4 Intensity measurements on diffraction arcs.
(a) The diffraction pattern.
(b) Radial intensity distributions for the sections (0, 1, 2,...) indicated. The arrows in curve 2 indicate the background level.

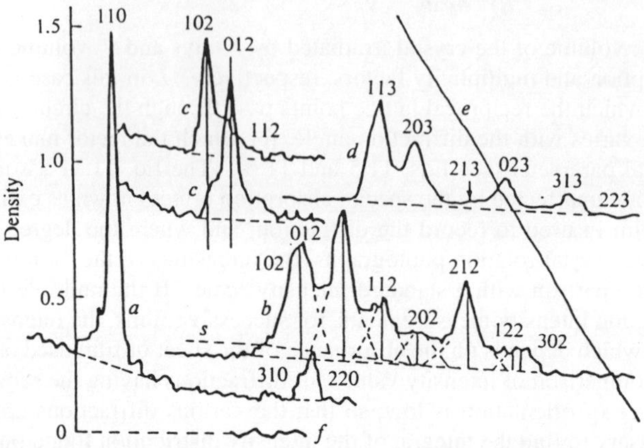

Fig. 11.5 Photometric intensity curves illustrating resolution of overlapping diffractions.
a: Equator; *b* and *d*: second and third layer lines; *c*: radial intensity distributions for 102 and 012 diffractions; *e*: standard intensity curve; *f*: unexposed film background.
[Reproduced with permission from R. Laugridge *et al.*, *J. Mol. Biol.*, **2**, 29, Academic Press (1960)]

mens so small that absorption may be neglected (*cf.* Section 8.4 and Eq. 1.4).

c) Lp correction. The corrections of the Lorentz and polarization factors are usually considered together and referred to as the *Lp* correction. The reciprocal of the Lorentz factor (1/*L*) for the rotating-crystal method (as well as the oscillating-crystal and Weissenberg methods) is as follows.

Equatorial diffractions, $1/L = \sin 2\theta$ (11.32)

Equi-inclination method, $1/L = \lambda\xi\cos\theta = \sin 2\theta(\sin^2\theta - \sin^2\mu)^{1/2}/\sin\theta$ (11.33)

The polarization factor is the factor $(1 + \cos^2 2\theta)/2$ given earlier in Eq. 4.2. With single crystal monochromators, however, the incident X-rays are already partly polarized, and the above expression should be replaced by $(1 + \cos^2 2\theta_M \cos^2 2\theta)/2$, where θ_M is the Bragg angle used with the monochromator. In practice, the *Lp* correction can be made by a computer. But, if X-ray film is being used, the *Lp* correction is found from charts[25] as $(Lp)^{-1}$. The *Lp* correction for precession photographs is complicated, but this correction can also be made with the aid of a computer or charts.[26]

d) Multiplicity correction. In the rotating crystal method (and in fiber photographs), the diffraction intensity is divided by the multiplicity[27] of the crystal planes.

C. Fixing the structure factor scale

The method described below requires the absolute value of the structure factor. This can be found experimentally,[28] but is usually determined by means of a statistical method proposed by Wilson.[29] Since any experimentally determined structure factor will be measured on a relative scale, it must be multiplied by a *scale factor c* to transpose it to the absolute scale.

$$cF_o(hkl) = \sum_j f_j \exp\{2\pi i(hx_j + ky_j + lz_j)\}\exp\{-B[(\sin\theta)/\lambda]^2\} \tag{11.34}$$

B is the temperature factor (*cf.* Section 5.2.3), with all the atoms executing identical isotropic thermal vibrations.

$$\begin{aligned} c^2 |F_o(hkl)|^2 &= \{\sum_j f_j \exp(hx_j + ky_j + lz_j)\} \\ &\qquad \times [\sum_j f_j \exp\{-2\pi i(hx_j + ky_j + lz_j)\}]\exp\{-2B[(\sin\theta)/\lambda]^2\} \\ &= [\sum_j f_j^2 + \sum\sum f_j f_j' \exp\{2\pi i[(h(x_j + x_j') + k(y_j + y_j') + l(z_j + z_j')]\}] \\ &\qquad \times \exp\{-2B[(\sin\theta)/\lambda]^2\} \end{aligned} \tag{11.35}$$

If $|F|^2$ is averaged over roughly equal intervals in $(\sin\theta)/\lambda$, the second terms of Eq. 11.35 for various *h*, *k*, and *l* cancel out, giving

$$c^2 < |F_o(hkl)|^2 > = \sum_j < f_j^2 > \exp\{-2B[(\sin\theta)/\lambda]^2\} \tag{11.36a}$$

This equation can be rewritten

$$\log_e[\sum_j < f_j^2 > / < |F_o(hkl)|^2 >] = \log_e(c^2) + 2B[(\sin\theta)/\lambda]^2 \tag{11.36b}$$

In this equation the average of $|F_o|^2$ is readily calculable for a series of ranges of $(\sin\theta)/\lambda$, while $\sum_j (f_j^2)$ can be calculated from tables of atomic scattering factors for the central points of the ranges. By plotting the term on the left-hand side of the equation as a function of $[(\sin\theta)/\lambda]^2$ a straight line should be obtained, the zero intercept and slope of which give the scale factor *c* and the temperature factor *B*, see Fig. 11.6.

D. Procedure for structure analyses

There are generally less than 100 independently observable diffractions for all layer lines In the X-ray diagram of a fibrous polymer. This clearly imposes limitations on the precision which can be achieved in polymer structure analysis, especially in comparison with the 2,000 or more diffractions observable for ordinary single crystals. However, the molecular chains of the high polymer usually possess some symmetry of their own, and it is often pos-

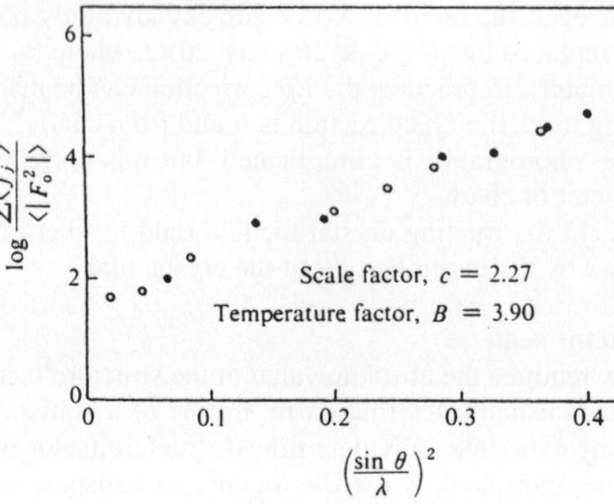

Fig. 11.6 Determination of scale and temperature factors.
(○) Equal ranges of (sin θ)/λ (●) Corresponding ranges of $(\sin\theta)^3/\lambda$

sible to devise a structural model of the molecular chain to interpret the fiber period in terms of the chemical composition by comparison with similar or homologous substances of known structure. Structural information from methods other than X-ray diffraction (*e.g.* infrared and NMR spectroscopy) is also sometimes helpful in devising a structural model of the molecular chain. The majority of the structure analyses which have so far been performed are based on models derived in this way. This is, of course, a trial and error method. The most credible models are arranged in the unit cell, and two-dimensional structure factors projected in the fiber direction (*i.e.* $F_c(hk0)$ if the c axis is the fiber axis) are usually calculated first. The model giving the best agreement with the observed $|F_o(hk0)|$ is then selected. In the preparation of this first model, and in fitting it into the unit cell, use is often made of the Patterson function, as will be described later. When this approximate analysis has been carried out, the analysis is naturally extended to the third dimension, and a more precise structure can be found by the usual Fourier-series method and refined by the least-squares procedure.

E. Calculation of the structure factor

For the structure factor (Section 11.2.3) we have in general, from Eq. 11.19.

$$\left.\begin{aligned} |F(hkl)|^2 &= A^2+B^2, \quad \alpha(hkl)=\tan^{-1}(B/A) \\ A &= \sum_0^n f_j \cos 2\pi(hx_j+hy_j+hz_j) \\ B &= \sum_0^n f_j \sin 2\pi(hx_j+hy_j+hz_j) \end{aligned}\right\} \quad (11.37)$$

where the summations are carried out over all the atoms in the unit cell. This is simplified in practice as a result of the symmetry of the crystal, and structure factor formulae for the various space groups are given in *International Tables for Crystallography*, Vol. B.[30] When the arrangement of the model of the molecular chain in the unit cell has been deduced, the coordinates (x_j, y_j, z_j) of the various atoms are determined, and structure factors are calculated for the various combinations of h, k, and l and compared with the experimental values.

The measure of agreement between the calculated $|F(hkl)|$ and the observed value is expressed by the reliability index or factor, or discrepancy index, R, in Eq. 11.38.

$$R = \frac{\sum \| F_o | - | F_c \|}{\sum | F_o |} \tag{11.38}$$

or by the weighted R index, R_w.

$$R_w = \left[\frac{\sum w(| F_o | - | F_c |)^2}{\sum w | F_o |^2} \right]^{1/2} \tag{11.39}$$

R or R_w approaches zero as the agreement between the observed structure factor F_o and the calculated value F_c improves. The results is usually accepted as reliable when the value is 0.15 or less.

The necessary values of the atomic scattering factors f may be obtained from the *International Tables for Crystallography*, Vol. C.[31] Equation and coefficients for the atomic scattering factor calculation are also given in the *International Tables for Crystallography*, Vol. C.[32]

11.2.5 Fourier transforms and syntheses and Patterson functions

A. Fourier transforms and syntheses

We have already seen in Sections 2.2.3 and 5.2.1 how the Fourier transform theory can be very successfully applied to the phenomenon of X-ray diffraction. This section presents some additional basic information necessary for successful application of the theory to various kinds of crystal structures and their diffraction patterns.

Equation 2.13 gives the amplitude of the diffracted wave for a continuous structure of infinite extension. When the scattering body is a crystal, A becomes $F(hkl)$ (*cf.* Eqs. 2.41 and 4.3), which is not continuous, but takes finite, non-zero values only at the positions of the reciprocal lattice points expressed by the indices hkl. Eq. 2.14, the Fourier transform of Eq. 2.13, shows that a continuous structure is given by the continuous integral over all space of its reciprocal space amplitude.

$$\rho(\boldsymbol{r}) = \int A(\boldsymbol{S}) \exp\{-2\pi i(\boldsymbol{S}\cdot\boldsymbol{r})\} dv_S$$

In the case of a crystal, from Eq. 2.38,

$$\begin{aligned} \rho_{\text{cryst}}(x,y,z) &= \int F(hkl) \exp\{-2\pi i(hx+ky+lz)\} dv_S \\ &= \frac{1}{V} \sum_h \sum_k \sum_l F(hkl) \exp\{-2\pi i(hx+ky+lz)\} \\ &= \frac{1}{V} \sum \sum \sum | F(hkl) | \exp\{-2\pi i(hx+ky+lz) + \alpha(hkl)\} \end{aligned} \tag{11.40}$$

The reason for the change in form from an integral to a summation in Eq. 11.40 is that the amplitude in reciprocal space for a crystal is not continuous over all space, but is concentrated only at the positions of the reciprocal lattice points, its value elsewhere being zero, so that it is sufficient to pick out the amplitudes for the positions of the lattice points hkl and sum them. This equation shows that the electron density at an arbitrary point (x, y, z) in the unit cell of the crystal can be calculated simply by determining the F for all the diffractions hkl and carrying out the above summation. The structure can thus be determined from the electron density and the positions of the atoms. Eq. 11.37 shows, however, that the F's are generally complex quantities, and the $|F_o|$ obtained experimentally cannot, unfortunately, be used directly for the calculation of Eq. 11.40. This equation is known as a Fourier series; its evaluation as a Fourier synthesis, and the method of analysis as Fourier analysis. The usual procedure for the evaluation of Fourier series in crystal analyses is first, as described in the previous section, to identify by some means the most credible model, and then to calculate the F_c for this model. Form the phase angles $\alpha(hkl)$ of these F_c for the observed F_o, the summation of the series is possible. The provisional models are initially based, as we have seen, upon various physical and chemical data. A powerful method of assessing their ultimate congruity has been developed with the use of the Patterson function, which is described below.

B. The Patterson function[33, 34]

If the scattering intensity $I(\boldsymbol{S})$ in Eq. 5.10 is replaced by the diffraction intensity $|F(hkl)|^2$ for a single crystal, then since $I(\boldsymbol{S}) = F(hkl)F^*(hkl) = |F(hkl)|^2$ this equation takes the following form.

$$Q(\boldsymbol{u}) = \int I(\boldsymbol{S})\exp\{-2\pi i(\boldsymbol{S}\cdot\boldsymbol{u})\}\mathrm{d}v_S = \int |F(hkl)|^2 \exp\{-2\pi i(hx+ky+lz)\}\mathrm{d}v_S \tag{11.41}$$

Since the right-hand side of Eq. 11.41 is the Fourier transform of FF^*, it is equal to the convolution of the Fourier transforms $\rho(\boldsymbol{r})$ and $\rho(\bar{\boldsymbol{r}})$ of F and F^*

$$Q(\boldsymbol{u}) = \int \rho(\boldsymbol{r})\rho(\boldsymbol{r}+\boldsymbol{u})\mathrm{d}v_r \tag{11.42}$$

As in Eq. 11.40 we can replace the integral by a summation over the reciprocal lattice points in the case of a single crystal, and we then define the Patterson function $P(u, v, w)$.

$$Q(\boldsymbol{u}) = P(u,v,w) = \frac{1}{V}\sum_h\sum_k\sum_l |F(hkl)|^2\exp\{-2\pi i(hu+kv+lw)\} \tag{11.43}$$

Unlike the Fourier series, Eq. 11.43 does not contain quantities that cannot be measured (*e.g.* phase angle, α *cf.* Eq. 11.40), and is therefore very useful in that it allows an unambiguous calculation using only the $|F(hkl)|^2$ found from the diffraction intensities. Moreover, whereas x, y, and z in Eq. 11.40 are coordinates in the unit cell, $P(u, v, w)$ is not the electron density at the point (u, v, w). On the contrary, from the very nature of such convolutions (as explained in connection with Eq. 5.9), it is the sum of the products of the electron densities at the two ends of a vector $\boldsymbol{u}(u, v, w)$ for parallel translations of this vector to all positions in the crystal (see Fig. 11.7). Thus, if there is an interatomic vector in the crystal that coincides with the vector $\boldsymbol{u}(u, v, w)$, $P(\boldsymbol{u})$ has a maximum. The value increases with the electron densities of the atoms at the two ends of $\boldsymbol{u}(u, v, w)$, and also with the number of interatomic vectors in the crystal that correspond to this vector. In setting up the model mentioned earlier and arranging it in the unit cell, the interatomic vectors for all the intramolecular and intermolecular pairs of atoms must coincide exactly with the maxima of $P(u, v, w)$. In this sense the use of the Patterson function is a most important means of determining structures by assessing structural models purely on the basis of the diffraction pattern without other assumptions. The reader is referred to accounts of the various other methods in the literature.†

† Methods devised for the determination of actual structures from $P(u, v, w)$ include the Harker method, superposition methods, and minimum function methods. For descriptions of the former, the reader is referred to D. Harker, *J. Chem. Phys.*, **4**, 381 (1936), and for the latter two, to M.J. Buerger, *Vector Space and its Application in Crystal Structure Investigation*, Chapter 10, John Wiley & Sons, Inc. N.Y. (1959).

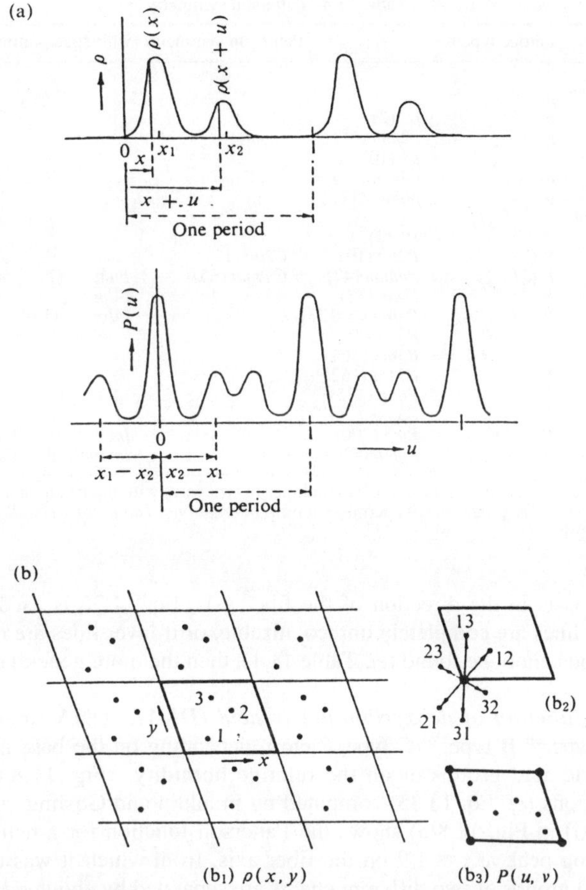

Fig. 11.7 Significance of the Patterson function.
(a) In this one-dimensional example $P(u)=\int_0^1 \rho(x)\rho(x+u)\mathrm{d}x$. Peaks appear in $P(u)$ at $u=\pm(x_2-x_1)$ corresponding to the peaks at x_1 and x_2 in $\rho(x)$.
(b) In this two-dimensional example each unit cell contains three point atoms (b_1), from which are derived an interatomic vector assembly (b_2) and the Patterson function $P(u,v,w)=\int_0^1\int_0^1 \rho(x,y)\rho(x+u,y+v)\mathrm{d}x\mathrm{d}y$.

The Patterson symmetry,[34] that is the space group of the Patterson function, $P(u, v, w)$ is given in Table 11.4.

C. Patterson function for fiber X-ray patterns

In the analysis of X-ray diffraction patterns where the orientation of the crystallites in the specimen or the aggregation of chain molecules is expected to exhibit a statistical density distribution with cylindrical symmetry, it is often convenient to use a cylindrical Patterson function rather than the normal Patterson function. Cylindrical Patterson functions will be discussed in detail in Section 14.6. If the layer lines are clear in the diffraction diagram, *i.e.*

Table 11.4 Patterson symmetry

Laue class	Lattice type	Patterson symmetry (with space-number)				
Two dimensions						
2	p					
2mm	p c	p2 (2)				
4	p	p2mm (6)	c2mm (9)			
4mm	p	p4 (10)				
6	p	p4mm (11)				
6mm	p	p6mm (17)				
Three dimensions						
$\bar{1}$	P	$P\bar{1}$ (2)				
2/m	P C	P2/m (10)	C2/m (12)			
mmm	P C I F	Pmmm (47)	Cmmm (65)	Immm (71)	Fmmm (69)	
4/m	P I	P4/m (83)		I4/m (87)		
4/mmm	P I	P4/mmm (123)		I4/mmm (139)		
$\bar{3}$	P R	$P\bar{3}$ (147)				$R\bar{3}$ (148)
$\bar{3}m1$	P R	$P\bar{3}m1$ (164)				$R\bar{3}m$ (166)
$\bar{3}1m$	P	$P\bar{3}1m$ (162)				
6/m	P	P6/m (175)				
6/mmm	P	P6/mmm (191)				
$m\bar{3}$	P I F	$Pm\bar{3}$ (200)		$Im\bar{3}$ (204)	$Fm\bar{3}$ (202)	
$m\bar{3}m$	P I F	$Pm\bar{3}m$(221)		$Im\bar{3}m$(229)	$Fm\bar{3}m$(225)	

Account is taken of the fact that Laue class $\bar{3}m$ combines in two ways with the hexagonal translation lattice, namely $\bar{3}m1$ as $\bar{3}1m$. [Reproduced with permission from *International Tables for Crystallography*, Teo Hahn ed. p. 15, IUCr. (1985)]

if periodicity exists in the direction of the fiber axis, Eq. 14.36 is used. If, on the other hand, the layer lines are completely unrecognizable, or if layer lines are observed but a distinct fiber period cannot be found (*cf.* Table 11.1), then the more general equation Eq. 13.21 is used.

a) Patterson function of deoxyribonucleic acid (DNA). DNA fibers assume various structures (A type,[35] B type,[36] C type,[37] etc.), depending on the base added to neutralize their phosphoric acid groups or on the relative humidity. Fig. 11.8 shows cylindrical Patterson functions (*cf.* Eq. 13.35) computed by Franklin and Gosling[38] for an A type Na-DNA. Curve (i) in Fig. 11.8(a) shows the Patterson function for a helix of radius 9.0 Å. There is a strong peak at $c = 1/2$ on the fiber axis, from which it was concluded that the phosphoric acid groups of two different chains are separated by about $c/2$, and a model consisting of two coaxial helices with a mutual displacement of $c/2$ in the axial direction was proposed. Curve (ii) in Fig. 11.8(a) is the Patterson function of the second strand of the helix. The principal peaks of the cylindrical Patterson function can be largely explained by these two curves. The well-resolved peaks situated at 5.7 Å from$(0, \frac{1}{2})$ are thought to correspond to the vector between the heavy atoms P–P in Na-DNA. The only diffraction observed near the meridian of the X-ray diagram belongs to the 11th layer line, and this can be interpreted as implying that each turn of the helix of radius 9 Å contains 11 nucleotides. In this case the intrahelix P–P vectors should appear in the positions marked × in the diagram. The Patterson diagram is obviously capable of this interpretation. Fig. 11.7(b) shows the positions of the interhelix P–P vectors when one molecule, *i.e.* a double molecular chain, is fitted into the unit cell ($a = 22.0$, $b = 39.8$, $c = 28.1$ Å, $\beta = 96.5°$). The weights for the positions marked ○ are 1/11 those of the positions marked × in diagram (a). The various diffractions were eventually indexed, and the analysis continued by synthesis and interpretation of the normal three-dimensional Patterson function.

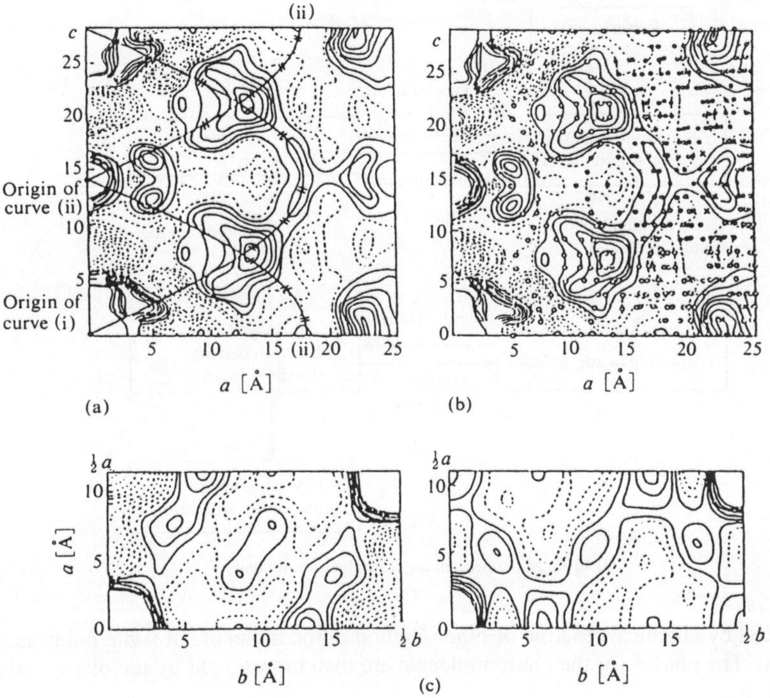

Fig. 11.8 Patterson function of crystalline Na-DNA.[38)]
(a) Cylindrical Patterson function. The curves (i) and (ii) are the Patterson functions of the two helices. The symbol ≠ denotes intrahelix P–P vectors.
(b) The same function as (a). The symbols ○ and × denote interhelix P–P vectors.
(c) Sections of the three-dimensional Patterson function. The left-hand side is at $c = 0$ and the right-hand side at $c = \frac{1}{2}$.
[Reproduced with permission from R.E. Franklin, R.G. Gosling, *Nature*, **172**, 156, Macmillan (1963)]

11.2.6 Determination of phases in Fourier syntheses

The procedure followed in structure analyses based on X-ray diffraction is, generally speaking, as shown in Fig. 11.9. The continuous arrows in the diagram show what can always be found experiment or by calculation. Thus there are two possible routes from the experimental value or $|F|^2$ to the desired electron density distribution. One of these, the determination of the electron density from the Patterson function, is the so-called imageseeking method. This method presents considerable practical difficulties. The corresponding step in the other path is the determination of the phase of the structure factor. The defects attendant upon both of these methods are indicated by the broken lines, which represent them in the diagram. If it were indeed a simple matter to determine the phases and carry out the calculation of Eq. 11.43, structures could be established almost as readily and unambiguously as by examining an object under a microscope. Since this is far from the actual situation, various experimental and analytical methods have been developed.

a) Heavy atom method.[39)] The phase contribution due to heavy atom (*e.g.* Cl, Br, I, Ag, or Hg, depending upon the molecular weight), which is originally included in the molecule

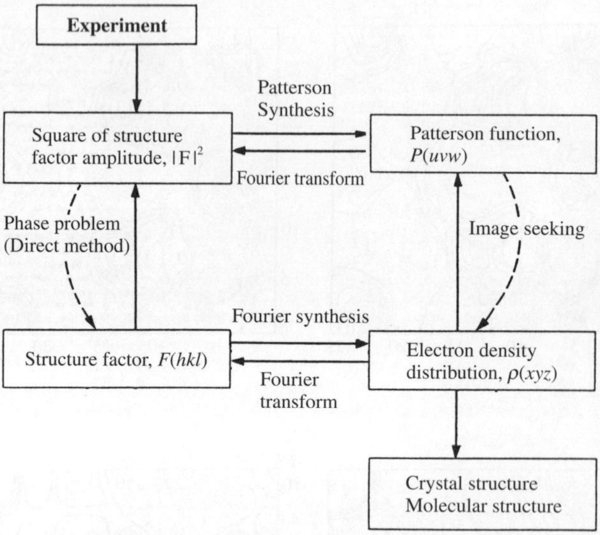

Fig. 11.9 Main steps in the crystal structure analysis.

or introduced by chemical reaction or other method to the molecule at some point, is determined first. The phases for the entire molecule are then represented by the phases calculated in this way.

b) Isomorphous replacement method.[40,41] If the crystal of unknown molecular structure is isomorphous to the crystal of a known molecular structure, and if the unknown molecule is supposed to have a structure similar to that of the known molecule except for a small part, phases calculated from the crystal of known molecular structure except the small part in question may be used to approximate the corresponding phases of the crystal of unknown structure to solve the structure. This isomorphous replacement method is often used to solve the structure of macromolecules in combination with the heavy atom method. (Section 12.4.1).

c) Direct method.[42-48] In this method the phases of various structure factors are found, without the use of semi-empirical procedures, by treating the intensities or phases of the structure factors as statistical groups and finding the phases from the probabilities of these quantitative relationships or phases. Very many analyses, including those of complicated substances, have been successfully carried out by such methods. They are described in detail in specialized textbooks[38] and will not be discussed here.

At present, the structures of organic and organometallic crystals of low molecular weight or even inorganic crystals are mostly solved by the direct method immediately after intensity data collection. Many convenient program systems have developed and are commercially available. These can be loaded on a computer attached to a single crystal diffractometer.

11.2.7 Refinement of the structure

In the case of organic and organometallic crystals of small molecules, the structure solved is usually refined by the least-squares procedure: the quantity minimized being

$\Sigma w(|F_o| - |F_c|)^2$, where w is the weight obtained from the counting statistics of diffraction intensities. The structure is first refined for non-hydrogen atoms with isotropic temperature factors, followed by the determination of the location of hydrogen atoms by difference Fourier synthesis and further refinement with anisotropic thermal parameters for non-hydrogen atoms and isotropic temperature factors for hydrogen atoms.

However, in the case of crystalline high polymers the number of intensity data is much less than that obtained from a single crystal. For instance, for Cu $K\alpha$ radiation ($\lambda = 1.5418$ Å), a single crystal of poly(ethylene terephthalate) theoretically could give 588 independent diffractions within $2\theta \fallingdotseq 100°$ (average spacing of 1 Å), while only 140 diffractions occur in the observable region for the crystalline fiber (the lowest spacing 1.1 Å). This lack of resolving power in the diffraction data has limited or precluded the use of powerful methods such as Fourier synthesis of electron density or least-squares refinement of atomic parameters since in the former, atomic positions would not be located, and in the latter the calculated diffraction amplitudes are insufficiently sensitive to changes in individual atomic coordinates even where there is a sufficient excess of data on parameters for the method to be undertaken. Consequently the usual method of structure refinement in such cases has been to incorporate knowledge of chemical sequence, bond lengths and bond angles in molecular models and then, by systematic trial and error, to adjust their conformations and crystal positions to obtain a good agreement between the structure factor observed, $|F_o|$, and that calculated, $|F_c|$, from the model system.[49)]

11.2.8 Crystal structure analysis of polyethylene

The fiber period c in the direction of the fiber axis is found with the aid of Eq. 8.1 from the spacing z_1 of the layer lines in the well oriented X-ray diagram given as Fig. 11.10(a). The value of c is also accurately determined from the $00l$ diffractions in rotation or Weissenberg photographs (Fig. 11.10(b)) for an axis normal to the fiber axis. The best value for c is 2.534 Å.

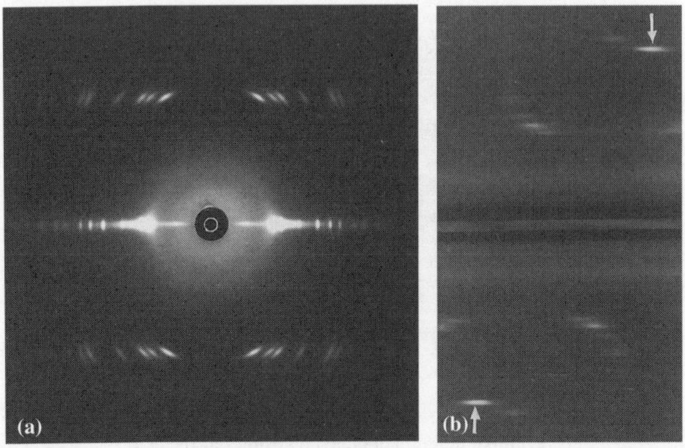

Fig. 11. 10 X-ray diffraction patterns of oriented polyethylene.
(a) Cu $K\alpha$ X-rays normal to vertical fiber axis.
(b) Equatorial Weissenberg photograph for axis normal to fiber axis.
The arrows indicate the 002 diffractions. The photograph yields an accurate measure of the fiber period and includes diffractions not observed in the fiber-axis rotation photograph.

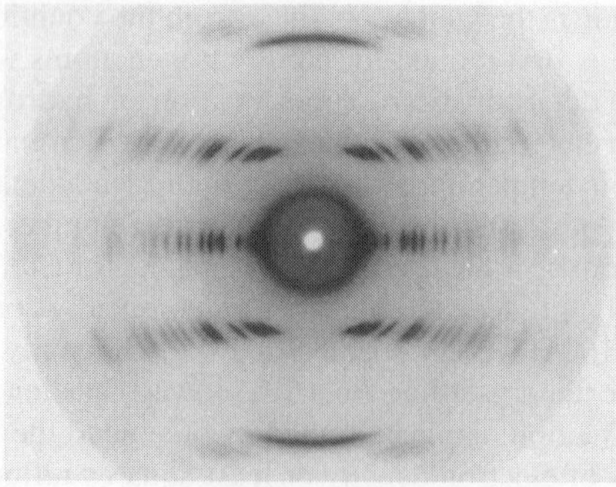

(c$_1$)

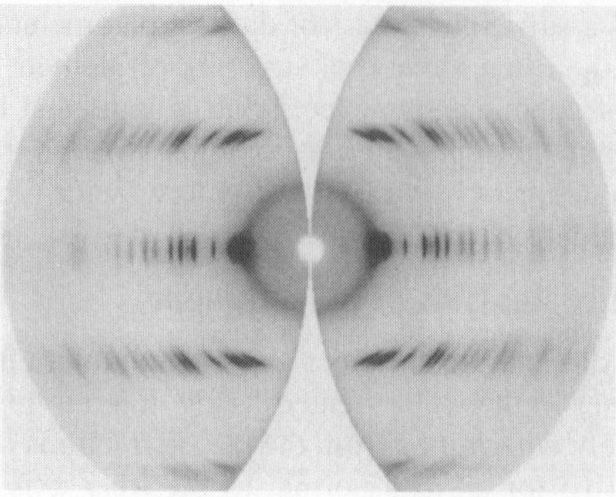

(c$_2$)

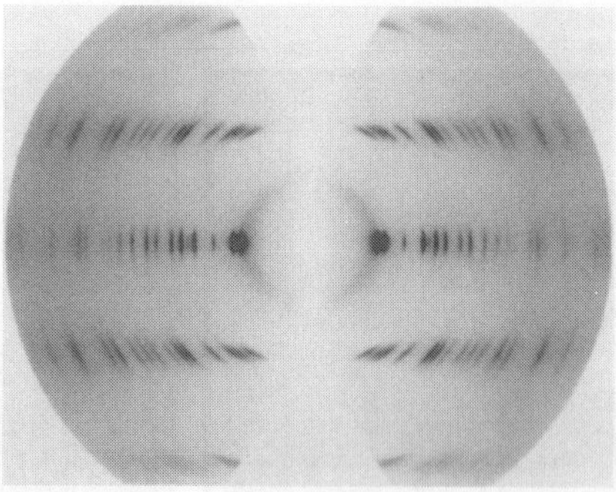

(c$_3$)

Fig 11.10 —*Continued*
(c$_1$) Fiber diagram obtained by using flat IP.
(c$_2$) $\xi\zeta$-Transformed diagram of (c$_1$).
(c$_3$) (c$_2$) After *Lp* correction.
(c$_1$), (c$_2$), (c$_3$) : [Reproduced with permission from K. Tashiro, *MAC Science*, **7**, 32-43, MAC Science (1994)]

Since the unoriented X-ray diagram is similar to that of saturated hydrocarbons[50)] and the crystal structure were also expected to be similar, Bunn determined the unit cell dimensions after indexing the equatorial and layer-line diffractions by trial and error based on the values for saturated hydrocarbons. The best values were $a = 7.40$ Å and $b = 4.93$ Å, with $\alpha = \beta = \gamma = 90°$. The results of the indexing are shown in Table 11.5,[50)] which also gives a comparison between the measured interplanar spacing d and the values calculated from the lattice constants obtained.

Table 11.5 Polyethylene diffraction data used in early analyses[†1]

Interplanar spacing [Å]		*hk*0 diffractions[†2]			Other diffractions[†2]		
Experimental	Calc.	Indices	Obs. intensity	Calc. intensity	Indices	Obs. intensity	Calc. intensity
4.106	4.102	110	4400	4400			
3.696	3.696	200	1160	1165			
2.964	2.956	210	35	48			
2.467	2.467	020	100	226			
2.346	2.340	120	(10)	18			
2.252	2.254				011	105	73
2.202	2.203	310	100	248			
2.162	2.156				111	75	48
2.088	2.089				201	140	94
2.063	2.050	220	70	175			
1.925	1.924				211	70	60
1.849	1.848	400	(5)	41			
1.720	1.720	320, 410	(5)	56	121	20	161
1.665	1.663				311	10	134
1.598	1.596	130	(2)	28	221	(3)	35
1.499	1.502	230	(1)	20			
1.434	1.435	510	(0)	9	321, 411	(3)	56
1.381	1.379	330	(0)	9	031	(2)	67
1.294	1.292				231	(2)	89
1.267	1.267	520	(0)	28	002	(3)	36
1.237	1.236	040, 600	(0)	1	511	(2)	103
1.210	1.210				112	(5)	108
1.123	1.127				022	(<1)	34
1.098	1.098	530			312	(1)	61

†1 Ref. 50 †2 The intensity values in parentheses were estimated visually, while the other values were measured with a photometer. The table lists only the diffractions for which measurements were made. [Reproduced from C.W. Bunn, *Trans. Faraday, Soc.*, **35**, 482, The Faraday Soc. (1939)]

Table 11.6 shows the indices of the observed diffractions in a form which readily enables the systematic extinctions to be identified. Reference to Table 11.5 indicates that the corresponding symmetry elements (lattice type, glide planes, and screw axes) are consistent with either *Pnam* or $Pna2_1$ space groups.

From Eq. 8.2 the number of repeat units $-CH_2-$ in the unit cell is

$$Z = \frac{7.40 \times 4.93 \times 2.534 \times 0.96}{14 \times 1.66} = 3.82 \cong 4$$

With $Z = 4$, the crystal density found from Eq. 8.3 is $D_X = 1.01$ g cm^{-3}(Mg m^{-3}) which is about 5–10% greater than the experimental values of 0.92–0.96 g cm^{-3}.

Table 11.6 Extinction rules space group for polyethylene

Diffractions	Occurrence characteristics (extinction rules)	Corresponding symmetry element (*cf.* Table 11.3)		Space group
hkl	All present	Lattice type	*P*	*Pnam* or $Pna2_1$
0kl	$k+l=2n$ present	Glide plane	n ($\perp$ *a* axis)	
h0l	$h=2n$ present	Glide plane	a ($\perp$ *b* axis)	
*hk*0	All present	Mirror plane present or no symmetry element	*m* or 1 ($\perp$ *c* axis)	
*h*00	($h=2n$, but this is included in the more general extinction rule for *h*0*l*)	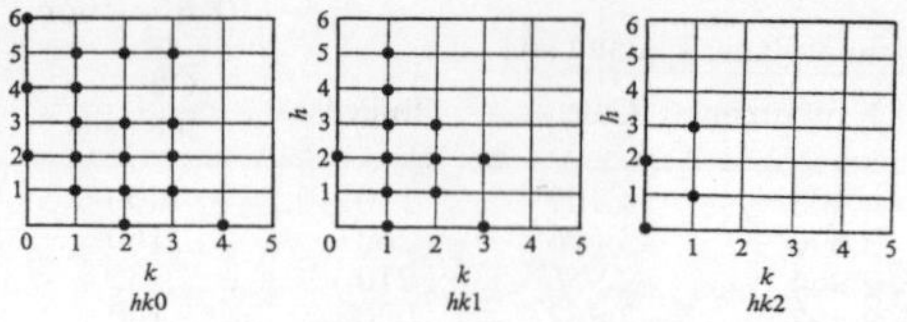		
0*k*0	($k=2n$, but this is included in the more general extinction rule for 0*kl*)			
00*l*	($l=2n$? Only 002 present. This is included in the more general extinction rule for (0*kl*)	Indices of the observed diffractions (marked with black circles)(*cf.* Table 11.5)		

The more highly symmetrical of the two possible space groups, *Pnam*, is listed in *International Tables for Crystallography*, Vol. A, as No. 62 D_{2h}^{16}-*Pnma* (*cf.* Fig. 3.11 and Table 3.4); we must therefore think of the *b* and *c* axes in the present example as being interchanged. Application of the symmetry operations for this space group gives eight general equivalent points. The fact that the unit cell was found only to contain four–CH_2–units implies that the carbon atoms must be located at the special points *c* in Table 3.4.

Pnma (standard, as appearing in *International Tables*) *Pnam* (present example)

$$\left.\begin{array}{lll} x, & \frac{1}{4}, & z; \\ \bar{x}+\frac{1}{2}, & \frac{3}{4}, & z+\frac{1}{2}; \\ \bar{x}, & \frac{3}{4}, & \bar{z}; \\ x+\frac{1}{2}, & \frac{1}{4}, & \bar{z}+\frac{1}{2}. \end{array}\right] \xrightarrow[\text{interchanged}]{b \text{ and } c \text{ axes}} \left[\begin{array}{lll} x, & \bar{y}, & \frac{1}{4}; \\ \frac{1}{2}-x, & \frac{1}{2}-y, & \frac{3}{4}; \\ \bar{x}, & y, & \frac{3}{4}; \\ \frac{1}{2}+x, & \frac{1}{2}+y, & \frac{1}{4}. \end{array}\right. \tag{11.44}$$

Since the *z* coordinates are fixed at $\frac{1}{4}$ and $\frac{3}{4}$, the positions of the carbon atoms can be found by determining the values of the coordinates *x* and *y*. The length of the *c* axis (the fiber period), 2.534 Å, is roughly equal to the calculated value of 2.52 Å for the distance between alternate carbons in the same molecular chain. Thus, since $Z=4$, two molecular chains

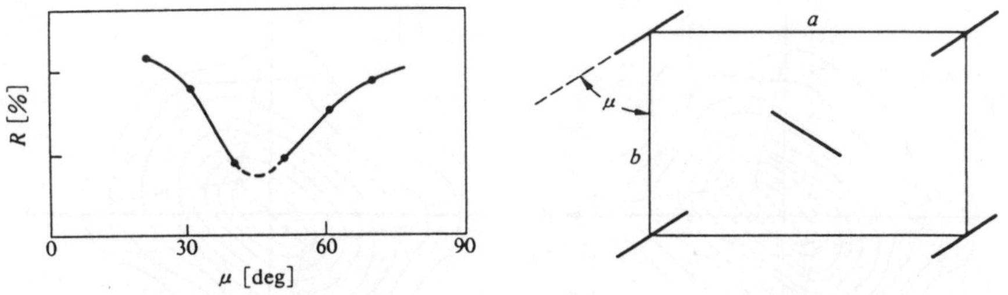

Fig. 11.11 R (see Eq, 11.41) for $hk0$ diffractions *vs.* μ (angle between the b axis and the plane containing the zigzag molecular chains).

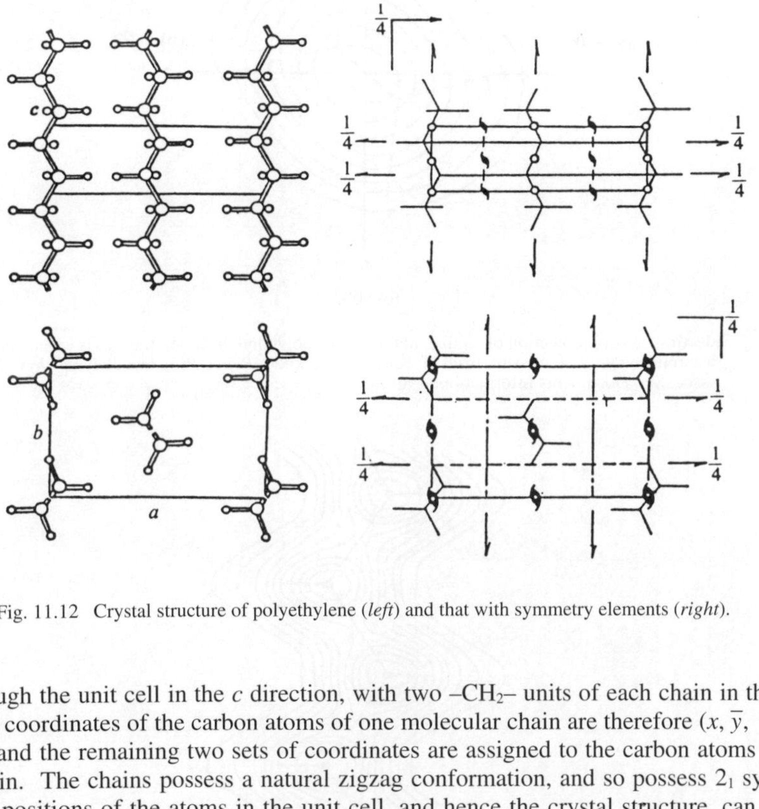

Fig. 11.12 Crystal structure of polyethylene (*left*) and that with symmetry elements (*right*).

pass through the unit cell in the c direction, with two $-CH_2-$ units of each chain in the unit cell. The coordinates of the carbon atoms of one molecular chain are therefore $(x, \bar{y}, \frac{3}{4})$ and $(\bar{x}, y, \frac{3}{4})$ and the remaining two sets of coordinates are assigned to the carbon atoms of the other chain. The chains possess a natural zigzag conformation, and so possess 2_1 symmetry. The positions of the atoms in the unit cell, and hence the crystal structure, can therefore be found by determining, for instance, the angle μ between the b axis and the plane in which the chains zigzag. Fig. 11.11 shows how the value of R(*cf.* Eq. 11.38) for the $hk0$ diffractions varies with μ. The value of $\mu \approx 40^\circ$, corresponding to the minimum of R, is close to that for the now accepted structure, and it is found by trial and error that the fractional coordinates of the carbon atoms are $x = 0.038$ and $y = 0.065$. The accepted crystal structure of polyethylene is given in Fig. 11.12. Calculation of the structure factors natural-

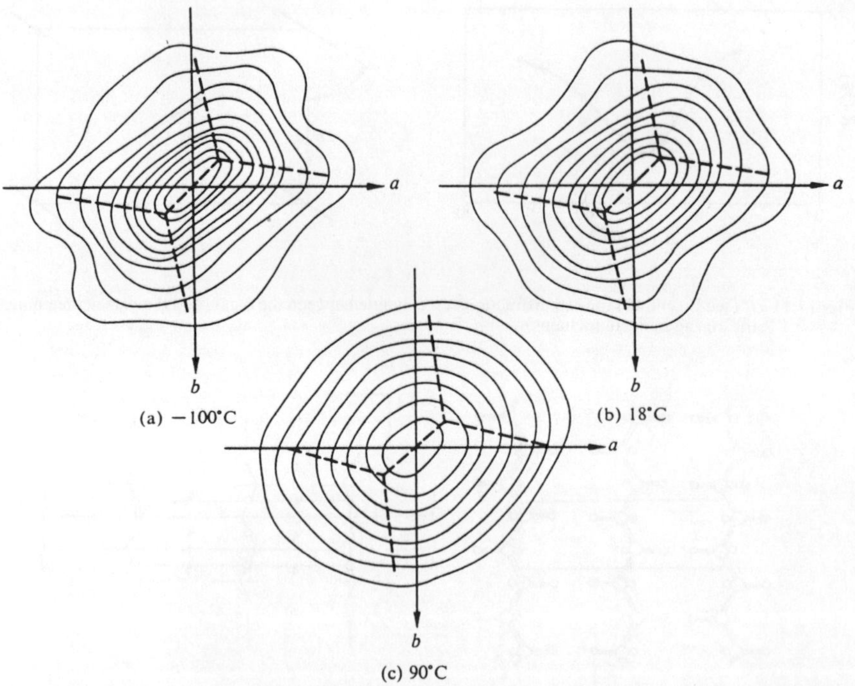

Fig. 11.13 Electron density projection of a molecular chain of polyethylene along the *c* axis (*c* axis down).[51)]
[Reproduced with permission from N. Kasai, M. Kakudo, *Repts. Progress Polym. Phys. Jpn.*, **11**, 145, Assoc. Sci. Documents Information (1968)]

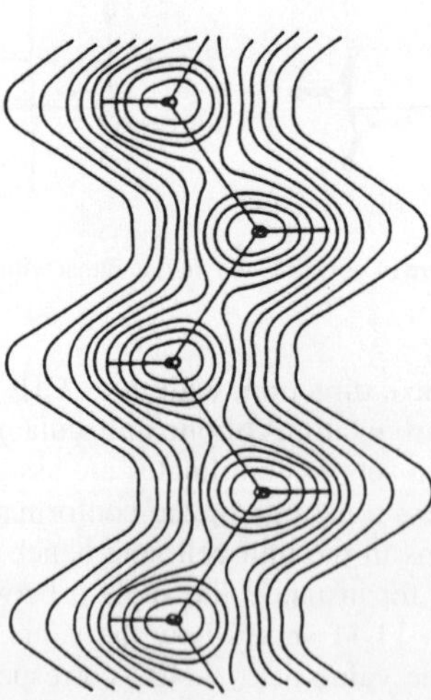

Fig. 11.14 Electron density projection in the plane of a zigzag molecular chain of polyethylene.[50)]
[Reproduced from C.W. Bunn, *Trans. Faraday. Soc.* **35**, 482, Faraday. Soc. (1939)]

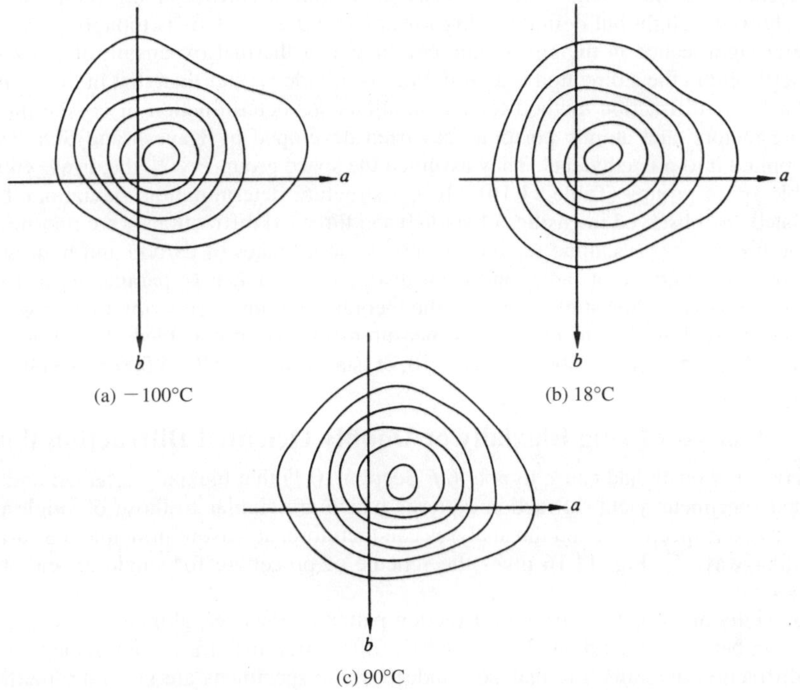

Fig. 11.15 Bounded projection, $B\,(x,y;\,0\rightarrow\frac{1}{2})$ of a molecular chain of polyethylene along the c axis (projection of a $-CH_2-$ unit) (c axis down).[51)]
[Reproduced with permission from N. Kasai, M. Kakudo, *Repts. Progress Polym. Phys. Jpn.*, **11**, 146, Assoc. Sci. Documents Information (1968)]

ly also gives the phases. By equating the phases of the $F_o(hk0)$ with those of the corresponding $F_c(hk0)$, we obtain the electron density projections $\rho(x, y)$ shown in Fig. 11.13,[50)] and by extension to three-dimensional data it is possible to obtain electron density distributions for a section containing a zigzag carbon chain (Fig. 11.14)[50)] and for a section containing a $-CH_2-$ unit (Fig. 11.15).

Kasai and Kakudo[51)] obtained a greater number of accurate diffraction data than Bunn by employing Mo $K\alpha$ radiation in conjunction with diffractometer measurements using a G.E. Single Crystal Orienter, and were able to refine the structure by the least-squares procedure. Diffraction intensity measurements were also carried out at − 100°C and at 90°C. The electron density projections shown in Fig. 11.13 and the bounded projections from $z=0$ to $z=\frac{1}{2}$, $B(x, y;\,0\rightarrow\frac{1}{2})^\dagger$ given in Fig. 11.15, were obtained from the results of these

† Projection of the electron density from $z=0$ to $z=\frac{1}{2}$ in the unit cell on the (001) plane; in the case of polyethylene the electron density distribution for one CH_2 group is projected.

measurements. The effect of thermal vibrations was observed in the rounding of the electron density distribution with rising temperature; the angle μ between the plane of the zigzag chain and the b axis simultaneously underwent a corresponding increase, and the plane showed a slight but definite swing towards the a axis. This is thought to be a factor of major significance in the anisotropic coefficient of thermal expansion of polyethylene; the coefficient in the a direction is an order of magnitude greater than that in the b direction.

An X-ray diffraction method for the simultaneous determination of crystallinity, temperature factors, and atomic positions has been developed by Kavesh and Schultz,[52)] and they applied it to polyethylene. They assumed the space group $Pna2_1$, the, lower of the two possible space groups (Table 11.16). In the structure determination, deviations between calculated and observed intensities of six *hk*0 and three *hk*l diffractions were minimized: the volume fraction of crystalline region, the atomic coordinates of carbon and hydrogen, and constants characteristic of radial and axial disorder were taken as parameters in the minimization process. In the study, errors in the theoretical atomic scattering factors were compensated by scaling the calculations to measurements on an extended chain polyethylene sample (Marlex 50: $\bar{M}_n = 8300$, $\bar{M}_w/\bar{M}_n = 18$; crystallized at 215°C, 5,000 atmospheric pressure; $x_{cr} = 1$).

11.3 Analyses Using Biaxially or Doubly Oriented Diffraction Patterns

We have previously had cause to note (*cf.* Section 10.1) that biaxially oriented and doubly oriented specimens yield diffraction patterns which are similar to those of single crystals and of twins respectively, and the analyses can therefore, at least in principle, be carried out in similar ways.[53)] Fig. 11.16 gives the scheme of procedure for single crystal structure analysis.

Biaxially and doubly oriented diffraction patterns, however, although they have better resolution between individual elements of the diffraction pattern, do not always give suitable diffraction diagrams for analysis. Indeed, these specimens are generally neither very significantly more crystalline nor more highly oriented than uniaxially oriented specimens. This means that their usefulness is limited to indexing, determination of the unit cell parameters and the space group, and measurement of the intensities of diffractions that overlap in uniaxially oriented diagrams, *i.e.* where the higher resolution is helpful.

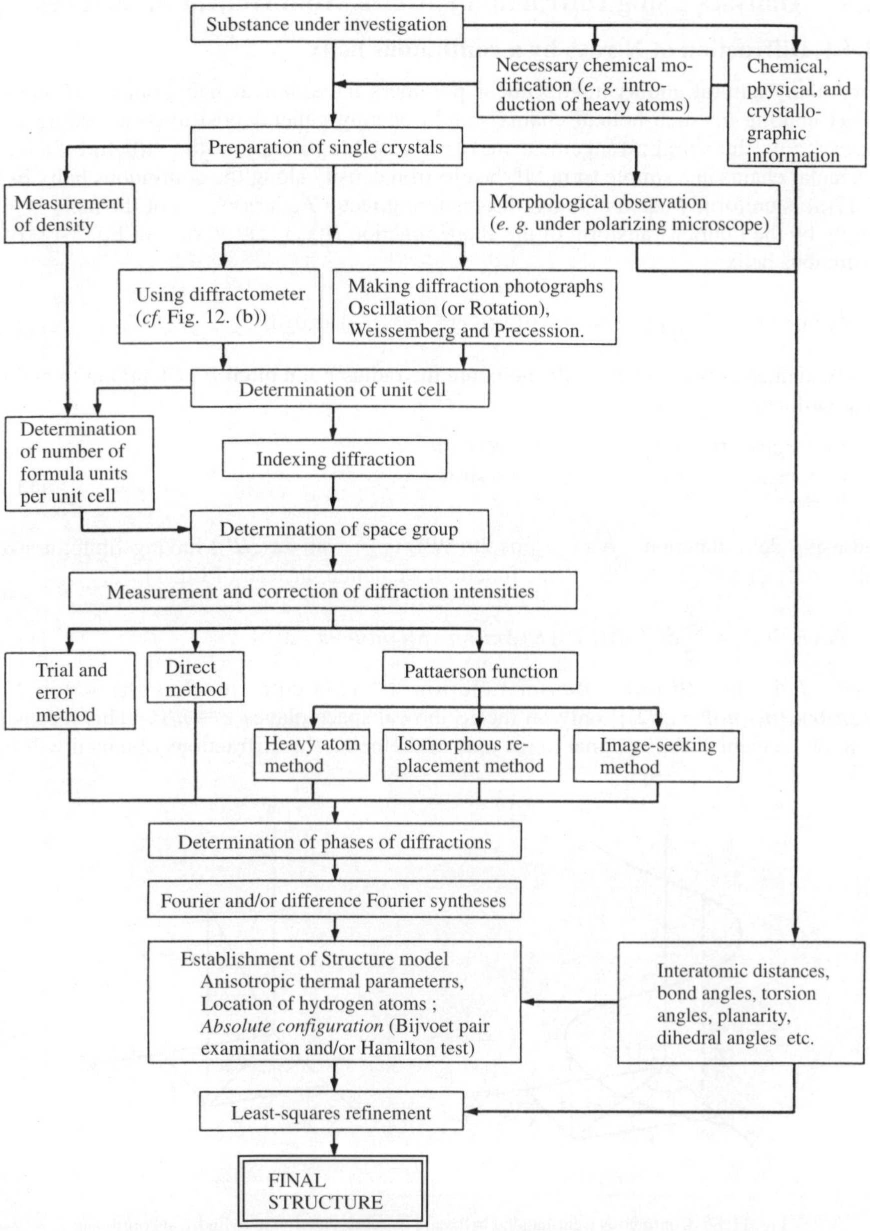

Fig. 11.16 Flow chart of the procedure for single crystal structure analysis.

11.4 Analyses Using Diffraction Patterns from Helical Structures

11.4.1 Diffraction of X-rays by a continuous helix

Very many natural and synthetic linear polymers have helical backbones. Though the atoms making up such helical chains should occupy general positions according to the space group, the spiral arrangement makes it possible to express the structure factors of molecular chains in a simple form. If the electron density along the continuous helix in Fig. 11.17(a) is uniformly equal to unity, the scattering factor $F_M(x^*, y^*, z^*)$ of the helix is given simply by the Fourier transform of the shape function $H(x, y, z)$ (*cf.* $\sigma(r)$ in Eq. 5.12) of the continuous helix:

$$F_M(x^*,y^*,z^*) = \iiint H(x,y,z)\exp\{2\pi i(xx^*, yy^*, zz^*)\}dxdydz \tag{11.45}$$

The fundamental parameters of the helix are the radius r and pitch P. Changing to cylindrical coordinates (Fig. 11.18)†

$$\begin{cases} x = r\cos 2\pi(z/P) \\ y = r\sin 2\pi(z/P) \\ z = z \end{cases} \qquad \begin{cases} x^* = R\cos\Phi \\ y^* = R\sin\Phi \\ z^* = Z \end{cases} \tag{11.46}$$

and using delta function[54] $\delta(x - r\cos 2\pi(z/P))\,(y - r\sin 2\pi(z/P))$ having finite, non-zero values only on the helix as the shape function, we obtain, instead of Eq. 11.45,

$$F_M(R,\Phi,Z) = \sum_{-\infty}^{\infty}\delta(Z-(n/P))J_n(2\pi Rr)\exp\{i[n(\Phi+\pi/2)\} \tag{11.47}$$

where J_n is the nth order Bessel function.[55] This equasion has the value $F_M = J_n(2\pi Rr)\exp\{i[n(\Phi + \pi/2)]\}$ only on the reciprocal space planes $Z = n/P$. This means that with the incident X-rays normal to the axis of the helix, the diffractions obtained will be in

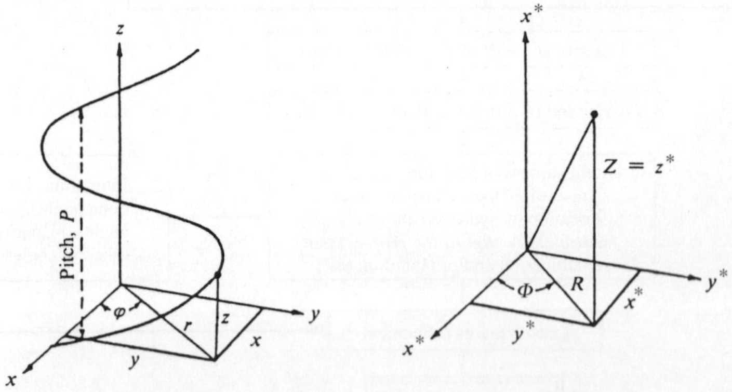

Fig. 11.17 Continuous right-handed helix and its representation in cylindrical coordinates. (a) Physical space; (b) Reciprocal space.

† R is equivalent to ξ and Z is equivalent to ζ in previous sections.

the form of layer lines, where n is the order number of the layer line. Calculation of $|F_M|^2$ from Eq. 11.47 gives a quantity that is proportional to the diffraction intensity. This is shown in Fig. 11.18(a), and the characteristics of the diffraction due to a continuous helix are shown in Fig. 11.18(b).[56)]

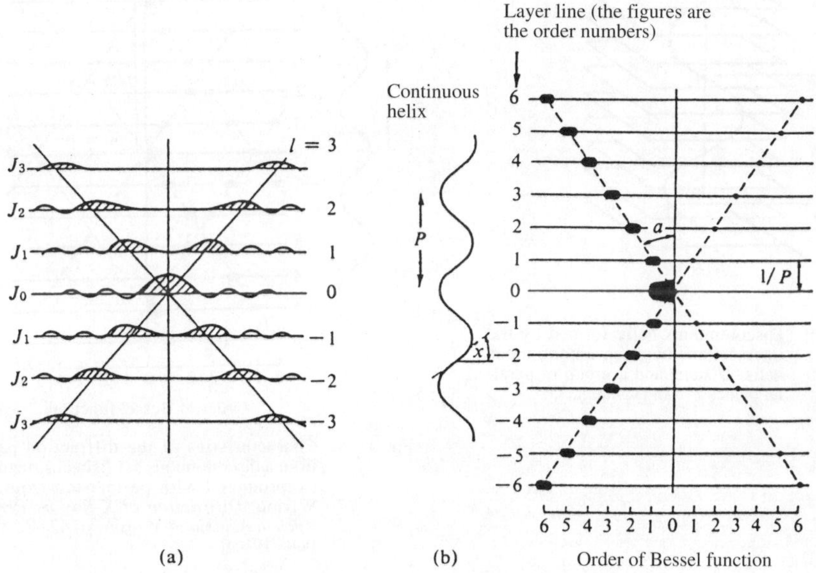

Fig. 11.18 Characteristics of the diffraction pattern from a continuous helical structure.
(a) Bassel function distribution for $n = 1$ (The intensities of various layer lines are influenced by J_n)
(b) The angle α is roughly equal to the pitch angle of the continuous helix[56)]: [Reproduced with permission from H. R. Wilson, *Diffraction of X-Ray by Protein, Nucleic Acids and Viruses*, p. 42, E. Arnold Pub. (1966)]

11.4.2 Diffraction of X-rays by a discontinuous helix

Consider a succession of points arranged with a constant interval p on a helix of pitch P. Each point can be represented as the intersection of the shape function H of the helix with a plane $z = kp$ (where k is an integer), see Fig. 11.19. This can be written with the aid of a delta function as $K(x, y, z) = \sum_{k=\infty} \delta(z - kp)$, and the Fourier transform F_M is the transform of $H(x, y, z) \cdot K(x, y, z)$. From the convolution theorem Eq. 5.7, we obtain

$$F_M(R, \Phi, Z) = \frac{1}{p} \sum_m \sum_n \delta(Z - (m/p) - (n/P)) J_n(2\pi Rr) \exp\{i[n(\Phi + \pi/2)]\} \tag{11.48}$$

Thus the structure factor for a discontinuous helix has finite values only when

$$Z = n/P + m/p \tag{11.49}$$

where n and m are integers (including zero), and F_M is given by summation over all values of n that satisfy this condition.

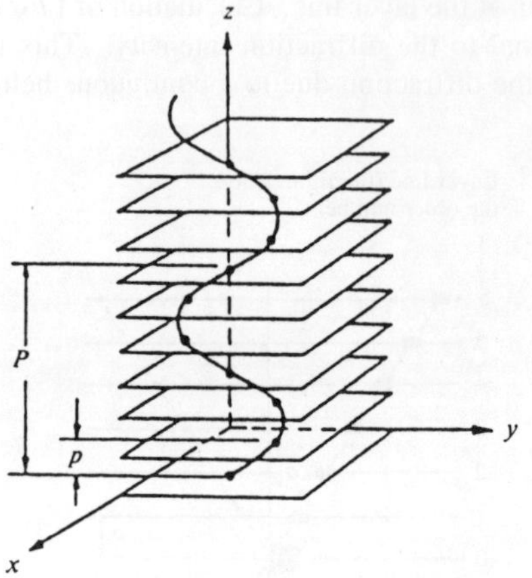

Fig. 11.19 Discontinuous helix formed by the intersections between a continuous helix, $H(x,y,z)$ and a group of parallel planes $z = kp$ (k: integer).

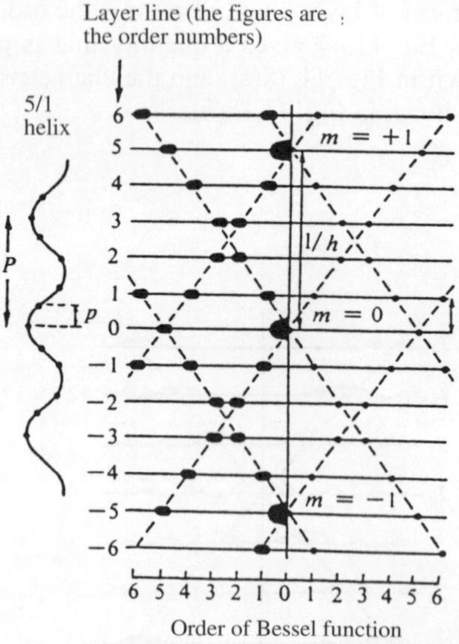

Fig. 11.20 Characteristics of the diffraction pattern from a discontinuous 5/1 helical structure.[56) [Reproduced with permission from H.R. Wilson, *Diffraction of X-Ray by Protein, Nucleic Acids and Viruses*, p. 42, E. Arnold Pub. (1966)]

A. When *P*/*p* is a whole number

For M points on N turns of a helix, the period in the z direction is

$$NP = Mp = c \tag{11.50}$$

Eq. 11.48 then becomes

$$F_{\mathrm{M}}(R,\Phi,Z) = \frac{1}{p}\sum_{n} J_n(2\pi Rr)\exp\{i[n(\Phi+\pi/2)]\} \tag{11.51}$$

and the condition for Eq. 11.49 becomes

$$l = nN + mM \tag{11.52}$$

in which the layer line number l is related to the period c (Fig. 11.20)[56)

B. When *P*/*p* is not a whole number

There is no period of the type described above, and the number of layer lines in the diffraction pattern becomes very large. However, the analysis can be carried out with aid of an approximate period found by taking layer lines relatively close in integral ratios, see Fig. 11.21[56)

C. Extension to real helical structures

If atoms of different kinds are arranged within the interval p along the z direction of the above discontinuous helix, so that there is an atom j having a scattering factor f_j at the point (r_j, φ_j, z_j), the structure factor of one helical molecule is as follows.

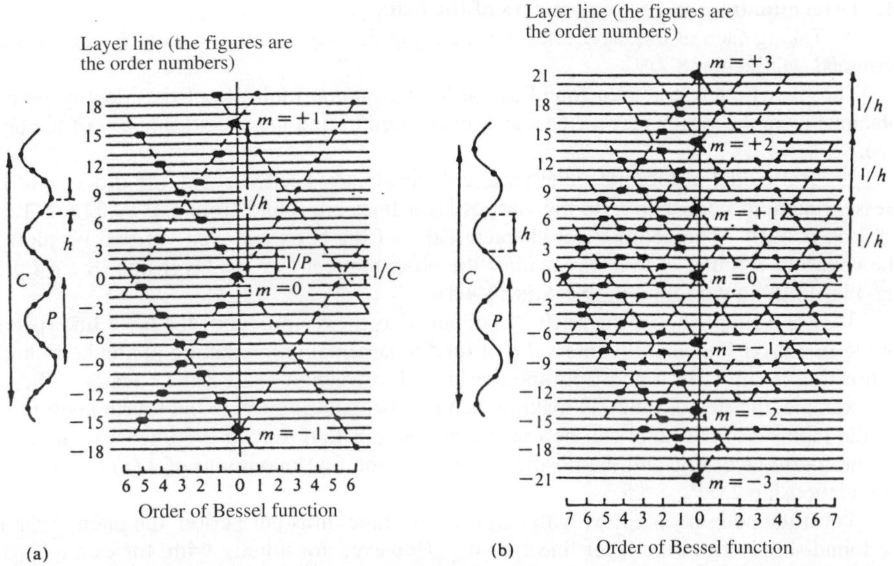

Fig. 11.21 Characteristics of the diffraction pattern from discontinuous helical structures with non-integral P/p.[56] (a) 16/3 helix. (b) 7/3 helix.
The black spots in the left-hand half of the diagrams correspond to the maxima of the Bessel functions. [Reproduced with permission from H. R. Wilson, *Diffraction of X-Ray by Protein, Nucleic Acids and Viruses*, pp.42.43, E. Arnold Pub. (1966)]

$$F_{\mathrm{M}}(R,\mathrm{F},l/c)=\frac{1}{p}\sum_{j}\sum_{n} f_j(2\pi R_c r_j)\exp\{i[n(\mathrm{F}-\varphi+(\pi/2))+2\pi l(z_j/c)\}] \tag{11.53}$$

When r such helical molecules are arranged parallel to one another to form a unit cell, the structure factor is obtained by summation over the r molecules using the coordinates (x_r, y_r, z_r) of standard points of the various helical molecules in the unit cell and the direction angles φ_r about the axis of the spiral.

$$F(hkl)=\sum_{r} F_{\mathrm{M}}(R,\Phi,Z)\exp\{2\pi(hx_r+ky_r+lz_r)\}\exp(-i[n\varphi_r]) \tag{11.54}$$

11.4.3 Interpretation of the diffraction pattern and structure analysis of helical polymers†

A. Indications of the presence of a helical structure

The presence of the following features in the X-ray diffraction pattern is a fairly reliable indication that the molecular chains have a helical structure.

1) The average intensities of the layer lines show a periodic variation from lower to higher order numbers.

2) There are diffraction sports near the meridian that are sufficiently intense to give a high average intensity, and the pattern has the characteristic X shape shown in Figs. 11.18, 11.20 and 11.21.

† This is not to be confused with the spiral orientation discussed in Chapter 10.

B. Determinations of the parameters of the helix

l) Taking each of the layer lines in turn as the first layer line, the identity period is determined (*cf.* Section 8.7).

The fiber period c is then found as the least common multiple of the identity periods obtained from the various layer lines, and this is used to find the order number l of the layer line.

2) If a model of the molecular structure has already been derived, the orders n of the Bessel functions contributing to the various layer lines are found from Eq. 11.52 (*cf.* Table 11.7 on p. 330). The geometrical characteristics of the helix are then verified by plotting the values of n found in this way against the order number l of the layer line (the n-l plot) (*cf.* Fig. 11.27 in the following Section 11.4.4).

3) For a simple helix, the pitch P is obtained as the reciprocal of the layer line spacing of the first layer line, and the interval p of the fundamental units making up the helix in the z direction is similarly found from the meridional diffraction nearest the equator. Then if, for example, the Bessel function making the principal contribution to the fourth layer line is J_4, the radius r of the helix can be found from the position R_{max} of the intensity maximum on the fourth layer line and the position $x = 5.3$ of the first maximum of $J_4(x)$ by means of the relationship $2\pi r R_{max} = 5.3$.

4) If the helix is complex, with (say) two or three turns per period, the pitch P cannot be found simply from the layer line spacing. However, for a helix with, for example, one period per three turns, N in Eq. 11.50 is 3, and $c = 3P$. The first, second, fourth, fifth,...layer lines thus correspond to high order Bessel functions of low intensity, whereas the third,

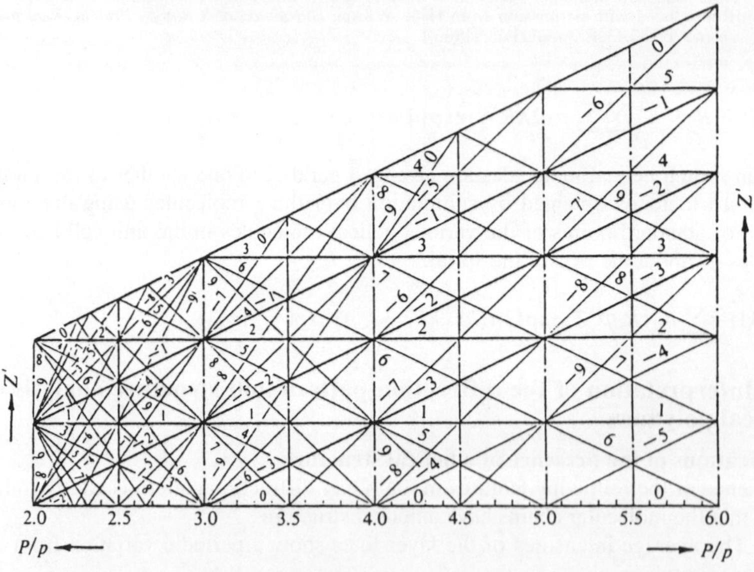

Fig. 11.22 Specially prepared diagram for determining the parameters $Z'(n)$ *vs.* P/p of Eq. 11.55.[57] Since m in independent of the layer line intensity (*cf.* Eq. 11.51), only n values (for $n \leq 9$) are shown. The positions of the various layer lines are recorded according to the Z' values, together with their average intensities, on a uniformly extensible tape. This tape is stretched out parallel to the ordinate at various P/p positions to obtain the best possible agreement between its Z' and the corresponding Z' values in the diagram, and the P/p for this point is then read off. Intense layer lines must naturally correspond to small n. [Reproduced with permission from Y. Mitsui, *Acta. Cryst.* **20**, 694, IUCr. (1966)]

sixth, etc. layer lines correspond to low-order Bessel functions, with the result that intense diffractions appear close to the meridian. The pitch P can therefore be found from the spacing of the layer lines of the latter type, and the number of turns N of the helix is obtained from P and the period c.

5) It is clear from Eq. 11.50 that when P/p is not an integer, N, M, and c are naturally large, and if P/p is slightly changed, N, M, and c undergo a large change. In such cases, therefore, it is preferable to find the parameters of the helix in terms of P and p instead of N, M, and c. From Eq. 11.49 we obtain

$$Z' = ZP = n + m\,(P/p) \tag{11.55}$$

$$Z'' = Zp = n\,(p/P) + m \tag{11.56}$$

The relative values of the Z' or Z'' coordinate of the various layer lines can thus be expressed by a first order equation in P/p or p/P. The parameters are determined with the aid of prepared diagrams as shown in Fig. 11.22 and 11.23[57] which give explicitly the relationships expressed in Eqs. 11.55 and 11.56. We first find P/p (or p/P), and then look for is smallest integral multiple M(the number of units per period). N is then found from Eq. 11.50, and the order number l of the layer line is obtained from Eq. 11.52. When the order numbers of the layer lines have been correctly found in this way, it is possible to calculate the fiber period c.

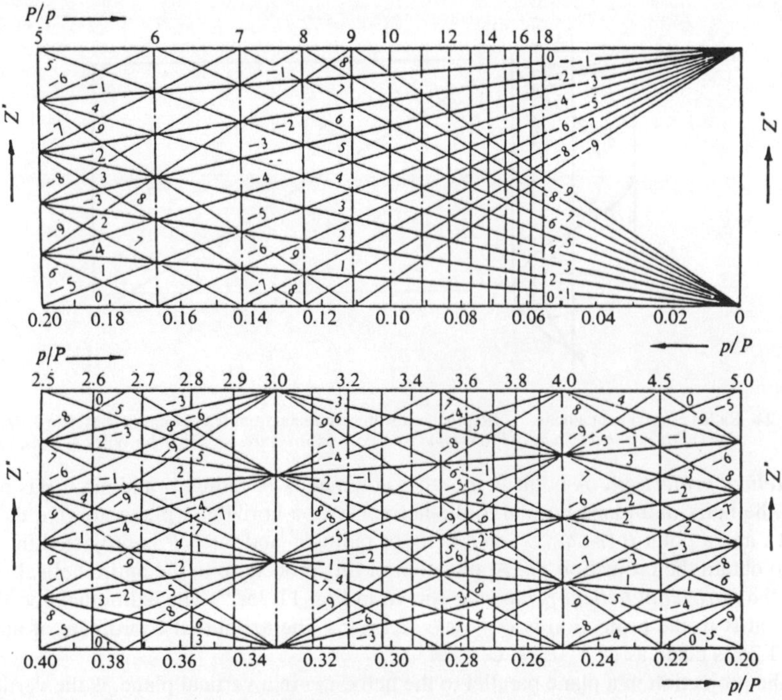

Fig.11.23 Specially prepared diagram for determining the parameters $Z'/(n)$ *vs.* p/P of Eq. 11.56.[57] Used in the same way as Fig. 11.22 this diagram is more convenient for meridional diffractions. [Reproduced with permission from Y. Mitsui, *Acta. Cryst.* **20**, 694, IUCr. (1966)]

C. Analysis of helical structures

Structure factors are calculated from Eqs. 11.33 and 11.34 for definite positions of the various atoms in the helical model found. Comparison between calculated and observed values enables refinement of the model by the trial and error method, the Fourier method, or, precisely, by the least-squares procedure. Since this stage is identical with the corresponding stage of the general crystal analysis of polymers, which has already been described in some detail, it will not be discussed again here.

The methods described by Shimanouchi and Mizushima,[58] Miyazawa,[59] and Nagai and Kobayashi[60] may be used to calculate the coordinates of the atoms in the helix. Another method of gaining an understanding of the structure is to examine the model in the form of a two-dimensional diagram, using the radial and helical projections[61] of the helix.

a) Radial projection. Consider a helix of uniform radius drawn on the surface of a cylinder. If this cylindrical surface is cut and spread out flat, inside uppermost, we obtain a two-dimensional picture of the helix (*cf.* Fig. 11.31). If the whole of the helical molecule does not lie on helices having the same radius, it may be projected on a suitable cylinder as shown in Fig. 11.24.[61] This is the radial projection of the helix. Klug, Crick, and Wyckoff[62] have shown that the Fourier transform of the lattice obtained in this way is the *n–l* plot. The relationship between the radial projection of the helix and the *n–l* plot is also useful for estimating, from the diffraction diagram, the magnitude of the modified Bessel term corresponding to a small modification of the model of the helical molecule during the analysis.

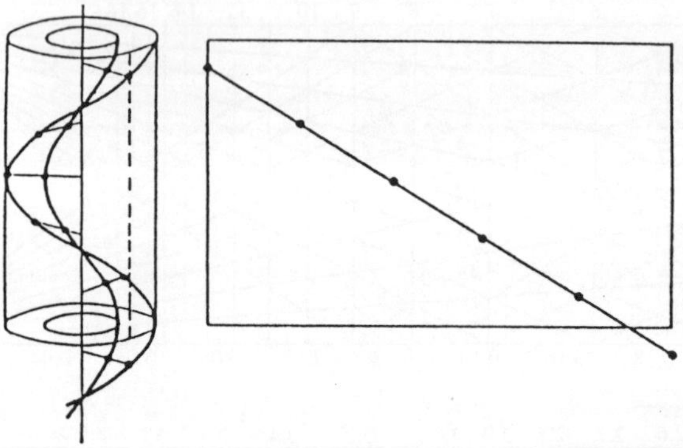

Fig. 11.24 Radial projection of a helix.[61] [Reproduced with permission from K.C. Holmes, D.M. Blow, *The Use of X-Ray Diffraction in the Study of Protein and Nucleic Acid Structures*, p. 221, John Wiley & Sons, Inc. (1966)]

b) Helical projection. A point A (r_1, φ_1, z_1) on a helix of radius r_1 and pitch P is projected along the helix in a plane normal to the helix, *i.e.* in a horizontal plane, to give the projection A_1, and a point B (r_2, φ_2, z_2) on a helix of radius r_2 and pitch P is projected in the same way to obtain the projection B_1. Application of this procedure to the entire helical molecule gives the *horizontal helical projection* shown in Fig. 11.25.[61] The azimuthal coordinate of the point B in the projection is $\varphi' = \varphi_2 - 2\pi z_2/P$. The azimuthal coordinate of atom A in Fig. 11.25 is taken as $\varphi' = 0$, *i.e.* $\varphi_1 = 2\pi z_1/P$.

The projection in a plane parallel to the helix, *i.e.* in a vertical plane, is the vertical helical projection. The point A is projected to A_1 and A_2, which are separated by a distance equal to the pitch P, and B is projected to B_1 and B_2 (see again Fig. 11.25). A_1 and A_2 are used to define the repeat distance of the projection, and the coordinate z' of B_2 measured

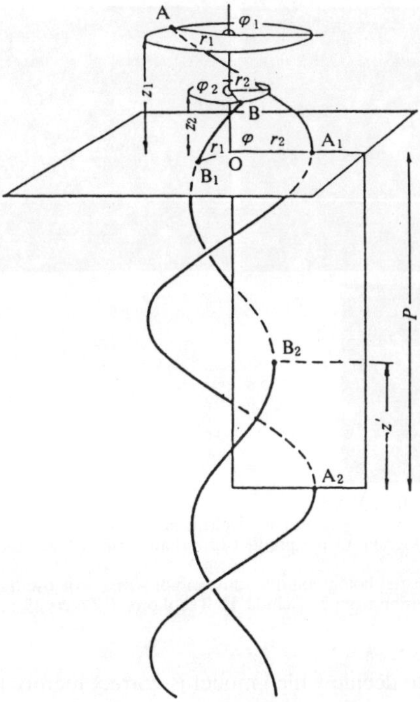

Fig. 11.25 Helical projection.[61] [Reproduced with permission from K.C. Holmes, D.M. Blow, *The Use of X-Ray Diffraction in the Study of Protein and Nucleic Acid Structures*, p. 223, John Wiley & Sons, Inc. (1966)]

from A_2 is given by $z' = z_2 - P\varphi_2/2\pi$. Vertical helical projections are used where the pitch is small. In general, however, the horizontal helical projection is used. Helical projections are useful for finding the order of the Bessel function on layer line l. Thus if the projections of various points for pitch $p = nc/l$ exhibits n-fold symmetry, the J_n term will be strong.

11.4.4 Determination of helical structures

A. Poly(oxymethylene) (POM) $(-CH_2-O-)_n$

The initial crystal structure analysis of POM was carried out using a c-axis rotation photograph[63] as shown in Fig. 11.26(a), and a more accurate analysis was then carried out on the basis diffraction data, Fig. 11.26(b) and (c), for highly oriented specimens obtained by polymerization of tetraoxane in the solid phase.[64] POM crystal belong to the trigonal system; the unit cell dimensions, expressed for a hexagonal lattice, are $a = 4.46$ Å and $c = 17.35$ Å, and the extinction rules indicate a C_s^2 or C_s^3 space group. The unit cell contains one molecular chain, and there are nine repeat units per fiber period (17.35 Å).

The average intensities of the layer lines (Table 11.7) are consistent with either a 9/4 or a 9/5 helix. Eq. 11.52 gives

$$l = 5n + 9m \text{ (9/5 model)} \tag{11.57}$$

$$l = 4n + 9m \text{ (9/4 model)} \tag{11.58}$$

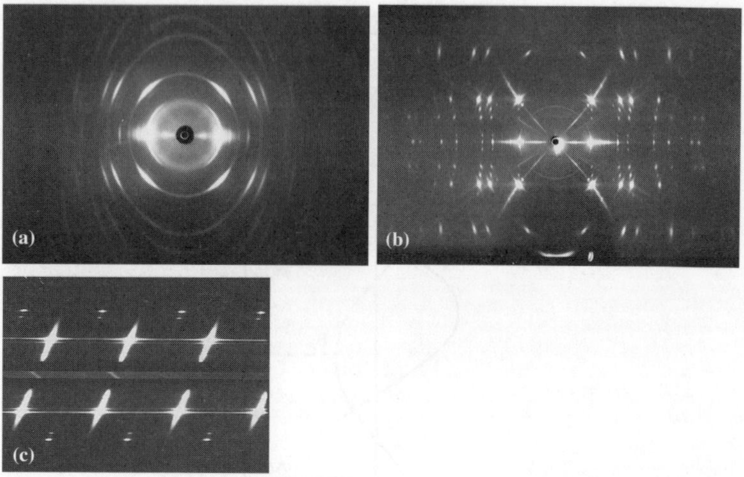

Fig, 11.26 X-ray diffraction patterns from POM.
(a) Fiber photograph of a uniaxially oriented Delrin specimen.[63)]
(b) *c*-Axis rotation photograph of the product of radiatin-induced polymerization of tetraoxane in the solid phase.[64)]
(c) Equatorial Weissenberg photograph (the same sample (b), *c* axis oscillation).[64)]
[Reproduced with permission from T. Uchida, H. Tadokoro, *J. Polym. Sci.*, 63, John Wiley & Sons, Inc. (1967)]

It is not possible in this case to decide which model is correct merely from the orders of the Bessel functions corresponding to the various layer lines (Table 11.7). However, whereas the Bessel functions corresponding to the fourth and fifth layer lines are both first order (though of opposite signs), the average intensities of these two layer lines are very different (see again Table 11.7). The molecular symmetry is such that two-fold axes normal to the molecular axis pass through the C and the O atoms. If the radii of the C and O helices are

Table 11.7 Average intensities of the layer lines and orders of the Bessel functions for POM

Order number *L* of layer line	Average intensity of layer line	9/5 helix: Order *n* of Bessel function	9/5 helix: Phase relation of C and O contributions to structure factor†	9/4 helix: Order *n* of Bessel function	9/4 helix: Phase relation of C and O contribution to structure factor†
0	Very strong	0	+	0	+
1	(Not observed)	2	−	−2	−
2	(Not observed)	4	+	−4	+
{ 3	Weak	−3	+	3	−
{ 4	Weak	−1	+	1	+
5	Strong	1	+	−1	−
6	(Not observed)	3	−	−3	+
7	(Not observed)	−4	−	4	−
8	Medium	−2	+	2	+
9	Medium	0	−	0	−
⋮		⋮	⋮	⋮	⋮
18	Medium	0	+	0	+

† + indicates that the C and O contributions have the same phase, and so reinforce each other, while − indicates that the contributions have opposite phases, and so tend to cancel each other out. See text.
[Reproduced with permission from T. Uchida, H. Tadokoro, *J. Polym. Sci.*, 63, John Wiley & Sons, Inc. (1967)]

taken to be equal, the structure factors of the fourth and fifth layer lines, considering only the first term of Eq. 11.58, are given by:

(for a 9/5 helix)

$$F_{\mathrm{M}}(R, \Phi, 4/c) = i \exp(-i\varphi)\{-f_{\mathrm{c}} J_{-1}(2\pi R r_{\mathrm{c}}) + f_{\mathrm{o}} J_{-1}(2\pi R r_{\mathrm{o}})\} \quad \text{(4th layer line)}$$

$$F_{\mathrm{M}}(R, \Phi, 5/c) = i \exp(i\varphi)\{f_{\mathrm{c}} J_{1}(2\pi R r_{\mathrm{c}}) + f_{\mathrm{o}} J_{1}(2\pi R r_{\mathrm{o}})\} \quad \text{(5th layer line)}$$

(for a 9/4 helix)

$$F_{\mathrm{M}}(R, \Phi, 4/c) = i \exp(i\varphi)\{f_{\mathrm{c}} J_{1}(2\pi R r_{\mathrm{c}}) + f_{\mathrm{o}} J_{1}(2\pi R r_{\mathrm{o}})\} \quad \text{(4th layer line)}$$

$$F_{\mathrm{M}}(R, \Phi, 5/c) = i \exp(-i\varphi)\{-f_{\mathrm{c}} J_{-1}(2\pi R r_{\mathrm{c}}) + f_{\mathrm{o}} J_{-1}(2\pi R r_{\mathrm{o}})\} \quad \text{(5th layer line)}$$

It follows, therefore, that the structure factors for the 9/5 helix involve the sum of the C and O contributions for the fifth layer line and the difference for the fourth, whereas the conditions are reversed in the case of a 9/4 helix. Examination of the other layer lines confirms that the 9/5 helix model is the more appropriate, as shown in Table 11.7.

Figure 11.27[64] shows the n–l plot which was constructed for POM using the order numbers n obtained from the solution of Eq. 11.57. The figure is symmetrical about the origin, indicating that a point (n, l) appears in company with the point $(-n, -l)$. Application of the formula given by Shimanouchi and Mizushima[58] gives a value of $r = 0.691$ Å for the radius of the helix. This value was used to calculate $J_n(2\pi R r_j)$ for $n = 0$–8, and it was confirmed that the contributions of the second and subsequent terms of the Bessel functions are very small (Fig. 11.28).[64] For this reason, only the first term was used

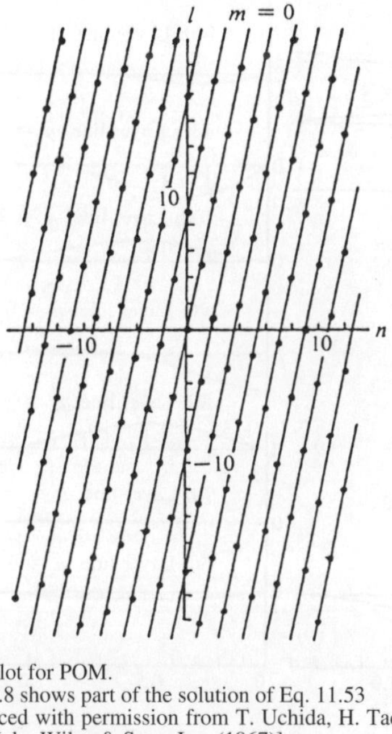

Fig. 11.27 The n–l plot for POM.
Table 11 .8 shows part of the solution of Eq. 11.53
[Reproduced with permission from T. Uchida, H. Tadokoro, *J. Polym. Sci.*, **A5**, 63, John Wiley & Sons, Inc. (1967)]

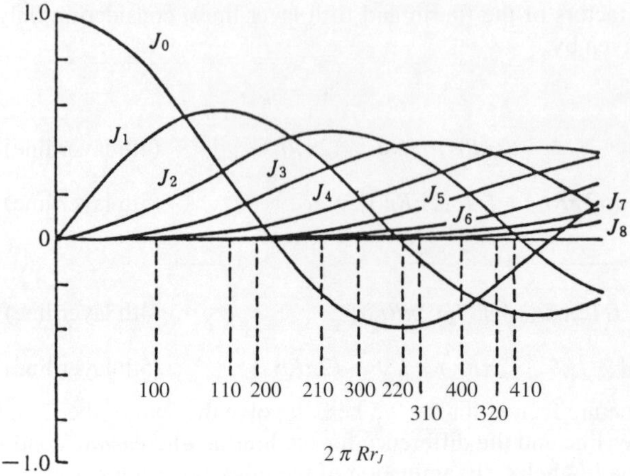

Fig. 11.28
Values of $J_n(2\pi Rr_j)$ for $n = 0$–8. The broken lines indicate the values of $2\pi Rr_j$ for the *hk*0 diffractions.

[Reproduced with permission from T. Uchida, H. Tadokoro, *J. Polym. Sci.*, **A5**, 63, John Wiley & Sons, Inc. (1967)]

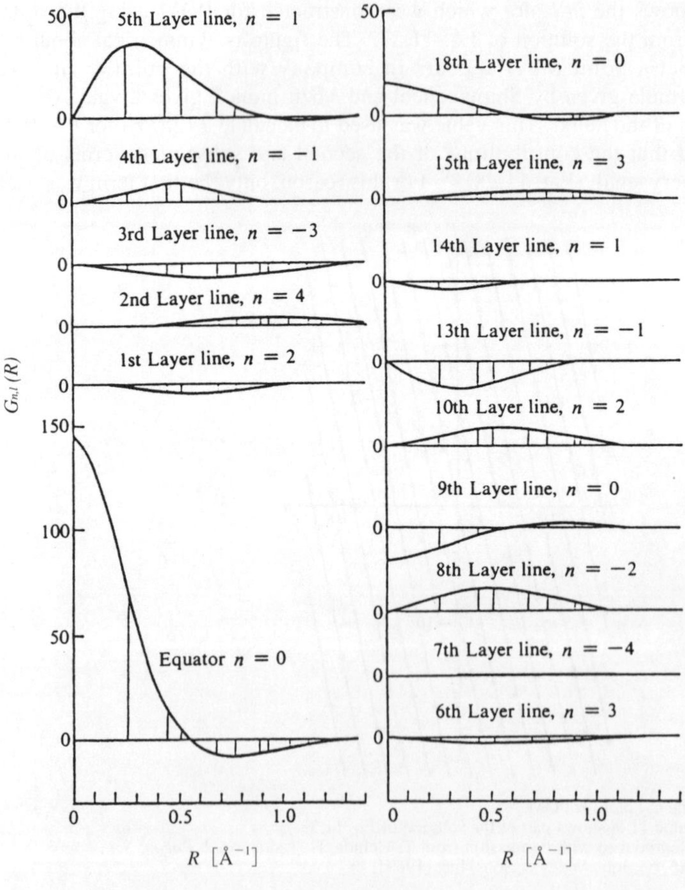

Fig. 11.29
Observed and calculated structure factors of POM.[64]

[Reproduced with permission from T. Uchida, H. Tadokoro, *J. Polym. Sci.*, **A5**, 63, John Wiley & Sons, Inc. (1967)]

in the remainder of the analysis. Refinement of the structure to an R value of 8.8% (starting with $r_c = r_o = 0.691$ Å and an isotropic temperature factor $B = 5.0$ Å^2, and finally using an anisotropic temperature factor) gave the parameters listed in Table 11.8.[64] Calculated structure factors and molecular structure are given in Fig. 11.29,[64] and Fig. 11.30, respectively.

Table 11.8 Atomic positional and thermal parameters of the POM molecule†

	r[Å]	φ	B_1	B_2	B_3
C	0.691	0°	5.3	4.3	0.0394
O	0.671	100°	3.9	2.8	0.0394
H	1.507	−28°48′			
H′	1.507	28°48′			

† Ref. 64. [Reproduced with permission from T. Uchida, H. Tadokoro, *J. Polym. Sci.*, **A5**, 63, John Wiley & Sons, Inc. (1967)]

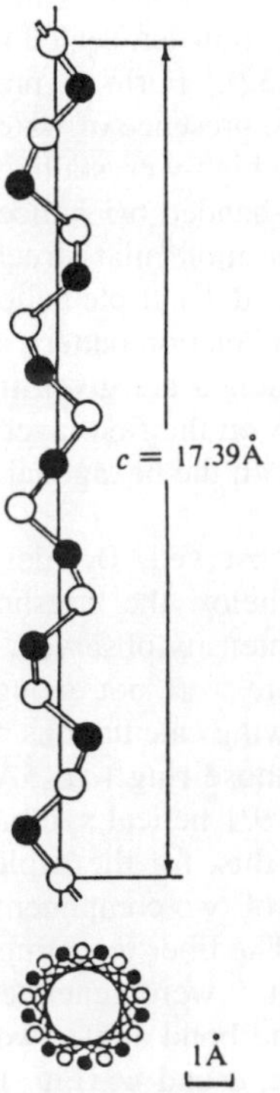

Fig. 11.30 The structure of POM.[64] ○, Oxygen atom; ●, Methylene group [Reproduced with permission from T. Uchida, H. Tadokoro, *J. Polym. Sci.*, **A5**, 63, John Wiley & Sons, Inc. (1967)]

B. Native curdlan, (*β*- (1→3)-D-glucan)

Curdlan is an extracellular, bacterial polysaccharide composed of *β*-(1→3) linked D-glucose residues (Fig. 11.31), and is produced from the mutant strain of *Alcaligenes faecalis* var. *myxogenes* 10C3. This polysaccharide was first isolated by T. Harada and coworkers,[65] and has several interesting physicochemical features, such as the ability to form a firm and resilient gel when heated in aqueous solution.

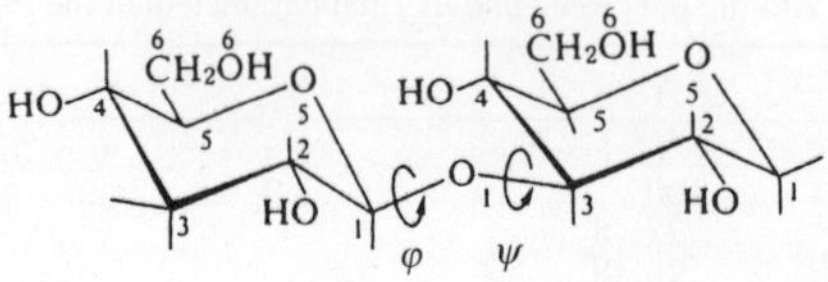

Fig. 11.31 *β*- (1→3)- Glucosidic linkage between two *β*-D-glucopyranese rings.

A number of X-ray diffraction patterns of oriented curdlan gel reported so far[66–71] are classified into three crystalline forms. Form I (the native form) is prepared from a 4% (w/v) aqueous suspension of curdlan powder, sealed in a glass tube, by heating in not water at about 70°C for 10 min (Fig. 11.32). Form II, prepared from form I by heat treatment above 120°C in a sealed bomb in the presence of water, is transformed into form III by drying in vacuum. Forms II and III are highly crystalline, the structures of which are shown to be a triple helix composed of right-handed 6/1 helices.[67–69] On the other hand, the form I allomorph is low crystalline and the molecular structure has not yet been made clear: 7/1 triple helix[66] mixture of 7/1-single and 7/1-triple helices,[66,68] or others.[68]

Figure 11.32 shows an X-ray diffraction pattern of the form I recorded with a cylindrical camera.[72] The crystal data obtained are given in Table 11.9. All diffractions except two (one with the spacing of 18.4 Å on the first layer line and the other with 6.13 Å on the third layer line) could be indexed with the hexagonal unit cell similar to those proposed by Fulton and Atkins.[70]

A total of 12 diffractions were observed. Besides this, 28 weaker diffractions were recognized but their intensities were below the threshold value. Their intensities were assumed to be one-half the weakest intensity observed, and they were used in the calculation only when $|F_o|<|F_c|$. Since there were not enough diffractions to determine the space group, *P*1 was assumed in the following calculations.

Considering the size of a pyranose ring (*ca.* 5Å), single-, double- and triple-helices composed of 5/1, 6/1, 7/1, 8/1 and 9/1 helical symmetry molecules were first adopted for molecular models. In addition to this, for the triple-helix model, shifts (1/3 and 2/3 or −1/3 and −2/3 of the unit height) of two component helices to the third component along the fiber axis must be considered. The fiber repeating period, 22.8 Å and pyranose rings of the standard $^4C_1$ chair conformation,[72] were generated using a linked-atom descrip-tion.[73] In these models all bond lengths and bond angles were held constant. Since two variable parameters at the glycosidic linkage, φ and ψ, (Fig. 11.31) are constrained by the fiber repeating period and the helical symmetry, there is no degree-of freedom in the main chain conformation (*e.g.* in the case of 6/1 helix, $\varphi = \angle C_2C_1O_1C_3 = 168.6°$ and $\psi = \angle C_1O_1C_3C_2 = -1\ 13.2°$). For the molecular conformation, therefore, the torsional angle $\chi = \angle O_6C_6C_5O_5$ is the only variable parameter (Fig. 11.31). In the calculations, the initial value of this angle was set to one of −60°, 60° and 180°. Since the right-handed helical

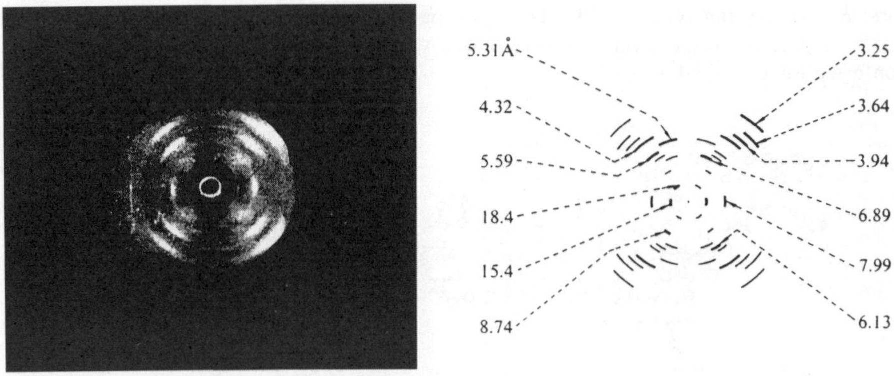

Fig. 11.32 X-ray diffraction pattern of curdlan, form I.[71] [Reproduced with permission from K. Okuyama *et al.*, *J. Carbohyd. Chem.*, **10**, 648, Marcel Dekker (1991)]

Table 11.9 Crystal data of curdlan, form I

	Okuyama *et al.*[71]	Takeba *et al.*[66]	Fluton and Atkins[70]
Sample condition	98%	Wet	100%
Crystal system	Orthorhombic	Orthorhombic	Hexagonal
a[Å]	28.8	26.4	17.01
b[Å]	16.6	16.4	17.01
c (fiber axis) [Å]	22.8	22.65	22.70
Density [g cm^{-3}]	1.27	1.17	1.24
Proposed helix model	6/1-single	7/1-single + 7/1-triple	7/1-triple

[Reproduced with permission from K. Okuyama *et al., J. Carbohyd. Chem.*, **10**, 648, Marcel Dekker (1991)]

sense in forms II and III, obtained from form I by solid state transformation, have been established,[66,68] only right-handed models were considered. All the double- and triple-helix models except those having 5/1 helical symmetry have O4-H...O5 intramolecular hydrogen bondings but have no unusual short intramolecular atomic contacts. The 6/1 helical symmetry molecules are connected by O2-H...O2 intermolecular hydrogen bondings.

Further, in forms II and III, all the curdlan molecules are packed in a parallel fashion, only the parallel packing models were examined. The 8/1 and 9/1 triple-helix models have too large radii of helices ($r > 10$Å) to be located in the unit cell. When the curdlan molecules is located in the unit cell, three parameters must be determined. They are the azimuthal angles μ_1 and μ_2 for curdlan molecules on the corner and the center of the unit cell, respectively (*cf.* Fig. 11.11), and the shift of the relative position along the fiber axis, u. The packing models were generated systematically by changing the values of these μ and u at 30° and 0.04 Å intervals, respectively.

Of 120 models, 11 which were found to be suitable in terms of interatomic contacts and agreement between observed and calculated structure factors were refined further. Of these 11 models, 6 converged to the same structure ($R = 0.14$) and the rest showed rather poor X-ray agreement ($R = 0.18 - 0.22$). Therefore, the former structure was refined further.

Finally, the 6/1-single helix model (Fig. 11.33) was found to be the most preferable in terms of interatomic contacts and agreement between observed and calculated structure factors: $R = 0.135$ and $R_w = 0.179$. The final packing parameters μ_1, μ_2 and u are 24.0°, 163.5° and 0.103, respectively. The only one variable dihedral angle, χ indicated *a trans* conformation (− 175.4°).

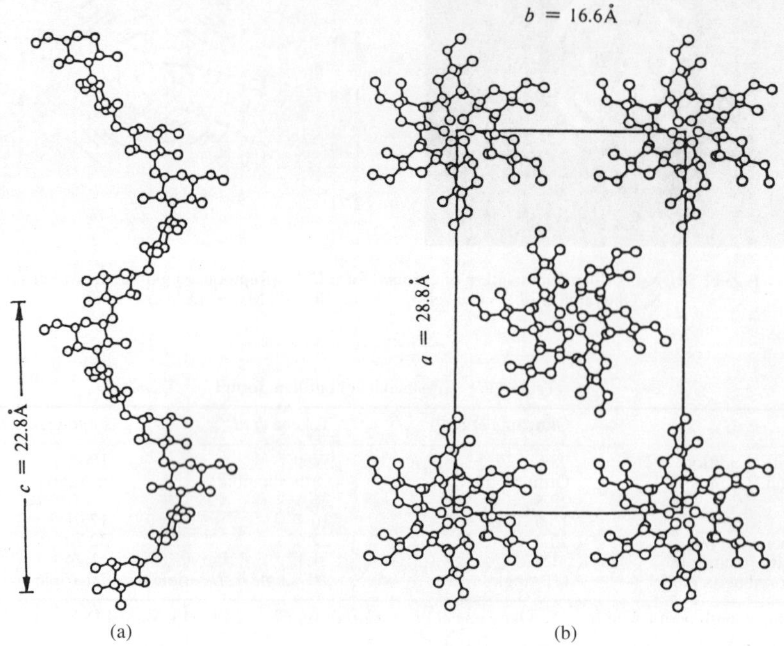

Fig. 11.33 Molecular and crystal structures of curdlan form I.[71)]
(a) Molecular structure.
(b) Cryotal structure.
[Reproduced with permission from K. Okuyama *et. al*, *J. Carbohyd. Chem.*, **10**, 649, Marcel Dekker, Inc. (1991)]

References

1. R.A. Young, E. Prince, R.A. Sparks, *J. Appl. Cryst.*, **15**, 357 (1982).
2. H.M. Rietveld, *J. Appl. Cryst.*, **2**, 65 (1969).
3. K.D. Rose, M.J. Cooper, E.J. York, A. Chakera, *Acta Cryst.*, **A26**, 682 (1970).
4. R.J. Hill, C.J. Howard, *J. Appl. Cryst.*, **20**, 467 (1987).
5. T.M. Sabine, R.B. von Dreele, J.E. Jørgensen, *Acta Cryst.*, **A44**, 374 (1988).
6. W.A. Dollase, *J. Appl. Cryst.*, **19**, 267 (1986).
7. M. Ahtee, M. Nurmela, P. Soutti, M. Jarvinen, *J. Appl. Cryst.*, **22**, 261 (1989).
8. R.A. Young, D.B. Wiles, *J. Appl. Cryst.*, **15**, 430 (1982).
9. D. Taupin, *J. Appl. Cryst.*, **6**, 266 (1973).
10. W.J. Mortier, M.I. Constenoble, *J. Appl. Cryst.*, **6**, 488 (1973).
11. W. Parrish, T.C. Huang, G.I. Ayers, *Trans. Am. Cryst. Assoc.*, **12**, 55 (1976).
12. G.S. Pawley, *J. Appl. Cryst.*, **14**, 357 (1981).
13. H. Toraya, *J. Appl. Cryst.*, **19**, 440 (1986).
14. a) M. Sakata, M.J. Cooper, *J. Appl. Phys.*, **12**, 554 (1979).
    b) E. Prince, *J. Appl. Phys.*, **14**, 157 (1981).
    c) H.G. Scott, *J. Appl. Phys.*, **16**, 159 (1983).
15. A. Immirzi, P. Iannelli, *Macromolecules*, **21**, 768 (1988).

16. H. Toraya, F. Marumo, T. Yamase, *Acta Cryst.*, **B40**, 145 (1984).
17. *International Tables for Crystallography*, Vol. C, 2nd. ed. (A.J.C. Wilson, E. Prince eds.), Mathematical, Physical and Chemical. Tables, pp. 770-804, Kluwer Academic, Dordrecht (1999).
18. Cambridge Structural Database. URL. http://www.ccdc.cam.ac.uk/prods/csd.html
19. R.L. Miller, *Crystallographic Data for Various Polymers*, *Polymer Handbook*, (J. Brundrup, E.H. Immergut eds.), III-1 to III-59, Interscience, N.Y. (1966).
20. J. Kakinoki, *Ann. Repts. Textile Res. Inst.* Osaka Univ., **5**, 19 (1950).
21. C.R. Bohnm J.R. Schaefgen and W.O. Statton, *J. Polymer Sci.*, **55**, 531 (1961).
22. E.W. Nuffield, *X-Ray Diffraction Methods*, p. 372, John Wiley & Sons, Inc., N.Y. (1966).
23. a) *International Tables for Crystallography*, Vol. A, Space-Group Symmetry, (T. Hahn ed.), p. 523, D. Reidel Pub., Dordrecht (1985).
b) *International Tables for X-Ray Crystallography*, Vol. I, (N.E M. Henry, K. Lonsdale eds.), Symmetry Groups, p. 292, Kynoch Press, Birmingham, 1952; 3rd ed. (1969).
24. Y. Takaki, *Memoirs of Osaka University of Liberal Arts and Education*, **10**, 161 (1962).
25. M.J. Buerger, *Crystal-Structure Analysis*, p. 178, John Wiley & Sons. Inc., N.Y. (1960).
26. M.J. Buerger, *The Precession Methbds in X-Ray Crystallography*, p. 191, John Wiley & Sons, Inc., N.Y. (1964).
27. In ref. 25, p. 187.
28. W.A. Wooster, A.J.P. Martin, *J. Sci. Instr.*, **17**, 83 (1940).
29. A.J.C. Wilson, *Acta Cryst.*, **2**, 378 (1949).
30. *International Tables for Crystallography*, Vol. B, Reciprocal Space (U. Shumueli ed.), D. Reidel Pub., Dordrecht, (1993).
31. In ref. 17, pp. 549-558.
32. In ref. 17, pp. 572-575.
33. a) A.L. Patterson, *Z. Krist.*, **A90**, 517 (1935).
b) H.L. Lipson, W. Cochran, *The Crystalline State*, Vol. III, The Determination of Crystal Structures, G. Bell & Sons, London, 1957; 3rd ed., pp. 12, 150, (1966).
34. G.H. Stout, L.H. Jensen, *X-Ray Structure Determination*, A Practical Guide, pp. 279-291, Macmillan, N.Y. (1965).
35. W. Fuller, M.H.F. Wilkins, H.R. Wilson, L.D. Hamilton, *J. Mol. Biol.*, **12**, 60 (1965).
36. S. Arnott, M.H.F. Wilkins, L.D. Hamilton, R. Langridge, *J. Mol. Biol.*, **11**, 391 (1965).
37. D.A. Marvin, M. Spencer, M.H.F. Wilkins, L.D. Hamilton, *J. Mol. Biol.*, **3**, 547 (1961).
38. R.E. Franklin, R.G. Gosling, *Nature* (*London*), **172**, 156 (1963).
39. In ref. 33. c), p. 308.
40. D. Harker, *Acta Cryst.*, **9**, 1 (1956).
41. M. F. Perutz, *Acta Cryst.*, **9**, 901 (1956).
42. D. Harker, J.S. Kasper, *J. Chem. Phys.*, **15**, 882 (1947).
43. D. Sayer, *Acta Cryst.*, **5**, 60 (1952).
44. H. Hauptmann, J. Karle, *The Solution of the Phase Problem*. I, The Centrosymmetric Crystal, Am. Cryst. Assoc. Monograph 3, Polycrystal Book Service, Boston (1953).
45. H. Hauptomann, J. Karle, *Acta Cryst.*, **11**, 149 (1958).
46. J. Karle, H. Hauptman, *Acta Cryst.*, **21**, 849 (1966).
47. M.M. Woolfson, *Direct Methods in Crystallography*, Oxford Univ. Press, Oxford (1961).
48. In ref. 33.c), p. 350.
49. a) S. Arnott, A. J. Wanacott, *Polymer*, **7**, 157 (1966).
b) P.J.C. Smith, S. Arnott, *Acta Cryst.*, **A34**, 3 (1978).
50. C.W. Bunn, *Trans. Faraday Soc.*, **35**, 482 (1939).
51. N. Kasai, M. Kakudo, Paper presented at the International Symposium on Macromolecular Chemistry, Prague, 1965; *Repts. Progr. Polymer Phys. Jpn.*, **11**, 145 (1968).
52. S. Kavesh, J. Schultz, *J. Polymer Sci., part A-2*, **8**, 243 (1970).
53. For example, ref. 33b) and ref. 34.
54. P.A. Dirac, *The Principle of Quantum Mechanics*, p. 58, Oxford Univ. Press, Oxford, (1947).
55. For example, H. Margenau and G.M. Murphy, *The Mathematics of Physics and Chemistry*, sect. 3-9, van Nostrand, N.Y. (1943).
56. H.R. Wilson, *Diffraction of X-Rays by Proteins, Nucleic Acids, and Viruses*, p. 42, E. Arnold, London (1966).
57. Y. Mitui, *Acta Cryst.*, **20**, 694 (1966).
58. T. Shimanouchi, S. Mizushima, *J. Chem. Phys.*, **23**, 707 (195).
59. T. Miyazawa, *J. Polymer Sci.*, **55**, 215 (1961).
60. K. Nagai, M. Kobayashi, *J. Chem. Phys.*, **36**, 1268 (1962).
61. K. C. Holmes, D.M. Blow, in: *Methods of Biochemical Analysis*, (D. Glick ed.), Vol. 13, p. 221, John Wiley & Sons, Inc., N.Y. (1965).

62. A. Klug, F.H.C. Crick, H.W. Wyckoff, *Acta Cryst.*, **11**, 199 (1958).
63. H. Tadokoro, T. Yasumoto, S. Murahashi, I. Nitta, *J. Polymer Sci.*, **44**, 266 (1960).
64. T. Uchida, H. Tadokoro, *J. Polymer Sci., part A-2*, **5**, 63 (1967).
65. T. Harada, M. Masada, K. Fujimori, I. Maeda, *Agric, Biol. Chem.*, **30**, 196 (1966).
66. H. Takeda, N. Yasuoka, N. Kasai, T. Harada, *Polymer, J.*, **10**, 365 (1978).
67. Y. Deslandes. R.H. Marchessault, A. Sarkao, *Macromolecules*, **13**, 1466 (1980).
68. N. Kasai, T. Harada, in: *Fiber Diffraction Methods* (A.D. French, H. Gardener eds.) ACS Symp. Ser. No. 141, p. 363, Am. Chem. Soc., Washington, D.C. (1980).
69. C.I. Chuah, A. Sarko, Y. Deslandes, R.H. Marchessault, *Macromolecules*, **16**, 1375 (1983).
70. W.S. Fulton, E.D.T. Atkins, in: *Fiber Diffraction Methods.* (A.D. French, H. Gardener eds.) ACS Symp. Ser. No. 141 , p. 385, Am. Chem. Soc., Washington, D.C. (1980).
71. K. Okuyama, A. Otsubo, Y. Fukuzawa. M. Ozawa, T. Harada, N. Kasai, *J. Carbohyd. Chem.*, **10**, 645 (1991).
72. S. Arnott, W.E. Scott, *J. Chem. Soc., Perkin Trans.*, **2**, 324 (1972).
73. P.J.C. Smith, S. Arnott, *Acta Cryst.*. **A34**, 3 (1978).

12. Crystal Structure Determination of Macromolecules

This chapter is a brief summary of the crystal structure determination[1,2] of macromolecular substances such as proteins, nucleic acids, viruses and others.[†] A flow chart of the structure determination of macromolecular substances is given in Fig. 12.1.

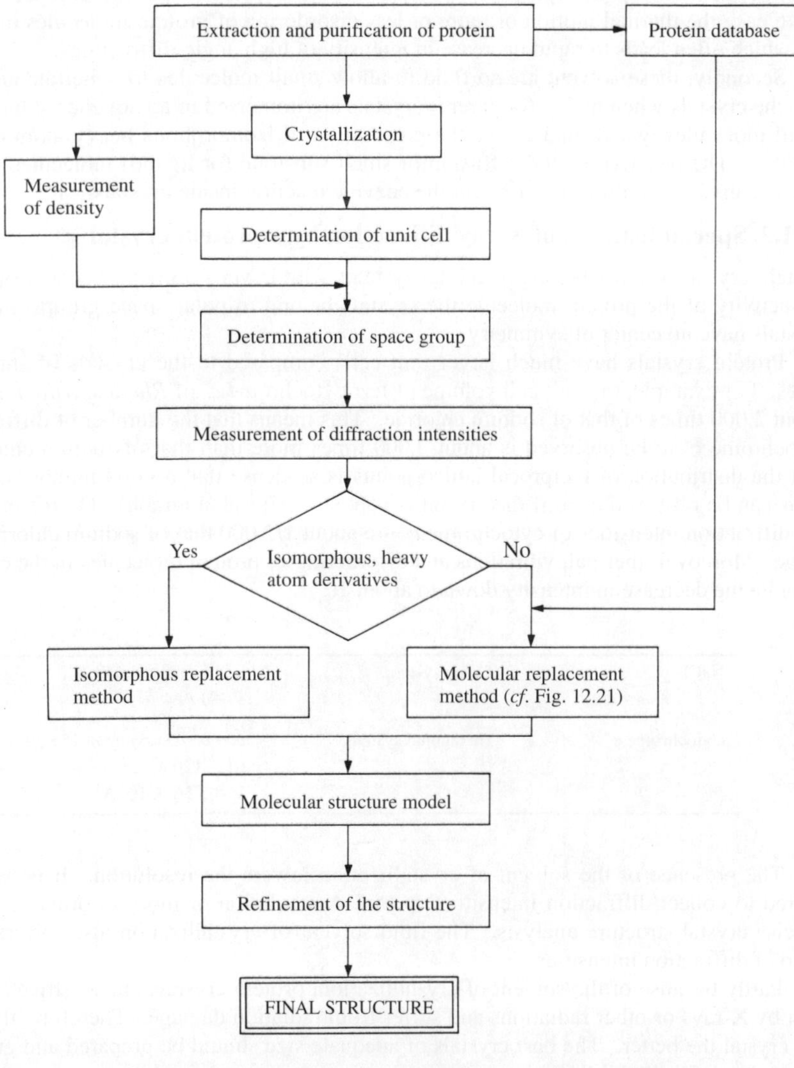

Fig. 12.1 Flow chart of the structure determination of macromolecules.

---

† These are simply called proteins in this chapter.

12.1 Characteristics of Protein Crystals

12.1.1 Solvent of crystallization

The protein crystals usually contain solvent of crystallization, mostly water: the amount of crystalline water being ordinary 30% to 60% or more. Some of them on the surface of the protein molecules are bound with the protein molecules by hydrogen bonding, whereas most of them fill up the spaces between protein molecules inside the crystal. This fact make easy the thermal motion or more or less disordering of protein molecules in the crystal, which often leads to rapid decrease in intensity of high angle diffractions.

Secondly, these solvent are so fluid to allow small molecules to penetrate and diffuse into the crystals when native (or parent) crystals are immersed in an aqueous solution of the small molecules, which make easy the preparation of isomorphous heavy atom derivative crystals. The penetration and diffusion of small substrate (or ligand) molecules into crystals also give possibilities to carry out the enzyme reaction inside crystals.

12.1.2 Special features of X-ray diffraction by a protein crystal

Protein crystals usually belong to crystal systems with lower symmetry because of the optical activity of the protein molecule the crystals belong to polar space groups, *i.e.* protein crystals have no center of symmetry.

Protein crystals have much larger unit cells compared to the crystals of small molecules. For example, the unit cell volume of ferricytochrome *c´* of *Rhodospirillum rubrum* is about 2,000 times of that of sodium chloride. This means that the number of diffractions of cytochrome *c´* to be observed is about 2,000 times more than that of sodium chloride and that the distribution of reciprocal lattice points is so dense that a good number of diffractions can be observed even if the crystal is kept still (still photograph). On the other hand, the diffraction intensities of cytochrome *c´* are about 1/2,000 that of sodium chloride on average. Moreover. thermal, vibrations and disordering of protein molecules in the crystal accelerate the decrease in intensity down to about 10^{-1}.

NaCl	Cubic system,	$a = b = c = 5.640$ Å $V = 179.4$ Å^3,
Cytochrome *c´*	Hexagonal system,	$a = b = 51.63,\ c = 155.39$ Å, $(\beta = 120°)$ $V = 3.56 \times 10^5$ Å^3

The presence of the solvent of crystallization lowers the resolution. It is usually required to collect diffraction intensities up to 2 Å resolution or more in order to carry out precise crystal structure analysis. The fluid solvent of crystallization also lowers the S/N ratio of diffraction intensities.

Partly because of the solvent of crystallization, protein crystals are sensitive to irradiation by X-rays or other radiations and suffer from radiation damage. Therefore, the fresher the crystal the better. The best crystals of adequate size should be prepared and grown just before intensity data collection.

12.2 Crystallization

12.2.1 Solubility of protein[1–4)]

Protein molecules in an aqueous solution ordinary take globular forms: molecular chains are folded to form a rather fixed shape. Among the amino acid residues in a polypeptide chain of a protein molecule, those that have a hydrophilic side chain are generally distributed on the surface whereas those having a hydrophobic side chain are found inside the globular molecule. Many hydrophilic amino acid residues on the surface have a carboxyl or amino group in their side chains; these have their own pK_a values and dissociate depending upon the pH value of the solution. Protein molecules are, therefore, considered to be polyvalent electrolytes. The solubility of a protein in water is affected by the ionic strength. The solubility dependence on ionic strength for a protein is shown in Fig. 12.2.[3)]

For the crystallization of a protein the principle is to decrease gradually the solubility of the protein in aqueous solution.[1,5)]

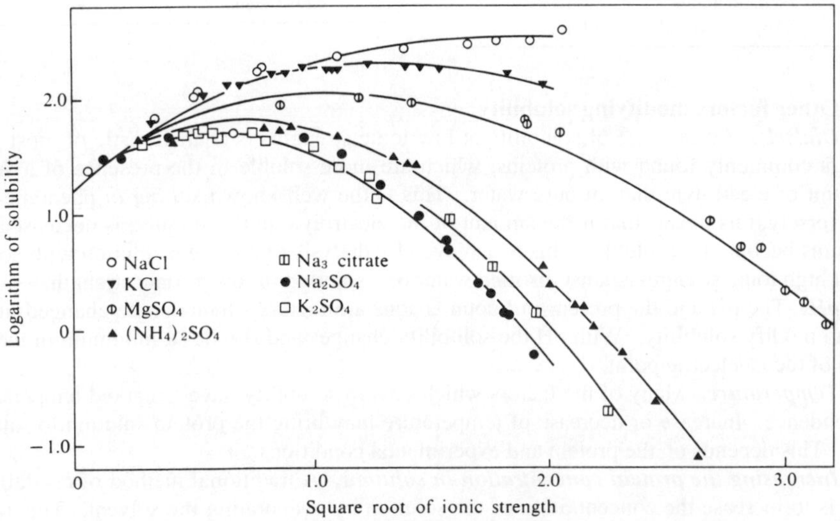

Fig. 12.2 The solubility of carbonmonoxyhemoglobin in various electrolytes at 25°C.[3)]
[Reproduced from A. A. Green, *J. Biol. Chem.* **95**, 47, Am. Chem. Soc. (1932)]

A. Precipitants and additives to control solubility

In order to decrease the solubility of protein precipitants are generally added to the protein solution. These include salts, organic solvents and high polymers.

a) Salts. Addition of a salt to an aqueous solution of protein makes the ionic strength of the solution higher, making the solubility of the protein lower (*salting out*). Commonly used salting out reagents are given in Table 12.1. Among them a typical example is ammonium sulfate, $(NH_4)_2SO_4$, which is the most frequently and successfully used. Sodium chloride, potassium or sodium phosphate and sodium citrate are also often used.

b) Organic solvents. A water soluble organic solvent can be used as an additive to produce marked change in solubility of proteins. This is partly the result of the lower dielectric constant of water-organic solvent mixture compared to water. The organic solvent may also decrease solubility through specific solvation of the protein and displacement of bound water. Commonly used organic precipitants are listed in Table 12.2. Among them 2-methyl-2,4-pentanediol (MPD or hexyleneglycol) has proved very useful. Ethanol is commonly used. Dioxane (1,4-dioxacyclohexane) is also useful.

c) High polymers. Poly(ethylene glycol) (PEG) is probably the only polymer additive that is used often. PEG 4,000, PEG 6,000 and PEG 8,000 (DP = 4,000, 6,000 and 8,000, respectively) are widely used. Sometimes PEG 20,000 or PEG 400 is also used.

Table 12.1 Commonly used salting-out reagents

Ammonium or sodium sulfate
Ammonium or sodium citrate
Potassium or sodium phosphate
Sodium chloride
Lithium chloride
Magnesium sulfate
Lithium sulfate
Sodium nitrate
Poly(ethylene glycol)(PEG)

Table 12.2 Commonly used organic precipitants

Ethanol
2-Methyl-2,4-pentanediol (MPD)
Acetone
Methanol
Dioxane
n-Propanol
i-Propanol
tert-Butanol

B. Other factors modifying solubility

a) Dialysis. An increase of solubility at low ionic strengths is characteristic of most ions and is commonly found with proteins, which are more soluble in the presence of a small amount of electrolyte than in pure water. This is the well-known *salting in* phenomenon. Conversely, this means that if the amount of the electrolyte in the solution is decreased the proteins become less soluble. This is achieved by dialysis of a protein solution with somewhat high ionic strength against distilled water or a solution of lower ionic strength.

b) pH. The pH and the presence of counter-ions are factors which can be changed in order to modify solubility. With pH the solubility changes and reaches a minimum in the region of the isoelectric point.

c) Temperature. Many of the factors which govern solubility have a marked temperature dependence. Increase or decrease of temperature may bring the protein solution to saturation. This depends on the protein and experimental conditions.

d) Increasing the protein concentration in solution. A traditional method of crystallization is to increase the concentration of the solution by evaporating the solvent. This technique is not generally useful with proteins as it is difficult to control and is conducive to crystallization of salts in the mother liquor.

Even after crystals have been obtained, it is often worthwhile to continue to explore other conditions which may produce a crystal form polymorph more suitable for X-ray study.

12.2.2 Techniques for crystallization[1, 6-16)]

Before growing good protein crystals with appropriate dimensions for X-ray diffraction it is necessary to take a wide survey to find crystallization conditions with a minimum amount of protein, say a few or favorably some tens of milligrams. In practice, crystallization conditions are examined for concentration of the protein, precipitants, ionic strength, pH, temperature and so on. For this purpose the following techniques are usually applied (Table12.3).

A. Methods of protein crystallization

a) Microdialysis, 1) Zeppezauer method,[7]
2) Button method,
b) Vapor diffusion,[8] 1) Hanging drop method,[9]
2) Sitting drop method,[10]
c) Seeding,[11–13]
d) Layering or free interface diffusion.[14,15]
These are summarized briefly in Table 12.3.

Techniques for automated survey for crystallization conditions of proteins have been reported[16] (*e.g.* Fig. 12.3) and automated instruments are commercially available.

Recently, studies on crystallization of proteins under a microgravity environment in space have been begun[17–19] in order to scrutinize the effects of gravity on crystal growth

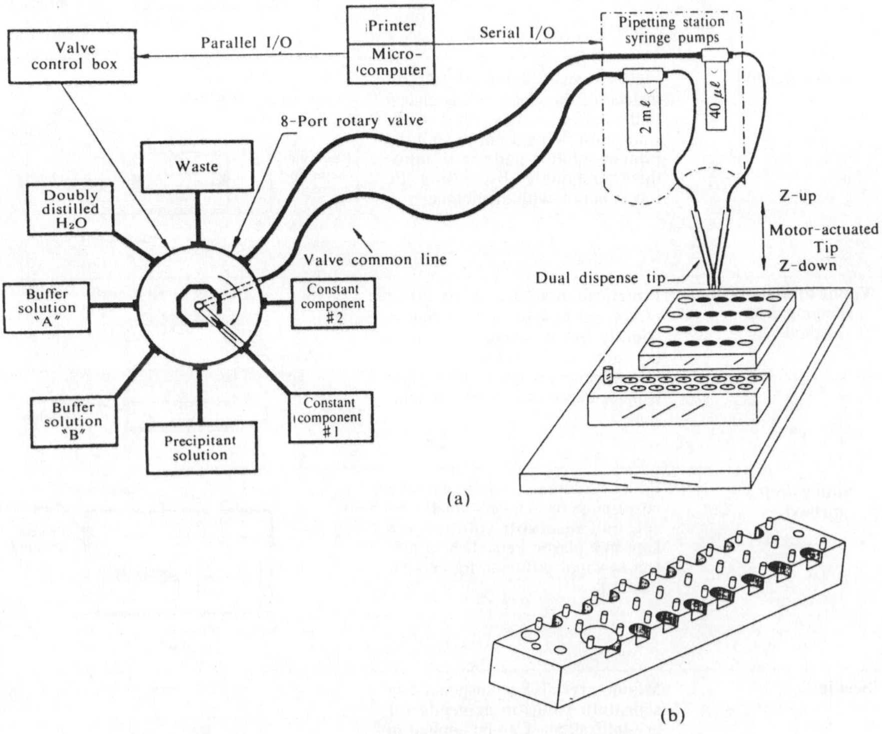

Fig. 12.3 An example of automated survey for crystallization conditions of proteins.
(a) A schematic drawing of hardware components in the liquid handling system. The system is operated by a microcomputer that controls movements of the pipetting station and rotary valve whose positions are relative to those fixed on initialization. The depression plate, shown as a 6×4 array of circles with the inner 4×4 filled, and the coverslip holder are fixed on the *x,y* translation table of the pipetting station. The coverslip holder, shown to scale in (b), was machined from plexiglass. Coverslips (one is shown) are positioned between three small pegs. Machined indentations facilitate manual pickup of the coverslips for inversion over the wells of the plate. Holes at either end of the holder are for set screws to attach the holder to the translation table. Additional holes position the protein-containing vials.
[Reproduced with permission from M.J. Cox, P.C. Weber, *J. Appl. Cryst.*, **20**, 366, IUCr. (1987)]

Table 12.3 Methods of protein crystallization

Batch crystallization	To a solution (~ 1 ml) of protein. dissolved at low ionic strength, precipitating reagent (salt or organic solvent) is added to bring a solution to a state of super-saturation to induce crystallization (classical technique).	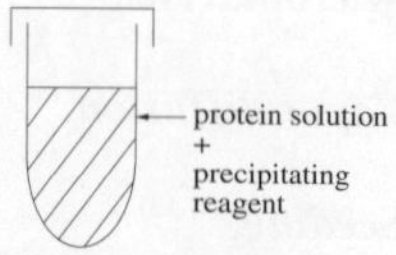
Microdialysis Zeppezauer method	Protein solution (20–100 ml) is injected in a glass microdiffusion cell, bottom of which is sealed with a semipermeable membrane, and the cell is introduced in a relevant buffer so that the buffer cover the dialysis membrane. pH of the protein solution can be modified easily.	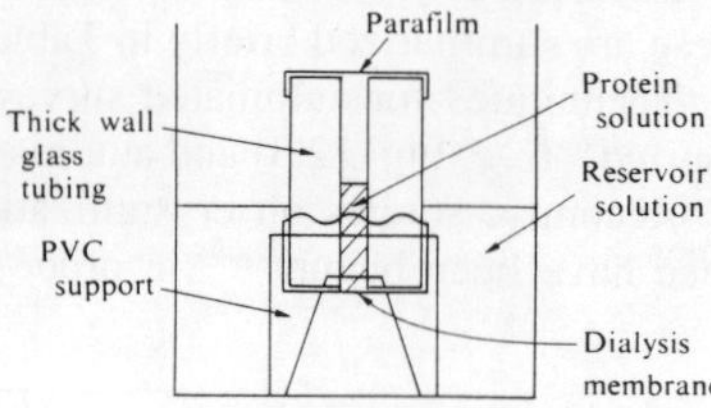
Button method	Plastic vessel (button, 20–100 μl), instead of the glass cell, is placed in the buffer for the microdialysis. Care must be paid not to spill the protein solution and not to introduce air bubbles by sealing the top of button with membrane.	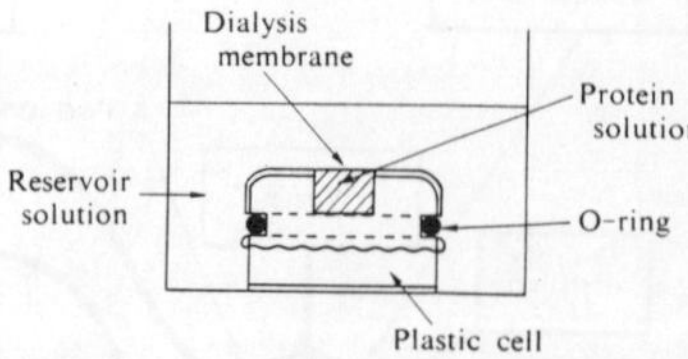
Vapor diffusion Hanging drop method	Protein is dissolved in 10–20 μl of solvent at a salt concentration slightly below where the protein crystallizes. A droplet of the solution hanged under a cover glass is placed above the reservoir solution.	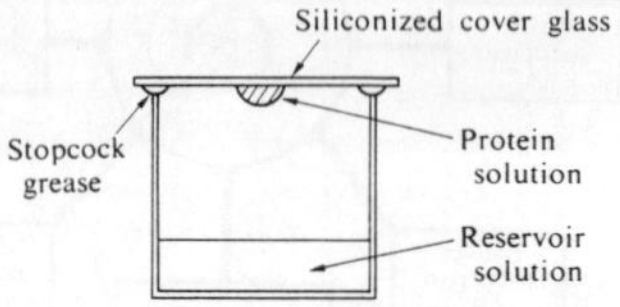
Sitting drop method	Protein solution, as a droplet on a cover glass or in a very small vessel, and reservoir solution are kept in a plastic Petri dish to subject to vapor diffusion for crystallization.	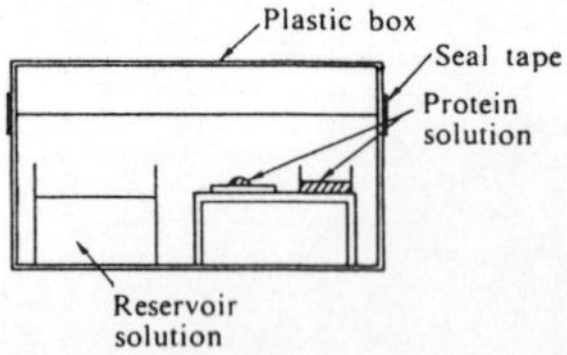
Seeding	Minute crystals are suspended in a protein solution as seeds for crystallization. Can be applied in all methods described above.	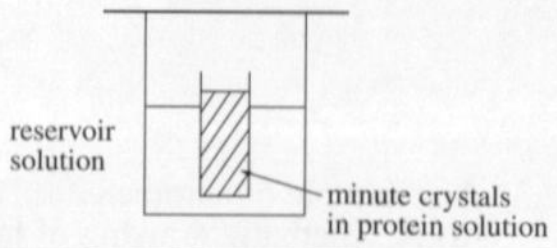
Layering or free interaction	Inside a capillary, protein solution is added gently on a layer of precipitating reagent solution.	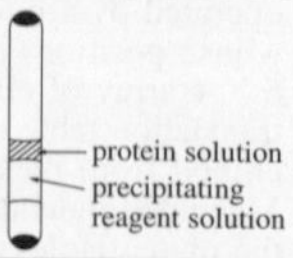

[Reproduced with permission from Y. Matsuura, *J. Jap. Assoc, Cryst. Growth*, **12**, 112, Jap. Assoc. Cryst. Growth (1985)] and to apply the information obtained on the crystal growth of proteins of better quality for X-ray diffraction studies on Earth.

B. Extraction, purification and crystallization of cytochrome *c´* from *Rhodopirillum rubrum* (*R. rubrum*)

The cytochrome *c´* was extracted from *R. rubrum* and purified by the method of Horio and Kamen with some modification (See scheme). The purified cytochrome *c´* in 50 mM-Tris-HCl buffer (pH 8.0) was precipitated by adding 50% (w/v) PEG 4,000. The resulting precipitate collected by centrifugation was dissolved in a minimum amount of 50 mM-Tris-HCl buffer (pH 8.0) containing 22% (w/v) PEG 4,000. Crystals are grown by dialyzing this protein solution against 50 mM-phosphate buffer (pH 7.0) using a Zeppezauer tube (Table 12.3). Long hexagonal, rod-shaped crystals shown in Fig. 12.4 grew within a few weeks.[20]

Scheme. Extraction and purification of cytochrome *c´* from *Rhodospirillum rubrum*†

Cells *R. rubrum* (carotenoid-less blue-green mutant strain G-9) (Wet weight 325 g)
↓
↓ Sonication: 20 kHz 50% duty cycle for 10 min.
↓ Ultra-centrifugation: 100,000 g for 90 min.
↓
Supernatant (*ca.* 1 ℓ (in 0.1 M tris-HCl, pH 8))
↓
↓ Hydrogen bond chromatography on DEAE-Sepharose CL-6B (bed volume 310 ml)
↓ Linear gradient of concentrations from 75 to 30% sat. $(NH_4)_2SO_4$
↓ Desalting and concentration
↓
Purified cytochrome *c´* (*ca.* 70mg)

† T. Horio, M.D. Kamen, *Biochim. Biophys. Acta.*, **48**, 266 (1961).

Fig. 12.4 A microphotograph of crystals of cytochrome *c´* from *R. rubrum*. Crystals were rod-shaped with a hexagonal section. Typical dimensions of the crystals were 0.1 to 0.4 mm in diameter and 0.5 to 2.0 mm in length. [Reproduced with permission from M. Yasui *et al.*, *J. Mol. Biol.*, **177**, 845, Academic Press (1984)]

12.2.3 Preparation of isomorphous heavy atom derivative crystals

For the phase determination of diffractions by the isomorphous replacement method, heavy atom derivatives crystals isomorphous to the native (or parent) crystals must be prepared. This is usually achieved by immersing the native protein crystals in the aqueous solution of appropriate heavy atom derivatives (*soaking*): heavy atom derivative ions penetrate the native crystal, spread inside the crystal by diffusion and bind preferentially at the special site of the protein molecule without destroying the isomorphism to the native crystal.

Among the heavy atom compounds reported[21] the following are frequently used: $HgAc_2$,[*1] $HgCl_2$, K_2HgI_4, MeHgCl, PCMBS,[*2] mersaryl,[*3] UO_2Ac_2, $K_2UO_2F_5$, $K_2Pt(CN)_4$, $AgNO_3$, and $SnCl_2$, etc.

A. Heavy atom derivative crystals of cytochrome c´ from *R. rubrum*

The general conditions for soaking surveyed are given below:[20]

Concentrations of heavy atom reagent:	1-20 mM in the mother liquor (Mother liquor: 30% PEG 4,000 in 50 mM potassium phosphate buffer)
Soaking time:	2-5 days

The soaked crystals were then examined for isomorphism with the native crystal by taking precession photographs: the quality and quantity of diffractions, *i.e.* changes in unit cell dimensions, diffraction intensities and resolutions. The results are summarized in Table 12.4. K_2PtCl_6 gave good derivative crystals and CH_3HgCl fairly good ones. Both of

Table 12.4 Results of survey for heavy atom derivatives for cytochrome *c´* from *R. rubrum*

Heavy atom reagent	Concentration (mM)	Soaking time (day)	Result
K_2PtCl_6	saturated ($<$ 3)	5	good
CH_3HgCl	20	2	fair
$K_3UO_2F_5$	0.8[†1]	1/24	bad[†2]
$SmCl_3$	1[†1]	2	bad[†3]
	1[†1]	1	bad[†3]
K_2IrCl_6	1[†1]	2	bad[†3]
	1[†1]	4	bad[†3]

†1 Mother liquor: 30% PEG4000 in 50 mM HEPES buffer pH7.
†2 Cracked.
†3 Neither isomorphous difference Patterson and Fourier map could be solved.

them were used for the preceding isomorphous replacement studies.

12.2.4 Crystal mounting[1,22,23]

Protein crystals are usually mounted in thin-walled (0.001 mm thick) glass or quartz capillary tubes.[22,23] A very small quantity of mother liquor should remain around the crystal so that the crystal is kept moist because a protein crystal contains much solvent of crystallization and some of them are fluid like a liquid, which will quickly distill out and cause the crystal structure to break.

Introduction of the crystals into the capillary tubes can be achieved by drawing up crystal by capillary action or gentle suction. A Pasteur pipette is also useful to pipette a crystal and a small amount of liquid into the capillary tube (Fig. 12.5).

*1 Ac = acetate, CH_3COO^-;
*2 PCMBS = *p*-chloromercuribenzene sulfonate, 

*3 mersary = Mersalate sodium mersaryl, 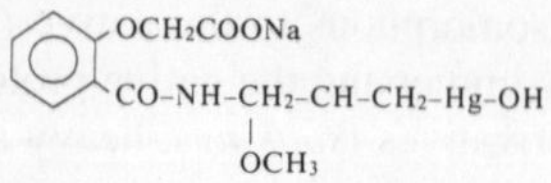

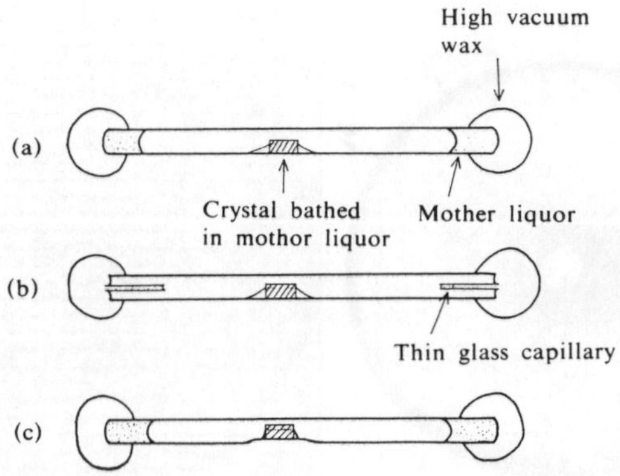

Fig. 12.5 Mounting protein crystals in thin-walled tubes.[la)]
(a) The solvent reservoir is held by capillary action of the mounting tube itself.
(b) The reservoir comprises two thin capillaries inside the mounting tube.
(c) The mounting tube has been flattened to prevent slippage.
[Reproduced with permission from T. L. Blundell, L. N. Jhonson, *Protein Crystallography*, p.81, Academic Press (1976)]

In order to improve X-ray diffraction pattern of a protein crystal sealed in capillary tube soft ablation of undesirable portion in the crystal by pulsed UV-laser processing has been reported useful.[24)]

12.3 Data Collection

12.3.1 Determination of preliminary crystallographic data

Prior to the systematic measurement of diffraction intensities, it is important to characterize the unit cell dimensions, V_m value, internal symmetry, and extent of diffraction from a crystal. This is most readily done on precession cameras.

The V_m (or V_M) is the crystal volume per unit of protein molecular unit, which is defined by Matthews as follows:[24)]

$$V_m = V/(N \cdot n \cdot M) \qquad (12.1)$$

where V is the volume of the unit cell, N the number of equivalent positions in the unit cell, n the number of molecules in an asymmetric unit and M the molecular weight. Usually, V_m has values of 1.68–3.53 Å^3/amu (av. 2.37 Å^3/ amu). Assuming the partial specific volume of proteins to be 0.74 ml/g the solvent content of a protein crystal V_{solv} is given as

$$V_{solv} \doteqdot 1 - (1.23\ V_m).$$

A. The *hk0* and *h0l* precession photographs[20)] of cytochrome *c´* of *R. rubrum*

These photographs shown in Fig. 12.6. revealed that the Laue symmetry of the crystal was $6/m$ and that the systematic absence of diffractions as $l = 6n$ for $00l$. The corresponding space group is either $P6_1$ or its enantiomorph $P6_5$. Approximate unit cell dimensions were determined from the precession photographs. They are given in Section 12.1.2. Assuming that one dimeric molecule of cytochrome *c´* is in an asymmetric unit, the volume of molecular mass, V_m calculated was 2.12 Å^3/Da, which is comparable to that normally found in protein crystals.[25)] The solvent content is 38%.

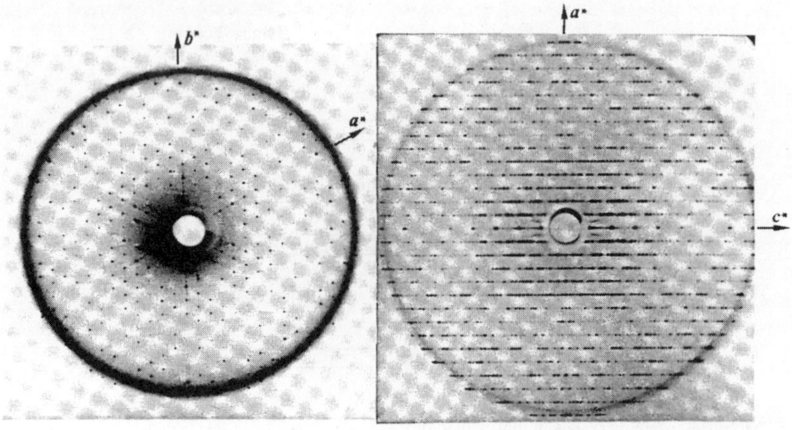

Fig. 12.6 Precession photographs of cytochrome *c′* from *R. rubrum*.[20]
(a) *hk*0 diffractions. (b) *h*0*l* diffractions. [Reproduced with permission from M. Yasui *et al.*, *J. Mol. Biol*, **177**, 845, Academic Press (1984)]

12.3.2 Collection of intensity data

For the collection of three-dimensional intensity data, the main techniques used are: automated diffractometry on a four-circle diffractometer (*cf.* Table 12.4), rotation or Weissenberg photography using imaging plate or electronic area-detector diffractometry with multi-wire proportional chamber (MWPC), X-ray CCD or X-ray TV system. Laue diffractometry using imaging plate (IP) with white X-rays is sometime useful.[26]

The number of diffractions, N_{diff}, to be measured will depend upon the resolution, d_m, the minimum interplanar spacing for which $|F|$'s are included in the Fourier summation of the electron density. The N_{diff} can be estimated by the following formula[1a]

$$N_{diff} = \frac{4}{3} \cdot \frac{\pi V}{d_m{}^3} \cdot \frac{1}{j}, \tag{12.2}$$

where $j = 1$ for primitive lattice, $j = 2$ for body-centered lattice or bottom-centered lattice and $j = 4$ for face-centered lattice. For a set of data in which each diffraction is measured only once, N_{ref} must be divided by m, where m is the multiplicity factor of the Laue group.

Accurate unit cell parameters and intensity data collection statistics of cytochrome *c′* and heavy atom derivatives are given in Table 12.5.

Table 12.5 Unit cell parameters and data collection statistics of cytochrome *c′* from *R. rubrum* and two isomorphous heavy atom derivatives[39a]

	Cell constants		No. of crystals	No. of observed diffractions	No. of independent diffractions†1 (∞ to 2.8 Å)	R_{merg} †2
	a[Å]	c[Å]				
Native	51.72	155.49	8	21,203	12,092	0.039
K_2PtCl_6	51.76	155.20	11	32,250	10,791	0.104
CH_3HgCl	51.65	154.92	6	14,291	5,851	0.036

†1 The native and platinum derivative data include Bijvoet pair diffractions (*hkl* and $\overline{hkl}$).
†2 $R_{merg} = \Sigma_h | \Sigma_i | \Sigma_{i,h} | - \langle | \Sigma_h | \rangle | / N\Sigma_h \langle | F_h | \rangle$, where $| F_{i,h} |$ is an individual measurement of diffraction *h* of the *i*th crystal, $\langle | F_h | \rangle$ the mean amplitude of that diffraction and *N* the numberof crystals.
[Reproduced with permission from M. Yasui *et al.*, *J. Biochem.*, **111**, 318, Japanese Biochem. Soc. (1992)]

12.4 Phase Determination

Three categories of problems arise, all pertinent to the design of modified proteins.[2b)]

1) If the crystal of a modified protein is nearly isomorphous with a parent crystal structure, then the difference Fourier method can be used with phases from the parent structure to show up the changes in structure.

2) If the crystal is novel but is composed of molecules that are similar to one of a known structure, then the molecular replacement method can often be used to build a model of properly positioned components from which phases can be computed by the structure factor equation.

3) If the molecule is the first of its class to be analyzed, then the crystallographic phase problem must be solved *ab initio*.

12.4.1 Isomorphous replacement[1, 27)]

The isomorphous replacement method played a central role in the diffraction analysis of nearly all truly new protein structures. In this method the small differences in the X-ray

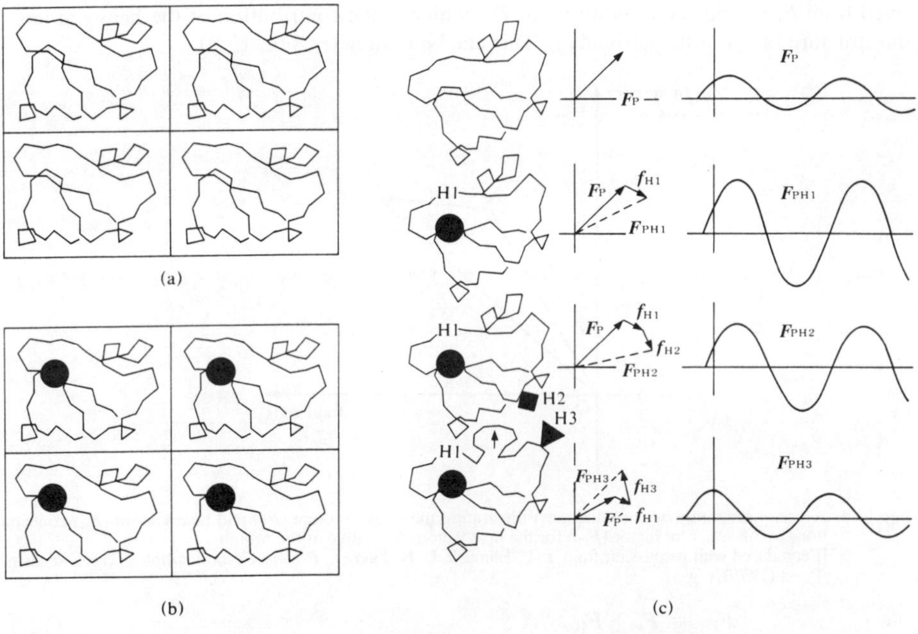

Fig. 12.7 Schematic illustration of the principle of the heavy atom isomorphous replacement method.[2, 28)]
(a) A single crystal of the native protein (only four unit cells are given).
(b) A single crystal of a heavy atom derivative which is isomorphous to the native crystal. Black circle represents the heavy atom group introduced.
(c) Phase relationships between various heavy atom derivatives. From top to bottom, illustrations for the native protein and derivative of heavy atom group 1 (H1), of heavy atom groups 1 and 2 (H1 and H2), and of heavy atom groups 1 and 3 (H1 and H3). In each row from left to right are given molecular structure, vector diagram showing the native protein (F_P) and heavy atom (f_{Hn}) for the heavy atom derivatives, and phase differences in the resultant structure factors. In the fourth row a part of the molecular chain (shown by an arrow) changed its shape by the introduction of heavy atom group 3. [Reproduced with permission from Alexander McPherson, *Preparation and Analysis of Protein Crystals*, pp.177, 260, Krieger Pub. (1982)]

diffraction intensities caused by the addition of a heavy atom to an otherwise unchanged crystal structure are used to determine the phases of diffractions. The essence of the method is shown in Fig. 12.7.[2, 28] It depends upon the preparation of heavy atom derivatives; this is usually possible by rational design or with reference to prior experience.

There are five stages in the isomorphous replacement:[1]

(1) The preparation of heavy atom derivatives (Section 12.2.3).
(2) The measurement of intensities of X-ray diffraction patterns for native and derivative crystals.
(3) The reduction and correction of intensity data.
(4) The determination of the heavy atom positions.
(5) The determination of phases.

The general theory of the isomorphous replacement method is described here.

A. Single isomorphous replacement[1]

Let $\boldsymbol{F}_P$ be the structure factor for the native protein. $\boldsymbol{F}_P$ is a vector and can be described in terms of structure factor magnitude, $|\boldsymbol{F}_P|$, and phase, α_P. The structure factor, $\boldsymbol{F}_{PH}$, of the heavy atom derivative is also a vector with magnitude $|\boldsymbol{F}_{PH}|$, and phase α_{PH}. $\boldsymbol{F}_{PH}$ can be derived from $\boldsymbol{F}_P$ by the vector addition of $\boldsymbol{F}_H$, which is the contribution of the heavy atoms to the structure factor of the derivative. This can be written (*cf.* Fig. 12.8)

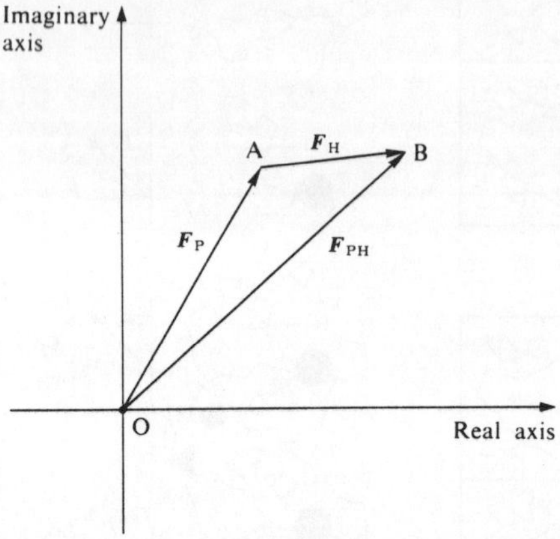

Fig. 12.8 A vector diagram (Argand diagram) illustrating the native protein ($\boldsymbol{F}_P$) and heavy atom ($\boldsymbol{F}_H$) contributions to the structure factor ($\boldsymbol{F}_{PH}$) for the heavy atom derivative of the protein.
[Reproduced with permission from T. L. Blundell, L. N. Johnson, *Protein Crystallography*, p.156, Academic Press (1976)]

$$\boldsymbol{F}_{PH} = \boldsymbol{F}_P + \boldsymbol{F}_H \tag{12.3}$$

Assuming that the $\boldsymbol{F}_{PH}$ and $\boldsymbol{F}_P$ are measured and that the position of heavy atoms in the unit cell is known, *i.e.* the vector $\boldsymbol{F}_H$ is known, from Fig. 12.9 using the cosine law

$$\alpha_P = \alpha_H + \cos^{-1}\left(\frac{|F_{PH}|^2 - |F_P|^2 - |F_H|^2}{2|F_P||F_H|}\right) = \alpha_H \pm \alpha' \tag{12.4}$$

This equation shows that there are two possible values for α_P which cannot be distinguished with one isomorphous derivative. Only when vectors $\boldsymbol{F}_P$ and $\boldsymbol{F}_H$ are colinear is there no ambiguity. This is illustrated graphically by the Harker construction for phase calculation[29] in Fig. 12.10. The vector $-\boldsymbol{F}_{\mathrm{H}}$ is drawn from the center O. The circle of radius $|\boldsymbol{F}_{\mathrm{PH}}|$ is drawn centered on the end of the vector at J. In general there are two points of intersection at G and H. The vectors OG and OH represents the two possibilities for $\boldsymbol{F}_{\mathrm{P}}$. It can be seen from Eq. 12.8 that the vectors OG and OH are symmetrically disposed about $\boldsymbol{F}_{\mathrm{H}}$. When $\boldsymbol{F}_{\mathrm{H}}$ and $\boldsymbol{F}_{\mathrm{P}}$ are colinear there is only one point where the circles touch and the solution is unambiguous.

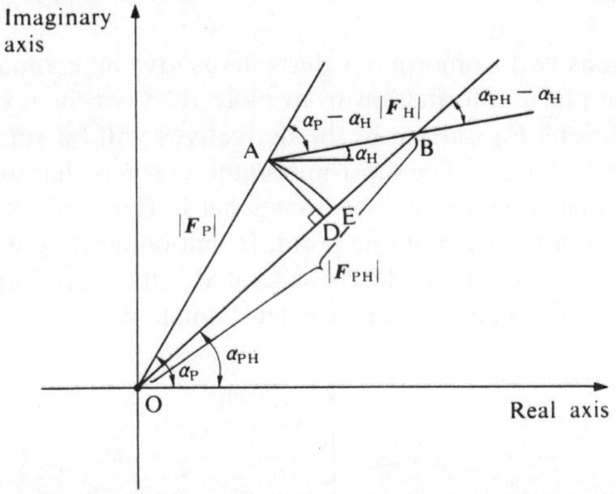

Fig. 12.9 A vector diagram defining the structure factor amplitudes and phases referred to in the text. [Reproduced with permission from T. L. Blundell, L. N. Johnson, *Protein Crystallography*, p.156, Academic Press (1976)]

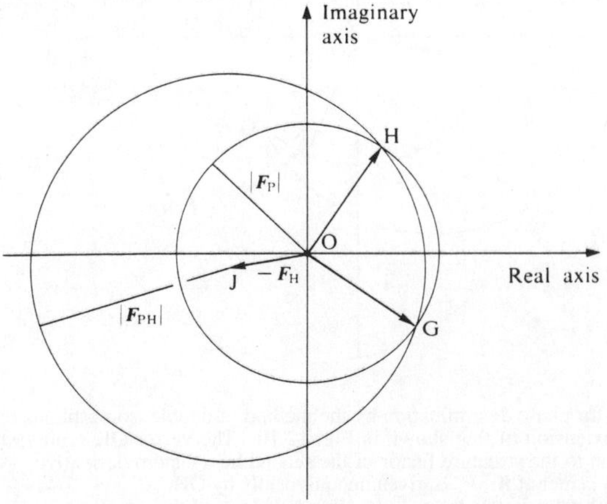

Fig. 12.10 Harker diagram for phase calculation by the method of single isomorphous replacement corresponding to the situation shown in Fig. 12.9. The vector OJ represents $-\boldsymbol{F}_{\mathrm{H}}$. Circles of radii $|\boldsymbol{F}_{\mathrm{P}}|$ and $|\boldsymbol{F}_{\mathrm{PH}}|$ are drawn with their centers at O and J respectively. The vectors OH and OJ represent the two possibilities for $\boldsymbol{F}_{\mathrm{P}}$.
[Reproduced with permission from T. L. Blundell, L. N. Johnson, *Protein Crystallography*, p.157, Academic Press (1976)]

B. Multiple isomorphous replacement (MIR)

The ambiguity mentioned above may be resolved by double isomorphous replacement in principle. The situation is shown in Fig. 12.11. Similar to Fig. 12.10 the first heavy atom derivative indicated that $\boldsymbol{F}_P$ may be one of two possible vectors given by OG and OH. The heavy atom contribution of the second derivative is drawn as $-\boldsymbol{F}_{H2}$. Two possible values of $\boldsymbol{F}_P$ are given by vectors OH and OL. The vector OH is indicated by both derivatives and this must be the correct choice for $\boldsymbol{F}_P$. From this graphic construction it is obvious that if $\boldsymbol{F}_{H2}$ is colinear with $\boldsymbol{F}_{H1}$ the second derivative will not help in sorting out the ambiguity in the phasing. Two derivatives with different heavy atoms or different occupancies (but the same relative occupancies for different sites) at the same atom positions will not be very useful.

For most diffractions two isomorphous derivatives give an estimate of α_P. However, it is advisable to include phase information from more than two heavy atom derivatives because for some diffractions $\boldsymbol{F}_H$ for one of the derivatives will be small, and for others $\boldsymbol{F}_{H1}$ and $\boldsymbol{F}_{H2}$ may colinear by chance. The most important reason is that the values used for $\boldsymbol{F}_{PH}$, $\boldsymbol{F}_P$ and α_H will be estimates which may be somewhat in error. Consequently the circle in Fig. 12.11 will usually not intersect at one point, H, but rather they will give rise to several intersections which may not be quite clearly separated if there are large errors. In fact it is necessary to estimate the errors for each phase determination.[30)]

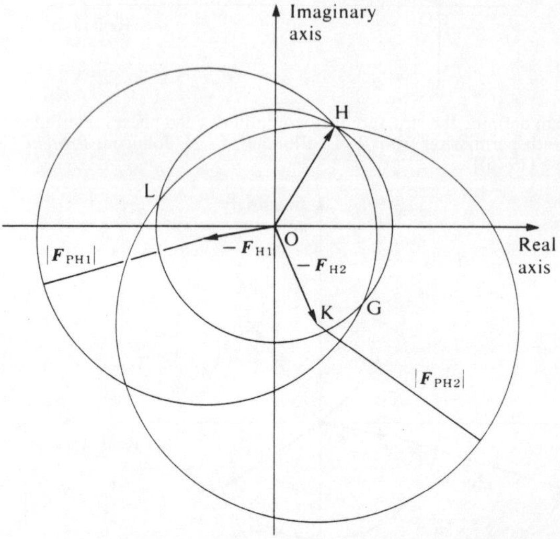

Fig. 12.11 Harker diagram for phase determination by the method of double isomorphous replacement. The construction is an extension of that shown in Fig. 12.10. The vector OK represents $-\boldsymbol{F}_H$ for the heavy atom contribution to the structure factor of the second heavy atom derivative. A circle of radius $\boldsymbol{F}_{PH2}$ is drawn with its center at K. $\boldsymbol{F}_P$ is given unequivocally by OH.
[Reproduced with permission from T. L. Blundell, L. N. Johnson, *Protein Crystallography*, p.161, Academic Press (1976)]

12.4.2 Anomalous scattering

In the calculation of phases of diffractions, small differences In the X-ray intensities may be caused by heavy atoms as a result of anomalous scattering (*cf.* Section 2.3.2)

A. Anomalous scattering and structure factor: Bijvoet pair

As shown earlier, if the X-ray wavelength is very close to the absorption edge of a heavy atom, the atomic scattering factor, f, must be represented as a complex quantity:

$$f = f_0 + \Delta f' + i\Delta f'' = f' + f''. \tag{12.5}$$

The imaginary part, f'', lags $\pi/2$ behind the primary wave, *i.e.* it is always $\pi/2$ in front of the scattered wave. The magnitude of f'' is proportional to the absorption coefficient.

In the protein structure which contains some heavy atoms the contribution of the heavy atom to the $F(hkl)$ diffractions of the derivative crystal are written as:

$$F_{\mathrm{H}}(hkl) = \sum_{j=0}^{H} f_j' \exp[2\pi i(hx_j + ky_j + lz_j)] \tag{12.6}$$

and

$$F_{\mathrm{H}}''(hkl) = \sum_{j=0}^{H} f_j'' \exp[2\pi i(hx_j + ky_j + lz_j)]. \tag{12.7}$$

for real and imaginary parts respectively, where H is number of heavy atoms in a unit cell. Corresponding expressions can be written for the inverse $F(\overline{h}\overline{k}\overline{l})$ diffractions. Then, the structure amplitudes of the direct and inverse diffractions, $\boldsymbol{F}^*(hkl)$ and $\boldsymbol{F}^*(\overline{h}\overline{k}\overline{l})$ (abbreviated as $\boldsymbol{F}(+)$ and $\boldsymbol{F}(-)$, respectively, hereafter) are expressed as

$$\boldsymbol{F}_{\mathrm{PH}}(+) = \boldsymbol{F}_{\mathrm{P}}(+) + \boldsymbol{F}_{\mathrm{H}}(+) + i\boldsymbol{F}_{\mathrm{H}}''(+) \tag{12.8}$$

$$\boldsymbol{F}_{\mathrm{PH}}(-) = \boldsymbol{F}_{\mathrm{P}}(-) + \boldsymbol{F}_{\mathrm{H}}(-) + i\boldsymbol{F}_{\mathrm{H}}''(-). \tag{12.9}$$

These relationships are shown in Fig. 12.12. The vectors $\boldsymbol{F}_{\mathrm{H}}$ and $\boldsymbol{F}_{\mathrm{H}}''$ are drawn at right angles to each other assuming that anomalous scattering atoms are all identical. This diagram shows that in general $\boldsymbol{F}_{\mathrm{PH}}(+)$ is no longer equal to $\boldsymbol{F}_{\mathrm{PH}}(-)$. Friedel's law is no longer obeyed. $\boldsymbol{F}(+)$ and $\boldsymbol{F}(-)$ are known as the Bijvoet pair.

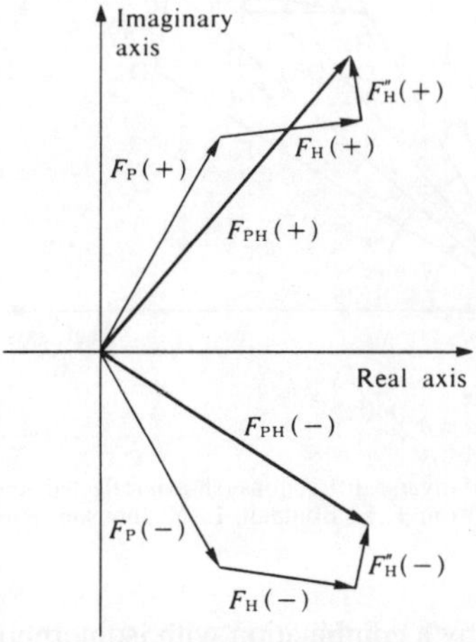

Fig. 12.12 Vector diagrams for the structure factors for the direct and inverse diffractions. Subscripts P and H express atoms which scatter normally and anomalously, respectively.
[Reproduced with permission from T. L. Blundell, L. N. Johnson, *Protein Crystallography*, p.172, Academic Press (1976)]

B. Anomalous scattering differences and phase determination[1)]

Bijvoet[32)] obtained the following useful results for the relation of the anomalous difference $\boldsymbol{F}_{\mathrm{PH}}(+)-\boldsymbol{F}_{\mathrm{PH}}(-)$ to the phases α_{PH} and α_{P} using the cosine rule (Fig. 12.13):

$$F_{\mathrm{PH}}(+)-F_{\mathrm{PH}}(-) \doteqdot \frac{2}{k} F_{\mathrm{H}} \cos\left(\alpha_{\mathrm{PH}}-\alpha_{\mathrm{H}}-\frac{\pi}{2}\right) \tag{12.10}$$

$$= \frac{2F_{\mathrm{H}}}{k} \sin(\alpha_{\mathrm{PH}}-\alpha_{\mathrm{H}}) \tag{12.11}$$

where $k=F_{\mathrm{H}}/F''_{\mathrm{H}}$.

Hence $\alpha_{\mathrm{PH}}=\alpha_{\mathrm{H}}+\pi/2+\theta$, where

$$\theta = \pm\cos^{-1}\left\{\frac{k[F_{\mathrm{PH}}(+)-F_{\mathrm{PH}}(-)]}{2F_{\mathrm{H}}}\right\} \tag{12.12}$$

Thus the phase of the heavy atom derivative α_{PH} cannot be determined unambiguously. There are two possible phases which are symmetrically disposed about $\alpha_{\mathrm{H}}+\pi/2$. These two possibilities are illustrated in Fig. 12.14. This diagram is constructed in the same way as that for isomorphous replacement by drawing circles of radii $F_{\mathrm{PH}}(+)$ and $F_{\mathrm{PH}}(-)$ with their centers at $-(\boldsymbol{F}_{\mathrm{H}}+\boldsymbol{F}''_{\mathrm{H}})$ (D´) and $-(\boldsymbol{F}_{\mathrm{H}}-\boldsymbol{F}''_{\mathrm{H}})$ (C´). The two points of intersection indicate the two possible vectors for F_{PH}, *i.e.* JH and JM. Thus two possibilities for F_{P} are given as OH and OM.

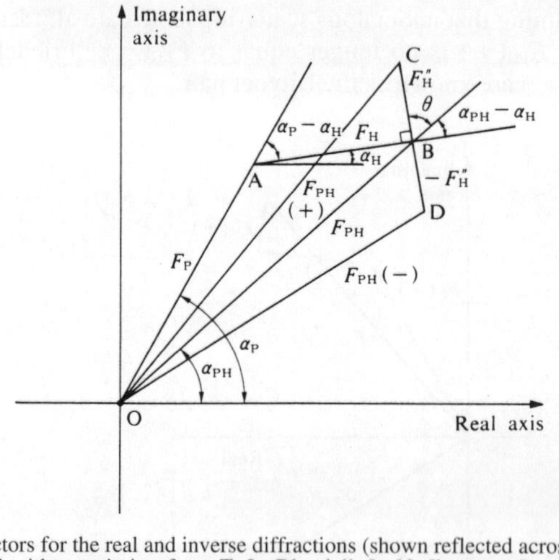

Fig. 12.13 Structure factors for the real and inverse diffractions (shown reflected across the real axis). [Reproduced with permission from T. L. Blundell, L. N. Jhonson, *Protein Crystallography*, p.178, Academic Press (1976)]

This difficulty may be solved by a combination with isomorphous replacement method (Fig. 12.15). The third circle of radius F_{P} added to the two circles of radii $F_{\mathrm{PH}}(+)$ and $F_{\mathrm{PH}}(-)$ indicate that F_{P} is given by the vector OH.

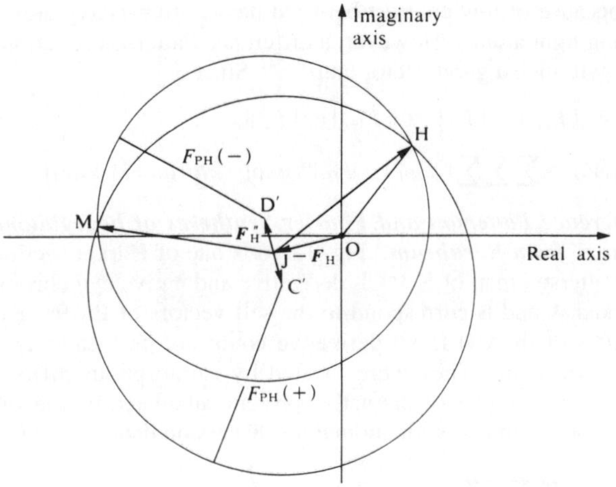

Fig. 12.14 Harker diagram to illustrate phase determination using anomalous scattering. The diagram is constructed in the same way as that for isomorphous replacement (Fig. 12.11). Circles of radii $F_{PH}(+)$ and $F_{PH}(-)$ are drawn with their centers at $-(\boldsymbol{F}_H+\boldsymbol{F}''_H)-(\boldsymbol{F}_H-\boldsymbol{F}''_H)$. The two points of intersection indicate the two possible vectors for $\boldsymbol{F}_{PH}$, *i.e.* JH and JM. Two possibilities for $\boldsymbol{F}_P$ are given as OH and OM.
[Reproduced with permission from T. L. Blundell, L. N. Jhonson, *Protein Crystallography*, p.178, Academic Press (1976)]

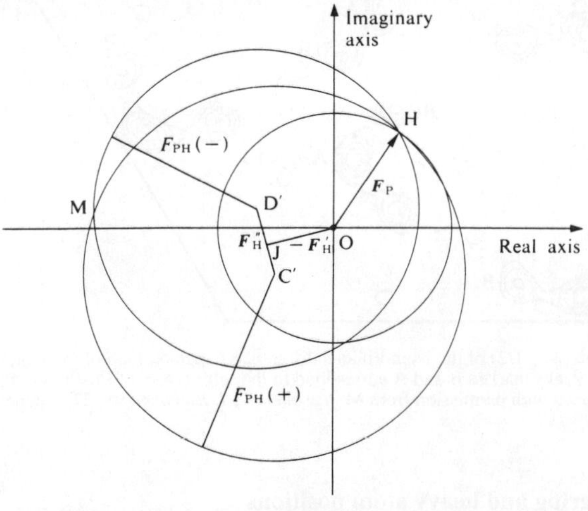

Fig. 12.15 Harker diagram to illustrate phase determination using anomalous scattering (Fig. 12.14) combined with isomorphous replacement for situation shown in Fig. 12.13. The intersection of the three circles of radii F_P, $F_{PH}(+)$ and $F_{PH}(-)$ at H indicates that $\boldsymbol{F}_P$ is given by the vector OH.
[Reproduced with permission from T. L. Blundell, L. N. Jhonson, *Protein Crystallography*, p.180, Academic Press (1976)]

12.4.3 Determination of the position of heavy atoms

A. Isomorphous difference Patterson function

Ordinary Patterson function with coefficient $|F_{PH}|^2$ may not be useful to locate the heavy atom in the unit cell because of low contrast between heavy atom-heavy atom vectors and other vectors concerning light atoms. However, a difference Patterson function with coefficients, $\| F_P | - | F_{PH} \|^2$ will give a good vector map.[33,34] Since

$$| F_H | = | F_{PH} | - | F_P | \fallingdotseq \| F_{PH} | - | F_P \|, \tag{12.13}$$

$$\Delta P(uvw) = \sum\sum\sum \| F_{\Pi H} | - |F_{\Pi} \|^2 \exp[-2\pi i(hu+kv+lw)] \tag{12.14}$$

a) Isomorphous difference Patterson and Fourier Syntheses of heavy atom derivatives and native cytochrome c´ from R. rubrum. Fig. 12.16 is one of Harker sections of an isomorphous difference Patterson map of K_2PtCl_6 derivative and native cytochrome *c´* from *R. rubrum*.[35] Peaks marked A and B correspond to the self vectors of Pt_1-Pt_1 and Pt_2-Pt_2, respectively. The Hg sites of the CH_3HgCl derivative could not be located by the isomorphous difference Patterson map. They were located by isomorphous difference Fourier synthesis using the best phase angles of the native protein calculated by the single isomorphous replacement method combined with anomalous dispersion data of K_2PtCl_6.

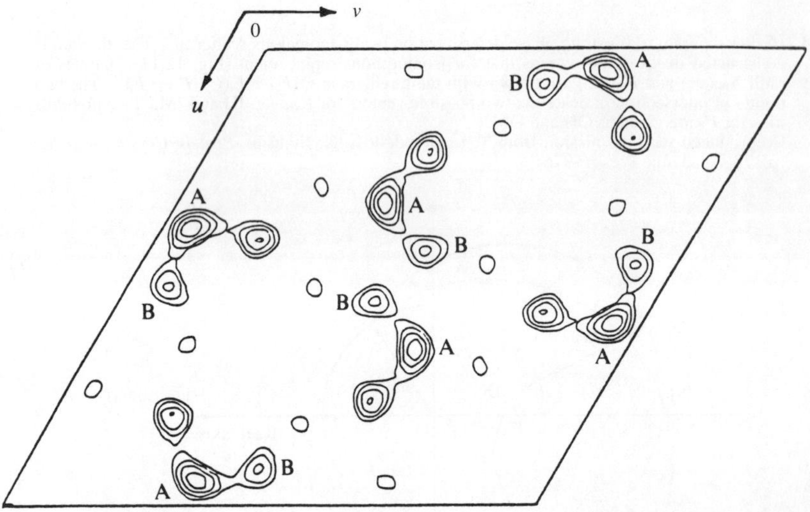

Fig. 12.16 A Harker section ($w = 1/2$) of the isomorphous difference Patterson map of the platinum derivative of cytochrome *c´*. Peaks marked B and A correspond to the self vectors of Pt_1-Pt_1 and Pt_2-Pt_2, respectively.[35] [Reproduced with permission from M. Yasui *et al*., *J. Biochem*, **98**, 77, Japanese Biochem. Soc. (1991)]

B. Anomalous scattering and heavy atom positions

As mentioned above, anomalous dispersion data may be combined with the isomorphous replacement method for the phase determination. If the protein contains two heavy atoms or more the atomic numbers of which are very close to each other, these heavy atoms may be identified by the use of anomalous dispersion.[36)]

a). Location of heavy atoms by anomalous scattering. Intensity data of cytochrome *c´* from *R. rubrum* were collected on a four-circle diffractometer using synchrotron X-rays (BL-14A) at the Photon Factory, National Laboratory for High Energy Physics, Tsukuba.[37] Since cytochrome *c´* contains iron in its heme, three wavelengths, $\lambda_1 = 1.077$, $\lambda_2 = 1.730$

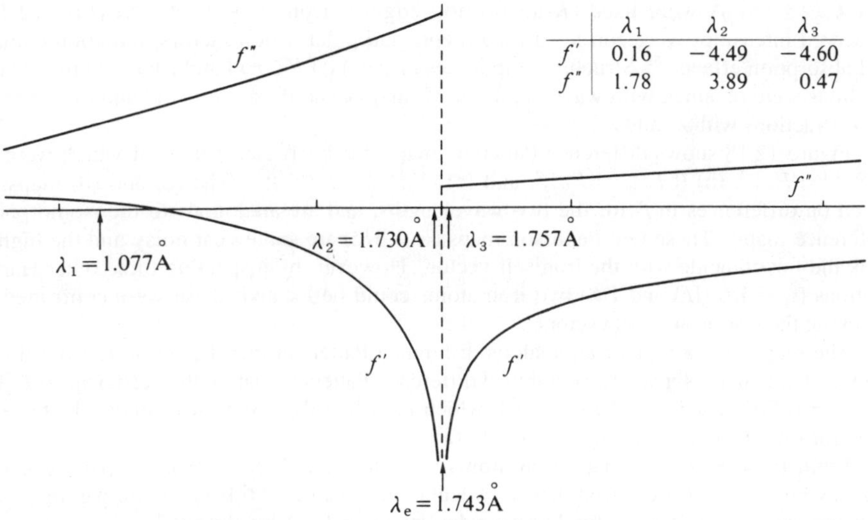

	λ_1	λ_2	λ_3
f'	0.16	−4.49	−4.60
f''	1.78	3.89	0.47

Fig. 12.17 Anomalous dispersion of the iron atom around the *K*-absorption edge.[37] Three sets of intensity data obtained with three wavelengths. λ_1, λ_2, and λ_3 were used to determine the positions of the iron atoms. [Reproduced with permission from S. Harada *et al., J. Cryst. Soc. Jpn.*, **29**, 272, Cryst. Soc. Jpn. (1987)]

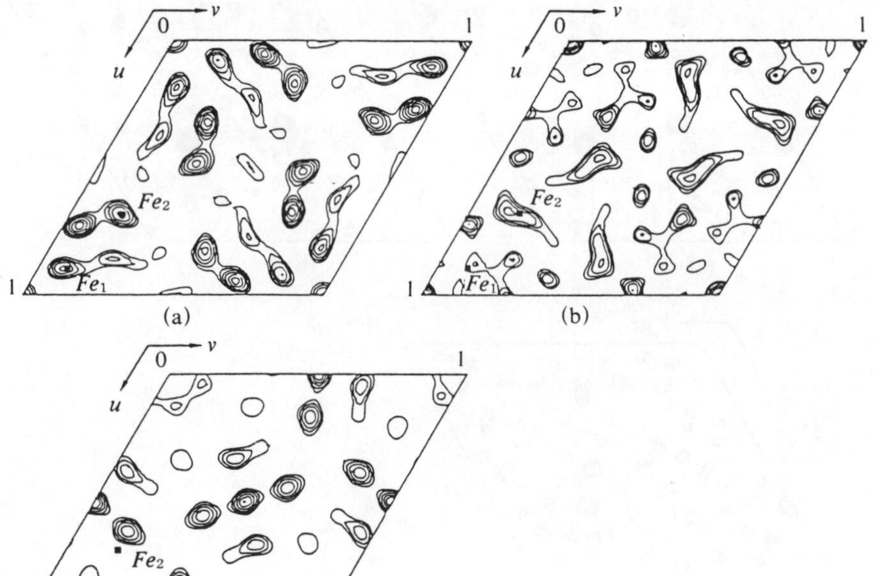

Fig. 12.18 Difference patterson maps of cytochrome c' ($w = 1/2$) with coefficient (a) $(|F_1| - |F_2|)^2$, (b) $(|F_1| - |F_3|)^2$, and (c) $(|F_2(+)| - |F_2(-)|)^2$. The positions of iron self vectors are marked.[37] [Reproduced with permission from S. Harada *et al., J. Appl. Cryst.*, **19**, 450. IUCr. (1986)]

and $\lambda_3 = 1.757$ Å were used (K-absorption edge of iron, $\lambda_e = 1.743$ Å) (Fig. 12.17). Observed intensities were corrected for Lorentz and polarization factors, radiation damage and absorption effects.[38] Structure amplitudes $|F_1(+)|$, $|F_2(+)|$ and $|F_3(+)|$ for *hkl* diffractions were obtained with wavelengths λ_1, λ_2 and λ_3 and those $|F_2(-)|$ and $|F_3(-)|$ for $\bar{h}\bar{k}\bar{l}$ diffractions with λ_2 and λ_3.

Figure 12.18 shows difference Patterson maps, the coefficients used of which were (a) $(|F_1| - |F_2|)^2$, (b) $(|F_1| - |F_3|)^2$, and (c) $(F_2(+) - F_2(-))^2$. The (a) and (b) maps are based on differences in f' for the two wavelengths, and are analogous to the isomorphous difference maps. These two Patterson maps ($w = 1/2$) are somewhat noisy and the highest peak did not coincide with the iron self vector. However, by inspecting three other Harker sections ($w = 1/6$, 1/3 and 1/6) two iron atoms could be located; these were confirmed by checking the iron-iron cross vectors.

The map (c) is a regular anomalous difference Patterson map based on f'',[2] but this is the worst among the three. The combined difference Patterson map with coefficient $[(|F_1| - |F_2|)^2 + (k/2)^2(|F_2(+)| - |F_2(-)|)^2]$, which is ordinarily expected to give a better estimate for iron than (a) or (c), was similar to (a).

From these Patterson maps iron atoms were located (Fig. 12.19), the parameters of which were refined for electron density calculations. Table 12.6 lists atomic parameters of iron and mercury used in the final phase calculations at 2.8 Å resolution.[39]

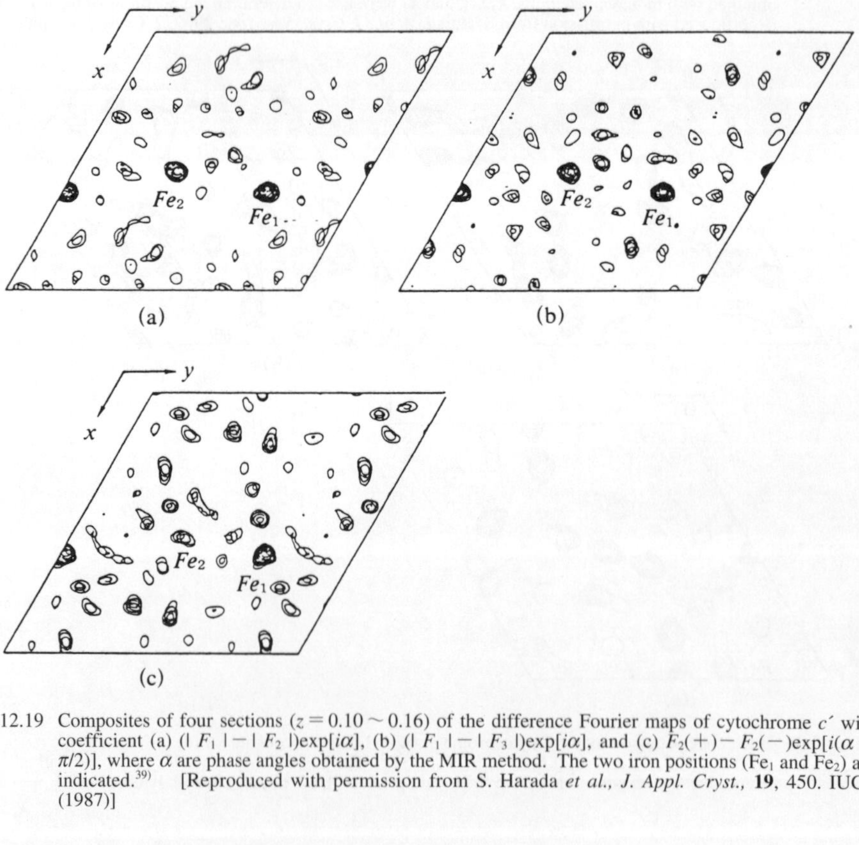

Fig. 12.19 Composites of four sections ($z = 0.10 \sim 0.16$) of the difference Fourier maps of cytochrome c' with coefficient (a) $(|F_1| - |F_2|)\exp[i\alpha]$, (b) $(|F_1| - |F_3|)\exp[i\alpha]$, and (c) $F_2(+) - F_2(-)\exp[i(\alpha - \pi/2)]$, where α are phase angles obtained by the MIR method. The two iron positions (Fe_1 and Fe_2) are indicated.[39] [Reproduced with permission from S. Harada *et al.*, *J. Appl. Cryst.*, **19**, 450. IUCr. (1987)]

Table 12.6 Heavy atom parameters of cytochrome *c*′ from *Rhodospirillum rubrum*[39)]

Derivative	site	*x* (in fractional unit of cell edges)	*y*	*z*	Occupancy (electrons)	*B* (Å^2)
K_2PtCl_6	1	0.253	0.535	−0.378	41.6	39.6
	2	0.162	0.186	0†	48.8	45.5
CH_3HgCl	1	0.900	0.600	0.094	12.5	28.8
	2	0.930	0.065	0.121	13.5	21.5

† fixed at 0 to define the origin of the Z axis.
[Reproduced with permission from M. Yasui, *Doctral Thesis*, Osaka Univ.]

12.5 Molecular Replacement Method[40)]

As the number of protein crystal structures analyzed has increased remarkably, a large amount of structural data has been and are being accumulated in the Protein Data Bank at Brookhaven.[41)] Looking at these data, the existence of structural similarity has been recognized in groups of proteins. Many proteins have been found that consist of the same subunits. In such cases, it may be possible to determine the molecular structure by limiting the phase angles of diffraction.

If two identical molecules differ only in orientation and position, it is possible to define a general operation involving only rotation and translations which bring the equivalent points of electron density in each molecule into coincidence.

If the equivalent points on the two molecules are defined by position vectors $\boldsymbol{X}_1$ and $\boldsymbol{X}_2$ in reference to a common orthogonal coordinate system, then

$$\boldsymbol{X}_2 = [\mathrm{C}]\boldsymbol{X}_1 + \boldsymbol{\tau} \tag{12.15}$$

where [C] is a rotation matrix and $\boldsymbol{\tau}$ a vector defining translation.

The Patterson function, which is the Fourier transform of $|F_P|^2$, contains intramolecular and intermolecular atomic vectors, and the intramolecular vectors are generally short and most of them are distributed near the origin. Picking up these vectors included in a sphere with radius *r*, which corresponds to the approximate size of the molecule, rotate them to search for the best rotation angle until they show good coincidence with the corresponding vectors of the known structure or with the self vectors (*rotation search*). Then, displacing the molecule in the crystal lattice to find the best translation vector until coincidence between the observed and calculated $|F_P|$'s or that between corresponding vectors is obtained (*translation search*); then the protein molecule is placed at the correct position with correct orientation. The electron density map is then computed using the phases determined with the orientation and positions of the protein molecule (*cf.* Fig. 12.20).

12.5.1 Structure solution of bacterial cytochrome c_2 from *Rhodopseudomonas viridis (Rps. viridis).*[42)]

The structure of cytochrome c_2 from *Rps. viridis* (trigonal rhombohedral system, space group $P3_121$ or $P3_221$, $a = b = 76.13$ and $c = 40.40$ Å; one protein molecule per asymmetric unit, $V_m = 2.70$ Å^3/Da) was solved on the basis of the structure of tuna cytochrome *c*[43)] as a search model by an automatic molecular replacement procedure.[44)] This procedure consists of four parts: structure factor calculation of the model molecule, fast-rotation function, multiple translation function and rigid-body refinement.

The model molecule, in which all the atoms of tuna cytochrome *c* together with those

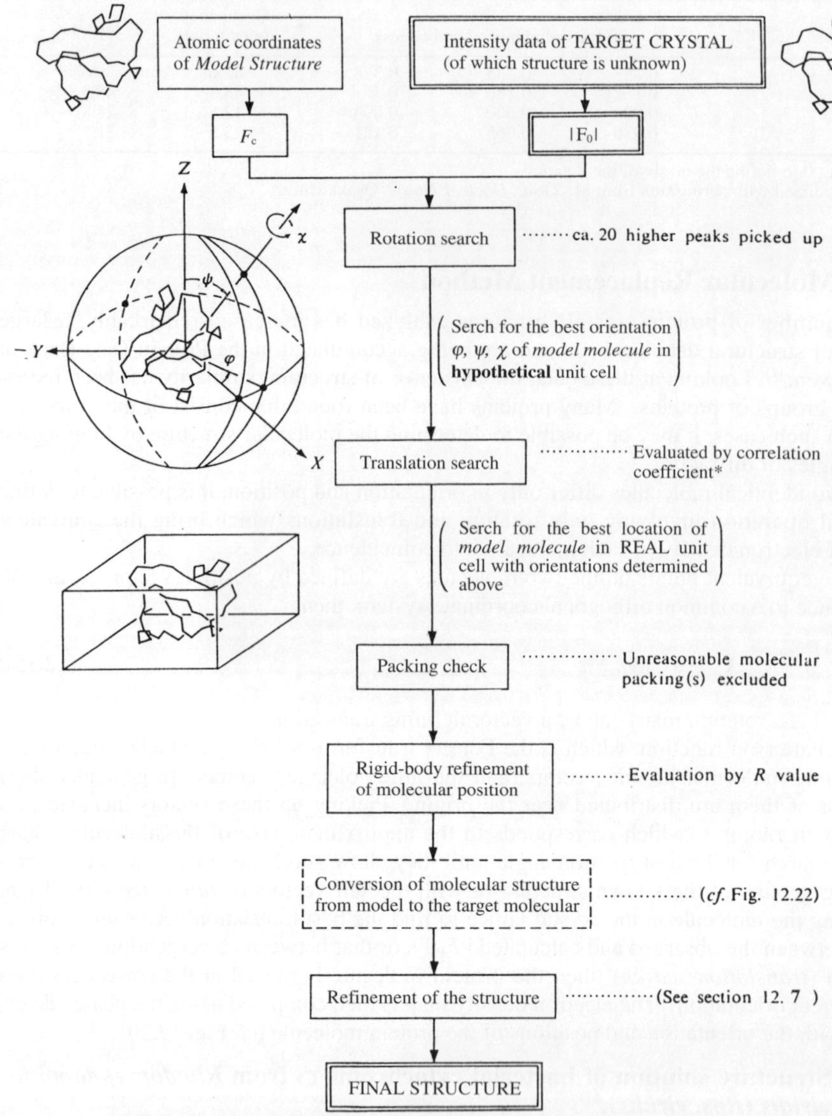

Fig. 12.20 Structure solution of a protein crystal by the molecular replacement method.

$$\text{* Correlation coefficient} = \frac{n\sum F_{\rm obs}F_{\rm calc} - \sum F_{\rm obs}F_{\rm calc}}{\{[n\sum F^2{}_{\rm obs} - (\sum F_{\rm obs})^2][n\sum F^2{}_{\rm calc} - (\sum F_{\rm calc})^2]\}^{1/2}}$$

where n is the number of diffractions

Figs.: [Reproduced with permission from Alexander MacPherson, *Preparation and Analysis of Protein Crystals*, p.260, Krieger Pub. (1982)]

of the side chains were included, was placed in a $P1$ cell with $a = b = c = 100$ Å and $\alpha = \beta = \gamma = 90°$. Triclinic structure factors were calculated between 10 and 6 Å resolution. Cross rotation function[45,46] was calculated as a function of the orientation angles ψ, φ, χ in the angular range, $\psi = 0$ to 90°, $\varphi = 0$ to 360°, $\chi = 0$ to 360° with 5° steps in each direction using the data between 10 and 6 Å resolution. The radius of the cut-off sphere in the Patterson function was set to 20 Å. A translation search was carried out for the 20 highest peaks in the rotation search. The agreement of observed and calculated structure factors was evaluated with a correlation coefficient between F_{obs} and F_{calc} at every position translated by about a 1 Å step. The molecular packing in the unit cell was checked and those cases giving packing with short intermolecular contacts were rejected. The model position was then improved by the rigid-body refinement[47] using the data between 10 to 6 Å resolution.

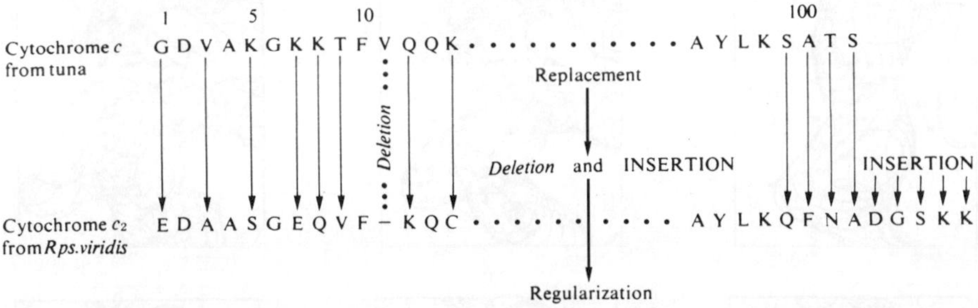

Fig. 12.21 Conversion from cytochrome *c* from tuna, adopted as a starting model to solve the structure by the molecular replacement method, to the structure of cytochrome c_2 from *Rhodopseudomonas viridis* for the refinement of the structure.
1) 50 different amino acid residues (indicated by arrows, only some of them are shown in this figure)
2) One amino acid residue was *deleted* and four new residues were INSERTED.
3) Regularization of the molecular chain with the reasonable bond lengths and bond angles etc. was carried out for further refinement.

Before starting refinement, the tuna cytochrome *c* model molecule used in the molecular replacement was modified so as to fit the sequence of the *Rps. viridis* cytochrome c_2 (Fig. 12.21). Refinement was carried out by both the restrained least-squares procedure and simulated-annealing method (Section 12.7): current $R = 0.219$ (3 Å resolution).[42]

12.6 Interpretation of Electron Density Maps: Model Building

Density map interpretation depends on two factors, the accuracy of phasing and the resolution limits (*cf.* Fig. 12.22).[2b]

In early times the Richard box was used to build an initial atomic model that fit the electron density map obtained. Minimaps, *i.e.* transparent stacks of contour sheets at a scale of 0.5 cm/Å are still used to trace the peptide chain.

Use of computer graphics systems to display the electron density superimposed on the structure model for chain tracing and also for detailed model building has become almost universal[48] (Fig. 12.23).[39]

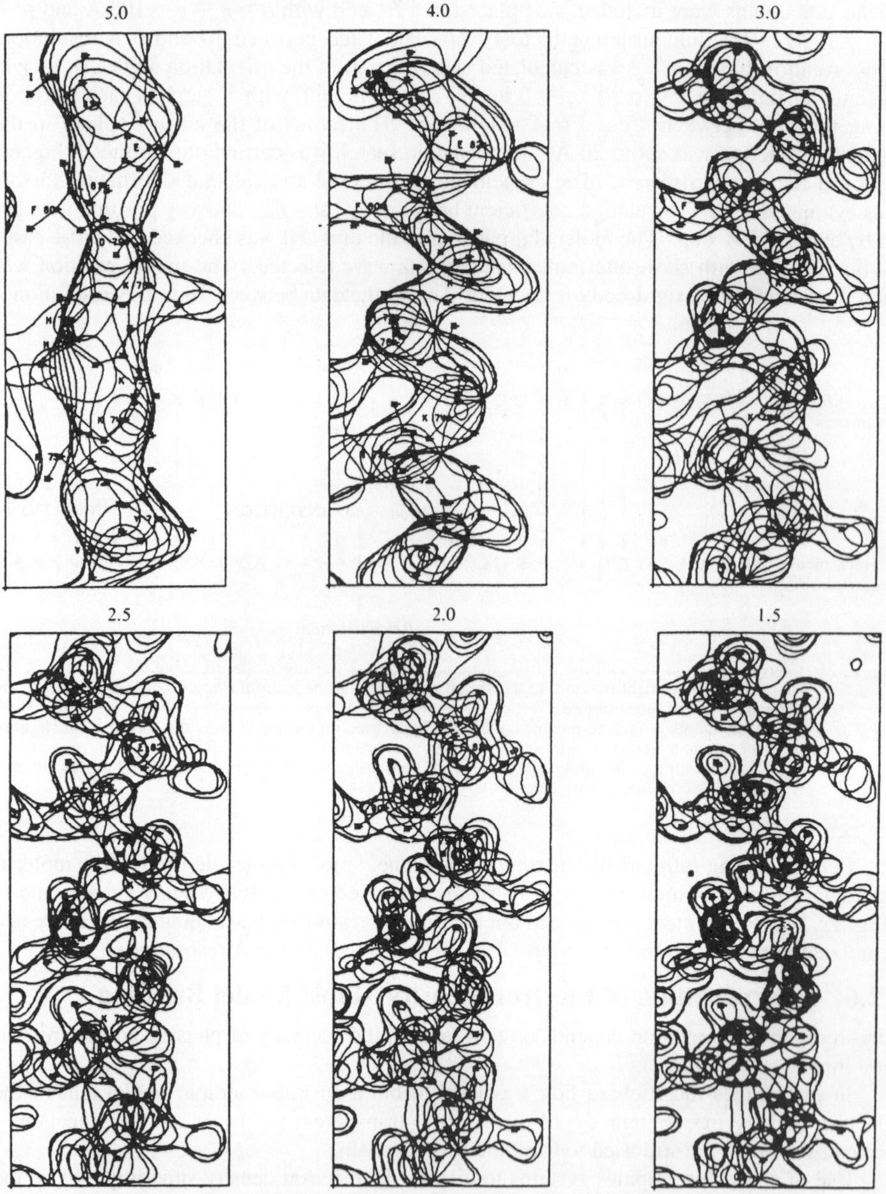

Fig. 12.22 Electron density distribution for the *C* helix of myohemerythrin at various resolutions. Each frame shows the projected electron density distribution from a Fourier synthesis truncated at the indicated nominal resolution. The number of terms included approximately doubles with each successive step in resolution: 638 unique diffractions contribute at 5.0 Å resolution, 1,208 at 4.0 Å, 2,699 at 3.0 Å, 4,531 at 2.5 A, 8,618 at 2.0 A and 16,747 at 1.5 A. In each case the final refined backbone model, including C_β atoms from the side chains, is shown superimposed on the density map. (Drawings by Steven Sheriff.) [Reproduced with permission from W. A. Hendrickson, *Protein Engineering*, p.11, Alan R. Liss, Inc. (1987)]

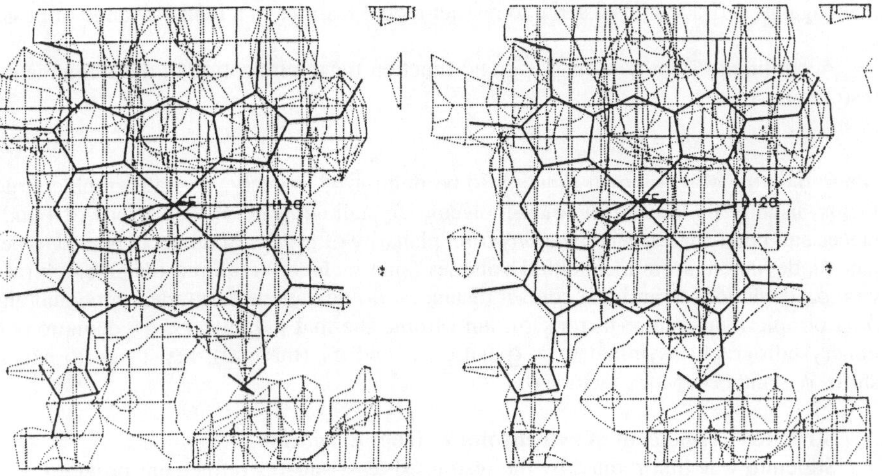

Fig. 12.23 Stereoscopic drawing of the improved electron density of the heme and the side-chain of His 120 of cytochrome *c′* from *Rhodospirillum rubrum* which was obtained after 5 cycles of phase refinement using the non-crystallographic 2-fold symmery.[39] The part of the model which was refined at 2.8 Å resolution is superposed on the electron density.
[Reproduced with permission from M. Yasui *et al.*, *J. Biochem.*, **111**, 318 Japanese Biochem. Soc. (1992)]

12.7 Refinement of the Structure

The refinement is usually carried out by a series of steps.

1) Each step consists of a few cycles of least-squares refinement with stereochemical and internal packing restraints.[49–55]

2) Next step is the re-fitting of the model structure to difference electron density maps with interactive computer graphics.[56]

3) At an appropriate stage of the final refinement, solvent molecules are usually included and alternative conformations for some atoms or residues in the protein may be introduced.

12.7.1 Restrained least-squares refinement

The aim of least-squares refinement is to minimize the difference between the observed and calculated structure amplitudes, $|F_o(hkl)|$ and $|F_c(hkl)|$. In the case of small molecules many strong reflections can be measured to carry out the least-squares refinement rigorously.

Macromolecules usually consist of a great many atoms and the number of atomic parameters to be refined is too many compared to the number of diffractions, even if hydrogen atoms are ignored. Secondly, relatively weak diffraction intensities of macromolecules limit the accuracy of intensities measured and the number of diffractions observed. Thirdly, initial structure models of macromolecules are usually quite inaccurate. These facts limit the rigorous structure refinement of macromolecular crystals. However, the wealth of prior knowledge about the stereochemistry of macromolecular structures can supplement the limited diffraction data. Several different methods have proposed to incorporate stereochemical knowledge into the refinement process.[57] These subsidiary conditions[58,59] on refinement against diffraction data serve to restrict model features to a realistic range of possibilities.

According to Hendrickson,[54] a grand function for minimization in the refinement with restraint is

$$\Phi = \Sigma\phi_j, \tag{12.16}$$

where ϕ_j is the observational function to be minimized, namely, crystallographic structure factors usually minimized in small-molecule crystallography, ϕ_1, deviations of bond distances and bond angles, ϕ_2, deviations from planarity of groups of atoms expected to be planar, ϕ_3, deviations from ideal chiral volumes (sign included) required for given chiral centers, ϕ_4, deviations from ideal contact distances, ϕ_5, deviations from ideal torsional angles, ϕ_6, isotropic temperature factors, ϕ_7, anisotropic thermal parameters, ϕ_8, deviations from noncrystallographic symmetry, ϕ_9 (positional) and ϕ_{10} (thermal), resistance to excessive shifts, ϕ_{11}, and occupancy factors, ϕ_{12}.

A. Structure refinement of cytochrome *c*´ from *R. rubrum*[39]

Structure was first refined by the restrained least-squares refinement program, PROLSQ.[59–61] Observed diffractions data of 4481 between 5.0 and 2.8 Å resolution ($|F_o| \geqq \sigma(F)$) were used. An overall temperature factor 15 Å^2 was adopted for all protein atoms at the initial stage. At the end of the 18th cycle the refined structure was checked against the multi-isomorphous replacement map, the averaged map and the omit map computed with coefficients ($2|F_o| - |F_c|$). After the rebuilding of the model, successive cycles of the refinement were carried out. The *R* factor decreased from 0.44 to 0.32.

Simulated annealing procedure of the refinement (See following Section) using the XPLOR program package.[63,64] Dynamics were carried out at 3,000 K (time steps 0.67 fs), then at 4,000 K, and slowly cooled down to 300 K in steps of 50 K. Final energy minimization and refinement of individual temperature factors gave the *R* factor of 0.22.

The atomic parameters obtained were further refined for 6 cycles with PROLSQ. This refinement did not reduce the *R* factor. However, stereochemical parameters such as bond lengths, bond angles and others showed improvement (root mean-squares (r.m.s.) deviation of bond lengths from ideal values: 0.022 Å and the r.m.s. error of atomic positions according to Luzatti:[62] *ca.* 0.25–0.35 Å).

12.7.2 Crystallographic refinement by simulated annealing[63–67]

In the case of macromolecules, the restrained least-squares refinement procedure mentioned before is easily trapped in a local minimum and does not correct the positions of residues that are misplaced by more than 1 Å so that manual adjustments of the structure, using a computer graphics system to display the electron density superimposed on the structural model, are necessary. It has recently been shown that the introduction of simulated annealing into the refinement procedure can reduce the need for manual intervention.[63] Monte Carlo and molecular dynamics simulations are both ways of generating conformations of the system that are consistent with a Boltzmann distribution appropriate to the specified temperature. For large biomolecular structures the molecular dynamics algorithm is generally more efficient at generating equilibrium structures,[68] and a molecular dynamics was therefore used to implement a simulated annealing procedure for refining protein structures.[63–65] The application of molecular dynamics to X-ray refinement is also a natural extension of the use of high-temperature molecular dynamics and cooling in searching for stable peptide conformations[69] and is closely related to the use of molecular dynamics to derive structures from nuclear magnetic resonance measurements .[70,71)]

The basic principle of this method is that the crystallographic discrepancy term, E_X, is added as a pseudo-energy term to the total potential energy, E_{MD}, of the system.

$$E_{MDX} = E_{MD} + E_X. \tag{12.17}$$

Then the total potential energy, E_{MDX}, is minimized during the refinement. This is achieved by solving the Newtonian equations of motion for all atoms using the forces derived from the energy terms, including the "X-ray energy," E_X. The temperature is kept constant by coupling to a heat bath.[72)] Owing to the presence of kinetic energy in the simulated system, energy barriers can be overcome in this process. This means that, for example, peptide planes can be flipped and loops can be rearranged by exploring large areas in conformational space. Thus, in principle, it is possible to reach a lower energy minimum than the nearest minimum.[67)]

The first term, the total potential energy of the system, E_{MD}, is given as

$$E_{MD} = E_i\ E_i^{crystal}. \tag{12.18}$$

Here, E_i, the empirical potential energy, is a function of all atomic positions of the system describing internal stereochemical interactions (bond lengths, bond angles, dihedral torsion angles, chiral centers, planarity of aromatic rings) as well as nonbonded (van der Waals and electrostatic) interactions.

$$\begin{aligned} E_i = \sum_{\text{bonds}} k_0(r-r_0)^2 + &\sum_{\text{angles}} k\theta(\theta - \theta_0) \\ &+ \sum_{\text{dihedrals}} k\phi\cos(n\phi + d) \\ &+ \sum_{\text{chiral,planar}} k\omega(\omega - \omega_0)^2 \\ &+ \sum_{\text{atom pairs}} (ar^{-12} + br^{-6} + cr^{-1}). \end{aligned} \tag{12.19}$$

The parameters of the empirical potential energy E_i are inferred from experimental as well as theoretical investigations.[73–79)] Since the system considered is a crystal but not a molecule, the influence of the crystal packing is important in crystallographic refinement, and the crystal symmetry interaction energy $E_i^{crystal}$ should be added to the empirical potential energy E_i.

$$E_i^{crystal} = \sum_s \sum_{i \geqq j} \text{NB}\{|\mathcal{F}^{-1}\,\text{MinG}(\mathcal{F} r_i - 2_s \mathcal{F} r_i + t_s)|\} \tag{12.20}$$

where NB{ } and MinG( ) are functions that are defined below, $\mathcal{F}$ is the matrix that converts orthogonal coordinates into fractional coordinates, the first sum extends over all symmetry operators ($\mathbb{O}_s$, $\boldsymbol{t}_s$) of the crystal, and the second sum extends over all pairs of atoms (i, j) for which the argument of the function NB{ } is less than a specified cutoff r_{out}, and r_i and r_j are coordinates of atoms i and j respectively. The function MinG($\boldsymbol{r}$) defines the minimum image distance in fractional coordinate space. It operates on each component of the three-dimensional vector r separately where the operation on each component x is given by

$$\text{MinG}(x) = \text{sign}(-x)\text{int}(|x| + 1/2) + x \tag{12.21}$$

The function int(y) is defined as the integer part of y, and sign(y) is defined as the sign of y. NB{ } has the form of a nonbonding interaction potential, *i.e.* it is the sum of van der Waals and electrostatic interactions,

$$\mathrm{NB}\{r\} = ar^{-12} - br^{-6} + cr^{-1} \tag{12.22}$$

The second term, "X-ray energy", E_X, is given as

$$E_X = \sum_h \frac{1}{\sigma_F{}^2(h)}(|F_o(h)| - k|F_c(h)|)^2, \tag{12.23}$$

where $F_o(h)$ corresponds to observed structure factors, $F_c(h)$ to calculated structure factors based on the current structure model, k is the scale factor between F_o and F_c, and $\sigma_P(h)$ (h substituted here for hkl) is a weighting term.

A. Structure refinement of cytochrome c_2 from *Rps. viridis* [42)]

Refinement by the simulated annealing was performed by the X-PLOR program.[1,2)] After 200 steps of preliminary energy minimization, 6,000 steps (time steps of 0.25 fs) of molecular dynamics were carried out at 6,000 K, followed by decreasing the temperature from 6,000 to 300 K in steps of 25 K. Tolerance by which any atomic coordinates from a position is specified as 0.2. Another 80 steps of energy minimization and individual temperature factor refinement completed the process.

12.7.3 Further refinement

At the final steps, water molecules and sulphate and other molecules, which were added as a precipitant or additive for the crystallization of the protein and which have migrated in the crystal during crystal growth, may be located on the difference Fourier maps. They are added in further cycles of the refinement.

12.7.4 Expression of the result

How to explain the result obtained effectively, attractively, easy to understand for the reader of your paper is very important. Currently used typical models are the following. There are of course many modifications and the reader can modify and improve them to express the results in the best way. At an initial stage, the balsa wood model may be used for the low resolution data just to obtain an idea of the whole shape and rough features of the macromolecule. For a detailed exhibition of the molecule in whole or part a solid ball and spoke model (including the ORTEP drawing) (Fig. 12.24) and space-filling model are used as well as in the case of small molecules. For simplicity, only the α-carbon atoms are connected by spokes to trace the feature of the folding of the main chain (Fig. 12.25). The others are the ribbon model or its modifications in which the α-helix part is expressed as a helically wound ribbon (or cylinder) and the β-pleated sheet part by a flat plate (or sometimes by a flat arrow showing the direction of the pleated sheet) (*cf.* Fig. 12.26). The accuracy of the results should also be given appropriately, if necessary.

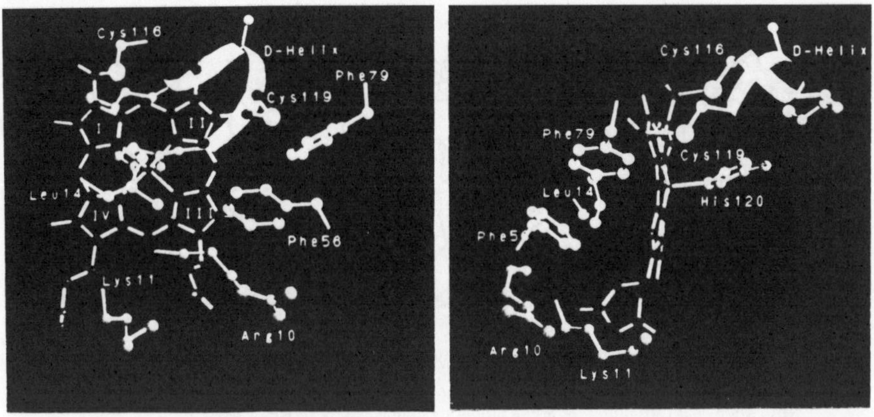

Fig. 12.24 View of the heme prosthetic group of cytochrome *c´* from *R. rubrum.*, from two different directions (ball and spoke model drawing).[39]
[Reproduced with permission from M. Yasui *et al.*, *J. Biochem.*, **111**, 320, Japanese Biochem. Soc. (1992)]

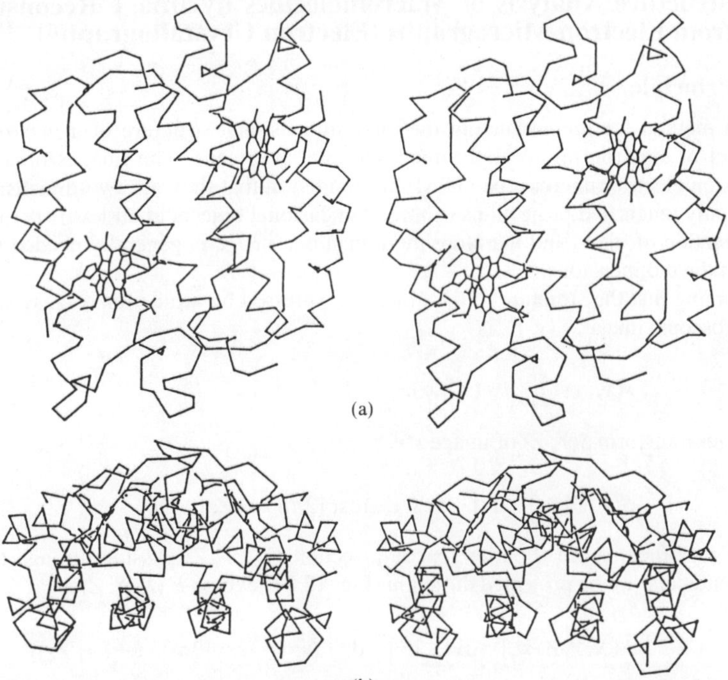

Fig. 12.25 Stereoscopic drawings of cytochrome *c´* from *R. rubrum* viewed from the direction of the non-crystallographic two-fold axis (a), and from the bottom of the helix bundle (b).[39] α-Carbon atoms are connected by spokes. Heme groups are also shown.
[Reproduced with permission from M. Yasui *et al.*, *J. Biochem.*, **111**, 320, Japanese Biochem. Soc. (1992)]

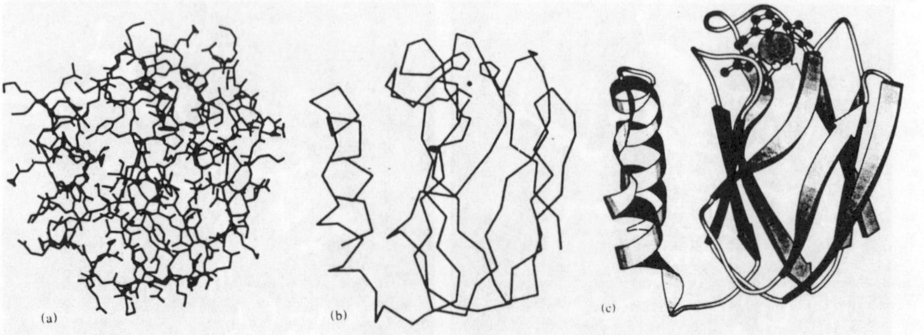

Fig. 12.26 Different drawings of the molecular structure of pseudoazurin from *Methylobacterium extorquens* AM1.
(a) All the non-hydrogen atoms are connected by spokes. Water molecules are not shown.
(b) Main chain structure; α-carbon atoms are connected by spokes.
(c) Ribbon model; α-helix part is expressed as helically wound ribbon and β-pleated sheets by flat arrows. Four ligands around the copper atom, expressed by a flat disc, are drawn by balls and spokes.
[Reproduced with permission from T. Inoue, Doctral Thesis, pp.68, 69, Osaka Univ. (1994)]

12.8 Structure Analysis of Macromolecules by Image Reconstruction from Electron Micrographs (Electron Crystallography)[80–83)]

12.8.1 Principle

This is a method for reconstructing the three-dimensional structure of a macromolecule from a set of electron microscopic images of the specimen. The images obtained by a transmission electron microscope (TEM) are limited in that they are two-dimensional, *i.e.*, they are only magnified projections of three-dimensional objects at all levels perpendicular to the direction of view, and therefore more than one image is generally needed to see the objects in three dimensions.

Assuming that the structure of the object is expressed by a function, $f(x, y, z)$, its electron microscopic image, $s(y, z)$, is

$$s(y,z) = \int_{-\infty}^{\infty} dx f(x,y,z). \tag{12.24}$$

The Fourier transform $S(Y, Z)$ of image $s(y, z)$

$$S(Y,Z) = \int_{-\infty}^{\infty} \mathrm{d}y \int_{-\infty}^{\infty} \mathrm{d}x\, s(y,z) \exp[2\pi i(Yy+Zz)] \tag{12.25}$$

is, in the reciprocal space (Fourier space), a section $F(0, Y, Z)$ passing through the origin and perpendicular to the projected direction (Fig. 12.27), where $F(X, Y, Z)$ is

$$F(X,Y,Z) = \int_{-\infty}^{\infty} \mathrm{d}x \int_{-\infty}^{\infty} \mathrm{d}y \int_{-\infty}^{\infty} \mathrm{d}z\, f(x,y,z) \exp[2\pi i(Xx+Yy+Zz)] \tag{12.26}$$

That is, an electron microscopic image gives the information about a section in the reciprocal space. If the structure of the object has adequate symmetry, a small number of sections can give a reasonable amount of information on $F(X, Y, Z)$, and the three-dimensional structure of the object $f(x, y, z)$ may be obtained by the inverse Fourier transformation of F

(*X*, *Y*, *Z*).

In the case of a two-dimensional crystal, if the crystal plane is assumed to be the *yz* plane, its Fourier transform *F* is discontinuous in the plane defined by $X =$ const. and continuous in the *X* direction. Information on *F* can, therefore, be obtained from many images of various electron beam incidence by Fourier transformation, the tilt method being often applied for this purpose (Fig. 12.27). If the crystal is one-dimensional with helical symmetry, a three-dimensional structure may be reconstructed from an image perpendicular to the screw axis. In the case of a zero-dimensional crystal such as spherical viruses having high point symmetry, a small number of images enable us to reconstruct the structure.

Electron microscopic images have a number of distortions due to various "noises": instability and discreteness of the beam, instability of the power supply of the lens, mechanical vibrations, the effect of extraneous fields, the inhomogeneity of photographic emulsion, and so on. In addition there are instrumental errors in measurements of electron micrographs. Quantitative data on the image in the digital or graphical form may be obtained using computer-controlled densitometers. Corrections of systematic instrumental effects, lens aberrations, defocusing, and phase constrast can be made.

Electron microscopic images can be improved experimentally by using optical diffraction and filtering technique or by using a computer.

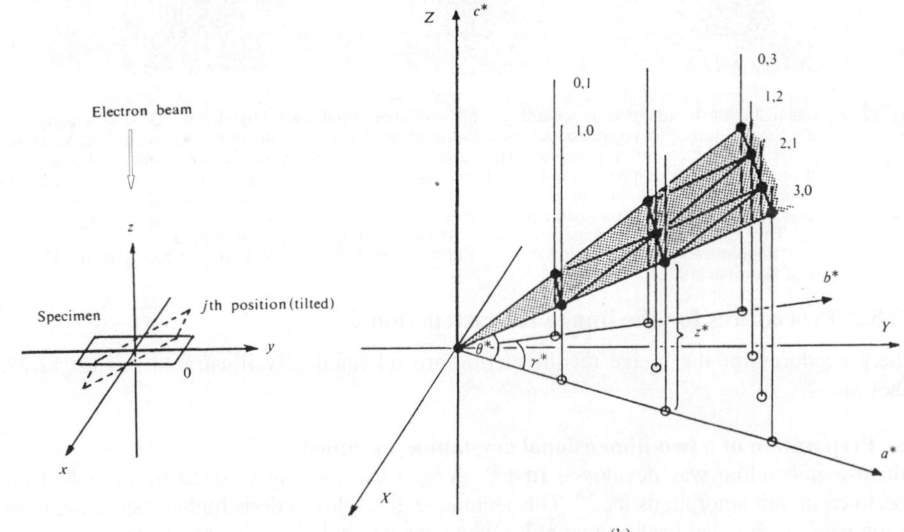

Fig. 12.27 Tilt method.[83)]
(a) The specimen is placed on the plane (defined by the *y* and *z* axes), and ordinarily the electron microscopic image is taken in this position with the electron bram perpendicular to this plane (0 direction).
(b) The Fourier transform of the image corresponds to the plane $X = 0$ in the reciprocal space (0 section).
Electron microscopic images from directions 1, 2,... and 1´, 2´,... respectively correspond to sections 1, 2,... and 1´, 2´,... in the reciprocal space. These provide information around the central section and enable us to reconstruct the three-dimensional structure in real space by inverse Fourier transformation.
(b): [Reproduced with permission from C. Toyoshima, *Protein, Nucleic Acid, Enzyme* (Tokyo), **37**, 1279, Kyoritsu (1992)]

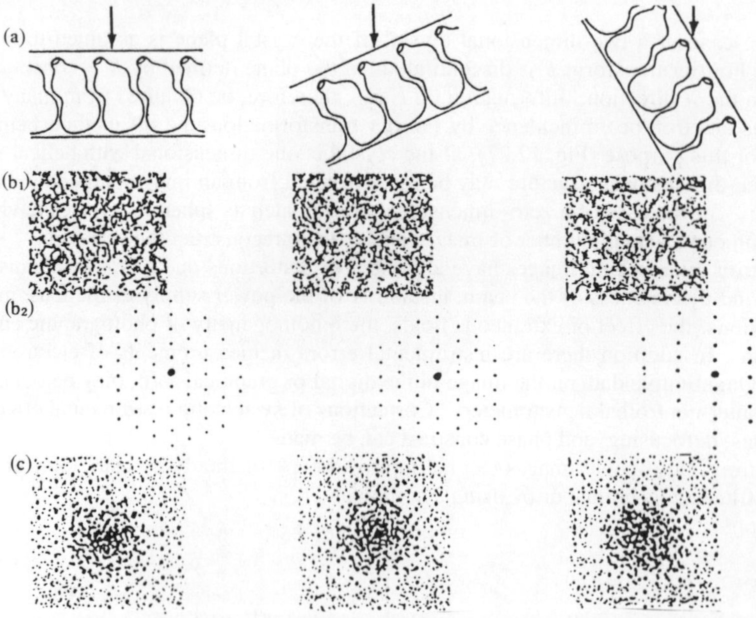

Fig. 12.28 Schematic illustration of structure analysis of two-dimensional crystal by electron crystallography.[6)]
(a) Preparation of two-dimensional crystal of macromolecules for electron microscopy: the specimen is tilted at three different angles. The arrow shows the direction of electron beam incidence.
(b_1) Electron micrographs of the two-dimensional crystal obtained with various incident angles of electron beam. (Because of quantum noise the S/N ratio of the images is generally not good).
(b_2) Corresponding electron diffraction patterns.
(c) The best part of each image is selected with the aid of optical transformation.
[Reproduced with permission from Y. Fujiyoshi, *Protein Nucleic Acid, Enzyme* (Tokyo), **37**, 561, Kyoritsu (1992)]

12.8.2 Procedures for the image reconstruction

The procedures for the image reconstruction are schematically illustrated in Fig. 12.28. They are:

A. Preparation of a two-dimensional crystalline specimen

Glucose embedding was developed first.[83)] Very rapid cooling enabled us to embed the specimen in thin amorphous ice.[84)] This technique gives diffractions higher than 20 Å resolution with much lower background noise than conventional glucose embedding.

In order to make the specimen visible to the TEM it is necessary to. add heavy metals as an electron dense contrasting medium by either ordinary or negative staining.

Bulk specimens such as muscle may be embedded, sectioned, and stained.

B. Electron micrographs and electron diffraction patterns

a) Electron micrographs. For the reconstruction of the three-dimensional structure of the specimen, numbers of electron microscopic images of the specimen with various tilt angles to the electron beam are needed. The necessary number of images will increase when the desired resolution of the structure to be reconstructed is high, the symmetry of the crystal is low, and the crystal is thick. For example, if the resolution desired is 20 Å and the thick-

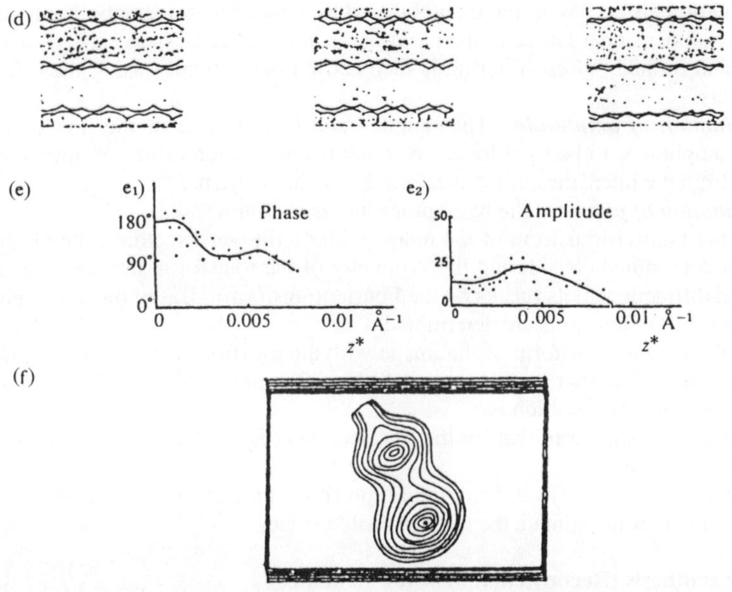

Fig. 12.28 (*Continued*)
(d) Density measurement of the selected part of each image, followed by Fourier transformation. refinement of the unit cell parameters, and noise-filtering.
(e) Phase determination.
The origin of the phase is first determined on the Fourier transforms of the image with zero-section. The determination proceeds in order from the Fourier transform of the images taken with the least tilt (1st section), the second least tilt (2nd section), the third least tilt (3rd section), etc. to those taken with the largest tilt angle. Amplitudes obtained are converted to the same scale. Better values of amplitudes may be obtained from the intensities of electron diffraction spots. Optimum phases(e_1) and amplitudes(e_2) are then determined plotting against z^*.
(f) The three-dimensional structure is obtained by a procedure similar to the Fourier synthesis in the case of crystal structure analysis.
[Reproduced with permission from Y. Fujiyoshi, *Protein Nucleic Acid, Enzyme*, **37**, 561, Kyoritsu (1992)]

ness of the specimen is *ca.* 100 Å, it is necessary to record images with tilt angles θ of every 10° from $-60°$ to $+60°$ along one direction. For the crystal with lower symmetry many more images such as another set of images along a different direction are needed.

As the electron microscopic image is a projection of the structure, the Fourier transform of an image corresponds to the section passing through the origin in the reciprocal space (Fourier space). The Fourier transform of the image with no tilt gives the Fourier components on the basal plane (zero section), and that of the image with any tilt angle θ gives those on the plane tilted with $\theta^*(= 90° - \theta)$ (Fig. 12.28).

For the symmetry of plane groups refer to the *International Tables for Crystallography*. Vol. A.

b) Electron diffraction patterns. At each tilt angle for the electron microscopic image the electron diffraction pattern of the corresponding part is recorded for auxiliary use.

C. Data processing

a) Selection of the best part in each electron microscopic image. The part of the image where the details of the structure show up best is selected by optical diffraction patterns.

b) Sampling of the optical density. Optical density of the selected part of the image is measured and digitalized. With the map obtained by Fourier transformation unit cell parameters of two-dimensional lattice in the basal plane in the Fourier space, a^* b^*, and γ^* are determined and refined. Noise-filtering is then carried out optically using the refined unit cell parameters.

c) Determination of amplitude. The sampled and filtered data of the image are transformed into amplitudes to be used for the reconstruction. Better values of amplitudes may be obtained from the intensities of the electron diffraction patterns.

d) Determination of phases. The basic procedure is as follows:

1) On the Fourier transform of the image with no tilt (zero section), the origin of the phase is first determined considering the symmetry of the transform and positions of symmetry-related diffraction spots (peaks on the Fourier transform). Based on this origin, phases of the other diffraction spots are determined.

2) On the Fourier transform of the image with the tilt (first section), the origin of this section is determined so that the phases of all the diffraction spots are close to the corresponding spots of the zero section.

3) A procedure similar to that for the first section is extended to the second, third, and higher sections in order.

4) As values of the phase and amplitude spread in some extent, both of them are plotted against z^* in order to estimate the most plausible values.

D. Fourier synthesis (Reconstruction of the structure)

A Fourier synthesis is finally carried out taking into account the symmetry of the object[†] and building up a three-dimensional map of the structure. The density at each point on this map is made up of all the different images of the object.

12.9 Structural Study of Macromolecules in Solution—NMR Investigations—[85,86)]

Recent development of the nuclear magnetic resonance (NMR) spectroscopy offers excellent information on the structure of macromolecules and also provides a powerful means of elucidating structure-function relationships in solution. NMR spectroscopy of small proteins has advanced to the point where one can routinely obtain site-specific information such as conformational changes, apparent pK_a values, hydrogen-bonding geometry, and side-chain mobility. These are mainly due to the progress of two-dimensional Fourier transform NMR, and especially the use of nuclear Overhauser effect (NOE) data and distance geometry algorithm.[87–91)] They are also supported by the development of multiple-quantum techniques, sequence-specific assignment of NMR data, and heteronuclear two-dimensional Fourier transform techniques, and others. Instead of distance geometry, application of molecular dynamics (*cf.* Section 12.7.2) is also carried out.[92,93)]

Figure 12.29 shown is an example of the molecular structure of *Tendusmit* determined by the use of NOE data and distance geometry algorithm,[94)] compared with the structure determined by X-ray diffraction.[95)] A good coincidence can be seen between the two structures with the exception of the N terminus.

† For example, for a helical particle the three-dimensional Fourier program takes the form of a Fourier-Bessel synthesis into which the appropriate helical parameters can be inserted.

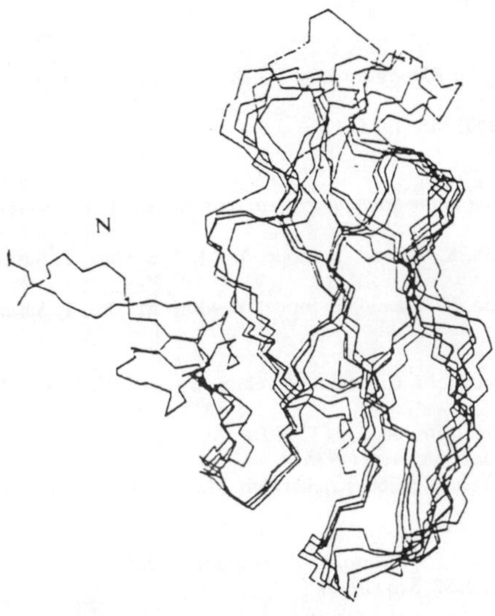

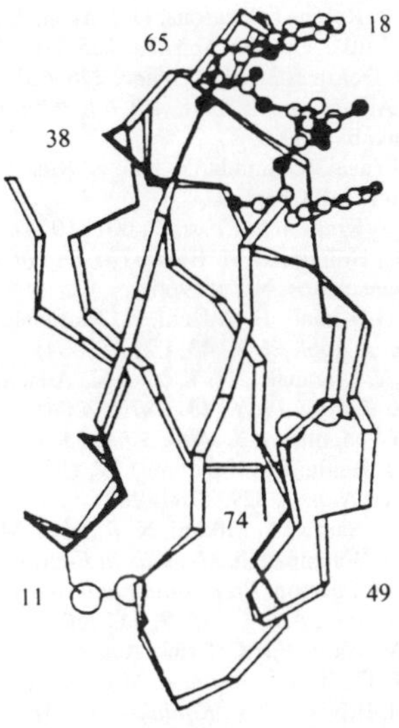

Fig. 12.29 Molecular structure of *Tendusmit*.
(a) Structure in solution obtained by NMR studies.[94] Structures obtained from four independent initial structures generated randomly. Convergence of the *N* terminus is not superimposed.
(b) Determined by means of X-ray diffraction.[95]
[Reproduced with permission from (a): A. D. Kleine *et al.*, *J. Mol. Biol.*, **189**, 377, Academic Press (1986), (b): J. W. Pflugrath e*t al., J. Mol. Biol.*, **189**, 383, Academic Press (1986)]

References

1. a) T.L. Blundell, L.N. Johnson, *Protein Crystallography*, Academic Press, London (1976).
   b) *International Tables for Cryotallography*, Vol. F, Cryotallography of Biological Macromolecules. (M. G. Rossmann, E. Arnold eds.), Kluwer Academic Pub., Dordrecht (2001).
2. a) A. McPherson, *Preparation and Analysis of Protein Crystals*, John Wiley & Sons, Inc., N.Y. (1982).
   b) W.A. Hendrickson. *X-Ray Diffraction in Protein Engineering*, pp.5–13. Alan R Liss. Inc. (1987).
3. a) A.A. Green *J. Biol. Chem.*, **93**, 495 (1931).
   b) A.A. Green *J. Biol. Chem.*, **95**, 47 (1932).
4. a) A.A. Green, W.L. Hughs, *Methods in Enzymology*, Vol. I. Academic Press, N.Y. (1955).
   b) E.T. Cohn, J.T. Edsall, *Proteins, Amino Acids and Peptides*, Rheinhold Pub., N.Y. (1943).
5. *J. Crystal Growth*, **76**, 535 (1986).
6. A. McPherson, *J. Cryst. Growth*, **110**, 1 (1991).
7. M. Zeppezauer, *Methods in Enzymology*, **22**, 253, Academic Press, N.Y. (1971).
8. D.R. Davis, D.M. Segal, *Methods in Enzymology*, **22**, 266, Academic Press, N.Y. (1971).
9. A. Wlodawer, K.O. Hodgson, *Proc. Natl. Acad. Sci. U.S.A.*, **72**. 398 (1975).
10. P. Kim, S.H. Vollmer, F.E. Freeman, *J. Biol. Chem.*, **259**, 3318 (1984)
11. N.E. Newcomwe, A. Liljas, J. Sundelin, L. Rask, P.A. Peterson. *J. Biol. Chem.*, **259**, 5230 (1984).
12. B. Shaanam, M. Shoham, A. Yonath, H. Lis, N. Sharon, *J. Mol. Biol.*, **174**, 723 (1984).
13. C. Thaller, L.H. Weaver, G. Eichele, E. Wilson, R. Karlsson, J.N. Jansonius, *J. Mol. Biol.*, **147**, 465 (1981).
14. F.R. Salemme, *Arch. Biochem. Biophys.*, **151**, 533 (1972).
15. D.W. Banner, G. Cesareni, D. Tsernoglou, *J. Mol. Biol.*, **170**, 1059 (1983).
16. a) M.J. Cox, P.C. Weber, *J. Appl. Cryst.*, **20**, 366 (1987).
    b) D.W. Morris, C.Y. Kim, A. McPherson, *Bioteçhniques*, **7**, 522 (1989).
    c) K.B. Ward, M.A. Prezzo, W.M. Zuk, *J. Cryst. Growth*, **90**, 325 (1988).
    d) W.M. Zuk, K.B. Ward, *J. Cryst. Growth*, **110**, 148 (1991).

e) B. Rubin, J. Talafous, D. Larson, *J. Cryst. Growth*, **110**, 156 (1991).
17. W. Littke, C. John, *Science*, **225**, 203 (1984).
18. L.J. DeLucas *et al.*, *Science*. **246**, 651 (1989).
19. S. Aibara *et al.*, *Abst. 4th International Conference of Biophysics and Synchrotron Radiation*. p. 184, Tsukuba (1992).
20. M. Yasui, S. Harada, Y. Kai, N. Kasai, *J. Mol. Biol.*, **177**, 845 (1984).
21. In ref. 1a), Chapt. 8.
22. M.V. King, *Acta Cryst.*, **7**, 601 (1954).
23. K.C. Holmes, D.M. Blow, *The Use of X-ray Diffraction in the Study of Protein and Nucleic Acid Structure*, Insterscience. N.Y. (1966).
24. a) H. Kitano, H. Adachi, A. Muranaka, H. Matsumura, K. Takano, T. Inoue, Y. Mori, S. Owa, T. Sasaki, *Jpn. J. Appl. Phys.*, **43**, L297 (2004).
e) A. Muranaka, H. Kitano, H. Adachi, H. Matsumura, K. Takano, T. Inoue, Y. Mori, M. Doi, T. Sasaki, *Jpn. J. Appl. Phys.*, **43**, L873 (2004).
25. B.W. Matthews, *J. Mlol. Biol.*, **33**, 491 (1968).
26. a) J. Haidu. P.A. Machin. J.W. Cambell. T.J. Greenhough, I.J. Clifton. S. Zurek. S. Gover, L.N. Jonson, M. Elder, *Nature*, **329**, 178 (1987).
b) Y. Satow, A. Mikuni, N. Kamiya, M. Ando, *Rev. Sci. Instr.*, **60**. 2394 (1989).
27. K.D. Watenpaugh. *Methods in Enzymology*, **115**, 3, Academic Press (1985).
28. A. McPherson, *Preparation and Analysis of Protein Crystals*, p.260, Krieger Pub. Malabar, Florida (1983).
29. D. Harker, *Acta Cryst.*, **9**, 1 (1956).
30. D.M. Blow, F.H.C. Crick, *Acta Cryst.*, **12**, 794 (1959).
31. A.F. Cullls. H. Muirhead, M.F. Perutz, M.G. Rossmann, A.C.T. North, *Proc. Roy. Soc.*, **A265**, 161 (1961).
32. J.M. Bijvoet, *Proc. Koninkl. Ned. Akad., Wetenschap (b)*, **52**, 313 (1949).
33. M.F. Perutz, *Acta Cryst.*, **9**, 867 (1956).
34. D.M. Blow, *Proc. Roy, Soc.*, **A247**, 302 (1958).
35. M. Yasui. M. Harada, Y. Kai, N. Kasai, J. *Biochem.*, **98**. 77 (1985).
36. a) H. Einspahr, K. Suguna, F.L. Suddath, G. Ellis, J.R. Helliwell, M.Z. Papiz, *Acta Cryst.*, **B41**, 336 (1985).
b) Y. Kitagawa, N. Tanaka, Y. Hata, Y. Katsube, Y. Satow, *Acta Cryst.*, **B43**, (1987).
37. S. Harada, M. Yasui, K. Murakawa, N. Kasai, Y. Satow, *J. Appl. Cryst.*, **19**, 448 (1986).
38. A.C.T. North, D.C. Phillips, F.S. Mathews, *Acta Cryst.*, **A24**, 351 (1968).
39. a) M. Yasui. S. Harada, Y. Kai, N. Kasai, M. Kusunoki. Y. Matsuura, *J. Biochem.*, **111**, 317 (1992).
b) M. Yasui, S. Harada, Y. Kai, N. Kasai, M. Kusunoki, Y. Matsuura, *J. Biochem.*, **111**, 317 (1992).
40. For example, *The Molecular Replacement Method*, (M.G. Rossmann ed.), Gordon & Breach, N.Y. (1972).
41. Protein Data Bank, Brookhaven National Laboratory, Upton, N.Y., U.S.A.
42. K. Miki, S. Sogabe, A. Uno, T. Ezoe, N. Kasai, M. Saeda, Y. Matsura, M. Mik, *Acta Cryst.*, **D50**. 271, (1994).
43. T. Takano, R.E. Dickerson. *J. Mol. Biol.*, **153**, 79, 95 (1981).
44. Y. Matsuura. *J. Appl. Cryst.*, **24**. 1063 (1991).
45. R.A. Crowther, In ref. 40, p. 173.
46. N. Tanaka, *Acta Cryst.*, **A33**, 191 (1977).
47. J.L. Sussman, S.R. Holbrook, G.M. Church, S.H. Kim, *Acta Cryst.*, **A33**, 800 (1977).
48. T.A. Jones, *J. Appl. Cryst.*, **11**, 268 (1978); T.A. Jones in *Computational Crystallography* (D. Sayer ed.) , p.303, Oxford Univ. Press, London (1982).
49. A.T. Brünger, M. Kaplus, G.A. Petsko, *Acta Cryst.*, **A45**, 50 (1989).
50. J.L. Sussman, S.R. Holbrook. S.R. Church, S.H. Kim, *Acta Cryst.*, **A29**, 183 (1973).
51. A. Jack, M. Levitt, *Acta Cryst.*, **A34**, 931 (1978).
52. J.H. Konnert, W.A. Hendrickson. *Acta Cryst.*, **A36**, 344 (1980).
53. D.S. Moss, A.J. Morffew, *Comput. Chem.*, **6**, 1 (1982).
54. W.A. Hendrickson, *Methods Enzymol.*, **115**, 252 (1985).
55. D.E. Tornued, L.F. Ten Eyck, B.W. Matthews, *Acta Cryst.*, **A43**. 489 (1987).
56. T.A. Jones, *J. Appl. Cryst.*, **11**, 268 (1978).
57. E. Prince, L.W. Finger, J.H. Kounert, in *International Tables for Crystallography*, Vol. C, 2nd ed. Mathematical, Physical, Chenical Tables (A.J.C. Wilson, E. Prince eds.), P.687, Kluwer Academic Pub., Dordrecht (1999).
58. W.A. Hendrickson, J.H. Konnert, in: *Computing in Crystallography* (R. Diamond, R. Rameseshan, K. Venkatesan eds.), p. 13.01 . Indian Acad. Sci., Bangalore, India (1980).
59. W.A. Hendrickson, J.H. Konnert, in: *Biomoleclular Structure, Function, Conformation and Evolution* (R. Srinivasan ed.), Vol. 1, p. 43, Pergamon, Oxford (1981).
60. A.T. Brünger, J. Kuriyan, M. Karlplus, *Science*, **235**, 458 (1987).
61. V. Shoemaker, J. Waser, R.E. Marsh, G. Bergman, *Acta Cryst.*, **12**, 600 (1959).

62. V. Luzatti, *Acta Cryst.*, **5**, 802 (1952).
63. A.T. Brünger, K. Kuriyan, M. Karplus, *Science*, **235**, 458 (1987).
64. A.T. Brünger, G.A. Petsko, M. Karplus, *Acta Cryst.*, **A45**, 50 (1989).
65. A.T. Brünger, *J. Mol. Biol.*, **203**, 803 (1988).
66. J. Kuriyan, A.T. Brünger, M. Karplus, W.A. Hendrickson, *Acta Cryst.*, **A45**, 396 (1989).
67. P. Gros, M. Fujinaga, B. Dijkstra, K.H. Kalk, W.G.J. Hol, *Acta Cryst.*, **B45**, 488 (1989).
68. S.H. Northrup, J.A. McCammon, *Biopolymers*, **19**, 1001 (1980).
69. For example, B.R. Brooks, R.W. Paster, F.W. Carson, *Proc. Natl. Acad. Sci. U.S.A.*, **84**, 4470 (1987).
70. R. Kaptein, E.R.P. Zuiderwerg, R.M. Scheek, R. Boelens, W.F. van Gunsteren, *J. Mol. Biol.*, **182**, 179 (1985).
71. A.T. Brünger, G.M. Clore. A.M. Gronenborn, M. Karplus, *Proc. Natl. Acad. Sci. U.S.A.*, **83**, 3801 (1986).
72. H.J.C. Berenden. J.P.M. Postma, W.F. van Gunsteren. A. Dinola, J.R. Haak, *J. Chem. Phys.*, **81**, 3684 (1984).
73. S. Lifson, P. Stern, *J. Chem. Phys.*, **77**, 4542 (1982).
74. M. Levitt, *J. Mol. Biol.*, **168**, 595 (1983).
75. B.R. Brooks, R.E. Bruccoleri, B.D. Olafson, D.J. State, S. Swaminathan, M. Karplus, *J. Comput. Chem.*, **4**, 187 (1983).
76. G. Nemethy, M.S. Pottie, H.A. Scherage, *J. Phys. Chem.*, **87**, 1883 (1983).
77. J. Hermans, H.J.C. Berendsen, W.F. van Gunsteren, J.P.M. Postma, *J. Mol. Biol.*, **23**, 1513 (1984).
78. L. Nilson, M. Karplus, *J. Comput. Chem.*, **7**, 591 (1986).
79. S.J. Weiner, P.A. Kollman, D.T. Guyen, D.A. Case, *J. Comput. Chem.*, **7**, 230 (1986).
80. D.J. de Rosier, A. Klug, *Nature* (London), **217**, 130 (1968).
81. R.A. Crowther, A. Klug, *Ann. Rev. Biochem.*, **44**, 161 (1975).
82. P.N.T. Unwin, R. Henderson, *J. Mol. Biol.*, **94**, 425 (1975).
83. Y. Fujiyoshi, *Protein, Nucleic Acid and Enzyme*, Supplement **20**, 560 (1993) (in Japanese).
84. R. Henderson, J.M. Baldwin, T.A. Ceska, F. Zemlin, E. Beckmann, K.H. Downing, *J. Mol. Biol.*, **213**, 899 (1990).
85. M. Adrian, J. Dubochet, J. Lepault,A.W. Dowall, *Nature*, **398**, 32 (1984).
86. For example, J.L. Markley, in: *Protein Engineering*, p. 15. Alan R. Liss, Inc. (1987).
87. T.F. Havel. I.D. Kuntz, G.M. Crippen, *Bull. Mathemat. Biol.*, **45**, 665 (1983).
88. G.M. Crippen, T.F. Havel, *Acta Cryst.*, **A34**, 282 (1978).
89. T.F. Havel, K. Würthrich, *J. Mol. Biol.*, **182**, 281 (1985).
90. W. Braun, *Quarterly Rev. Biophys.*, **19**, 115 (1987).
91. W. Braun, N. Go, *J. Mol. Biol.*, **186**, 611 (1985).
92. R. Kaptein, E.R.D. Zunderweg, R.E. Scheek, R. Boeleus, W.F. von Gunsteren, *J. Mol. Biol.*, **182**, 179 (1985).
93. G.M. Clore, A.T. Brünger, M. Karplus, A.M. Gronenborn, *J. Mol. Biol.*, **191**, 523 (1986).
94. A.D. Kline. W. Braun, K. Würthrich, *J. Mol. Biol.*, **189**, 377 (1986).
95. J.W. Pflugrath, E. Wiegand, R. Huber, L.J. Vertesy, *J. Mol. Biol.*, **189**, 383 (1986).

13. Analysis of the Breadth and Shape of Diffraction Patterns

The broadening of X-ray diffraction patterns observed consists of instrumental broadening (broadening due to observation and recording conditions) and true broadening. As shown schematically in Fig. 13.1, the true broadening of the diffraction patterns is determined by two factors: 1) broadening due to the crystallite[†] size (*cf.* Fig. 2.21) and/or 2) broadening due to the lattice distortion inside the crystallites. Fig. 13.2 gives an example of powder diffraction photographs.[1] Line profiles show remarkable changes due to crystallite size.

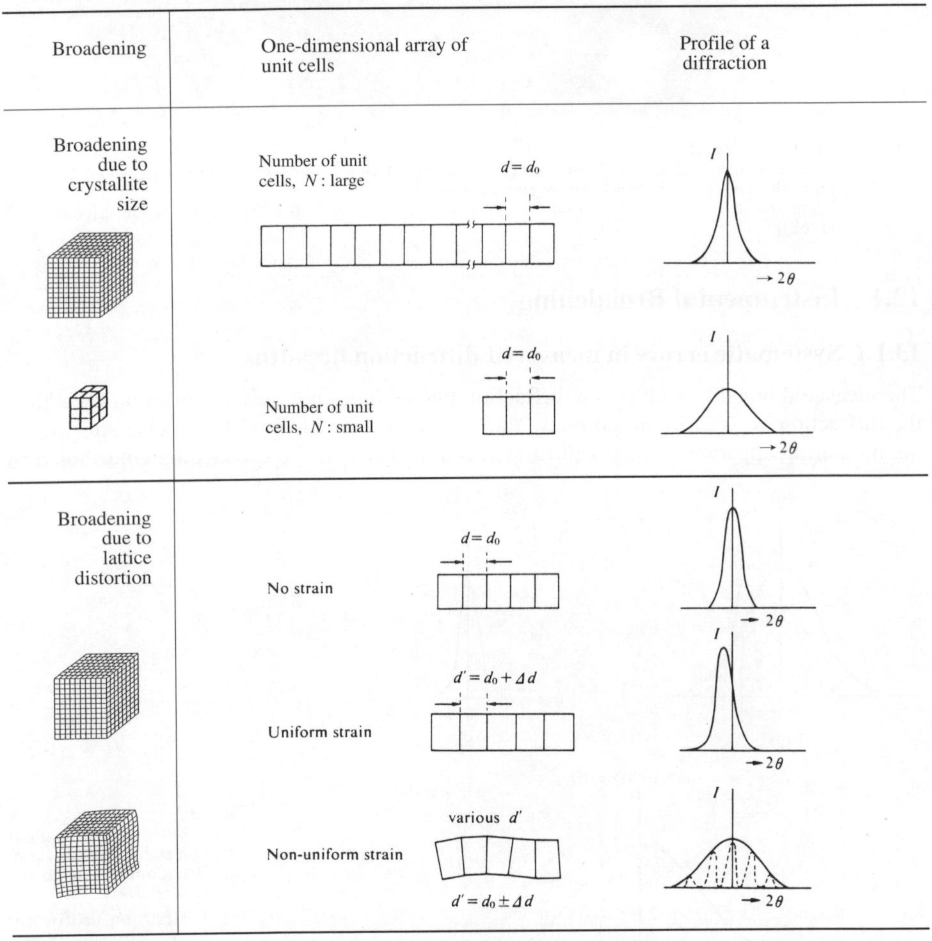

Fig. 13.1 Schematic illustration of the broadening of the diffraction pattern. [Reproduced from B.D. Cullity, *Elements of X-Ray Diffraction*, p.266, Addison-Wesley Pub. (1956)]

† In this chapter, as in Chapter 10, "crystallite" means "crystallite" and "crystalline region" (or "crystalline part").

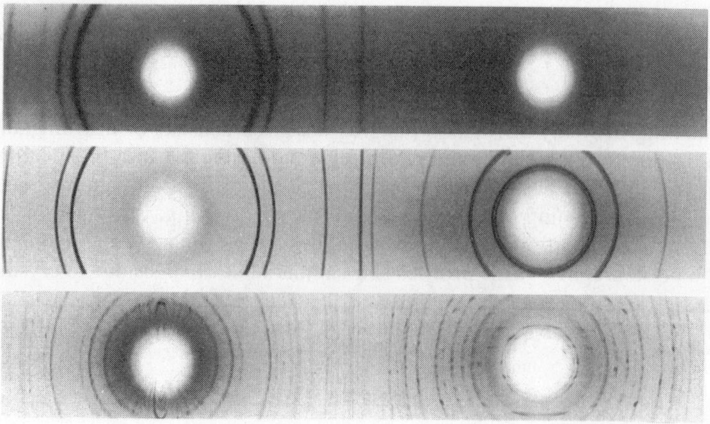

Fig. 13.2 Crystallite-size effects.[1)]
(a) Raney nickel, about 10^{-6} cm average diameter; (b) nickel powder of "ideal" size, about 10^{-3} cm; (c) rutile, TiO_2 powder insufficiently ground, about 10^{-2} cm
[Reproduced with permission from L.V. Azároff, *Elements of X-Ray Diffraction*, p.550, McGraw-Hill (1968)]

13.1 Instrumental Broadening

13.1.1 Systematic errors in measured diffraction breadths

The measured breadth (width) of a diffraction line includes not only the inherent breadth of the diffraction $G(S)$ (or more correctly $I_{th}(S)$), but also broadening that is characteristic of the measuring apparatus. This is illustrated in Fig. 13.3. If the broadening contributed by

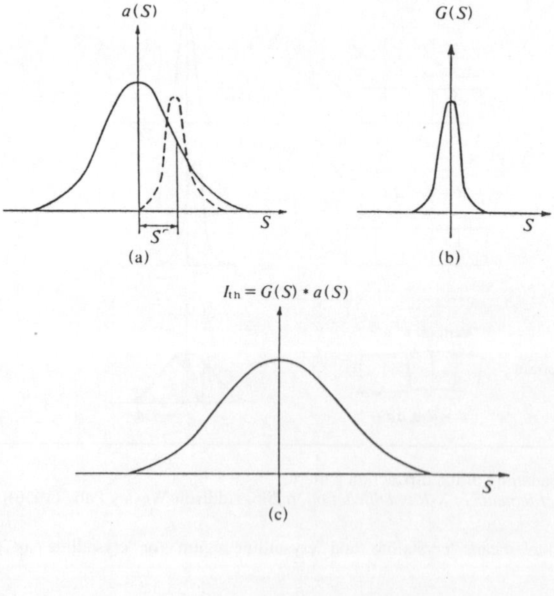

Fig. 13.3
Contributions to the measured breadth of a diffraction.
(a) Broadening due to the apparatus.
(b) Diffraction profile for ideal pinhole X-ray beam.
(c) Actual broadening of the diffraction.

the apparatus alone is of the form $a(S)$ (the instrumental broadening function), and that due only to the very small but otherwise perfect crystals is of the form $G(S)$, the overall broadening (for the given crystals with the given apparatus) will be of the form $G(S) * a(S)$. This is an application of the convolution operator described in Section 5.2.1, which is here used to average a function $G(S)$ over a function $a(S)$ for all values of $G(S)$. The intensity is given by

$$I_{\text{th}}(S) = \int_{-\infty}^{\infty} G(S-S')a(S')dS' = G(S) * a(S) \tag{13.1}$$

13.1.2 Methods of correcting the line profile

A. Reference method (including $K\alpha_1$, $K\alpha_2$ doublet correction)

The instrumental broadening function $a(S)$ is determined from a reference line of a standard specimen which gives no broadening due to particle size (*i.e.* which consists of particles of effectively infinite size). Debye-Scherrer lines of the standard specimen are produced under the same experimental conditions as the specimen under study. If the integral breadths of $I(S)$, $a(S)$, and $G(S)$ are

$$B = \frac{\int I(S)\mathrm{d}S}{I(0)}, \qquad b = \frac{\int a(S)\mathrm{d}S}{a(0)}, \qquad G = \frac{\int G(S)\mathrm{d}S}{G(0)}, \tag{13.2}$$

the correction curve, β/B *vs.* b/B can be constructed (Fig. 13.4).[2)] Then, if $a(S)$ and $G(S)$ can be approximated by Gaussian functions,

$$B^2 = \beta^2 + b^2 \tag{13.3}$$

while if $a(S)$ and $G(S)$ are Cauchy (Lorentz) functions†

$$B = \beta + b \tag{13.4}$$

Before this correction is applied, however, a further correction must be made: the monochromatic X-rays normally used contain both $K\alpha_1$ and $K\alpha_2$ radiations, and this presents

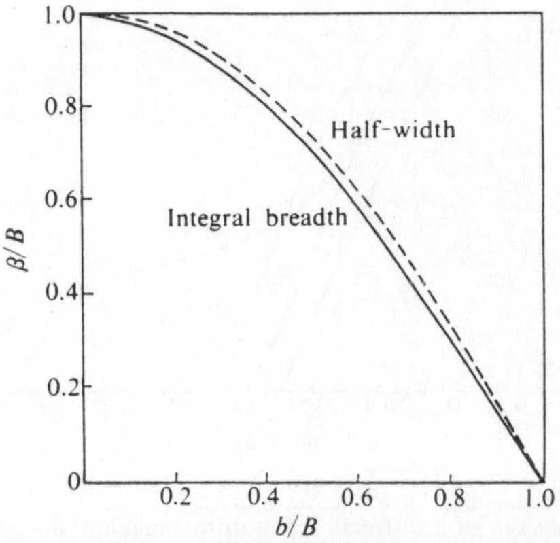

Fig. 13.4 Correction curve for the broadening due to the apparatus.[2)]
[Reproduced from F.W. Jones, *Proc, Roy, Soc.*, **A156**, 16, Royal Society (1938)]

† These B's are not to be confused with those expressing isotropic temperature factors.

a further source of broadening for which correction is necessary. The separation δ of the $K\alpha_1$ and $K\alpha_2$ diffractions increases with the diffraction angle 2θ as shown in Fig. 13.5.[2)]

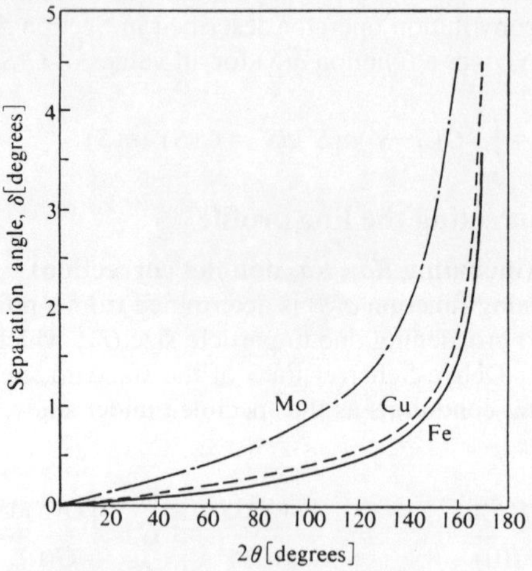

Fig. 13.5 Angular separation between $K\alpha$ doublet components for the principal X-rays *vs*. diffraction angle.[2)] [Reproduced from H.P. Klug, L.E. Alexander, *X-Ray Diffraction Procedures for Polycrystalline and Amorphous Materials*, p.505, John Wiley & Sons, Inc. (1954)]

Postulating that the $K\alpha_1$ and $K\alpha_2$ profiles are usually similar figures having an intensity ratio of 2 : 1, Jones showed that the sum of the two profiles divided by the height of the composite maximum is the observed integral breadth.[2)] Fig. 13.6 shows the correction curve

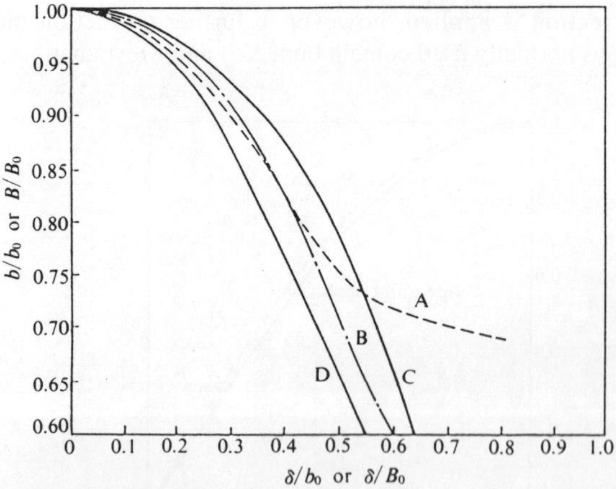

Fig. 13.6 Correction curve for the broadening due to the $K\alpha$ doublet.[2)]
A, Using the integral breadth of a diffraction due to α_1 radiation alone, for θ close to 80° (Jones' curve). B, Profile intermediate between C and D.
C, Profile approximates to Gaussian function.
D, Profile approximates to Cauchy (Lorentz) function between B and C.
[Reproduced from H.P. Klug, L.E. Alexander, *X-Ray Diffraction Procedures for Polycrystalline and Amorphous Materials*, p.504, John Wiley & Sons, Inc. (1954)]

obtained in this way. The values labeled 0 are the uncorrected experimental breadths.

B. Fourier method

The instrumental broadening function can be represented by a convolution

$$a(S) = a_1(S) * a_2(S) * \ldots * a_j(S) * \ldots \tag{13.5}$$

where $a_j(S)$ is a line broadening function due to a single special cause, such as finite width of the X-ray source, width of the counter slit, height of the source and of the counter slit, horizontal and vertical divergences, absorption, specimen orientation, misalignment of the slits, etc.[2)] The Fourier transform of Eq. 13.5 gives

$$A = \mathcal{F}[a(S)] = A_1 A_2 \ldots A_n = \prod_{j=1}^{n} A_j \tag{13.6}$$

from which an unknown broadening function can be obtained by inverse Fourier transformation.

$$a_k = \mathcal{F}^{-1}[A / \prod_{j \neq 1}^{n} A_j] \tag{13.7}$$

C. Variance method

If we denote the variance of the broadening function $a_j(S)$ by W_j,

$$W_j = \frac{\int_{-\infty}^{\infty} (S - \langle S \rangle)^2 a_j(S) \mathrm{d}S}{\int_{-\infty}^{\infty} a_j(S) \mathrm{d}S} \tag{13.8}$$

From Eq. 13.6 we obtain

$$W = \sum_{j=1}^{n} W_j$$

This also permits the isolation of the broadening due to a single cause. Some of the approximate functions (*e.g.* Lorentz function) encountered have infinite variances, which may be avoided by using truncated profiles. However, in the Fourier method, the truncated profile leads to a meaningless oscillation of the broading function.[3)]

13.2 Relationship between the Size and Shape of an Ideal Crystal and the Broadening of Its Diffraction Pattern

13.2.1 Broadening due to the Laue function

As mentioned in Section 2.9, the shape of the X-ray diffractions due to ideal crystals depends on the product of the structure factor and the Laue diffraction function. From Eq. 2.38, therefore

$$G(S) = \frac{\sin^2\{\pi N_1(\boldsymbol{S}\cdot\boldsymbol{a})\}}{\sin^2\{\pi(\boldsymbol{S}\cdot\boldsymbol{a})\}} \cdot \frac{\sin^2\{\pi N_2(\boldsymbol{S}\cdot\boldsymbol{b})\}}{\sin^2\{\pi(\boldsymbol{S}\cdot\boldsymbol{b})\}} \cdot \frac{\sin^2\{\pi N_3(\boldsymbol{S}\cdot\boldsymbol{c})\}}{\sin^2\{\pi(\boldsymbol{S}\cdot\boldsymbol{c})\}} \tag{13.9}$$

where G is the broadening of the reciprocal lattice point in reciprocal space, *i.e.* the broadening of the diffraction pattern (*cf.* Fig. 2.21). The broadening of the reciprocal lattice points can thus be made to yield a measure of the number of unit cells along the direction normal to the planes producing the diffraction in question (that is, the direction from the origin to the reciprocal lattice point for the planes) and subsequently an estimate of the corresponding crystallite size. Eq. 13.9 is the analytical expression of the function for a crystal in the form of a parallelepiped with N_1, N_2, and N_3 unit cells respectively in the directions of the principal axes. The geometrical characteristics of crystals with other, specific shapes (*e.g.* spheres, ellipsoids, etc.) influence the summation of the exponential terms in Eq. 2.38 in the Fourier synthesis of the lattice points, and lead to functions which differ in form from *G*, In these cases *G* takes a form similar to the scattering function (see Appendix, Table 5). The reader is referred to other publications[4,5] for further details, but with the exception of high polymers, which in almost every case exhibit significant lattice distortion (see Section 13.6.1), the function *G* is generally a satisfactory approximation.

13.2.2 Variation in the shape of diffractions with $|F|^2 \cdot G$

We now turn to another problem that arises in the analysis of the shapes of diffractions, particularly when considerable broadening is caused by imperfect or very small crystals. The shape of the diffractions is determined ultimately by the product of *G* and $|F|^2$, although hitherto in our treatment the dominant factor has been *G*. This dominance is effective as long as the variation of $|F|^2$ with diffraction angle is much slower than that of *G*.

It must not be forgotten, however, that $|F|^2$ may significantly affect the product (and therefore the shape of the diffractions) if, when *G* is fairly broad, it varies at all rapidly with diffraction angle.

A further complication, even in the absence of any rapid variation in $|F|^2$, can arise whenever the number of unit cells (*N* in Fig. 2.21) is very small and *G* correspondingly broad, for then the secondary maxima between the principal maxima at the reciprocal lattice points may be clearly distinguishable, and will appear to be diffractions from a separate reciprocal lattice. This effect will be discussed in Section 13.3.4 in connection with the diffraction pattern of Nylon 6,6.

13.3 Calculation of Crystallite Size from the Broadening of the Diffraction Pattern

13.3.1 The Scherrer formula

Assuming that the crystallite is a cube of edge length $L = Na$, Scherrer and Bragg gave the following approximation, based on the Laue diffraction function, for the broadening of diffractions:

$$\beta = \frac{K\lambda}{L\cos\theta} \tag{13.10}$$

where λ is the wavelength of the X-rays, and θ the Bragg angle. *K* is known as the Scherrer constant (Table 13.1).[6]

Table 13.1 Values for the Scherrer constant *K*[6]

Form of β	*K*
For the half-breadth $\beta_{\frac{1}{2}}$	0.9
For the integral breadth β (general)	1.05
Case of a disordered layer lattice (For irregularity in the drection of the *c* axis, the 00*l* dffractions are sharp, while the unsymmetrical *hk* diffractions are broadened. This value is for the *hk* diffractions.)	1.84

[Reproduced from B.E. Warren, *Phys. Rev.*, **59**, 693, Am. Inst, Phys. (1941)]

A. Jones'method[2)]

The background is first eliminated from the intensity distribution curves of the various diffractions, and any overlapping diffractions are resolved (*cf.* Fig. 13.3). The integral breadth B_0 of the various diffractions is then found from these corrected intensity curves. The integral breadths b_0 are also found for the various diffractions of a standard specimen. It is often necessary to convert the resulting integral breadths into the breadths due only to the $K\alpha_1$ radiation. This is achieved by finding the $K\alpha_1$, $K\alpha_2$ separation angle δ from Fig. 13.5 for the appropriate Bragg angle, and calculating δ/B_0 and δ/b_0. B/B_0 is found from δ/B_0 and b/b_0 from δ/b_0 with the aid of Fig. 13.6, and it is then possible to find the breadths B and b of the diffractions due only to the $K\alpha_1$ radiation, and to calculate b/B. A $b/B - \beta/B$ correction curve (*cf.* Fig. 13.4) for the conditions under which the apparatus operates is used to obtain β/B, from which the true breadth β of the diffraction is found.

B. Crystallite size of polythylene

An X-ray powder diffractometer, using Cu $K\alpha$ radiation (with Ni-Co balanced filters) at 30 kV and 20 mA, was used to estimate the dimensions of polyethylene crystallites in an undrawn specimen of polyethylene, assuming that the broadening of the diffraction lines might be taken as entirely attributable to small crystallite size. Breadth standards were derived from measurements on a standard quartz specimen under identical experimental conditions. The integral breadths of the diffraction lines were used. The analysis proceeded as follows.

1) The $K\alpha_1$, $K\alpha_2$ separation angles δ for the various diffraction angles were read off from Fig. 13.5 (Table 13.2, (a) and (b) column 4).

2) The values of b/b_0 and B/B_0 were found for δ/b_0 and δ/B_0 respectively from Fig. 13.6 (Table 13.2, column 6).

3) Using the b values for the standard specimen in the vicinity of the appropriate 2θ value, β/B corresponding to the b/B for the specimen under examination was found from Fig. 13.4 and used to calculate β (Table 13.2(b), columns 8–11).

Table 13.2 Estimation of crystallite size: typical calculation

(a) Standard quartz specimen

hk·l	2θ[deg]	b_0	δ	δ/b_0	b/b_0	b
10·0	20.8	0.21	0.07	0.333	0.88	0.185
10·1	26.5_5	0.21	0.07_2	0.343	0.88	0.185
11·0	36.4	0.21	0.11_7	0.557	0.73	0.15_3
20·1	45.7	0.21	0.14_3	0.681	0.70_5	0.14_3
11·2	50.0	0.21_5	0.15_9	0.739	0.69_8	0.15

(b) Polyethylene specimen

hkl	2θ[deg]	B_0	δ	δ/B_0	B/B_0	B
110	21.6_5	1.022	0.07_1	0.065	0.99_5	1.01_7
200	23.8	1.014	0.07_1	0.070	0.99_5	1.00_9
220	44.0	1.40_5	0.14_1	0.100	0.99_0	1.39_1
400	49.1_8	1.31_1	0.15_9	0.121	0.98_5	1.29_1

hkl	b/B	β/B	β[deg]	β[rad]	$\cos\theta$	L [Å]
110	0.18_2	0.96_4	0.980	0.0171	0.9822	96
200	0.18_3	0.96_4	0.973	0.0171	0.9785	96
220	0.10_6	0.98_4	1.36_9	0.0239	0.9272	73
400	0.11_6	0.97	1.25_2	0.0218	0.9092	81

4) The crystallite dimension L was calculated from Eq. 13.10. Since the integral breadth was used, the value used for K was 1.05 (*cf.* Table 13.1).

13.3.2 Effect of crystallite size distribution

When there is a distribution of crystallite sizes in the specimen, the broadening of the diffraction profile results from the combination of the profiles due to the various size fractions, and is generally of the form

$$I(S) = \int \frac{\sin^2\{\pi N(\boldsymbol{S}\cdot\boldsymbol{a})\}}{\sin^2\{\pi(\boldsymbol{S}\cdot\boldsymbol{a})\}} W(N)\boldsymbol{d}N \qquad (13.11)$$

where $[\sin^2\{\pi N(\boldsymbol{S}\cdot\boldsymbol{a})\}]/[\sin^2\pi(\boldsymbol{S}\cdot\boldsymbol{a})]$ is the one-dimensional diffraction function, and $W(N)\mathrm{d}N$ is the number of crystallites with sizes between N and $N + \mathrm{d}N$, The experimental $I(\boldsymbol{S})$ is therefore resolved by some suitable method into the $I(\boldsymbol{S}_j)$ due to the various crystallite size fractions (*cf.* Fig. 13.22(a)), and the resulting profile curves are analysed (*cf.* Section 13.6.2D).

13.3.3 Effect of crystallite shape

Powder diffraction diagrams caused by different crystal habits are pictured in Figs. 13.7 and 13.8[7] Diffractions in Fig. 13.7 give saw-toothed line shape due to one-dimensional array such as random layer lattices, regular interposition of amorphous materials, and others. Needle-like or plate-like crystals may give a diffraction pattern, schematically shown in Fig. 13.8. In the case of a needle crystal (needle axis: c axis), for example, among the broad hkl diffractions only the $00l$ diffraction will give sharp diffractions, whereas in the

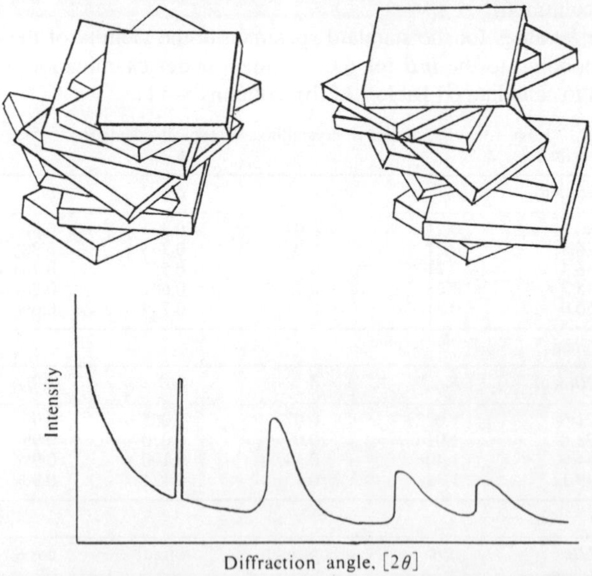

Fig. 13.7 Schematic illustration of a diffraction diagram with saw-toothed line shape caused by two-dimensional crystals, random layer lattices, regular interposition of amorphous material, or displacement of whole lattice layers.[7]
[Reproduced from F. Schossberger, *Adv. X-ray Anal.*, **1**, 85, Plenum Press (1957)]

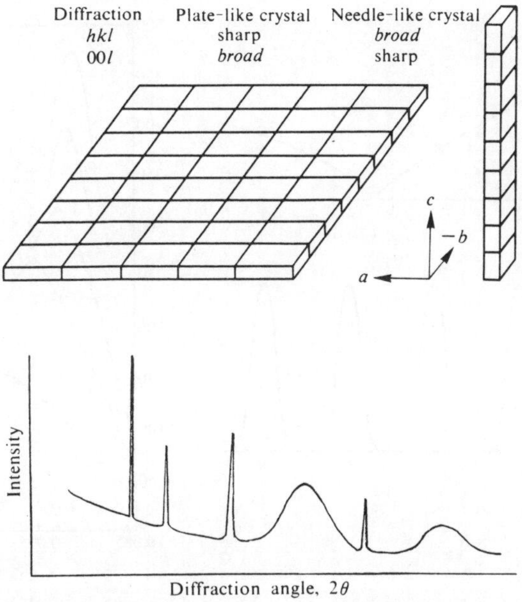

Fig. 13.8 Schematic illustration of selective line broadening in a diffraction diagram, caused by needle-like or plate-like crystals having large dimensions in one direction and small in the other.[7)]
[Reproduced from F. Schossberger, *Adv. X-ray Anal.*, **1**, 85, Plenum Press (1957)]

case of a plate-1ike crystal in which the *a* and *b* axes are in the plane of plate, among the sharp *hkl* diffractions broad 00*l* diffractions will be observed. Distinction between these needle- and plate-like crystals is possible because of selective broadening of the lines if the small dimensions are in the range of Scherrer broadening (20–1,500 Å).

13.3.4 Application to very small crystallites

When the crystallites are very small, consisting of only a few unit cells, the principal maxima of the Laue function (Fig. 2.16) exhibit considerable broadening. If $|F|^2$ varies at all rapidly with diffraction angle in the region of these maxima, the shape of the diffraction changes. The effect of the change is such that the position of the principal maximum of $I_{th} = |F|^2G$ is displaced in the direction of higher $|F|^2$ and *vice versa* (Fig. 13.9), which affects values of the cell constants determined.

A. 020 Diffraction of Nylon 6 (Perlon L)[8)]

An example is given in Fig. 13.10 of the broadening of 020 diffraction of Nylon 6 (Perlon L). Fig. 13.10 gives the the calculated variation of the position and approximate shape of the 020 diffraction of Nylon 6 as the number of unit cells in the direction of the *b* axis (fiber axis) varies from one to five. Wallner[8)] found that the maximum of the 020 diffraction moved from $\sin\theta = 0.095$ for a normal specimen to $\sin\theta = 0.092$ for a heat-treated specimen. On comparison with the calculated positions of the maxima in Fig. 13.10, he deduced that the crystallite size in the normal specimen corresponded to about two unit cells in the

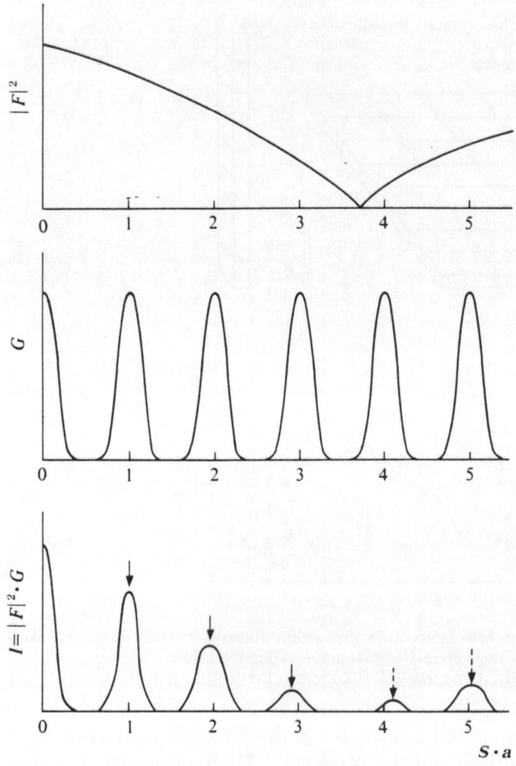

Fig. 13.9 Shifts of diffraction peakes.
When G is fairly broad, each diffraction peak shifts inside or outside from the peak position of G depending upon rapid decrease or increase of $|F|^2$.

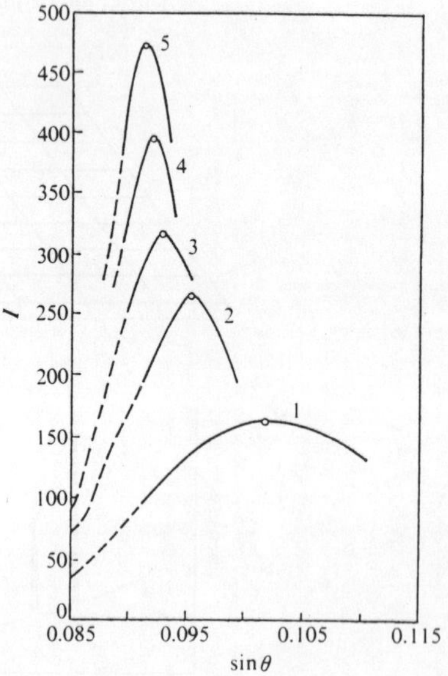

Fig. 13.10
Variation in the intensity of the 020 diffraction from Nylon 6 (Perlon L) for one to five unit cells in the b direction (fiber axis).[8)]
[Reproduced from L.G. Wallner, *Monatsh. Chem.*, **79**, 271, Springer Verlag (1949)]

fiber direction, while the heat treatment had resulted in crystal growth giving a crystallite dimension of about four unit cells.

B. 0k0 Diffraction of Nylon 12

Table 13.3 gives a result of a discrepancy of the fiber period (b axis) of Nylon 12 determined from various 0k0 diffractions at different temperatures. Values of the fiber period determined showed an increase for the higher order diffractions. A schematic representation of diffraction profiles at different drawing temperatures is given in Fig. 13.11. In Fig. 13.11, as a measure of the crystallite size the half breadth of each diffraction in 2θ, $\beta_{\frac{1}{2}}$ is shown by an arrow at the peak position, and also as a measure of the degree of crystallites orientation, the half breadth of each diffraction arc along the azimuthal (ψ) direction, $H_{\frac{1}{2}}$ is given by a vertical arrow. Fig. 13.12 depicts changes in diffraction profiles of Nylon 12 with draw ratio. With increase of the draw ratio slight shifts of the 0k0 diffraction peak and slight increase of crystallite size were noted, while improvement in the degree of crystallite orientation was observed.

Table 13.3 Fiber period determined of a Nylon 12 specimen

Drawing temp. [°C]	Draw ratio	Fiber period [Å]					
		020	040	060	080	0,10,0	0,12,0
room temp.	1.0	31.7_2					
50	4.0	29.2_6	30.3_4	30.6_5			
70	4.0	29.6_3	30.5_8	30.9_3			
100	4.0	29.9_7	30.9_4	31.2_6			
130	4.0	30.2_7	31.2_1	31.4_5	31.7_6	31.7_2	31.7_1
140	2.4	30.8_2	31.4_0				
140	3.0	30.7_8	31.4_0	31.3_9			
140	3.4	30.6_5	31.3_3	31.4_8	31.8_6		
140	4.0	30.6_7	31.3_4	31.4_9	31.8_2		
150	4.0	30.7_2	31.4_0	31.5_6	31.8_7	31.7_8	31.7_8
160	4.0	31.1_1	31.5_7	31.5_8			

[Reproduced with permission from T. Ishikawa *et al.*, *J. Chem. Soc. Jpn.*, p.1747, Chem. Soc. Jpn. (1973)]

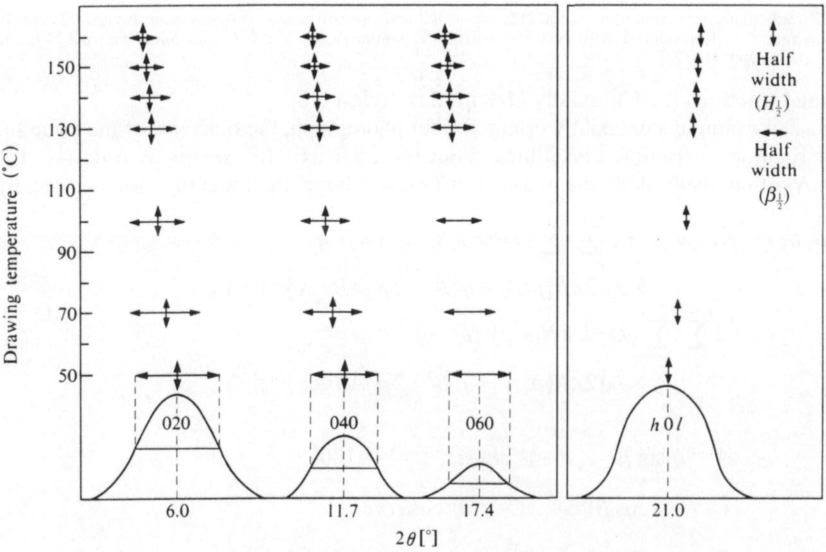

Fig. 13.11 Schematic representation of the change in 0*k*0 and *h*0*l* diffraction profiles from Nylon 12 with drawing temperature.[8] Vertical and horizontal arrows respectively show half widths, $H_{\frac{1}{2}}$ and $\beta_{\frac{1}{2}}$ measured along the ψ and 2θ directions of 0*k*0 and *k*0*l* diffractions.
[Reproduced with permission from T. Ishikawa, *et. al.*, *J. Chem. Soc. Jpn.*, p.1749, Chem. Soc. Jpn. (1973)]

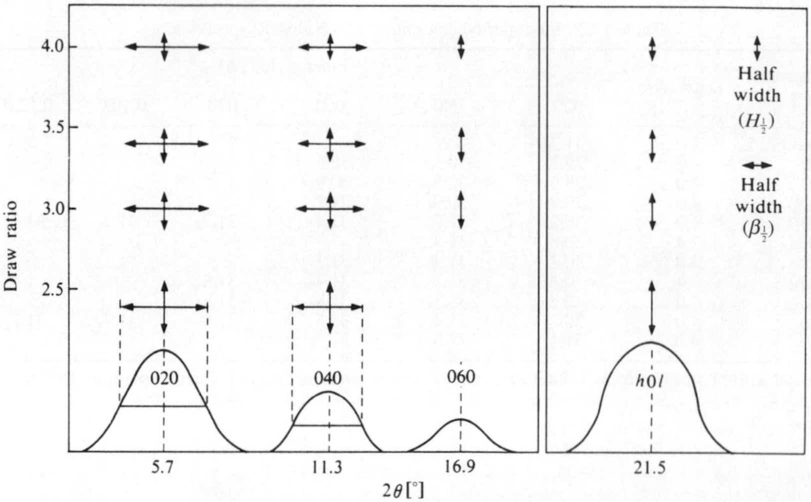

Fig. 13.12 Schematic representation of the change in $0k0$ and $h0l$ diffraction profiles from Nylon 12 with draw ratio.[8) [Reproduced with permission from T. Ishikawa, *et. al.*, *J. Chem. Soc. Jpn.*, p.1749, Chem. Soc. Jpn. (1973)]

C. Laue Functions for Uniaxially Oriented α-Nylon 6,6

Taking as an example a uniaxially oriented fiber photograph, the behavior of the Laue function for rotation of triclinic crystallites about the fiber axis (c axis) is as follows. If the number N_3 of unit cells along the c axis is sufficiently large, the Laue function is given by

$$\begin{aligned} G(R, l/c) = N_3{}^2[N_1N_2 &+ 2\sum_{p=0}^{N_1-1}\sum_{q=0}^{N_2-1}(D+E)(N_1-p)(N_2-q) \\ &\times J_0(2\pi R\{p^2A^2+q^2B^2-2pqAB\cos\gamma^*\}^{1/2}) \\ &+2\sum_{p=0}^{N_1-1}\sum_{q=0}^{N_2-1}(D-E)(N_1-p)(N_2-q) \\ &\times J_0(2\pi R\{p^2A^2+q^2B^2+2pqAB\cos\gamma^*\}^{1/2}) \end{aligned} \tag{13.12}$$

where

$$A = a\sin\beta, \quad B = b\sin\alpha, \quad \gamma^* = 180° - \gamma$$

$$C_1 = la(\cos\beta)/c, \quad C_2 = lb(\cos\alpha)/c$$

$$D = \cos 2\pi pC_1 \cos 2\pi qC_2, \quad E = \sin 2\pi pC_1 \sin 2\pi qC_2$$

a, b, c, α, β, and γ are the unit cell constants; N_1, N_2, and N_3 the number of unit cells in the directions of the a, b, and c axes respectively; l the order number of layer line in the direction of the fiber axis (the c axis); and R the distance from the fiber axis in the cylindrical coordinate system of the reciprocal lattice.

Figure 13.13 shows the Laue functions $G(R, l/c)$ found by Keller and Maradudin[9) with the aid of Eq. 13.12 for the second layer line of uniaxially oriented Nylon 6,6 (α form). Fig. 13.13(a) and (b) show the Laue functions for two-dimensional net planes with $N_1 = 1 \sim 7$, $N_2 = 1$, and $N_1 = 1$, $N_2 = 1 \sim 7$. These net planes correspond to the 010 and 100 planes respectively. Calculations were also carried out for three-dimensional lattices with

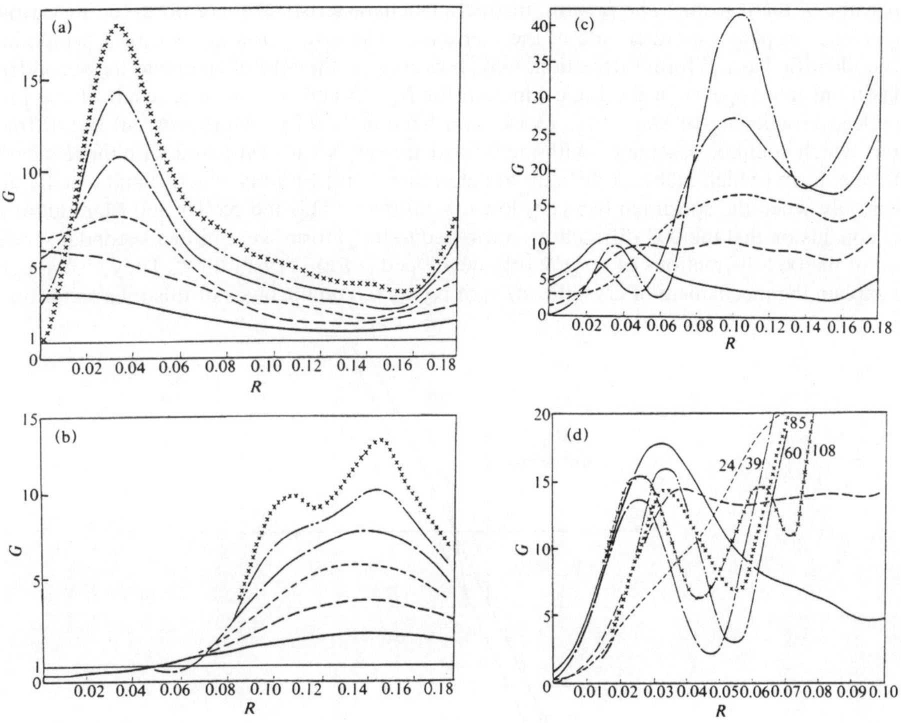

Fig. 13.13 Laue functions $G(R, l/c)$ for uniaxially oriented α form Nylon 6,6.[9]

(a) Two-dimensional net planes ($N_2 = 1$, $N_2 = 1 \sim 7$)
———, $N_1 = 1$ (*i.e.* G = const. = 1); ———, $N_1 = 2$; - - - - - -, $N_1 = 3$; ------, $N_1 = 4$;
—·—, $N_1 = 5$; —··—··—, $N_1 = 6$; ××××, $N_1 = 7$

(b) Two-dimensional net planes ($N_1 = 1$, $N_2 = 1 \sim 7$)
———, $N_2 = 1$ (*i.e.* G = const. = 1); ---------, $N_2 = 2$; - - - - - -, $N_2 = 3$; ·········, $N_2 = 4$;
—·—, N_2;$= 5$; —··—··—, $N_2 = 6$; ××××, $N_2 = 7$

(c) Three-dimensional lattice ($N_1 = 5$). The maximum at $R \approx 0.10$ corresponds to the 002 diffraction. Note the secondary maximum at $R \approx 0.03$
———, $N_2 = 1$; --------, $N_2 = 2$; ·········, $N_2 = 3$; —·—, $N_2 = 4$; —··—··—, $N_2 = 5$

(d) Three-dimensional lattice ($N_1 = 7$), the extension of (b) to the tnird dimension. The figures in the diagram give the height of the maximum for $R \approx 0.01$. The secondary maximum at $R \approx 0.027$ is the same as in (c)
———, $N_2 = 1$; --------, $N_2 = 2$; ·········, $N_2 = 3$; —·—, $N_2 = 4$; —··—··—, $N_2 = 5$;
××××, $N_2 = 6$; –×—×–, $N_2 = 7$
[Reproduced from A. Keller, A. Maradudin, *J. Phys. Chem. Solids.*, **2**, 301, Pergamon Press (1957)]

$N_1 = 5$, $N_2 = 1 \sim 5$ and $N_1 = 7$, $N_2 = 1 \sim 7$. It is evident that the Laue function varies in a complex manner at small values of R (Fig. 13.13(c) and (d)). Fig. 13.13 (d) shows the Laue functions for the three-dimensional analogues of the net planes of Fig. 13.13 (b); a maximum appears close to $R = 0.10$ for $N_2 \leqq 3$. This maximum corresponds to the 002 diffraction ($R = 0.104$) of well-developed crystals of Nylon 6,6 (α form) as illustrated in Fig. 13.14.[9] As this maximum appears, the small peak at $R \approx 0.03$ disappears, only to reappear at $N_2 = 4 \sim 5$.

The β form of Nylon 6,6 has also been reported.[10] The presence of several diffractions that cannot be indexed in accordance with the α form has been interpreted as implying the presence of the β form. The β form, diffractions characteristically are observed in various layer lines in pairs on either side of the meridian. The diffraction at $R \approx 0.027$, which has been identified as a β form diffraction, may, however, be thought of as due to the secondary maximum that appears in the Laue function for $N_1 = 5$ and $N_2 = 4$ as a result of the pronounced broadening of G(*cf.* Fig. 13.14[9] and Section 13.2.2). Moreover, all the diffractions which could be associated with the β form are very weak compared with those clearly of the α form (which makes it difficult to determine their positions exactly) and usually appear only when the specimen has very low crystallinity. This led Keller and Maradudin to the conclusion that the 002 diffractions attributed to the β form were in fact secondary maxima of the 002 diffractions of very poorly developed α form crystallites. They also sought to explain the mechanism of crystallization of Nylon 6,6 on the basis of this interpretation.

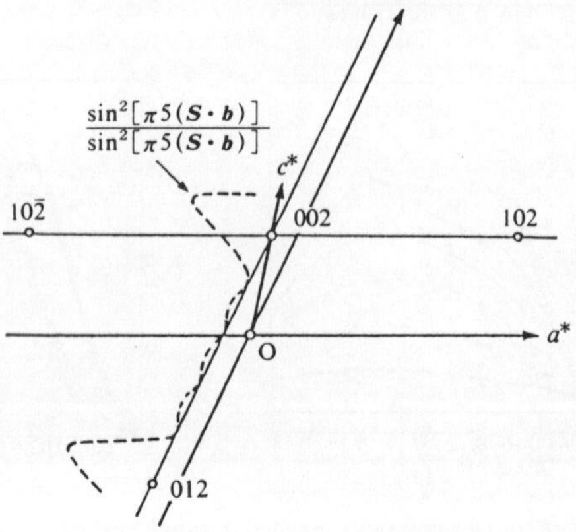

Fig. 13.14 Diagrammatic representation of the maximum in R at ≈ 0.01 corresponding to the 002 diffraction from α form Nylon 6,6.[9] [Reproduced from A. Keller, A. Maradudir, *J. Phys. Chem. Solids.* **2**, 301, Pergamon Press (1957)]

13.4 Estimation of Lattice Distortion from Line Broadening

If the crystal has lattice distortions due to stress by mechanical or physical treatments or if the atomic density of some component in crystalline substance is not uniform, the interplanar spacing d is not constant throughout the crystal, and different parts of the crystal diffract at different angles (Fig. 13.1).

The broadening of a diffraction pattern of such distorted crystals may be estimated from the Bragg equation. If the maximum strain in the crystal is $g(=\Delta d/d)$, the spacing d will vary from $d(1+g)$ to $d(1-g)$, and θ will vary over a range numerically equal to $2g \tan\theta$. Since the deviation of the rays is 2θ, the range of angle over which the diffraction is appreciable is twice this,[11] or

$$\beta^{\mathrm{D}} \fallingdotseq 4g \tan\theta \qquad (13.13)$$

Eq. 13.13 can be written as

$$\beta^{\mathrm{D}} \fallingdotseq 4\bar{g} \tan\theta \qquad (13.13a)$$

where $\bar{g}$ is the weight averaged strain. For a Gaussian strain profile, it follows that

$$\bar{g} = (1/2)(2\pi)^{1/2} < g^2 >^{1/2}$$

where $< g^2 >^{1/2}$ is the mean-square strain of the lattice.

As shown earlier, the integral breadth of the line broadening due to the volume-weighted crystallite size $< L >_V$ is given by the Scherrer equation

$$\beta_C = \frac{K\lambda}{<L>_V \cos\theta} \tag{13.10a}$$

If at least two orders of diffractions are observed and if only one source of the broadening is operative (either crystallite size or distortion, but not both), it is possible in principle to distinguish which factor is present by ascertaining whether the $\beta/\tan\theta$ *vs.* θ or the $\beta\cos\theta$ *vs.* θ plot is approximately constant (Fig. 13.15).

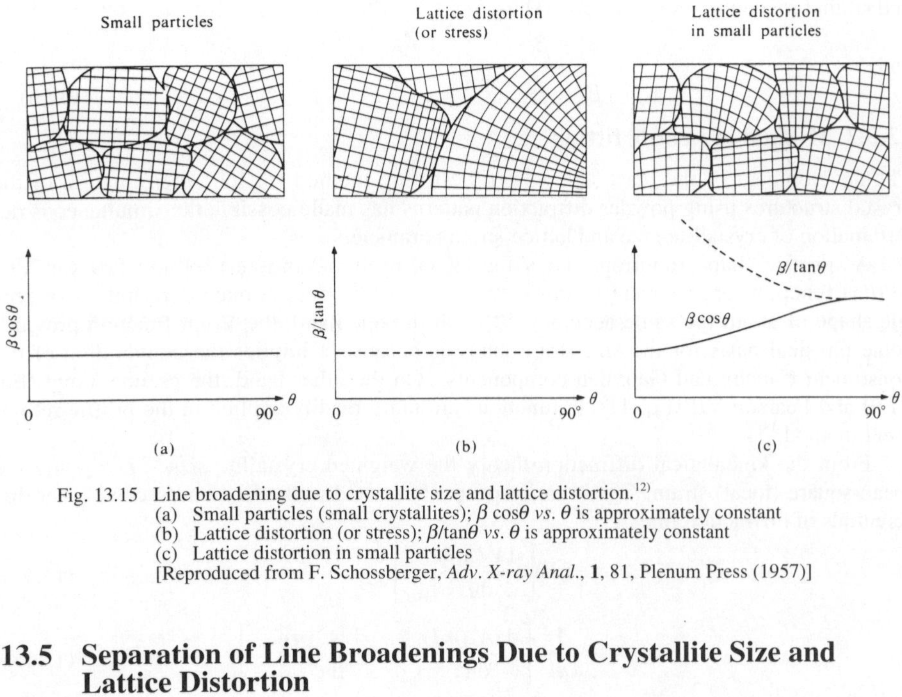

Fig. 13.15 Line broadening due to crystallite size and lattice distortion.[12)]
(a) Small particles (small crystallites); $\beta\cos\theta$ *vs.* θ is approximately constant
(b) Lattice distortion (or stress); $\beta/\tan\theta$ *vs.* θ is approximately constant
(c) Lattice distortion in small particles
[Reproduced from F. Schossberger, *Adv. X-ray Anal.*, **1**, 81, Plenum Press (1957)]

13.5 Separation of Line Broadenings Due to Crystallite Size and Lattice Distortion

13.5.1 Method of integral breadths

If both size and distortion broadening are operative, we can estimate the magnitude of two contributing factors by approximating the shapes of two contributing diffraction profiles appropriately. Frequently they are assumed to be either Cauchy (Lorentzian) or Gaussian function. If both diffraction profiles are approximated as being of the Cauchy type, the net integral breadth, β, of the pure diffraction profile will be the sum of the size and distortion breadths, β^S and β^D

$$\beta = \beta^{\mathrm{S}} + \beta^{\mathrm{D}} \qquad \text{(Cauchy)}, \tag{13.14}$$

whereas for two Gaussian profiles

$$\beta^2 = (\beta^{\mathrm{S}})^2 + (\beta^{\mathrm{D}})^2 \text{ (Gaussian)}. \tag{13.15}$$

This method is convenient if the diffraction peaks are isolated or if the overlapped peaks can be resolved to individual peaks correctly. The labor of evaluating Fourier transforms can be avoided, although at the cost of reduced information and accuracy, by restoring to integral breadths in place of the complete profiles of the diffractions.

It has been experimentally verified that crystallite size broadening can often be approximated by a Cauchy function, whereas distortion broadening is better described by a Gaussian function.[2–14] A Voigt function (Eq. 11.7) is often adopted as the profile-shape function in the profile fitting method. In this case the Cauchy and Gaussian components of the structurally broadened profile (integral breadths β_{C} and β_{G}, respectively) are due to size and strain broadening, respectively. Thus

$$\beta^{\mathrm{S}} = \beta_{\mathrm{C}}$$

and

$$\beta^{\mathrm{D}} = \beta_{\mathrm{G}} \tag{13.16}$$

13.5.2 Method of profile fitting

The development of the least-squares profile-fitting method for the determination of the crystal structures using powder diffraction patterns has made possible the simultaneous determination of crystallite-size and lattice-strain parameters.

As profile shape functions the Voigt, $V(x)$, pseudo-Voigt, $pV(x)$ and Pearson VII, $pVII(x)$ functions are frequently considered; these afford approximate descriptions of profile shape of about the same accuracy.[15-18] On the one hand, the Voigt function provides some physical basis for the size-strain analysis because it implies the convolution of the constituent Cauchy and Gaussian components. On the other hand, the pseudo-Voigt (Eq. 11.9) and Pearson VII (Eq. 11.11) functions are more readily applied in the profile-refinement method.[18]

From the kinematical diffraction theory the weighted crystallite size $< L >_{\mathrm{a}}$, and the mean-square (local) strain, $< g^2 >$ are respectively related to the first and second order differentials of diffraction lines.[19,20]

$$<L>_{\mathrm{a}} = -(\mathrm{d}l)\left[\left.\frac{\mathrm{d}A^{\mathrm{f}}(n,l)}{\mathrm{d}n}\right|_{n\downarrow 0}\right]^{-1} \tag{13.17}$$

$$<g^2> = \frac{1}{4\pi l^2}\left[\left.\frac{\mathrm{d}^2A^{\mathrm{f}}(n,l)}{\mathrm{d}n^2}\right|_{n\downarrow 0} - \left.\frac{\mathrm{d}^2A^{\mathrm{S}}(n)}{\mathrm{d}n^2}\right|_{n\downarrow 0}\right] \tag{13.18}$$

where A^{f} and A^{S} are the normalized cosine Fourier coefficients of the pure only structurally broadened profile and of the size profile, respectively, n is the harmonic number, d is the interplanar spacing of the diffracting planes and l is the order of the diffraction.

According to Eq. 13.17, $< L >_{\mathrm{a}}$ can be determined from a single-1ine, whereas $< g^2 >$ [and $\mathrm{d}^2A^{\mathrm{S}}(n)/\mathrm{d}n^2$] can be determined from two orders of a diffraction. However, because of experimental difficulties, Eq. 13.17 and 13.18 have not been used as the basis of a multiple-line method for size-strain analysis: Owing to truncation of profiles and background levels estimated too high, the Fourier coefficients of small harmonic number become unreliable ("hook" effect).[19] Because the profile-shape assumptions are of different strength for the various orders of a diffraction, the reliability of multiple-line method on the basis of Eqs. 13.17 and 13.18 is very limited and even nonsense may result (negative size and/or negative strain values). However, if the size estimates, according to Eq. 13.17 from the separated orders coincide, a valuable strain estimate, according to Eq. 13.18, may be obtained.[19–21)]

In single-line analysis an additional assumption is required. From a recent review on single-line methods[19] it appears that the following assumptions may be appropriate:

(Assumption 1)

$$\left.\frac{d^2A^S(n)}{dn^2}\right|_{n\downarrow 0} = 0, \tag{13.19}$$

which includes the approximative description of the size-broadened profile by a Laue function, or

(Assumption 2)

$$\left.\frac{d^2A^S(n)}{dn^2}\right|_{n\downarrow 0} = \frac{4}{a^2}(\beta_C{}^1)^2 \tag{13.20}$$

where the size-broadened profile has been approximated by a Cauchy function (The period a is expressed in the same units as the integral breadth $\beta_C{}^f$ of the structurally broadened profile).

For the case of a

Voigt profile-shape function:

By Assumption 1

$$L_a = \lambda/(2\beta_C{}^f \cos\theta) \tag{13.21}$$

and

$$<g^2>^{1/2} = \{(\beta_G{}^f)^2 - (2/\pi)(\beta_G{}^f)^2\}^{1/2}/\{2(2\pi)^{1/2}\tan\theta\} \tag{13.22}$$

and for a meaningful result

$$\beta_G{}^f \geq (2\pi)^{1/2}\,\beta_C{}^f. \tag{13.23}$$

Alternatively (by Assumption 2)

$$<g^2>^{1/2} = \beta_G{}^f/\{2(2\pi)^{1/2}\tan\theta\}. \tag{13.24}$$

Similarly, for the case of a pseudo-Voigt profile-shape function, useful equations are also given.[19]

13.5.3. Method of Fourier transforms[22,23]

The experimental diffraction profile, corrected for instrumental broadening, $G(S)$ is the convolution of the crystallite size profile, $G_S(S)$ and distortion profile, $G_D(S)$,

$$G(s) = G_S(S) * G_D(S). \tag{13. 25}$$

By the convolution theorem, $A_S(t)$, the Fourier transforms of $G_S(S)$ and $A_D(t)$, that of $G_D(S)$ are related to the transform, $A(t, S)$, of the corrected experimental profile, $G(S)$

$$A(t, S) = A_S(t) \cdot A_D(t, S) \tag{13.26}$$

where, $S = (2\sin\theta)/\lambda$ and t is the length of vector $\boldsymbol{t}$ vertical to the diffraction plane.

If we denote the mean square lattice distortion over the domain t by $<g_t{}^2>$, according to Bertaut[24], Warren and Averbach,[25] and Warren,[26] $A_D(t, S)$ can be expressed for small values of g, and low orders of diffraction as

$$A_D(t, S) = \exp[-2\pi^2 S^2 t^2 < g_t^2 >]. \tag{13.27}$$

If we substitute Eq. 13.27 into Eq. 13.26 and take logarithms, the result is

$$\log A(t, S) = \log A_S(t) - 2\pi^2 S^2 t^2 < g_t^2 > \tag{13.28}$$

If several orders of diffraction can be observed, plotting the log $A(t, S)$ against S^2 for any desired values of t, $A_S(t)$ and $< g_t^2 >$ can be obtained from the ordinate intercept and slopes at low angles, respectively.

For details and practical applications the reader is referred to papers of Warren and Averbach (Cu-Si single crystal)[27] and Buchanan and Miller[28] (isotactic polystyrene).

13.6 Analyses Including Background Scattering Due to Imperfect Crystals

In Chapter 5 we discussed Hosemann's theory of X-ray diffraction by paracrystal lattices. According to this theory, in addition to the broadening due to the Laue function (G or L^2 *cf.* Sections 2.9 and 13.2), there is a significant contribution to the X-ray diffraction broadening from lattice distortions of the second kind (function g, *cf.* Section 13.6.1). The measurements of crystallite size mentioned in Section 13.3 are based on broadening due to the Laue function, and those in Section 13.4 are based on lattice distortion, but these are obviously inadequate where the specimen has a typically paracrystalline structure (*e.g.* a high polymer crystal). On the other hand, it is doubtful on theoretical grounds whether the essentially one-dimensional statistics of Hosemann are entirely adequate to deal with this phenomenon. It is, however, in the authors' view, a major step forward beyond analyses which treat only "pure" diffraction breadths toward a soundly-based general treatment for the diffraction patterns of distorted crystals. As such it yields results which can be meaningfully compared with those of the earlier methods. The practical difficulties of drawing a rather arbitrary distinction between crystalline and amorphous scattering in the diffraction pattern or scattering curve which treatment involves, are nevertheless considerable. A few examples of analytical methods based on this theory will now be presented. The reader is referred to Chapter 5 for a statement of the basic principles.

13.6.1 Broadening of diffraction patterns from paracrystalline structures

The diffraction patterns of paracrystalline structures are often represented by a continuous X-ray intensity curve as shown in Fig. 13.16. Eq. 5.44 (*cf.* Eq. 5.38) is expressed in general by the following equation.

$$I(\boldsymbol{S}) = N[<A_{cell}{}^2> - <A_{cell}>^2] + N<A_{cell}>^2(1-D_t{}^2) + \frac{1}{v}<A_{cell}>^2 D_t{}^2 Z(\boldsymbol{S}) * | S(\boldsymbol{S}) |^2 \tag{13.29}$$

The crystalline diffraction peak is represented by the third term, in which the main factor determining the shape and breadth of the diffraction pattern is $Z(\boldsymbol{S}) * | S(\boldsymbol{S}) |^2$. Where the value of $< A_{cell} >^2$ changes abruptly within an angular range smaller than the breadth of $Z(\boldsymbol{S}) * | S(\boldsymbol{S}) |^2$, the shape of the diffraction pattern is drawn out under the influence of $< A_{cell} >^2$ and is consequently governed by $< A_{cell} >^2 Z(\boldsymbol{S}) * | S(\boldsymbol{S}) |^2$. If the change in $< A_{cell} >^2$ is assumed to be slow in relation to $Z(\boldsymbol{S})$ and $| S(\boldsymbol{S}) |^2$ the shape of the diffraction curve may in general be regarded as given by $Z(\boldsymbol{S}) * | S(\boldsymbol{S}) |^2$. For cases where the variation

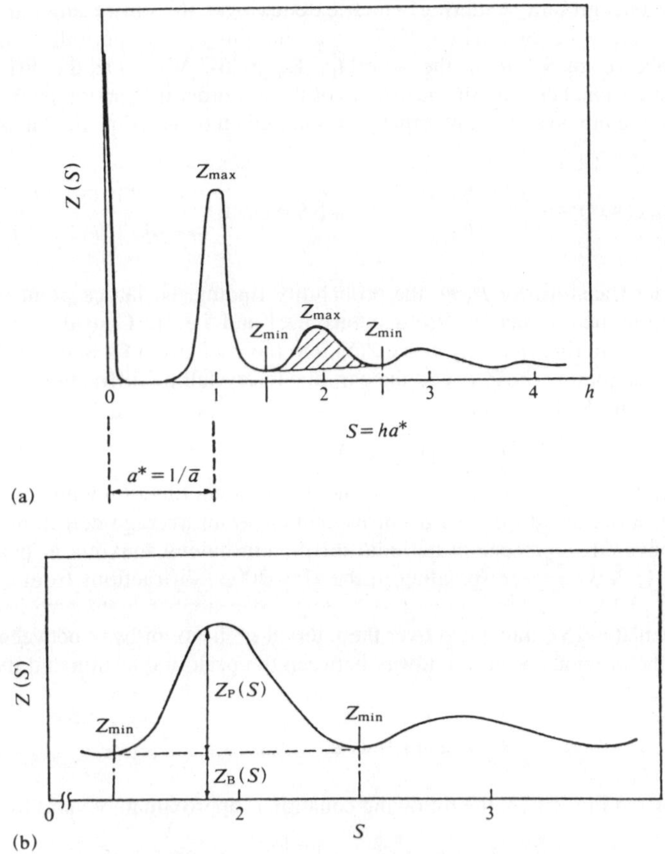

Fig. 13.16 Continuous intensity distribution for X-ray from a paracrystalline substance.
(a) Crystalline peaks superimposed on background scattering.
(b) Enlarged section of (a) (see text).

of $< A_{\mathrm{cell}} >^2$ has an important effect, the reader is referred to Section 2.9.1 (*cf.* Fig. 5.12(b)). The influence of $Z(\boldsymbol{S}) * | S(\boldsymbol{S}) |^2$ on the shape of the diffraction depends upon the following factors.

1) If the breadth of $Z(\boldsymbol{S})$ is much greater than that of $| S(\boldsymbol{S}) |^2$, *i.e.* if both the degree and extent of lattice distortions are large, the breadth of the diffraction pattern is dominated by the influence of $Z(\boldsymbol{S})$, the paracrystalline lattice factor.

2) If the breadth of $| S(\boldsymbol{S}) |^2$ is greater than the breadth of $Z(\boldsymbol{S})$, *i.e.* if the regularity of the lattice is high but the regions themselves are small, the shape of the diffraction pattern is more strongly influenced by the profile of $| S(\boldsymbol{S}) |^2$ (the shape factor or the small angle scattering function).

3) If the breadths of $| S(\boldsymbol{S}) |^2$ and of $Z(\boldsymbol{S})$ are comparable, the shape of the diffraction is affected by both, depending upon their convolution $Z(\boldsymbol{S}) * | S(\boldsymbol{S}) |^2$.

We now consider the two elements which constitute Fig. 13.13, *i.e.* the crystalline peaks and the background scattering which extends over the entire angular range. The paracrystalline lattice factor $Z(\boldsymbol{S})$ ($= K_k(\boldsymbol{S})$) is a one-dimensional periodic function with a reciprocal lattice vector $\boldsymbol{S} = \boldsymbol{a}^*$ as the period (*cf.* Eq. 5.26). With h as the diffraction index for the one-dimensional crystal, the maximum of the hth order diffraction peak and the minimum between the hth and $(h+1)$th order peaks are given by (*cf.* Eqs. 5.32 and 5.33)

$$Z_{\max}(ha^*) = \frac{1+|\mathcal{F}(ha^*)|}{1-|\mathcal{F}(ha^*)|}, \qquad Z_{\min}[(h+\tfrac{1}{2}a^*)] = \frac{1-|\mathcal{F}[(h+\frac{1}{2})a^*]|}{1+|\mathcal{F}[(h+\frac{1}{2})a^*]|} \tag{13.30}$$

$\mathcal{F}$ is the Fourier transform of $H_k(\boldsymbol{r})$, the probability finding the lattice point with vector $\boldsymbol{r}$ from the origin in the $\boldsymbol{a}$ direction (See Sections 5.3.1 and 5.3.2). Consider now the resolution of the total scattering intensity curve $Z(\boldsymbol{S})$ into the background scattering Z_B, which we take to be the line joining the $Z_{\min}$, and the part due to crystalline diffraction, which projects above this line (Fig. 13.13(b)), *i.e.*

$$Z(\boldsymbol{S}) = Z_P(\boldsymbol{S}) + Z_B(\boldsymbol{S}) \tag{13.31}$$

The fact that $K(\boldsymbol{S})$ in Eq. 5.26 (*i.e.* $Z(\boldsymbol{S})$ is a one-dimensional function with a period equal to the reciprocal lattice length $a^* = 1/\bar{a}$ corresponding to an average cell dimension $\bar{a}$ was mentioned above. It represents a periodic diffraction having maxima at points $S = ha^*$ such that $h = 1, 2, 3,...$, (corresponding to the 100, 200,... diffractions from crystals). As shown in Fig. 13.13, for a given period a^* having order number h, the area under the total diffraction intensity $Z(\boldsymbol{S})$, integrated over the interval a^* from midway between the peak in question and the previous peak to midway between the peak in question and the next peak, is constant, *i.e.*

$$\int_{(h-\frac{1}{2})a^*}^{(h+\frac{1}{2})a^*} Z(\boldsymbol{S})\boldsymbol{dS} = 1/\bar{a} \tag{13.32}$$

From Eq. 5.33 and Fig. 13.13, the following equation is approximately valid for $Z(\boldsymbol{S})$.

$$\int_{(h-\frac{1}{2})a^*}^{(h+\frac{1}{2})a^*} Z_B(\mathrm{S})\mathrm{d}S \cong a^* \frac{1-|\mathcal{F}|}{1+|\mathcal{F}|} = Z_B(\mathrm{S})/\bar{a} \tag{13.33}$$

The area of the crystalline diffractions is therefore

$$\int_{(h-\frac{1}{2})a^*}^{(h+\frac{1}{2})a^*} Z_P(\mathrm{S})\mathrm{d}S = \frac{1}{\bar{a}}\{1-Z_B(\mathrm{S})\} \tag{13.34}$$

Since $Z_{\max} = 1/Z_{\min}$, the height $Z_{P,\max}$ above Z_B at the maximum $Z_{\max}$ is

$$Z_{P,\max}(\boldsymbol{S}) = 1/Z_B(\boldsymbol{S}) - Z_B(\boldsymbol{S}) \tag{13.35}$$

The integral breadth $\beta_P(h)$ of the crystalline diffraction emerging above the background level is therefore the area (Eq. 13.34) divided by the expression in Eq. 13.35.

$$\beta_P(h) = \frac{1}{\bar{a}}\frac{Z_B(\mathrm{S})}{1+Z_B(\mathrm{S})} = \frac{a^*}{2}\{1-\mathcal{F}(\mathrm{S})\} \tag{13.36}$$

The actual form of $\mathcal{F}(\boldsymbol{S})$ for the hth order diffraction, *i.e.* for $\boldsymbol{S} = h\boldsymbol{a}^*$, is given by Eq. 5.35, and the integral breadth of the diffraction due to the paracrystalline lattice is:

$$\beta_P(h) \cong \frac{1}{2\bar{a}}\{1-\exp(-2\pi^2 g^2 h^2\} \quad (13.37)$$

Of the three possibilities, 1), 2), and 3) mentioned at the beginning of this section, 1) (where $Z(\boldsymbol{S})$ is dominant) has broadening which is clearly that due to Eq. 13.37. In case 3), on the other hand, where neither $Z(\boldsymbol{S})$ nor $|S(\boldsymbol{S})|^2$ predominates, the correct procedure in principle is to calculate $|S(\boldsymbol{S})|^2 * Z(\boldsymbol{S})$, but this is difficult in practice. It may therefore be acceptable experimental procedure to take the root mean square of the breadths due to the two effects as an approximation to the measured breadth of the diffraction. Thus if the integral breadth of the broadening due to the small-angle scattering accompanying the hth order diffraction is β_S, the breadth $\beta(h)$ of the hth order diffraction due to the two effects is

$$\beta(h) = [\beta_P^2(h) + \beta_S^2]^{1/2} \quad (13.38)$$

In our one-dimensional model, let the scattering body consist of a row of N units with average cell dimension $\bar{a}$, so that its total length is $N\bar{a} = L$. Determination of the integral breadth from the small angle scattering function (scattering function of a rod in Table 6 of the Appendix) gives

$$|S(\boldsymbol{S})|^2 = \left(\frac{\sin \pi SL}{\pi S}\right)^2, \qquad \beta_S = 1/L = 1/(N\bar{a}) \quad (13.39)$$

The combination of Eqs. 13.37 and 13.39 with Eq. 13.38 gives

$$\beta(h) = \frac{1}{2\bar{a}}\left[\frac{4}{N^2} + \{1-\exp(-2\pi^2 g^2 h^2)\}^2\right]^{1/2} \quad (13.40)$$

Eq. 13.40 gives the diffraction breadth due to both effects. From Eqs. 13.38 and 13.40 we obtain

$$\beta^2(h) - 1/L^2 = \{1/(4\bar{a})\}\{1 - \exp(-2\pi^2 g^2 h^2)\}^2$$

$$g = (0.342/h)[-\log(1 - 2\bar{a}\{\beta^2(h) - 1/L^2\}^{1/2})]^{1/2} \quad (13.41)$$

where $h = 1, 2, 3,...$ Since $1/L^2$ is common to this set of diffractions, it can be eliminated and g can be found. This is a useful method of finding g without eliminating the effect of $|S(\boldsymbol{S})|^2$ form β to obtain the diffraction breadth due purely to the paracrystal structure (*cf.* Section 13.6.2).

13.6.2 Analysis of the broadening of diffractions from paracrystal structures

A. Calculation of the factor D_1 for distortions of the first kind

We shall now examine in detail the effect of the factor D_1 (Debye factor) for distortions of the first kind upon the diffraction intensity. As is evident from Eq. 13.29, this factor D_1 so acts on the crystalline diffraction terms as to reduce the intensity. All the energy lost from the crystalline diffraction in this way is dissipated as continuous background scattering in the form of the second term in the equation. Thus if the intensity of the scattering due to D_1 in the continuous scattering cannot be correctly estimated, crystallinity measurements based on the background scattering are meaningless. D_1 is totally disregarded in the simple method given in Section 14.2 for the measurement of crystallinity. If D_1 is referred purely to distortions due to the temperature factor, its effect might cancel out when the measurements were carried out at a constant temperature. In reality, however, D_1 arises from distortions of the first kind due to various causes. Consequently, differences in the structures of different specimens will also influence the values of D_1, and there are many cases where this factor cannot be disregarded. In this section we shall examine in detail the significance of D_1 particularly for paracrystalline theory.

For a paracrystal, if the size of the crystalline regions, *i.e.* N, is very large (say $N \gtrsim 5$), $|S(\mathbf{S})|^2$ is point-like with respect to $Z_B(\mathbf{S})$. Thus, from Eqs. 13.29 and 13.31,

$$I = I_P + I_B; \qquad I_B = I_{B_1} + I_{B_2} + I_{B_3} \tag{13.42}$$

$$I_{B_1} = N[< A^2_{cell} > - < A_{cell} >^2] \tag{13.43}$$

$$I_{B_2} = N < A_{cell} >^2 [1 - D_1^2(\mathbf{S})] \tag{13.44(a)}$$

$$I_{B_3} = N < A_{cell} >^2 D_1^2(\mathbf{S}) Z_B(\mathbf{S}) \tag{13.44(b)}$$

$$I_P = \frac{1}{\bar{a}} < A_{cell} >^2 D_1^2(\mathrm{S}) Z_P(\mathrm{S}) * |S(\mathrm{S})|^2 \tag{13.45}$$

The diffuse scattering I_{B_1} is usually negligible. The integral intensity of I_P for the hth order diffraction is given by the product of the integral value of $|S(\mathbf{S})|^2$ and Eq. 13.34 multiplied by $(1/\bar{a}) < A_{cell} >^2 D^2$

$$\begin{aligned} \int_h I_P(\mathrm{S})\mathrm{d}S &= \frac{1}{\bar{a}} < A_{cell} >^2 D_1^2(\mathrm{S}) \int_h Z_P(\mathrm{S})\mathrm{d}S \int_h |S(\mathrm{S})|^2 \,\mathrm{d}S \\ &= \frac{1}{\bar{a}} < A_{cell} >^2 D_1^2(h)\{1 - Z_B(h)\} \end{aligned} \tag{13.46}$$

Division by the interpolated background intensity $I_B(h)$ at $\mathbf{S} = h\mathbf{a}^*$ gives the new variable

$$C(h) = \frac{\int_h I_P \mathrm{d}S}{I_B(h)} = D_1^2(1 - Z_B) / \bar{a}[1 - D_1^2(1 - Z_B)] \tag{13.47}$$

Thus the distortion of the first kind, $D_1^2(h)$, is obtained by the calculation of Eq. 13.47, which can also be expressed in the form

$$\begin{aligned} D_1^2(h) &= \frac{\bar{a}C(h)}{1 + \bar{a}C(h)} \frac{1}{1 - Z_B(h)} \\ \frac{1}{1 - Z_B(h)} &= \frac{1 + |\mathcal{F}|}{2|\mathcal{F}|} = \frac{1 + \exp(-2\pi^2 g^2 h^2)}{2\exp(-2\pi^2 g^2 h^2)} \end{aligned} \tag{13.48}$$

B. Determination of the crystallinity of a paracrystalline specimen

Ruland[29-32] and Kilian[33,34] have proposed a method of finding the crystallinity by application of the paracrystal theory. This is an important method of paracrystal analysis, and gives the distortion factor as well as an estimate of the crystallinity. The scattering energy of the coherent X-rays scattered over all space by the specimen is expressed by the following equation.

$$\int_0^\infty I(S)\mathrm{d}v_S = 4\pi \int_0^\infty S^2 I(S)\mathrm{d}v = 4\pi \int_0^\infty S^2 < f^2 > \mathrm{d}S \tag{13.49}$$

$$< f^2 > = \sum n_j f_j^2 / \sum n_j \tag{13.50}$$

f_j is the scattering factor of the atom j, n_j is the number of atoms j in the unit cell, and $<f^2>$ is the mean scattering intensity of the atoms.

If the proportion by weight of the crystalline regions of the specimen is x_{cr}, then

$$\int_0^\infty I_{cr}(S)\mathrm{d}v_S = 4\pi\int_0^\infty S^2 I_{cr}(S)\mathrm{d}S = x_{cr}\cdot 4\pi\int_0^\infty S^2 <f^2> D_t^2\mathrm{d}S \tag{13.51}$$

D_t^2 in this equation is a distortion factor, but unlike D_1, it includes distortions of both the first and second kinds. From Eqs. 13.49 and 13.51 we obtain

$$x_{cr} = \frac{\int_0^\infty S^2 I_{cr}\mathrm{d}S}{\int_0^\infty S^2 I\mathrm{d}S}\,\frac{\int_0^\infty S^2 <f^2>\mathrm{d}S}{\int_0^\infty S^2 <f^2> D_t^2\mathrm{d}S} \tag{13.52}$$

This equation should be valid even if the limits of integration are not 0 to ∞, provided that the range (say $S_1 \sim S_2$) is sufficiently wide. We can therefore write

$$x_{cr} = \frac{\int_{S_1}^{S_2} S^2 I_{cr}(S)\mathrm{d}S}{\int_{S_1}^{S_2} S^2 I(S)\mathrm{d}S}\,\frac{\int_{S_1}^{S_2} S^2 <f^2>\mathrm{d}S}{\int_{S_1}^{S_2} S^2 <f^2> D_t^2\mathrm{d}S} = x'_{cr}\cdot K \tag{13.53}$$

$$x'_{cr} = \frac{\int_{S_1}^{S_2} S^2 I_{cr}(S)\mathrm{d}S}{\int_{S_1}^{S_2} S^2 I(S)\mathrm{d}S} \tag{13.54}$$

$$K(S_1, S_2, D_t^2, <f^2>) = \frac{\int_{S_1}^{S_2} S^2 <f^2>\mathrm{d}S}{\int_{S_1}^{S_2} S^2 <f^2> D_t^2\mathrm{d}S} \tag{13.55}$$

Since the crystallinity x_{cr} is independent of the limits of integration, x'_{cr} is found from Eq. 13.51 for various ranges of integration, and K and D_t are adjusted so that $x'_{cr}\cdot K = x_{cr}$ is independent of the limits of integration $S_1 \sim S_2$. The resulting value of x_{cr} is a true measure of crystallinity. From Eqs. 13.42 to 13.48, the background scattering I_B is

$$\begin{aligned} I_B &\cong I_{B_2} + I_{B_3} \\ &= N<A_{cell}>^2[1 - D_1^2(S)] + N<A_{cell}>^2 D_1^2(S) Z_B(S) \\ &= N<A_{cell}>^2\left[1 - D_1^2(S)\frac{2|\mathscr{F}|}{1+|\mathscr{F}|}\right] \\ &= N<A_{cell}>^2[1 - D_t^2(S)] \end{aligned} \tag{13.56}$$

The distortion factor is therefore

$$D_t^2(S) = D_1^2(S)\frac{2|\mathscr{F}|}{1+|\mathscr{F}|} \tag{13.57}$$

Among the distortions of the first kind, the component due to thermal vibrations can be approximated by the Debye-Waller temperature factor $\exp(-B_T S^2)$, and the effect of lattice imperfections by the Gaussian factor $\exp(-B_1 S^2)$. If the component due to distortions of the second kind is also assumed to be Gaussian it can be expressed as $\exp(-B_2 S^2)$. Thus, as a first approximation, we can write

$$D_t^2 = \exp(-BS^2)$$

$$B \cong B_T + B_1 + B_2 \tag{13.58}$$

This is a useful equation which shows the possibility of finding the factors for distortions of the first and of the second kinds independently of the total distortion factor D_t.

C. Crystallitinity and lattice distortion of some polymers

a) Crystallinity of polypropylene.[30] Fig. 13.17[30] shows the $S^2I(S)$ *vs.* S curve obtained

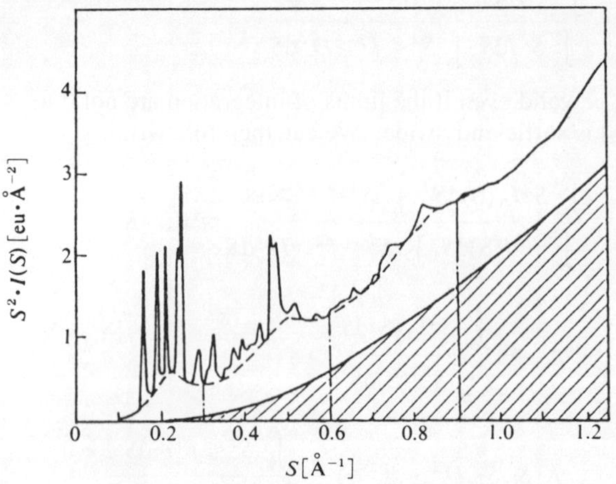

Fig. 13.17 Diffraction intensity distribution ($S^2 I(S)$ *vs.* S) for a polypropylene specimen.[30] [Reproduced with permission from W. Ruland, *Acta. Cryst.*, **14**, 1184, IUCr. (1961)]

from the diffraction intensity distribution of a specimen of polypropylene. The broken line is the curve joining the minima between the crystalline diffractions. Above this is the $S^2I_{cr}(S)$ *vs.* S curve, and below it is the background scattering. The values of x'_{cr} in Eq. 13.54 for the case in which there is no distortion are found by graphical integration of the $S^2I(S)$ *vs.* S and $S^2I_{cr}(S)$ *vs.* S curves from S_1 $(=0.1)$ to the points indicated by the vertical dot-dash lines (see Table 13.4).[30] For $D_t^2(S)$, a nomogram showing the relationship between K and S_2 with $S_1 = 0.1$ and for various values of B must also be prepared (see Fig. 13.18). This is used to find the value of K such that x_{cr} is constant and independent of the integration range S_1–S_2.

It is evident from Table 13.4 that although the crystallinity naturally varies with the thermal history of the specimen, the value of the distortion function is the same. It is thought, however, that the lattice distortion generally varies with the treatment of the specimen. Ruland[30] therefore concluded that the effect of lattice distortion is small in comparison with that of thermal vibration in the case of polypropylene.

b) Crystallinity and lattice distortion factor of Nylon 6.[31,32] The crystallinity and the B values are found in exactly the same way as for polypropylene (see Table 13.5).[32] The B value, unlike that of polypropylene, varies considerably with the thermal history. This B value includes contributions due to lattice distortions of the first and of the second kinds as well as to thermal vibrations, as shown in Eq. 13.58.

The measurements on Nylon 6 for Table 13.5 were all carried out at room temperature,

Table 13.4 The crystallinities of polypropylene speciments for $B = 0$ and 4[†1]

Range $S_1 \sim S_2$	Specimen I[†1]		Specimen II[†2]		Specimen III[†3]		Specimen IV[†4]	
	x'_{cr} ($B=0$)	x_{cr} ($B=4$)	x'_{cr} ($B=0$)	x_{cr} ($B=4$)	x'_{cr} ($B=0$)	x_{cr} ($B=4$)	x'_{cr} ($B=0$)	x_{cr} ($B=4$)
0.1 ~ 0.3	0.270	0.329	0.353	0.431	0.546	0.666	0.120	0.146
0.1 ~ 0.6	0.159	0.294	0.222	0.411	0.333	0.616	0.078	0.144
0.1 ~ 0.9	0.105	0.305	0.145	0.421	0.220	0.638	0.044	0.128
0.1 ~ 1.25	0.067	0.315	0.095	0.447	0.145	0.682	0.029	0.136
$\langle x_{cr} \rangle$		0.31		0.43		0.65		0.14

†1 Isotactic polypropylene heated to its melting point and quenched in water at room temperature. †2 as above, heated for 1 h at 105°C. †3 As above, heated for 30 min at 160°C. †4 Atactic polypropylene.
[Reproduced with permission from W. Ruland, *Acta. Cryst.*, **14**, 1184, IUCr. (1961)]

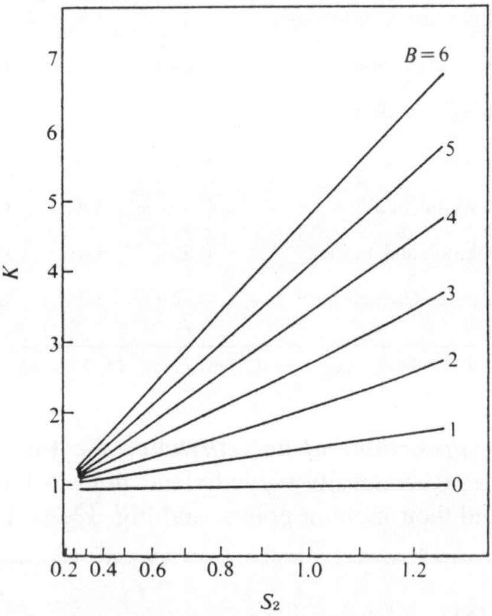

Fig. 13.18 Nomogram giving K for various values of B and S_2 for polypropylene specimens (with $S_1 = 0.1$).[30)] [Reproduced with permission from W. Ruland, *Acta. Cryst.*, **14**, 1184, IUCr. (1961)]

and since the various specimens cannot belong to different crystal systems, it may be assumed that the thermal vibration contribution B_T to B does not vary from specimen to specimen. The variation of B must therefore reflect the degree of lattice distortion. B_T must be equal to or smaller than the smallest B value in Table 13.5, *i.e.* 3.0. If we take ($B - 3.0$) as the contribution due to lattice distortions, it varies between 0 and 2.6. Thus

$$B_T \leqq 3.0, \qquad B_1 + B_2 \geqq B - 3.0$$

The contributions B_1 and B_2 can be resolved by finding B_2 with the aid of the broadening of the diffraction pattern. If it is assumed that the crystalline regions are very large and the term containing N can be disregarded in Eq. 13.40 (which gives the breadth of the diffraction), expansion of the exponential term gives an approximation to the lattice distortion

$$g \cong \bar{a}_j \beta(h)/\pi h \qquad (13.59)$$

The contribution B_2 of lattice distortions the second kind to B can be found from the broadening of the diffraction by means of the following equation.

$$B_2 = 0.7 \times 2\pi^2 \overline{\Delta}^2 = 1.4\pi^2 g^2 \overline{a}_j^2 \tag{13.60}$$

In Eq. 13.59, $\beta(h)$ is the integral breadth of the diffraction, $\overline{a}_j$ is the average spacing of the crystal planes giving this diffraction, and Δ in Eq. 13.60 is the variation of $\overline{a}_j$. The B_2 found in this way are shown in Table 13.5. The g_{am} in Table 13.5 are the g obtained by considering parts of the amorphous regions that exhibit regularity as a type of paracrystal that is different from that in the crystalline regions, and treating the maximum at the beginning of the diffraction intensity curve as a primary interference due to this type of paracrystal.

Table 13.5 The crystallinities and lattice distortions of Nylon 6 specimens[32)]

Specimen	x_{cr}	B	B_2	g_{cr}	g_{am}
1. Bulk specimen heated for 10 min at 195°C in nitrogen	0.33	3.0	0.39	0.042	0.18
2. Bulk specimen heated for 6 h in boiling water	0.24	4.2	0.30	0.037	0.17
3. Bulk specimen powdered and heated for 10 min at 195°C in nitrogen	0.27	3.0	0.34	0.039	0.17
4. Bulk specimen powdered and heated for 6 h in boiling water	0.33	5.6	0.45	0.045	0.17
5. Specimen 4 heated for a further 15 min at 200°C in nitrogen	0.31	3.9	0.37	0.041	0.17
6. Precipitated from formic acid solution and washed with boiling water	0.35	4.4	0.45	0.045	0.17
7. Precipitated from formic acid soln. and heated for 20 min at 2000°C in nitrogen	0.32	3.7	0.35	0.040	0.18

[Reproduced with permission from W. Rulland, *Faserforsch. Textiltech.*, **15**, 532, Akademie Verlag (1964)]

c) Crystallinity and lattice distortions of polyethylene.[33,34)] Fig. 13.19 shows how the crystallinities of polyethylenes (low density polyethylene and Marlex 6015) vary with temperature between − 150°C and their melting points, and Fig. 13.20 shows the variation of B

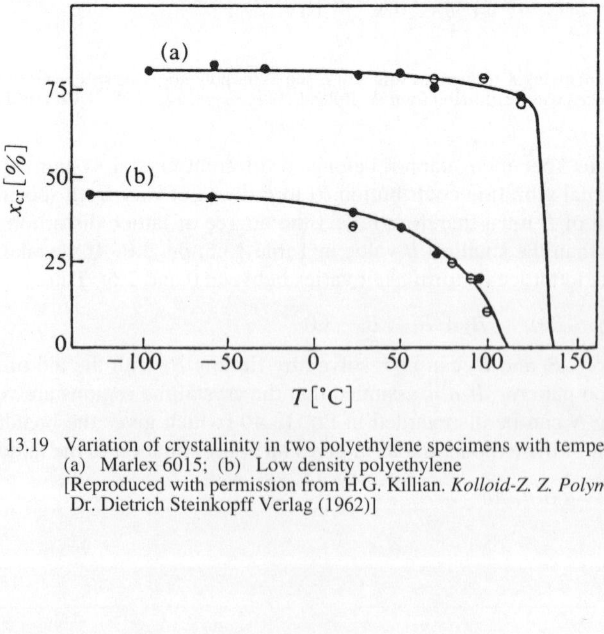

Fig. 13.19 Variation of crystallinity in two polyethylene specimens with temperature.[33, 34)]
(a) Marlex 6015; (b) Low density polyethylene
[Reproduced with permission from H.G. Killian. *Kolloid-Z. Z. Polymere*, **183**, 9, Dr. Dietrich Steinkopff Verlag (1962)]

over the same range. The crystallinities were estimated in the same way as in Example a) above. The B values found at the same time show a linear decrease with falling temperature, until at very low temperatures they become practically constant.

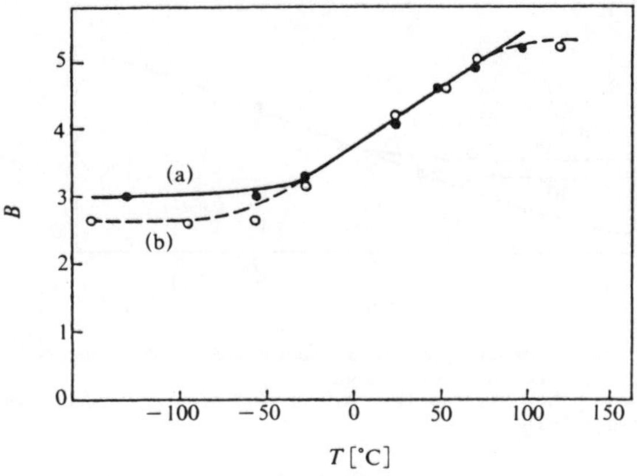

Fig. 13.20 Variation of B with temperature.[33, 34)]
(a) Marlex 6015; (b) Low density polyethylene.
[Reproduced with permission from H.G. Killian. *Kolloid-Z. Z. Polymere*, **183**, 9, Dr. Dietrich Steinkopff Verlag (1962)]

According to the Debye theory, the temperature factor B_T for isotropic crystals in general is expressed as follows[34)]

$$B_T = [6h^2/(M\kappa\Theta)][\Phi(x)/x + 1/4][(\sin\theta)/\lambda]^2 \tag{13.61}$$

$$x = \Theta/T, \quad \Phi(x) = (1/x)\int_0^x \xi/(\exp\xi - 1)\delta\xi$$

M is the molecular weight, κ is the Boltzman constant, and Θ is the characteristic temperature.

The B value obtained from the distortion factor found in the crystallinity measurements also contains the effect of lattice distortions.

$$B = B_T + B_1 + B_2 = B_T + B_P \tag{13.62}$$

In general, however, the effect of the thermal vibrations predominates at high temperatures, and B_T, *i.e.* the expression in Eq. 13.61, may be taken as a good approximation to B. At low temperatures, on the other hand, the effect of thermal vibrations is small and the effect of lattice distortions (B_P) predominates. The effect of lattice distortions (B_P) is therefore given by the difference between the B' curve based on the plot of the measured B value against temperature and the theoretical B' curve obtained on the assumption that the high-temperature B value approximately follows Eq. 13.61, the curves obtained being extrapolated to 0 K (see Fig. 13.21). Distortion of the lattice distortion factor g from

$$g \cong (\overline{\Delta}^2)^{1/2}/\bar{a} = (1/\pi)[(B_P/2)^{1/2}/\bar{a}] \tag{13.63}$$

which closely resembles Eq. 13.59, gives $g = 0.06$ for low density polyethylene and $g = 0.04$ for Marlex 6015.

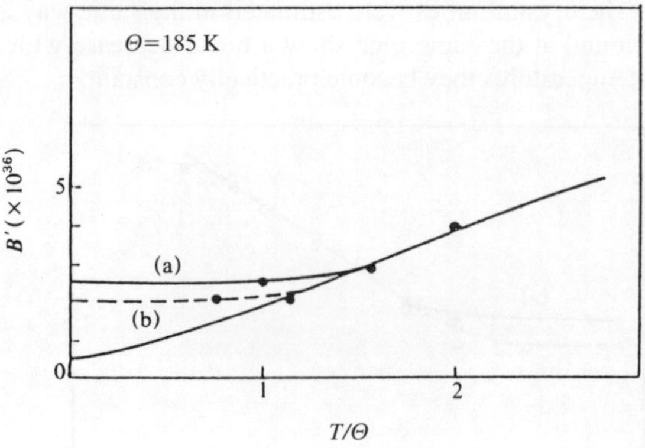

Fig. 13.21 Variation of B' with temperature according to the equation $B' = (6h^2/\kappa)\{\Phi(x)/x + 1/4\} = 2BM\Theta$.[33, 34)]
(a) Marlex 6015; (b) Low density polyethylene
[Reproduced with permission from H.G. Killian. *Kolloid-Z. Z. Polymere*, **183**, 10, Dr. Dietrich Steinkopff Verlag (1962)]

D. Determination of *g* from line breadth $\beta(h)$

The factor $g(=\Delta a/\bar{a})$, which expresses the degree of distortion of the lattice points (distortion of the second kind), was dealt with in some detail in Chapter 5, and examples of analytical methods have also been given. We shall now describe a method of calculating g directly from the broadening of the X-ray diffraction pattern.

The first step is to record the diffraction pattern and the intensity distribution (including the background intensity) over the entire range of angles with a diffractometer (or photometer). A one-dimensional intensity curve is then constructed for some fixed direction radiating from the origin of the reciprocal lattice and passing through a succession of reciprocal lattice points with indices which are simple multiples of each other, *e.g.* the 100, 200, 300, etc. or the 110, 220, 330, etc. points. The integral breadths $\beta(h)$ of the parts of these diffractions which emerge above the background level are then measured. The greater the number of diffractions the better, but a minimum of three (those with $h = 1$, 2, and 3) are necessary. The equation used is Eq. 13.41, and $\bar{a}$ is the interplanar spacing found from the first order diffractions. Thus, when the 100, 200, 300, etc. diffractions are used, $\bar{a} = d_{(100)}$. Substitution of $\bar{a}$ and $\beta(h)$ for the various order numbers h and elimination of L enables g to be found.

Alternatively, if we square both sides of Eq. 13.40, expanding the bracket containing the exponential term, and take the first two terms of the expansion, we obtain

$$\beta^2(h) \cong (1/\bar{a}^2)(1/N^2 + \pi^4 g^4 h^4) = 1/L^2 + \pi^4 g^4 h^4/\bar{a}^2 \qquad (13.64)$$

This shows that a linear relationship holds between the square of the integral breadth $\beta^2(h)$ of the diffraction and h^4. N and g can therefore be found from the gradient and the intercept on the ordinate, respectively, of the plot of $\beta^2(h)$ against h^4. Moreover, since $N\bar{a} = L$, the crystal dimension L can also be found.

E. Analysis in cases where the crystallite size is not uniform

If the crystallite size is not uniform, there will obviously no longer be a linear relationship between $\beta^2(h)$ and h^4. If the diffraction profile is assumed to be Gaussian, a method for the resolution of Gaussian functions having different breadths is applied, as in the Jellinek-Solomon-Fankuchen analysis for small-angle scattering[35] (*cf.* Section 6.2.5). A graph of log I against $|S - S_h|^2$ is drawn (Fig. 13.22(a))[36] for a given diffraction, using the values of $S - S_h(S = (2\sin\theta)/\lambda))$ measured from the maximum S_h of the intensity profile of this dif-

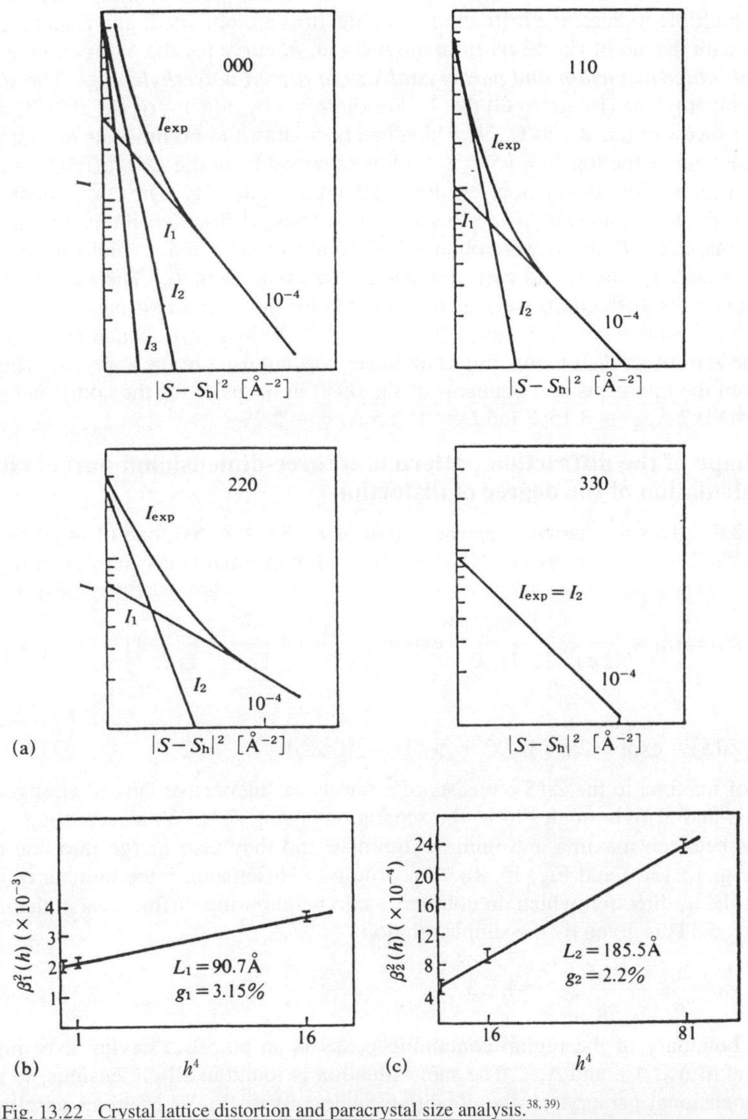

Fig. 13.22 Crystal lattice distortion and paracrystal size analysis.[38, 39]
(a) Log I *vs.* $|S - S_h|^2$ for *hh*0 diffractions of polyethylene with draw ratio 60 : 1.
(b) Plot of $\beta^2(h)$ *vs.* h^4 for component I_1(using $\beta(h)$ of the 000, 110, and 220 diffractions).
(c) Plot $\beta^2(h)$ *vs.* h^4 for component I_2 (using 000, 110, 220, and 330 diffractions).
[Reproduced with permission from F.J. Balta-Calleja *et al.*, *Makromol. Chem.*, **92**, 25, Hüthig & Wepf Verlag (1966)]

fraction. The component I_1 is found by drawing a tangent to the lower end of this curve. A second component I_2 is obtained by drawing a tangent to the lower end of the curve derived by subtraction of the diffraction intensity of the first tangent, *i.e.* of the component I_1, from the original diffraction intensity curve. Successive repetitions of this operation give the components I_1, I_2, I_3,..., and in this way the diffraction can be resolved into the component diffractions due to the crystals of different sizes. Using the I_1, I_2, I_3,... for various diffractions with multiple indices, the diffraction breadths $\beta(h)$ are obtained, and L and g can then be derived with the aid of Eq. 13.64 from the $\beta^2(h)$ *vs.* h^4 curve for the h direction used.

a) Crystal lattice distortion and paracrystal size in drawn polyethylene.[36] The specimen used was polyethylene (Hostalen G) that had been drawn by a factor of six at 70°C and then by a further factor of ten at 120°C, *i.e.* which had been drawn to 60 times its original length. Fig. 13.22(a) shows the log I *vs.* $|S - S_h|^2$ plots obtained from the $hh0$ diffraction profiles, including the zero order diffraction, *i.e.* the small angle scattering. The components I_1 were obtained by drawing tangents to the lower ends of these diffraction intensity curves, and then the components I_2 and I_3 were obtained. The log I *vs.* $|S - S_h|^2$ plot for the 330 diffraction was already linear, and corresponded to the component I_2. The integral breadths $\beta(h)$ of the various diffractions were then calculated for the resulting groups of I_1, I_2,..., and the $\beta^2(h)$ *vs.* h^4 plots were constructed (see Fig. 13.22(b) and (c)). Since I_3 was obtained only for the zero order diffraction, this component was not used in the analysis. The results derived from the intercepts and gradients of the $\beta^2(h)$ *vs.* h^4 plots for the components I_1 and I_2 were $L_1 = 90.7$ Å, $g_1 = 3.15\%$ and $L_2 = 185.5$ Å, $g_2 = 2.2\%$.

13.6.3 Shape of the diffraction pattern of a three-dimensional paracrystal and calculation of the degree of distortion

Figure 13.23[37] gives a schematic representation of $K_1(\boldsymbol{S})(=Z_1(\boldsymbol{S}))$, one of the three K factors of Eq. 5.32 for a given $H_1(x)$ $(=H_{100})$, a three-dimensional Gaussian function (*cf.* Fig. 5.11):

$$H_1(x+\bar{a}_1)=\frac{1}{(2\pi)^{3/2}\mathrm{D}_{11}\mathrm{D}_{21}\mathrm{D}_{31}}\exp\left[-\frac{1}{2}\left(\frac{x^2}{\mathrm{D}_{11}{}^2}+\frac{y^2}{\mathrm{D}_{21}{}^2}+\frac{z^2}{\mathrm{D}_{31}{}^2}\right)\right] \tag{13.65}$$

and

$$Z_1(\boldsymbol{S})=\exp[-2\pi^2(\Delta_{11}{}^2X^2+\Delta_{21}{}^2Y^2+\Delta_{31}{}^2Z^2)] \tag{13.66}$$

The zone of maxima in the $Z_1(\boldsymbol{S})$ consists of a family of 'biconcave lenses' arranged along the $\bar{a}_1{}^*$ axis and at right angles to it, the separations being $\bar{a}_1{}^*$. With increasing $|\boldsymbol{S}|$, the differences between maxima and minima diminish and they also merge into one another (see also Fig. 13.13(a) and Fig. 15.30). According to Hosemann,[38] the number of diffractions n_{i1} in the a_1 direction which do not merge into neighboring diffractions in the a_i direction (*cf.* Fig. 5.11) is given by the simple relation

$$g_{i1}=\frac{\Delta_{xi1}}{\bar{a}_1}=\frac{0.35}{n_{i1}},\quad i=1,2,3 \tag{13.67}$$

The outer boundary of the region containing peaks is an ellipsoid having axes inversely proportional to Δ_{11}, Δ_{21}, and Δ_{31}. The same situation is found in all dimensions, so that for a three-dimensional paracrystal, the condition analogous to Eq. 13.67 gives an ellipsoidal surface (a limiting ellipsoid of diffraction) in reciprocal space. This condition gives $|\mathcal{F}| \approx 0.1$ in Eqs. 5.30 and 13.30.

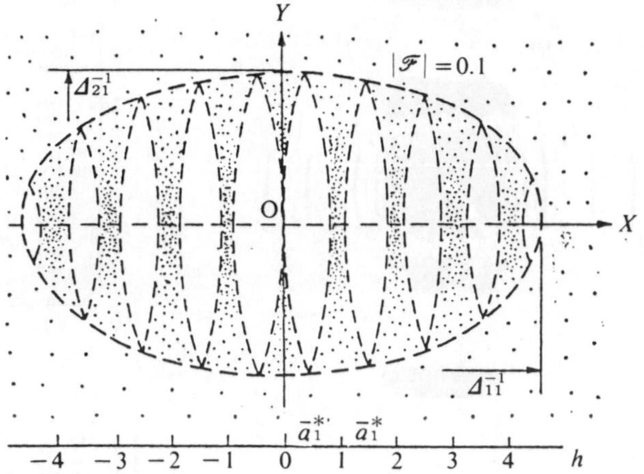

Fig. 13.23 Schematic representation of a diffraction pattern from a one-dimensional paracrystal.[37) [Reproduced with permission from B.K. Vainshtein, *Diffraction of X-rays by Chain Molecules*, p.233, Elsevier (1966)]

A. X-ray diffraction from β-keratin fibers from seagull's feathers[39)

Figure 13.24 shows the analysis of the X-ray diffraction from β-keratin fibers from seagulls' feathers in light of the above results. Diagram (a) shows the X-ray diffraction pattern, and diagram (b) shows a representation of (a) by means of reciprocal lattice coordinates in a plane containing the fiber direction. The line –·–·– delineates the boundary $|\mathscr{F}| = 0.1$ for the h_2 direction, and the line –··–··– delineates that for the h_1 direction; the two together give an ellipse. The average lattice constants $\bar{a}_1$, $\bar{a}_2$, and $\bar{a}_3$ can be found from the diffraction angles, and g_1, g_2, g_3, and $\Delta_i x_k$ can be obtained from the diffraction orders n_{ik} on either side of the boundary.

For example, in the range $h_2 \leqq 4$ in diagram (b) of Fig. 13.24, the diffraction pattern is indistinguishable at $h_1 = 7$ to 8, and in the meridional direction, the boundary passes through (2, 15) and (3, 14) to meet the axis at $h_2 = 17$. Thus from Eq. 13.67 we find

$$n_{21} = 17 = 0.35\bar{a}_2/\Delta_1 x_2, \qquad \Delta_1 x_2 = 0.35 \times 95/17 = 1.8\ \text{Å}$$

Similarly, from $n_{11} = 5$ we obtain $\Delta_1 x_1 = 0.35\bar{a}_1/5 = 2.5$ Å. The various constants found in this way are given in (c) of Fig. 13.24. H_1, H_2, and H_3 can then be calculated from the degrees of distortion. Diagram (d) shows H_1, H_2, and $H_1 * H_2$. Figure 13.24(e) shows a model of the paracrystalline lattice constructed with the aid of (d), and (f) shows the diffraction pattern obtained by its optical Fourier transformation (*cf.* Section 5.2.4). The broadening of this pattern is very similar to that of the X-ray diffraction pattern in (a). This provides a good example of one fruitful application of Hosemann's theory. In real structures, however, the atoms are not so point-like, hence the structure factor $< A_{\text{cell}} >^2$ weakens the intensity at larger values of $|\boldsymbol{S}|$. In this case the limiting ellipsoid of diffraction is actually given by $|\mathscr{F}| \approx 0.3$ and the numerical factor Eq. 13.67, 0.35, becomes 0.24.[39)

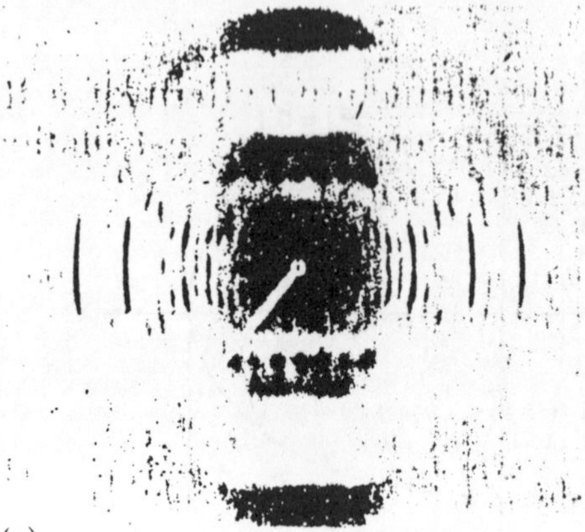

(a)

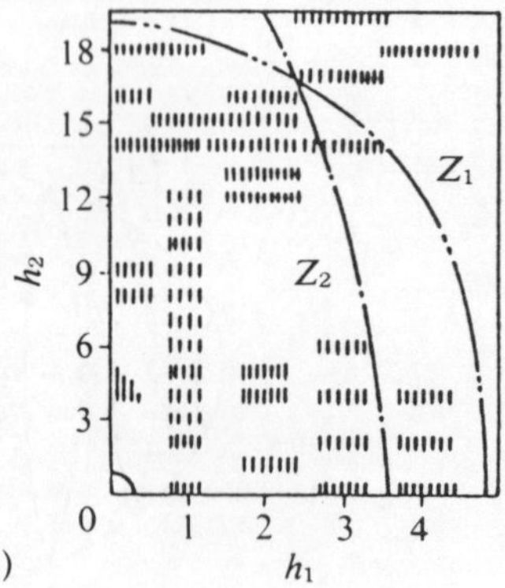

(b)

Paracrystal parameter	Result of analysis
$\bar{a}_1$	34[Å]
$\bar{a}_2$	95
$\bar{a}_3$	?
$\Delta_2 x_2$	1.4
$\Delta_2 x_1$	3.7
$\Delta_1 x_1$	2.5
$\Delta_1 x_2$	1.8
Long period, L_1	130
L_2	950
L_3	?

(c)

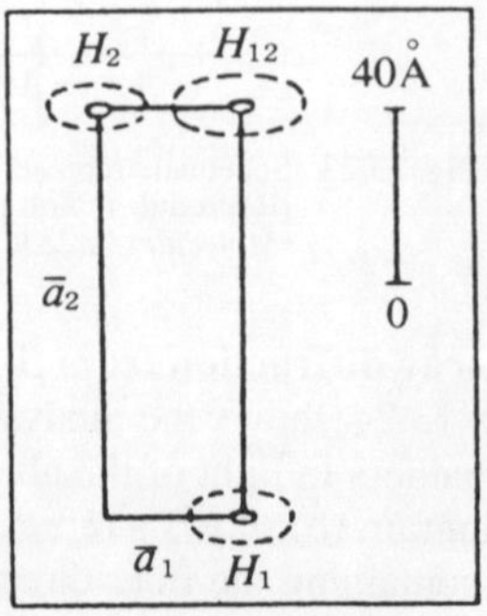

(d)

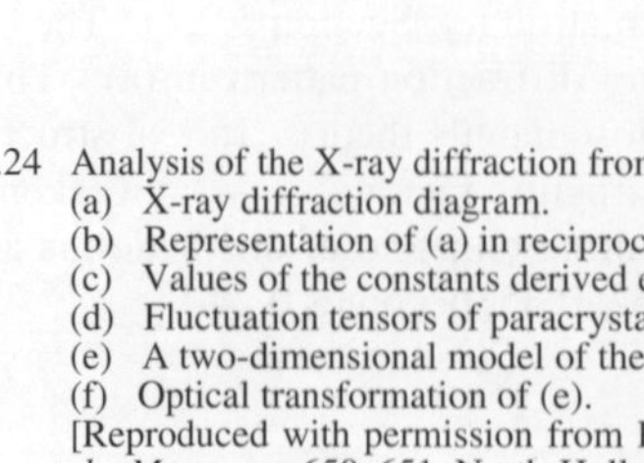

(e)

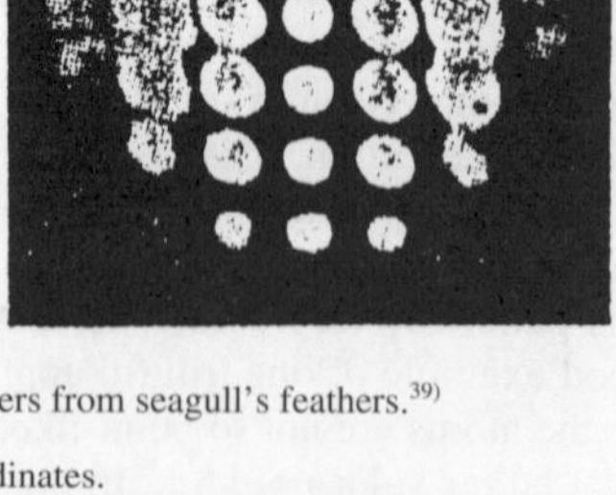

(f)

Fig. 13.24 Analysis of the X-ray diffraction from β-keratin fibers from seagull's feathers.[39]
(a) X-ray diffraction diagram.
(b) Representation of (a) in reciprocal lattice coordinates.
(c) Values of the constants derived experimentally.
(d) Fluctuation tensors of paracrystalline superlattice cell.
(e) A two-dimensional model of the paracrystalline super lattice constructed from (d).
(f) Optical transformation of (e).
[Reproduced with permission from R. Hoseman, S.N. Bagchi, *Direct Analysis of Diffraction by Matter*, pp.650, 651, North-Holland Pub. (1962)]

References

1. L.V. Azároff, *Elements of X-ray diffraction*, McGraw-Hill, N.Y. (1968).
2. a) H.K. Klug, L.E. Alexander, *X-Ray Diffraction Procedures for Plycrystalline and Amorphous Materials*, John Wiley & Sons, Inc., N.Y. (1954).
   b) F.W. Jones, *Proc. Roy. Soc.*, (*London*), **A 166**. 16 (1938).
3. R.A. Young, R.J. Gierder, A.J.C. Wilson, *Act. Cryst.*, **22**, 155 (1969).
4. A.J.C. Wilson, *Elements of X-Ray Crystallography*, Chapter 10, Addison-Wesley Reading (1969).
5. *X-Ray Crystallography*, (I. Nitta ed.), Maruzen, Tokyo, Vol. I (1959) and Vol. II (1961).
6. B.E. Warren, *Phys. Rev.*, **59**, 693 (1941).
7. F. Schossberger, *Adv. X-Ray Analysis*, **1**, 73 (1957).
8. T. Ishikawa, A. Sugihara, T. Harada, S. Nagai, N. Yasuoka, N. Kasai, *J. Chem. Soc. Jpn., Chem. and Ind. Chem.*, 1744 (1973).
9. A. Keller, A. Maradudin. *Phys. Chem. Solids*, **2**, 301 (1957).
10. C.W. Bunn, E.V. Garner, *Proc. Roy. Soc.* (*London*), **A 189**, 39 (1947).
11. A.J.C. Wilson, *X-Ray Optics*, p. 4, Methuen, London (1949).
12. N.C. Halder, C.N.J. Wagner, *Adv. X-ray Analysis*, **9**, 91 (1966).
13. R.K. Nandai, S.P. Sen Gupta, *J. Appl. Cryst.*, **11**, 6 (1978).
14. Th.H. de Keijser, J.L. Langford, E.J. Mittemeijer, A.B.P. Vogels, *J. Appl. Cryst.*, **15**, 308 (1982).
15. J.I. Langford, *J. Appl. Cryst.*, **11**, 10 (1978).
16. A.M. Hindeleh, D.J. Johnson, *Polymer*, **13**, 423 (1972).
17. M.M. Hall Jr., V.G. Veeraraghaven, H. Rubin, P.G. Winchell, *J. Appl. Cryst.*, **10**, 66 (1977).
18. R.A. Young, D.B. Wiles, *J. Appl. Cryst.*, **15**, 430 (1982).
19. R. Delhez, Th.H. de Keijser, E.J. Mittemeijer, *Fresenius Z. Anal. Chem.*, **312**, 1 (1982).
20. J.N. Eastbrook, A.J.C. Wilson, *Proc. Phys. Soc.*, (*London*), **B 65**, 67 (1952).
21. Th.H. de Keijser, E.J. Mittemeijer, H.C.F. Rosendaal, *J. Appl. Cryst.*, **16**, 309 (1983).
22. In ref. 2a) (Sect. 7.3.1).
23. L.E. Alexander, *X-Ray Diffraction Methods in Polymer Science*, pp. 437-452, John Wiley & Sons, Inc., N.Y. (1969).
24. E.F. Bertaut, *Compt. rend.*, **228**, 492 (1949).
25. B.E. Warren, B.L. Averbach, *J. Appl. Phys.*, **21**, 595 (1950).
26. B.E. Warren, *Acta Cryst.*, **8**, 483 (1955).
27. B.E. Warren, Averbach, *J. Appl. Phys.*, **23**, 497 (1952).
28. D.B. Buchanan, R.L. Miller, *J. Appl. Phys.*, **37**, 4003 (1966).
29. W. Ruland, *Acta Cryst.*, **13**, 1059 (1960).
30. W. Ruland, *Acta Cryst.*, **14**, 1180 (1961).
31. W. Ruland, *Polymer*, **5**, 89 (1964).
32. W. Ruland, *Faserforsch. Textiltech.*, **15**, 532 (1964).
33. H.G. Kilian, *Koiloid-Z. Z. Polymere*, **183**, 1 (1962).
34. H.G. Kilian, *Kolloid-Z. Z. Polymere*, **185**, 13 (1962).
35. M.H. Jellinek, E. Solomon, I. Fankuchen, *Ind. Eng. Chem.*, **37**, 158 (1945).
36. F.J. Balta-Calleja, R. Hosemann, W. Wilke, *Makromol. Chem.*, **92**, 25 (1966).
37. B.K. Vainshtein, *Diffraction of X-Rays by Chain Molecules*, p. 233, Elsevier, Amsterdam (1966).
38. R. Hosemann, S.N. Bagchi, *Direct Analysis of Diffraction by Matter*, p. 334, North-Holland, Amsterdom (1962).
39. In Ref. 38, p. 333.

14. Analyses Using the Total Diffraction Intensity Distribution Curves of High Polymers

In the method of paracrystal analysis described in the previous chapter, and in the method for the determination of crystallinity from the intensity of scattering due to the amorphous component to be described in Section 14.2 below, as with many other methods of X-ray diffraction analysis, it is necessary to deal with the background scattering (as distinct from the crystalline diffractions) over the entire angular range of measurable intensities. Adequate precision can usually be obtained, in such cases, with relative measurements of intensities. However, just as the absolute intensity is essential for the analysis of crystal structure, so the determination of the absolute intensities gives access to more precise information when the background diffraction (which arises from interatomic coherent scattering) is to be used for structural analysis. The determination of $P(r)$ in Sections 2.5 to 2.8 is an example of this kind of analysis.

We commence this chapter with an account of methods for eliminating systematic errors from the continuous diffraction intensity background as distinct from the crystalline diffractions, not only so that overall quantitative factors such as crystallinity can be determined, but also in order to extend the analysis to the details of atomic structure.

14.1 Correction for Coherent Background Scattering

14.1.1 Correction of the measured intensity for the effect of polarization

As was mentioned earlier in connection with crystal structure analyses, polarization is present in the atomic scattering intensities themselves, and a correction is normally made for this by multiplying by the reciprocal of the polarization factor $(1+\cos^2 2\theta)/2$ over the entire scattering intensity curve. When a monochromator is used, the polarization factor becomes $(1+\cos^2 2\theta_M \cos^2 2\theta)/2$, where θ_M is the Bragg angle of the monochromator.

14.1.2 Normalization of the scattering intensity

After correction for the effect of polarization, the scattered intensity still includes an element due to Compton scattering. This has an intensity given by Eq. 14.1, where c is a scaling constant.

$$I(s) = cI_{obs}(s) \times 1/(\text{polarization factor}) = I_a(s) + I_m(s) + I_{inco}(s) \qquad (14.1)$$

I_{inco} is the Compton scattering intensity, and $s = (4\pi \sin\theta)/\lambda$.

The coherent scattering has been resolved into I_a(the first term of Eq. 2.28′, p. 35) and I_m(the second term of Eq. 2.28′) (*cf.* Fig. 2.13). When the scattering angle 2θ is very large, the effect of $I_m(s)$ is negligible, and we obtain

$$I(s) \cong I_a(s) + I_{inco}(s) \qquad (14.2)$$

In this equation,

$$I_a(s) \cong \sum_{j=1}^{n} f_j^2(s) \qquad (14.3)$$

where f_j is the atomic scattering factor of the jth atom.

The incoherent scattering term is given by Eq. 2.18.

$$I_{inco} = I_e R_{BD}[Z - \sum |f_{mm}|^2 - \sum_{m \neq n}\sum |f_{mm}|^2]$$

If the number and nature of the atoms in the structure are known, $I_a(s)$ and $I_{inco}(s)$ can be calculated from Eqs. 14.3 and 2.18 after the appropriate atomic scattering factors[1)] and Compton scattering factors[2)] have been found from *International Tables for Crystallography*, Vol. C. The right-hand side of Eq. 14.2 is then calculated, and the scaling constant c is found by comparison at sufficiently large scattering angles with the $I_{cor}(s)$ obtained by correction of the observed scattering intensity for the polarization, absorption, etc. The $I(s)$ of Eq. 14.1 is then obtained by using this constant c to normalize $I_{inco}(s)$. The coherent diffraction $I_{co}(s)$ can be found in crystallinity analyses by elimination of $I_{inco}(s)$, and the molecular interference term $I_m(s)$ can also be found, *e.g.* in the analysis of the radial distribution function $P(r)$, by subtracting $[I_a(s) + I_{inco}(s)]$ from $I(s)$.

14.2 Determination of Crystallinity

As illustrated in Fig. 14.1, for the determination of crystallinity, *unoriented diffraction diagram* should be used in order to measure the intensity distribution along the radial direction.

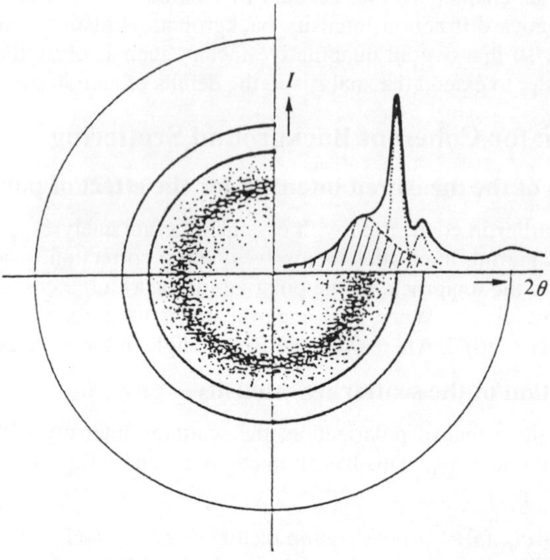

Fig. 14.1 Schematic illustration of an diffraction diagram of an unoriented crystalline polymer. Radial intensity distribution is given in the first quadrant (*cf.* Fig. 14.4).

Fig. 14.2 shows intensity distribution of a crystalline polyethylene measured at different temperatures.[3)] Sharp crystalline 110 and 200 diffractions were observed overlapped on the first peak of amorphous halo at 27°C. With increase of temperature from 60° to 100°C, the crystalline diffraction peaks showed remarkable decrease in intensity, whereas the amorphous peak showed intensity increase. Finally, at 120°C the crystalline peaks disappeared completely and only the amorphous halo increased in intensity. These results have made easier the separation of the first peak of amorphous halo and two crystalline diffraction peaks after the correction for incoherent scattering, polarization effect, and other background scattering.

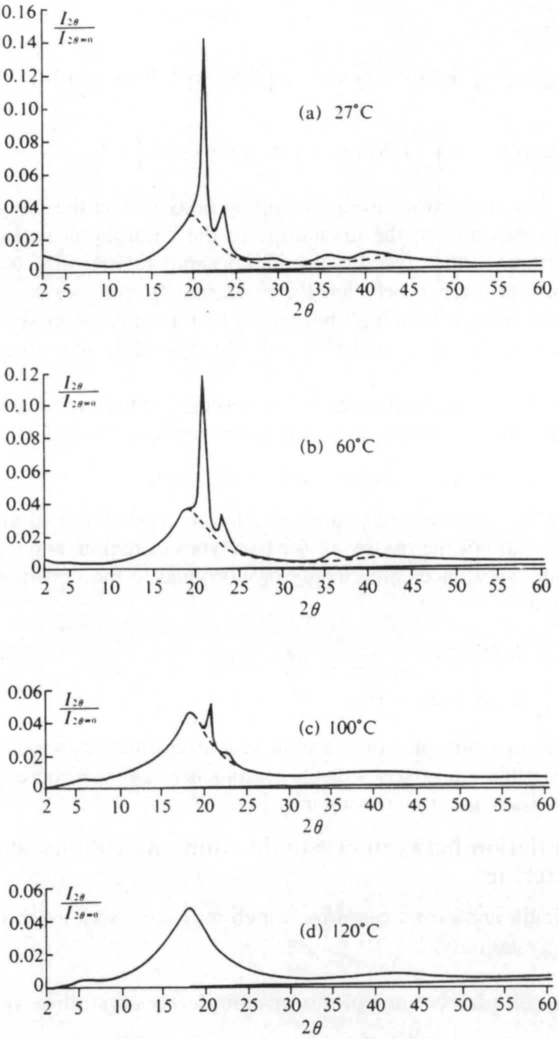

Fig. 14.2 X-ray diffraction intensity curves of an unoriented polyethylene specimen at four different temperatures.[3] [Reproduced from S. Krimm, A.V. Tobolsky, *J. Polym. Sci.*, **7**, 52, 53, John Wiley & Sons, Inc. (1951)]

14.2.1 Principles of the measurement of crystallinity

Under conditions of constant incident X-ray intensity, the total intensity of the coherent scattering from a scattering body of fixed mass should be constant, irrespective of the steric structure of the atoms inside the body.

Striking with a structure of N electrons, whose scattering amplitude is given by Eq. 2.10, the Q function has at $x = 0$ a hump, which consists of N self-convolution products of each electron density $s_k(x)$. Since the integral intensity of an electron is given by (*cf.* Eq. 2.3)

$$\int I_e dv_S$$

one obtains from Eqs. 5.5 and 5.9

$$Q(x) = \int \rho(y)\rho(x+y)\mathrm{d}v_y = \int I(S)\exp\{2\pi i(S\cdot x)\}\mathrm{d}v_S \tag{14.4}$$

$$\lim_{x\to 0} Q(x) = \int_{-\infty}^{\infty} I(S)\mathrm{d}v_S = \int \rho^2(y)\mathrm{d}v_y = N\int I_e \mathrm{d}v_S, \tag{14.5}$$

N is the total number of electrons in the scattering body so that the integral intensity will be constant whatever the ratio of the crystalline to the amorphous regions in the scattering body. If the structures of the crystalline and amorphous regions are clearly distinguishable, and if the mutual interference between the scattering from these two regions may be ignored, the total scattering from a high polymer which remains after subtraction of the incoherent background scattering is separated into the crystalline scattering I_{cr} and amorphous scattering I_{am}.

The crystallinity of a high polymer, x_{cr}, is defined in terms of its total mass, M, and the respective masses of the crystalline and amorphous regions, M_{cr} and M_{am}, by

$$x_{cr} = M_{cr}/M, \qquad \text{where} \qquad M_{cr} + M_{am} = M \tag{14.6}$$

Thus, the ratio of the scattering intensities due to the crystalline and amorphous regions is nearly equal to the ratio of the masses of the two types of region, since the average electron densities of organic substances are closely proportional to the densities of the substances. Hereupon we obtain

$$\frac{\int_0^{\infty} I_{cr}(S)S^2\mathrm{d}S}{\int_0^{\infty} I_{am}(S)S^2\mathrm{d}S} = \frac{N_{cr}}{N_{am}} = \frac{M_{cr}}{M_{am}} \tag{14.7}$$

Thus the principle that the ratio of the total scattering intensities of these two regions is equal to the ratio of the total masses of the regions is established here. This is the basis of the crystallinity measurement in general use.

14.2.2 Differentiation between crystalline and amorphous scattering in coherent scattering

This is very difficult, and errors can arise which may have serious consequences upon the measurement of crystallinity.

A. Cases where completely amorphous or completely crystalline specimens cannot be obtained

Since there are no *a priori* indications of the true apportionment, a smooth curve is drawn as in Fig. 14.3[4] (not to be confused with Fig. 14.5) to connect the minima between the crystalline peaks on the assumption that the intensity above it is contributed by the crystalline region, I_{cr}, and that below it by the amorphous region, I_{am}. This method gives an excessively high value for I_{am}, with a crystallinity x_{cr} which is therefore smaller than the true value. The principal sources of error are as follows.

1) The smooth curve will inevitably mask weak crystalline diffractions (including paracrystalline diffractions) and transfer their contribution to amorphous scattering, I_{am}, effectively reducing the measured crystallinity.

2) The density of packing of the molecular chains does not vary greatly from the crystalline to the amorphous regions, so that the strong crystalline peaks are usually superim-

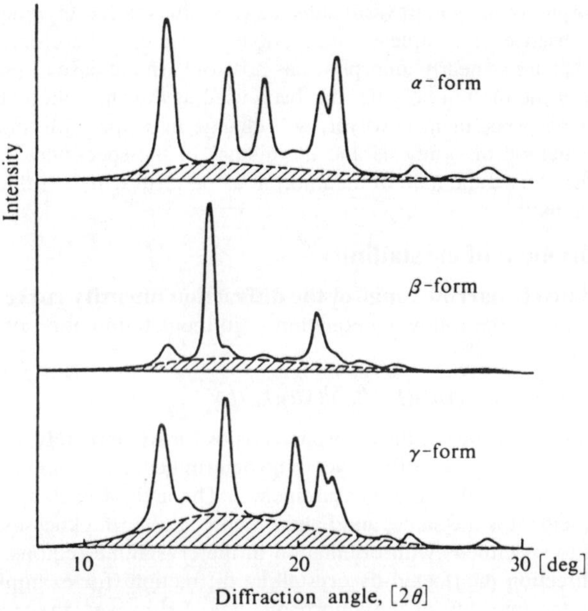

Fig. 14.3 Resolution of the diffraction pattern of isotactic crystalline forms of polypropylene.[4)] Shaded area: amorphous scattering. Background due to incoherent scattering is not shown. [Reproduced with permission from A. Turner-Jones *et al.*, *Macromol. Chem.*, **75**, 134, Hüthig & Wepf Verlag (1964)]

posed on the maxima of amorphous halos, and the base of the crystalline peak is included in I_{am}, again with an effective reduction in the measured crystallinity.

3) Diffuse scattering due to crystal lattice imperfections, and in particular the background scattering due to crystal lattice distortions of the first kind, cannot be properly distinguished from the amorphous halo. It is generally impossible to assess what effect the contribution of distortions of the first kind may have upon the measured crystallinity.

It is obvious from the foregoing that the crystallinity values obtained in this way are suspect, depending as they do upon how the amorphous scattering curve is drawn, and that too much credence cannot be given to their absolute values. In this respect, the crystallinity determination of Section 13.6.2B, which involved the application of the paracrystal theory, is probably to be preferred; it does take into account formally the structural distortion that is a characteristic feature of high polymers.

B. Cases where completely amorphous or completely crystalline specimens can be obtained

The basic problem of resolving the diffraction pattern into its crystalline and amorphous elements still remains in these cases, but it is possible, for instance, to determine the amorphous fraction by comparing the intensity of the halo for a completely amorphous specimen over a certain angular range with that for the unknown specimen under identical conditions. The crystallinity can thus also be obtained.

In practice, there are few so-called crystalline polymers which can be obtained in the completely amorphous state apart from substances such as poly(ethylene terephthalate) and rubber. In the absence of completely amorphous specimens, therefore, it is customary to use specimens that are as nearly amorphous as possible. In the case of polyethylene, for example, the halo of the molten polymer has been used as the amorphous halo.[3,5] The amorphous halo of a real specimen, however, will change in shape with temperature, and also varies with the method of synthesis and the history of the specimen.[5] Matthews, Peiser, and Richards[6] have used the halo of an amorphous paraffin with a similar structure instead of the halo of the melt.

14.2.3 Measurement of crystallinity

A. Using a relatively narrow range of the diffraction intensity curve

It is convenient to use the following equation[7–9] in the determination of crystallinity using the amorphous halo.

$$1 - x_{cr} = (1/t')(I_{am}'/I_0')/(1/t)(I_{am}/I_0) \tag{14.8}$$

I_{am} and I_{am}' are the intensities of the amorphous halos for a completely amorphous specimen and for the specimen with crystallinity yet to be determined, and I_0 and I_0' are the intensities of the incident X-rays used in the measurements. The area of each specimen upon which the X-rays are incident is the same, and t and t' here are their thicknesses. This method can also be applied to specimens with orientation in the crystalline regions, in which case the intensity in a direction unaffected by crystalline diffraction (for example, the meridional) may be used as the intensity of the amorphous halo. However, the use of the method is open to question when orientation is also present in the amorphous halo.

The number of empirical equations which have been proposed for the estimation of crystallinity of individual substances is very great. The reader is referred to other textbooks[10–12] and reviews[13] for details.

a) Crystallinity of polyethylene determined by an empirical equation. The correction for the background scattering due to incoherent scattering and other reasons is first carried out. Based on the changes in diffraction intensity distribution curves at different temperatures shown in Fig. 14.2, the amorphous scattering was assumed to be similar to that of the molten sample (Fig. 14.2(d)), and it was first subtracted from the corrected intensity distribution curve. The resolution of the remaining two crystalline diffraction patterns is based on the symmetric nature of the 110 and 200 diffractions (Fig. 14.4). The areas under the three component curves, A_{am}, A_{110} and A_{200}, were determined planimetrically (or by graphic integration), which should be corrected for Lorentz, polarization and thermal factors. The crystallinity can be determined by the empirical equation,

$$\frac{1-x_{cr}}{x_{cr}} = K\frac{I_{am}}{I_{110}+I_{200}} = K\frac{Q_{am}A_{am}}{Q_{110}A_{110}+Q_{200}A_{200}} \tag{14.9}$$

or

$$x_{cr} = \frac{Q_{110}A_{110}+Q_{200}A_{200}}{Q_{am}A_{am}+Q_{110}A_{110}+Q_{200}A_{200}} \tag{14.9a}$$

where the combined correction factors are $Q_{am} = 0.75$, $Q_{110} = 1.0$ and $Q_{200} = 1.46$, respectively.[14] Mathews and coworkers gave values of these factors as 0.74, 1.0 and 1.43, respectively, whereas Bryant *et al.* gave 0.69, 1.0 and 1.43[15].

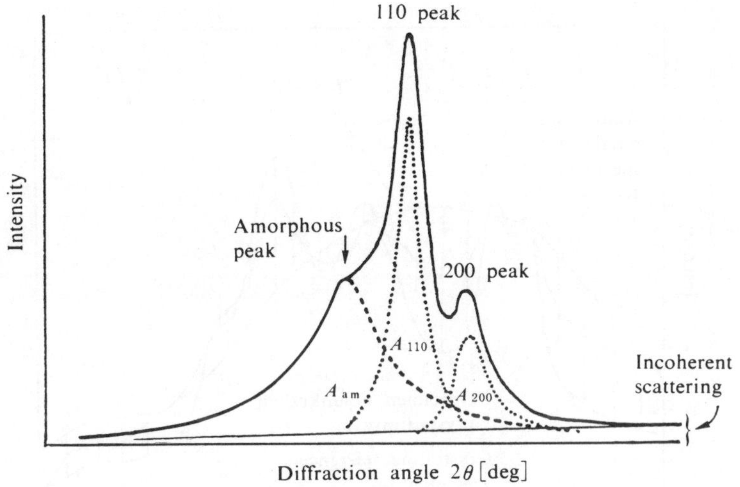

Fig. 14.4 Resolution of the diffraction pattern of polyethylene (See text).

B. Using a wide range of the diffraction intensity curve[16,17] (Crystallinity Index)
Instead of seeking to establish an absolute crystallinity for a given substance, it may be enough to find a relative crystallinity (a crystallinity index) with respect to reference specimens which exhibit the highest degree of observable crystallinity and, on the other hand, which approach most closely the completely amorphous. Let the diffraction intensity of the specimen with unknown crystallinity take the values I_u over a wide angular range, and let the corresponding intensities of the most crystalline and the most nearly amorphous specimens of the same substance over the same wide range of angles be I_{cr} and I_{am} respectively. The crystallinity x_{cr} is then given by the following equation.

$$x_{cr} = [(I_u - I_{am})_j - K]/(I_{cr} - I_{am})_j \tag{14.10}$$

and the analytical procedure is as follows.

1) The diffraction intensity differences $(I_u - I_{am})_j$ and $(I_{cr} - I_{am})_j$ are determined at various diffraction angles j.

2) The results are plotted on a graph with $(I_u - I_{am})_j$ as the ordinates and $(I_{cr} - I_{am})_j$ as the abscissae.

3) The straight line that best fits the plotted points is drawn. The slope of this straight line is the crystallinity x_{cr}. It takes the value unity for a 100% crystalline specimen, and approaches zero for a completely amorphous specimen. If the scales of the various diffraction intensity curves are well matched, the intercept K is zero.

a) Crystallinity index of poly(ethylene terephthalate).[17] Figure 14.5(a) shows the diffraction intensity profiles of crystalline and amorphous reference specimens of unoriented polyethylene terephthalate, together with that of the specimen with undetermined crystallinity. The plot of intensity differences gives the graph shown in Fig. 14.5(b). From the slope of the straight line, the crystallinity index $x_{cr} = 0.695$.

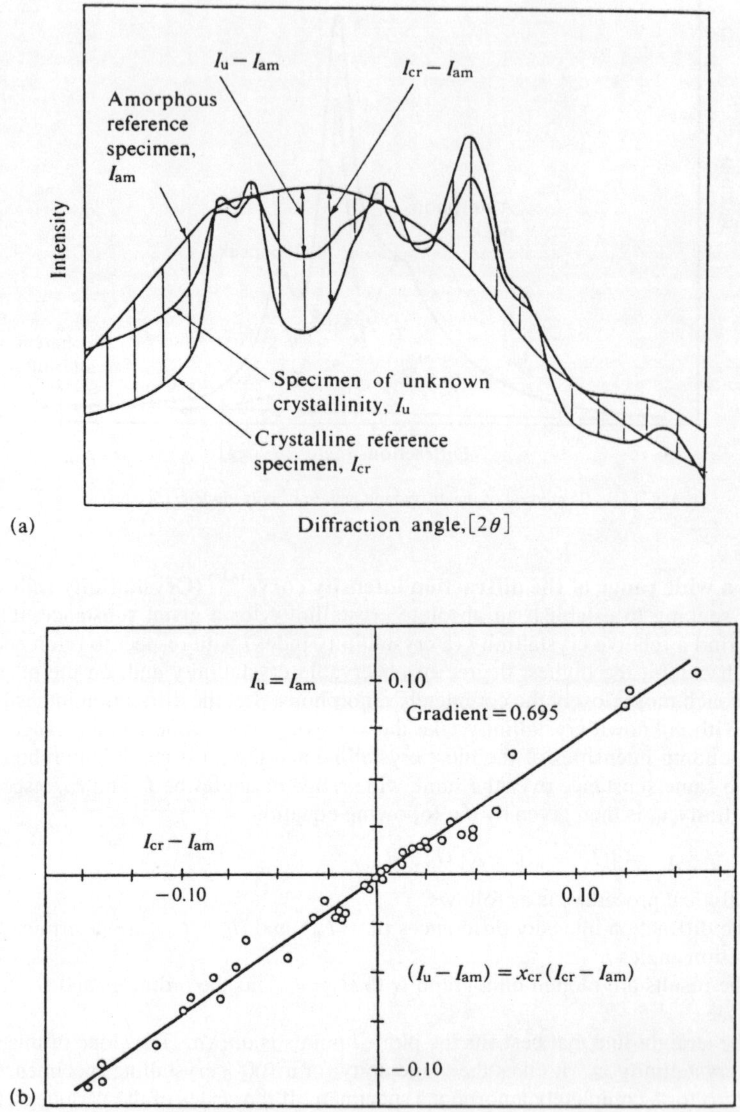

Fig. 14.5 Derivation of the crystallinity index x_{cr} from the diffraction intensity curves for polyethylene terephthalate.[16,17]
(a) Diffraction intensity curves for the specimen, and for crystalline and amorphous reference specimens.
(b) Plot of the intensity differences derived from (a), see text.
[Reproduced with permission from W.O. Statton, *J. Appl, Polym. Sci.*, **7**, 803, John Wiley & Sons, Inc. (1963)]

C. Determination of the crystallinity of a paracrystalline specimen
(See Section 13.6.2A)

D. Other methods for the assessment of crystallinity

The above methods are based on th assumption that the crystalline regions have an ideal crystal lattice. The measurement of crystallinity for paracrystalline structures with lattice distortions has been described in Section 13.3.2. and methods of analysis based on small-angle scattering will be discussed in Section 15.4.

E. Comparison with methods of crystallinity measurement not involving X-rays

Crystallinity can be measured by many methods other than those described above. The most important of these are summarized in Table 14.1.

Apart from the extreme case, where perfect crystals are mixed with completely amorphous material, high polymers are generally thought to contain regions in which the distortion of the arrangement of the molecular chains shows a continuous, gradual variation from the regular arrangement of crystalline regions to the completely random arrangement of the amorphous regions. If a line is drawn through such regions of gradual variation such that everything on the side with the more regular arrangement is regarded as crystalline and everything on the other side as amorphous, the "crystallinity" will vary according to where

Table 14.1 Principal methods of crystallinity measurement

Method		Crystallinity (x_{cr})	
X-ray	Wide-angle diffraction Small-angle scattering	Intensity of crystalline diffraction I_{cr} Intensity of amorphous halo I_{am} etc. See text Long period L and crystallite size t	$x_{cr} \approx t/L$
IR spectroscopy	Bands that change with crystallinity[†] Dichroism	Extinction of crystalline band E_c Extinction of amorphous band E_a	$\frac{x_{cr}}{1-x_{cr}} = \frac{E_c}{E_a} \cdot k$
NMR spectrpscopy		Width and shape of the absorption curve	
Density		Density of crystalline fraction ρ_c $(=\rho_x)$ Density of amorphous fraction ρ_a Density of specimen ρ	$\frac{1}{\rho} = \frac{x_{cr}}{\rho_c} - \frac{1-x_{cr}}{\rho_a}$
Thermal methods	Measurement of specific heat Differential thermal analysis Heat of combustion	Heat of fusion ΔH_c of a 100% crystalline specimen Enthalpy H_a of the amorphous material Enthalpy H of the specimen	$x_{cr} = \frac{H_a - H}{\Delta H_c}$
Adsorption	Latent heat of freezing of adsorbed water Adsorption ratio (water, iodine) Hailwood-Horrobin method	Ratio of quantity adsorbed by test specimen to that of reference specimen Adsorption isotherm	Measurement of accessibility a: accessibility of specimen a_{cr}: accessibility of crystalline fraction $a = a_{cr}x_{cr} + (1-x_{cr})$
Other chemical methods	Heavy hydrogen exchange Hydrolysis rate Acetal formation		

Other methods tried: comparison of dichroism and birefringence, comparison of X-ray and optical orientation, recrystallization of amorphous specimens, moisture permeability, dielectric constant, gas diffusion and density, modulus of elasticity in torsion, elongation. etc.

[†] In addition to the bands mentioned here, bands that are unaffected by crystallization and bands that depend on both the crystalline and the amorphous fractions have been used.

the line is drawn. Each method of measurement will, in effect, have its own characteristic means of settling the position of this boundary line. Thus, it is the exception rather than the rule for the values given by the methods listed in Table 14.1 to be in agreement for the same specimen.

These very differences, however, and the characteristics of the methods of measurement which give rise to them, can be exploited to obtain further information about the molecular structure, crystal structure, or fine texture. For example, copolymers of styrene and *o*-fluorostyrene are always classified as crystalline under X-ray analysis, but the IR spectra indicate that copolymers with *o*-fluorostyrene contents from 30 to 70% are *amorphous*,[18] as shown in Table 14.2. This is because the sequence of styrene and *o*-fluorostyrene units is completely random, with the result that the so-called crystalline bands in the IR spectrum, which faithfully reflect the partial structures, absorb only very weakly, and are therefore not observed. On the other hand, the molecular chains have the same fiber period as the homopolymers (6.63 Å), and the presence or absence of a fluorine atom attached to the benzene ring makes little difference to the cross section normal to the molecular chains, so that the packing of the chains is practically unaffected; consequently the copolymers always crystallize. The marked difference between the crystallinities found by X-ray methods and by IR spectroscopy is therefore highly significant.

Table 14.2 Comparison between the X-ray and IR crystallinities of copolymers[18]

Polymer	Crystallinity	
	X-ray	IR
Isotactic polystyrene	crystalline	crystalline
Styrene/ *o*-fluorostyrene copolymer	crystalline	*amorphous* (for compositions between 30 and 70%)
Isotactic poly-*o*-fluorostyrene	crystalline	crystalline

Polymer	M.p. [°C]	Unit cell dimensions [Å]
Isotactic polystyrene	235	$c = 6.63,\ a = 21.90$
Styrene/ *o*-fluorostyrene copolymer	↕ continuous variation	$c = 6.63$, ↕ continuous variation
Isotactic poly-*o*-fluorostyrene	270	$c = 6.63,\ a = 21.10$

14.3 Analysis of the Radial Distribution Function *P*(*r*)

14.3.1 Calculation of the radial distribution function

In the most commonly occurring cases, *i.e.* isotropic amorphous solids and liquids, the radial distribution function *P*(*r*) is given by Eq. 2.35 (Section 2.8).

$$4\pi r P(r) = (2/\pi)\int_0^\infty sI'(s)\sin sr\,ds \qquad (2.35)$$

where $s = (4\pi \sin\theta)/\lambda$.

Thus the distribution function *P*(*r*) can be calculated from the Fourier transform of the reduced scattering intensity $I'(s)$, Eq. 2.35.

The radial distribution function obtained in this way is used as a basis for the investigation of the average macromolecular structure in isotropic amorphous specimens. The statistical arrangements of the nearest neighbors and of the next-nearest neighbors of a given atom, the short-range order of the structure, can be found from the radial distribution function of an amorphous solid or liquid. Particularly, the contribution of atoms in the same molecule (intramolecular radial distribution function $P_{intra}(r)$) can be estimated by comparison with the radial distribution functions of similar amorphous polymers, from the temperature dependence of the radial distribution function of the same polymer,[19] or from the distribution of heavy atoms introduced into the polymer molecules.[20] Subtraction of $P_{intra}(r)$ gives the intermolecular radial distribution function $P_{inter}(r)$, from which we can obtain information about the arrangement of the molecular chains in relation to one another.

A. *Termination-effect correction*

Whereas the angular range of the Fourier integral giving the radial distribution function $P(r)$ is $0-\infty$, the experimental scattering intensity curve is recorded between the measurement limits s_1 and s_2. The lower limit s_1 is close to zero, while the upper limit s_2 is the highest value at which measurements are possible, and depends on the apparatus and the wavelength of the X-rays. Since the integrated function $sI'(s)$ in Eq. 2.33 tends to zero as $s \to 0$, the termination (truncation) effect in the region where s is smaller than s_1 is negligible. Where s is large, however, $sI'(s)$ is also large. This terminal effect leads to broadening of the peak of the distribution curve $P(r)$, and a secondary-maximum appears on the flank of the peak, which results in large errors. It is usual to induce an artificial convergence of $sI'(s)$ by multiplying by a factor such as $\exp(-as^2)$.

14.3.2 The radial distribution function of Nylon 6,6

Figure 14.6(a) shows the radial distribution function $P(r)$ obtained from the amorphous

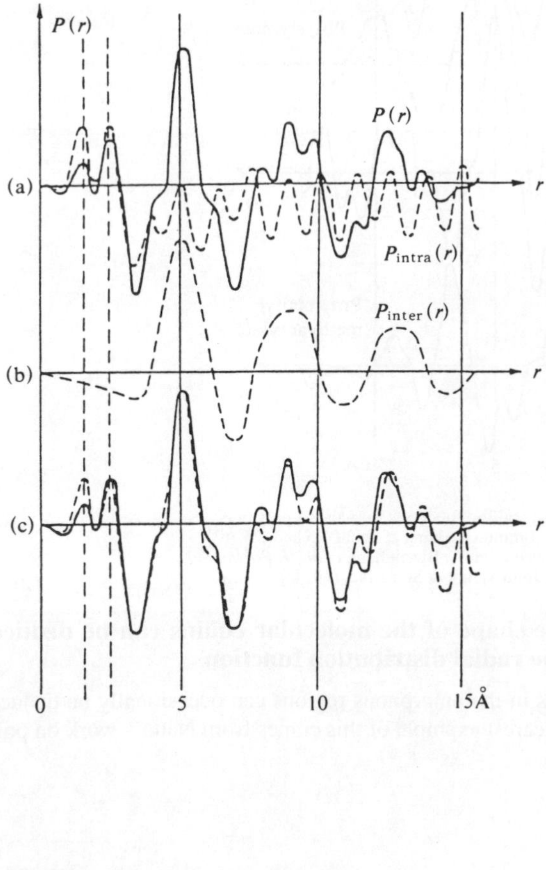

Fig. 14.6
Radial distribution functions of Nylon 6,6.[21]
(a) Continuous curve: $P(r)$ derived from observed intensity distribution.
Broken curves: $P_{intra}(r)$ for an infinite planar zigzag molecular chain.
(b) $P_{inter}(r)$ derived by subtracting the broken curve in (a) from the continuous curve
(c) Continuous curve: as in (a).
Broken curve: synthesis of the broken curve in (a) with that in (b), showing good agreement with the experimental distribution function.
[Reproduced from A. Bjørunhang *et al.*, *J. Polym. Sci.*, **12**, 691, John Wiley & Sons, Inc. (1954)]

scattering of Nylon 6,6 using Eq. 2.35. Comparison of the radial distribution functions of other polymers, such as poly(vinyl acetate), poly(vinyl alcohol), and polystyrene (given for comparison in Fig. 14.7)[21] reveals the presence of common maxima at about 1.5 Å and 2.5 Å. These maxima are thought to be due to a partial structure that is common to these high polymers and consists mainly of pairs of successive carbon atoms and pairs of alternate carbon atoms in the molecular backbone. $P(r)$ thus appeared to be strongly dependent on the arrangement of the molecular chains with respect to one another, and the intramolecular interatomic vector contribution $P_{\text{intra}}(r)$ was obtained (see broken curve in Fig. 14.6(a)). Fig. 14.6(b) shows the intermolecular radial distribution function $P_{\text{inter}}(r) = P(r) - P_{\text{intra}}(r)$ derived by subtracing the broken curve in Fig. 14.6(a) from the continuous curve and smoothing out minor irregularities. The first maximum in this curve shows that the average distance between adjacent chains is about 5 Å.

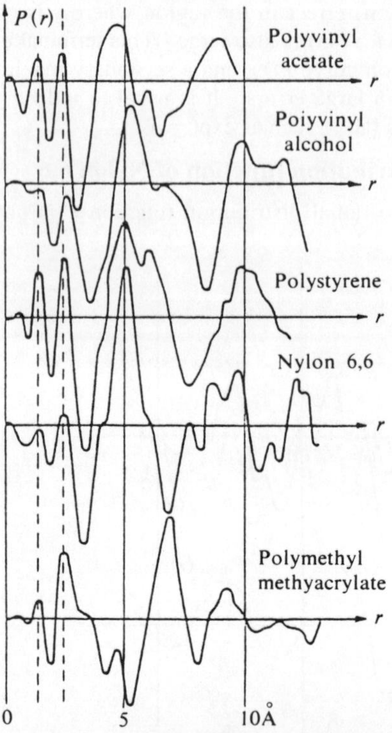

Fig. 14.7 Radial distribution functions for various polymers. Note the common maxima at about 1.5 and 2.5 Å.[21] [Reproduced from A. Bjørunhang *et al.*, *J. Polym. Sci.*, **12**, 691, John Wiley & Sons, Inc. (1954)]

14.3.3 Special cases where the shape of the molecular chains can be deduced without determining the radial distribution function

The shapes of the molecular chains in the amorphous regions can occasionally be deduced by special methods. Perhaps the clearest example of this comes from Natta's work on poly-

p-chlorostyrene.[22] Natta obtained amorphous poly(*p*-chlorostyrene) by polymerization of *p*-chlorostyrene in the presence of stereospecific catalysts. Hydrogenation and dechlorination of this amorphous poly(*p*-chlorostyrene) with Raney nickel as the catalyst gave a crystalline product which was identified from its unoriented X-ray diffraction pattern and IR spectrum as isotactic poly(vinylcyclohexane). It was therefore concluded that the amorphous poly(*p*-chlorostyrene) had also been isotactic.

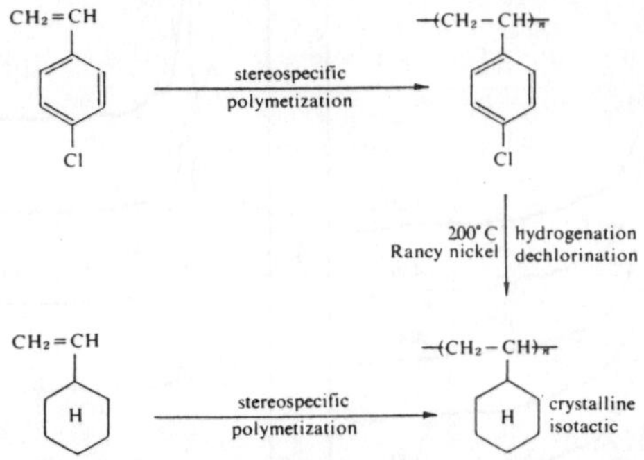

Scheme
Polymerization of *p*-chlorostyrene and dehydrogenation/dechlorination of the product to poly(vinylcyclohexane).[22]
[Reproduced with permission from G. Natta, *Makromol. Chem.*, **35**, 94, Hüthig & Wepf Verlag (1960)]

14.4 Recognition of Oriented Diffraction Mixed with Unoriented Amorphous Scattering

The discussion in the previous section was concerned with the spherically symmetrical radial distribution function *P*(*r*) (*cf.* Eq. 2.35, quoted above). This may also be regarded as a two-dimensional *Q* function (*cf.* Eq. 5.9) involving the radius *r* and the angle θ with respect to a fixed direction in space, averaged with respect to θ. θ here is a spherical polar coordinate as in (r, θ, φ), and is not be confused with the Bragg angle.

$$P(r) \equiv Q_0(r) = \langle Q(r,\theta) \rangle_\theta = (1/2)\int_0^\pi Q(r,\theta)\sin\theta d\theta$$
$$= \sum_{n=0}^{\infty} P_{2n}(\cos\theta)\sin\theta d\theta \qquad (14.11)$$

where P_{2n} is Legendre's polynomial. Thus the contribution which the diffraction intensity due to the oriented part makes to the total observed intensity is given by the Fourier transform of the difference between the Fourier transform $Q(r, \theta)$ of the observed intensity and the spherically symmetrical $Q_0(r)$ of Eq. 14.11.

$$\int_0^\pi \{Q(r,\theta) - Q_0(r)\}\exp\{2\pi i(Rr)dr\} = I(R,\Theta) - I_0(R) = I'(R,\Theta) \qquad (14.12)$$

Subtraction of the average of the observed diffraction intensities $I(R, \Theta)$ (*cf.* Fig. 14.12) over all directions

$$I_0(R) = (1/2)\int_0^\infty I(R,\Theta)\sin\Theta d\Theta \qquad (14.13)$$

from the observed diffraction intensity gives the divergence $I'(R, \Theta)$ of the diffraction intensity from spherical symmetry. It is thus possible to ascertain the presence and determine the intensity distribution of oriented diffraction even when it is mixed with a predominantly

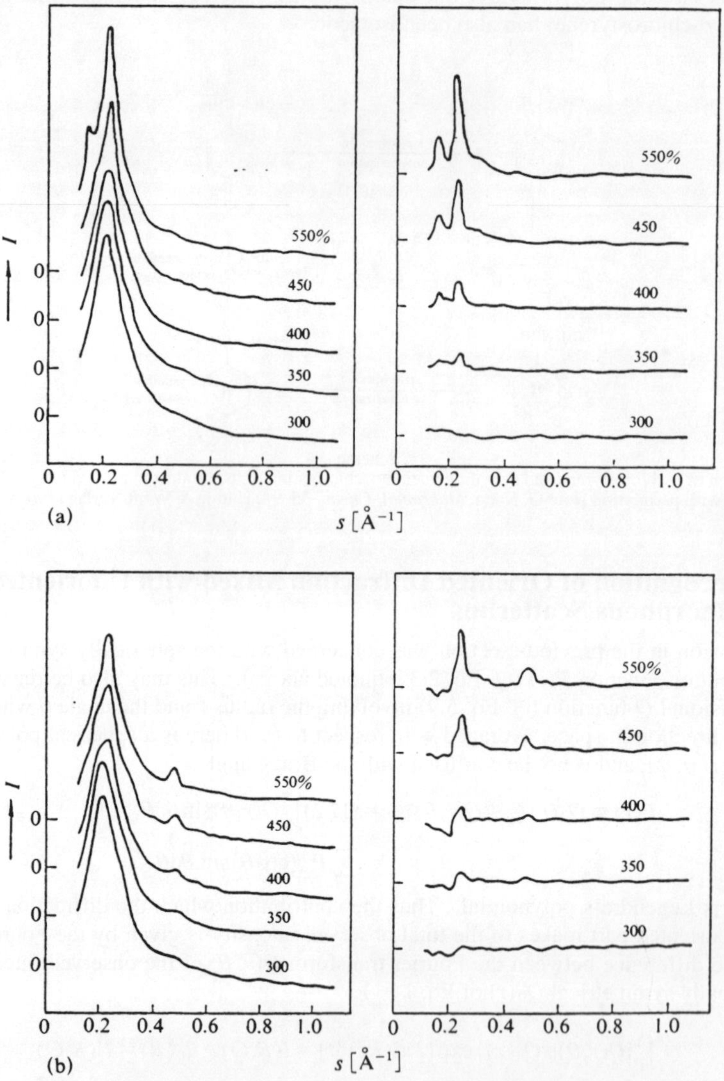

Fig. 14.8 The effect of elongation on the diffraction intensity distribution of natural rubbers.[23)]
(a) Equatorial intensity distribution as observed (left-hand diagram) and after eliminating the contribution due to unoriented amorphous scattering (right-hand diagram).
(b) As in (a) for the meridional intensity distributions.
[Reproduced with permission from M.E. Milberg, *J. Polymer Sci.*, **A 4**, 801, John Wiley & Sons, Inc. (1966)]

unoriented amorphous scattering and would not normally be accessible to direct measurement.[23)]

14.4.1 Resolution of oriented diffraction masked by unoriented amorphous scattering.[23)]

The left-hand diagram of Fig. 14.8(a)[22)] shows the equatorial diffraction intensity distributions $I\,(R, \Theta)$ of stretched natural rubber, while the right-hand diagram shows the difference $I'(R, \Theta) = I\,(R, \Theta) - I_0\,(R)$ between these intensity distributions and the average $I_0\,(R)$ of such diffraction intensity distributions over all directions. The oriented crystalline diffraction, which is completely masked by the very strong unoriented amorphous diffraction in the original intensity distribution, appears in these diagrams at an elongation of only 350%. This is even more clearly evident in the meridional diffraction intensity distributions in Fig. 14.8 (b).[23)] Fig. 14.9 shows the elimination of unoriented amorphous scattering with the aid of Eq. 14.12 in the case of atactic polystyrene fibers to leave the oriented amorphous scattering (*cf.* Section 14.5).

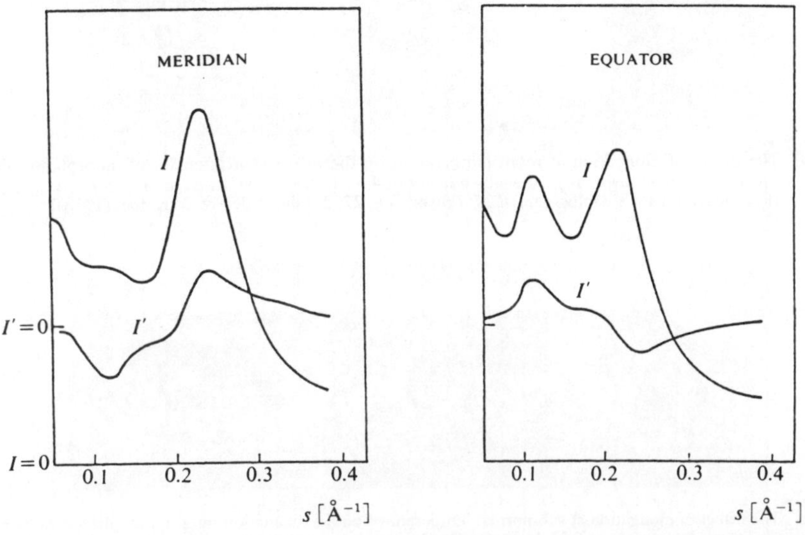

Fig. 14.9 Comparison between meridional and equatorial diffraction intensity distributions of atactic polystyrene.[23)] I' is the broad amorphous scattering halo that remains after elimination of the effect of unoriented amorphous scattering from I. The difference between meridional and equatorial distributions indicates intrinsic orientation. [Reproduced with permission from M.E. Milberg, *J. Polym. Sci.*, **A 4**, 801, John, Wiley & Sons, Inc. (1966)]

14.5 Analysis of the Orientation of Molecular Chains in Amorphous Regions

14.5.1 Orientation of molecular chains in amorphous regions

Just as drawing, rolling, compression, etc., lead to the orientation of crystallites, so the molecular chains in amorphous regions can acquire a degree of orientation under the influence of drawing, rolling, compression, etc. The presence of this orientation can be recognized by the appearance of intensity maxima in the originally uniform amorphous halos in the equatorial or meridional regions. Fig. 14.10(a) shows the diffraction pattern of an amorphous silicone rubber; (b) and (c) respectively show the patterns obtained when this specimen was elongated by factors of 4.5 and 8.75 at room temperature.[24] The drawing direction corresponded to the vertical direction in the diagram. In contrast with the behavior of drawn specimens in general, no crystalline diffractions appear in this case. However, the amorphous scattering shows a distinct concentration of intensity around the equator, suggesting partial orientation of the molecular chains parallel to the drawing direction. Despite the partial orientation of the molecular chains, the drawing was not sufficient to cause crystallization of the specimen. In Fig. 14.11, on the other hand, the molecular chains become so highly oriented as a result of drawing that crystallization occurs, and oriented crystalline diffractions appear with equatorial concentration of the amorphous halo.[24]

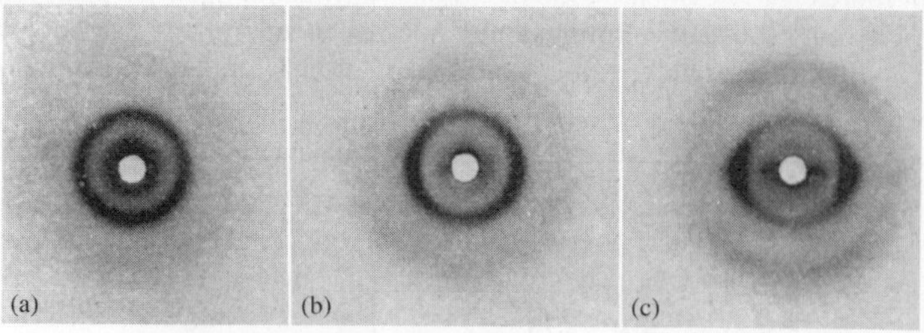

Fig. 14.10 The effect of elongation at room temperature on the diffraction intensity of amorphous silicone rubber.[24] (a) Undrawn; (b) Drawn by a factor of 4.5 : 1; (c) Drawn by a factor of 8.75 : 1 [Reproduced from S. Ohlberg, *et al.*, *J. Polym. Sci.*, **27**, 2, John Wiley & Sons, Inc. (1958)]

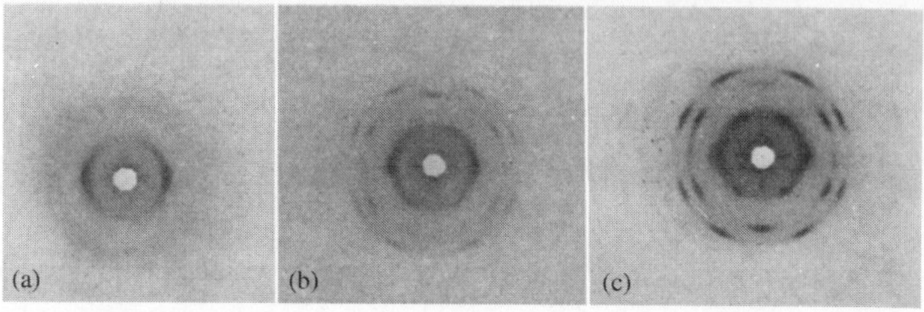

Fig. 14.11 The effect of elongation at sub-normal temperatures on the diffraction intensity of silicone rubber.[24] (a) Drawn by a factor of 6.5 : 1 at − 20°C; (b) As (a), but at − 30°C; (c) As (a), but at − 60°C [Reproduced from S. Ohlberg, *et al.*, *J. Polym. Sci.*, **27**, 10, John Wiley & Sons, Inc. (1958)]

14.5.2 Degree of orientation of the molecular chains; practical measure of parallelism of amorphous chains

A. Practical measure of parallelism of amorphous chains

Corresponding to the practical measure of parallelism used in the case of crystals (Section 10.3.1), a criterion of orientation based on the half-width H° of the azimuthal intensity distribution of the amorphous halo is given by (*cf.* Eq. 10.13).

$$\Pi_{\text{amorph}} = \frac{180^\circ - H^\circ}{180^\circ} \times 100 \tag{14.14}$$

Alternatively, however, the $<\varphi>$ discussed below may be determined and employed.

B. Mean orientation angle of molecular chains

The average of the orientation angles φ of the molecular chains is found from the azimuthal intensity distribution of amorphous scattering with the aid of the following equation (Fig. 14.12[24] and *cf.* Fig. 10.19).

$$<\varphi> = \int_0^{\pi/2} \psi I(\psi) \sin\psi d\psi \, / \int_0^{\pi/2} I(\psi) \sin\psi d\psi \tag{14.15}$$

Here, φ is the angle of orientation of the molecular chains with respect to the reference axis (*e.g.* drawing direction), and ψ is the angle made with the equator (*cf.* Fig. 10.19). $I(\psi)$ is the intensity of the amorphous scattering in direction ψ. It is clear from Eq. 14.15 that $<\varphi>$ has a value of 57.3°(= 1 rad) when $I(\psi)$ is constant, *i.e.* when the arrangement of the molecular chains is completely random.

a) Oriention of molecular chains in drawn amorphous silicone rubber.[24] Figure 14.12 shows the directional intensity distribution in the amorphous halo of silicone rubber drawn by a factor of 6.15, which exhibits a distinct concentration of intensity towards the equator ($\psi = 0°$). Determination of the mean orientation angle $<\varphi>$ of the molecular chains from this intensity distribution with the aid of Eq. 14.15 gives $<\varphi> = 45.0°$.

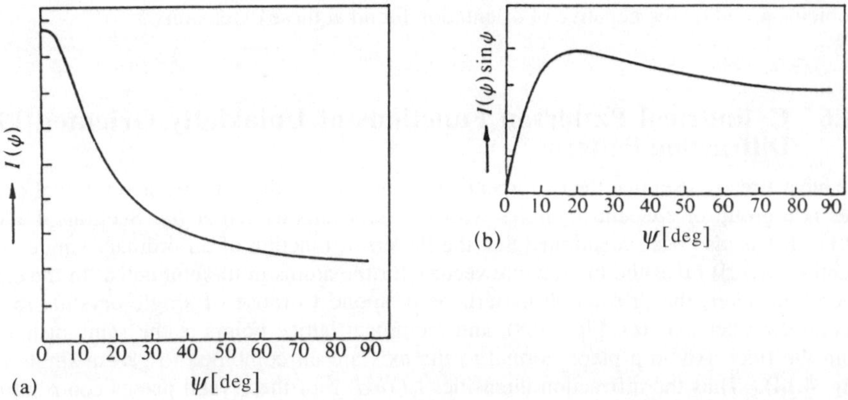

Fig. 14.12 Directional intensity distribution of the halo of amorphous silicone rubber in various directions. (a) $I(\psi)$ *vs.* ψ; (b) $I(\psi)\sin\psi$ *vs.* ψ
[Reproduced from S. Ohlberg, *et al., J. Polym. Sci.*, **27**, 2, John Wiley & Sons, Inc. (1958)]

14.5.3 Estimation of the degree of orientation of molecular chains in amorphous regions by methods other than X-ray methods

Methods based on optical birefringence and infrared dichroism are currently under investigation. It should be possible in principle to find the degree of orientation from the values of these quantities. At present the results are being evaluated in comparison with the crystallinities and degrees of crystallite orientation found by X-ray methods. As is evident from Fig. 14.13,[25] the variations of the orientation with elongation merely exhibit the same overall trend, with very considerable differences in detail. However, it may be that these methods will prove useful at small extensions (where determination by X-ray methods is diffi-

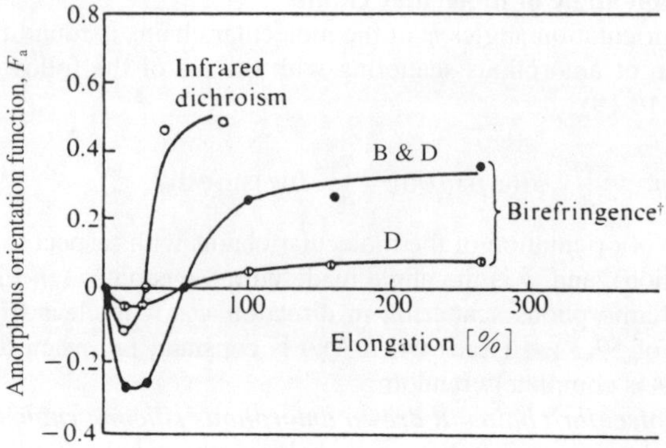

Fig. 14.13 Variation of the orientation of amorphous molecular chains with elongation.[25] Determined by using †: bond polarizabilities of Bun & Daubeny (B & D) and of Denbigh (D) . [Experimental data supplied by R.S.Stein in private communication (1957)]

cult), provided, of course, that detailed investigation is able to solve the interpretational problems posed by the negative of orientation found at these extensions.

14.6 Cylindrical Patterson Functions of Uniaxially Oriented Fiber Diffraction Patterns

We have already seen that the reciprocal lattice of a uniaxially oriented assembly of crystallites is a group of concentric circles with the fiber axis as center (*cf.* Sections 4.4.4 and 10.1). It has also been mentioned that the Patterson function of an ordinary single crystal (Section 11.2.5) gives the interatomic vectors for the atoms in the unit cells. In the case of fibers, however, the diffraction patterns correspond to those of single crystals rotating around the fiber axis (*cf.* Fig. 10.8), and reciprocal lattice points at the same distance R' from the fiber axis in a plane normal to the axis are all combined to give a single circle (Fig. 4.14). Thus the diffraction intensities $|F(hkl)|^2$ of the crystal planes contributing to this circle cannot be separated, and consequently Patterson functions cannot be calculated on the basis of the individual $|F(hkl)|^2$. However, although the Patterson function cannot be employed, the fiber diffraction pattern itself will necessarily reflect the atomic arrangement within the fiber structure. Moreover, in distinction from the case of single crystals, where only the reciprocal lattice points *hkl* can be measured, fiber specimens give rise to an additional, background intensity. It must therefore be possible to derive some structural information from the continuous total scattering intensity distribution. A method has been proposed for the direct Fourier transformation of the rotationally averaged intensity (corresponding to the normal X-ray fiber diagram) of a fiber structure. Since this method, which is described in more detail below, makes use of the *two-dimensional* continuous intensity over the entire range of diffraction angles, it can even be applied to diffraction patterns that lack crystalline diffractions, such as those of nematic or smectic assemblies of chain molecules.

14.6.1 The cylindrical distribution function

The convolution $Q(\boldsymbol{r})$ of Eq. 5.9 is generally expressed as the Fourier transform of the X-ray intensity distribution $I(\boldsymbol{S})$:

$$Q(\boldsymbol{r}) = \int \rho(\boldsymbol{u})\rho(\boldsymbol{u}+\boldsymbol{r})\mathrm{d}\boldsymbol{u} = \int I(\boldsymbol{S})\exp\{-2\pi i(\boldsymbol{S}\cdot\boldsymbol{r})\}\mathrm{d}\boldsymbol{S} \tag{14.16}$$

Where the atomic density distribution in the specimen is on average cylindrically symmetrical about the fiber axis, whether the specimen is an assembly of crystallites or of parallel chain molecules, it is convenient to express the electron density $\rho(r)$ in terms of cylindrical coordinates (r, φ, z) (Fig. 14.14). We have, of course, already found that the reciprocal lattice of a fiber structure is such that it is more convenient to give the continuous intensity distribution in terms of cylindrical coordinates (R, Φ, Z) than the indices hkl. Calculation of the Fourier transform of the intensity (Eq. 14.16) using $I(R, \Phi, Z)$ for $I(\boldsymbol{S})$ gives

$$\begin{aligned} Q(\boldsymbol{r}) = Q(r,z) = \int_0^{\infty}\int_{-\infty}^{\infty} & RI(R,Z)\exp(-2\pi iZz) \\ & \times\left[\int_0^{2\pi}\exp\{-2\pi iRr\cos(\Phi-\varphi)\}\mathrm{d}\varphi\right]\mathrm{d}R\mathrm{d}Z \end{aligned} \tag{14.17}$$

The integral in φ becomes the Bessel function $J_0(2\pi Rr)$, and we can write

$$Q(r,z) = 2\pi\int_0^{\infty}\int_{-\infty}^{\infty} RI(R,Z)J_0(2Rr)\cos(2\pi Zz)\mathrm{d}R\mathrm{d}Z \tag{14.18}$$

This is the cylindrical distribution function.

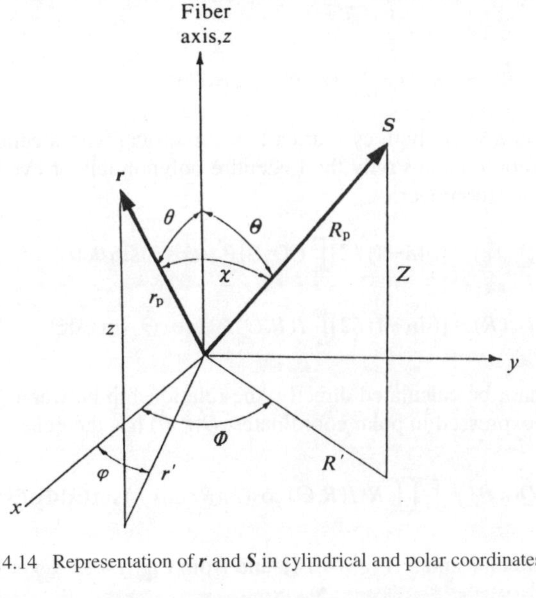

Fig. 14.14 Representation of $\boldsymbol{r}$ and $\boldsymbol{S}$ in cylindrical and polar coordinates.

Since $I(S)$ and $Q(\boldsymbol{r})$ are both cylindrically symmetrical in this case, Eq. 14.18 is applicable whenever 1) the polymer chains themselves are arranged parallel to the fiber axis and have a statistical density distribution, $\rho(r)$, which is cylindrically symmetrical about the fiber axis (nematic or smectic states) or 2) the distribution of the orientations of the crystallites in the specimen is cylindrically symmetrical (uniaxial, spiral, ring, or tilted orientations, *cf.* Section 10.1). For example, in the case of a uniaxially oriented group of crystallites in a fiber, *i.e.* where the reciprocal lattice can be regarded statistically as a group of concentric circles of uniform density around the fiber axis as in Fig. 4.14, the diffraction intensity generally depends only on the reciprocal lattice axis Z in the fiber direction and the radii R of the various reciprocal lattice circles in a plane normal to the fiber direction. It is independent of the angle Φ between the radius R and a reference line (*cf.* Fig. 14.14). Thus, although direct determination of the unit cell dimensions and indexing of the diffractions is not possible for specimens with cylindrically symmetrical orientation of the crystallites (case 2) above), the cylindrical distribution function is obtainable. Moreover, Eq. 14.32 below can be used where periodicity exists in the direction.

14.6.2 Representation of $Q(r)$ in polar coordinates

When the statistical density distribution in the specimen is cylindrically symmetrical, but the divergence from spherical symmetry is not very large, it is convenient to express the density distribution function in polar coordinates. If the spherical polar coordinates (r_p, θ, φ) and (R_p, Θ, Φ)† are used in place of the cylindrical coordinates (r, φ, z) and (R, Φ, Z) in real and reciprocal space (*cf.* Fig. 14.14), since the density distribution in the specimen is cylindrically symmetrical, it varies with the radius r of the spherical surface and the colatitude θ, and the density distribution function $Q(\mathbf{r})$ can be expanded as follows with the aid of Legendre polynomials $P_n(\cos\theta)$.[26,27]

$$Q(\mathbf{r}) = Q(r,\theta) = \sum_{n=0}^{\infty} Q_{2n}(r) P_{2n}(\cos\theta) \tag{14.19}$$

$$I(\mathbf{S}) = I(R,\theta) = \sum_{n=0}^{\infty} I_{2n}(R) P_{2n}(\cos\Theta) \tag{14.20}$$

Since $Q(\mathbf{r})$ and $I(\mathbf{S})$ are both cylindrically symmetrical with a center of symmetry at the origin, the expansion contains only the Legendre polynomials of even order $P_{2n}(\cos\theta)$. In the above equations, moreover,

$$Q_{2n}(r) = [(4n+1)/2]\int_0^{\pi} Q(r,\theta) P_{2n}(\cos\theta)\sin\theta \mathrm{d}\theta \tag{14.21}$$

$$I_{2n}(R) = [(4n+1)/2]\int_0^{\pi} I(R,\Theta) P_{2n}(\cos\Theta)\sin\Theta \mathrm{d}\Theta \tag{14.22}$$

Since $Q(r, \theta)$ cannot be calculated directly, the relationship between $Q_{2n}(r)$ and $I_{2n}(R)$ must be found. When expressed in polar coordinates, $Q(r, \theta)$ has the general form

$$Q(r,\theta) = \int_0^{2\pi}\int_0^{\pi}\int_0^{\infty} R^2 I(R,\Theta)\cos(2\pi Rr\cos\chi)\sin\Theta \mathrm{d}\Phi \mathrm{d}\Theta \mathrm{d}R \tag{14.23}$$

χ is the angle between r and S (Fig. 14.14), and from Fig. 14.12

$$\cos\chi = \cos\theta\cos\Theta + \sin\theta\sin\Theta\cos(\Phi - \varphi) \tag{14.24}$$

Using the Bessel function addition theorem and the Legendre polynomial addition theorem, we obtain

† The suffix p of r_p and R_p is omitted in the description which follows in this section.

$$Q(r,\theta)=\int_0^{\pi}\int_0^{\infty} I(R,\Theta)\sum_{n=0}^{\infty}(-1)^n(4n+1)J_{2n}(2\pi Rr)\sum_{m=-n}^{n}\frac{(n-m)!}{(n+m)!}P_{2n}{}^{m}(\cos\Phi)$$
$$\times P_{2n}{}^{m}(\cos\theta)\left[\int_0^{2\pi}\exp\{im(\Phi-\varphi)\}\mathrm{d}\Phi\right]R^2\sin\Theta\mathrm{d}\Theta\mathrm{d}R \qquad (14.25)$$

Because of the integral inside the brackets in Eq. 14.25 , the terms with $m \neq 0$ disappear, and

$$Q(r,\theta)=\sum_{n=0}^{\infty}(-1)^2 P_{2n}(\cos\theta)\int_0^{\infty}4\pi R^2 J_{2n}(2\pi Rr)$$
$$\times\left[\{(4n+1)/2\}\int_0^{\pi} I(R,\Theta)P_{2n}(\cos\Theta)\sin\Theta\mathrm{d}\Theta\right]\mathrm{d}R \qquad (14.26)$$

Comparison of Eq. 14.26 and Eq. 14.22 shows that the expression in square brackets is equal to $I_{2n}(R)$, and comparison with Eq. 14.19 leads to the desired relationship

$$Q_{2n}(r)=(-1)^n 4\pi\int_0^{\infty}R^2 I_{2n}(R)J_{2n}(2\pi Rr)\mathrm{d}R \qquad (14.27)$$

Thus $I_{2n}(R)$ is found from the observed diffraction intensity $I(R, \Theta)$ with the aid of Eq. 14.22, and $Q_{2n}(r)$ is then obtained from Eq. 14.27. $Q(r, \theta)$ can then be found from Eq. 14.19. When the specimen exhibits perfect spherical symmetry, Eq. 14.19 reduces to the radial distribution function $P(r)$ (*cf.* Eqs. 2.35 and 14.11).

In practice, instead of Eqs. 14.18 and 14.19, we frequently use the forms obtained after subtraction of the average electron density in the specimen, $Q^0(r', z)$ or $Q^0(r, \theta)$.

$$Q(r',z)-Q^0(r',z)=2\pi\iint R'I(R',Z)J_0(2\pi R'r')\cos(2\pi Zz)\mathrm{d}R'\mathrm{d}Z \qquad (14.28)$$

$$Q(r,\theta)-Q^0(r,\theta)=\sum_{n=0}^{\infty}\{Q_{2n}(r)-Q_{2n}{}^0(r)\}P_{2n}(\cos\theta) \qquad (14.29)$$

$$Q_{2n}(r)-Q_{2n}{}^0(r)=(-1)^n(1/2\pi^2)\int_0^{S_0}s^2 I_{2n}{}'(R)J_{2n}(sr)\mathrm{d}s \qquad (14.30)$$

where $s=2\pi S=(4\pi\sin\theta)/\lambda$

$$I_{2n}{}'(R)=\{(4n+1)/2\}\int_0^{\pi}I(R,\Theta)P_{2n}(\cos\Theta)\sin\Theta\mathrm{d}\Theta \qquad (14.31)$$

$I(R, \Theta)$ is naturally the intensity obtained on subtraction of the incoherent scattering and of the atomic coherent scattering.

A. Analysis of the diffraction pattern of ramie[26,27)]

The diffraction intensities used were obtained mainly from a Weissenberg photograph for an axis normal to the fiber axis, and secondarily from a fiber-axis rotation photograph. The latter was necessary because of the blind region due to the shadow of the specimen holder in the Weissenberg photograph. Diagrams (a) and (b) in Fig. 14.15 show the results obtained after application of the corrections for polarization and absorption to the diffraction intensities. Fig. 14.15(c) shows the diffraction intensity distribution $I(R, Z)$ derived from these. Fig. 14.15(a) shows the distribution function $Q(r, z)-Q^0$ derived from Eq. 14.28;

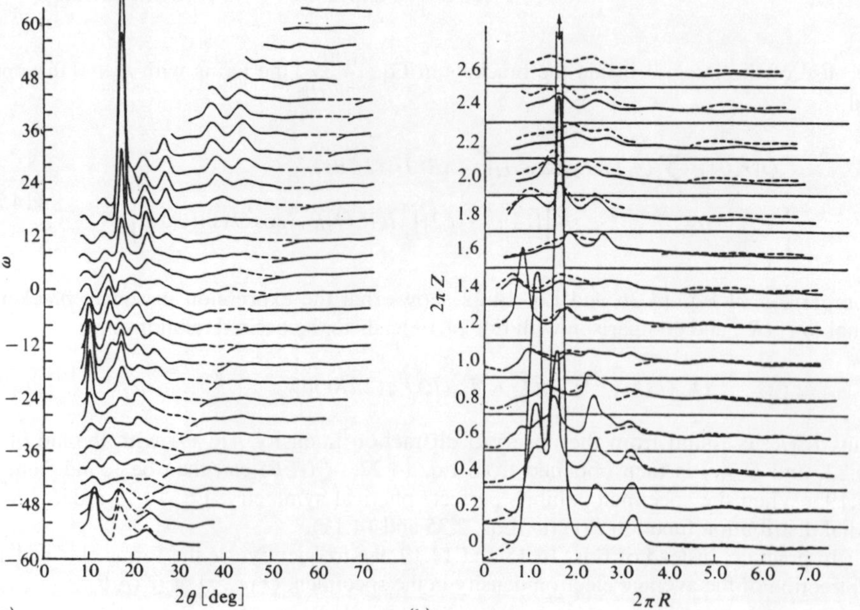

(a) (b)

Fig. 14.15 Diffraction intensities of ramie.[26,27)]
(a) Data from a Weissenberg photograph (axis normal to the fiber axis). The dotted sections are the "blind" regions due to the shadow of the specimen holder.
(b) Data from a rotation photograph: used to "fill in" the blind regrons in (a). The dotted sections are derived from the Weissenberg photograph data of (a).
(c) Diffraction intensity distribution $I(R, Z)$ derived from (a) and (b). The contour lines are at intervals of 20 in arbitrary intensity units. The broken lines indicate contours of 100 and 200, and areas with less than 20 units are shaded.

[Reproduced from N. Norman, *Doctoral Thesis*, pp.57, 59, Univ. Oslo (1954)]

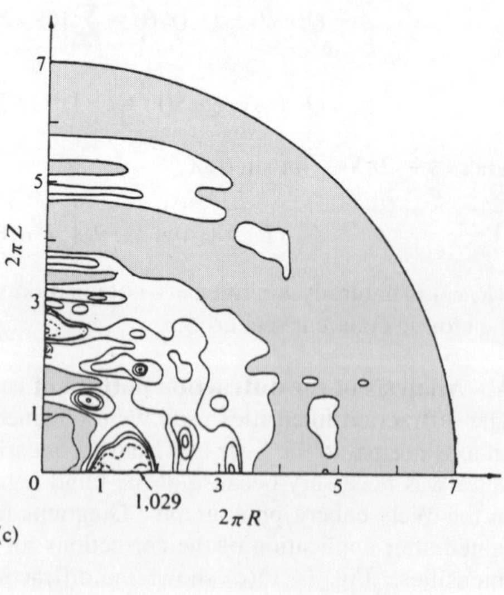

(c)

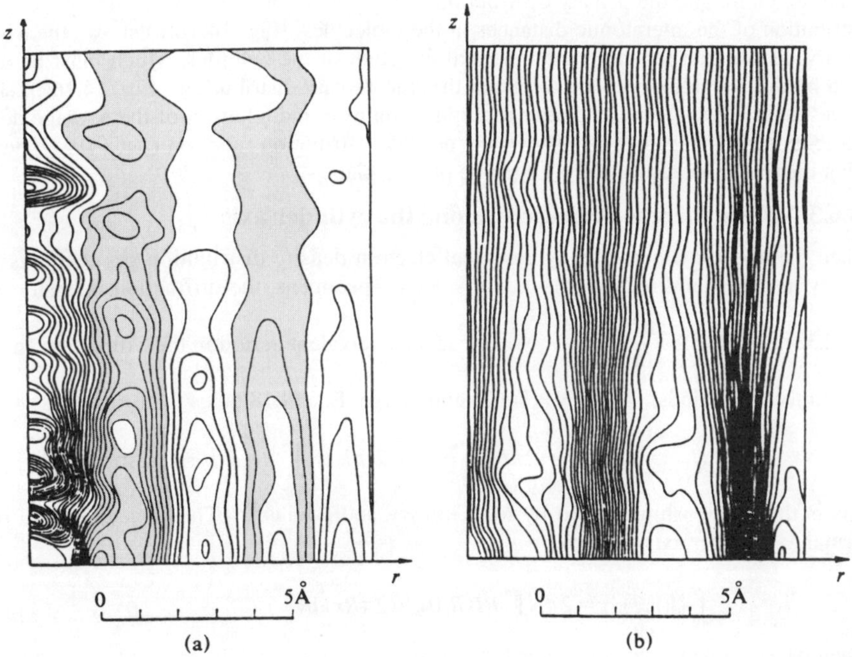

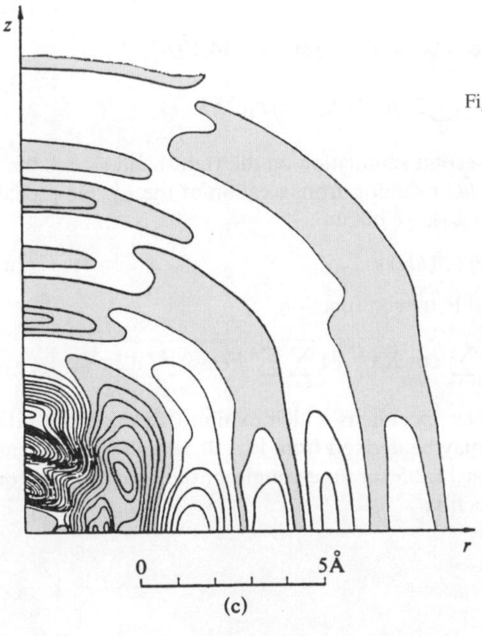

Fig. 14.16 Cylindrical distribution functions for ramie.
(a) $Q(r, z) - Q^0$ (Shaded areas have negative values).
(b) $rQ(r, z) + rQ^0$.
(c) $Q(r, \theta) - Q^0$ (Shaded areas have negative values).
[Reproduced from N. Norman, *Doctoral Thesis*, pp.61, 67, Univ. Oslo (1954)]

the packing of the molecular chains is found from the density distribution in the circumferential direction, and the density distribution for $r = 0$, *i.e.* on the z axis, corresponds to the distribution of the interatomic distances in the molecules. Fig. 14.16(a) shows the average density distribution for the circumferential direction of the cylinder, which must be multiplied by the weighting factor r to obtain the true density distribution. Fig. 14.16(b) shows the resultant $rQ(r, z)$ obtained after multiplication by r and addition of the average density distribution rQ^0. Fig. 14.16(c) shows the density distribution function $Q(r, \theta) - Q^0$ derived using the first three terms of the Legendre polynomials.

14.6.3 Where there is periodicity along the cylinder axis

When the above cylindrically symmetrical electron density distribution $\rho(r)$ exhibits periodicity in the axial direction (z), *i.e.* for smectic specimens, the diffraction intensity (from Eq. 2.38) is $I(R, Z)\cdot\dfrac{\sin^2(N\pi cZ)}{\sin^2(\pi cZ)}$, where c is the equivalent period in the z direction and N is the number of periods. When N is sufficiently large, Eq. 14.18 takes the form

$$Q(r) = Q(r,z) = (2\pi N/c)\sum_{l=-\infty}^{\infty}\cos 2(\pi lz/c)\int_0^{\infty} RI(R,l/c)J_0(2\pi Rr)\,\mathrm{d}R \quad (14.32)$$

This is the relationship derived by MacGillavry and Bruins.[28] The projection in a plane normal to the fiber axis is given by

$$\int_0^c Q(r,z)\mathrm{d}z = 2\pi N\int_0^{\infty} RI(R,0)J_0(2\pi Rr)\mathrm{d}R \quad (14.33)$$

When the specimen is crystalline and the crystallites are uniaxially oriented (with the c axis as the fiber axis), the diffracted intensity is expressed with the aid of δ functions[29] by

$$I = I(R, Z)\delta(R - Z)\delta(Z - Z_l) \quad (14.34)$$

Since $Z_l = l/c$, the distribution function can be derived from Eq. 14.19.

$$Q(r,z) = \sum_{l=-\infty}^{\infty}\cos 2(\pi lz/c)\sum_{hk} 2\pi R_{hk} I(R_{hk}, l/c) J_0(2\pi R_{hk} r) \quad (14.35)$$

Again, since the $2\pi R_{hk}I(R_{hk}, l/c)$ of the second summation on the right-hand side is the integral intensity of the circle with indices *hkl* resulting from rotation of the *hkl* reciprocal lattice points about the fiber axis (*cf.* Fig. 4.14), we obtain

$$2\pi R_{hk}I(R_{hk}, l/c) = (N/v)\,|F(hkl)|^2 \quad (14.36)$$

$Q(r, z)$ is therefore simply the cylindrical Patterson function.

$$P(r,z) = Q(r,z) = (N/V)\sum_{l=-\infty}^{\infty}\cos 2(\pi lz/c)\sum\sum_{h,k=-\infty}^{\infty} |F(hkl)|^2 J_0(2\pi R_{hk} r) \quad (14.37)$$

Thus for smectic specimens or crystalline specimens which exhibit (say) uniaxial orientation, the cylindrical Patterson function may be derived from Eq. 14.32 rather than from Eq. 14.18. The reader is referred to Section 11.2.5 for an example of the analysis and for the significance of the general Patterson function.

References

1. *International Tables for Crystallography*, Vol. C. Mathematical, Physical and Chemical Tables 2nd ed., (A.J.C. Wilson, E. Prince eds.) pp.477-486 (free atoms) and pp.487-499. (chemically significant ions), Kluwer Academic Pub., Dordrecht (1999). *cf.* Table 2A of Appendix (free atoms).
2. In ref. 1; p. 652; *cf.* Table 3 of Appendix.
3. S. Krimm, A.V. Tobolsky, *J. Polymer Sci.*, **7**, 57 (1951).
4. A. Turner Jones, J.M. Aizlewood, D.R. Beckett, *Makromol. Chem.*, **75**, 134 (1964).
5. M. Kakudo, R. Ullman, *J. Polymer Sci.*, **45**, 91 (1960).
6. J.L. Matthews, H.S. Peiser, R.B. Richards, *Acta Cryst.*, **2**. 85 (1949).
7. P.H. Hermans, R. Platzek, *Kolloid-Z. Z. Polymere*, **88**, 68 (1939).
8. J.J. Arlman, *Appl. Sci. Res.*, **Al**, 347 (1949); **A2**, 1 (1948).
9. J.M. Goppel, *Appl. Sci. Res.*, **Al**, 3 (1947): J.J. Arlman, J.M. Goppel, *Appl. Sci. Res.*, **Al**, 462 (1949).
10. *X-Ray Crystallography* (I. Nitta ed.) Vol. I, pp. 587–598, Maruzen, Tokyo (1961).
11. L.E. Alexander, *X-Ray Diffraction Methods in Polymer Science*, pp. 137–197, John Wiley & Sons, Inc., N.Y. (1969).
12. G. Bodor, *Structural Investigation of Polymers*, pp. 300–340, Akadémiai Kiadó, Budapest (1991).
13. V.D. Gupta, R.B. Beevers, *Chem. Rev.*, **62**, 665 (1962).
14. S.L. Aggarwal, G.P. Tiley, *J. Polymer Sci.*, **18**. 17 (1955).
15. W. M.D. Bryant, C.T. Tordella, R.H.H. Pierce, Jr., Paper presented at the 118th National Meeting Am. Chem, Soc.. Chicago, September 3–8, 1950.
16. J.H. Wakelin, H.S. Virgin, E. Crystal, *J. Appl. Phys.*, **30**, 1654 (1959).
17. W.O. Statton, *J. Appl. Polym. Sci.*, **7**, 803 (1963).
18. G. Natta, *Makromol. Chem.*, **35**, 94, (1959); G. Natta, P. Corradini, D. Slanesi, D. Morero, *J. Polym. Sci.*, **51**, 527 (1961).
19. H.G. Kilian, K. Buoeke, *J. Polymer Sci.*, **58**, 311 (1962).
20. K. Katada, *Acta Cryst.*, **16**, 290 (1963).
21. A. Bjørnhaug, Φ. Ellefsen, B.A. Tønnesen, *J. Polymer Sci.*, **12**, 691 (1954).
22. G. Natta, *Makromol. Chem.*, **35**, 94 (1960).
23. M.E. Milberg, *J. Polymer Sci.*, *A-1*, **4**, 801 (1966).
24. S.M. Ohlberg, L.E. Alexander, E.L. Warrick. *ibid.*, **17**, 1 (1958).
25. R.S. Stein, *ONR Technical Report*, Nonr. 2141 (00), NR 356–378.
26. N. Norman, *Acta Cryst.*, **7**, 462 (1954).
27. N. Norman, Doctoral Thesis, Univ. Oslo, pp. 57–67 (1954).
28. C.H. MacGillavry, E.M. Bruins, *Acta Cryst.*, **1**, 156 (1948).
29. In ref. 10, Vol. I, pp. 941–951.

15. Analysis of X-ray Small-angle Scattering

15.1 Preparative Procedure

15.1.1 Detection and recording of the small-angle scattering

It was shown in Chapter 6 that X-ray small-angle scattering generally occurs within an extremely narrow range of angles. Further, the intensity must be measured continuously over a range that can exceed $1 : 10^5$. This is far wider than the range of densities that can be recorded on one film, so that it is necessary to build up the total scattering intensity curve from a number of emulsions which have received different exposures. It is also necessary to use suitable attenuators or to reduce the X-ray tube current in order to keep the measured intensities within the optimum range for the detector. The scattering profile for the entire intensity range is then obtained by combining the measured values. (*cf.* Fig. 15.1).[1,2]

However, the development of the imaging plate (IP) saves the scientist from this troublesome procedure: only one or two exposures are enough to record the whole small-angle scattering pattern because of its excellent dynamic range (*cf.* Section 8.2.2.C.b)). The posi-

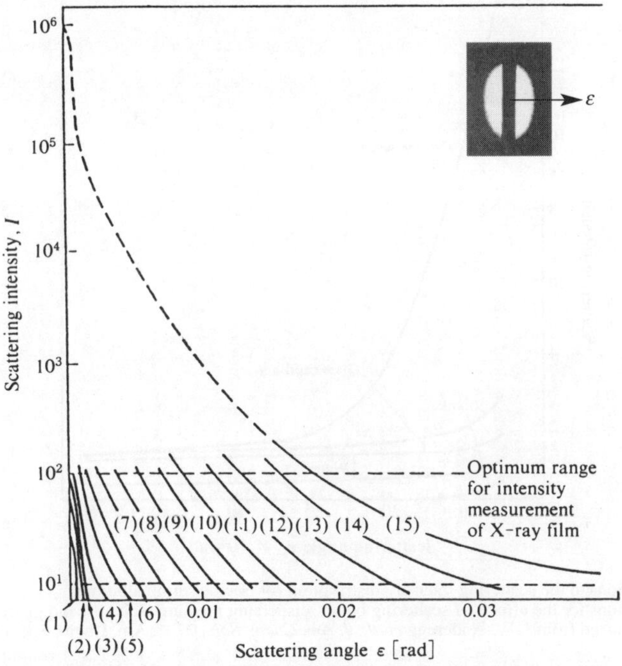

Fig. 15.1 Derivation of the complete scattering intensity curve from a number of films exposed within their linear range of density *vs.* intensity (cf. Fig. 6.1.a).[1,2]
Cu $K\alpha$ radiation (pentaerithrytol crystal monochromator) was used with the silica gel specimen (Santocel C.S.). x (distance from the center to the observed point on the film) or s $(= 2\pi\varepsilon/\lambda)$ may substitute for ε ($\varepsilon \fallingdotseq x/R_F$, R_F: specimen-to-film distance).
[Reproduced from M. Kakudo *et al.*, *J. Chem. Soc., Jpn.Chem. Sect.*, **78**, 823, Chem. Soc. Jpn. (1957)]

tion-sensitive proportional counter (PSPC) and multi-wire proportional counter (MWPC) are also useful for measuring the intensity distribution precisely for a wide range of scattering angles. These procedures make time-resolved measurements of the small-angle scattering diagram possible (say) by the order of milliseconds or less.

15.1.2 Corrections to the scattering intensity distribution

A. The polarization factor

Since the scattering angle ε ($=2\theta$) is extremely small, $\cos 2\theta \fallingdotseq 1$. The polarization factor $(1+\cos^2 2\theta)/2$ is therefore also approximately equal to unity, and generally no correction is required.

B. Elimination of scattering caused by air and by the collimator edges

Measurements of the scattering caused by air, by the edges of the collimator, by dispersing media (solvents), by glass capillaries, etc. are carried out under conditions identical with those used in the measurements on the specimen, and the results are deduced from the scattering intensity obtained with the specimen in position, see Fig. 15.2.[3)]

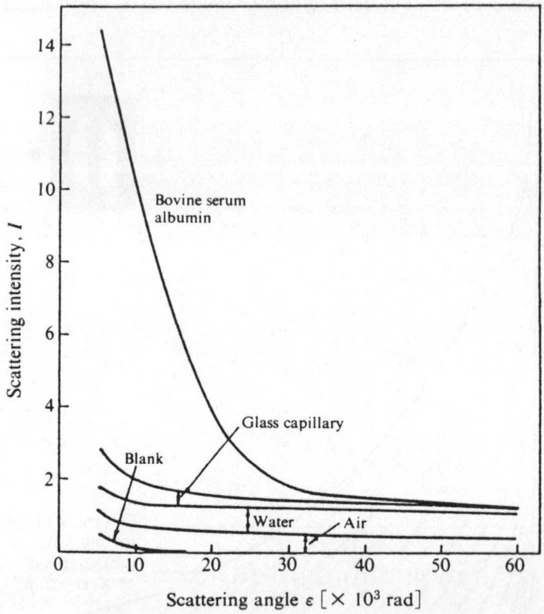

Fig. 15.2 Corrections to the scattering intensity distribution for bovine serum albumin.[3)] Corrections for the effects of scattering by air, dispersing medium (solvent) and glass capillary. [Reproduced from J. W. Anderegg *et al.*, *J. Am. Chem. Soc.*, **77**, 1, Am, Chem, Soc. (1951)]

C. Correction for errors caused by the size of the collimator

Except where the collimator system is a very fine pinhole, the intensity curve obtained with a long, narrow slit, after correction as indicated above, must be further corrected by reduction to the profile that would be produced by an ideal fine pinhole X-ray beam.

a) Correction of the observed intensity curve.

1) Correction for the effect of slit width: this is necessary when the collimator is wide, and it is normal to carry this correction out before correcting for the effects of slit height. If the Fourier transforms of the intensity distribution $I_{PB}(s)$ of the primary X-ray beam in the direction of the slit width, the diffraction $\tilde{I}_H(s)$ free from the effect of the slit width, and the diffraction $\tilde{I}_{WH}(s)$ as affected by the slit width are $J_{PB}(u)$, $J_H(u)$ and $\tilde{J}_{WH}(u)$ respectively, then

$$\tilde{I}_{wh}(s) = \int_{-\infty}^{\infty} I_{PB}(t)\,\tilde{I}_H(s-t)\mathrm{d}t,$$

where $s = 2\pi\varepsilon/\lambda$, and t is a variable of integration of no physical meaning.
Hence

$$\tilde{J}_{WH}(u) = \tilde{J}_H(u)\, J_{PB}(u) \tag{15.1}$$

Thus the diffraction $\tilde{I}_H(s)$ without the effect of the slit width can be found by inverse transformation of the quotient $\tilde{J}_H(u)$ obtained by dividing the Fourier transform $\tilde{J}_{WH}(u)$ of the measured diffraction intensity curve $\tilde{I}_{WH}(s)$ by that of the intensity distribution of the primary X-ray beam $J_{PB}(u)$.

2) Correction for the slit height: if the slit is assumed to be extremely narrow and of infinite height, the true scattering intensity $I(s)$ is calculated from the following equation[4,5] using the gradient $\tilde{I}_H{}'$ (the differential coefficient of the intensity curve) of the measured intensity $\tilde{I}_H$.

$$I(s) = (-1/\pi c)\int_0^{\infty} [\tilde{I}_H{}'\{(s^2+v^2)^{1/2}\}/(s^2+v^2)^{1/2}]\mathrm{d}v \tag{15.2}$$

where v is a variable of integration of no physical significance. This correction is not necessary in the Guinier plot method (Section 15.2.1, *cf.* Eq. 15.8).

If the slit is assumed to be extremely narrow, but with a finite height, the equation takes the form[5,6]

$$I(s) = C\Big\{-\int_0^{h} [\tilde{I}_H{}'\{(s^2+v^2)^{1/2}\}/(s^2+v^2)^{1/2}]\mathrm{d}v \\ -2\int_h^{h\sqrt{2}} [\tilde{I}_H{}'\{(s^2+v^2)^{1/2}\}/(s^2+v^2)^{1/2}]\mathrm{d}v + \tilde{I}\{(s^2+2h^2)^{1/2}\}\Big\} \tag{15.3}$$

where h is the height of the cross section of the direct beam in the plane of observation measured from the center.

In a generalization of Guinier's approximation, Hossfeld represented the scattering functions by series of orthogonal Hermite functions, which allowed for a closed treatment of the geometric errors of the size of the collimator, and also of the distortion effect caused by non-monochromatic radiation. This method leads to simple recurrence formulae.[7]

These corrections can be easily carried out with electronic computers.

b) Use of modified theoretical scattering intensity curves. Whereas the above methods enable the measured intensity $\tilde{I}_H$ to be corrected to the intensity that would be obtained with an ideal collimator, there is another method by which a theoretical curve is converted into the intensity curve that would be expected with any given collimator.[8] This method, however, can only be applied to small-angle scattering from particles with certain shapes.

15.2 Analysis of Particle Size and Shape

Some structural parameters, such as the radius of gyration, R_g, the molecular weight, M_r (using the scattering intensity at zero angle, $I(0)$), the hydrated volume, the maximum intra-particle distance from the distance distribution function, $p(r)$, etc. can be computed directly from the small-angle scattering data.

The methods of analysis can be classified into three:[9]

a) Analysis in reciprocal space. Small-angle scattering curves measured with point collimation (slit error free) or those corrected for slit errors (desmeared curve) is used for the analysis. This is the most common method. Several special plots,

1) $\log I(s)$ *vs.* s^2 (Guinier plot) and its analogues, $\log (s \cdot I)$ *vs.* s^2 and $\log (s^2 \cdot I)$ *vs.* s^2 plots (Section 15.2.1),

2) $\log I$ *vs.* s plot (curve fitting method, Section 15.2.2), and others are applied for interpretation.

b) Analysis in experimental space. Smeared small-angle scattering curves $\tilde{I}$ are used for the analysis, and this method does not require any slit error correction process. If the small-angle scattering is measured with a long slit camera, such as a Kratky U-slit camera, in which the slit height can be approximated as infinitely high compared to the narrow slit width, and if we can derive theoretically the relationship between the smeared scattering functions, $\tilde{\Psi}$, and the observed (smeared) scattering intensity, $\tilde{I}$, we can interpret the observed (smeared) scattering curves, $\tilde{I}$ without slit-error correction. For example: $\log \tilde{I}$ *vs.* s^2 plot can be used instead of $\log (s \cdot I)$ *vs.* s^2 plot, the scattering power $\tilde{Q} = \int s \cdot \tilde{I}(s) \mathrm{d}s$ instead of $Q = \int s^2 \cdot I(s) \mathrm{d}s$ and s^{-3} rule can be applied for $\tilde{I}$ instead of s^{-4} rule, etc.

c) Analysis in real space. Distance distribution functions, $p(r)$ or $p_L(r)$ and maximum dimensions, D_{max} are mainly used for the analysis of the small-angle scattering curves obtained.

15.2.1 The Guinier plot

A. Dilute system

The small-angle scattering intensities for highly dispersed, uniform particles (as in an aerosol), or for colloidal particles of uniform size in dilute solution (*e.g.* solutions of homogeneous proteins) are given for various particle shapes by the scattering functions in Table 5 of the Appendix. In most instances, however, the particle shape is unknown. As a first approximation, the radius of gyration is calculated by applying a Gaussian scattering function, which utilizes the radius of gyration R_g with respect to the particle's center of gravity, and Eq. 6.13:

$$I = I_e M n^2 \exp(-s^2 R_g^2/3) = I(0) \exp(-4\pi^2\varepsilon^2 R_g^2/3\lambda^2) \qquad (15.4)$$

where

$$I(0) = I_e M n^2 \quad \text{and} \quad s = 2\pi\varepsilon/\lambda = 4\pi\theta/\lambda$$

$$\log I = \log I(0) - s^2 R_g^2/3 = \log I(0) - (4\pi^2/3\lambda^2)\varepsilon^2 R_g^2$$

A plot of $\log I$ against s^2 or ε^2 is linear, and the radius of gyration R_g may be found from the gradient (see Fig. 6.7). In the case of a graph of $\log I$ *vs.* ε^2 if the gradient is α (the value of which must be appropriately converted if $\log_{10}$ semilogarithmic graph paper has been used), then

$$R_g = (\lambda/\pi)(-3\alpha/4)^{1/2} \tag{15.5}$$

In practice, however, the curve departs radically from a Gaussian curve at large scattering angles. This may be due to the fact that the Gaussian curve is an approximation, or it may derive from experimental error, or the effects of electron density variations within the particles. Accordingly, the gradient for the smallest possible scattering angles will give the most accurate value (see Fig. 15.3).

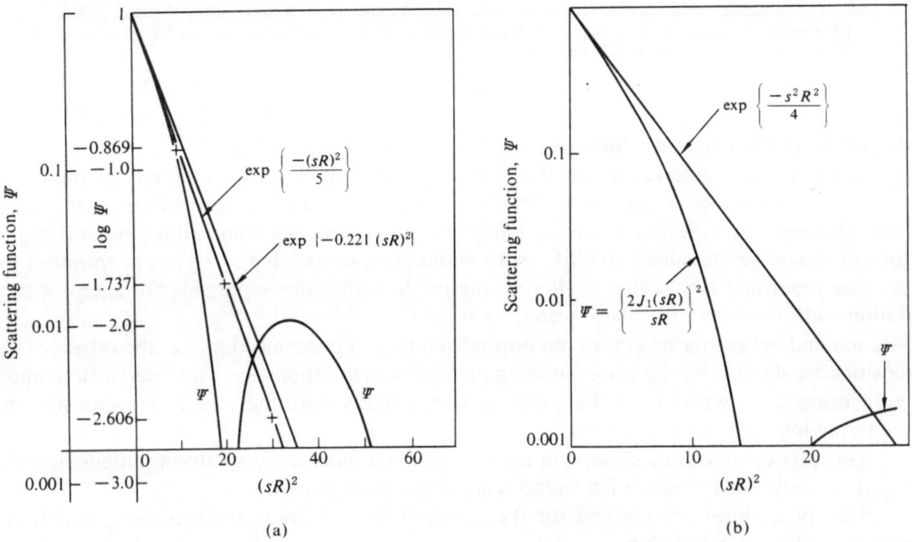

Fig. 15.3 Comparison between the scattering function, Ψ, and the approximate (Gaussian) functions.
(a) Spherical particle The log Ψ *vs.* $(sR)^2$ (or log I *vs.* ε^2) plot may be taken as practically linear for $0 < (sR)^2 < 10$. The slope at $sR = 1.5$ gives the diameter (about 5% below the actual diameter) of the sphere.[10)]
(a): [Reproduced from J. Turkevitch, H. H. Hubell, *J. Am. Chem. Soc.*, **71**, Am. Chem. Soc. (1951)]
(b) a fixed cylindrical particle.
(b) The error in the radius of the cylinder as derived from the gradient where $sR = 1.5$ is about 5%, and increases to 15% at $sR = 2.5$. The departure from linearity is increasingly marked for $(sR)^2 > 2$.

If the mass and density of the particles are known, it is possible to calculate the radius of gyration R_{min} for spherical particles. The ratio of the measured value to R_{min}, *i.e.* $R/R_{min} = f$, may then be used as a criterion of the anisotropy in particle shape. Moreover, if the particle has a simple geometrical shape the radius of gyration can be calculated. Table 15.1[10)] shows the radii of gyration of particles with simple shapes. In such cases, a number of possible particle shapes that correspond to the measured volume and radius of gyration can be found. The analysis may then be repeated for these shapes using the scattering functions given in Table 6 of the Appendix to refine the results (*cf.* Section 15.2.2).

Table 15.1 Radii of gyration of particles having simple shapes[11]

Shape	Radius of gyration R_g
Sphere (radius *R*)	$(3/5)^{\frac{1}{2}}R$
Spherical shell (external radius *R*, internal radius *cR*)	$(3/5)^{\frac{1}{2}}R[(1-c^5)/(1-c^3)]^{\frac{1}{2}}$
Ellipsoid of revolution (semiaxes *a*, *a*, *wa*)	$a[(2+w^2)/5]^{\frac{1}{2}}$
Cylinder (height 2*H*, radius *R*)	$(R^2/2+H^2/3)^{\frac{1}{2}}$
Thin disc (radius *R*)	$2^{-\frac{1}{2}}R$
Fiber (length 2*H*)	$3^{-\frac{1}{2}}H$
Rectangular Prallelepiped (length 2*a*, width 2*b*, height 2*c*)	$[(a^2+b^2+c^2)/3]^{\frac{1}{2}}$
Cube (edge length 2*a*)	a

[Reproduced with permission from C. H. MacGillavry, G. D. Rieck eds., *International Tables for X-Ray Crystallography*, Vol.III, 327. IUCr.(1962)]

B. Effect of the concentration (See also Sections 6.2.5 and 15.2.3)
It is said that, for example, for globular proteins concentration effects can only be neglected for concentrations below 1 mg/ml. Therefore, as a rule, small-angle scatterings of four or five solutions with different concentrations in the range between 5 and 30 mg/ml and that of solvent should be measured, and the results should be extrapolated to zero concentration.

The procedure is as follows: after averaging the intensities separately for each concentration and for solvent, the blank scattering is subtracted from the scattering of each protein solution and the scattering curves are normalized to unit concentration, *i.e.* the observed intensities are divided by the corresponding protein concentrations, *c*. The interparticle interference increases with increasing concentration, usually with the effect of decreasing the scattering intensity at small angles.

This procedure can be done with the smeared and unsmeared scattering intensities.[12,13] Fig. 15.4 shows an example of smeared scattering curves.

Two procedures are applied for the extrapolation to zero concentration, which are shown in Figs. 15.5 and 15.6

a) Extrapolation to zero concentration.

1) The upper thick line in Fig. 15.5 is the extrapolated curve: the extrapolation to zero concentration is carried out in the normal plot, $\tilde{I}/c$ *vs.* ε at each measured scattering angle.

2) The second procedure shown in Fig. 15.6 is a Zimm plot:[14] this procedure is familiar in light scattering studies of high polymer solutions.

3) A third possible way is to calculate the radius of gyration $\tilde{R}_g$ form the Guinier plot for each concentration (Fig. 15.7) and to plot the $\tilde{R}_g$ values against the concentration *c*, which are shown in Fig. 15.8. The value of $\tilde{R}_g$ extrapolated to zero concentration $\tilde{R}_{g,c\leftarrow 0}$ is 65.6 Å. The distance distribution function $p(r)$ for the scattering curve extrapolated to zero concentration may be used as a reference to confirm the disappearance of the concentration effect (Fig. 15.9) (*cf.* Section 15.2.3).

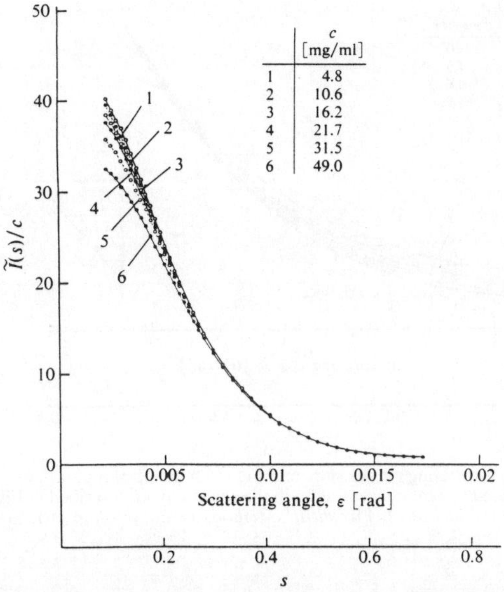

Fig. 15.4 Slit-smeared scattering curves of hemocyanin (*Astacus leptodactylus*) for the indicated concentrations c with Cu $K\alpha$ radiation., The curves are normalized to $c = 1$ by plotting $\tilde{I}/c$; $\tilde{I}$ slit-smeared scattering intensity.[12,13)]
[Reproduced with permission from I. Pilz *et al.*, *Methods in Enzymology*, **61**, 208, Academic Press (1980)]

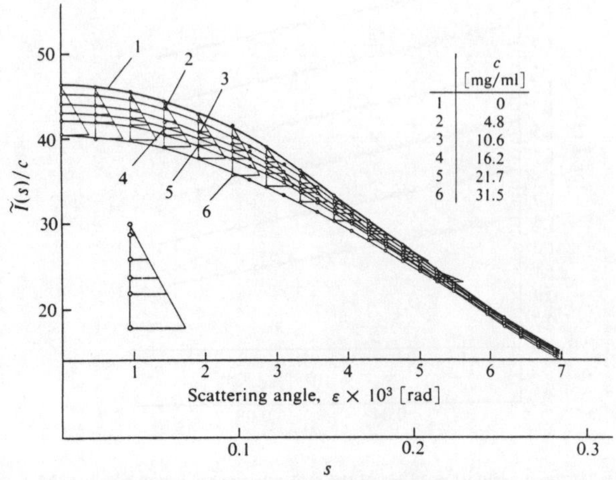

Fig. 15.5 Innermost portions of the scattering curves shown in Fig. 15.4.
The curves are extrapolated to zero concentration (curve 1) by plotting the corresponding concentration at different scattering angles parallel to the abscissa in arbitrary units, as indicated.[12,13)]
[Reproduced with permission from I. Pilz *et al.*, *Methods in Enzymology*, **61**, 209, Academic Press (1980)]

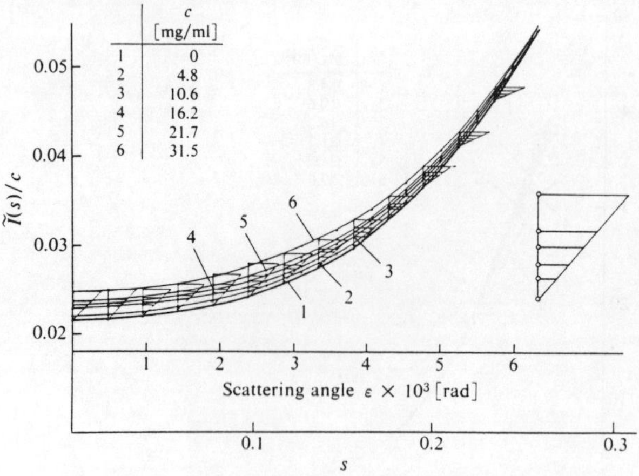

Fig. 15.6 Innermost portions of the scattering curves shown in Fig. 15.5 in a Zimm plot. The curves are extrapolated to zero concentration in the same way as described in Fig. 15.5.[12,13] [Reproduced with permission from I. Pilz *et al.*, *Methods in Enzymology*, **61**, 209, Academic Press (1980)]

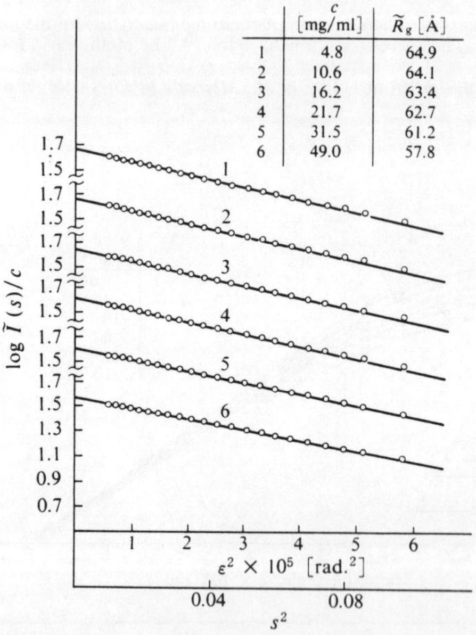

Fig. 15.7 Guinier plots of the innermost portions of the scattering curves shown in Fig. 15.4. The values of the slit-smeared radii of gyration $\tilde{R}_g$ obtained for the different concentrations are indicated.[12,13]
[Reproduced with permission from I. Pilz *et al.*, *Methods in Enzymology*, **61**, 210, Academic Press (1980)]

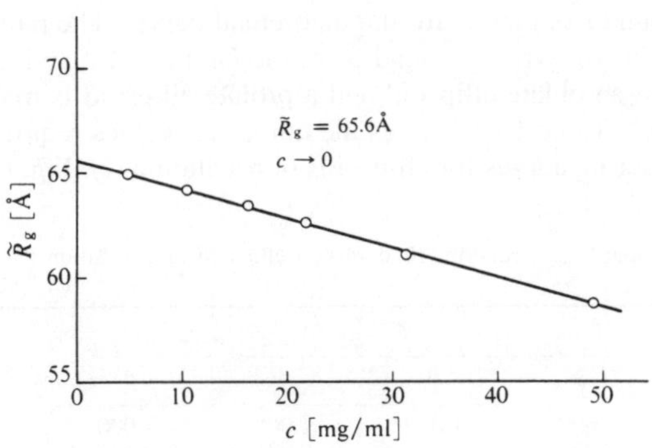

Fig. 15.8 Slit-smeared radii of gyration $\tilde{R}_g$ calculated from Fig. 15.7 plotted against the protein concentration c.[12,13) [Reproduced with permission from I. Pilz *et al.*, *Methods in Enzymology*, **61**, 211, Academic Press (1980)]

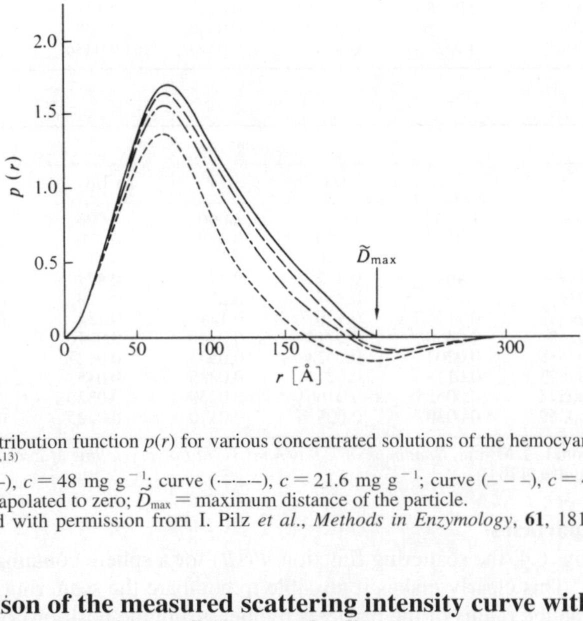

Fig. 15.9 Distance distribution function $p(r)$ for various concentrated solutions of the hemocyanin of *Astacus leptodactylus*.[12,13)
Curve (– – –), $c = 48$ mg g^{-1}; curve (·——·), $c = 21.6$ mg g^{-1}; curve (– – –), $c = 4.8$ mg g^{-1}, curve (——), c extrapolated to zero; $\tilde{D}_{max}$ = maximum distance of the particle.
[Reproduced with permission from I. Pilz *et al.*, *Methods in Enzymology*, **61**, 181, Academic Press (1980)]

15.2.2 Comparison of the measured scattering intensity curve with the theoretical curve (Curve fitting method)

When theoretical scattering intensity curves have been derived for particles having relatively simple shapes, it is possible to determine the desired parameters by trial and error involving appropriate variation of the parameters and direct comparison of the overall shapes of the theoretical and measured curves, or by looking for the theoretical profile that conforms best to the measured curve in the graph of log Ψ *vs.* log ε.

Figure 6.6(b) shows log Ψ *vs.* log ε graphs of theoretical scattering curves for ellipsoids of revolution with equal radii of gyration (axial ratio 1 : 1 : w). If any one of these curves is modified by moving the coordinates until it conforms to the measured curve, the shape of the particles can be found from the w for the theoretical curve. The particle size can be determined from the radius of gyration found from the Guinier plot and the value of w. The differentiation between an oblate ellipsoid and a prolate ellipsoid is made at the tail of the curve (see Fig. 15.10).[3] Table 15.2[15] gives the numerical values required for the construction of theoretical scattering curves for ellipsoids of revolution (*cf.* Fig. 6.6(b)).

Table 15.2 Scattering functions for ellipsoids of revolution[15]

(a) Prolate ($w > 1$)

sR	w						
	1.5	2	3	4	6	10	∞
0.0	1.000	1.000	1.000	1.000	1.000	1.000	1.000
0.5	0.920	0.920	0.921	0.921	0.922	0.922	0.920
1.0	0.713	0.718	0.726	0.730	0.735	0.737	0.738
1.5	0.461	0.477	0.505	0.522	0.537	0.546	0.551
2.0	0.242	0.274	0.327	0.359	0.389	0.407	0.418
2.5	0.0990	0.138	0.206	0.250	0.292	0.318	0.334
3.0	0.0293	0.0616	0.128	0.176	0.225	0.259	0.278
3.5	0.00638	0.0240	0.0760	0.122	0.176	0.214	0.239
4.0	0.00333	0.00789	0.0421	0.0827	0.138	0.180	0.210
4.5	0.00447	0.00294	0.0216	0.0546	0.108	0.153	0.187
5.0	0.00450	0.00270	0.0101	0.0347	0.0836	0.131	0.168

(b) Oblate ($w < 1$)

sR	w						
	2/3	1/2	1/3	1/4	1/6	1/10	0
0.0	1.000	1.000	1.000	1.000	1.000	1.000	1.000
0.5	0.920	0.920	0.920	0.920	0.920	0.920	0.920
1.0	0.712	0.714	0.716	0.717	0.717	0.718	0.718
1.5	0.457	0.464	0.472	0.475	0.477	0.479	0.479
2.0	0.236	0.250	0.267	0.274	0.278	0.284	0.284
2.5	0.0931	0.112	0.136	0.148	0.154	0.162	0.166
3.0	0.0262	0.0450	0.0714	0.0846	0.0916	0.102	0.106
3.5	0.00688	0.0203	0.0428	0.0555	0.0627	0.0738	
4.0	0.00523	0.0113	0.0274	0.0385	0.0453	0.0562	
4.5	0.00525	0.00625	0.0166	0.0259	0.0320	0.0426	
5.0	0.00369	0.00307	0.00934	0.0170	0.0225	0.0328	

[Reproduced from L. I. Mirkin, *Handbook of X-Ray Analysis of Polycrystalline Materials*, pp. 671, 672, Consultants Bureau (1964)]

A. Globular particles

As shown in Fig. 6.4, the scattering function $\Psi(sR)$ for a sphere contains a large number of weak maxima. This clearly makes it possible to compare the scattering curves as a whole, and to determine the radius of the particles by measuring the positions of the maxima. The maxima of the scattering function $\Psi(sR)$ are situated at $s_{max}R = k\pi - 3/k\pi$, where k is a positive integer greater than unity, and the corresponding maximum values are $9/(k\pi)^4 + 81/(k\pi)^6$. The minima are situated at $\tan(s_{min}R) \cong s_{min}R$, *i.e.* $s_{min}R \cong (2j+1)/4 - 1/(2j+1)\pi^2$, where j is a positive integer greater than zero, and have a minimum value of zero. It should also be possible to use the number of observed maxima as a criterion of sphericity. Particles of polystyrene latex are well known to be sufficiently uniform in shape and size to be used as a standard of particle size in electron microscopy. A correspondingly large number of peaks have been measured in the small-angle scattering of this material.[16,17]

B. Very long rod-like or very broad disc-shaped particles (*cf.* Section 6.2.2 B and C) In general, for particles of special shapes, such as rods of infinite length or plates of infinite breadth, it is possible to separate the integrations for specified dimensions such as length and thickness in the Fourier transform of $Q_s(\boldsymbol{r})$ (the convolution of the shape function). For example, the scattering functions of an unoriented assembly of M rectangular parallelepipeds with thickness T, breadth B, and length L, and having uniform electron density ρ, can be resolved into the functions Ψ_T, Ψ_B, Ψ_L for the three fixed directions.

$$I(s) = I_e M\rho^2 V^2 < \Psi_T \Psi_B \Psi_L >, \tag{15.9}$$

In Eq. 15.9, the sign $< \ >$ represents the angle averaged with respect to azimuth. If the length L is very long, Ψ_L should have a much greater angular variation than the other two functions, and Ψ_L may therefore be averaged separately from the other two. Further, when B and T are very large in relation to L, it should also be possible to average Ψ_B and Ψ_T separately. Since the respective averages will be

$$< \Psi_B > = \pi/sB, \quad < \Psi_T > = \pi/sT, \quad < \Psi_L > = \pi/sL \tag{15.10}$$

the scattering functions for long rod-like particles will be

$$I(s) \cong < \Psi_L > < \Psi_B \Psi_T > \cong (\pi/sL) I_{BT}(s) \tag{15.11}$$

and for very broad disc-shaped particles

$$I(s) \cong < \Psi_L > < \Psi_B \Psi_T > \cong I_L(s)(\pi/sB)(\pi/sT) \tag{15.12}$$

or

$$\text{(long rod-like particles)} \qquad I(s)s \cong I_{BT}(s) \tag{15.13}$$

and

$$\text{(very broad disc-shaped particles)} \qquad I(s)s^2 \cong I_L(s) \tag{15.14}$$

Thus if the scattering intensity profile $I(s)$ for such a particle is multiplied by s (or ε), the curve should correlate with the intensity I_{BT} due only to the shape of the cross section of a rod, and multiplication by s^2 (or ε^2) will give a curve that correlates with the intensity due only to the thickness of a broad disc. The respective shapes and sizes can then be found by comparison of I_{BT} or I_L with the theoretical intensity. It is possible, for example, to calculate the radius of gyration R_{gs} of the cross section from the gradient of the Guinier plot $\log I_{BT}(= \log (I \cdot \varepsilon))$ *vs.* ε^2 using I_{BT}[18–20] (see Fig. 6.8)

$$\log I_{BT} = \log (I \cdot \varepsilon) = \log (I \cdot \varepsilon)_0 - K R_{gs}^2 \varepsilon^2, \quad \text{and} \quad K = 2\pi^2/\lambda^2 \tag{15.15}$$

It is also possible to deduce the shape of the cross section from the plot of $\log I_{BT}$ against $\log \varepsilon$ (see explanation below in connection with Fig. 15.10(c)).

a) Small-angle scattering of γ-globulin.[21] This is an example of small-angle scattering of a long protein molecule. The Guinier plot of the scattering intensity curve departs radically from linearity as shown by Fig. 15.10(a). Determination of the radius of gyration of the molecule as a whole from a tangent to the small-angle portion of the curve gave $R_g = 70$ Å. Since it had previously been deduced from other types of measurements that the molecule was rod-like, $\log (I \cdot \varepsilon)$ was plotted against ε^2 to determine the size of the cross section of the molecule (Fig. 15.10(b)). The disparity at very small angles is attributable to the finite length of the rod. The radius of gyration of the cross section asfound from the gradient of the tangent is $R_{gs} = 15$ Å. If it is assumed that the molecule is cylindrical, then the height H is given by

$$H^2 = 12(R_g{}^2 - R_{gs}{}^2) = 12(70^2 - 15^2) \quad \therefore \quad H = 236\ \text{Å}$$

It was deduced from the log $(I \cdot \varepsilon)$ *vs.* log ε plot (see Fig. 15.10(c)) that the cross section has an axial ratio of 1 : 3, and hence

$$R_{gs} = (a^2 + b^2)/4 \ (a \text{ and } b \text{ are the semiaxes of the ellipse})$$

The values obtained were $a = 19$ Å, and $b = 57$ Å. Calculation of the molecular weight from these values gives $M_r = 1.57 \times 10^5$, which agrees well with the value of 1.56×10^5 obtained by the ultracentrifuge method.

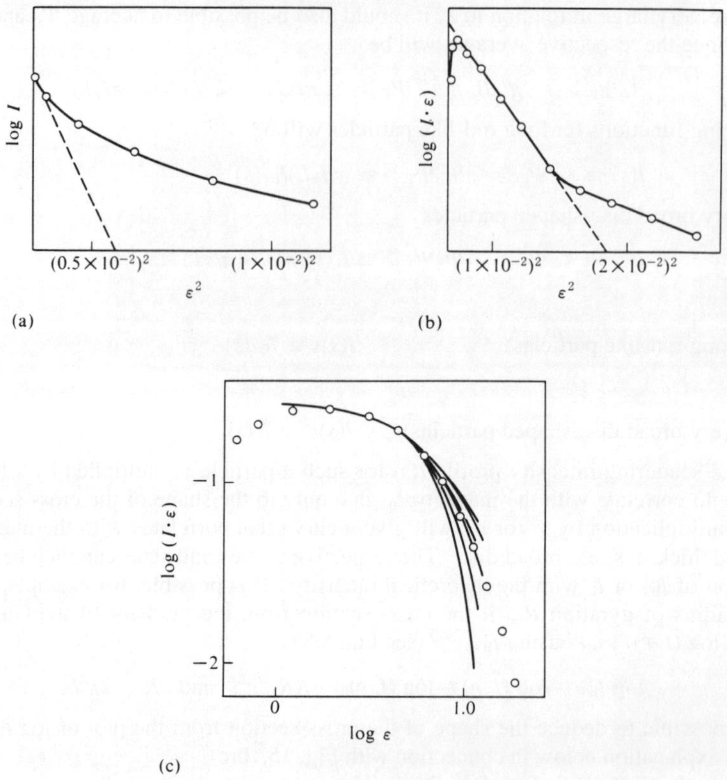

Fig. 15.10 Analysis of small-angle scattering by an aqueous solution of γ-globulin.[21)]
(a) Guinier plot of log I *vs.* ε^2 (markedly non-linear).
(b) Guinier plot of log $(I \cdot \varepsilon)$ *vs.* ε^2.
(c) log $(I \cdot \varepsilon)$ *vs.* log ε (for cross-section axial ratio).
[Reproduced from O. Kratky *et al.*, *J. Polym. Sci.*, **16**, 163, John Wiley & Sons, Inc. (1955)]

15.2.3 The distance distribution function

As mentioned before (Section 6.2.3) the distance distribution, $p(r)$ or $p_L(r)$ can be determined by Eq. 6.51 and Eq. 6.53 from the small-angle scattering intensity curves measured and corrected for slit errors.

$$p(r) = \gamma_0(r) \cdot r^2 \tag{6.53}$$

$$\gamma_0(r) = \frac{1}{2\pi^2\rho^2 Vr}\int_0^\infty s \cdot I(s) \sin sr ds \tag{6.51a}$$

A. The formation of dimer[9,22,23]

Figure 15.11 shows the distance distribution function of dimer models. The models are built from two uniform, prolate ellipsoids of revolution ($w = b/a = 2$, *i.e.* axial ratio $2a : 2a : 2b = 1 : 1 : 2$). In the figure the distance distribution of the monomer (solid line), that of the dimer (broken line) and also the difference of the two distance distribution functions (thick solid line) are shown. The difference between the parallel and linear dimers is obvious. The two rectangular arrangements, T and L types, lie between the parallel and linear dimers.

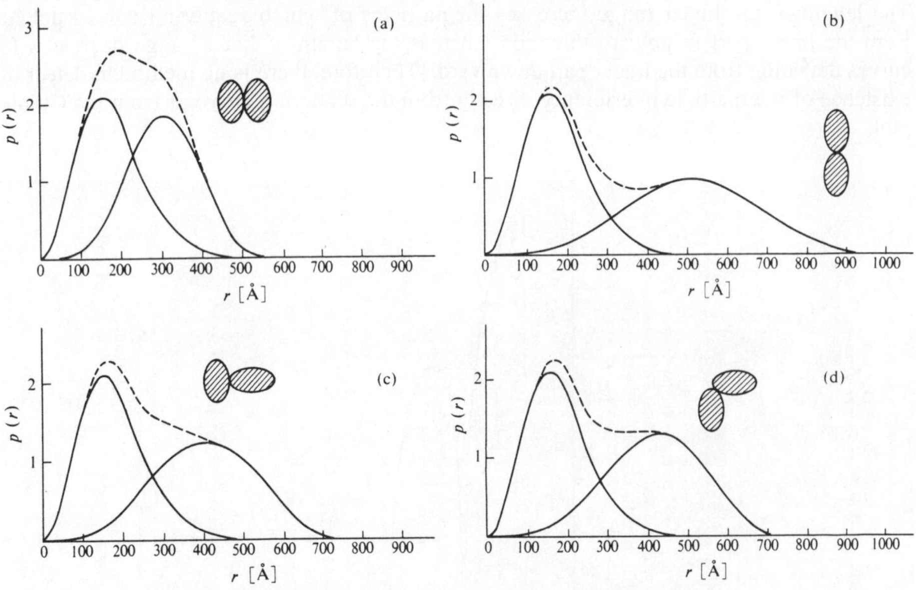

Fig. 15.11 Distance distribution function $p(r)$ from dimer models built from prolate ellipsoids.[22,23] Solid line, monomers; broken line, dimers; thick solid line, difference between dimers and monomers. (a) Parallel arrangement. (b) linear arrangement, (c) T type, (d) L type.
[Reproduced with permission from I. Pilz *et al.*, *Methods in Enzymology*, **61**, 176, Academic Press (1980)]

B. The interparticle interference[24,25)]

A rough approximation is the model consists of N identical hard spheres with radius R. The scattering intensity of this model is given by Guinier and Fournet* based on theories of Born and Green[26)]

$$I(s) = I_e \langle N \rangle F^2(sR) \frac{1}{1+(8v_0 / v_1)\varepsilon F(2sR)} \tag{15.16}$$

where $\Phi^2(sR) = \Psi(sR)$ and v_0/v_1 is the packing parameter, which is equal to 0.72 in the case of hexagonal and cubic closest packings, and ε is a constant approximately equal to unity. The approximation for small s values

$$I(s) = \text{const.} \cdot F^2(sR) \frac{1}{1+(8v_0 / v_1)} \tag{15.17}$$

and for low concentrations

$$I(s) = \text{const.} \cdot \Phi^2(sR)(1 - 8v_0/v_1) \tag{15.18}$$

A linear approximation to zero concentration discussed above is based on Eq. 15.18, which is an approximation of Eq. 15.17.

The scattering curves for several values of $8v_0/v_1$ is already given in Fig. 6.16. The apparent radius of gyration decreases with increasing concentration (Figs. 15.12 and 15.13). The length of the linear range increases for particles of which scattering curves deviate from the linear part in upward direction whereas the length of linear range decreases for curves deviating from the linear part downward. Therefore, there is no method to detect the existence of interparticle interference directly from the scattering curve or from the Guinier plot.

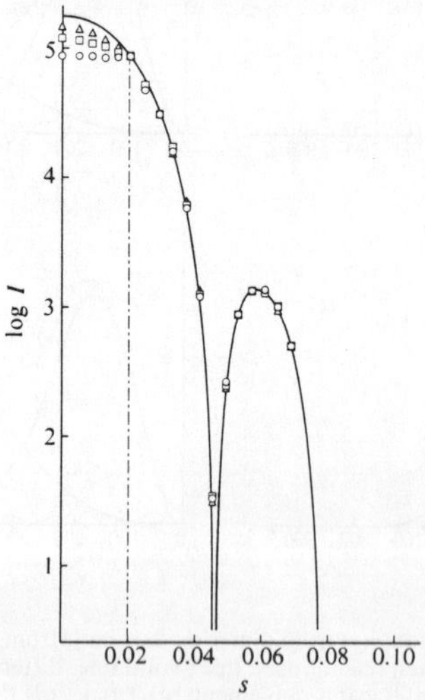

Fig. 15.12 Interparticle interference effect. (Hard sphere model, $R = 100$ Å)
Scattering functions for different volume concentrations.
Solid curve, $v_0/v_1 = 0$; Δ, $v_0/v_1 = 1/32$; □, $v_0/v_1 = 1/16$; ○, $v_0/v_1 = 1/8$. Dashed vertical line; termination for indirect transformation (first data point at $sR = 2.0$)
[Reproduced with permission from I. Pilz *et al.*, *Methods in Enzymology*, **61**, 177, Academic Press (1980)]

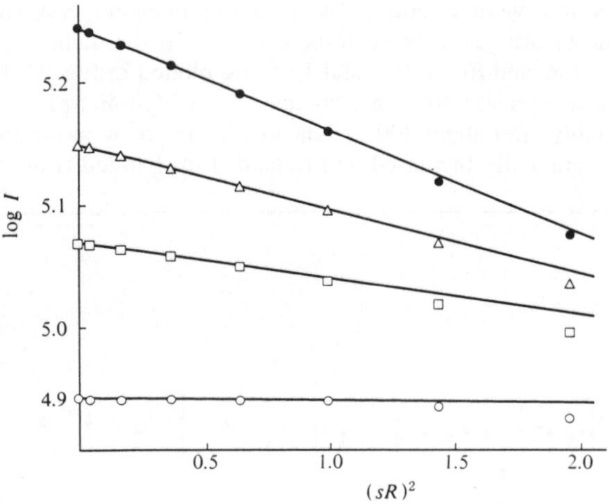

Fig. 15.13 Interparticle Interference effect, hard sphere model as Fig. 15.12. Guinier plot: ●, $v_0/v_1 = 0$, △, $v_0/v_1 = 1/32$; □, $v_0/v_1 = 1/16$; ○, $v_0/v_1 = 1/8$. [Reproduced with permission from I. Pilz *et al.*, *Methods in Enzymology*, **61**, 178, Academic Press (1980)]

The distance distribution is affected considerably by interparticle interferences. It is lowered with increasing distance *r*, goes to a negative minimum in the region of the maximum distance of the particle and the oscillations vanish at larger *r* values. This has been shown for the hard sphere model in Fig. 15.9.

A. Acid denaturation of Taka-Amylase A[27]

The structural change of Taka-Amylase A by acid denaturation has been studied by a stopped-flow small-angle X-ray scattering. Taka-Amylase A (EC 3.2.1.1, α-1,4-glucan-4-glucano hydrolase) is monomeric, composed of a single polypeptide chain of 478 amino residues. The three-dimensional molecular structure, determined at a 3.0 Å resolution by an X-ray crystal structure analysis is nearly ellipsoidal with approximate dimensions of 35 × 45 × 80 Å.[28] It may be approximated as a prolate ellipsoid of revolution with axial lengths of 40 × 40 × 80 Å ($w = b/a = 2$).

The aqueous solution of protein (20.2 mg ml^{-1}) was mixed with a 0.1 M glycine/NaOH–HCI (1 M = 1 mol dm^{-3}) buffer in a stopped-flow mixing device. The pH of the mixed solution was adjusted to 3.3. Nickel-filtered Cu $K\alpha$ radiations were collimated with a 0.3 × 10 mm slit, and reflected and focused by nickel-coated glass mirror through two limiting slits: one located immediately behind a mirror holder to collimate reflected X-rays, the other located just before a stopped-flow mixing device. The scattered X-rays were recorded on a 50 mm-long one-dimensional position sensitive proportional counter (delay-line type). The flow gas (11 mm in thickness) used was a mixture of 90% argon and 10% methane. The width of the detector window was 10 mm. The detector length was 50 mm and divided into 256 channels on a multi-channel analyzer. The sample-to-detector distance was 310.0 mm. The scattered X-rays were collected in the range of scattering from 3.8×10^{-3} to 7.7×10^{-2} rads.[29]

The scattering intensities, $\tilde{I}(s)$, recorded on both sides of the primary beam were averaged at equivalent points after subtracting background intensities. The coincidences of intensities on both sides, R_{sym} were all good. The scattering intensity, $I(s)$ was then obtained by a deconvolution (desmearing) for the beam-height effect by the method of Glatter.[30]

The time courses of R_g and $I(0)$ at 0°C and 15°C are plotted in Fig. 15.14. Radii of gyration, R_g and zero-angle intensity, $I(0)$ were obtained from a Guinier plot. At 15°C, R_g and $I(0)$ increased appreciably after about 800 s (indicated by an arrow in the figure, within the time during which R_g gradually increased and remained unchanged (Fig. 15.14(b)). This

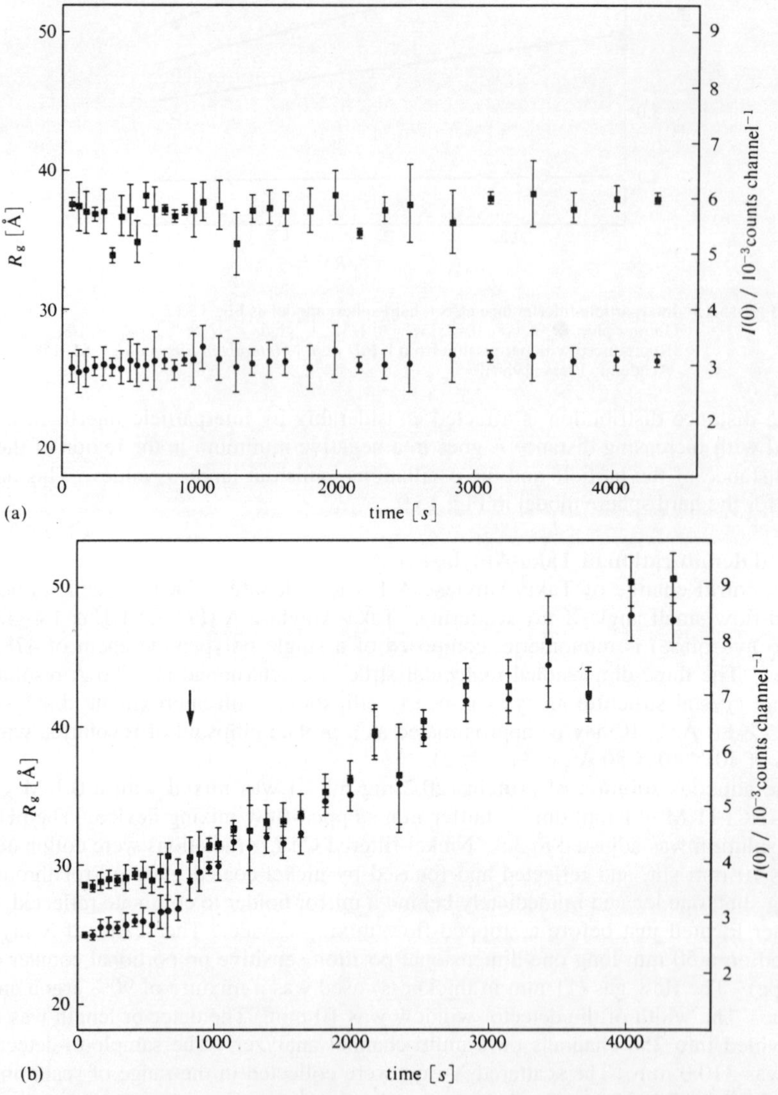

Fig. 15.14 Time courses of radius of gyration, R_g(●) and zero-angle scattering intensity, $I(0)$(■) (Taka-Amylase A).[27] (a) at 0°C and (b) at 15°C. Error bar is drawn by 3σ, where σ is the standard deviation obtained from the least-squares method. [Reproduced with permission from T. Hozaki *et al.*, *Bull. Chem. Soc. Jpn.*, **59**, 3748, Chem, Soc. Jpn. (1986)]

phenomenon was also observed at 10°C; however, the time at which the large increase of R_g and $I(0)$ observed was retarded to about 3,400 s. At 5°C, only the phenomenon of a gradual increase of R_g and an almost constant $I(0)$ was observed over the whole range of the time course. Neither R_g nor $I(0)$ changed at 0°C (Fig. 15.14(a)).

These results show that the acid-denaturation of Taka-Amylase A can be explained by two steps of a structural change from the time course of R_g and $I(0)$. The arrow in Fig. 15.14(b) indicates the time between the first and second steps of the acid denaturation observed.

1) First step: The log I *vs.* log s plot of the scattering intensity for frame 1 and 13 at 15°C is compared with the theoretical scattering functions of prolate ellipsoids (Fig. 15.15). This comparison was performed in a time range in which $I(0)$ was nearly constant, because the increase of $I(0)$ corresponds to that of molecular mass.[31] The structural parameters of the initial and final stages in the first step are summarized in Table 15.3. Those obtained for the initial stage, $a = 20$ Å, $b = 40$ Å ($R_g = 22$ Å) are in good agreement with those obtained by the crystal structure analysis taking into the account the hydration in solution

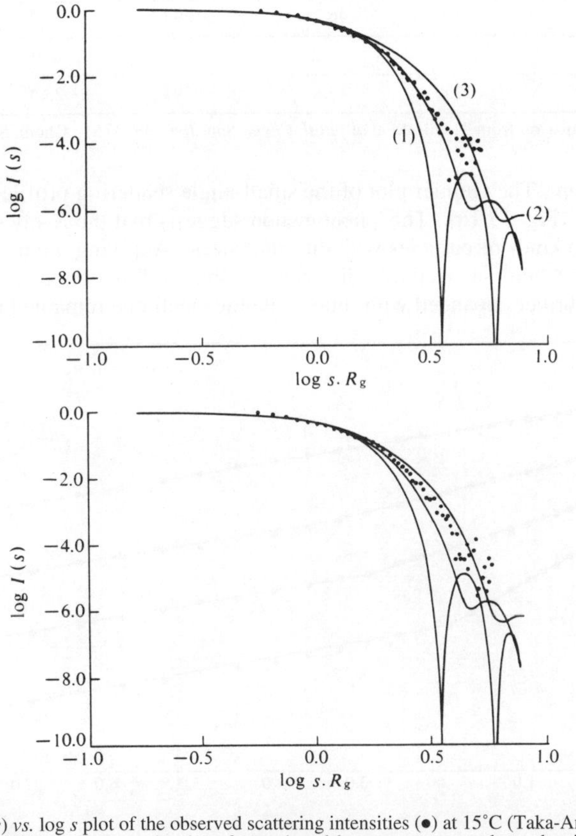

Fig. 15.15 Log $I(s)$ *vs.* log s plot of the observed scattering intensities (•) at 15°C (Taka-Amylase A).[27] The top corresponds to the time frame 1 and bottom corresponds to the time frame 13 at 15°C. Superimposed are the theoretical scattering functions (solid lines) of the prolate ellipsoids with the eccentricities, w of (1) $w = 1.0$, (2) $w = 2.0$ and (3) $w = 3.0$. [Reproduced with permission from T. Hozaki *et al.*, *Bull. Chem. Soc. Jpn.*, **59**, 3750 , Chem. Soc. Jpn. (1986)]

and/or experimental errors. As seen in Fig. 15.15 the overall molecular structure changed with time; the eccentricity w of the ellipsoid increased from 2.0 to 2.4; however, the value of a did not show any change. This indicates that the Taka-Amylase A molecule elongates with time by acid denaturation.

Table 15.3 Structural parameters of the initial stage (Time 1) and the final stage (Time 2) in the first step[27)]

t[°C]	Time 1 [s]			Time 2 [s]		
	R_g[Å]	a[Å]	b[Å]	R_g[Å]	a[Å]	b[Å]
0		60			4,320	
	25.9	23.7	47.3	26.2	23.1	48.5
5		60			4,320	
	25.5	23.3	46.6	30.2	23.7	56.8
10		60			3,420	
	25.6	23.4	46.7	29.6	23.3	53.7
15		60			810	
	25.0	22.8	45.6	29.0	23.3	55.9

[Reproduced with permission from T. Hozaki *et al.*, *Bull. Chem. Soc. Jpn.*, **59**, 3751 , Chem. Soc. Jpn. (1986)]

2) Second step: The Guinier plot of the small-angle scattering profile at 15°C became biphasic with time (Fig. 15.16). This phenomenon suggests that the X-ray scatterer in solution comprises two kinds of particles with different sizes. Applying the method of Jellinek, Solomon and Fankuchen[32)] as well as a distance distribution function, it was elucidated that the large particle further expanded with time while the small one remained unchanged. The

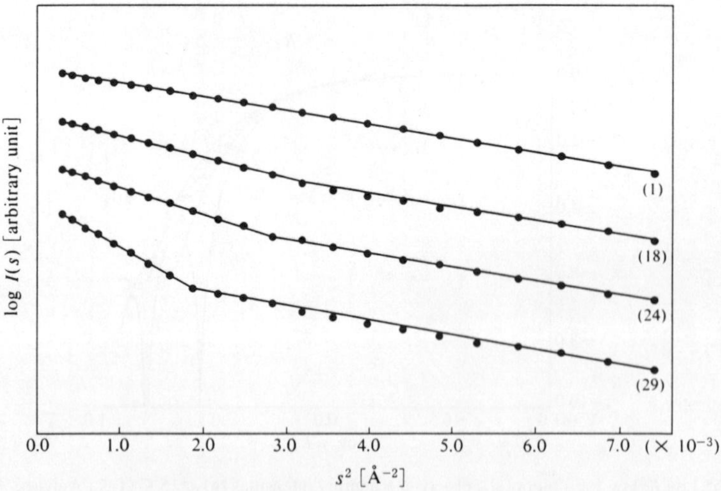

Fig. 15.16 Guinier plots of the small-angle scattering profiles at 15°C (Taka Amylase A).[27)]
Time frames 1, 18, 24, and 29 are plotted.
[Reproduced with permission from T. Hozaki *et al.*, *Bull. Chem. Soc. Jpn.*, **59**, 3751 (1986)]

R_g of the small particle showed a good coincidence with that of the Taka-Amylase A monomer. These facts suggest that Taka-Amylase A molecules aggregates with each other and the X-ray scatterer in solution consists of large aggregate and small elongated molecules transformed from native Taka-Amylase A by a partial unfolding of the polypeptide chain.

B. Structure of a cockroach lipophorin[33,34)]

Lipophorin is a diglyceride-carrying lipoprotein found in insects. Fig. 15.17 shows small-angle X-ray scattering intensities of lipophorin in aqueous solution of sucrose at various concentrations, ρ_2. Scattering intensities varied widely with sucrose concentration. An aqueous solution of 0% sucrose concentration showed a broad peak at $s = 0.01$ Å^{-1} (channel 35 in Fig. 15.17). This peak moved toward the smaller angle with increase of sucrose concentration, and the zero angle intensity, $I(0)$, also decreased. However, at sucrose concentrations over 30% $I(0)$ showed increase with increase in sucrose concentration and the small-angle scattering curves changed to decrease monotonically.

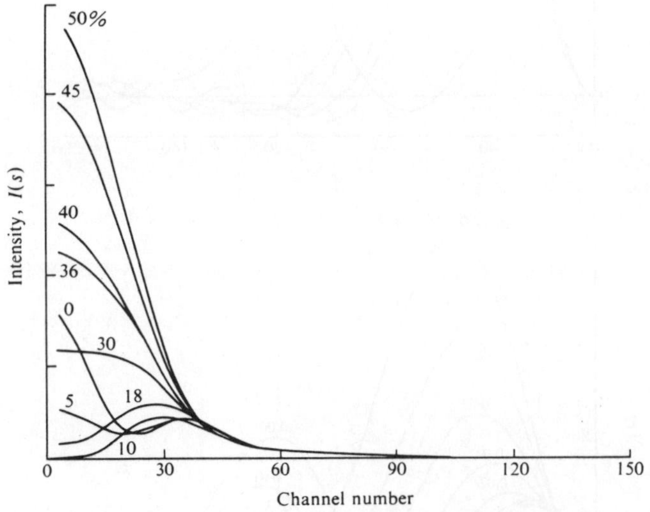

Fig. 15.17 Changes in small-angle scattering intensities of cockroach lipophorin in aqueous solutions of sucrose of various concentrations (%).[33,34)]
Abscissa is channel number of multichannel analyzer which correspond to scattering parameter, *s*.
[Reproduced with permission from M. Sato, *Protein, Nucleic Acid, Enzyme (Tokyo)*, **35**, 2883, Kyoritsu Pub. (1990)]

At 0% sucrose concentration the $p(r)$ function (Section 6.2.3.B–D) gave two positive peaks at $r = ca.$ 30 and 110 Å and a small negative peak at $r = ca.$ 70 Å, which suggests that the distance distribution between regions of positive and negative constrast to the water may be clear (Fig. 15.18(a)). With increase of sucrose concentration the peak at $r = ca.$ 30 Å became remarkably larger, but the other two peaks decreased into ripples.

Since $[I(0)]^{1/2} = \langle \Delta\rho \rangle | S_1(0) |$ and $\langle \Delta\rho \rangle = \bar{\rho}_1 - \bar{\rho}_2$ (Section 6.2.2.D), the average electron density of a lipophorin particle, $\bar{\rho}_1$ could be determined by plotting values of $[I(0)]^{1/2}$ against ρ_2 : $\bar{\rho}_1 = 0.355\ e/\text{Å}^3$ (Fig. 15.19(a)). Values of the contrast $<\Delta\rho>$ at various sucrose concentrations were then calculated. The R_g^2 *vs.* $1/<\Delta\rho>$ plot was approximately linear (Fig. 15.19(b)). Finally, obtained by the least-squares procedure were $a = 38.4\ e/\text{Å}$, $b = 0.03\ e^2/\text{Å}^4$, and $R_g = 49.1$ Å. This result, *i.e.* $a > 0$ and $b \doteqdot 0$, suggests that

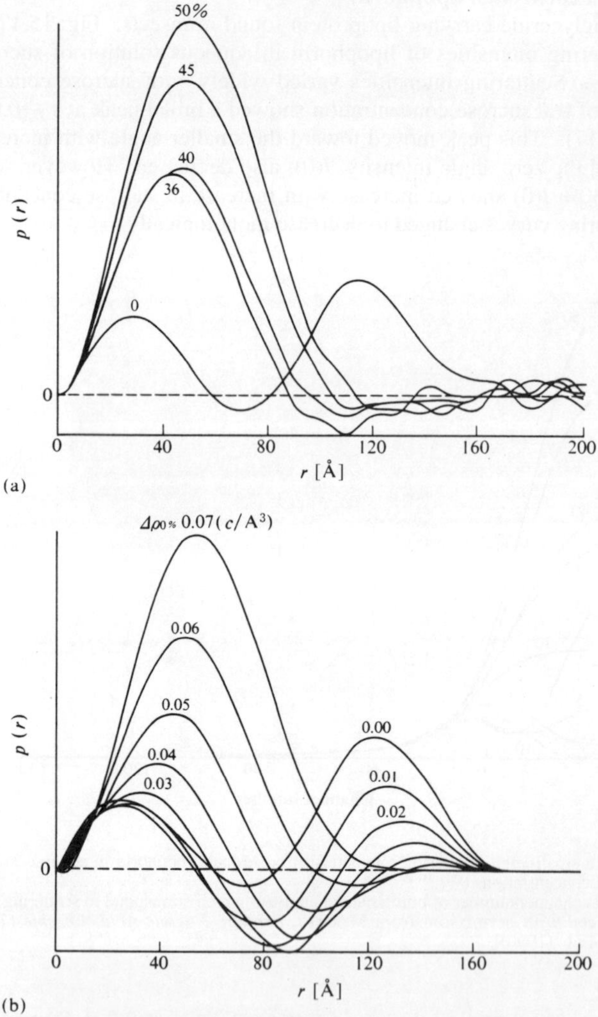

Fig. 15.18 Changes in distance distribution function, $p(r)$ of lipophorin in aqueous solutions of sucrose of various concentrations (%).[33,34]

(a) $P(r)$'s determined from the scattering intensity distributions observed.

(b) $P(r)$'s calculated from the electron density distribution assumed. $\Delta\rho = (\rho_2 - \rho_0)$, where ρ_0 (%), is the electron density of the solvent at 0% sucrose concentration.

[Reproduced with permission from M. Sato, *Protein, Nucleic Acid, Enzyme (Tokyo)*, **35**, 2883, Kyoritsu Pub. (1990)]

the center of gravity of regions which have different electron densities coincides approximately with the center of the gravity of the particle shape (V_1) and that the higher region in electron density locates on the surface and the lower region occupies the inner side of the particle.

It was concluded that a spherical lipophorin molecule has a three-layered inner struc-

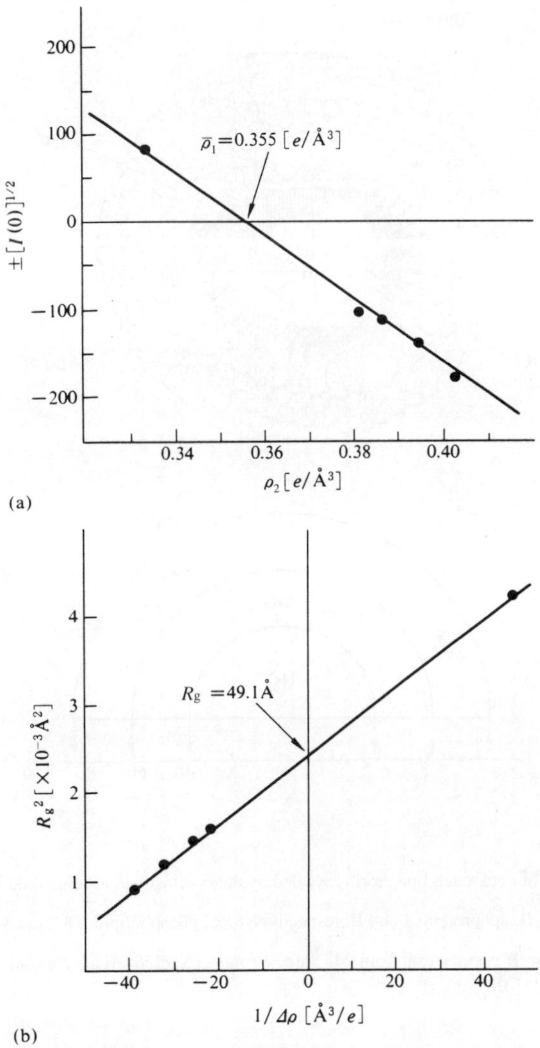

Fig. 15.19 Effects of contrast variation (cockroach lipophorin).[33,34]
(a) ρ_2 dependency of zero angle intensity, $[I(0)]^{1/2}$ *vs.* ρ_2;
(b) Contrast dependency of radii of gyration, R_g^2 *vs.* $1/\Delta\rho$
[Reproduced with permission from M. Sato, *Protein, Nucleic Acid, Enzyme (Tokyo)*, **35**, 2883, Kyoritsu Pub. (1990)]

ture: inside the apoprotein layer there exists a lipid layer that may be divided into two layers. That is: the innermost core portion is occupied by hydrocarbon molecules, *n*-pentacosane, 3-methylpentacosane, and 6,9-heptacosadiene, which are much less polar than the middle layer lipids (Fig. 15.20).

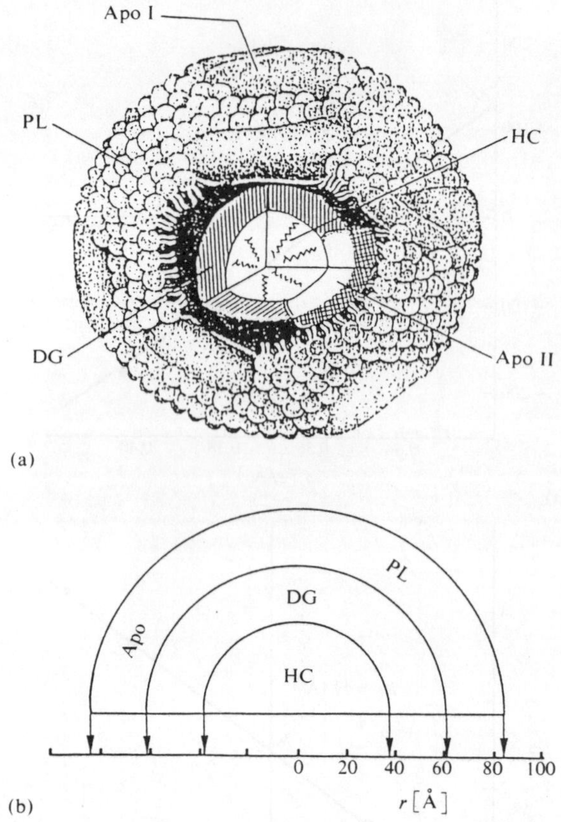

Fig. 15.20 Inner structure of cockroach lipophorin revealed by the contrast variation method (a) and its schematic representation (b).[33,34)]
Apo I and Apo II. apoproteins I and II, respectively; PL, phospholipid; DG, diacyl glyceride; HC, hydrocarbon.
[Reproduced with permission from M. Sato, *Protein, Nucleic Acid, Enzyme (Tokyo)*, **35**, 2883, Kyoritsu Pub. (1990)]

15.2.4 Other analytical methods

In addition to the analytical methods described in Sections 15.2.2 through 15.2.4, there are many other methods of small-angle scattering analysis for dilute or dense systems of colloidal particles, such as Warren's total scattering intensity measurement method,[35] Porod's method, which utilizes absolute intensity measurements[21] and the methods described by Debye,[36,37] Luzzati,[38,39] and others.[40–43] The reader is referred to other sources[5,49–55] for details of these method. Reference should also be made to other books for the analysis of small-angle scattering for stacked lamellar micelles and for parallel assemblies of cylindrical micelles.[3,18,20,43–53]

15.3 Analysis of Small-angle Scattering for Solutions of Chain Macromolecules

15.3.1 Persistence of polymer chain

A. Direction persistence

When chain macromolecules are distributed in a liquid in the form of tightly packed particles, their small-angle scattering can be analyzed by the methods described in earlier sections using the normal particle shape factors. Dependent upon the solvent or on the characteristics of the polymer chains themselves, the molecular chains may be dispersed in the solvent as loosely tangled "random coils," see Fig. 15.21.[56] Small-angle scattering from

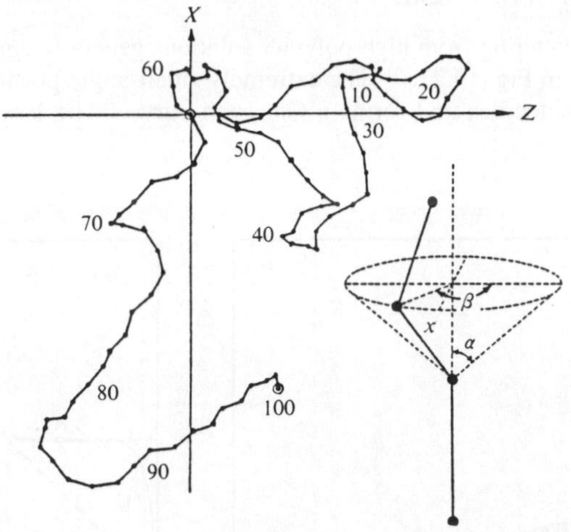

Fig. 15.21 A projection of a molecular chain model consists of 100 straight links onto the *XZ* plane.[56] ($\cos\alpha = 0.8$, where α is the direction angle; β, azimuth; x, length of a link)

systems of this type cannot be analyzed by the simple application of methods such as those described in Section 15.2. Instead of the small-angle scattering intensity equation, which is calculated from the total volume at a constant density, we begin with the general equation (Eq. 2.23) for the intensity of the X-rays diffracted from molecules. If the scattering terms of the individual atoms are omitted, the determination of the scattering function $\Psi(s)$, which is represented by

$$I(s) \propto \psi(s) = \sum_j \sum_k < \frac{\sin s r_{jk}}{s r_{jk}} >, \tag{15.19}$$

where $s = (4\pi \sin\theta)/\lambda \cong 4\pi\theta/\lambda = 2\pi\varepsilon/\lambda$, corresponds to calculation of the sum of the phases of the groups of individual pairs of atoms r_{jk} in the molecule. Expansion of $\Psi(s)$ into a series and omission of the terms below the mean square $<r^2_{jk}>$ of r_{jk} gives

$$\begin{aligned}\Psi(s) &= 1 - s^2 < r_{jk}^2 > /3! + s^4 < r_{jk}^4 > /5! - s^6 < r_{jk}^6 > /7! + \dots \\ &= 1 - s^2 R_g^2/3 \end{aligned} \tag{15.20}$$

Since R_g in this equation is the radius of gyration of the molecular chain, $2R_g = < r^2_{jk} >$. From statistical theory, the relationship between R_g and the mean square of the end-to-end distance $< r_{\mathrm{end}}^2 >$ for an extremely long molecular chain is

$$R_g^2 = < r_{\mathrm{end}}^2 > /6 \tag{15.21}$$

Further, since the overall length L_c of the molecular chain is very long, the characteristic value that determines the linearity of the molecule in solution, *i.e.* the persistence length p, can be defined for r_{end}^2 by

$$\lim_{L_c \to 0} < r_{\mathrm{end}}^2 > = 2\pi L_c \tag{15.22}$$

The small-angle scattering from high polymer solutions generally consists of four parts as shown schematically in Fig. 15.22.[56] The extremely small-angle portion ① is related to the overall size of the molecules and forms a Gaussian curve. The large angle portion ③ is

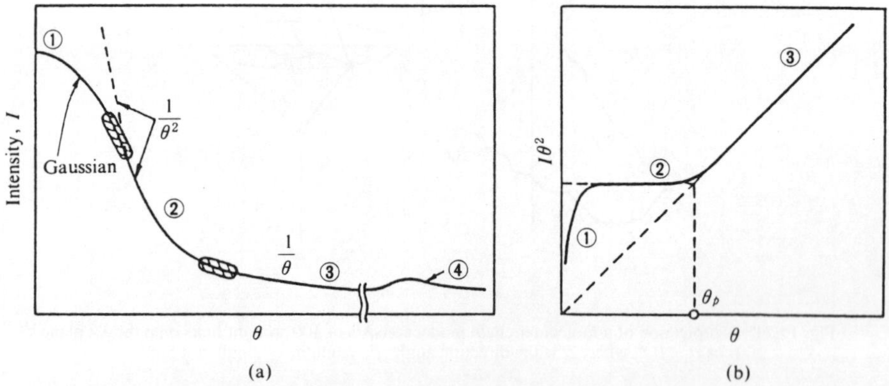

Fig. 15.22 Schematic representation of the small-angle scattering intensity distribution for solution of loosely tangled "random coil" polymer.[56]
(a) The plot of I *vs.* θ generally consists of the four portions shown (see text for explanation).
(b) The plot of $I\theta^2$ *vs.* θ illustrating the place of θ_p (see text).
[Reproduced with permission from O. Kratky, *Angew. Chem.*, **72**, 475, Verlag Chemie (1960)]

caused by interference between the atoms in the localized linear sections of the molecular chain, and the intensity is approximately proportional to $1/\theta$. The portion ② is caused by interference intermediate between these, and the intensity is proportional to $1/\theta^2$. In addition, scattering due to the period of the monomer units, the solvent molecules, and other short-range interatomic interferences generally appears at very large scattering angles.

In the plot of $I\theta^2$ *vs.* θ for this scattering curve, the portion ② is an almost horizontal straight line, and the portion ③ is a straight line passing through the origin, with a smooth transition between them. The Gaussian portion gives a curve that decreases from the highest point in the center, as shown in Fig. 15.22(b).

The point θ_p at which the two linear portions intersect (the point of inflection) depends upon p. According to Porod, the intensity in portion ② is $6/(ps^2)$, while that in portion ③ is π/s. At the point of inflection between ② and ③, therefore, we have

$$6/(ps^2) = \pi/s \tag{15.23}$$

Thus

$$p = 6/(\pi s_p) = 3\lambda/2\pi^2 \tag{15.24}$$

where ($s_p = 4\pi\theta_p/\lambda$) and p can be found from θ_p from Eq. 15.24

$$u_p = ps_p = 6/\pi \cong 1.91 \tag{15.25}$$

The above results were obtained using only the $< r_{jk}^2 >$ term in Eq. 15.20. Despite continued efforts by Peterlin[57,58] and Kratky, Porod *et al.*[19,59] to improve the approximation by adding the $< r_{jk}^4 >$ and higher terms, no simple conclusions have been reached as to which approximation will give accurate results. Peterlin,[60] followed by Heine, Kratky, and Rappert,[61] estimated the improvement of the approximation by numerical calculations. These results generally agreed with the results given by Porod's early theory, but differ in the following particulars.

1) Except in the case of short molecular chains, the point of inflection between the $1/\theta^2$ portion and the $1/\theta$ portion corresponds to $u_p = 2.3$.

2) Even for long molecules, the $I\theta^2$ plot of the $1/\theta^2$ portion is not horizontal as in Fig. 15.22(b), but ascends slightly with increasing angle.

It is therefore better, as far as the degree of approximation is concerned, to use the result obtained from the numerical calculation

$$u_p = (p4\pi\theta_p)/\lambda = 2.3 \tag{15.26}$$

rather than Eq. 15.25 to calculate p from the plot of $s^2 I$ *vs.* s.

a) Small-angle scattering from an acetone solution of nitrocellulose.[59] The measurements were carried out with a U camera and a counter. Fig. 15.23 shows the plot of $x^2 I$ *vs.* x (instead of $s^2 I$ *vs.* s),[59] where x is the displacement of the counter tube and $\varepsilon = x/R_F$. A persistence length of $p = 55.0$ Å was found from the point of inflection θ_p, using Eq. 15.26. Eq. 15.25 gives $p = 45.7$ Å.†

† The original report gives $p = 38.5$ Å, but this is an error for $p = 45.7$ Å.

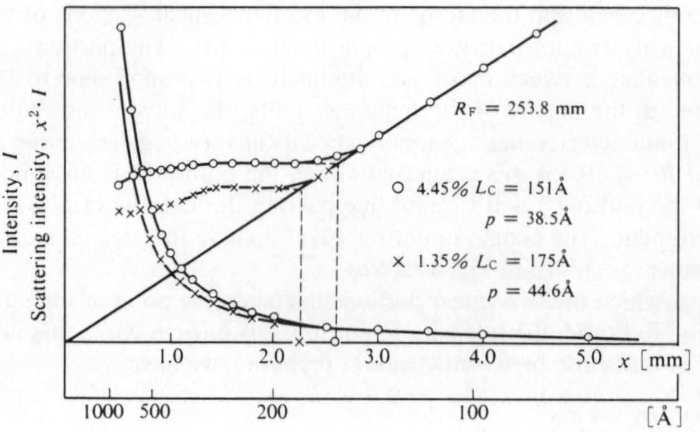

Fig. 15.23 Small-angle scattering intensity curve from a 4.45% acetone solution of nitrocellulose ($\widetilde{DP}_n = 1{,}640$).[59] [Reproduced with permission from S. Heine *et al. Makromol. Chem.*, **44**, 692, Hüthig & Wepf Verlag (1961)]

B. Curvature persistence

The above treatment only dealt with the persistence where the position of one bond with respect to an adjacent bond is locally *trans* or nearly so, *i.e.* where the molecular chain is nearly linear. In some cases, however, *e.g.* when the arrangement of successive bonds is locally always *gauche* (there is no *G´ gauche* in counter rotation), the molecular chain will be twisted into a helix. This is another type of persistence, which Kirste and Kratky call curvature persistence,[62] while they identify the above-mentioned linear type as direction persistency.

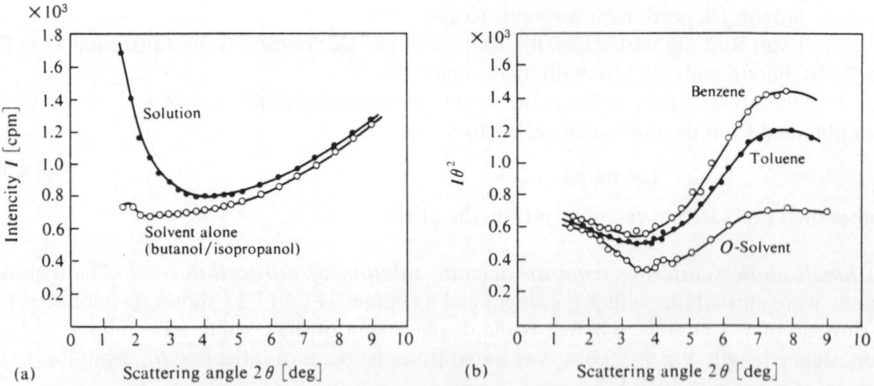

Fig. 15.24 Small-angle scattering from a 2% solution of poly(methyl methacrylate).[62]
(a) Scattering intensities for the solution and for the solvent. showing the small difference at large angles. Accurate intensity measurements are very difficult in the region where the maximum appears in the plot of $I\theta^2$ *vs.* θ
(b) Plot of $I\theta^2$ *vs.* θ which, in the region normally linear, has a maximum due to the helical parts of the molecule
[Reproduced with permission from O. Kratky, *Kolloid-Z Z. f. Polymere*, **182**, 20, 21, Dr. Dietrich Steinkopf Verlag (1962)]

a) Small-angle scattering of an atactic polymethyl methacrylate (PMMA) solution.[62] Unlike the plots for solutions of ordinary chain polymers, the plot of s^2I *vs.* s (or $\theta^2 I$ *vs.* θ) the small-angle scattering from a 2% atactic PMMA solution has a maximum in the linearly ascending large angle portion as shown in Fig. 15.24.[56] This is caused by the contribution due to localized regions of helical structure.

15.3.2 Scattering intensity from stiff chain molecules†

As stated above, the scattering function of a stiff chain (or worm-like) molecule can be expressed for very small and very large scattering angles by the forms of the Debye's scattering factor for random coil chain and by that for rigid rod molecule, respectively.

Assume a linear continuous string (or chain) which has so-called *stiffness* but *no elasticity*. In this stiff chain an arbitrary partial chain segment with its two ends at contour lengths t and t' from one end of a polymer chain is considered, and here its end-to-end distance is denoted as $\boldsymbol{r}$ (instead of r_{end} as in the previous section) (Fig. 15.25). A normalized distribution function of a scattering element at $\boldsymbol{u}$ from its center of distribution t is defined as $g(\boldsymbol{u}|\boldsymbol{v})$: $\boldsymbol{v}$ is the unit vector to define the chain direction, which may be approximated to the tangent at a point t. The $f(\boldsymbol{r}, \boldsymbol{v}, \boldsymbol{v}')$ is defined as the normalized distribution function of the end-to-end vector $\boldsymbol{r}$ and the chain directions $\boldsymbol{v}$ and $\boldsymbol{v}'$ at both ends of a partial chain segment contained between two points t and t'.[66]

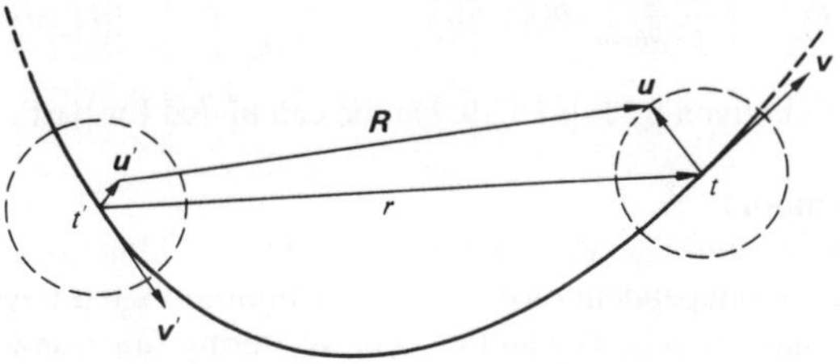

Fig. 15.25 Distribution of scattering elements around the central axis of a polymer chain.

Then the distribution function of the relative coordinates R of the two scattering elements found around the points t and t' can be given as

$$H(\boldsymbol{R}) = \int\cdots\int f(\boldsymbol{r},\boldsymbol{v},\boldsymbol{v}')g(\boldsymbol{u}\,|\,\boldsymbol{v})g(\boldsymbol{u}'\,|\,\boldsymbol{v}')\delta(\boldsymbol{r}+\boldsymbol{u}-\boldsymbol{u}'-\boldsymbol{R})\mathrm{d}\boldsymbol{r}\mathrm{d}\boldsymbol{u}\mathrm{d}\boldsymbol{u}'\mathrm{d}\boldsymbol{v}\mathrm{d}\boldsymbol{v}', \quad (15.27)$$

where δ is the three-dimensional delta function. Assuming that $g(\boldsymbol{u}|\boldsymbol{v})$ and $g(\boldsymbol{u}'|\boldsymbol{v}')$ are independent of $\boldsymbol{r}$, and using the Fourier transform of $H(\boldsymbol{R})$, the scattering function of this chain molecule can be written as

$$P(\varepsilon) = \frac{I(\varepsilon)}{I(0)} = \frac{1}{T^2}\int_{-\infty}^{+\infty}\int_0^T\int_0^T H(\boldsymbol{R})\exp[-i(\boldsymbol{s}\cdot\boldsymbol{R})]dtdt'd\boldsymbol{R}, \quad (15.28)$$

where $I(0)$ is the scattering intensity at zero angle, T is the total contour length of polymer chain, ε is the scattering angle, and $s = |\,\boldsymbol{s}\,| = (4\pi/\lambda)\sin(\varepsilon/2) = (4\pi/\lambda)\sin\theta$. Assuming that $g(\boldsymbol{u}|\boldsymbol{v})$ and $g(\boldsymbol{u}'|\boldsymbol{v}')$ are independent of $\boldsymbol{r}$, from these equations the Fourier transform of $H(\boldsymbol{R})$ can be written in the form

† after R. Koyama[63–65]

$$\Phi(s) = \int_{-\infty}^{+\infty} H(\boldsymbol{R}) \exp[-i(\boldsymbol{s} \cdot \boldsymbol{R})] \mathrm{d}\boldsymbol{R}, \tag{15.29}$$

$$= \int \Phi_{\mathrm{f}}(s, \nu, \nu') \Phi_{\mathrm{g}}(s, \nu) \Phi_{\mathrm{g}}(s, \nu') \mathrm{d}\nu \mathrm{d}\nu', \tag{15.30}$$

where $\Phi_{\mathrm{f}}(s, \boldsymbol{\nu}, \boldsymbol{\nu}')$, $\Phi_{\mathrm{g}}(s, \boldsymbol{\nu})$ and $\Phi_{\mathrm{g}}(s, \boldsymbol{\nu}')$ and the Fourier transforms of $f(\boldsymbol{r}, \boldsymbol{\nu}, \boldsymbol{\nu}')$, $g(\boldsymbol{u}|\boldsymbol{\nu})$ and $g(\boldsymbol{u}'|\boldsymbol{\nu}')$ given by

$$\Phi_{\mathrm{f}}(s, \nu, \nu') = \int f(\boldsymbol{r}, \nu, \nu') \exp[-i(\boldsymbol{s} \cdot \boldsymbol{R})] \mathrm{d}\boldsymbol{r}, \tag{15.31}$$

$$\Phi_{\mathrm{g}}(s, \nu) = \int g(\boldsymbol{u} \mid \nu) \exp[-i(\boldsymbol{s} \cdot \boldsymbol{u})] \mathrm{d}\boldsymbol{u}, \tag{15.32}$$

$$\Phi_{\mathrm{g}}(s, \nu') = \int g(\boldsymbol{u}' \mid \nu') \exp[-i(\boldsymbol{s} \cdot \boldsymbol{\nu}')] \mathrm{d}\boldsymbol{u}'. \tag{15.33}$$

The particle scattering function $P(\varepsilon)$ of this polymer can be given by the equation

$$P(\varepsilon) = \frac{1}{T^2} \int_{-\infty}^{+\infty} \int_0^T \int_0^T H(\boldsymbol{R}) \exp[-i(\boldsymbol{s} \cdot \boldsymbol{R})] \mathrm{d}t \mathrm{d}t' \mathrm{d}\boldsymbol{R}. \tag{15.34}$$

When Eq. 15.28 is used we can write this $P(\varepsilon)$ in the form,

$$P(\varepsilon) = \frac{1}{T^2} \int_0^T \int_0^T \Phi(s) \mathrm{d}t \mathrm{d}t'. \tag{15.35}$$

In this equation, $\Phi(s)$, given by Eq. 15.29, can be calculated for particular chain models.

A. Simple chain model

As the simplest chain model the distribution of scattering elements around the axis of polymer chain is assumed independent of the chain configuration, namely $g(\boldsymbol{u}|\boldsymbol{\nu})$ and $g(\boldsymbol{u}'|\boldsymbol{\nu}')$ do not depend on the chain directions $\boldsymbol{\nu}$ and $\boldsymbol{\nu}'$. The scattering function is given by

$$P(\varepsilon) = \left(1 - \frac{1}{2} \langle u^2 \rangle s^2\right) P_0(\varepsilon) + \cdots. \tag{15.36}$$

In particular, the distribution function $g(u)$ can be approximated by a Gaussian function,

$$P(\varepsilon) = \exp\left(-\frac{1}{2} \langle u^2 \rangle s^2\right) P_0(\varepsilon). \tag{15.37}$$

$P_0(\varepsilon)$ is the particle scattering function of an optically ideal thin polymer which has no cross section.

B. More constrained model

a) Scattering element around the point t is distributed only on the plane perpendicular to the central axis of polymer chain.

1) Limiting case of very large flexibility. The chain direction at the end of a chain segment is considered not to depend on its integral configuration:

$$P(\varepsilon) = \exp\left(1 - \frac{1}{3} \langle u^2 \rangle s^2\right) P_0(\varepsilon). \tag{15.38}$$

These two equations, Eqs. 15.36 and 15.38, have the same form but different numerical constants, 1/2 and 1/3, in the cross section factors.

2) Case of zero Flexibility. The chain is completely nonflexible and rod-1ike, the end-to-end distance of chain segment is equal to its contour length, and the chain directions at both ends v and v' coincide with r.

$$P(\varepsilon)=\exp\left(1-\frac{1}{2}<u^2>s^2\right)P_0(\varepsilon). \tag{15.39}$$

This equation has the same form as Eq. 15.37.

b) Long stiff chain molecules with arbitrary flexibility.

The scattering function $P(\varepsilon)$ is given as

$$P(\varepsilon)=\exp\left(1-\frac{1}{3}<u^2>s^2\right)P_0(\varepsilon) \\ -\frac{1}{3}<u^2>s^2\frac{1}{sT}\left\{\left(1+\frac{16}{3q^2}\right)\tan^{-1}\left(\frac{3}{4}s\right)-\frac{4}{s}\right\}, \tag{15.40}$$

and $q=s/(2\mu)=(4\pi/\lambda)\sin(\varepsilon/2)/(2\mu)=(4\pi/\lambda)\sin\theta/(2\mu)$, where μ is a parameter of the flexibility of the chain.

$$P_0(\varepsilon)=\frac{2}{L^2}\int_0^L(L-x)\exp\left[1-\frac{1}{3}q^2xf(x)\right]\cdot\frac{\sin\{q\cdot xg(x)\}}{q\cdot xg(x)}\mathrm{d}x, \tag{15.41}$$

$$xf(x)=\frac{1}{2}(2\mu)^2<r^2>\left\{1-\frac{1}{2}\cdot10^{1/2}\left(1-\frac{3}{5}\mathrm{K}\right)^{1/2}\right\}, \tag{15.42}$$

$$x^2g^2(x)=\frac{1}{2}(2\mu)^2<r^2>10^{1/2}\left\{1-\frac{3}{5}\mathrm{K}\right\}^{1/2}, \tag{15.43}$$

$$K=<r^4>/<r^2>^2 \quad \text{and} \quad L=2\mu T. \tag{15.44}$$

L is a reduced contour length of polymer and here $x=t/T$. $<r^2>$ and $<r^4>$ are the second and fourth moments of $\boldsymbol{r}$, respectively.

$$<r^2>=\frac{1}{\mu}\left\{t-\frac{1}{2\mu}(1-\exp[-2\mu t])\right\}, \tag{15.45}$$

$$<r^4>=\frac{1}{\mu^2}\left\{\frac{5}{3}t^2-\frac{26}{9}\cdot\frac{t}{\mu}-\frac{1}{54}\cdot\frac{1}{\mu^2}\left(1-\exp[-6\mu t]\right) \\ +\frac{2}{\mu^2}\left(1-\exp[-2\mu t]\right)-\frac{t}{\mu}\exp[-2\mu t]\right\}. \tag{15.46}$$

Besides Koyama's theory introduced above, Yoon and Flory[67] presented another scattering function based on their theory.

C. A stiff chain molecule expanded by intramolecular interaction[65]

The mean square end-to-end distance of a partial chain segment with contour length t is already given as

$$<r^2>=(1+2\mu^2)(2\mu t-1+\exp[-2\mu t]). \tag{15.45}$$

For sufficiently long contour length this leads to

$$<r^2> \approx t/\mu. \tag{15.47}$$

Denoting by $\alpha(t)$ the expansion coefficient of partial chain segments this mean quantity for the expanded polymer chain should be written as

$$<r^2> \approx t\alpha^2(t)/\mu. \tag{15.48}$$

Comparing Eq. 15.48 with Eq. 15.47, it can be supposed that each chain segment is a simple stiff chain which has the flexibility $\mu/\alpha^2(t)$ and contour length t, since its contour length can be assumed as constant independently of the interaction. Therefore, the mean values of $<r^2>$ and $<r^4>$ in this expanded polymer chain can be assumed to be represented by the same equations in cases where there is no interaction, but replacing, μ by $\mu/\alpha^2(t)$.

$$<r^2>=\frac{\alpha^4(t)}{2\mu^2}\left\{\frac{2\mu t}{\alpha^2(t)}-1+\exp\left[\frac{-2\mu t}{\alpha^2(t)}\right]\right\}, \tag{15.49}$$

$$\begin{aligned}<r^4>=\frac{5\alpha^8(t)}{3\mu^4}&\left\{\frac{\mu^2t^2}{\alpha^4(t)}-\frac{26}{5}\cdot\frac{\mu t}{\alpha^2(t)}-\frac{1}{90}\left(1-\exp\left[\frac{-6\mu t}{\alpha^2(t)}\right]\right)\right.\\&\left.+\frac{6}{5}\left(1-\exp\left[\frac{-2\mu t}{\alpha^2(t)}\right]\right)-\frac{3}{5}\cdot\frac{\mu(t)}{\alpha^2(t)}\exp\left[\frac{-2\mu t}{\alpha^2(t)}\right]\right\}.\end{aligned} \tag{15.50}$$

For sufficiently large t Eq. 15.49 reproduces Eq. 15.48.

Since a stiff chain segment having sufficiently long contour length can be treated as a Gaussian coil, $\alpha(t)$ for this chain can be determined by the excluded volume theory of random coil chains. From the perturbation theory of this chain it can be written

$$\alpha^2(t)=1+C(t)z-\cdots, \qquad (|z|\ll 1), \tag{15.51}$$

where z is the interaction parameter independent of t. The function $C(t)$ depends not only on the length of the partial segment but also on its position along the polymer chain. For simplicity, averaging it over this position, it becomes

$$\begin{aligned}C(t)=\frac{32}{9}x^{1/2}&\left[1+\frac{3}{8}x^{1/2}+(1-x)^{-1}x^{-3/2}\right.\\&\left.\times\left\{\frac{4}{5}-2x+\frac{6}{5}x^{5/2}-\frac{4}{5}(1-x)^{5/2}\right\}\right].\end{aligned} \tag{15.52}$$

$$=\frac{32}{9}x^{1/2}\left[1-\frac{9}{8}x^{1/2}-\cdots\right] \qquad x\ll 1. \tag{15.53}$$

Further, to determine $\alpha(t)$ for higher interaction the following equation is assumed, which conforms to Eq. 15.51,

$$\alpha^5(t)-\alpha^3(t)=C(t)z. \tag{15.54}$$

Thus using the function $\alpha(t)$ determined by Eq. 15.54, two equations Eq. 15.49 and Eq. 15.50 give an approximate chain configuration in the expanded state.

Figure 15.26 shows $(L/6)q^2P(q)$ *vs.* q plots of X-ray solution scattering of poly(p-chlorostyrene) ($<M>=6\times 10^5$) in propylbenzene, an endothermic θ solvent and that in t-butyl acetate an exothermic θ solvent: both plots show an good coincidence with respec-

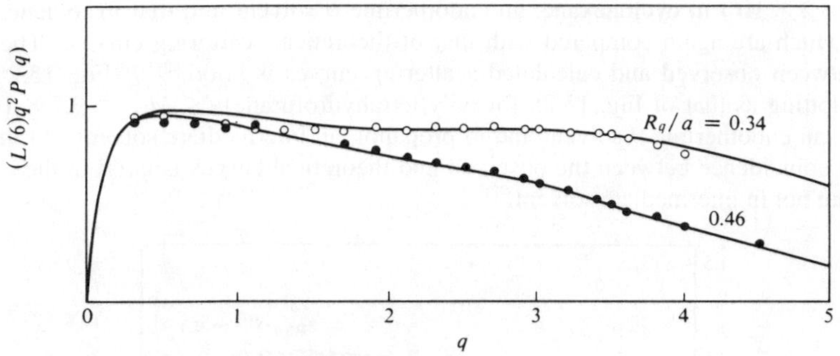

Fig. 15.26 X-ray small-angle scattering of poly(*p*-chlorostyrene) (PPCS) solutions of two different θ solvents.[65)]
$q=s/(2\mu)$, M : parameter of the flexibility of the chain.
Solid line —, theoretical curve obtained by Koyama's scattering theory. $R_q/a \equiv 2\mu < u^2 >^{1/2}$.
Black circles ●, observed values for PPCS in *t*-butyl acetate ($\theta = 61.5$°C, 2%).
White circles ○, observed values for PPCS in propylbenzene ($\theta = 43.1$°C, 2%).
[Reproduced with permission from Y. Izumi *et al.*, *Colloid. Polym. Sci.*, **256**, 1, Dr. Dietrick Steinkopf Verlag (1978)]

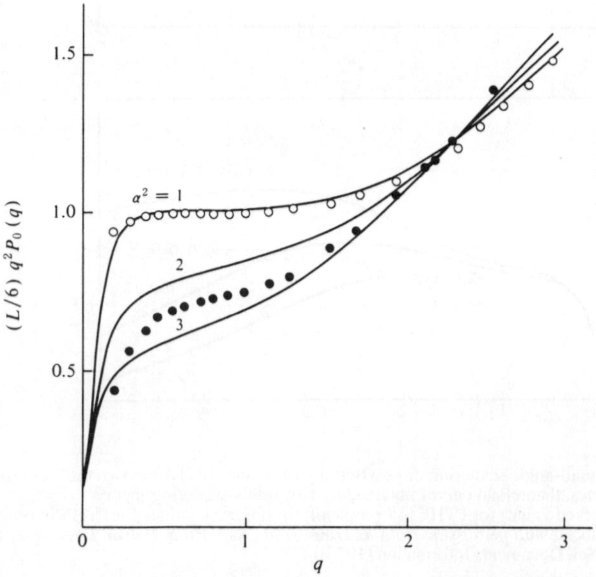

Fig. 15.27 X-ray small-angle scattering of polystyrene (PS) in two different solvents.[66–70)]
Solid lines —, theoretical curves obtained by Koyama's scattering theory.
Black circles ●, observed values for PS in toluene (30°C, extrapolated to zero concentration after corrected for the effect of cross section).
White circles ○, observed values of cyclohexane ($\theta = 35$°C, 1.48%).
[Reproduced with permission from Y. Taru *et al.*, *Rep. Prog. Polym. Phys. Jpn.*, **15**, 37, Assoc. Sci. Documents Information (1972)]

tive theoretical curves.[68)] The whole structure of poly(*p*-chlorostyrene) chain molecule in these two θ solvents can be described with the $P_0(\varepsilon)$ (Eq. 15.41), whereas the difference in molecular shape is explained by the difference in radii of gyration of the cross section.

Figure 15.27 depicts the similar plots of X-ray solution scattering of polystyrene

($<M> = 2 \times 10^6$) in cyclohexane, an endothermic θ solvent and that in toluene, a good solvent, which are again compared with that of theoretical scattering curves. The coincidence between observed and calculated scattering curves is good.[69–71] Fig. 15.28 shows similar plotting as that of Fig. 15.26 for poly(tetrahydrofurane) ($<M> = 9.7 \times 10^4$) in *i*-propanol, an endothermic θ solvent and in propanol, an intermediate solvent. In these two cases the coincidence between the observed and theoretical curves is good in the case of θ solvent but not in intermediate solvent.[72)]

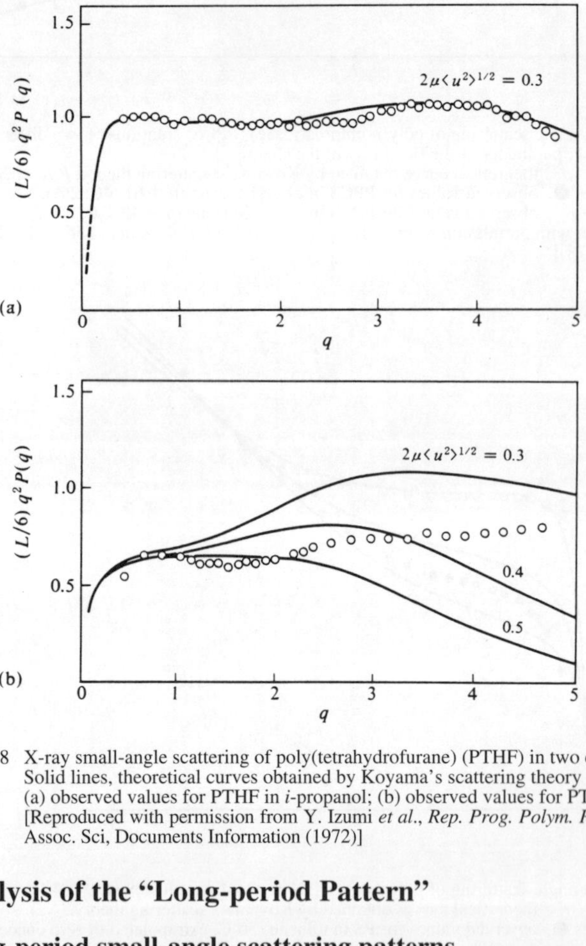

Fig. 15.28 X-ray small-angle scattering of poly(tetrahydrofurane) (PTHF) in two different solvents. Solid lines, theoretical curves obtained by Koyama's scattering theory
(a) observed values for PTHF in *i*-propanol; (b) observed values for PTHF in propanol.
[Reproduced with permission from Y. Izumi *et al.*, *Rep. Prog. Polym. Phys. Jpn.*, **15**, pp.41, 45, Assoc. Sci, Documents Information (1972)]

15.4 Analysis of the "Long-period Pattern"

15.4.1 Long-period small-angle scattering patterns

Figure 15.29 shows a schematic illustration of wide- and small-angle scattering diagrams of drawn poyethylene,[73)] their intensity distribution along the equator and meridian, and the corresponding fine structure and fine texture (cf. Fig. 6.1 (c)).

Various types of long-period small-angle scattering patterns have been observed, as shown In Table 15.4 (pp.452, 453). These differ in the position, intensity, and shape of the diffraction elements from the X-ray diffraction patterns at the usual wider angles. The influ-

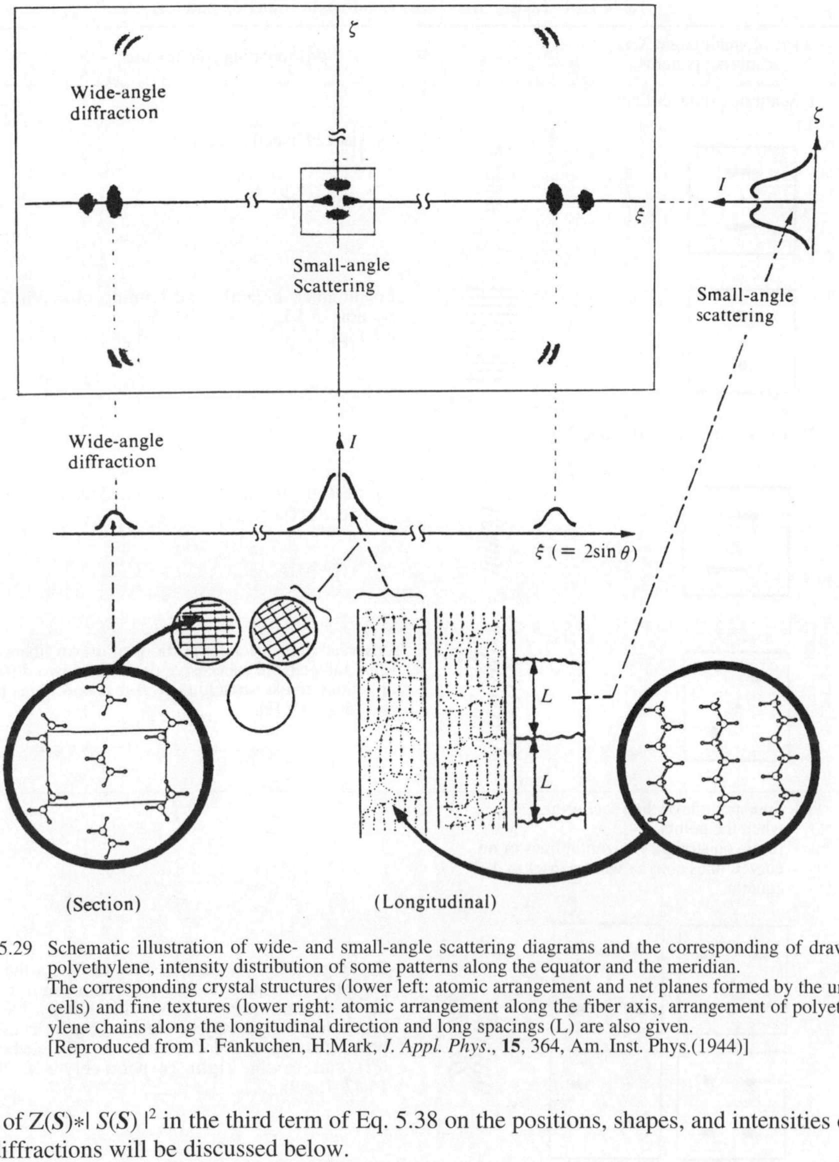

Fig. 15.29 Schematic illustration of wide- and small-angle scattering diagrams and the corresponding of drawn polyethylene, intensity distribution of some patterns along the equator and the meridian.
The corresponding crystal structures (lower left: atomic arrangement and net planes formed by the unit cells) and fine textures (lower right: atomic arrangement along the fiber axis, arrangement of polyethylene chains along the longitudinal direction and long spacings (L) are also given.
[Reproduced from I. Fankuchen, H.Mark, *J. Appl. Phys.*, **15**, 364, Am. Inst. Phys.(1944)]

ence of $Z(\boldsymbol{S})*|S(\boldsymbol{S})|^2$ in the third term of Eq. 5.38 on the positions, shapes, and intensities of the diffractions will be discussed below.

A. The magnitude of the long period and the order number at which it appears

The form of the function $Z(\boldsymbol{S})$ depends upon the degree of ordering in the arrangement of the particles within the specimen— "particle" here being taken in the broadest sense to include regions of discontinuous density, for example the crystalline regions. In the two extremes of a perfectly regular crystal on the one hand, and a completely irregular amorphous substance on the other, $Z(\boldsymbol{S})$ takes the form of the Laue function $G(L_1{}^2$ in Eq. 2.45) and the

Table 15.4 Long period X-ray small-angle scattering patterns

Type of small-angle X-ray scattering patterns	Corresponding fine texture
I. Scattering in layer lines	
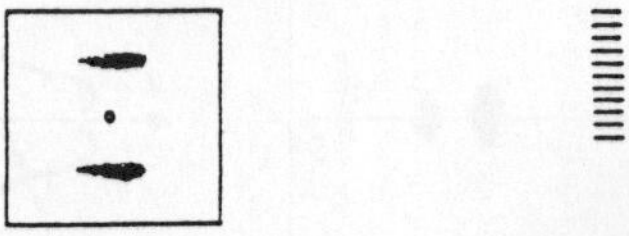	*e.g.* stacked lamellar crystals.
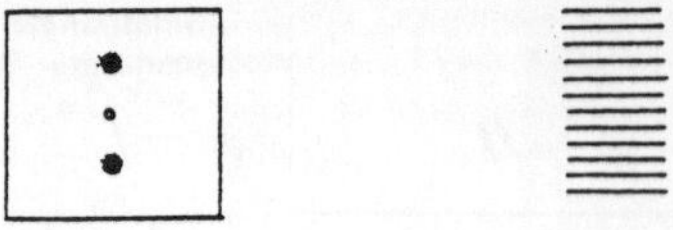	For number, breadth, and fanning of maxima, see Section 15.3.1. (*cf.* Figs. 15.31, 15.35, 15.37, 15.38 and 15.39).
II. Two-point layer line scattering	
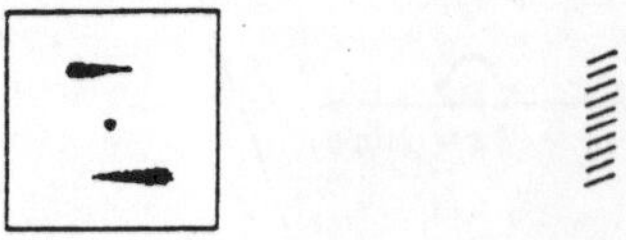	*e.g.* stacked lamellar crystals inclined with respect to the fiber direction.
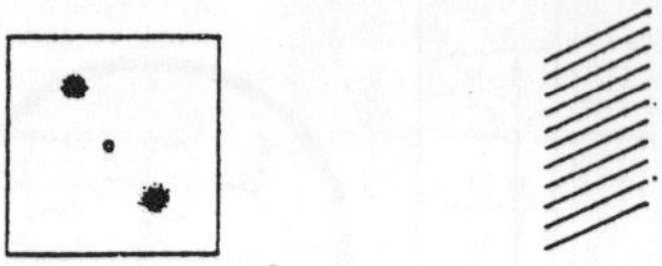	Observed in the necking portions of drawn fibers or in films subjected to successive drawing in two different directions (cross stretching) (*cf.* text and Figs. 15.36 and 15.47–15.49).
III. Four-point layer line scattering when the points (a) lie on straight horizontal lines or on curved lines convex with respect to the equator	
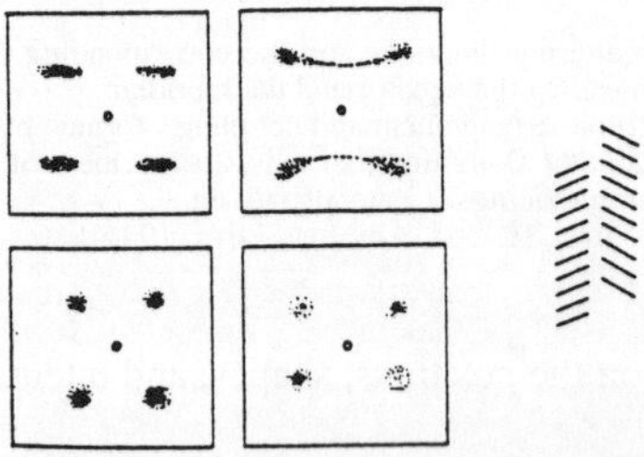	Best regarded as the superposition of two of the two-point layer line patterns of type II with a mirror symmetry relation between them. Accordingly, the corresponding texture may be regarded *e.g.* as an assembly of two types of stacked structures inclined to the left and to the right respectively (*cf.* Figs. 15.47–15.49).

(*Continued on page* 453)

radial distribution function $P(r)$ (Eq. 2.35) respectively. In the general, intermediate cases, $Z(\boldsymbol{S})$ varies with the value of the lattice distortion factor g (*cf.* Section 13.4) as shown in Fig. 15.30.[74] In cases such as that of collagen, a very large number of small-angle layer line diffractions appear in the meridional direction and g is extremely small, so that the long

Table 15.4 — (*Continued*)

(b) lie on a curve concave with respect to the equator 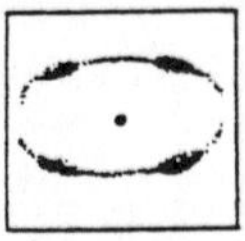		This is treated as elliptical scattering of type V in which four maxima have developed, and is explained as due to a cylindrically symmetrical structure corresponding to slight distortion of the spherical symmetry of type V (*cf.* Figs. 15.40).
(c) lie on two oblique straight lines 		*e.g.* doubly oriented structures. This may be regarded as a superposition of the layer line scattering of type I (*cf.* Fig 15.49).
IV. Scattering in a ring 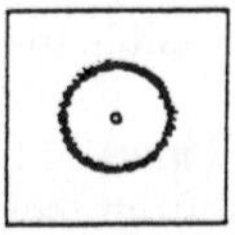		(a) 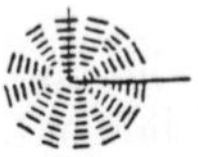 (b)  Statistically spherically symmetrical assemblies of crystallites, *e.g.* spherulites (a) or unoriented assemblies of stacked lamellar crystals (b) (*cf.* Fig. 15.35). Radial distribution functions can be obtained over a very wide range in the specimen as described in Section 15.3.
V. Scattering in an ellipse 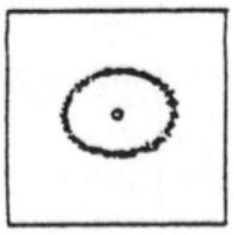		Statistically cylindrically symmetrical assemblies of crystallites, *e.g.* deformed spherulites or stacked lamellar crystals with cylindrically symmetrical orientation. Often observed at intermediate stages of deformation of the specimen by drawing or compression. Cylindrical distribution functions can be obtained over a very wide range in the specimen, as described in Section 14.6, and can be used for the study of the fine texture (*cf.* Fig. 15.40 and Table 15.5).

period L can be derived from Eq. 8.2 (which gives the identity period in the direction of the axis of rotation from the layer-line spacing in the case of single crystals) by applying it to each of the layer-line diffractions. Since the angle is extremely small, however, $\tan\varepsilon \cong \sin\varepsilon \cong \varepsilon$ ($\varepsilon = 2\theta$, the scattering angle), and instead of Eq. 8.2 we can use

$$L\varepsilon = n\lambda \tag{15.55}$$

The values found for the long period from the diffractions of various orders agree within the limits of error, and it is usually found that the period is some multiple of an axial length of the unit cell period of the specimen. Diffractions as high as the fourth or fifth order can be observed under favorable conditions, for example with many synthetic fibers. On the other hand, where the value of g is high, as in Fig. 15.30(b) and (c), only the first- or at most second-order diffraction will be even vaguely discernible for long-period determination. The positions of the intensity maxima are located by Eq. 5.30. Since the maxima

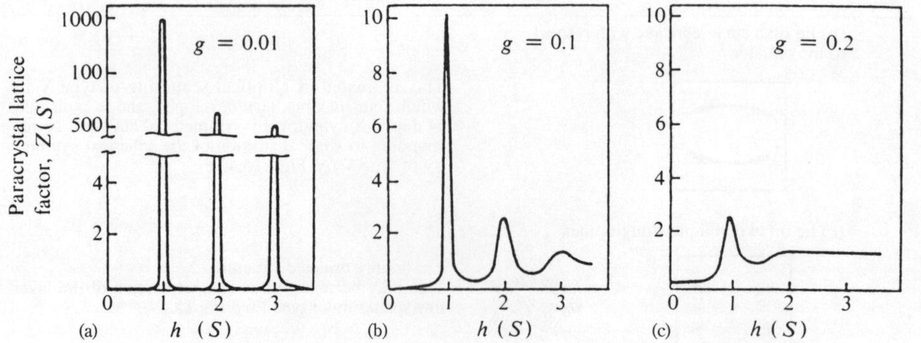

Fig. 15.30 The effect of the distortion factor, g, on the paracrystal lattice factor, $Z(S)$ [Reproduced with permission from P. H. Lindenmeyer, R. Hosemann, *J. Appl. Phys.*, **34**, 43, Am. Inst. Phys. (1963)]

appear at approximately $S = n/\bar{L}$ for an average long period $\bar{L}$, it is possible to find the average long period $\bar{L}$ by application of Eq. 15.55.

As shown for the case of polyethylene in Fig. 15.31,[75] there is a marked difference between the L_1 found from the first-order diffraction and the L_2 found from the second-order diffraction. This is attributed to the fact that the lattice point function $H(\boldsymbol{r})$ described in Section 5.3.1 (a superlattice with crystal lattices as single lattice points) is not symmetrical with respect to the maxima, as is the Gaussian function or the Cauchy function in Eqs. 13.3 and 13.4[76]

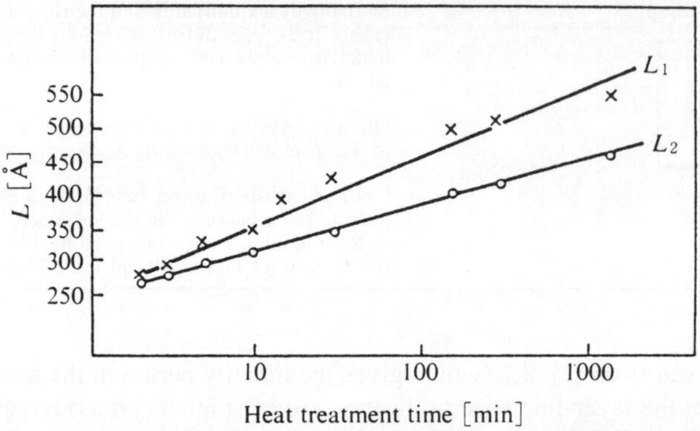

Fig. 15.31 Differences between the long period of a bulk polyethylene specimen heat-treated at 130°C as derived from first- and second-order diffractions.[75] [Reproduced with permission from E. W. Fischer, G. F. Schmidt, *J. Appl. Phys.*, **35**, 71, Am. Inst. Phys. (1964)]

B. Breadth of the long-period diffractions

This is determined largely by the shape factor $|S(\boldsymbol{S})|^2$. As described in Section 13.6.1, the breadth of the long-period diffractions in the vertical direction (the meridional direction) decreases as the number of particles N contributing to the long period lattice increases, and increases with decreasing N. The number of diffractions that can be observed accordingly also decreases with decreasing N.

In the horizontal direction, as is clear from the example of the line lattice in Fig. 15.32, the long-period diffractions are narrow when the particles are large, whereas broad diffractions in the form of layer lines appear when the particles are small.

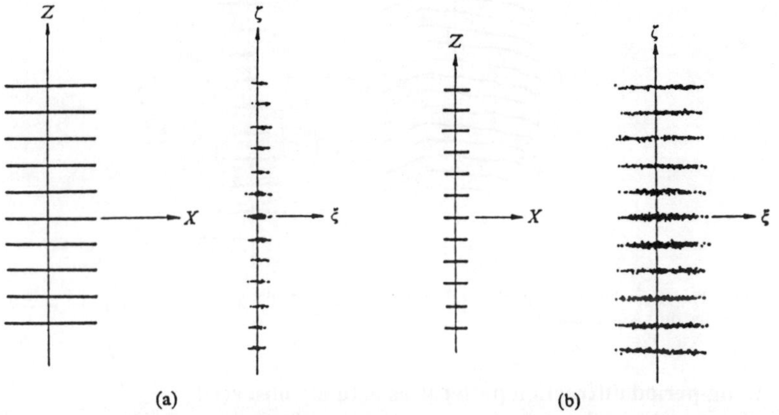

Fig. 15.32 The effect of the breadth of a line lattice on the diffraction pattern.
(a) Great horizontal breadth gives rise to small horizontal spread in diffractions.
(b) Small horizontal breadth gives rise to great horizontal spread in diffractions.

Considering next the effects of particle shape, again for a line lattice, when the particles are inclined to the right as in Fig. 15.33, the diffractions appear on a straight line inclined to the left.[77] Moreover, when the particles are curved or undulating, an intensity distribution develops in the horizontal direction, and the pattern exhibits the phenomenon known as fanning shown in Fig. 15.34).[77]

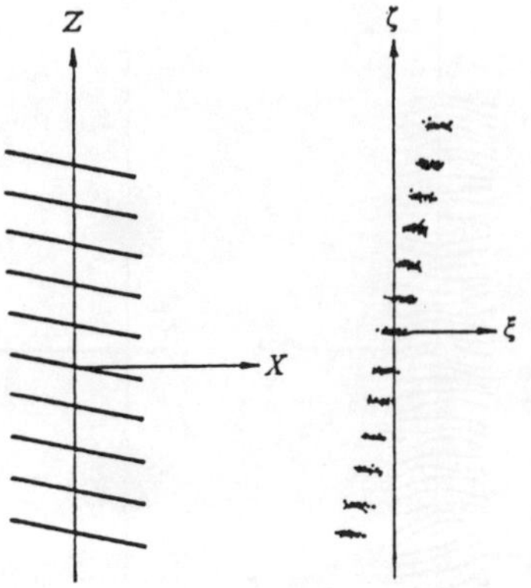

Fig. 15.33 The effect of a non-perpendicular line lattice on the diffraction pattern.

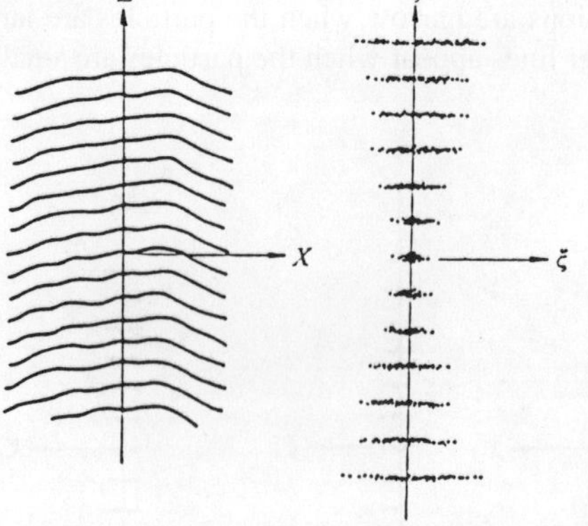

Fig. 15.34 The effect. known as fanning, of an undulating line lattice (constant spacing in Z direction) on the diffraction pattern.

C. Long-period diffraction patterns as actually observed

The patterns obtained in practice are not so simple as the examples cited representatively above. The diffraction intensity is in fact proportional to $< A_p(\boldsymbol{S}) >^2 Z(\boldsymbol{S}) * \mid S(\boldsymbol{S}) \mid^2$, where $A_p(\boldsymbol{S})$ is the particle structure factor. The examples were, for simplicity, based on the assumption that $< A_p(\boldsymbol{S}) >^2 = 1$, but $< A_p(\boldsymbol{S}) >^2$ does vary, and positions and intensities of the maxima vary with it. Though the factors discussed are two-dimensional, the extension to three dimensions and their effect on the real diffraction pattern can to some extent be grasped by analogy with the relationship between the structural models proposed by Hosemann and Bonart and their optical diffraction patterns[78,79] (see Fig. 15.35).

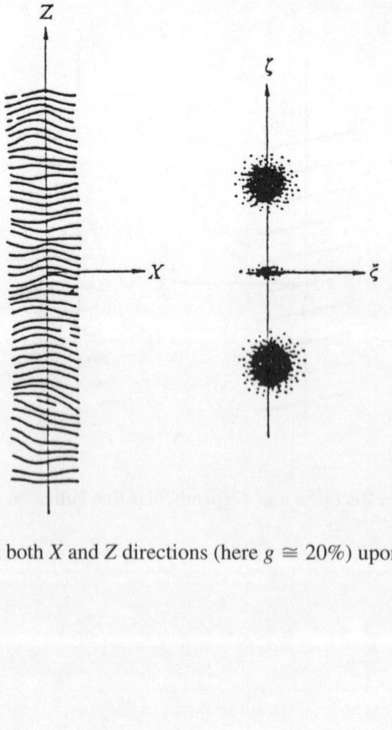

Fig. 15.35 The effect of distortion in both X and Z directions (here $g \cong 20\%$) upon the diffraction pattern: first-order visible only.

Table 15.4 (pp.452–453) shows a classification of long-period diffraction patterns as they have actually been observed. The main points to be considered in interpreting these small-angle scattering patterns, irrespective of the long period diffractions and central diffuse scattering, are the patterns obtained with X-rays incident from various directions and their correspondence with the wide-angle diffraction patterns (Fig. 15.36).[80)]

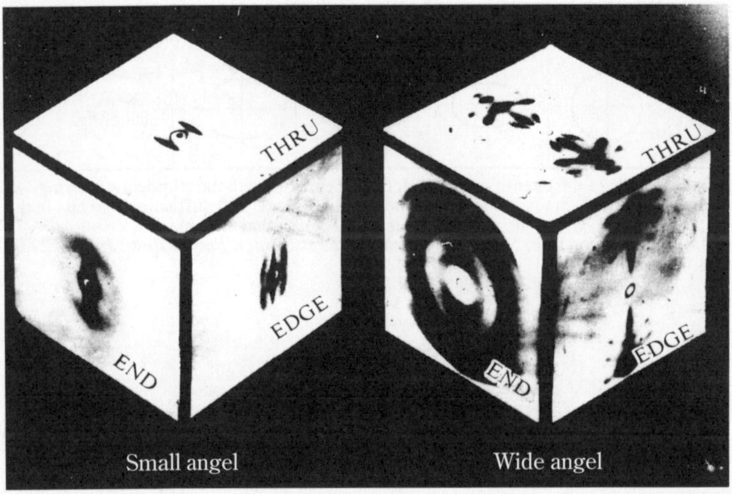

Fig. 15.36 X-ray diffraction patterns of rolled poly(ethylene terephthalate), with small-angle scattering on the left and wide-angle diffraction in the right. The directions of the incident X-rays as follows: THRU, Perpendicular to the rolled surface; EDGE, Parallel to the rolled surface and perpendicular to the machine direction; END, Parallel to both the rolled surface and the machine direction, *cf.* Fig. 8.37 for the preparation of the specimen.[80)] [Reproduced from W. O. Statton, G. M. Goddazd, *J. Appl. Phys.* **28**, 1112, Am. Inst. Phys. (1957)]

a) Variation of the X-ray small-angle scattering patterns for a poly(vinyl alcohol) film with elongation.[81,82)] Table 15.5 shows the variation of the small-angle scattering pattern of poly(vinyl alcohol) (PVA) film on drawing after heat treatment in air for 10 min at 160°C. As the draw ratio increases, the scattering pattern changes from a ring to an ellipse, and then to layer lines. The intensity gradually decreases until no small-angle scattering is observed apart from the central diffuse scattering; after heat treatment, however, small-angle scattering reappears. On the other hand, the wide-angle diffraction pattern initially shows spiral orientation, which changes to uniaxial orientation (b axis parallel to the fiber axis) with increasing draw ratio.

Figure 15.37(a) shows the intensity distribution of the small-angle scattering for a draw ratio of 1.5 :1. The ellipse is broken in the meridional direction and intensity maxima appear at angles of about 40° to the meridian. Figure 15.37(b) shows the cylindrical distribution function $Q(r, z)$ obtained for this type of scattering intensity distribution from Eq. 14.21. The distribution function is elongated in the drawing direction (the z direction).

Table 15.5 The effects of increasing draw ratio on poly(vinyl alcohol) film

Draw ratio [%]		0	20	50	200	300	300 plus heat-treatment
D_L [Å]	*meridional*	134	155	168	165	weak scattering	L 135
	equatorial		128	134	?		
δ' [deg]			—			≈38	
ψ [deg]			—	30	14	0	<2
Scattering pattern (schematic)		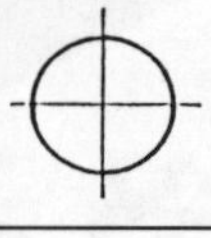	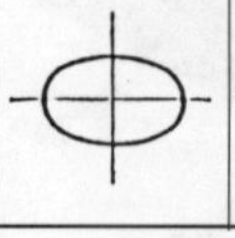	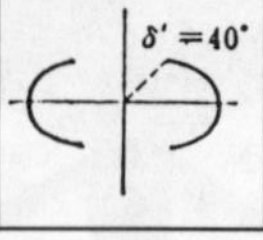 $\delta'=40°$	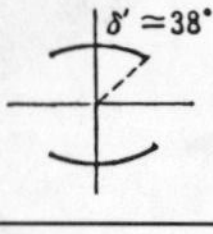 $\delta'=38°$	weak scattering	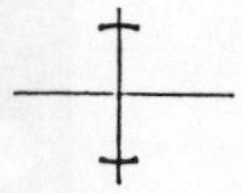

δ' is the azimuthal angle of the intensity maximum from the meridian in the elliptical scattering. ψ is the corresponding azimuthal angle, with respect to the equator, of the 100 and 200 diffraction maxima in the wide-angle fiber pattern: $\delta = 90° - \psi$ is thus the angle between the fiber axis and the a axis.
[Reproduced with permission from N. Kasai, M. Kakudo, *Repts. Progr. Polym. Phys. Jpn.*, **9**, 293,Assoc. Sci. Documeents Intormation (1966)]

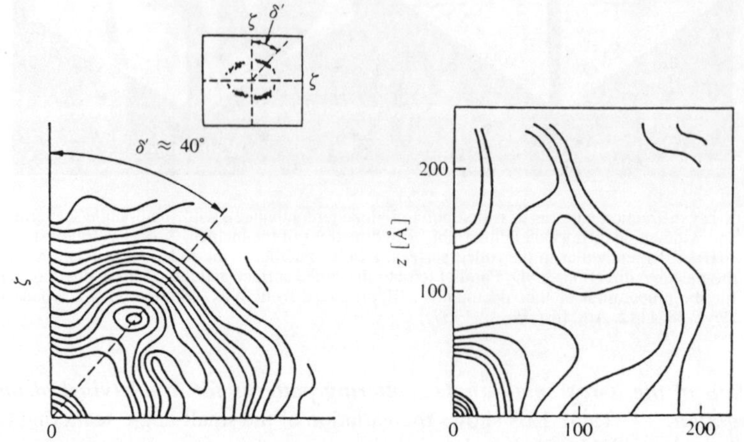

Fig. 15.37 Analysis of the ellipsoidal small-angle scattering from poly(vinyl alcohol).
(a) Intensity distribution (draw ratio 1.5 : 1).
(b) Fourier transform of (a) (cylindrical Patterson function $Q(r, z)$).
Near the center contour lines are omitted.
[Reproduced with permission from N. Kasai, M. Kakudo, *J. Polym. Sci.*, **A2**, 1958, John Wiley & Sons, Inc. (1964)]

b) Fine texture in the necking portions of cold drawn polyethylene.[83–86] In the investigation of the continuous variation of the structure and texture in the necking portions of a cold drawn specimen of an unoriented polyethylene rod molded from the melt, practically all the typical X-ray small-angle scattering patterns listed in Table 15.4 were observed. Wide-angle and small-angle scattering measurements were carried out simultaneously on the necking portions of a flat specimen about 0.3 mm thick, which had been cut out in such a way as to include the central axis (Fig. 15.38), at the points shown in Fig. 15.39 (see also Fig.

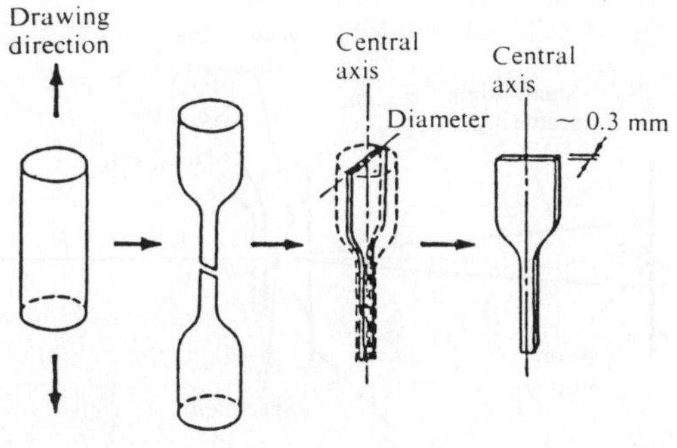

Fig. 15.38 Preparation of the specimen for the investigation of the fine texture of cold-drawn polyethylene (Figs. 15.39–15.46). [Reproduced with permission from N. Kasai *et al.*, *J. Chem. Soc. Jpn. Ind. Chem. Section*, **64**, 56, Chem. Soc. Jpn. (1961)]

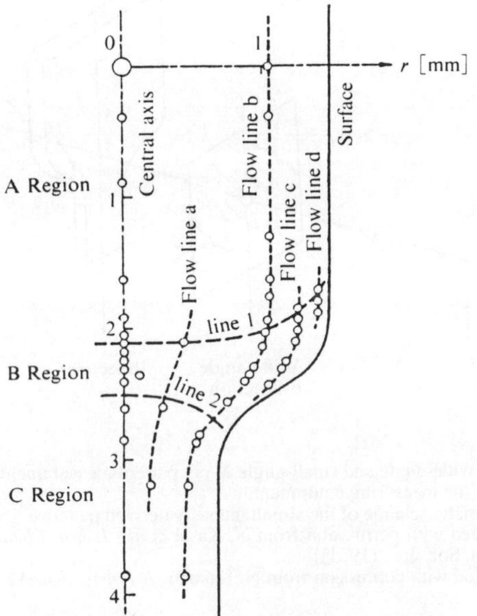

Fig. 15.39 Diagram locating the points for which wide-angle diffraction and small-angle scattering measurements were made. [Reproduced with permission from N. Kasai *et al.*, *J. Jpn. Chem. Soc. Ind. Chem. Section*, **64**, 56, Chem. Soc. Jpn. (1961)]

15.40).[84] The changes in the wide- and small-angle diffraction patterns are shown in Figs. 15.41 and 15.42, respectively.

The wide-angle diffraction pattern first exhibits preferred orientation (spiral orientation) at line 1 in Fig. 15.39. The *c* axis gradually begins to move into the fiber direction, while the *a* axis becomes perpendicular to the fiber direction. The small-angle scattering

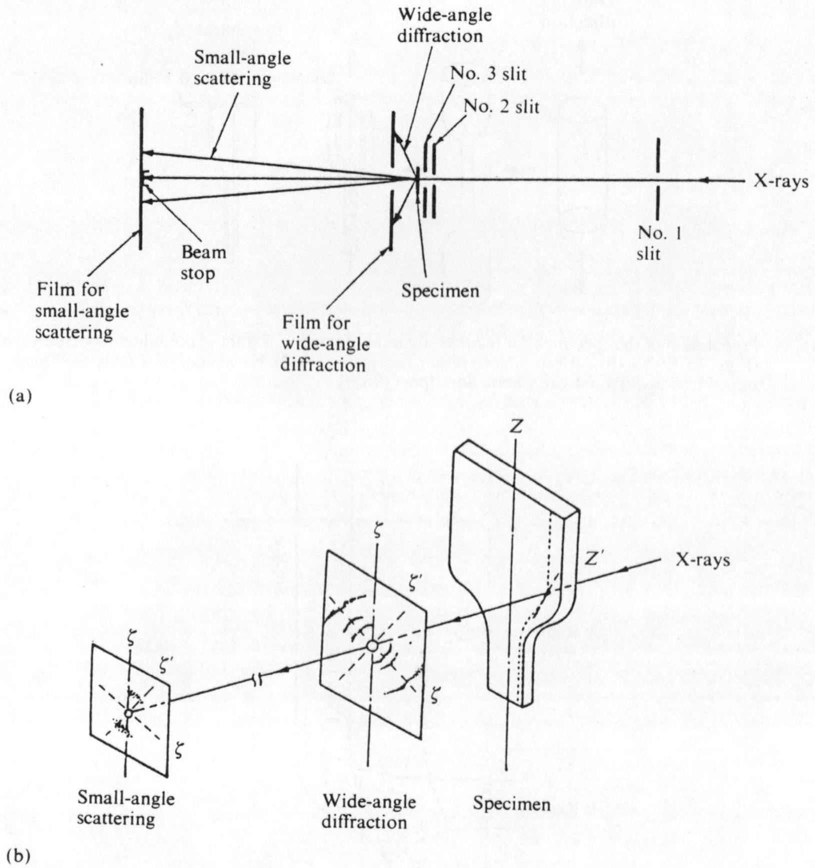

Fig. 15.40 Simultaneous wide-angle and small-angle X-ray pattern measurements.
(a) Sketch of the measuring equipment.
(b) Diagrammatic scheme of the simultaneously derived patterns.
(a): [Reproduced with permission from N. Kasai *et al.*, *J. Jpn. Chem. Soc., Ind. Chem. Section*, **64**, 57, Chem. Soc. Jpn. (1961)]
(b): [Reproduced with permission from N. Kakudo, *J. Polym. Sci.*, **A2**, 1958, John Wiley & Sons, Inc. (1964)]

pattern changes from ring scattering to anisotropic elliptical scattering. These changes appear almost simultaneously with the changes in the external shape near the surface of the specimen, but their appearance is progressively delayed with increasing proximity to the center.

The changes in the narrow band between line 1 and line 2 (within about 0.2 mm at the central axis and about 0.7 mm at the surface for a specimen about 3 mm in diameter) are extremely complex. At the central axis, the small-angle scattering ellipse fades in the meridional direction, and layer line scattering simultaneously appears superimposed outside the ellipse. The ellipse ultimately disappears, and only the layer line scattering remains.

With increasing distance from the central axis, the meridian of the small-angle diffrac-

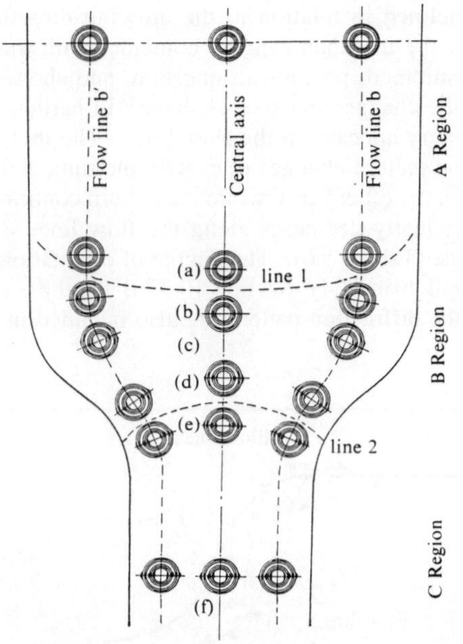

Fig. 15.41 Schematic diagram of the variation in the wide-angle pattern for the necking portions of cold drawn polyethylene.

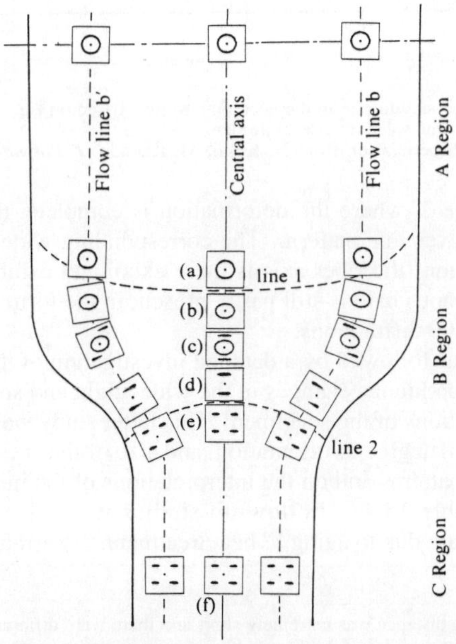

Fig. 15.42 Schematic diagram of the variation in the small-angle pattern for the necking portions of cold drawn polyethylene. [Reproduced with permission from N. Kasai, M. Kakudo, *J. Polym. Sci.*, **A2**, 1963, John Wiley & Sons, Inc. (1964)]

tion pattern becomes inclined in relation to the drawing direction of the specimen as a whole. The direction of the meridian roughly coincides with the tangent to the flow line (Fig. 15.39) at the measurement position in question, and shows changes along the flow line that correspond to the changes in external shape. In particular, a two-point layer line pattern with point symmetry appears on the flow lines. The inclination of the meridian of the wide-angle diffraction pattern changes in exactly the same way as that of the small-angle scattering pattern. On the other hand, while the pattern continues to show spiral orientation, the spiral angle gradually decreases along the flow line, so that the orientation approaches the uniaxial type (Table 15.6). The degree of orientation (represented here by the practical measure of parallelism Π, *cf.* Section 10.3.1B†) and the crystallite size as estimated from the breadth of the diffraction pattern are also included in Table 15.6, see also Fig. 15.43.[84)]

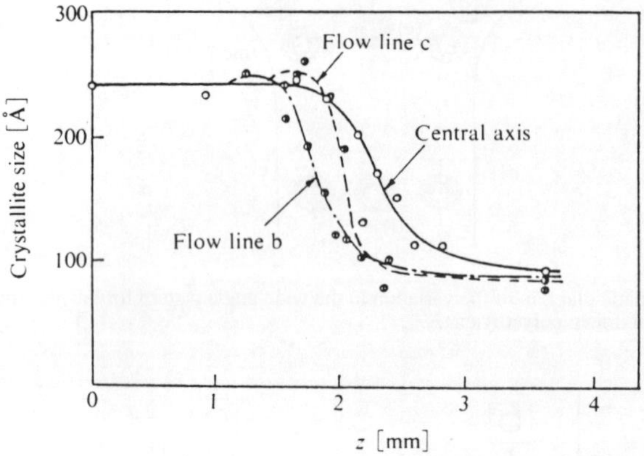

Fig. 15.43 Variation in the crystallite size in the necking portions, based on Eq. 13. 10 and the assumption that line broadening is due solely to crystallite size.
[Reproduced with permission from N. Kasai, M. Kakudo, *J. Polym. Sci.*, **A2**, 1967, John Wiley & Sons, Inc. (1964)]

Finally, beyond line 2, where the deformation is complete, the small-angle scattering becomes a four-point layer line pattern. The corresponding wide-angle diffraction pattern shows uniaxial orientation (the fiber axis is the *c* axis), and exhibits the highest degree of orientation. The amorphous halo is still partly present in the form of a ring, mainly concentrated around 110 and 200 diffractions.

The above study was followed by a detailed investigation of the changes in fine texture that accompanied the continuous changes in the wide-angle and small-angle diffraction patterns in the necking portions of the specimen. This latter study was based on the changes in the preferred orientation, degree of orientation, and size of the crystallites as found from the wide-angle diffraction patterns and on the interpretations of the individual small-angle scattering patterns (*cf.* Table 15.4). Follow-up studies were also made of the secondary changes in the fine texture due to aging,[85)] heat treatment,[86)] γ-irradiation,[85)] etc.

† Since specimen-to-film distance was extremely short and there were differences in the positions of the diffraction patterns on the two sides of the film (which was coated with emulsion on both surfaces), the emulsion was scraped off one side after exposure. For this reason, the accuracy of diffraction intensity was inadequate and the determination of the degree of orientation was confined to the determination of Π.

Table 15.6 Analysis of the necked portion of cold-drawn polyethylene

(a) On the central axis

z [mm]	Wide-angle diffraction: Orientation: Type of orientation	Spiral angle φ[deg]	Degree of orientation Π[%]	Crystallite size t [Å]	Small-angle scattering: Scattering pattern	Long period D_L [Å]	Long period L[Å]	δ' [deg]	Remarks
1.66	Unoriented		0	245	Circular	130			
1.93			0	230	(same)	130			
									line 1
2.19	Spiral orientation	42.5	80	200	Ellipse	170 (m) 130 (eq)			
2.27	(a axis perpendicular to central axis)	35	80	165	Ellipse plus layer line	170 (m) 130 (eq)	(110)		
2.47		18.5	85	150	Layer line		(140)		
									line 2
2.61	(as above)	< 15	86	110	Four-point layer line		125 125	±42 ±40	
2.85	Uniaxial orientation or spiral orientation with an extremely	Extremely small	90	110	(same)				
3.87	small spiral angles	(same)	93	90	(same)		125	±40	

(b) On a flow line (flow line b)

z [mm]	Wide-angle diffraction: Orientation: Type of orientation	Spiral angle φ[deg]	Degree of orientation Π[%]	Crystallite size t [Å]	Small-angle scattering: Scattering pattern	Long period D_L[Å]	Long period L[Å]	δ' [deg]	Remarks
1.66	Unoriented		0	245	Circular	130			
									line 1
1.98	Spiral orientation	25.5	45	220	Ellipse	167 (m) 125 (eq)			
2.15	(a axis perpendicular to tangent to flow line at measurement point)	23.5	47		Ellipse plus two-point layer	167 (m) 140 (eq)	(120)	(−40)	
2.31		18.7	69	75	Two-point layer line		115	−40	
2.35		13.5	83		(same)		125	−37	
2.38		< 10	88		(same)		120	−37	
									line 2
2.41	Uniaxial orientation or spiral orientation	Extremely small	89	70	Four-point layer line		125	±42	
2.85	with an extremely	(same)	91	75	(same)		123	±38	
3.87	small spiral angles	(same)	93	70	(same)		125	±38	

D_L replaces L for the long period with circular and elliptical scattering. Under D_L, (m) and (eq) are the long-period values in the meridian and equatorial directions respectively in elliptical scattering. Further, the signs + and − with δ' indicate right inclination and left inclination respectively of the two-point layer line patterns (*cf.* Fig. 15.28).

± denotes a four-point layer line pattern.

[Reproduced with permission from N. Kasai, M. Kakudo, *J. Polym. Sci.*, **A2**, 1961, 1962, John Wiley & Sons, Inc. (1964)]

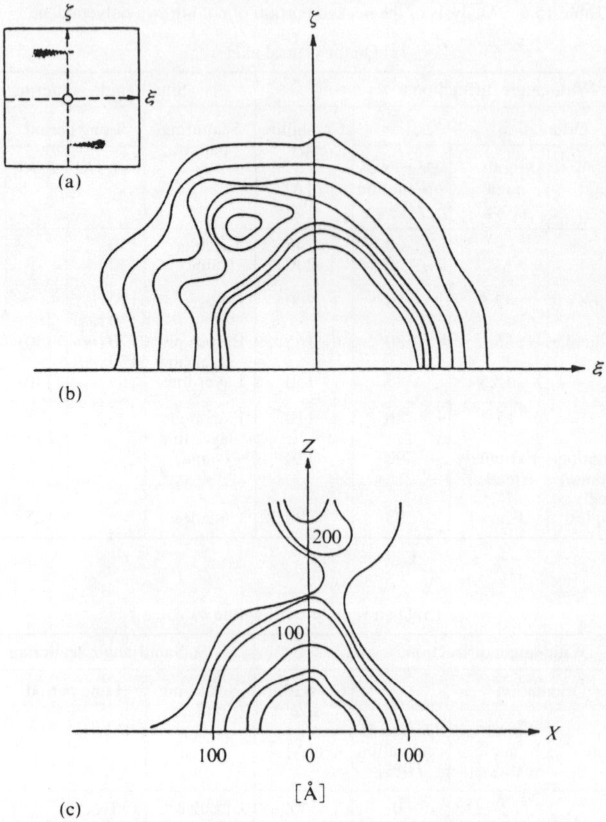

Fig. 15.44 Analysis of a two-point layer line pattern with point symmetry (1) (See also (2) and (3) in Figs. 15.45 and 15.46).
(a) Schematic representation of small-angle scattering pattern.
(b) Intensity distribution of (a); central contour lines omitted.
(c) Fourier transform of (b); central contour lines omitted.
[Reproduced with permission from N. Kasai, M. Kakudo, *J. Polym. Sci.*, **A2**, 1971, John Wiley & Sons, Inc. (1964)]

Figure 15.44(a) shows a sketch of one of the two-point layer line patterns mentioned above and (b) shows its intensity distribution (intensities close to scattering angle zero are omitted).[84] In this case the scattering pattern has point symmetry. Since the cylindrically symmetrical distribution function given earlier for PVA cannot be used, Fig. 15.44(c) was obtained by Fourier transformation in rectangular coordinates as described below.[82]

$$Q(x,z) = (2/A)\int_0^{\infty}\int_{-\infty}^{\infty} I(\xi,\zeta)\cos\{2\pi(\xi x+\zeta z)\}\mathrm{d}\xi\mathrm{d}$$

The comparative lack of detail of Fig. 15.44(c) makes interpretation difficult. However, because the long period appears to be in the meridional direction and there is a slight broadening in the second quadrant, a model similar to that in Fig. 15.45[84] was proposed. This fine texture model is also consistent with the changes in small-angle scattering that occur before

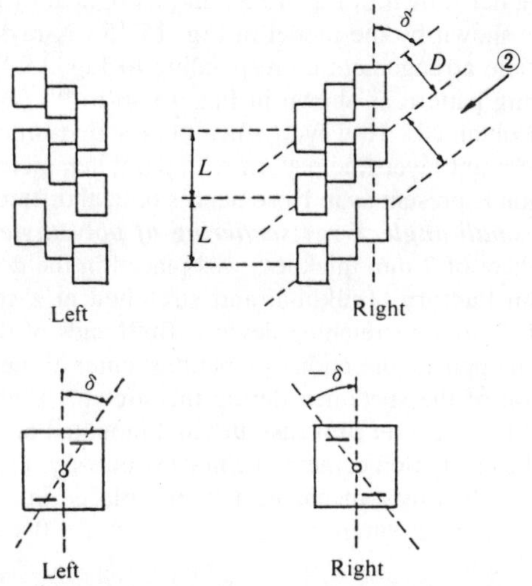

Fig. 15.45 Analysis of a two-point layer line pattern with point symmetry (2). The model of the fine texture is derived from Fig. 15.44(c) and supplementary information (see text). [Reproduced with permission from N. Kasai, M. Kakudo, *J. Polym. Sci.*, **A2**, 1972, John Wiley & Sons, Inc. (1964)]

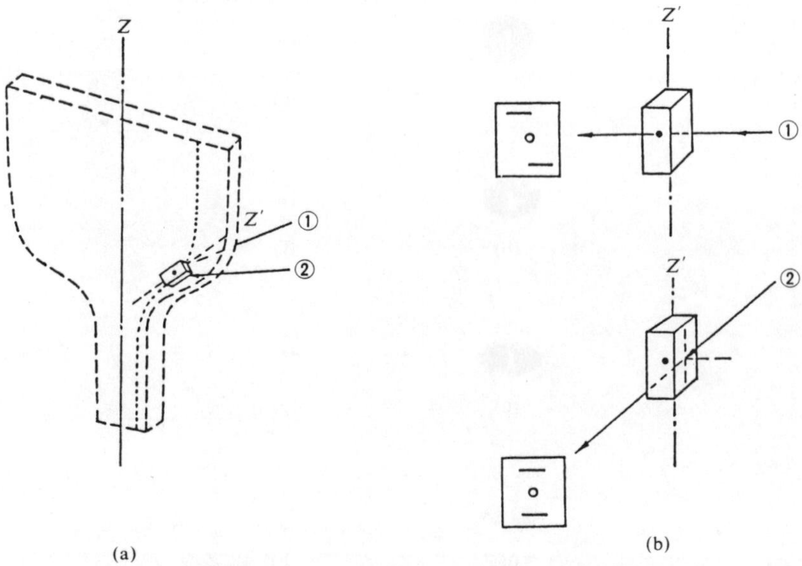

Fig. 15.46 Analysis of a two-point layer line pattern with point symmetry (3). [Reproduced with permission from N. Kasai, M. Kakudo, *J. Polym. Sci.*, **A2**, 1973, John Wiley & Sons, Inc. (1964)]

and after this pattern appears. This is consistent with the subsequently published relationship between the model shown in Fig. 15.33 and its optical diffraction pattern. If the fine texture is in fact as shown by the model in Fig. 15.45. X-rays incident from direction ② will indicate a particle arrangement corresponding to Fig. 15.27 and should give rise to a small-angle scattering pattern as shown in Fig. 15.46(b).[84] Such a small-angle scattering pattern was in fact obtained. However, since only a first-order scattering maximum was observed in the two-point layer line pattern with point symmetry, it may be concluded that appreciable distortion is present as in Hosemann's optical diffraction model.[78]

c) Time-resolved small-angle X-ray scattering of polyethylene sheet during stretch.[87] The polyethylene sheet of 2 mm thickness was placed in the doubly focusing camera (Fig. 8.47) at the Photon Factory, Tsukuba, and stretched at a speed of 12 mm/min (32% Stretch/min) by a DC-driven stretching device. Both ends of the specimen were stretched at equal velocities in opposite directions so that the center of the X-ray beam could impinge on the same position of the specimen during the stretch. Eight 126×126 mm^2 imaging plates (IP) were held in separate film cassettes and mounted on a remote control turn-table, which is a part of the oscillation camera of Enraf-Nonius (*cf.* Fig. 8.10). The exposure time for each IP was controlled by a mechanical shutter placed in front of the specimen. Each small-angle X-ray scattering pattern was recorded with a 1.0 s exposure time at 70-s intervals.

The dynamic changes of small-angle X-ray scattering patterns during the stretch (Fig 15.47) can be classified into four phases:

1). First from (a) to (b) a central diffuse isotropic (or ring) pattern changed to become elliptical.

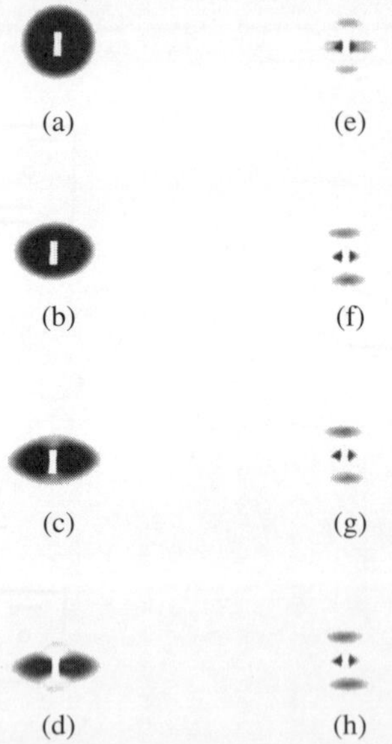

Fig. 15.47 Time-resolved small-angle X-ray scattering patterns from polyethylene sheet recorded on IP during stretch (12 mm/min, 32% stretch/min, strech direction vertical).[87]
Exposure time for each pattern was 1 s. Intervals between exposures were 70 s. X-ray wavelength was 1.55 Å.
[Reproduced with permission from Y. Amemiya *et al.*, *Topics in Current Chemistry*, **147**, 137, Springer Verlag (1988)]

2). Secondly, from (c) to (d) the elliptical pattern changed into a dipolar shape along the equator and the scattering intensity decreased. At the same time, a weak meridional scattering in layer line began to appear at 1/90–1/95 $Å^{-1}$.

3). The third change, from (e) to (h), is the disappearance of the dipolar-shape diffuse scattering and the intensity increase of the meridional scattering.

4). The meridional scattering in a layer line was finally split into two, and thus changed into a four-point scattering pattern (not shown in Fig. 15.47, but *cf.* Fig. 15.36 and see Table 15.6).

These results agree well with those obtained by the *static experiment* from the necking portions of the cold drawn polyethylene specimen[84)] schematically shown in Fig. 15.41.

15.4.2 Anisotropy in the small-angle scattering pattern and in orientation and particle distribution

As shown in Tables 15.4 and 15.5, several types of anisotropic small-angle scattering patterns have been observed. Since the small-angle scattering corresponds to a structure or fine texture with very large dimensions, the anisotropy of the scattering pattern corresponds either 1) to the crystallite orientation distribution in crystalline specimens with extremely large unit cells such as proteins (just as does the anisotropy of the wide-angle diffraction pattern) or 2) to the anisotropy of the distribution of the crystallites or regions of discontinuous density in the specimen.

Case 1) can be treated in exactly the same way as ordinary oriented diffraction patterns. The relationship between an anisotropic scattering pattern and the corresponding fine texture for case 2) was discussed in Section 15.4.1. There is no one-to-one correspondence between the anisotropy of the small-angle scattering pattern and the orientation of crystallite or crystalline regions; however, some examples of the determination of crystallite orientation from small-angle scattering patterns are given below.

Figure 15.48(a) shows examples of the small-angle scattering observed for a number of

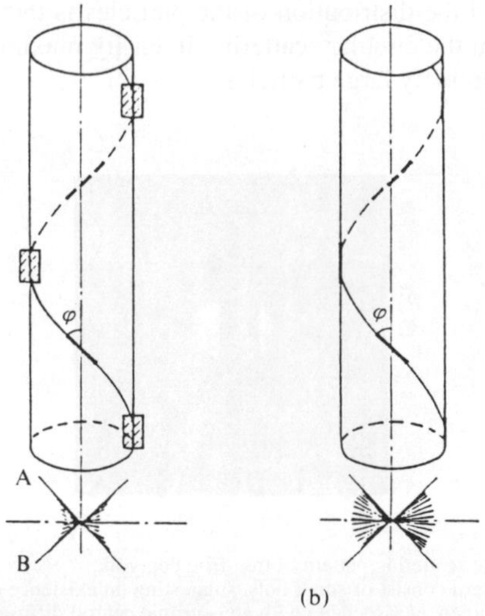

Fig. 15.48 The small-angle scattering patterns from spirally oriented micelles.[88,89)]
(a) Lamellar micelles; (b) Cylindrical micelles.
[Reproduced from A. N. J. Heyn, *J. Am. Chem. Soc.*, **72**, 2284, Am. Chem. Soc. (1950)]

vegetable fibers,[88,89] which indicates an orientation in which the cellulose micelles are inclined with respect to the fiber direction (spiral orientation). The spiral angle φ can be determined from the angle between the two intersecting diffuse scattering patterns. Fig. 15.48(a) is that obtained for lamellar micelles. When the micelles are cylindrical, the scattering pattern has a continuous fan-like appearance as shown in Fig. 15.48(b).

Figure. 15.49 shows the small-angle scattering pattern of a Nylon 6 specimen prepared by polycondensation of ε-aminocaproic acid in the solid phase by heating single crystals under vacuum at a temperature about 30°C below the melting point.[90] In this diagram, the two layer-line scattering patterns intersect at an angle of 110°, which is twice the angle of inclination of the b axis (the fiber axis) found from fiber photographs. The two results are thus in good agreement.

Fig. 15.49 Small-angle scattering pattern of doubly oriented Nylon 6 obtained by polymerization in the solid phase.
Mutually crossed two long periods are overlapped with central diffuse small-angle scattering in elliptical shape.

Figure 15.50 shows the small-angle scattering pattern of tree-frog egg yolk.[91] Orientation is absent, and the distribution of the particles is therefore isotropic. The presence of a fine structure in the circular scattering intensity maximum suggests that this is an isotropic assembly of extremely large particles.

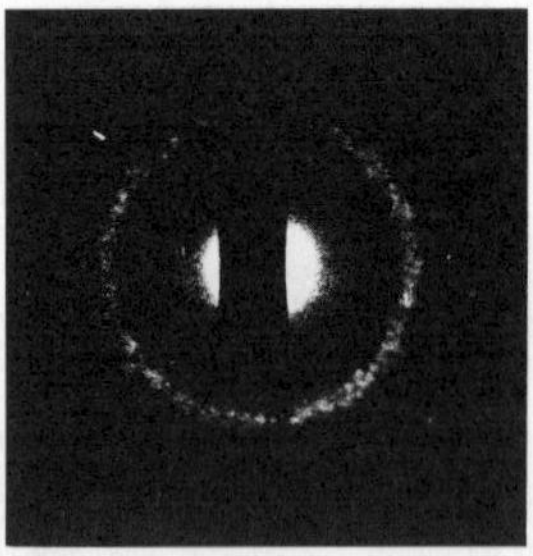

Fig. 15.50 Small-angle scattering pattern of tree-frog egg yolk.
A ring pattern consist of small dots, suggesting an existence of higher order structure in the specimen, is associated with an isotropic central diffuse scattering.

15.5 Analysis of Crystallinity from Small-angle Scattering

15.5.1 Analysis using the long-period pattern

As stated in Section 15.4, the long period L(Å) is thought to be associated with a periodic structure of alternating regions: ···–crystalline–amorphous–crystalline–amorphous–···as shown in Fig. 15.51. If the crystallite size is t(Å) (as found, say, from the breadth of wide-angle diffractions, *cf.* Section 13.3), the crystallinity can be estimated very approximately from the following formula

$$x_{cr} = t/L \tag{15.56}$$

If the paracrystal size L(Å) found from Eq. 13.64 is substituted here, the value for crystallinity obtained should correspond to that described in Section 14.2.3.

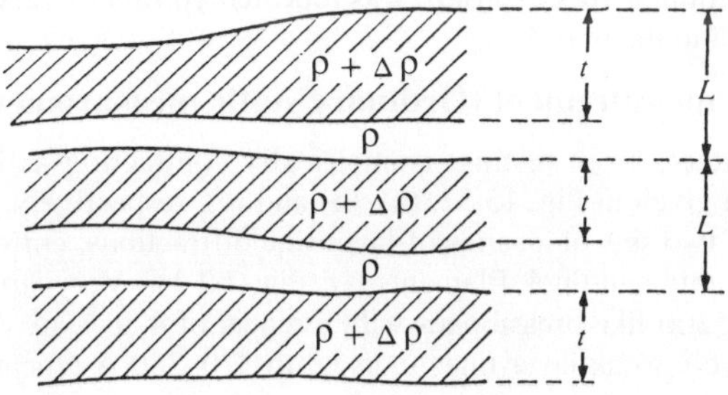

$$x_{cr} \cong t/L$$

Fig. 15.51 Schematic diagram illustrating the structure thought to be associated with the long period L.

15.5.2 Analysis using the central diffuse scattering

As stated in Section 6.2, the small-angle scattering intensity is proportional to the difference between the electron densities of the specimen particles and of the dispersion medium. Hermans and Weidinger[92)] have proposed a method of determining crystallinity by finding the total scattered energy in the central diffuse scattering region from absolute intensity measurements, and then applying Debye's small-angle scattering theory[39)] or, what is essentially the same, Porod's theory.[37,38)]

15.6 Analysis of Well-oriented Small- and Wide-angle Diffractions

In principle there is not much difference from the crystal structure analysis of well-oriented high polymers (Chapter 11) except for the existence of long spacings and higher order structures. However, rather weak and broad diffractions are concentrated in the small-angle region, so there may be some difficulty in obtaining precise measurements of spacing, intensity, indexing, and others.

If the diffuse small-angle scattering is associated with these diffractions it can be analyzed in a manner similar to those methods described in Sections 15.2–15.5. The corresponding structure should be elucidated by considering the diffractions obtained through both small and wide angles and also diffuse small-angle scattering.

15.6.1 X-ray diffraction patterns from contracting muscle[93,94]

X-ray diffraction patterns from contracting striated muscle of the bullfrog (*Rana catesbeiana*) were measured and the corresponding structural changes examined.

A freshly dissected sartorius muscle of the bullfrog was held vertically in a specimen chamber by clamping the pelvic bone at the bottom end and connecting the tibial end to a force transducer. Oxygenated Ringer solution (115 mM NaCl, 2.5 mM KC1, 1.8 mM $CaC1_2$, pH adjusted to 7.2 with $NaHCO_3$) was continuously perfused through the specimen chamber kept at 6°C. The muscle was isometrically stimulated at a sarcomere length of 2.4 μm for 1.3 s through a multi-electrode assembly in the chamber with trains of supramaximal rectangular pulses (3 ms duration at 25 Hz). A mechanical shutter placed in front of the specimen was synchronously opened for 1.0 s when a steady tetanic tension had developed to let the X-rays pass through only during the plateau phase of isometric tension. The plateau tension of the 10th contraction was 83% of the initial tension. To record the resting pattern before contraction, a 10-s exposure was repeated 10 times every 2 s by opening the shutter without stimuli to the muscle.

A. Time-averaged measurement (Ordinary static measurements)

X-ray diffraction patterns from resting, isometrically contracting and rigor states of the same frog muscle are given in Fig. 15.52, (a), (b) and (c), respectively. These patterns (*cf.* Fig. 15.53) consist of two sets of a series of layer line diffractions, corresponding to the helical structures of the thin and thick filaments (*cf.* Fig. 7.2 1). Very strong layer line series appearing in Fig. 15.52(a) like breastbones with a repeat of 429 Å are due to the thick filaments, and the relatively weak layer lines indexed to 370–380 Å repeat are due to the thin filaments. Since the thin and thick filaments have different periodicities from each other, each layer line series generally does not overlap. For convenience, the two sets of layer lines are considered separately, although separate indexing of the two sets is not always reasonable.

The equatorial diffractions appear as distinct Bragg diffractions up to about the 3,0 diffraction ($d =$ *ca.* 120 Å), and beyond this they become broad and seem to overlap, showing that the hexagonal lattice is in a paracrystalline state. In the resting state, the 1,0 diffraction is much stronger than the 1,1 diffraction (cf. Fig.15.53) the intensity ratio ($I(1,0)/I(1,1)$) being *ca.* 2.5.

A particular feature of the layer lines with 429 Å repeat is that the layer lines of $3n$ order have a strong intensity on the meridian, whereas all other layer lines have intensity off the meridian. They could be well indexed by the hexagonal lattice, indicating that the axial and azimuthal orientations of the thick filaments are well defined in the resting state. The 429 Å repeat of the entire thick filament was explained by the basic structure of the thick filament: three-stranded helices of myosin molecules, each with a pitch 3×429 Å, 9 subunits per one turn in each helix (9/1 helix) with an axial separation of 143 Å (Fig. 7.21(c)).

Two series of helix crosses that extend diagonally from the origin and from the 54.6 Å on the meridian are determined mainly by the F-actin helix in the thin filament. Each layer-line diffraction corresponds to the pitch of a particular helix passing through the actin monomers; for example, the 590 Å diffraction corresponds to the pitch of a left-handed genetic helix, the 510 Å diffraction to that of a right-handed genetic helix and the first layer line to the half-pitch of each of two long-strand right-handed helices. Most of these layer lines are axially diffuse and accurate determination of the helical parameters is particularly difficult.

When the muscle went into the rigor state, the layer lines due to the thick filaments

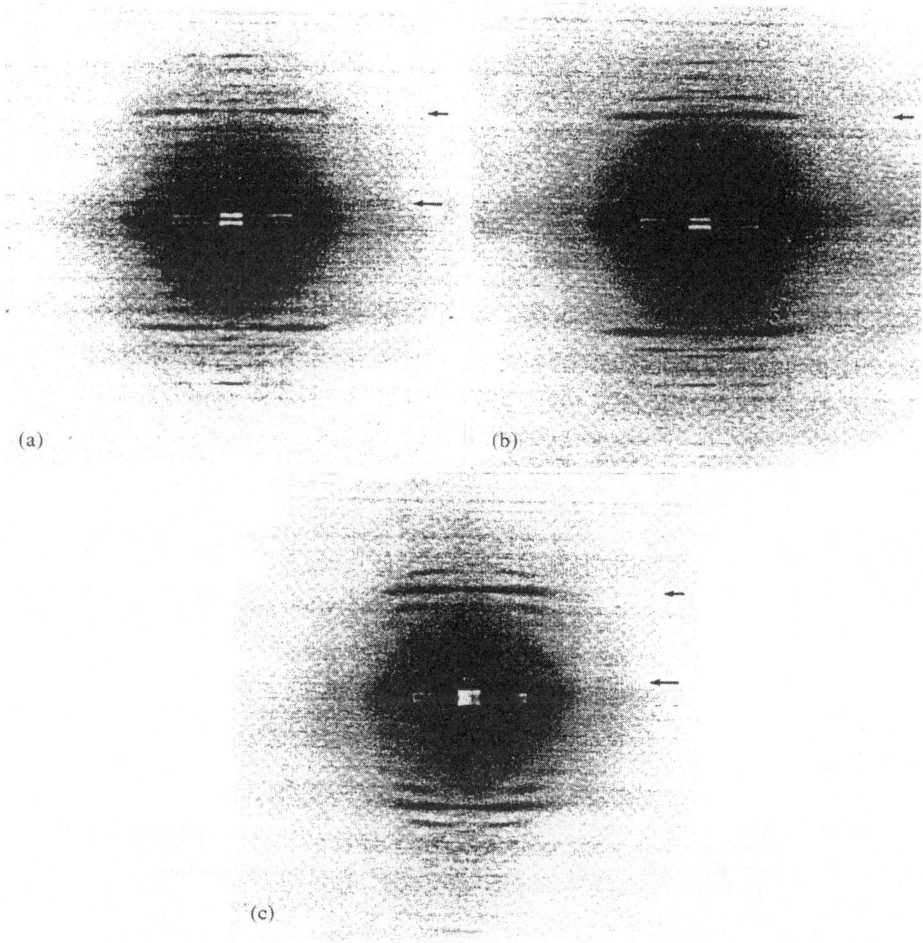

Fig. 15.52 X-ray diffraction photographs of a frog skeletal muscle.[94)]
Recorded with imaging plate using synchrotron radiation. Exposure time 1.5 s. Fiber axis vertical.
(a) Resting state,
(b) isometrically contracting state.
(c) rigor state.
Short arrow indicates the 590 Å layer line; the long arrow, the first layer line of thin filament. The patterns were taken using the same specimen.
[Reproduced with permission from K. Wakabayashi, Y. Amemiya, *Handbook on Synchrotron Radiation*, **4**, 629, Elsevier (1991)]

were generally weakened whereas those due to the thin filaments were markedly intensified (Fig. 15.52(c) and Fig. 15.53). A large intensity change also occurred in the equatorial pattern: the 1,1 diffraction became much stronger than the 1,0 diffraction, the ratio ($I(1,0)/I(1,1)$) being *ca.* 0.25.

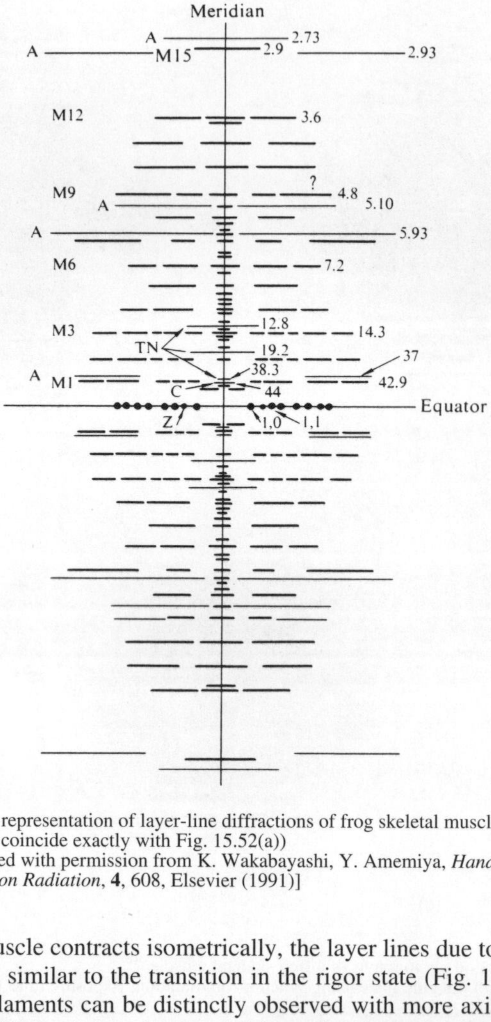

Fig. 15.53 Schematic representation of layer-line diffractions of frog skeletal muscle (resting state).[94) (Does not coincide exactly with Fig. 15.52(a))
[Reproduced with permission from K. Wakabayashi, Y. Amemiya, *Handbook on Synchrotron Radiation*, **4**, 608, Elsevier (1991)]

When a resting muscle contracts isometrically, the layer lines due to thick filaments are remarkably weakened, similar to the transition in the rigor state (Fig. 15.52(b)). The layer lines due to the thin filaments can be distinctly observed with more axial sharpness than in the other state (Fig. 15.54). A large equatorial intensity change also took place similar to that in the rigor state; the intensity of the 1,0 diffraction decreased and that of the 1,1 diffraction increased by a factor of *ca*. 2, the ratio ($I(1,0)/I(1,1)$) being 0.3.

Figure 15.55(a) depicts a perspective of intensity distribution of the diffraction pattern shown as Fig. 15.52(b), in which a very strong and high level of diffuse small-angle scattering is superimposed on the layer line diffractions. After removal of the peak values, the entire distribution of the diffuse scattering was derived (Fig. 15.55(b)). The Guinier plots made on the meridian and in the 45° direction between the equator and the meridian showed approximately straight lines except for the very small angle regions. The radii of gyration obtained from the slope of the linear parts was *ca*. 30 Å, being similar to that of the S1 par-

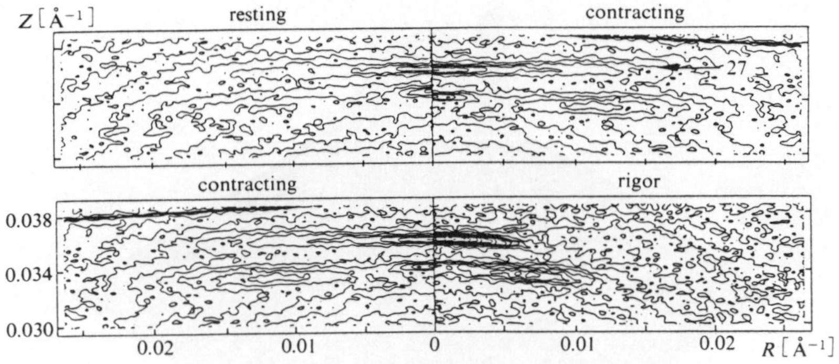

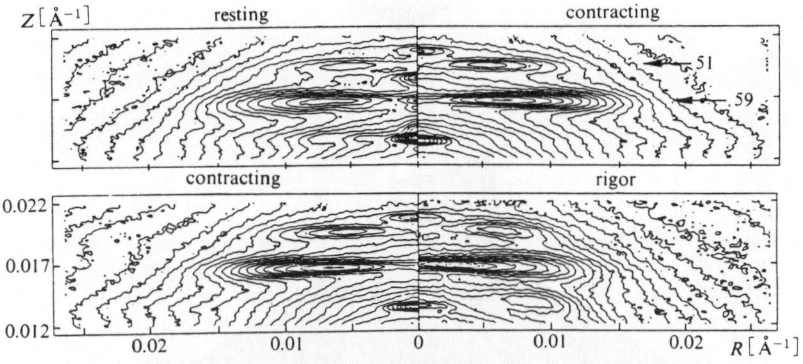

Fig. 15.54 Comparisons of intensity contour map.[94]
(a) Around the 27 Å meridional diffraction
(b) Around the 51 and 59 Å actin layer lines
[Reproduced with permission from K. Wakabayashi, Y. Amemiya, *Handbook on Synchrotron Radiation*, **4**, 641, Elsevier (1991)]

ticle in solution[95] which suggests that a large part of the diffuse small-angle scattering comes from the myosin heads[96] (*cf.* Fig.7.21).

In order to obtain clues as to the cause of the thin filament layer lines, the difference cylindrical Patterson function,[97]

$$\Delta Q(r,z) = 2\sum_{l=1}^{\infty}\left\{\int_0^{\infty} 2\pi R I_l(R) J_0(2\pi R r)\mathrm{d}r\right\}\cos(2\pi l z / c) \tag{15.57}$$

was computed using the intensity data of 15 layer lines except for the equatorial diffractions. In the difference cylindrical Patterson function, r and z are the radial and axial coor-

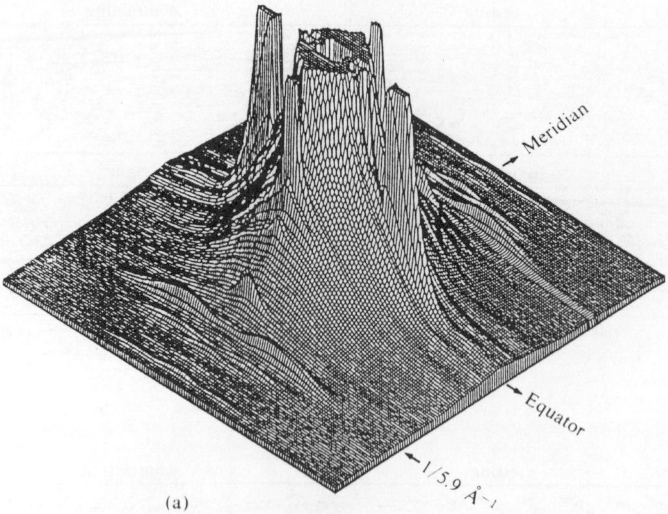

(a)

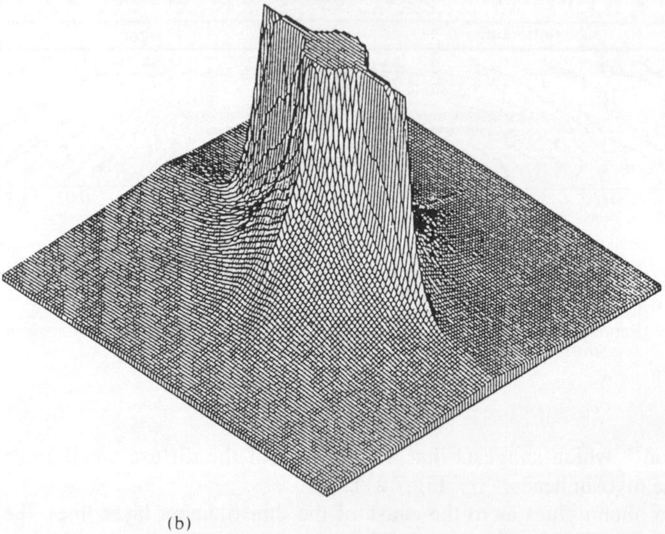

(b)

Fig. 15.55 A perspective of intensity distribution of the X-ray diffraction pattern from an isometrically contracting muscle of Fig. 15.52(b).[94]
(a) The original pattern,
(b) the diffuse small-angle scattering after removal of the layer-line diffractions.
[Reproduced with permission from K. Wakabayashi, Y. Amemiya, *Handbook on Synchrotron Radiation*, **4**, 644, Elsevier (1991)]

dinates in real space, R and Z the same coordinates in reciprocal space, and $J_0(2\pi Rr)$ denotes the zero order Bessel function (*cf.* Section 14.6). The $\Delta Q(r, z)$ is the autocorrelation function of $\Delta\rho(\boldsymbol{r})$ in the cylindrical coordinates defined by

$$\Delta\rho(\rho) = \rho(\rho) - <\rho(x,y)>\delta(z), \qquad <\rho(x,y)> = \int \rho(\rho)\delta z / c, \qquad (15.58)$$

where $\rho(\boldsymbol{r})$ is the electron density of the thin filament, $\boldsymbol{r}$ the position vector and c the axial repeat.[97] Fig. 15.56(b) shows a contour map of $\Delta Q(r, z)$ of the contracting state, in which the solid dots stand for the heads of vectors connecting centers of actin monomers along the 59 Å pitch genetic helix with a radius of 25 Å. The positive contour lines extend along the curve passing through these dots.

Precise examination of Fig. 15.56, referring the Eq. 15.58, suggests that during the contraction 1) myosin heads are not bound physically to action, and 2) if binding take place, the coherent contribution of the bound portions of myosin heads would be small and their location would be randomly distributed along the actin filament since the myosin periodicity is incommensurate with the actin periodicity.

It is assumed that the intensity increase of the layer lines during contraction. *i.e.*, the increase in $\Delta\rho(\boldsymbol{r})$ of the thin filament, is due to a structural change within the thin filament, which could be caused in a cooperative way along the whole filament by interaction with the myosin heads. By referring to three-dimensional reconstructions of the thin filament using electron micrographs, models of the thin filament in the resting and contracting states are shown in Fig. 15.57.[98]

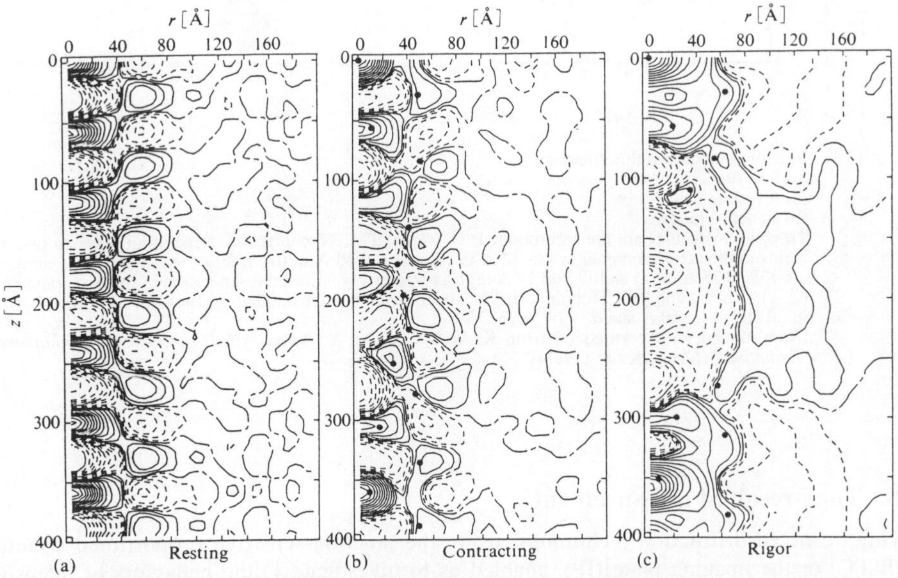

Fig. 15.56 Difference cylindrical Patterson functions of muscle in resting, isometrically contracting and rigor states.[94] [Reproduced with permission from K. Wakabayashi, Y. Amemiya, *Handbook on Synchrotron Radiation*, **4**, 647, Elsevier (1991)]

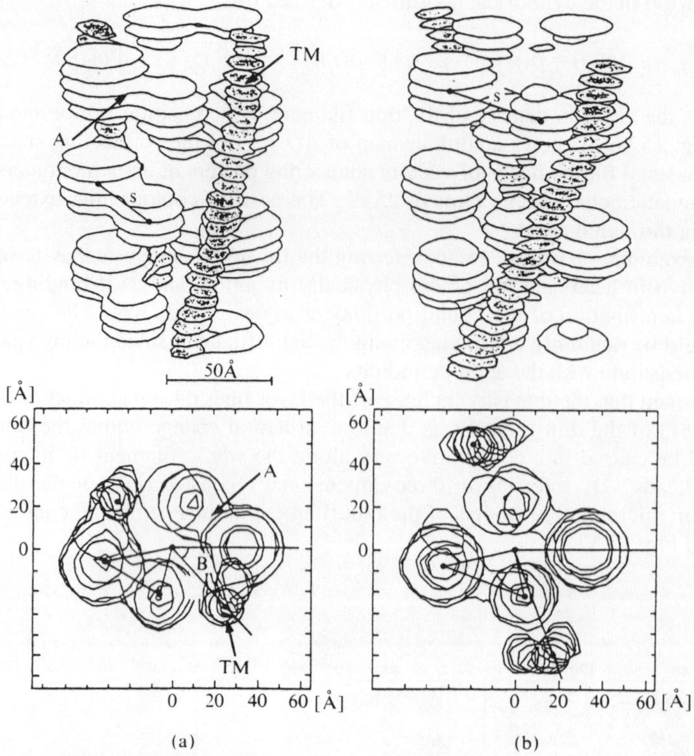

Fig. 15.57 Structural models of thin filament.[98)]
Surface (upper) and cross-section (lower).
(a) A resting model,
(b) an active model.
Troponin molecules are not incorporated. A, Actin; TM, Tropomyosin. Separation between two domains in the actin monomer is $s = 32$ Å in (a) and $s = 40$ Å in (b); radii of small and large domains are 13 and 18 Å in (a) and 11 and 17 Å in (b), respectively. The axial separation of the two domains is *ca.* 11 Å in (a) and (b). Tropomyosins locate at a radial and azimuthal positions of 35 Å and $\theta = 45°$ in (a) and at *ca.* 50 Å and $\theta = 70°$ in (b).
[Reproduced with permission from K. Wakabayashi, Y. Amemiya, *Handbook on Synchrotron Radiatim*, **4**, 651, Elsevier (1991)]

B. Time-resolved measurements

Time-resolved diffraction techniques, using the position-sensitive proportional counter (PSPC) or the imaging plate (IP), enabled us to investigate 1) the behaviors of intensity changes in the pattern in relation to the onset and decay of tension during isometric contraction, and 2) the effects of length perturbations on the diffraction intensities when applied to an isometrically contracting muscle.

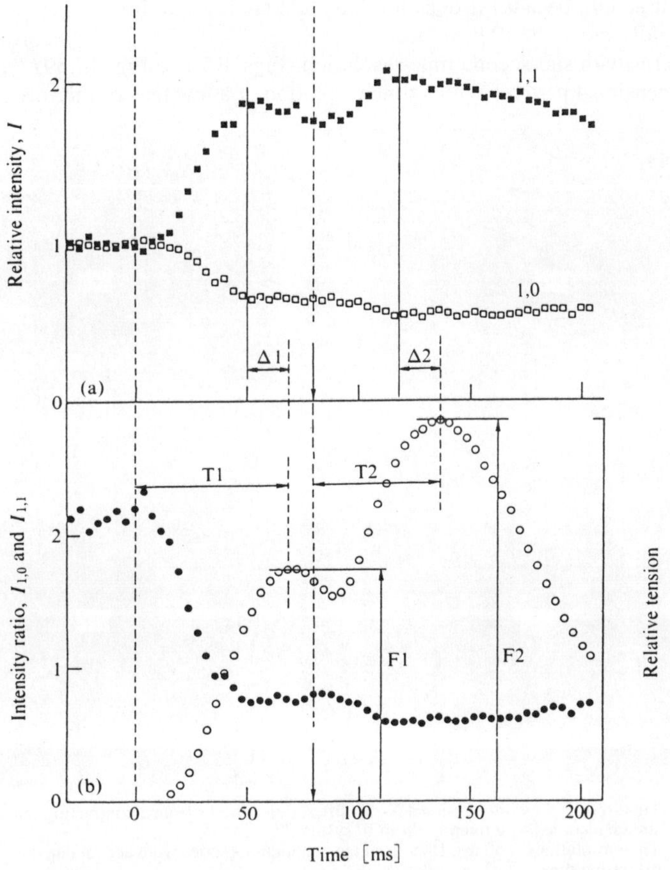

Fig. 15.58 Time courses of changes in the 1,0 and 1,1 equatorial diffractions when stimulated by two successive pulses at an interval of 80 ms at 13°C.[99)]
The first stimulus was given at $t = 0$ and the second at the time shown by vertical arrow with broken line.
(a) Intensity changes of the 1,0 and 1,1 diffractions. Intensities are normalized at resting phase as 1.0.
(b) Intensity ratio (●) of the 1,0 and 1,1 diffractions. ○denotes the tension. T1, T2 are the times to peak tensions at each twitch measured from each onset of stimulus. △1 (19.2 ms) and△2 (19.7 ms) are the time lags between the peak tension and the maximum intensity changes. Time resolution is 4 ms.
[Reproduced with permission from H. Tanaka *et al.*, *Biophys. Chem.*, **25**, 161, Elsevier (1986)]

a) Use of one-dimensional PSPC. A block diagram of the data acquisition system using a PSPC and the principles of time-resolved X-ray diffraction data collection are given in Fig. 8.6.[94)]

The two inner equatorial diffractions, 1,0 and 1,1, exhibit very large changes in intensity during the isometric contraction (Fig. 15.58).[99)] This change has been interpreted as a lateral movement of myosin heads away from the thick filament backbones and towards the thin filaments at the trigonal positions in the hexagonal unit cell.

b) Use of IP. The IP system has enormous potential for dynamic structural studies of muscle contraction, because it does not have a counting rate limit. Two different devices that can rapidly move the IP have been constructed.[100–103] One is for recording two-dimensional patterns with sub-second time resolution (Figs. 8.7 and Fig. 15.59)[94] and the other is for one-dimensional patterns with a time resolution of a few tens of microseconds.

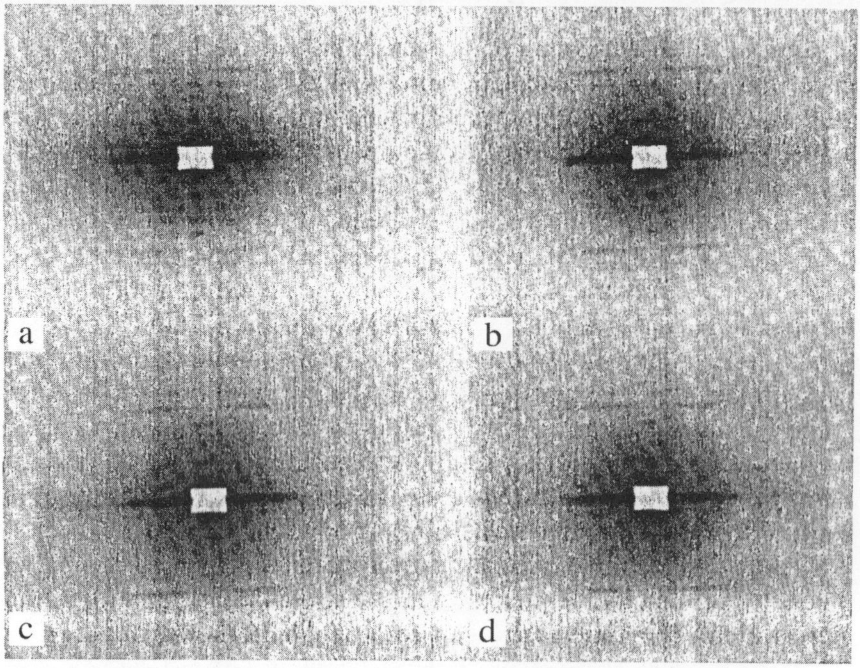

Fig. 15.59 Time-resolved two-dimensional X-ray diffraction patterns from a contracting muscle during slow stretch recorded by a rotating-drum IP system.[94]
Time resolutions is 50 ms. Data from ten contractions series were accumulated.
(a) resting time
(b) plateau phase of isometric contraction
(c) during stretch
(d) after stretch
[Reproduced with permission from K. Wakabayashi, Y. Amemiya, *Handbook on Synchrotron Radiation*, **4**, 670, North-Holland Pub. (1991)]

References

1. M. Kakudo, N. Kasai, M. Kimura, Y. Kubota, T. Watase, *J. Chem. Soc. Jpn., Pure Chem. Sec.*, **78**. 821 (1957).
2. N. Kasai, M. Kakudo, T. Watase, *Technol. Repts. Osaka Univ.*, **8**, 443 (1958).
3. J.W. Anderegg, W.W. Beeman, S. Sullivan, P. Kaesberg, *J. Am. Chem. Soc.*, **7**, 2927 (1955).
4. A. Guinier, G. Fournet, *J. Phys. Radium*, **8**. 345 (1947).
5. A. Guinier, G. Fournet, *Small-Angle Scattering of X-Rays*, pp. 101–120, Wiley, N.Y. (1955).
6. O. Kratky, G. Porod, L. Kahovec, *Z. Elektrochem.*, **55**, 53 (1951).
7. F. Hossfeld, *Acta Cryst.*, **A24**, 643 (1968).
8. R.E. Buerge, J.C. Draper, *Acta Cryst.*, **22**, 6 (1967).
9. I. Pilz, O. Glatter, O. Kratky, in *Methods in Enzymology*, Vol. 61, Part H. p. 164, Academic Press, London (1979).

10. J. Turkevitch, H.H. Hubbel, *J. Am. Chem. Soc.*, **73**, 1 (1951).
11. *International Tables for X-Ray Crystallography*, Vol. III, p. 327, Kynoch Press, Birmingham (1962).
12. In ref. 9, pp. 207–211.
13. I. Pilz, in: O. Glatter, O. Kratky, eds., *Small-angle X-Ray Scattering*, pp. 245–249, Academic Press, London (1982).
14. B.H. Zimm, I. Meyerson, *J. Am. Chem. Soc.*, **68**, 911 (1946).
15. *Handbook of X-Ray Analysis of Polycrystalline Materials* (L.I. Mirkin ed.), pp. 671–672, Consultant Bureau, N.Y. (1964).
16. K.L. Yudowitch, *J. Appl. Phys.*, **22**, 214 (1951).
17. W.E. Danielson, L. Schenfil, J.W.M. DuMond, *J. Appl. Phys.*, **23**, 860 (1952).
18. O. Kratky, *Angew. Chem.*, **72**, 467 (1960).
19. O. Kratky, G. Porod, *Acta Phys. Austriaca*, **2**, 133 (1948).
20. *X-Ray Crystallography* (I. Nitta ed.), Vol. II, p. 560. Maruzen, Tokyo (1961).
21. O. Kratky, G. Porod, B. Paletta, A. Sekora, *J. Polymer Sci.*, **16**, 163 (1955).
22. In ref. 9, pp. 175–176.
23. In ref. 13, pp. 178–179.
24. In ref. 9, pp. 176–181.
25. In ref. 13, pp. 244–249.
26. M. Born, H.S. Green, *Proc. Rcy. Soc. (London)*, **A188**, 10 (1946); **A190**, 455; **A191**, 168 (1947).
27. T. Houzaki, M. Kato, M. Sato, N. Tanaka, Y. Morimoto, Y. Hata, Y. Katsube, N. Kasai, *Bull. Chem. Soc. Jpn.*, **59**, 3747 (1986).
28. Y. Matsuura, M. Kusunoki, W. Date, S. Harada, S. Bando, N. Tanaka, M. Kakudo, *J. Biochem.*, **86**, 1773 (1979).
29. M. Sato, M. Kato, N. Kasai, Y. Hata, N. Tanaka, M. Kakudo, M. Tanaka, M. Ozawa, *Biochem. International*, **5**, 595 (1982).
30. O. Glatter, *J. Appl. Cryst.*, **7**, 147 (1974).
31. In ref. 5, pp. 24–28.
32. M.H. Jellinek, E. Solomon, I. Fankuchen, *Ind. Eng. Chem.*, **18**, 172 (1946).
33. a) C. Katagiri, M. Sato, N. Tanaka, *J. Biol. Chem.*, **262**, 15857 (1987).
    b) M. Sato, *Protein, Nucleic Acid, Enzyme* (Tokyo), **35**, 2883 (1990).
34. a) C. Katagiri, M. Sato, S. de Kont, Y. Katsube, *Biochem.*, **30**, 9675 (1991).
    b) M. Sato, *Protein, Nucleic Acid and Enzyme*, **38**, 1223 (1993) and Private communication.
35. B.E. Warren, *J. Appl. Phys.*, **20**, 96 (1949).
36. P. Debye, A.M. Bueche, *J. Appl. Phys.*, **20**, 518 (1949).
37. P. Debye, H.R. Anderson, H. Brumberger, *J. Appl. Phys.*, **28**, 679 (1957).
38. V. Luzzati, *Acta Cryst.*, **10**, 643 (1957).
39. R. Baro, V. Luzzati, *Acta Cryst.*, **12**, 144 (1959).
40. R. Hosemann, *Kolloid-Z.*, **117**, 13 (1950).
41. G. Porod, *Kolloid-Z.*, **124**, 83 (1951).
42. G. Porod, *Kolloid-Z.*, **125**, 51 (1951).
43. W.W. Beeman, P. Kaesberg, J.W. Anderegg, M.B. Webb, Size of Particles, Lattice Defects, in *Handbuch der Physik* (S. Fusse ed.), Springer Verlag, Leipzig (1957).
44. O. Kratky, *Naturwissenschaften*, **42**, 237 (1955).
45. O. Kratky, *Progress Biophys. Biophys. Chem.*, **13**, 105 (1963).
46. G. Porod, *Fortsch. Hochpolymer. Forsch.*, **2**, 263 (1961).
47. R. Hosemann, S.N. Bagchi, *Direct Analysis of Diffraction by Matter*, North-Holland, Amsterdam (1962).
48. W.O. Statton, Small-Angle Scattering X-Ray Studies of Polymers in Newer Methods of Polymer Characterization, in: *Handbook of X-Rays*, (B. Kaeibe, ed.), Interscience, N.Y. (1963).
49. *Small-Angle X-Ray Scattering* (H. Brumberger ed.), Gordon & Breach (1967).
50. L.E. Alexander, *X-Ray Diffraction Methods in Polymer Science*, pp. 280–356, Wiley-Interscience, N. Y. (1969).
51. C.R. Cantor, P.R. Schimmel, *Biophysical Chemistry*, Part II, Techniques for the Study of Biological Structure and Function, pp. 811–817, W.H. Freeman & Co., San Francisco 1980.
52. O. Glatter, X-ray techniques in Small-angle techniques, in: *International Tables for Crystallography*, Vol. C, (A.J.C. Wilson and ed.), Mathematical. Physical and Chemical Tables, pp. 89–103, Kluwer Academic, Dordrecht (1999).
53. In ref. 15, pp. 673–675.
54. In ref. 20, Vol. II, pp. 517–577.
55. In ref. 9 and ref 13.
56. O. Kratky, Possibilities of X-Ray Small-Angle Analysis in the Investigation of Dissolved and Solid High Polymer Substances, in: *Macromolecular Chemistry-2*, p. 488, Butterworths, London (1966).

57. A. Peterlin, *Makromol. Chem.*, **9**, 244 (1952).
58. A. Peterlin, *J. Polymer Sci.*, **10**, 425 (1953).
59. S. Heine, O. Kratky, G. Porod, J. Schmitz, *Makromol. Chem.*, **44–46**, 682 (1952).
60. A. Peterlin, *J.Polymer Sci.*, **47**, 403 (1960).
61. S. Heine, O. Kratky, J. Rappert, *Makromol. Chem.*, **56**, 150 (1952).
62. R. Kirste, O. Kratky, *Z. Physik. Chem. Neuge Folge*, **81**, 363 (1962).
63. R. Koyama, J. *Phys. Soc. Jpn.*, **34**, 1029 (1973).
64. R. Koyama, *J. Phys. Soc. Jpn.*, **36**, 1409 (1974).
65. R. Koyama, *J. Phys. Soc. Jpn.*, **41**, 1077 (1976).
66. J.J. Hermans, R. Ullman, *Phvsica*, **18**, 951 (1952).
67. D.Y. Yoon, P.J. Flory, *Macromolecules*, **9**, 294, 299 (1976).
68. Y. Izumi, K. Shinbo, M. Fujii, Y. Miyake, *Colloid Polymer Sci.*, **256**, 1 (1978).
69. T. Taru, E. Wada, S. Tatsumiya, *Repts. Prog. Polymer Phys. Jpn.*, **15**, 37 (1972).
70. E. Wada, Y. Taru, S. Tatsumiya, *Repts. Prog. Polymer Phys. Jpn.*, **15**, 41 (1972).
71. H. Hiramatsu, *Repts. Progr. Polymer Phys. Jpn.*, **19**, 39 (1976).
72. Y. Izumi, M. Fujii, K. Shinbo, Y. Miyake, *Repts. Progr. Polymer Phys, Jpn.*, **18**, 45 (1975).
73. I. Fankuchen, H. Mark, *J. Appl. Phys.*, **15**, 364 (1944).
74. P.H. Lindenmeyer, R. Hosemann, *J. Appl. Phys.*, **34**, 42 (1963).
75. C. Reinhold, E.W. Fischer, A. Peterlin, *J. Appl, Phys.*, **35**, 71 (1964).
76. E.W. Fischer, G.F. Schmidt, unpublished data cited in ref. 74.
77. S.G. Tomlinson, L.G. Erickson, *Acta Cryst.*, **13**, 395 (1960).
78. a) R. Bonart, R. Hosemann, *Makromol. Chem.*, **39**, 105 (1960).
    b) L. Loboda-Cackovic, R. Hosemann, H. Catkovic, F. Ferrero, E. Feracini, *Polymer*, **17**, 303 (1976).
    c) K. Kaji, T. Mochizuki, A. Akiyama, R. Hosemann, *J. Material Sci.*, **13**, 972 (1978).
79. In ref. 47, pp. 646–652.
80. W.O. Statton, G.M. Goddard, *J. Appl. Phys.*, **28**, 1111 (1957).
81. M. Kakudo, N. Kasai, in: *Poly(vinyl alcohol)* (Soc. High Polymer Sci. Jpn., ed.), pp. 223–231 (1955); N. Kasai, M. Kakudo, T. Watase, *J. Chem. Soc. Jpn., Ind. Chem. Sect.* **59**, 786 (1956).
82. N. Kasai, M. Kakudo, *Repts. Progr. Polymer Phys. Jpn.*, **9**, 239 (1966).
83. N. Kasai, M. Kakudo, *Repts. Progr. Polymer Phys. Jpn.*, **6**, 319 (1963).
84. N. Kasai, M. Kakudo, *J. Polymer Sci.*, **A2**, 1955 (1964).
85. N. Kasai, M. Kakudo, T. Watase, *Repts. Progr, Polymer Phys. Jpn.*, **9**, 243 (1966).
86. N. Kasai, K. Ogawa, M. Kakudo, Paper presented at the 11th Annual Meeting of the Soc. High Polymer Soc. Jpn. (1962).
87. a) Y. Amemiya, T. Kojima, T. Nomura, H. Chikashi, K. Satow, Y. Matsushita, *Photon Factory Activity Report*, **4**, 306 (1986).
    b) Y. Amemiya, Y. Satow, T. Matsushita, J. Chikawa, K. Wakabayashi, J. Miyahara, *Topics in Current Chemistry*, **147**, 122–144 (1988).
88. In ref. 5., p. 178.
89. A.N.J. Heyn, *J. Am. Chem. Soc.*, **72**. 2284 (1950).
90. N. Kasai, G. Bodor, H. Moravez, B. Post., IUPAC International Symposium on Macromolecular Chemistry, Tokyo-Kyoto (1966).
91. R. Honjin, T. Nakamura, Private communication.
92. P.H. Hermans, A. Weidinger, *Makromol. Chem.*, **39**, 67 (1960).
93. Y. Amemiya, K. Wakabayashi, H. Tanaka, Y. Ueno, J. Miyahara, *Science*, **237**, 164 (1987).
94. K. Wakabayashi, Y. Amemiya, Progress in X-ray synchrotron diffraction studies of muscle contraction, in: *Handbook on Synchrotron Radiation* (S. Ebashi, M. Koch, R. Rubenstein ed.), Vol. 4, pp. 597–678, North-Holland, Amsterdam (1991).
95. K.M. Kretzschmal, R.A. Mandelson, M.F. Morals, *Biochem.*, **17**, 2314 (1978).
96. F.R. Poulsen, J. Lowy, *Nature*, **303**, 146 (1983); J. Lowy, F.R. Poulsen, *J. Mol. Biol.*, **194**, 595 (1987).
97. K. Namba, K. Wakabayashi, T. Mitsui, *J. Mol. Biol.*, **138**, 1 (1980).
98. K. Wakabayashi, Y. Ueno, Y. Amemiya, H. Tanaka, In: *Molecular Mechanism of Muscle Contractions*, (H. Sugi, G.H. Pallock ed.), p. 353, Plenum Press, N.Y. (1988).
99. Y. Amemiya, *J. Cryst. Soc. Jpn.*, **17**, 119 (1985).
100. H. Tanaka, T. Kobayashi, Y. Amemiya, K. Wakabayashi, *Biophys. Chem.*, **25**. 161 (1986).
101. Y. Amemiya, Y. Satow, T. Matsushita. J. Chikawa. K. Wakabayashi, J. Miyahara, *Topics in Current Chemistry*, **147**, 121 (1988).
102. Y. Amemiya, T. Matsushita, A. Nakagawa, Y. Satow, J. Miyahara, J. Chikawa, *Nucl. Inslrum. Methods*, **A266**. 645 (1988).
103. Y. Amemiya, S. Kishimoto, T. Matsushita, Y. Satow, M. Ando, *Rev. Sci. Instrum.*, **60**. 1552 (1989).

Appendix

Appendix- Table 1 Mass absorption (Mass attenuation) coefficients, μ_m [cm^2 g^{-1}] for selected radiations [Å][†]

Absorber	Cr $K\bar{\alpha}$ 2.2909	Cr $K\beta_1$ 2.0848	Fe $K\bar{\alpha}$ 1.9373	Fe $K\beta_1$ 1.7565	Cu $K\bar{\alpha}$ 1.5418	Cu $K\beta_1$ 1.3922	Mo $K\bar{\alpha}$ 0.7107	Mo $K\beta_1$ 0.6323	Ag $K\bar{\alpha}$ 0.5608	Ag $K\beta_1$ 0.4970
1 H	0.412	0.405	0.400	0.396	0.391	0.388	0.373	0.370	0.367	0.363
2 He	0.498	0.425	0.381	0.335	0.292	0.268	0.202	0.197	0.193	0.189
3 Li	1.30	1.01	0.839	0.663	0.500	0.412	0.198	0.187	0.179	0.172
4 Be	3.44	2.59	2.09	1.58	1.11	0.853	0.256	0.229	0.209	0.195
5 B	7.59	5.69	4.55	3.39	2.31	1.73	0.368	0.309	0.267	0.237
6 C	15.0	11.2	8.99	6.68	4.51	3.33	0.576	0.458	0.374	0.315
7 N	24.7	18.6	14.9	11.0	7.44	5.48	0.845	0.645	0.503	0.404
8 O	37.8	28.4	22.8	17.0	11.5	8.42	1.22	0.908	0.685	0.329
9 F	51.5	38.9	31.3	23.3	15.8	11.6	1.63	1.19	0.879	0.660
10 Ne	74.1	56.1	45.2	33.8	22.9	16.9	2.35	1.69	1.23	0.906
11 Na	94.9	72.1	58.2	43.7	29.7	22.0	3.03	2.17	1.56	1.13
12 Mg	126	96.2	77.8	58.5	40.0	29.6	4.09	2.92	2.09	1.50
13 Al	155	118	95.9	72.3	49.6	36.8	5.11	3.64	2.59	1.85
14 Si	196	151	122	92.7	63.7	47.5	6.64	4.73	3.35	2.38
15 P	230	177	144	109	75.5	56.4	7.97	5.67	4.01	2.84
16 S	281	217	177	135	93.3	69.8	9.99	7.11	5.02	3.55
17 Cl	316	244	200	152	106	79.5	11.5	8.20	5.79	4.09
18 Ar	342	266	218	167	116	87.5	12.8	9.14	6.46	4.56
19 K	421	328	270	207	145	109	16.2	11.6	8.19	5.78
20 Ca	490	382	314	242	170	129	19.3	13.8	9.79	6.92
21 Sc	516	403	332	256	180	137	20.8	14.9	10.6	7.47
22 Ti	590	444	358	277	200	152	23.4	16.8	11.9	8.43
23 V	74.7	479	399	309	219	166	26.0	18.7	13.3	9.42
24 Cr	86.8	67.0	492	385	247	185	29.9	21.5	15.4	10.9
25 Mn	97.5	75.3	61.6	375	270	207	33.1	23.8	17.0	12.1
26 Fe	113	86.9	71.0	54.3	302	232	37.6	27.1	19.4	13.8
27 Co	124	96.0	78.5	60.0	321	248	41.0	29.6	21.2	15.1
28 Ni	144	112	91.3	69.8	48.8	279	46.9	34.0	24.4	17.4
29 Cu	153	118	96.8	74.0	51.8	39.2	49.1	35.7	25.6	18.3
30 Zn	171	132	108	82.7	57.9	43.8	54.0	39.3	28.2	20.2
31 Ga	183	142	116	88.6	62.1	47.0	57.0	41.5	29.8	21.4
32 Ge	199	155	127	96.9	67.9	51.4	61.2	44.6	32.1	23.1
33 As	219	170	139	106	74.7	56.5	66.1	48.2	34.8	25.0
34 Se	234	182	149	114	80.0	60.5	69.5	50.8	36.8	26.5
35 Br	260	202	165	127	89.0	67.4	75.6	55.5	40.3	29.1
36 Kr	277	215	176	135	95.2	72.1	79.3	58.4	42.5	30.7
37 Rb	303	236	193	148	104	79.0	85.1	63.0	45.9	33.2
38 Sr	328	256	210	161	113	85.9	90.6	67.2	49.1	35.6
39 Y	358	279	229	176	124	94.0	97.0	72.1	52.9	38.4
40 Zr	386	300	247	191	139	101	16.3	76.1	55.9	40.7
41 Nb	416	325	267	205	145	110	17.7	81.0	59.8	43.6
42 Mo	442	345	284	219	154	117	18.8	13.8	72.0	52.5
43 Tc	474	370	305	235	166	126	20.4	14.9	66.0	48.4
44 Ru	501	392	323	249	176	134	21.7	15.8	11.4	50.6
45 Rh	536	420	346	267	189	144	23.3	17.0	12.3	53.5
46 Pd	563	441	363	281	199	151	24.7	18.0	13.0	55.5
47 Ag	602	472	389	301	213	163	26.5	19.4	14.0	10.1
48 Cd	626	490	405	313	222	169	27.8	20.2	14.6	10.6
49 In	663	519	428	332	236	180	29.5	21.6	15.6	11.3
50 Sn	691	524	447	347	247	188	31.0	22.6	16.4	11.8

Appendix- Table 1 Mass absorption (Mass attenuation) coefficients, μ_m [cm^2 g^{-1}] for selected radiations [Å]† (*Continued*)

Absorber	Cr		Fe		Cu		Mo		Ag	
	$K\bar{\alpha}$ 2.2909	$K\beta_1$ 2.0848	$K\bar{\alpha}$ 1.9373	$K\beta_1$ 1.7565	$K\bar{\alpha}$ 1.5418	$K\beta_1$ 1.3922	$K\bar{\alpha}$ 0.7107	$K\beta_1$ 0.6323	$K\bar{\alpha}$ 0.5608	$K\beta_1$ 0.4970
51 Sb	723	570	471	365	259	198	32.7	23.9	17.3	12.5
52 Te	740	585	483	374	267	204	33.8	24.7	17.9	12.9
53 I	796	631	522	408	288	220	36.7	26.8	19.4	14.0
54 Xe	721	652	540	422	299	229	38.2	28.0	20.2	14.6
55 Cs	760	686	569	446	317	243	40.7	29.8	21.5	15.6
56 Ba	570	645	586	461	325	252	42.3	31.0	22.4	16.2
57 La	225	744	618	483	348	266	44.9	32.9	23.8	17.2
58 Ce	238	494	561	510	368	282	47.7	34.9	25.3	18.3
59 Pr	238	188	448	539	390	299	50.7	37.2	26.9	19.5
60 Nd	251	198	455	492	404	310	53.0	38.8	28.1	20.4
61 Pm	294	232	194	588	426	328	56.3	41.3	29.9	21.7
62 Sm	279	221	204	163	434	335	57.8	42.4	30.8	22.3
63 Eu	309	244	203	408	434	352	60.9	44.7	32.4	23.5
64 Gd	298	235	195	153	403	360	62.6	46.0	33.4	24.2
65 Tb	332	263	219	171	321	376	65.8	48.3	35.1	25.5
66 Dy	325	257	214	168	362	387	68.3	50.2	36.5	26.5
67 Ho	347	272	228	178	129	402	71.3	52.4	38.1	27.7
68 Er	352	278	232	182	132	417	74.4	54.8	39.9	29.0
69 Tm	386	305	253	196	140	108	77.9	57.4	41.8	30.4
70 Yb	387	304	251	196	142	108	80.4	59.3	43.2	31.4
71 Lu	431	339	280	218	156	121	84.0	61.9	45.1	32.8
72 Hf	425	334	277	216	155	120	86.9	64.1	46.7	34.0
73 Ta	432	339	283	220	158	122	90.4	66.7	48.7	35.4
74 W	457	361	301	234	168	130	93.8	69.2	50.5	36.8
75 Re	501	394	327	257	187	143	97.4	71.9	52.5	38.3
76 Os	499	392	327	255	184	142	100	74.1	54.1	39.5
77 Ir	520	411	340	265	191	148	104	77.0	56.3	41.1
78 Pt	541	423	357	261	188	145	107	79.7	58.3	42.6
79 Au	551	434	361	279	210	155	112	82.9	60.7	44.4
80 Hg	541	416	339	260	188	141	115	85.4	62.6	45.8
81 Tl	597	487	403	314	226	175	118	87.9	64.5	47.2
82 Pb	643	507	420	327	235	181	122	90.8	66.6	48.8
83 Bi	666	524	434	339	244	189	126	94.1	69.1	50.6
84 Po	691	544	452	354	254	196	132	98.3	72.3	53.0
85 At	680	533	444	345	248	186	117	102	75.1	55.1
86 Rn	734	576	477	373	267	205	108	101	72.1	54.5
87 Fr	758	597	493	384	277	213	87.0	104	77.0	56.7
88 Ra	743	585	487	380	273	210	88.0	108	79.3	58.4
89 Ac	739	618	530	444	317	285	90.8	110	82.4	60.7
90 Th	768	509	485	389	306	219	96.5	98.7	83.9	61.9
91 Pa	738	582	482	375	271	208	101	119	87.8	64.8
92 U	766	617	528	400	288	222	102	74.9	88.6	65.5

†[Reproduced with permission from, Source: *International Tables for Crystallography*, Vol. C, *Mathematical, Physical and Chemical Tables*, 2nd ed. (A.J.C. Wilson, E. Prince eds.), pp. 230-235, IUCr, D. Reidel Pub., Dordrecht (1999)]

Appendix-Table 2A Mean atomic scattering factors in electrons for free atoms†

Z / Element / $(\sin\theta)/\lambda$[Å^{-1}]	1 H	2 He	3 Li	4 Be	5 B	6 C	7 N	8 O	9 F	10 Ne
0.00	1.000	2.000	3.000	4.000	5.000	6.000	7.000	8.000	9.000	10.000
0.05	0.947	1.957	2.708	3.707	4.724	5.749	6.776	7.798	8.815	9.830
0.10	0.811	1.837	2.215	3.065	4.060	5.107	6.180	7.245	8.302	9.351
0.15	0.641	1.663	1.904	2.463	3.316	4.311	5.385	6.472	7.560	8.643
0.20	0.481	1.460	1.742	2.060	2.699	3.560	4.563	5.623	6.709	7.805
0.25	0.350	1.254	1.626	1.828	2.263	2.949	3.825	4.808	5.851	6.928
0.30	0.251	1.060	1.513	1.692	1.979	2.494	3.219	4.089	5.054	6.079
0.35	0.180	0.887	1.393	1.600	1.799	2.171	2.747	3.489	4.353	5.302
0.40	0.130	0.738	1.270	1.520	1.681	1.948	2.393	3.006	3.759	4.617
0.45	0.095	0.613	1.149	1.443	1.596	1.794	2.132	2.629	3.270	4.029
0.50	0.071	0.509	1.033	1.362	1.526	1.685	1.942	2.338	2.874	3.535
0.55	0.053	0.423	0.924	1.279	1.463	1.603	1.802	2.115	2.559	3.126
0.60	0.040	0.353	0.823	1.195	1.402	1.537	1.697	1.946	2.309	2.517
0.65	0.031	0.295	0.732	1.112	1.339	1.479	1.616	1.816	2.112	2.517
0.70	0.024	0.248	0.650	1.030	1.276	1.426	1.551	1.714	1.956	2.296
0.80	0.015	0.177	0.512	0.876	1.147	1.322	1.445	1.568	1.735	1.971
0.90	0.010	0.129	0.404	0.740	1.020	1.219	1.353	1.463	1.588	1.757
1.00	0.007	0.095	0.320	0.622	0.900	1.114	1.265	1.377	1.482	1.609
1.10	0.005	0.072	0.255	0.522	0.790	1.012	1.172	1.298	1.398	1.502
1.20	0.003	0.055	0.205	0.439	0.690	0.914	1.090	1.221	1.324	1.418
1.30	0.003	0.042	0.165	0.369	0.602	0.822	1.004	1.145	1.254	1.346
1.40	0.002	0.033	0.134	0.311	0.524	0.736	0.921	1.070	1.186	1.280
1.50	0.001	0.026	0.110	0.263	0.457	0.659	0.843	0.997	1.120	1.218
1.60		0.021	0.091	0.223	0.398	0.588	0.769	0.926	1.055	1.158
1.70		0.017	0.075	0.190	0.347	0.525	0.700	0.857	0.990	1.099
1.80		0.014	0.063	0.163	0.304	0.468	0.636	0.792	0.928	1.041
1.90		0.011	0.053	0.139	0.266	0.418	0.578	0.731	0.868	0.984
2.00		0.010	0.044	0.120	0.233	0.373	0.525	0.674	0.810	0.929

Z / Element / $(\sin\theta)/\lambda$[Å^{-1}]	11 Na	12 Mg	13 Al	14 Si	15 P	16 S	17 Cl	18 Ar	19 K	20 Ca
0.00	11.000	12.000	13.000	14.000	15.000	16.000	17.000	18.000	19.000	20.000
0.05	10.568	11.507	12.439	13.434	14.458	15.484	16.511	17.536	18.204	19.091
0.10	9.760	10.472	11.230	12.134	13.138	14.177	15.234	16.298	16.733	17.331
0.15	9.027	9.502	10.059	10.769	11.629	12.583	13.597	14.647	15.243	15.723
0.20	8.335	8.735	9.158	9.673	10.327	11.109	11.991	12.949	13.728	14.304
0.25	7.618	8.078	8.465	8.859	9.335	9.927	10.633	11.441	12.268	12.961
0.30	6.881	7.446	7.873	8.231	8.600	9.039	9.576	10.216	10.977	11.705
0.35	6.156	6.817	7.316	7.698	8.029	8.376	8.782	9.272	9.908	10.590
0.40	5.471	6.194	6.766	7.202	7.547	7.856	8.181	8.558	9.061	9.650
0.45	4.848	5.595	6.222	6.719	7.103	7.417	7.706	8.011	8.403	8.885
0.50	4.293	5.034	5.692	6.240	6.674	7.017	7.305	7.575	7.889	8.275
0.55	3.811	4.520	5.186	5.769	6.250	6.633	6.941	7.207	7.474	7.788
0.60	3.398	4.059	4.713	5.312	5.829	6.254	6.595	6.875	7.125	7.392
0.65	3.048	3.652	4.277	4.878	5.418	5.877	6.254	6.560	6.814	7.057
0.70	2.754	3.297	3.883	4.470	5.020	5.505	5.915	6.252	6.523	6.762
0.80	2.305	2.729	3.221	3.750	4.284	4.790	5.245	5.639	5.961	6.228
0.90	1.997	2.317	2.712	3.164	3.649	4.138	4.607	5.036	5.406	5.717
1.00	1.784	2.022	2.330	2.702	3.122	3.570	4.023	4.460	4.859	5.209
1.10	1.634	1.812	2.049	2.346	2.698	3.092	3.509	3.931	4.337	4.710
1.20	1.524	1.660	1.841	2.076	2.364	2.699	3.070	3.462	3.855	4.233
1.30	1.438	1.546	1.687	1.872	2.104	2.384	2.704	3.056	3.423	3.791
1.40	1.367	1.459	1.571	1.717	1.903	2.133	2.405	2.713	3.045	3.391
1.50	1.304	1.387	1.481	1.598	1.747	1.935	2.162	2.427	2.722	3.039
1.60	1.247	1.326	1.408	1.505	1.626	1.779	1.967	2.192	2.450	2.733
1.70	1.191	1.270	1.346	1.430	1.530	1.655	1.811	2.000	2.221	2.470
1.80	1.137	1.219	1.292	1.367	1.453	1.557	1.686	1.844	2.033	2.250
1.90	1.084	1.169	1.243	1.313	1.389	1.477	1.585	1.717	1.876	2.063
2.00	1.032	1.120	1.195	1.264	1.333	1.411	1.502	1.614	1.748	1.908

Appendix-Table 2A Mean atomic scattering factors for free atoms (*Continued*)

Z / Element / $(\sin\theta)/\lambda[\text{Å}^{-1}]$	21 Sc	22 Ti	23 V	24 Cr	25 Mn	26 Fe	27 Co	28 Ni	29 Cu	30 Zn
0.00	21.000	22.000	23.000	24.000	25.000	26.000	27.000	28.000	29.000	30.000
0.05	20.131	21.171	22.208	23.329	24.274	25.304	26.331	27.356	28.448	29.401
0.10	18.356	19.410	20.474	21.789	22.611	23.678	24.744	25.807	27.084	27.927
0.15	16.645	17.635	18.661	20.022	20.764	21.829	22.900	23.973	25.370	26.124
0.20	15.135	16.044	17.003	18.260	19.012	20.046	21.093	22.150	23.540	24.283
0.25	13.732	14.572	15.465	16.561	17.364	18.354	19.364	20.390	21.687	22.478
0.30	12.423	13.198	14.026	14.965	15.806	16.744	17.709	18.696	19.869	20.720
0.35	11.244	11.949	12.705	13.513	14.353	15.233	16.145	17.084	18.133	19.027
0.40	10.226	10.852	11.530	12.227	13.031	13.845	14.695	15.576	16.514	17.421
0.45	9.377	9.920	10.515	11.118	11.858	12.598	13.379	14.196	15.034	15.926
0.50	8.687	9.148	9.660	10.180	10.840	11.502	12.209	12.956	13.707	14.559
0.55	8.132	8.518	8.952	9.400	9.973	10.557	11.188	11.862	12.533	13.328
0.60	7.682	8.007	8.373	8.756	9.245	9.753	10.309	10.909	11.507	12.235
0.65	7.312	7.588	7.898	8.227	8.639	9.077	9.561	10.090	10.621	11.276
0.70	6.996	7.240	7.506	7.791	8.137	8.512	8.930	9.392	9.861	10.442
0.80	6.460	6.676	6.892	7.118	7.368	7.645	7.955	8.301	8.663	9.108
0.90	5.975	6.200	6.406	6.606	6.808	7.023	7.259	7.519	7.799	8.132
1.00	5.501	5.752	5.972	6.172	6.359	6.545	6.738	6.944	7.166	7.417
1.10	5.030	5.310	5.553	5.768	5.962	6.143	6.318	6.495	6.681	6.879
1.20	4.570	4.872	5.139	5.372	5.586	5.775	5.950	6.118	6.285	6.453
1.30	4.131	4.445	4.730	4.982	5.215	5.420	5.601	5.776	5.939	6.096
1.40	3.722	4.038	4.333	4.597	4.849	5.070	5.270	5.451	5.617	5.775
1.50	3.352	3.660	3.956	4.226	4.490	4.725	4.939	5.133	5.308	5.473
1.60	3.023	3.316	3.604	3.874	4.144	4.388	4.611	4.819	5.005	5.180
1.70	2.733	3.006	3.281	3.545	3.814	4.062	4.295	4.511	4.705	4.892
1.80	2.485	2.734	2.992	3.244	3.506	3.753	3.989	4.211	4.413	4.610
1.90	2.271	2.496	2.733	2.971	3.221	3.463	3.697	3.922	4.128	4.332
2.00	2.090	2.290	2.506	2.727	2.963	3.195	3.424	3.647	3.855	4.063

Z / Element / $(\sin\theta)/\lambda[\text{Å}^{-1}]$	31 Ga	32 Ge	33 As	34 Se	35 Br	36 Kr	37 Rb	38 Sr	39 Y	40 Zr
0.00	31.000	32.000	33.000	34.000	35.000	36.000	37.000	38.000	39.000	40.000
0.05	30.308	31.276	32.274	33.280	34.291	35.304	35.948	36.802	37.816	38.847
0.10	28.675	29.534	30.473	31.449	32.450	33.467	33.907	34.458	35.364	36.356
0.15	26.783	27.504	28.307	29.175	30.095	31.055	31.681	32.171	32.904	33.758
0.20	24.935	25.567	26.235	26.962	27.749	28.590	29.368	29.988	30.631	31.363
0.25	23.174	23.791	24.386	25.001	25.658	26.364	27.148	27.863	28.485	29.141
0.30	21.481	22.136	22.724	23.288	23.857	24.453	25.158	25.875	26.483	27.092
0.35	19.847	20.560	21.185	21.751	22.288	22.820	23.432	24.090	24.673	25.243
0.40	18.278	19.047	19.725	20.328	20.874	21.388	21.934	22.522	23.071	23.606
0.45	16.794	17.598	18.326	18.977	19.558	20.087	20.605	21.141	21.660	22.166
0.50	15.410	16.227	16.989	17.682	18.307	18.870	19.391	19.902	20.404	20.892
0.55	14.142	14.947	15.721	16.444	17.107	17.709	18.252	18.764	19.263	19.745
0.60	12.996	13.770	14.535	15.269	15.958	16.594	17.167	17.696	18.204	18.693
0.65	11.974	12.702	13.440	14.166	14.865	15.524	16.125	16.678	17.203	17.706
0.70	11.073	11.745	12.442	13.145	13.837	14.504	15.126	15.702	16.246	16.767
0.80	9.604	10.151	10.741	11.362	12.001	12.645	13.272	13.872	14.443	14.996
0.90	8.510	8.937	9.411	9.928	10.480	11.057	11.645	12.230	12.798	13.361
1.00	7.702	8.028	8.396	8.809	9.262	9.752	10.270	10.806	11.339	11.883
1.10	7.099	7.348	7.631	7.952	8.312	8.711	9.147	9.612	10.088	10.588
1.20	6.633	6.830	7.050	7.299	7.580	7.898	8.252	8.640	9.046	9.486
1.30	6.254	6.419	6.597	6.795	7.016	7.266	7.548	7.863	8.200	8.574
1.40	5.926	6.076	6.231	6.395	6.574	6.773	6.996	7.249	7.523	7.833
1.50	5.627	5.774	5.917	6.063	6.216	6.380	6.562	6.764	6.985	7.238
1.60	5.342	5.493	5.636	5.775	5.913	6.056	6.210	6.376	6.554	6.760
1.70	5.065	5.224	5.372	5.511	5.645	5.778	5.913	6.055	6.205	6.375
1.80	4.792	4.961	5.117	5.262	5.398	5.528	5.656	5.785	5.914	6.059
1.90	4.523	4.702	4.867	5.020	5.162	5.295	5.420	5.544	5.662	5.790
2.00	4.260	4.447	4.621	4.782	4.932	5.071	5.200	5.323	5.440	5.558

Appendix-Table 2A Mean atomic scattering factors for free atoms (*Continued*)

Z	41	42	43	44	45	46	47	48	49	50
Element	Nb	Mo	Tc	Ru	Rh	Pd	Ag	Cd	In	Sn
$(\sin\theta)/\lambda$[Å^{-1}]										
0.00	41.000	42.000	43.000	44.000	45.000	46.000	47.000	48.000	49.000	50.000
0.05	39.970	41.003	41.945	43.061	44.088	45.232	46.139	47.085	47.980	48.934
0.10	37.606	38.656	39.480	40.770	41.834	43.172	43.964	44.797	45.534	46.361
0.15	34.916	35.907	36.658	37.959	39.015	40.357	41.157	41.923	42.603	43.309
0.20	32.305	33.195	33.936	35.088	36.086	37.286	38.154	38.930	39.639	40.302
0.25	29.881	30.665	31.409	32.356	33.267	34.283	35.192	36.007	36.774	37.462
0.30	27.692	28.382	29.104	29.866	30.675	31.523	32.416	33.251	34.059	34.794
0.35	25.760	26.368	27.042	27.662	28.370	29.077	29.910	30.725	31.538	32.303
0.40	24.077	24.620	25.229	25.749	26.363	26.961	27.707	28.468	29.247	30.011
0.45	22.615	23.109	23.651	24.106	24.640	25.153	25.805	26.492	27.209	27.938
0.50	21.336	21.796	22.280	22.696	23.167	23.617	24.181	24.784	25.425	26.096
0.55	20.195	20.638	21.080	21.476	21.900	22.307	22.795	23.320	23.881	24.482
0.60	19.156	19.595	20.012	20.403	20.798	21.177	21.607	22.063	22.552	23.081
0.65	18.187	18.635	19.042	19.438	19.820	20.186	20.575	20.978	21.405	21.868
0.70	17.268	17.732	18.142	18.551	18.932	19.296	19.661	20.027	20.408	20.815
0.80	15.533	16.036	16.477	16.922	17.326	17.711	18.069	18.405	18.736	19.073
0.90	13.915	14.448	14.925	15.405	15.845	16.266	16.651	17.000	17.329	17.646
1.00	12.427	12.968	13.466	13.968	14.440	14.893	15.316	15.698	16.053	16.384
1.10	11.098	11.621	12.116	12.620	13.107	13.580	14.035	14.451	14.840	15.201
1.20	9.945	10.430	10.900	11.385	11.866	12.342	12.813	13.253	13.670	14.062
1.30	8.972	9.404	9.833	10.282	10.740	11.200	11.669	12.116	12.548	12.962
1.40	8.169	8.542	8.919	9.323	9.743	10.173	10.623	11.060	11.492	11.913
1.50	7.516	7.831	8.154	8.506	8.880	9.270	9.687	10.101	10.518	10.933
1.60	6.969	7.251	7.521	7.823	8.148	8.492	8.869	9.249	9.639	10.034
1.70	6.564	6.780	7.004	7.258	7.535	7.833	8.165	8.505	8.860	9.227
1.80	6.216	6.397	6.582	6.794	7.028	7.282	7.569	7.867	8.184	8.516
1.90	5.927	6.080	6.234	6.412	6.608	6.824	7.069	7.326	7.603	7.897
2.00	5.680	5.813	5.946	6.097	6.262	6.443	6.651	6.871	7.110	7.367

Z	51	52	53	54	55	56	57	58	59	60
Element	Sb	Te	I	Xe	Cs	Ba	La	Ce	Pr	Nd
$(\sin\theta)/\lambda$[Å^{-1}]										
0.00	51.000	52.000	53.000	54.000	55.000	56.000	57.000	58.000	59.000	60.000
0.05	49.915	50.906	51.911	52.917	53.527	54.345	55.351	56.385	57.439	58.468
0.10	47.250	48.174	49.142	50.125	50.603	51.122	51.982	53.047	54.281	55.342
0.15	44.056	44.848	45.702	46.588	47.291	47.839	48.523	49.579	50.957	52.022
0.20	40.955	41.623	42.340	43.088	43.888	44.586	45.212	46.233	47.610	48.660
0.25	38.100	38.709	39.333	39.967	40.713	41.456	42.078	43.042	44.323	45.336
0.30	35.465	36.079	36.675	37.251	37.904	38.598	39.212	40.104	41.256	42.210
0.35	33.016	33.663	34.279	34.850	35.440	36.063	36.659	37.474	38.496	39.379
0.40	30.745	31.424	32.075	32.671	33.241	33.818	34.397	35.139	36.042	36.854
0.45	28.663	29.352	30.030	30.656	31.236	31.798	32.370	33.051	33.849	34.596
0.50	26.784	27.458	28.141	28.785	29.382	29.948	30.523	31.154	31.862	32.553
0.55	25.113	25.748	26.412	27.054	27.661	28.238	28.817	29.409	30.040	30.683
0.60	23.646	24.226	24.851	25.470	26.072	26.652	27.231	27.791	28.358	28.960
0.65	22.366	22.885	23.459	24.038	24.619	25.189	25.759	26.289	26.803	27.367
0.70	21.253	21.711	22.228	22.758	23.303	23.851	24.401	24.901	25.370	25.899
0.80	19.424	19.783	20.193	20.618	21.072	21.547	22.031	22.469	22.867	23.325
0.90	17.958	18.262	18.599	18.943	19.310	19.701	20.106	20.481	20.824	21.214
1.00	16.696	16.986	17.293	17.591	17.900	18.224	18.561	18.881	19.182	19.513
1.10	15.537	15.841	16.150	16.438	16.722	17.008	17.300	17.583	17.854	18.139
1.20	14.429	14.759	15.090	15.390	15.676	15.953	16.227	16.491	16.745	17.003
1.30	13.355	13.712	14.072	14.396	14.700	14.988	15.265	15.526	15.776	16.024
1.40	12.321	12.698	13.082	13.432	13.759	14.067	14.362	14.633	14.888	15.138
1.50	11.341	11.726	12.125	12.494	12.845	13.175	13.489	13.776	14.042	14.303
1.60	10.431	10.811	11.214	11.592	11.956	12.305	12.636	12.939	13.218	13.493
1.70	9.602	9.966	10.360	10.736	11.104	11.461	11.807	12.123	12.414	12.704
1.80	8.861	9.201	9.576	9.940	10.303	10.661	11.009	11.333	11.631	11.932
1.90	8.208	8.518	8.868	9.212	9.558	9.907	10.253	10.576	10.878	11.185
2.00	7.642	7.921	8.239	8.556	8.881	9.213	9.550	9.868	10.166	10.473

Appendix-Table 2A Mean atomic scattering factors for free atoms (*Continued*)

Z / Element / $(\sin\theta)/\lambda[\text{Å}^{-1}]$	61 Pm	62 Sm	63 Eu	64 Gd	65 Tb	66 Dy	67 Ho	68 Er	69 Tm	70 Yb
0.00	61.000	62.000	63.000	64.000	65.000	66.000	67.000	68.000	69.000	70.000
0.05	59.497	60.525	61.552	62.557	63.603	64.627	65.627	66.673	67.696	68.717
0.10	56.403	57.463	58.521	59.427	60.634	61.689	62.591	63.798	64.852	65.904
0.15	53.091	54.163	55.228	56.005	57.366	58.437	59.212	60.578	61.648	62.718
0.20	49.720	50.786	51.847	52.588	53.985	55.059	55.803	57.213	58.292	59.371
0.25	46.364	47.406	48.444	49.209	50.549	51.611	52.390	53.750	54.825	55.903
0.30	43.186	44.180	45.176	45.965	47.208	48.240	49.059	50.329	51.384	52.444
0.35	40.289	41.221	42.160	42.951	44.087	45.073	45.908	47.081	48.099	49.127
0.40	37.694	38.559	39.433	40.207	41.241	42.171	42.996	44.075	45.046	46.029
0.45	35.370	36.169	36.980	37.726	38.665	39.536	40.337	41.327	42.246	43.178
0.50	33.269	34.008	34.761	35.479	36.329	37.143	37.914	38.822	39.686	40.565
0.55	31.349	32.036	32.737	33.428	34.199	34.958	35.699	36.531	37.342	38.169
0.60	29.581	30.222	30.877	31.543	32.243	32.953	33.664	34.425	35.187	35.964
0.65	27.948	28.547	29.161	29.802	30.438	31.103	31.786	32.483	33.198	33.929
0.70	26.442	27.002	27.576	28.192	28.772	29.394	30.049	30.688	31.359	32.045
0.80	23.796	24.281	24.781	25.335	25.822	26.366	26.958	27.497	28.086	28.690
0.90	21.616	22.030	22.459	22.940	23.353	23.821	24.343	24.800	25.311	25.837
1.00	19.853	20.202	20.565	20.970	21.323	21.721	22.167	22.556	22.995	23.447
1.10	18.430	18.728	19.035	19.372	19.675	20.011	20.385	20.718	21.089	21.474
1.20	17.262	17.523	17.789	18.072	18.338	18.623	18.934	19.221	19.535	19.860
1.30	16.266	16.507	16.747	16.995	17.234	17.483	17.746	17.998	18.266	18.542
1.40	15.378	15.613	15.841	16.072	16.296	16.522	16.753	16.980	17.215	17.454
1.50	14.551	14.790	15.020	15.247	15.465	15.680	15.895	16.107	16.321	16.536
1.60	13.755	14.005	14.245	14.477	14.697	14.913	15.123	15.329	15.533	15.735
1.70	12.980	13.243	13.494	13.741	13.968	14.190	14.406	14.612	14.815	15.013
1.80	12.220	12.497	12.763	13.022	13.259	13.491	13.718	13.929	14.137	14.338
1.90	11.481	11.767	12.044	12.317	12.564	12.808	13.047	13.267	13.483	13.691
2.00	10.773	11.064	11.345	11.631	11.886	12.141	12.392	12.621	12.847	13.064

Z / Element / $(\sin\theta)/\lambda[\text{Å}^{-1}]$	71 Lu	72 Hf	73 Ta	74 W	75 Re	76 Os	77 Ir	78 Pt	79 Au	80 Hg
0.00	71.000	72.000	73.000	74.000	75.000	76.000	77.000	78.000	79.000	80.000
0.05	69.707	70.723	71.745	72.767	73.788	74.810	75.832	76.914	77.936	78.899
0.10	66.789	67.757	68.758	69.778	70.799	71.832	72.872	74.086	75.135	76.018
0.15	63.478	64.326	65.229	66.172	67.126	68.107	69.108	70.343	71.380	72.198
0.20	60.103	60.870	61.675	62.519	63.378	64.269	65.189	66.317	67.296	68.088
0.25	56.683	57.434	58.189	58.961	59.742	60.548	61.380	62.344	63.241	64.029
0.30	53.282	54.054	54.799	55.536	56.270	57.013	57.773	58.582	59.395	60.177
0.35	49.998	50.796	51.548	52.274	52.986	53.692	54.401	55.094	55.835	56.600
0.40	46.906	47.717	48.479	49.205	49.910	50.596	51.274	51.897	52.581	53.318
0.45	44.038	44.849	45.615	46.344	47.048	47.726	48.387	48.977	49.622	50.326
0.50	41.398	42.197	42.962	43.691	44.396	45.072	45.726	46.308	46.929	47.601
0.55	38.970	39.752	40.508	41.236	41.940	42.617	43.269	43.860	44.469	45.113
0.60	36.733	37.494	38.238	38.960	39.662	40.340	40.994	41.601	42.207	42.829
0.65	34.666	35.404	36.132	36.846	37.544	38.222	38.878	39.502	40.110	40.718
0.70	32.752	33.465	34.175	34.878	35.569	36.244	36.901	37.539	38.153	38.753
0.80	29.334	29.992	30.658	31.327	31.993	32.654	33.305	33.958	34.581	35.176
0.90	26.413	27.008	27.618	28.238	28.865	29.495	30.125	30.766	31.387	31.980
1.00	23.950	24.473	25.016	25.576	26.148	26.732	27.323	27.930	28.530	29.112
1.10	21.902	22.352	22.823	23.313	23.821	24.345	24.882	25.437	25.998	26.554
1.20	20.219	20.598	20.998	21.418	21.856	22.314	22.789	23.281	23.789	24.303
1.30	18.842	19.159	19.494	19.847	20.219	20.610	21.019	21.445	21.892	22.354
1.40	17.709	17.975	18.256	18.552	18.864	19.194	19.541	19.902	20.287	20.692
1.50	16.759	16.988	17.228	17.478	17.742	18.019	18.312	18.616	18.943	19.290
1.60	15.939	16.145	16.356	16.575	16.801	17.038	17.287	17.545	17.821	18.116
1.70	15.208	15.403	15.598	15.796	15.998	16.206	16.422	16.644	16.880	17.131
1.80	14.534	14.727	14.916	15.104	15.293	15.483	15.678	15.875	16.081	16.298
1.90	13.894	14.091	14.282	14.469	14.653	14.835	15.018	15.202	15.388	15.581
2.00	13.277	13.481	13.679	13.871	14.057	14.239	14.418	14.595	14.770	14.949

Appendix-Table 2A Mean atomic scattering factors for free atoms (*Continued*)

Z / Element / $(\sin\theta)/\lambda[Å^{-1}]$	81 Tl	82 Pb	83 Bi	84 Po	85 At	86 Rn	87 Fr	88 Ra	89 Ac	90 Th
0.00	81.000	82.000	83.000	84.000	85.000	86.000	87.000	88.000	89.000	90.000
0.05	79.798	80.750	81.700	82.669	83.654	84.649	85.286	86.104	87.077	88.085
0.10	76.773	77.607	78.438	79.312	80.226	81.169	81.666	82.202	82.985	83.867
0.15	72.912	73.645	74.365	75.117	75.908	76.737	77.430	77.990	78.605	79.294
0.20	68.830	69.530	70.203	70.871	71.553	72.258	73.035	73.728	74.326	74.922
0.25	64.807	65.523	66.204	66.852	67.481	68.109	68.841	69.576	70.208	70.798
0.30	60.970	61.712	62.425	63.093	63.725	64.332	64.996	65.696	66.345	66.951
0.35	57.383	58.138	58.875	59.575	60.236	60.862	61.489	62.140	62.785	63.405
0.40	54.072	54.820	55.563	56.283	56.974	57.631	58.256	58.879	59.517	60.143
0.45	51.041	51.766	52.495	53.215	53.921	54.604	55.242	55.862	56.497	57.127
0.50	48.276	48.969	49.669	50.373	51.075	51.767	52.420	53.048	53.684	54.317
0.55	45.753	46.411	47.077	47.752	48.435	49.119	49.777	50.413	51.050	51.684
0.60	43.442	44.069	44.700	45.343	45.997	46.659	47.310	47.948	48.580	49.211
0.65	41.313	41.914	42.517	43.127	43.750	44.384	45.017	45.646	46.268	46.889
0.70	39.337	39.921	40.501	41.085	41.678	42.281	42.891	43.504	44.110	44.716
0.80	35.755	36.322	36.879	37.430	37.980	38.533	39.095	39.664	40.229	40.795
0.90	32.561	33.127	33.680	34.220	34.751	35.277	35.804	36.335	36.863	37.391
1.00	29.687	30.252	30.805	31.344	31.872	32.389	32.900	33.408	33.912	34.413
1.10	27.109	27.662	28.208	28.744	29.271	29.787	30.292	30.790	31.283	31.770
1.20	24.824	25.350	25.875	26.397	26.915	27.426	27.926	28.418	28.906	29.387
1.30	22.827	23.313	23.804	24.298	24.794	25.291	25.779	26.263	26.744	27.219
1.40	21.110	21.546	21.992	22.446	22.909	23.379	23.845	24.312	24.779	25.244
1.50	19.652	20.034	20.429	20.836	21.256	21.689	22.123	22.564	23.008	23.454
1.60	18.424	18.754	19.097	19.453	19.826	20.215	20.608	21.014	21.427	21.846
1.70	17.394	17.674	17.969	18.277	18.602	18.944	19.295	19.660	20.036	20.421
1.80	16.524	16.764	17.017	17.281	17.562	17.859	18.165	18.488	18.823	19.170
1.90	15.780	15.989	16.207	16.435	16.677	16.934	17.199	17.481	17.776	18.083
2.00	15.131	15.317	15.510	15.711	15.922	16.143	16.377	16.623	16.880	17.149

Z / Element / $(\sin\theta)/\lambda[Å^{-1}]$	91 Pa	92 U
0.00	91.000	92.000
0.05	89.144	90.180
0.10	85.066	86.130
0.15	80.563	81.595
0.20	76.115	77.080
0.25	71.829	72.712
0.30	67.810	68.607
0.35	64.121	64.838
0.40	60.758	61.409
0.45	57.679	58.283
0.50	54.836	55.410
0.55	52.191	52.748
0.60	49.719	50.268
0.65	47.405	47.950
0.70	45.241	45.784
0.80	41.333	41.869
0.90	37.930	38.454
1.00	34.946	35.458
1.10	32.292	32.794
1.20	29.897	30.391
1.30	27.714	28.199
1.40	25.720	26.192
1.50	23.905	24.360
1.60	22.266	22.699
1.70	20.807	21.207
1.80	19.518	19.886
1.90	18.394	18.723
2.00	17.423	17.713

†[Reproduced with permission from *International Tables for Crystallography*, Vol. C, *Mathematical, Physical and Chemical Tables*, 2nd ed. (A. J. C. Wilson, E. Prince eds.), pp. 549–558, IUCr, D. Reidel Pub., Dordrecht (1999)]

Appendix-Table 2B Coefficients for analytical approximation to the scattering factors of Table 2A

	a_1	b_1	a_2	b_2	a_3	b_3	a_4	b_4	c
1 H	0.489918	20.6593	0.262003	7.74039	0.196767	49.5519	0.049879	2.20159	0.001305
2 He	0.873400	9.10370	0.630900	3.35680	0.311200	22.9276	0.178000	0.982100	0.006400
3 Li	1.12820	3.95460	0.750800	1.05240	0.617500	85.3905	0.465300	168.261	0.037700
4 Be	1.59190	43.6427	1.12780	1.86230	0.539100	103.483	0.702900	0.542000	0.038500
5 B	2.05450	23.2185	1.33260	1.02100	1.09790	60.3498	0.706800	0.140300	-0.19320
6 C	2.31000	20.8439	1.02000	10.2075	1.58860	0.568700	0.865000	51.6512	0.215600
7 N	12.2126	0.005700	3.13220	9.89330	2.01250	28.9975	1.16630	0.582600	-11.529
8 O	3.04850	13.2771	2.28680	5.70110	1.54630	0.323900	0.867000	32.9089	0.250800
9 F	3.53920	10.2825	2.64120	4.29440	1.51700	0.261500	1.02430	26.1476	0.277600
10 Ne	3.95530	8.40420	3.11250	3.42620	1.45460	0.230600	1.12510	21.7184	0.351500
11 Na	4.76260	3.28500	3.17360	8.84220	1.26740	0.313600	1.11280	129.424	0.676000
12 Mg	5.42040	2.82750	2.17350	79.2611	1.22690	0.380800	2.30730	7.19370	0.858400
13 Al	6.42020	3.03870	1.90020	0.742600	1.59360	31.5472	1.96460	85.0886	1.11510
14 Si	6.29150	2.43860	3.03530	32.3337	1.98910	0.678500	1.54100	81.6937	1.14070
15 P	6.43450	1.90670	4.17910	27.1570	1.78000	0.526000	1.49080	68.1645	1.11490
16 S	6.90530	1.46790	5.20340	22.2151	1.43790	0.253600	1.58630	56.1720	0.866900
17 Cl	11.4604	0.010400	7.19640	1.16620	6.25560	18.5194	1.64550	47.7784	-9.5574
18 Ar	7.48450	0.907200	6.77230	14.8407	0.653900	43.8983	1.64420	33.3929	1.44450
19 K	8.21860	12.7949	7.43980	0.774800	1.05190	213.187	0.865900	41.6841	1.42280
20 Ca	8.62660	10.4421	7.38730	0.659900	1.58990	85.7484	1.02110	178.437	1.37510
21 Sc	9.18900	9.02130	7.36790	0.572900	1.64090	136.108	1.46800	51.3531	1.33290
22 Ti	9.75950	7.85080	7.35580	0.500000	1.69910	35.6338	1.90210	116.105	1.28070
23 V	10.2971	6.86570	7.35110	0.438500	2.07030	26.8938	2.05710	102.478	1.21990
24 Cr	10.6406	6.10380	7.35370	0.392000	3.32400	20.2626	1.49220	98.7399	1.18320
25 Mn	11.2819	5.34090	7.35730	0.343200	3.01930	17.8674	2.24410	83.7543	1.08960
26 Fe	11.7695	4.76110	7.35730	0.307200	3.52220	15.3535	2.30450	76.8805	1.03690
27 Co	12.2841	4.27910	7.34090	0.278400	4.00340	13.5359	2.34880	71.1692	1.01180
28 Ni	12.8376	3.87850	7.29200	0.256500	4.44380	12.1763	2.38000	66.3421	1.0341
29 Cu	13.3380	3.58280	7.16760	0.247000	5.61580	11.3966	1.67350	64.8126	1.19100
30 Zn	14.0743	3.26550	7.03180	0.233300	5.16520	10.3163	2.41000	58.7097	1.30410
31 Ga	15.2354	3.06690	6.70060	0.241200	4.35910	10.7805	2.96230	61.4135	1.71890
32 Ge	16.0816	2.85090	6.37470	0.251600	3.70680	11.4468	3.68300	54.7625	2.13130
33 As	16.6723	2.63450	6.07010	0.264700	3.43130	12.9479	4.27790	47.7972	2.53100
34 Se	17.0006	2.40980	5.81960	0.272600	3.97310	15.2372	4.35430	43.8163	2.84090
35 Br	17.1789	2.17230	5.23580	16.5796	5.63770	0.260900	3.98510	41.4328	2.95570
36 Kr	17.3555	1.93840	6.72860	16.5623	5.54930	0.226100	3.53750	39.3972	2.82500
37 Rb	17.1784	1.78880	9.64350	17.3151	5.13990	0.274800	1.52920	164.934	3.48730
38 Sr	17.5663	1.55640	9.81840	14.0988	5.42200	0.166400	2.66940	132.376	2.50640
39 Y	17.7760	1.40290	10.2946	12.8006	5.72629	0.125599	3.26588	104.354	1.91213
40 Zr	17.8765	1.27618	10.9480	11.9160	5.41732	0.117622	3.65721	87.6627	2.06929
41 Nb	17.6142	1.18865	12.0144	11.7660	4.04183	0.204785	3.53346	69.7957	3.75591
42 Mo	3.70250	0.277200	17.2356	1.09580	12.8876	11.0040	3.74290	61.6584	4.38750
43 Tc	19.1301	0.864132	11.0948	8.14487	4.64901	21.5707	2.71263	86.8472	5.40428
44 Ru	19.2674	0.808520	12.9182	8.43467	4.86337	24.7997	1.56756	94.2928	5.37874
45 Rh	19.2957	0.751536	14.3501	8.21758	4.73425	25.8749	1.28918	98.6062	5.32800
46 Pd	19.3319	0.698655	15.5017	7.98929	5.29537	25.2052	0.605844	76.8986	5.26593
47 Ag	19.2808	0.644600	16.6885	7.47260	4.80450	24.6605	1.04630	99.8156	5.17900
48 Cd	19.2214	0.594600	17.6444	6.90890	4.46100	24.7008	1.60290	87.4825	5.06940
49 In	19.1624	0.547600	18.5596	6.37760	4.29480	25.8499	2.03960	92.8029	4.93910
50 Sn	19.1889	5.83030	19.1005	0.503100	4.45850	26.8909	2.46630	83.9571	4.78210
51 Sb	19.6418	5.30340	19.0455	0.460700	5.03710	27.9074	2.68270	75.2825	4.59090
52 Te	19.9644	4.81742	19.0138	0.420885	6.14487	28.5284	2.52390	70.8403	4.35200
53 I	20.1472	4.34700	18.9949	0.381400	7.51380	27.7660	2.27350	66.8776	4.07120
54 Xe	20.2933	3.92820	19.0298	0.344000	8.97670	26.4659	1.99000	64.2658	3.71180
55 Cs	20.3892	3.56900	19.1062	0.310700	10.6620	24.3879	1.49530	213.904	3.33520

Appendix-Table 2B Coefficients for analytical approximation to scattering factors (*Continued*)

	a_1	b_1	a_2	b_2	a_3	b_3	a_4	b_4	c
56 Ba	20.3361	3.21600	19.2970	0.275600	10.8880	20.2073	2.69590	167.202	2.77310
57 La	20.5780	2.94817	19.5990	0.244475	11.3727	18.7726	3.28719	133.124	2.14678
58 Ce	21.1671	2.81219	19.7695	0.226836	11.8513	17.6083	3.33049	127.113	1.86264
59 Pr	22.0440	2.77393	19.6697	0.222087	12.3856	16.7669	2.82428	143.644	2.05830
60 Nd	22.6845	2.66248	19.6847	0.210628	12.7740	15.8850	2.85137	137.903	1.98486
61 Pm	23.3405	2.56270	19.6095	0.202088	13.1235	15.1009	2.87516	132.721	2.02876
62 Sm	24.0042	2.47274	19.4258	0.196451	13.4396	14.3996	2.89604	128.007	2.20963
63 Eu	24.6274	2.38790	19.0886	0.194200	13.7603	13.7546	2.92270	123.174	2.57450
64 Gd	25.0709	2.25341	19.0798	0.181951	13.8518	12.9331	3.54545	101.398	2.41960
65 Tb	25.8976	2.24256	18.2185	0.196143	14.3167	12.6648	2.95354	115.362	3.58324
66 Dy	26.5070	2.18020	17.6383	0.202172	14.5596	12.1899	2.96577	111.874	4.29728
67 Ho	26.9049	2.07051	17.2940	0.197940	14.5583	11.4407	3.63837	92.6566	4.56796
68 Er	27.6563	2.07356	16.4285	0.223545	14.9779	11.3604	2.98233	105.703	5.92046
69 Tm	28.1819	2.02859	15.8851	0.238849	15.1542	10.9975	2.98706	102.961	6.75621
70 Yb	28.6641	1.98890	15.4345	0.257119	15.3087	10.6647	2.98963	100.417	7.56672
71 Lu	28.9476	1.90182	15.2208	9.98519	15.1000	0.261033	3.71601	84.3298	7.97628
72 Hf	29.1440	1.83262	15.1726	9.59990	14.7586	0.275116	4.30013	72.0290	8.58154
73 Ta	29.2024	1.77333	15.2293	9.37046	14.5135	0.295977	4.76492	63.3644	9.24354
74 W	29.0818	1.72029	15.4300	9.22590	14.4327	0.321703	5.11982	57.0560	9.88750
75 Re	28.7621	1.67191	15.7189	9.09227	14.5564	0.350500	5.44174	52.0861	10.4720
76 Os	28.1894	1.62903	16.1550	8.97948	14.9305	0.382661	5.67589	48.1647	11.0005
77 Ir	27.3049	1.59279	16.7296	8.86553	15.6115	0.417916	5.83377	45.0011	11.4722
78 Pt	27.0059	1.51293	17.7639	8.81174	15.7131	0.424593	5.78370	38.6103	11.6883
79 Au	16.8819	0.461100	18.5913	8.62160	25.5582	1.48260	5.86000	36.3956	12.0658
80 Hg	20.6809	0.545000	19.0417	8.44840	21.6575	1.57290	5.96760	38.3246	12.6089
81 Tl	27.5446	0.655150	19.1584	8.70751	15.5380	1.96347	5.52593	45.8149	13.1746
82 Pb	31.0617	0.690200	13.0637	2.35760	18.4420	8.61800	5.96960	47.2579	13.4118
83 Bi	33.3689	0.704000	12.9510	2.92380	16.5877	8.79370	6.46920	48.0093	13.5782
84 Po	34.6726	0.700999	15.4733	3.55078	13.1138	9.55642	7.02588	47.0045	13.6770
85 At	35.3163	0.685870	19.0211	3.97458	9.49887	11.3824	7.42518	45.4715	13.7108
86 Rn	35.5631	0.663100	21.2816	4.06910	8.00370	14.0422	7.44330	44.2473	13.6905
87 Fr	35.9299	0.646453	23.0547	4.17619	12.1439	23.1052	2.11253	150.645	13.7247
88 Ra	35.7630	0.616341	22.9064	3.87135	12.4739	19.9887	3.21097	142.325	13.6211
89 Ac	35.6597	0.589092	23.1032	3.65155	12.5977	18.5990	4.08655	117.020	13.5266
90 Th	35.5645	0.563359	23.4219	3.46204	12.7473	17.8309	4.80703	99.1722	13.4314
91 Pa	35.8847	0.547751	23.2948	3.41519	14.1891	16.9235	4.17287	105.251	13.4287
92 U	36.0228	0.529300	23.4128	3.32530	14.9491	16.0927	4.18800	100.613	13.3966

†[Reproduced with permission from *International Tables for Crystallography*, Vol. C, *Mathematical, Physical and Chemical Tables*, 2nd ed. (A. J. C. Wilson, E. Prince eds.), pp. 572–574, For the range $2.0 < (\sin\theta)/\lambda < 6.0$ Å^{-1} see pp. 559, 575. IUCr., D. Reidel Pub., Dordrecht (1999)]

Appendix-Table 3 The incoherent scattering function for elements up to $Z = 55^{\dagger}$

Element	(sinθ)/λ [Å^{-1}] 0.10	0.20	0.30	0.40	0.50	0.60	0.70	0.80	0.90	1.00	1.50	2.00
1 H	0.343	0.769	0.937	0.983	0.995	0.998	0.994	0.999	1.000	1.000	1.000	1.000
2 He	0.296	0.881	1.362	1.657	1.817	1.902	1.947	1.970	1.983	1.990	1.999	2.000
3 Li	1.033	1.418	1.795	2.143	2.417	2.613	2.746	2.834	2.891	2.928	2.989	2.998
4 Be	1.170	2.121	2.471	2.744	3.005	3.237	3.429	3.579	3.693	3.777	3.954	3.989
5 B	1.147	2.531	3.190	3.499	3.732	3.948	4.146	4.320	4.469	4.590	4.895	4.973
6 C	1.039	2.604	3.643	4.184	4.478	4.690	4.878	5.051	5.208	5.348	5.781	5.930
7 N	1.08	2.858	4.097	4.792	5.182	5.437	5.635	5.809	5.968	6.113	6.630	6.860
8 O	0.977	2.799	4.293	5.257	5.828	6.175	6.411	6.596	6.755	6.901	7.462	7.764
9 F	0.880	2.691	4.347	5.552	6.339	6.832	7.151	7.376	7.552	7.703	8.288	8.648
10 Ne	0.812	2.547	4.269	5.644	6.640	7.320	7.774	8.085	8.312	8.490	9.113	9.517
11 Na	1.503	2.891	4.431	5.804	6.903	7.724	8.313	8.729	9.028	9.252	9.939	10.376
12 Mg	2.066	3.444	4.771	6.064	7.181	8.086	8.784	9.304	9.689	9.975	10.766	11.229
13 Al	2.264	4.047	5.250	6.435	7.523	8.459	9.225	9.830	10.296	10.652	11.592	12.083
14 Si	2.293	4.520	5.808	6.903	7.937	8.867	9.667	10.330	10.864	11.286	12.408	12.937
15 P	2.206	4.732	6.312	7.435	8.419	9.323	10.131	10.827	11.411	11.888	13.209	13.790
16 S	2.151	4.960	6.795	8.002	8.960	9.829	10.626	11.336	11.952	12.472	13.990	14.641
17 Cl	2.065	5.074	7.182	8.553	9.539	10.382	11.158	11.867	12.499	13.050	14.750	15.487
18 Ar	1.956	5.033	7.377	8.998	10.106	10.967	11.726	12.424	13.061	13.629	15.489	16.324
19 K	2.500	5.301	7.652	9.405	10.650	11.568	12.329	13.014	13.645	14.220	16.212	17.152
20 Ca	3.105	5.690	7.981	9.790	11.157	12.163	12.953	13.635	14.256	14.830	16.921	17.970
21 Sc	3.136	5.801	8.169	10.071	11.561	12.648	13.545	14.256	14.885	15.460	17.630	18.782
22 Ti	3.114	5.860	8.312	10.304	11.901	13.140	14.093	14.856	15.509	16.095	18.334	19.585
23 V	3.067	5.858	8.375	10.454	12.156	13.514	14.574	15.413	16.111	16.721	19.032	20.379
24 Cr	2.609	5.577	8.206	10.415	12.264	13.770	14.960	15.902	16.670	17.323	19.730	21.168
25 Mn	2.949	5.791	8.380	10.604	12.486	14.062	15.346	16.376	17.211	17.910	20.411	21.938
26 Fe	2.891	5.781	8.432	10.733	12.687	14.343	15.716	16.831	17.737	18.488	21.097	22.704
27 Co	2.832	5.764	8.469	10.844	12.867	14.596	16.050	17.249	18.229	19.039	21.777	23.462
28 Ni	2.772	5.726	8.461	10.894	12.980	14.780	16.317	17.602	18.664	19.543	22.445	24.211
29 Cu	2.348	5.455	8.310	10.778	12.942	14.847	16.494	17.885	19.043	20.002	23.107	24.957
30 Zn	2.654	5.631	8.388	10.901	13.094	15.020	16.709	18.163	19.395	20.427	23.745	25.683
31 Ga	2.791	5.939	8.599	11.082	13.290	15.233	16.947	18.445	19.734	20.831	24.370	26.400
32 Ge	2.839	6.229	8.912	11.338	13.536	15.486	17.215	18.741	20.074	21.224	24.983	27.109
33 As	2.793	6.365	9.236	11.658	13.828	15.775	17.511	19.056	20.420	21.612	25.583	27.810
34 Se	2.799	6.589	9.601	12.033	14.168	16.098	17.835	19.391	20.778	22.003	26.171	28.504
35 Br	2.771	6.748	9.940	12.440	14.552	16.456	18.185	19.747	21.149	22.399	26.747	29.190
36 Kr	2.703	6.760	10.157	12.828	14.969	16.849	18.562	20.123	21.535	22.804	27.313	29.870
37 Rb	3.225	7.062	10.431	13.206	15.410	17.282	18.974	20.526	21.940	23.221	27.871	30.543
38 Sr	3.831	7.464	10.746	13.576	15.860	17.745	19.420	20.956	22.367	23.654	28.423	31.210
39 Y	3.999	7.700	11.010	13.899	16.279	18.215	19.891	21.416	22.820	24.110	28.970	31.870
40 Zr	4.064	7.879	11.236	14.176	16.658	18.672	20.373	21.895	23.294	24.583	29.517	32.522
41 Nb	3.672	7.684	11.213	14.317	16.949	19.081	20.844	22.386	23.787	25.077	30.067	33.167
42 Mo	3.625	7.690	11.260	14.444	17.196	19.455	21.300	22.877	24.288	25.581	30.620	33.808
43 Tc	3.987	7.984	11.512	14.653	17.456	19.816	21.748	23.370	24.797	26.093	31.173	34.447
44 Ru	3.559	7.857	11.531	14.782	17.685	20.150	22.172	23.855	25.312	26.621	31.740	35.081
45 Rh	3.499	7.863	11.591	14.883	17.858	20.428	22.557	24.318	25.819	27.148	32.309	35.715
46 Pd	3.103	7.725	11.441	14.824	17.943	26.653	22.904	24.756	26.316	27.677	32.888	36.349
47 Ag	3.362	7.785	11.598	14.969	18.082	20.858	23.212	25.162	26.792	28.195	33.465	36.983
48 Cd	3.700	7.980	11.812	15.185	18.263	21.064	23.501	25.546	27.252	28.705	34.046	37.618
49 In	3.852	8.297	12.083	15.444	18.489	21.288	23.779	25.906	27.691	29.203	34.634	38.255
50 Sn	3.917	8.615	12.415	15.746	18.760	21.541	24.059	26.252	28.113	29.687	35.226	38.894
51 Sb	3.871	8.811	12.777	16.088	19.067	21.823	24.349	26.590	28.518	30.157	35.822	39.536
52 Te	3.097	9.076	13.171	16.466	19.407	22.134	25.655	26.927	28.912	30.613	36.422	40.181
53 I	3.903	9.287	13.564	16.876	19.227	22.471	24.980	27.269	29.298	31.056	37.024	40.827
54 Xe	3.841	9.340	13.892	17.307	20.175	22.833	25.324	27.619	29.680	31.488	37.628	41.477
55 Cs	4.320	9.615	14.217	17.753	20.612	23.228	25.691	27.981	30.064	31.914	38.232	42.129

†[Reproduced with permission from *International Tables for Crystallography*, Vol. C, *Mathematical, Physical and Chemical Tables,* 2nd ed. (A.J.C. Wilson, E. Prince eds.), p. 652, IUCr., D. Reidel Pub., Dordrecht (1999)]

Appendix-Table 4 Real and imaginary dispersion corrections, $\Delta f'$ and $\Delta f''$ (f' and f'') for atomic scattering factors[†] for selected radiations [Å]

Elements	Cr 2.289620		Fe 1.935970		Cu*K*α 1.540520		Mo*K*α 0.709260		Ag*K*α 0.559360	
	$\Delta f'$	$\Delta f''$	$\Delta f'$	$\Delta f''$	$\Delta f'$	$\Delta f''$	$\Delta f'$	$\Delta f''$	$\Delta f'$	$\Delta f''$
3 Li	0.0023	0.0008	0.0015	0.0006	0.0008	0.0003	-0.0003	0.0001	-0.0004	0.0000
4 Be	0.0083	0.0033	0.0060	0.0023	0.0038	0.0014	0.0005	0.0002	0.0001	0.0001
5 B	0.0190	0.0094	0.0140	0.0065	0.0090	0.0039	0.0013	0.0007	0.0004	0.0004
6 C	0.0364	0.0213	0.0273	0.0148	0.0181	0.0091	0.0033	0.0016	0.0015	0.0009
7 N	0.0606	0.0416	0.0461	0.0293	0.0311	0.0180	0.0061	0.0033	0.0030	0.0019
8 O	0.0928	0.0731	0.0716	0.0518	0.0492	0.0322	0.0106	0.0060	0.0056	0.0036
9 F	0.1324	0.1192	0.1037	0.0851	0.0727	0.0534	0.0171	0.0103	0.0096	0.0061
10 Ne	0.1793	0.1837	0.1426	0.1318	0.1019	0.0833	0.0259	0.0164	0.0152	0.0098
11 Na	0.2295	0.2699	0.1857	0.1957	0.1353	0.1239	0.0362	0.0249	0.0218	0.0150
12 Mg	0.2778	0.3812	0.2309	0.2765	0.1719	0.1771	0.0486	0.0363	0.0298	0.0220
13 Al	0.3260	0.5212	0.2774	0.3807	0.2130	0.2455	0.0645	0.0514	0.0406	0.0313
14 Si	0.3647	0.6921	0.3209	0.5081	0.2541	0.3302	0.0817	0.0704	0.0522	0.0431
15 P	0.3898	0.8984	0.3592	0.6628	0.2955	0.4335	0.1023	0.0942	0.0067	0.0580
16 S	0.3899	1.1410	0.3848	0.8457	0.3331	0.5567	0.1246	0.1234	0.0826	0.0763
17 Cl	0.3508	1.4222	0.3920	1.0596	0.3639	0.7018	0.1484	0.1585	0.0998	0.0984
18 Ar	0.2609	1.7458	0.3696	1.3087	0.3843	0.8717	0.1743	0.2003	0.1191	0.1249
19 K	0.0914	2.1089	0.3068	1.5888	0.3868	1.0657	0.2009	0.2494	0.1399	0.1562
20 Ca	-0.1987	2.5138	0.1867	1.9032	0.3641	1.2855	0.2262	0.3064	0.1611	0.1926
21 Sc	-0.6935	2.9646	-0.0120	2.2557	0.3119	1.5331	0.2519	0.3716	0.1829	0.2348
22 Ti	-1.6394	3.4538	-0.3318	2.6425	0.2191	1.8069	0.2776	0.4457	0.2060	0.2830
23 V	-4.4818	0.4575	-0.8645	3.0644	0.0687	2.1097	0.3005	0.5294	0.2276	0.3376
24 Cr	-2.1308	0.5468	-1.9210	3.5251	-0.1635	2.4439	0.3209	0.6236	0.2496	0.3992
25 Mn	-1.5980	0.6479	-3.5716	0.4798	-0.5229	2.8052	0.3368	0.7283	0.2704	0.4681
26 Fe	-1.2935	0.7620	-2.0554	0.5649	-1.1336	3.1974	0.3463	0.8444	0.2886	0.5448
27 Co	-1.0738	0.8897	-1.5743	0.6602	-2.3653	3.6143	0.3494	0.9721	0.3050	0.6296
28 Ni	-0.9005	1.0331	-1.2894	0.7671	-3.0029	0.5091	0.3393	1.1124	0.3147	0.7232
29 Cu	-0.7338	1.1930	-1.0699	0.8864	-1.9646	0.5888	0.3201	1.2651	0.3240	0.8257
30 Zn	-0.6166	1.3712	-0.9134	1.0193	-1.5491	0.6778	0.2879	1.4301	0.3242	0.9375
31 Ga	-0.4989	1.5674	-0.7701	1.1663	-1.2846	0.7763	0.2307	1.6083	0.3179	1.0589
32 Ge	-0.3858	1.7841	-0.6412	1.3291	-1.0885	0.8855	0.1547	1.8001	0.3016	1.1903
33 As	-0.2871	2.0194	-0.5260	1.5069	-0.9300	1.0051	0.0499	2.0058	0.2758	1.3314
34 Se	-0.1919	2.2784	-0.4179	1.7027	-0.7943	1.1372	-0.0929	2.2259	0.2367	1.4831
35 Br	-0.1095	2.5578	-0.3244	1.9140	-0.6763	1.2805	-0.2901	2.4595	0.1811	1.6452
36 Kr	-0.0316	2.8669	-0.2303	2.1472	-0.5657	1.4385	-0.5574	2.7079	0.1607	1.8192
37 Rb	0.0247	3.1954	-0.1516	2.3960	-0.4688	1.6079	-0.9393	2.9676	0.0068	2.0025
38 Sr	0.1037	3.6029	-0.0489	2.7060	-0.3528	1.8200	-1.5307	3.2498	-0.1172	2.2025
39 Y	0.1263	3.9964	0.0138	3.0054	-0.2670	2.0244	-2.7962	3.5667	-0.2789	2.4099
40 Zr	0.1338	4.4226	0.0659	3.3301	-0.1862	2.2449	-2.9673	0.5597	-0.5364	2.6141
41 Nb	0.1211	4.8761	0.1072	3.6768	-0.1121	2.4826	-2.0727	0.6215	-0.8282	2.8404
42 Mo	0.0801	5.3484	0.1301	4.0388	-0.0483	2.7339	-1.6832	0.6857	-1.2703	3.0978
43 Tc	-0.0025	5.8597	0.1314	4.4331	0.0057	3.0049	-1.4390	0.7593	-2.0087	3.3490
44 Ru	-0.1091	6.4069	0.1220	4.8540	0.0552	3.2960	-1.2594	0.8363	-5.3630	3.6506
45 Rh	-0.2630	6.9820	0.0861	5.2985	0.0927	3.6045	-1.1178	0.9187	-2.5280	0.5964
46 Pd	-0.4640	7.5938	0.0279	5.7719	0.1215	3.9337	-0.9988	1.0072	-1.9556	0.6546
47 Ag	-0.7387	8.2358	-0.0700	6.2709	0.1306	4.2820	-0.8971	1.1015	-1.6473	0.7167
48 Cd	-1.1086	8.9174	-0.2163	6.8017	0.1185	4.6533	-0.8075	1.2024	-1.4396	0.7832
49 In	-1.5975	9.6290	-0.4165	7.3594	0.0822	5.0449	-0.7276	1.3100	-1.2843	0.8542
50 Sn	-2.2019	10.3742	-0.6686	7.9473	0.0259	5.4591	-0.6537	1.4246	-1.1587	0.9299
51 Sb	-3.0637	11.1026	-0.9868	8.5620	-0.0562	5.8946	-0.5866	1.5461	-1.0547	1.0104
52 Te	-4.2407	11.8079	-1.4022	9.2067	-0.1759	6.3531	-0.5308	1.6751	-0.9710	1.0960
53 I	-5.6353	12.6156	1.9032	9.8852	-0.3257	6.8362	-0.4742	1.8119	-0.8919	1.1868
54 Xe	-8.1899	11.7407	-2.6313	10.5776	-0.5179	7.3500	-0.4205	1.9578	-0.8200	1.2838
55 Cs	-10.3310	12.8551	-3.5831	11.2902	-0.7457	7.9052	-0.3680	2.1192	-0.7527	1.3916

Appendix-Table 4 Real and imaginary dispersion corrections, $\Delta f'$ and $\Delta f''$ (f' and f'') for atomic scattering factors[†] for selected radiations [Å] (*Continued*)

Elements	Cr 2.289620		Fe 1.935970		Cu*K*α 1.540520		Mo*K*α 0.709260		Ag*K*α 0.559360	
	$\Delta f'$	$\Delta f''$	$\Delta f'$	$\Delta f''$	$\Delta f'$	$\Delta f''$	$\Delta f'$	$\Delta f''$	$\Delta f'$	$\Delta f''$
56 Ba	-11.0454	10.0919	-4.6472	12.0003	-1.0456	8.4617	-0.3244	2.2819	-0.6940	1.5004
57 La	-12.8190	3.5648	-6.3557	12.8927	-1.4094	9.0376	-0.2871	2.4523	-0.6411	1.6148
58 Ce	-9.3304	3.8433	-8.0962	11.8734	-1.8482	9.6596	-0.2486	2.6331	-0.5890	1.7358
59 Pr	-7.9841	4.1304	-10.9279	9.2394	-2.4164	10.2820	-0.2180	2.8214	-0.5424	1.8624
60 Nd	-7.1451	4.4278	-10.5249	9.9814	-3.1807	10.9079	-0.1943	3.0179	-0.5012	1.9950
61 Pm	-5.0783	4.7422	-13.2062	3.6278	-4.0598	11.5523	-0.1753	3.2249	-0.4626	2.1347
62 Sm	-6.0570	5.0744	-9.3497	3.8839	-5.3236	12.2178	-0.1638	3.4418	-0.4287	2.2815
63 Eu	-5.6630	5.4178	-7.9854	4.1498	-8.9294	11.1857	-0.1578	3.6682	-0.3977	2.4351
64 Gd	-5.3778	5.7756	-7.1681	4.4280	-8.8380	11.9157	-0.1653	3.9035	-0.3741	2.5954
65 Tb	-5.0951	6.1667	-6.5583	4.7292	-9.1472	9.1891	-0.1723	4.1537	-0.3496	2.7654
66 Dy	-4.8149	6.5527	-6.0597	5.0280	-9.8046	9.8477	-0.1892	4.4098	-0.3302	2.9404
67 Ho	-4.5587	6.9619	-5.6628	5.3451	-14.9734	3.7046	-0.2175	4.6783	-0.3168	3.1241
68 Er	-4.4106	7.3910	-5.3448	5.6776	-9.4367	3.9380	-0.2586	4.9576	-0.3091	3.3158
69 Tm	-4.2698	7.8385	-5.0823	6.0249	-8.0393	4.1821	-0.3139	5.2483	-0.3084	3.5155
70 Yb	-4.1523	8.2969	-4.8591	6.3813	-7.2108	4.4329	-0.3850	5.5486	-0.3157	3.7229
71 Lu	-4.0630	8.7649	-4.6707	6.7484	-6.6179	4.6937	-0.4720	5.8584	-0.3299	3.9377
72 Hf	-4.0564	9.2832	-4.4593	7.1518	-6.1794	4.9776	-0.5830	6.1852	-0.3548	4.1643
73 Ta	-3.9860	9.8171	-4.3912	7.5686	-5.7959	5.2718	-0.7052	6.5277	-0.3831	4.3992
74 W	-3.9270	10.3696	-4.2486	8.0005	-5.4734	5.5774	-0.8490	6.8722	-0.4201	4.6430
75 Re	-3.9052	10.9346	-4.1390	8.4435	-5.2083	5.8923	-1.0185	7.2310	-0.4693	4.8944
76 Os	-3.9016	11.5251	-4.0478	8.9067	-4.9801	6.2216	-1.2165	7.6030	-0.5280	5.1558
77 Ir	-3.9049	12.1453	-3.9606	9.3923	-4.7710	6.5667	-1.4442	7.9887	-0.5977	5.4269
78 Pt	-3.9435	12.7910	-3.8977	9.8985	-4.5932	6.9264	-1.7033	8.3905	-0.6812	5.7081
79 Au	-3.9908	13.4551	-3.8356	10.4202	-4.4197	7.2980	-2.0133	8.8022	-0.7638	5.9978
80 Hg	-4.1029	14.1373	-3.8228	10.9650	-4.2923	7.6849	-2.3894	9.2266	-0.8801	6.2989
81 Tl	-4.2233	14.8643	-3.8103	11.5300	-4.1627	8.0900	-2.8358	9.6688	-1.0117	6.6090
82 Pb	-4.4167	15.5987	-3.8519	12.1106	-4.0753	8.5060	-3.3944	10.1111	-1.1676	6.9287
83 Bi	-4.6533	16.3448	-3.9228	12.7017	-4.0111	8.9310	-4.1077	10.2566	-1.3494	7.2566
84 Po	-4.9604	17.1410	-4.0267	13.3329	-3.9670	9.3834	-5.1210	11.0496	-1.5613	7.5986
85 At	-5.3399	17.9390	-4.1781	13.9709	-3.9588	9.8433	-7.9122	9.9777	-1.8039	7.9509
86 Rn	-5.7275	18.7720	-4.3331	14.6313	-3.9487	10.3181	-8.0659	10.4580	-2.0847	8.3112
87 Fr	-6.2180	19.6009	-4.5387	15.3016	-3.9689	10.8038	-7.2224	7.7847	-2.4129	8.6839
88 Ra	-6.7502	20.4389	-4.7764	15.9778	-4.0088	11.2969	-6.7704	8.1435	-2.8081	9.0614
89 Ac	-7.4161	21.3053	-5.0617	16.6687	-4.0794	11.7994	-6.8494	8.5178	-3.2784	9.4502
90 Th	-8.2118	22.2248	-5.3692	17.4018	-4.1491	12.3296	-7.2400	8.8979	-3.8533	9.8403
91 Pa	-9.4459	23.1548	-5.7337	18.1406	-4.2473	12.8681	-8.0334	9.2807	-4.6067	10.2413
92 U	-9.9362	23.1239	-6.1485	18.8728	-4.3638	13.4090	-9.6767	9.6646	-5.7225	10.6428

[†][Reproduced with permission from *International Tables for Crystallography*, Vol. C, *Mathematical, Physical and Chemical Tables*, Second ed. (A.J.C. Wilson, E. Prince eds.), pp. 255-257, IUCr, D. Reidel Pub., Dordrecht, (1999)]

Appendix-Table 5 Conversion formulae for reciprocal lattice and real lattice parameters

Crystal system		Conversion formula
Triclinic		$a=\frac{\sin\alpha^*}{a^*\phi},\ b=\frac{\sin\beta^*}{b^*\phi},\ c=\frac{\sin\gamma^*}{c^*\phi}$ $\phi^2=1-\cos^2\alpha^*-\cos^2\beta^*-\cos^2\gamma^*+2\cos\alpha^*\cos\beta^*\cos\gamma^*$ $\cos\alpha=\frac{\cos\beta^*\cos\gamma^*-\cos\alpha^*}{\sin\beta^*\sin\gamma^*},\ \cos\beta=\frac{\cos\alpha^*\cos\gamma^*-\cos\beta^*}{\sin\alpha^*\sin\gamma^*},$ $\cos\gamma=\frac{\cos\alpha^*\cos\beta^*-\cos\gamma^*}{\sin\alpha^*\sin\beta^*}$
Monoclinic	(*b* as unique axis)	$a=1/(a^*\sin\beta^*),\ b=1/b^*,\ c=1/(c^*\sin\beta^*),\ \alpha=\gamma=90^\circ,\ \beta=180^\circ-\beta^*$
Orthorhombic		$a=1/a^*,\ b=1/b^*,\ c=1/c^*,\ \alpha=\beta=\gamma=90^\circ$
Hexagonal	(*c* as unique axis)	$a=b=\frac{2}{\sqrt{3}a^*},\ c=\frac{1}{c^*},\ \alpha=\beta=90^\circ,\ \gamma=120^\circ$
Rhombohedral (Trigonal)	(hexagonal lattice)	As given above for the hexagonal system
	(rhombohedral lattice)	$a=b=c=\frac{\sin\alpha^*}{a^*\sqrt{1-3\cos^2\alpha^*+2\cos^3\alpha^*}},\ \cos\alpha=\frac{\cos^2\alpha^*-\cos\alpha^*}{\sin^2\alpha^*}$
Tetragonal		$a=b=1/a^*,\ c=1/c^*,\ \alpha=\beta=\gamma=90^\circ$
Cubic		$a=b=c=1/a^*,\ \alpha=\beta=\gamma=90^\circ$

Appendix-Table 6 Scattering functions ($s = 2\pi S = 2\pi\varepsilon/\lambda$)

Shape of particle	Scattering function, Ψ	Notes
General form	$\exp(-s^2R_g^2/3)$	R_g is the radius of gyration with respect to the center of gravity, Eq. (2.51)[†1]
Sphere	$\Phi^2 = \left\{\dfrac{3(\sin sR - sR\cos sR)}{(sR)^3}\right\}^2 = \dfrac{9}{2}\pi\left\{\dfrac{J_{3/2}(sR)}{(sR)^{3/2}}\right\}^2$	R is the radius of the sphere. $J_{3/2}$ is a Bessel function of order 3/2[†1-3], $J_{3/2}(z) = \dfrac{\sqrt{2}}{\pi z}\left\{\dfrac{\sin z}{z} - \cos z\right\}$
	(Approximation) $\exp(-0.221s^2R^2)$ $\exp(-s^2R^2/5)$	Accuracy slightly poorer than that of the above
Ellipsoid of revolution	$\int_0^{\pi/2} \Phi^2(sa\sqrt{\cos^2\theta + w^2\sin^2\theta})\cos\theta d\theta$	a and b are the semiaxes, $w = b/a$; b is the axis of rotation[†1]
stationary	(Approximation) $\exp\{-s^2(a^2/4)\}$ (Equator) $\exp\{-s^2(b^2/5)\}$ (Meridian)	X-ray incidence normal to the b axis;
Cylinder of revolution	$(\pi R^2H)^2\left\{1 - \dfrac{1}{6}sR\left(1 + \dfrac{2}{3}\alpha^2\right) + \cdots\right\}$	Cylinder of radius R and length H, $\alpha = \dfrac{2R}{H}$ [†1]
Rod-like (radius negligible)	$\dfrac{S_i(2sH)}{sH} - \dfrac{\sin^2 sH}{(sH)^2}$	S_i is an integral sine[†3, 5]. $S_i(x) = \int_0^x \dfrac{\sin z}{z}dz$
Disc-like (length negligible)	$\dfrac{2}{s^2R^2}\left(1 - \dfrac{1}{sR}J_1(2sR)\right)$	J_1 is a first order Bessel function[†3] $J_1(z) = \dfrac{z}{2} - \dfrac{z^3}{2^2 1!2!} + \dfrac{z^5}{2^5 2!3!} - \cdots$
stationary	$\dfrac{2J_1(sR)}{sR}$	$H \to \infty$ and H can be neglected[†5]
	(Approximation) $\exp\{-s^2(R^2/4)\}$ (Equator) $\exp\{-s^2(H^2/3)\}$ (Meridian)	X-ray incidence normal to H axis;
Layer of infinite extent	$\dfrac{1}{sT}\left\{\dfrac{\sin sT/2}{sT/2}\right\}^2$	Independent of the shape of the layer; T is the thickness[†3]

†1 A. Guinier, *Ann. Phys. (Paris)*, **12**, 161 (1939).
†2 B. E. Warren, *J. Appl. Phys.*, **13**, 364 (1942).
†3 O. Kratky, G. Porod, *J. Colloid Sci.*, **4**, 36 (1949).
†4 D. L. Dexter, *Phys. Rev.*, **90**, 1007 (1953).
†5 G. Fournet, *Doctoral Thesis, Paris Univ.*, série A, no. 3356 (1950).

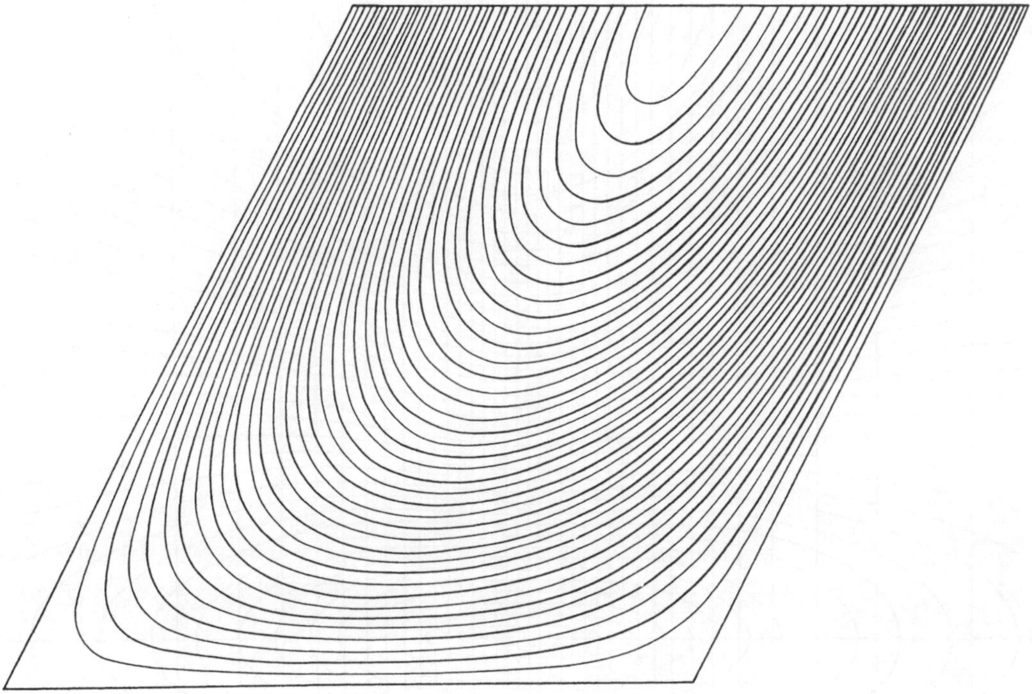

Appendix–Fig. 1 Wooster chart.[†]
For standard camera of radius 28.65 mm. Reciprocal lattice points on equidistant parallel lines in a section through the reciprocal lattice lie on the same curve in the chart. The chart is drawn with a spacing $d/\cos \nu = 0.05$ for the reciprocal lattice coordinates (d is the spacing of the crystal planes, and ν is the angle of inclination in the equi-inclination method).
[†][Reproduced with permission from M. J. Buerger, *X-Ray Crystallography*, p. 275, John Wiley & Sons, Inc. (1942)]

Appendix–Fig. 2 Bernal chart for reading ξ- and ζ-coordinates for diffraction on a rotation or oscillation photograph. For standard cylindrical camera of radius 28.65 mm.[†]
[†][Reproduced with permission from M. J. Buerger, *X-Ray Crystallography,* p. 148, John Wiley & Sons, Inc. (1942)]

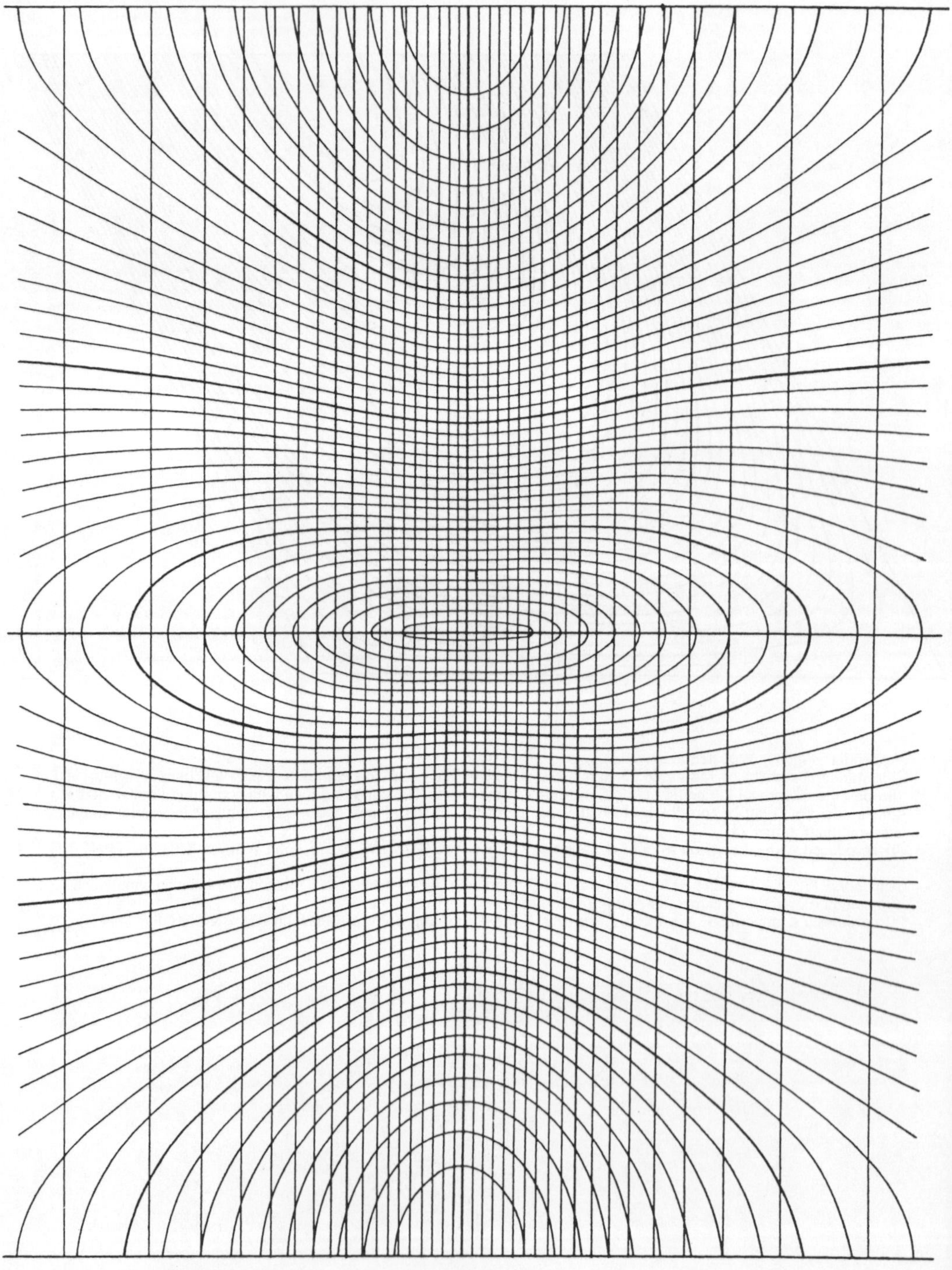

Index

E

F

G

H

M

N

O

P

Q

R

S

T